THERMAL INSULATION: MATERIALS AND SYSTEMS

A conference
sponsored by
ASTM Committee C-16
on Thermal Insulation
Dallas, TX, 2–6 Dec. 1984

ASTM SPECIAL TECHNICAL PUBLICATION 922

Frank J. Powell, National Bureau of Standards,
and Stanley L. Matthews, consultant, editors

ASTM Publication Code Number (PCN)
04-922000-10

1916 Race Street, Philadelphia, PA 19103

Library of Congress Cataloging-in-Publication Data

Thermal insulation: materials and systems/a conference sponsored by
ASTM Committee C-16 on Thermal Insulation, Dallas, TX, 2–6 Dec. 1984;
Frank J. Powell and Stanley L. Matthews, editors.
(ASTM special technical publication 922)
"ASTM publication code number (PCN) 04-922000-10."
Includes bibliographies and index.
ISBN 0-8031-0493-6
1. Insulating materials—Congresses. 2. Insulation (Heat)—
Congresses. I. Powell, Frank J., 1923- . II. Matthews, Stanley L.
III. ASTM Committee C-16 on Thermal Insulation. IV. Series.
TH1715.A1T47 1987 87-27045
620.1′1296—dc19 CIP

Library of Congress Catalog Card Number: 87-27045

NOTE

The Society is not responsible, as a body,
for the statements and opinions
advanced in this publication.

Printed in Baltimore, MD
December 1987

Stanley L. Matthews

In Memoriam

This volume is dedicated to the memory of Stan Matthews, coeditor of this book, who passed away before the book was in print. Stan was for many years a tireless worker for ASTM and, particularly, for ASTM Committee C-16 on Thermal Insulation. In Committee C-16, he was constantly involved at every level, from task group commitment to serving as a leading officer. His dedication to ASTM and to professional associations, the government, and the entire thermal insulation industry was admired and respected by all who actively worked with him. Stan's whole life was one of vigor, from his days as a Harvard M.B.A. graduate student, through his career as a U.S. Army colonel, and for the past 25 years as a champion for the thermal insulation industry, including eleven terms as president of the Mineral Insulation Manufacturers Association. The honors accorded Stan have been numerous, including ASTM's highest recognition, the Award of Merit. He will be missed.

Foreword

The conference on Thermal Insulation: Materials and Systems was held in Dallas, Texas, on 2–6 Dec. 1984. ASTM Committee C-16 on Thermal Insulation was sponsor of the event. Frank J. Powell, National Bureau of Standards, and Stanley L. Matthews, consultant, presided as chairmen of the conference and also served as editors of this publication.

Related ASTM Publications

Guarded Hot Plate and Heat Flow Meter Methodology, STP 879 (1985), 04-879000-10

Building Applications of Heat Flux Transducers, STP 885 (1985), 04-885000-10

Corrosion of Metals Under Thermal Insulation, STP 880 (1985), 04-880000-27

Thermal Insulations, Materials, and Systems for Energy Conservation in the '80s, STP 789 (1983), 04-789000-10

Thermal Insulation Performance, STP 718 (1980), 04-718000-10

Thermal Transmission Measurements of Insulation, STP 660 (1979), 04-660000-10

Thermal Insulations in the Petrochemical Industry, STP 581 (1975), 04-581000-10

Heat Transmission Measurements in Thermal Insulations, STP 544 (1974), 04-544000-10

A Note of Appreciation to Reviewers

The quality of the papers that appear in this publication reflects not only the obvious efforts of the authors but also the unheralded, though essential, work of the reviewers. On behalf of ASTM we acknowledge with appreciation their dedication to high professional standards and their sacrifice of time and effort.

ASTM Committee on Publications

ASTM Editorial Staff

Helen Mahy
Janet R. Schroeder
Kathleen A. Greene
William T. Benzing

Contents

Materials

Walls and Windows

Air and Moisture

Research

Reports

Indexes

Introduction

Since ASTM Committee C-16 on Thermal Insulation was founded in March 1938, its policy has been to sponsor programs periodically that provide the latest information regarding test methods and insulation materials. As a result of these programs, more than a dozen ASTM Special Technical Publications have been published on thermal insulation materials and technology, most of them in the last decade, based on symposia sponsored by the committee.

Activity and interest in thermal insulation have increased as a result of the space program in the 1960s and energy conservation in the 1980s. In 1978, over 500 people participated in a symposium when the topic was enlarged to include not only technical measurement but also all the parameters related to thermal insulation for energy conservation. The trend has continued with sponsorship of conferences and symposia alternating between ASTM and the American Society of Heating, Refrigerating, and Air-Conditioning Engineers (ASHRAE), with additional sponsorship by both the U.S. Department of Energy and Oak Ridge National Laboratory. Proceedings of the ASHRAE-sponsored conferences are available as ASHRAE Special Publications. The ASTM-sponsored conferences and symposia continue to be published as ASTM Special Technical Publications. ASTM Committee C-16 is now sponsoring a major conference in December 1987.

The conference published in this volume, Thermal Insulation: Materials and Systems, took as its theme "Controlling Fuel Costs Through the Use of Thermal Insulation." This topic was intended to cover both industrial and residential types of thermal insulation. Fuel prices do fluctuate, as witnessed by records over the past ten years. Therefore, it becomes doubly important that engineers, scientists, and economists, as well as managers, have available the latest and best tools to use in making current technical and business decisions. Energy is still expensive and continues to be a major factor in building technology and in industry. The editors hope that the information published by this conference will add to and sharpen the means by which profitable progress can be made.

The manufacture and use of thermal insulation materials continues to be the primary method for optimizing the use of energy in both the residential and industrial sectors. Traditional materials have been improved and new materials and systems are regularly being developed.

ASTM Committee C-16 on Thermal Insulation is pleased to continue to provide the format for discussion and documentation of this complex subject. Also, this conference provides excellent input to Committee C-16 in discharging its responsibility for promulgating standards for insulation materials, systems, and test methods. New and improved testing equipment and methods and the development and production of thermal insulation that is effective, durable, and safe provide assistance in the increasingly difficult task of controlling fuel costs.

The contributions of the organizing committee, the authors, and the session chairmen; the untiring efforts of the reviewers; and the cooperation of people from the many organizations involved are greatly appreciated. The editors wish specifically to acknowledge the conference organizing committee members: S. H. Cady, E. Bales, R. Baumgardner, M. Bomberg, J. Boulin, E. Freeman, W. Gerken, F. Govan, P. Greene, J. Kimpflen, T. Lundy, J. D. McAllister, D. L. McElroy, W. McHale, R. Morgan, and I. Seigler. Special gratitude is also extended to the session chairmen, who presided during the conference sessions: J. Kimpflen, F. A. Govan, R. L. Baumgardner, E. Freeman, D. McElroy, S. H. Cady, M. Hollingsworth, J.

Boulin, H. Trechsel, T. Kusuda, R. P. Tye, M. Van Geem, C. Bankvall, and D. Ober. The cooperation and the many hours of work on the part of the ASTM staff have made this book possible and are gratefully acknowledged.

Frank J. Powell

National Bureau of Standards, Gaithersburg, MD 20899; symposium cochairman and editor.

Stanley L. Matthews

Consultant, Colorado Springs, CO 80906; symposium cochairman and editor.

Controlling Fuel Costs

Wilbur T. Coyle[1]

Thermal Envelope Case History for a School District

REFERENCE: Coyle, W. T., **"Thermal Envelope Case History for a School District,"** *Thermal Insulation: Materials and Systems, ASTM STP 922,* F. J. Powell and S. L. Matthews, Eds., American Society for Testing and Materials, Philadelphia, 1987, pp. 5-7.

ABSTRACT: This paper discusses the actual selection and application of thermal envelope energy conservation measures for the 102 223 m^2 (1 100 000 ft^2) of Weld County School District No. 6 buildings, which are located in the cities of Evans and Greeley, Colorado. Since 1976, the district has followed an intensive energy conservation program in which mechanical, electrical, and structural retrofits have been prioritized on a cost/benefit basis, and measures have been financed by school funds or by 50-50 grants from the Institutional Grants Program of the U.S. Department of Energy. The thermal envelope measures discussed include replacement windows, storm windows, vestibules, overlay of thermal insulation on glass block, attic insulation, and rigid roof insulation. These measures, along with the usual retrofitting of controls for mechanical and electrical systems, have resulted in a cost savings of $305 342 for the district in 1983, in comparison with 1976. Finally, mention is made in this paper of the systemizing of the district's total energy conservation efforts in order to recognize the need for motivational measures aimed at the building occupants. This effort complements the ongoing thermal envelope and hardware retrofit measures.

KEY WORDS: cost savings, energy conservation, school building energy conservation, thermal envelope, retrofits, federal energy conservation programs, thermal insulation

The 102 230 m^2 (1 100 000 ft^2) of buildings in Weld County School District No. 6 are located on 23 sites in the cities of Evans and Greeley, Colorado. All the buildings are connected to a central processor for the monitoring of fire, security, and energy conservation. Renovations or retrofitting may be performed by either outside contractors or in-house maintenance personnel. Utility bills are reviewed promptly, and monthly cost savings reports are issued to the principals of the buildings.

The average annual number of heating degree-days for Greeley is 6442. The student population is 10 500, and the annual utility budget is $1.1 million. In spite of the ambitious energy conservation program, the utility budget has risen to 4% of the 1984 annual budget, in comparison with 1½% of the 1976 budget. Any dollars that can be saved on utilities can be used for educational materials and teachers' salaries. The 1983 versus 1976 energy conservation report estimates $305 342 in cost savings for the district. A more detailed calculation would have to take into account weather patterns and varying building usage. Also, the report shows that the energy used annually by the district has been reduced from 1 339 351 kJ/m^2 (118 000 Btu/ft^2) in 1976 to 907 955 kJ/m^2 (79 993 Btu/ft^2) in 1983.

Funding Priorities

Proposed energy conservation projects have to compete with other remodeling projects for the same funds. Cost/benefit studies, based on simple payback analysis, are prepared in order to

[1]Supervisor of energy conservation, Weld County School District No. 6, Greeley, CO 80631.

establish priorities. Appropriate energy-conserving projects are selected for applications to be made for the U.S. Department of Energy (DOE) Institutional Grants Program. Starting in 1979 and continuing through 1983 the school district has participated in Cycles I to V for a total of $187 322 in federal grants. Maintenance funds and personnel are used for minor energy-savings projects.

Energy Savings Measures

The energy conservation program for the district includes controls, equipment conversions, operations, changes, and thermal envelope modifications. The control changes include installing electronic hot water reset controls, closing relief vents, using dead-band thermostats, installing time clocks, and replacing outmoded pneumatic controls with solid-state controls. The equipment replacement program consists of replacing old or obsolete mechanical and electrical equipment with better designed and more efficient equipment, which reduces the energy consumed on utilities. The operations of the district have been governed by energy conservation guidelines adopted by the School Board as the result of the 1973 fuel crisis.

Thermal Envelope Retrofits

Windows

When the district has the opportunity to remodel its buildings, the exposed amount of glass is reduced and energy-efficient, wood, double-hung windows with double glazing are used. The panels above the new windows are covered with an insulated filler. New schools built in the district also use the same type of window to provide fresh air and natural light with minimum heat loss. One of the districts federal grants was for the installation of inside magnetic fastener storm windows for seven schools. The bids came in too high for the district to accept, but a decision was made to proceed with a trial installation. The interior storm windows were installed, but the integral adhesive used to bond the magnetic strip to the window frames failed. The contractor removed the inside storm windows, and no others are planned at this time.

Vestibules

Vestibules have been installed in five buildings in the last two years. These vestibules are inside the exterior doors, and the surrounding area is built of concrete block with an epoxy paint finish. Hollow metal door frames were especially fabricated for the vestibules, with transom panels and solid core wood doors. The vestibule doors swing out of the way when not needed for energy savings. Acceptance of the vestibules by all the tenants has been very good.

Glass Block

Glass block was designed for daylight lighting of classrooms in schools built in the 1950s and 1960s. However, the glass block was not energy efficient and generally had to be covered with drapes by the teachers when films or slides were being shown to their classes. This study of the best and most economical way to reduce the glass block heat losses considered options that varied from (1) completely removing the glass block and replacing it with solid masonry to (2) covering the glass block, in place, with some more energy-efficient materials. Our decision was to use a commercial system that combined 5 mm (2.0 in.) of polystyrene insulation glued to the glass block, which was then covered by a layer of fiberglass reinforcing fabric; the fabric was then covered by a portland cement Type I primer/adhesive; and, finally, an acrylic waterproof stucco with color was applied as a finish. The interior surface of the glass block was left exposed. The glass block retrofit projects have been in place for less than two years, so there are no conclusive energy savings results yet.

Attic Insulation

Most of the roofs in the district have low-pitch slopes with rigid insulation incorporated within the roof deck system. However, the old high school auditorium area had only 15 mm (6.0 in.) of loose-fill, mineral wool ceiling insulation. Under a DOE federal grant, 30 mm (12.0 in.) of loose-fill, mineral wool insulation was added. Other applications of loose-fill insulation in this district have been minimal.

Rigid Roof Insulation

Most district buildings were constructed before energy conservation became a serious concern. Roof decks consisted of board sheathing with a minimal cane fiber insulation board for insulation. Now the district requires an *R* value of 4.3 $K \cdot m^2/W$(25°F $ft^2 \cdot h/Btu$) for remodeling and 5.7 $K \cdot m^2/W$ (33°F $ft^2 \cdot h/Btu$) for new construction. The roofs are replaced on about a 20 to 25-year life cycle. During the replacement of any roof, the old roofing is torn down, and then the old insulation board material is removed down to the bare roof deck. The old roof deck is repaired, and all the roof penetrations, flashings, and fascias are extended in order to accommodate at least 10 mm (4.0 in.) of new insulation. Usually, the new roof is an ethylene-propylene-diene monomer (EPDM) ballasted membrane roof system.

Interestingly, the district was able to accelerate its roof insulation program because of a severe hail storm in 1983. Insurance coverage was obtained to replace over 46 465 m^2 (500 000 ft^2) of roofing, and the School Board had the foresight to provide funds for about 23 233 m^2 (250 000 ft^2) of roof insulation board to bring the hail-damaged roofs up to our standard *R* value of 4.3 $K \cdot m^2/W$ (25°F $ft^2 \cdot h/Btu$). Some of the roofs were insulated with one layer of fiberglass insulation plus one layer of polyurethane insulation, and other roofs were insulated with two layers of polyurethane insulation. The roofs were completed by using either a white-finished EPDM membrane or a gravel-ballasted EPDM membrane.

Systems Approach

Most institutions spend little effort in solving the motivational aspects of their energy conservation program. It is easy to quantify "hardware" activities such as the retrofitting of mechanical, electrical, and structural changes in buildings. However, the tenants or users of the district's buildings do have an influence on the amount of energy used. School District No. 6 of Weld County, Colorado, has developed a systems approach to control the interactions between the hardware activities and the motivational activities.

The systems approach is used by the district's Energy Conservation Committee to manage the energy conservation program. This committee is composed of representatives of all sectors of the district, has the full support of the superintendent, and meets monthly during the school year to manage the program. An annual energy conservation report is issued to the School Board to inform them on activities of the past year and plans for the future. The district's systems approach to energy conservation has been recognized by the State of Colorado with its 1984 Energy Conservation Pacesetter Award and by the U.S. Department of Energy with its 1984 National Award for Energy Innovation.

Helmuth E. Worbs[1]

Utilization of Energy Code Compliance Procedures for the Prediction of Commercial Building Annual Fuel Consumption

REFERENCE: Worbs, H. E., **"Utilization of Energy Code Compliance Procedures for the Prediction of Commercial Building Annual Fuel Consumption,"** *Thermal Insulation: Materials and Systems, ASTM STP 922,* F. J. Powell and S. L. Matthews, Eds., American Society for Testing and Materials, Philadelphia, 1987, pp. 8–20.

ABSTRACT: Many state model energy codes provide, as an optional alternative to building design by performance standards, a numerical computational procedure to predict a commercial building's annual energy fuel consumption. This annual energy fuel consumption value determines if the building complies with the code energy efficiency. The physical constraints of the building are considered in the calculation—that is, the building envelope and internal heat gains involving people, lighting, equipment, and ventilation. In addition, the efficiency of the building's heating and cooling mechanical plant are also considered.

This paper will first describe the computational procedures, together with the imposed assumptions and limitations, and then describe how this procedure can be adapted to a computer. It is then suggested that this computer program would be valuable in determining the changes one could expect in annual fuel usage by varying certain building design parameters.

An actual representative commercial office building was then selected using the actual building design parameters as the reference building. Variation in building orientation, wall and roof *U* values, changes in shading and glazing values, changes in allowable ventilation levels, and changes in the efficiency of the heating and cooling mechanical heating, ventilating, and air-conditioning system are all presented to show which ones exhibited a large sensitivity to overall building energy usage. The results will allow the design engineer to make some general conclusions as to which factors are most intensive from an energy conservation point of view.

KEY WORDS: air conditioning, cooling load, heat load, thermal insulation

Nomenclature

Symbols

- *A* Area, m^2
- *CM* Overall glass *U* value, $W/m^2 \cdot °C$
- *FA* Total plan floor area, m^2
- *GA* Glass area, m^2
- *GM* Glass conduction multiplier, $W/m^2 \cdot °C$
- *Q* Building design ventilation rate, L/s
- *RA* Roof area, m^2
- *RM* Roof multiplier, dimensionless
- *SCM* Glass shading coefficient, dimensionless

[1]Associate professor, College of Engineering, University of Central Florida, Orlando, FL 32816.

SM Solar multiplier
U Overall heat transfer coefficient, $W/m^2 \cdot °C$
VM Ventilation multiplier, m/s
WA Wall area, m^2
WM Wall multiplier factor, dimensionless
X Building total glass area, m^2

Subscripts

d Door
g Glass
i Building wall orientation
t Code standard value, as dictated by specified wall or roof grouping
w Wall

Superscripts

r Roof value
w Wall value

Model energy code compliance procedures fall into three basic categories:

(*a*) the point system method,
(*b*) design standards, and
(*c*) building performance criteria.

The first method involves assigning points for certain characteristics of the building. Certain point values are assigned for insulation, percentage of glazing, lighting level, and so on. The point values may be either maximized or minimized for increased energy efficiency. Currently, the Florida Model Energy Efficiency Code for Building Construction [*1*] encourages minimum total points for maximum energy savings.

The second method relates mainly to minimum standards for insulation, percentage of glazing, allowable levels of ventilation, and efficiency of the mechanical heating and cooling plant, to name a few. The first method is used largely in residential construction while most designers of heating, ventilating, and air-conditioning (HVAC) equipment and architects opt for the second method.

A third method, which is equally acceptable by most state energy efficiency building codes, involves calculating or predicting the building's annual energy usage and comparing this value with some code-designated limiting budget value. If the computer energy usage is less than the code-budgeted value, then, of course, the building complies with the code. This procedure essentially involves calculating the building's heating or cooling load on an hour-by-hour basis for the entire climatic design year and adding up these values. Certain simplifying assumptions can be made, such as that the hourly variation in outdoor air temperature is the same for each day in a particular month and also that the solar fenestration, though varying on an hourly basis, is the same for each day in the month. While this considerably reduces the number of calculations, the process, if accomplished with paper and a pocket calculator, is still tedious and time-consuming. The "bin" method just described can, of course, be programmed on a microprocessor computer with the hourly variation in outdoor air temperature and solar conditions stored on disk for a specified climatic zone. The HVAC design engineer can then rapidly look at the effects of varying different parameters in the mechanical design of the building and also have a rapid means of obtaining cost data for various design concepts. In essence, he can perform

"What If?" analysis and look at the effect of different building design concepts on the overall annual energy consumption.

This paper will describe the computerized calculation procedure and then the physical characteristics of an office building recently constructed in central Florida. It will next present the results of the computations showing which parameters most affect the energy consumption of the building.

Program Description

In actuality the cooling and heating load parameters are grouped for a 3-h period. These parameters essentially involve the outdoor ambient temperature and, in the case of the wall transmission, a reference U value. The procedure for computing the heat gain due to wall transmission is as described in the 1981 *ASHRAE Handbook of Fundamentals* and also the *ASHRAE Cooling and Heating Load Calculation Manual* [2,3], which employs the cooling load temperature difference method (CLTD). This method includes the thermal capacity of the wall system and accounts for the fact that the peak cooling load may occur some time after the peak instantaneous heat gain. Variation in building orientation is also considered in the program, since data for different wall orientations are available in the American Society of Heating, Refrigeration, and Air Conditioning Engineers (ASHRAE) guides. The permitted wall orientations are for 45° increments from true north, as are shown in Table 1.

This table is an actual representation of the cathod-ray-tube (CRT) display screen and illustrates the interactive nature of the program. Since the assigned values for Orientations 1 through 4 are arbitrary, it is obvious that the building can be rotated through 360° in increments of 45° to determine the effect of building orientation on the heat gain and the annual energy consumption.

Wall transmission is determined by evaluating the wall multiplier factors, as shown in Eq 1

$$WM_i = \frac{U_i^w (WA)_i}{U_t^w (FA)} \tag{1}$$

TABLE 1—*Building orientation referenced from true north.*

Characteristic	Value
Wall area	
Orientation 1[a], m^2	113.3
Orientation 2, m^2	147.7
Orientation 3, m^2	61.3
Orientation 4, m^2	80.8
Total gross conditioned floor area	1384.2
Wall U value, W/m^2 · °C	0.34
Wall weight per unit area, kg/m^3	248
Construction lightweight frame	
Energy table U value, W/m^2 · °C	0.46
Lightweight frame, W/m^2 · °C	0.46

[a]Key to orientation:
1 = north/northeast.
2 = east/southeast.
3 = south/southwest.
4 = west/northwest.

where

U_w = actual wall U value,
WA = wall area,
U_t^w = reference U value, and
FA = total plan floor area.

The total plan floor area in this equation must be equal to the floor area on which the energy budget is based. Also, the actual wall U value, is a weighted average, which accounts for the heat transmission through doors and other glass areas that are not accounted for in the fenestration calculation. Namely, for $i = 1$ to 4 (wall orientation)

$$U_i^w = \frac{U_w A_w + U_g A_g + U_d A_d}{A} \tag{2}$$

where $A = A_w + A_g + A_d$.

Wall thermal capacity effects are accounted for by having separate data tables in the program for lightweight curtain walls, lightweight frame walls, medium-weight walls, heavy-weight walls, and extra-heavy-weight walls.

Heat gain through roofs is handled in a similar fashion except that, for roofs, 18 separate and distinct configurations are in the program, each with its own reference roof U value. The program user is free to vary the U value for the actual building configuration. The roof multiplier equation is

$$RM = \frac{U^r(RA)}{U_t^r(FA)} \tag{3}$$

The energy calculation due to solar fenestration is accomplished by matrix multiplication tables in the program for each month of the year. These tables contain the solar heat gain factors as a function of orientation for that particular month. Separate tables are employed for different overhangs and side fins. The solar multiplier factor is

$$SM_i = \frac{(GA)_i(SCM)}{(FA)} \tag{4}$$

for $i = 1$ to 4 (glass orientation).

GA is the glass area and SCM is essentially the shading coefficient. Similar to their treatment in air-conditioning load calculations, the glass conduction cooling loads for the four building orientations are added together. The equation accounting for glass conduction is

$$GM = \frac{(CM)(X)}{(FA)} \tag{5}$$

Obviously CM in the equation is the overall glass U value.

The ventilation parameter is handled by

$$VM = \frac{(Q)}{(FA)} \tag{6}$$

The simplicity of the *VM* parameter should clearly indicate the nature of the energy usage calculation obtained from the program. Clearly, the ventilation load is the difference between the total enthalpy of the outside air evaluated at the wet bulb temperature and the room air, also evaluated at the wet bulb temperature. The program uses the 3-h average variation of the outside dry and wet bulb temperature and assumes that the indoor design conditions are as dictated by energy code guidelines. It is obvious from these constraints that the program cannot calculate energy usage figures for conditions other than code design values. As previously pointed out, the use of this program is to obtain rapid estimates of changes in annual energy usage as a function of variations in design parameters. Variations in building energy usage as a function of operating procedures is beyond the scope of this program.

The lighting loads are input into the program as the total load in watts due to lighting and are factored thus into the calculation. The output also contains the wattage per square foot for direct comparison to code limitations. Each building type, by its nature, will have a representative occupancy density, and, in addition, the activities of the people in the building will also vary according to the building use. Representative building types available in the program are shown in Table 2. The variation in internal heat gains due to occupants as a function of building usage is immediately obvious if we examine a typical office building, a theater, and a bowling alley. According to Ref *3,* one can present data as shown in Table 3. The program contains a data table for each building type, which incorporates an average value for the population density and a representative value for the occupancy heat gains as suggested by the Table 3. The program user may adjust this value by using a "people multiplier" other than 1, which the user may infer as an increase or decrease in occupancy from the suggested value or as a variation in occupancy heat gain from the program value. Clearly, the "people multiplier" can also be thought of as varying both occupancy levels and heat gain from those occupants.

The program will also take into consideration reduced air-flow requirements as a result of venting the return air through the light fixtures. This will be examined and described subsequently in the paper from the aspects of venting the return air to a ceiling plenum and venting the return air through the ceiling light fixture, which is tied to a return air duct system.

Nationally recognized fuel energy program calculation procedures are designed primarily for climates where heating is more important than cooling. This program is based on the Florida Model Energy Code, which is climate specific for Florida. Therefore, some of the assumptions built into the program may not be appropriate for northern climates. However, it is the author's contention that the program can be adapted to be of service in estimating annual fuel consumption for commercial buildings in northern climates. It is worth mentioning here that the subsequent building description study and conclusions derived from the study may be appropriate only for semitropical locations, as represented by central Florida.

In keeping with these considerations, the model energy code includes a morning heating start-up factor for office buildings even though the building will, in all likelihood, require cooling much of the day, even in the cool winter months. The start-up heat energy for office buildings in Florida is given in Table 4.

An additional factor that contributes to the overall annual energy usage is the efficiency of the cooling and heating/air-conditioning system. Obviously, a more efficient mechanical plant, or one that uses a heat pump as opposed to electric strip heat, will use less energy. The bottom line is, how much less? We can factor this into an economic analysis to determine if the proposed design modifications are cost-effective.

Building Description

The three-story office building used for this study was recently constructed in central Florida. Most of the space is utilized for engineering consulting services, with the remaining space being leased out to other professionals such as attorneys and accountants. As far as the engineering

TABLE 2—*Index to building type.*

Code	Description
ZA	place of assembly, auditorium
ZB	bank or savings and loan
ZC	clinic
ZD	drugstore
ZE	school
	classroom
	gymnasium (conditioned)
	office (same as ZO)
	laboratory
	auditorium
	dining
	kitchen
ZG	supermarket
ZH	hotel, motel
ZL	library
ZM	mercantile
	strip shopping center
	department store
	mall (conditioned)
	storage (same as ZS)
ZN	nursing home
ZO	office building
ZP	hospital
	autopsy/morgue
	central supply
	operating suite
	emergency department
	intensive care unit
	laboratory
	general patient care
	dining
	kitchen
	office
ZR	restaurant
ZS	storage, warehouse (conditioned)
ZT	theater
ZV	air terminal
	commercial
	concourse
	storage (same as ZS)
	dining
	kitchen
ZW	place of worship
ZX	bowling alley
ZZ	special (any building not listed)

TABLE 3—*Heat gain from occupants per 92.9 m^2 (1000 ft^2).*

Building type	Adjusted Group Total, kJ/h/Person	Persons/100 m^2	Heat Gain/100 m^2, kJ/h
Office building	443	10.8	4 789.4
Theater	369	161.5	59 594
Bowling alley	1 012	75.3	76 204

TABLE 4—*Start-up heat energy.*

Zone	kJ/m^2
Northern Florida	29 500
Central Florida	15 880
Southern Florida	5 670

utilization of the space is concerned, the central core on the second and third floors is used for an open bay drafting and design work area. The first floor is used primarily for private executive offices and also provides space for reception and the executive staff.

The wall construction consists of 1.9-cm stucco on metal lath with 1.27-cm exterior plywood sheathing. The insulation consists of 15.2-cm batt fiberglass. The interior finish consists of 1.37-cm drywall. The insulation thicknesses vary somewhat throughout the building, but for purposes of analysis and design, an average wall U value of 0.34 W/m^2 · °C was assumed.

The roof construction consists of a normal built-up roof over 7.6-cm lightweight concrete decking. The underside of the decking has had a 15.2-cm foil-backed insulation installed, resulting in a reflective air space gap between the insulation and the 1.7-cm gypsum plaster ceiling. This results in a U value for the roof system of 0.19 W/m^2 · °C.

Glazing considerations for the building have resulted in an average glass factor of 33% for the building. The glazing consists of a 0.69-cm single-pane bronze tinted glass throughout, which results in a shading coefficient equal to 0.7. These design parameters are fully described in Table 5 and are referred to as the actual building design in the table.

The actual building design occupancy levels at the time of the study were 137 people approximately evenly distributed over the three floors. The design ventilation rate was taken to be 2.4 L/s per person. This resulted in a design ventilation rate of 323 L/s for the building. The computer program has a standardized value of 9.3 m^2/person for occupancy levels for office buildings. Since the objective of this study was to investigate the variation in building design parameters on the annual energy fuel usage, the author decided to arrive at a standardized set of parameters and vary each one, in turn, to obtain its effect on energy consumption. The "code reference building" was therefore established, in which many of the building design parameters were arbitrarily set at the nominal value dictated by the computer program calculations. Specifically, the occupancy was taken to be 9.3 m^2/person, and this resulted in a people density of 149 people for the building. The ventilation rate was also increased to 352 L/s. The nominal reference values for the wall and roof U value are also used in the code reference building calculation, and it is noted that these values are $U_t^{r/w}$, as defined by Eqs 1 and 3.

Variation in the treatment between the actual building design and the code reference building also occurred in the treatment of the heat dissipation due to the air handler fan motors. Each floor is serviced by a separate air handler with its individual split-system condensing unit air-to-air heat pump installed on the roof. The first-floor air handler is installed in a mechanical equipment room so that the return air enters the room and then enters the evaporative coils through a return air filter unit on the side of the air handler. In this specific case, the motor, as well as the fan, is considered to be in the conditioned space. Thus, one must account for the heat generated by the fan horsepower and also the inefficiencies in the electric motor. The second and third-floor air handlers are installed in the ceiling plenum space, as shown in Fig. 1. These systems have a ducted return air system, and the design selected has resulted in the fan motors being out of the airstream. Therefore, only the heat dissipation generated by the fan horsepower is considered to be included in the overall building heat gain. Since there is no assurance that the specified air handler type would actually be purchased and installed, the code reference building has a worst-case condition of the motor and the driven machine being in the condi-

TABLE 5—*Florida model energy code.*

Characteristics	Actual Building Design	Code Reference Building
Wall area		
Orientation 1, m^2	113.3	113.3
Orientation 2, m^2	147.9	147.9
Orientation 3, m^2	61.3	61.3
Orientation 4, m^2	80.8	80.8
Conditioned floor area, gross, m^2	1 384.2	1 384.2
Wall U value, $W/m^2 \cdot °C$	0.34	0.46
Roof 12	0.19	0.51
Roof area, m^2	418	418
Glass area		
Orientation 1, m^2	50.4	50.4
Orientation 2, m^2	82.7	82.7
Orientation 3, m^2	45.9	45.9
Orientation 4, m^2	29.3	29.3
Shading coefficient, SC	0.7	1.0
Conduction multiplier, $W/m^2 \cdot °C$	5.67	5.67
Ventilation air rate, L/s	323	352
People multiplier	0.919	1.0
Lighting, W	40 230	40 230
Fixtures		
vented	yes	yes
ducted	yes	yes
Electric heat dissipation, hp[a]	8.72	9.933
Heating system multiplier	0.38	0.38
Cooling system multiplier	0.45	0.45
Exhaust fan, hp[a]	0.16666	0.16666
Operating hours/day	8	8
Electrical cost, $/kWh	0.05	0.05

[a] 1 hp = 0.746 kW.

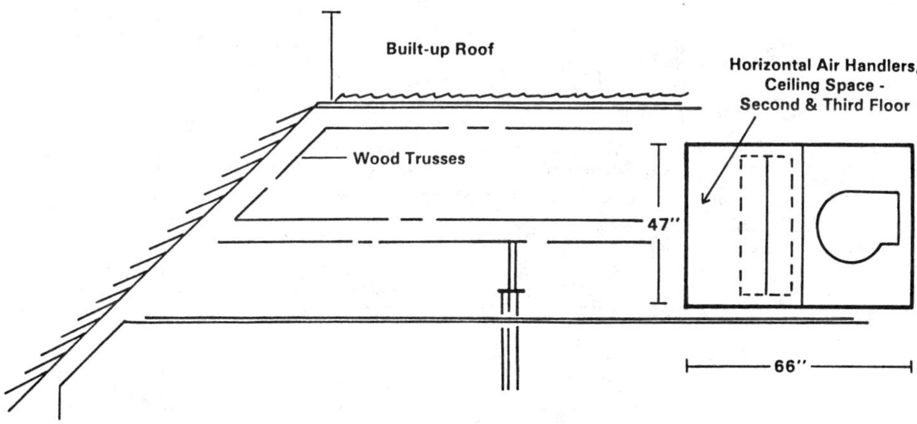

FIG. 1—*Ceiling plenum air handler installation.*

tioned space for each of the air handlers, as shown in Table 5. A 0.12-kW ($^1/_6$-hp) exhaust fan was also assumed to operate for the normal 8-h workday.

The actual design lighting level for the building was taken from the code-recommended value of 29 W/m^2. This value was also used in the code reference building. The only variation presented in this study is that the annual energy consumption is shown as a function of lighting level. The annual energy budget value dictated by the energy code was 132 kWh/year/m^2 at the time this study was undertaken. It is of interest to note that this figure has been reduced to 126 kWh/year/m^2 due to the availability of more efficient heating and cooling plants. Note, however, that the actual design building came in well under the budget figure—specifically, at 102.4 kWh/year/m^2. Since the building orientation numbering system is arbitrary, it is well to begin by noting that in the actual building orientation, 1 is north (refer to Table 1).

The results of rotating the building in successive 45° increments clockwise are shown in Table 6. These results were calculated as in subsequent parametric variations for the code reference building.

Since the building has a fairly uniform glass-to-total-wall-area ratio on all four sides and is well insulated, large variations in annual energy usage as a function of building orientation, as demonstrated by the results, are not to be expected.

The effect of wall insulation was next examined by arbitrarily varying the U value from the code reference value up to 5.67 W/m^2 · °C. These results are shown in Table 7. Before trying to

TABLE 6—*Annual energy usage for different building orientations.*

Azimuth, deg	Annual Energy Usage, kWh/year/m^2
0	112.5
45	112.8
90	112.5
135	111.8
180	111.5
225	111.2
270	111.2
315	111.8
Total variation	<0.6%

TABLE 7—*Wall* U *values and annual energy usage.*

U Value, W/m^2 · °C	Annual Energy Usage, kWh/year/m^2
0.46	112.5
0.57	112.5
0.85	112.8
1.13	113.1
1.70	113.4
2.84	114.0
4.54	115.0
5.67	115.8
Total variation	<2.2%

explain why such a small unexpected variation in annual energy consumption occurs for such a large variation in wall transmission factors, it may be appropriate to examine the two limiting conditions on a monthly basis. This is done in Table 8, which explains the apparent anomaly. This table reveals an interesting situation which is substantiated by the computer calculations. The building is actually in the cooling mode much of the winter months even though November through February are the coldest design months. This situation results from the fact that net cooling is required because of the occupancy load and the heat of lights. The higher U values permit a greater heat loss due to transmission, which reduces the net cooling load during these months. This condition is indicated by the footnote in the table. Actually the higher transmission losses assist in cooling the building during the colder winter months, and, hence, the integrated effect over the twelve-month period is not as large as one might expect. Table 9 shows the variation in annual energy usage as a function of the roof U value. A similar argument can be made for the roof thermal conductance to show the net cooling benefit during the cool winter months. It is important to remember that these conclusions are valid only for the cases studied in this model for the semitropical region under consideration. This effect would be expected to be nullified in northern Florida regions. This effect would, however, show a distinctly more pronounced pattern in the southern Florida region.

Glazing considerations produce a more distinct variation in building annual fuel consumption, as is shown in Table 10. Table 10 also shows the variation in fuel usage for the actual

TABLE 8—*Building monthly heat gain for the code reference building.*

Month	$U = 0.46$ W/m² · °C, kWh/m²	$U = 5.67$ W/m² · °C, kWh/m²
January	6.3[a]	4.3[a]
February	7.4[a]	6.5[a]
March	8.6	9.2
April	9.1	10.1
May	10.0	11.4
Summer	42.5	49.7
October	8.9	9.6
November	8.3[a]	8.5[a]
December	7.4[a]	6.5[a]
Total	108.5	115.7

[a]Indicates those months when wall transmission actually produces a heat loss.

TABLE 9—*Roof* U *values.*

U Value, W/m² · °C	Annual Energy Usage, kWh/year/m²
0.51	112.5
0.85	112.8
1.13	113.1
2.27	114.3
4.54	116.9
Total variation	<3.9%

TABLE 10—*Fenestration annual energy usage.*

Glass Type[a]	Code Reference Building, kWh/year/m^2	Actual Design Building, kWh/year/m^2
0.64 cm, $SC = 1.0$	112.5	112.1
0.64-cm bronze tint, $SC = 0.7$	108.9	108.7
0.64-cm bronze reflective, $SC = 0.43$	105.8	105.5
Koolshade (external shading), $SC = 0.25$	103.6	103.3
	Total reduction, 8.1%	
	Bronze reflective, 5.9%	

[a]SC = glass shading coefficient.

design building, since this was one of the significant parameters addressed by the architectural and engineering consultants on the project. The shading coefficient used for each type of glazing is also shown in the table. Although an external venetian-type shading device such as Koolshade was never considered on the project, its theoretical value in reducing energy consumption is shown. Maximum energy savings of approximately 6% occur as a result of using a 0.64-cm bronze reflective glass instead of ordinary 0.64 clear plate glass. The energy savings are more than 8% when an external shading device is used.

A significant parameter, one which greatly affects the annual energy usage, is the fresh-air ventilation required by a building. This paper will not go into the altruistic virtues of reduced ventilation requirements nor discuss the optional energy-conserving alternatives available to the building designer. Needless to say, the 2.4-L/s limitation imposed by the designer is a minimum value at best and should probably be supplemented with high-efficiency filtration or the use of charcoal filters. Table 11 shows the rapid increase in annual energy consumption as the ventilation requirement is increased. The calculation did not include 16.5 L/s per person, since this value exceeded the building's annual energy budget of 132 kWh/year/m^2. Note that Refs *2* and *3* suggest a recommended ventilation range of 7.5 to 12.5 L/s per person, with 7.5 L/s for each person for office buildings as an absolute minimum.

As was previously discussed, venting return air through the ceiling light fixtures will definitely reduce the cooling load to the conditioned space. This will usually result in lower fan operating costs because of the reduced supply air-flow requirements imposed on the system. The cooling load presented by the lights is still accountable to the mechanical plant refrigeration system. What happens is that this heat load is added to the return air, thus changing the coil entering air conditions. Should the light fixtures be vented, only some of this heat may escape the refrigeration system through natural or forced ventilation of attic spaces. This conclusion is substantiated by the results shown in Table 12. The results, which show that the values for fixtures that are neither vented nor ducted are somewhat less, remain a bit obscure. The total variation is less than 2½%. Although it is agreed that venting and ducting light fixtures may reduce the installed system capacity and may, indeed, reduce the annual energy consumption significantly in certain climatic regions, the significant reductions expected from this consideration were not achieved for this typically semitropical office building design.

The increases in annual energy usage were perhaps most dramatic in the case of varying the lighting level from the prescribed 29 W/m^2. Former lighting levels have been on the order of 54 to 75 W/m^2, and, as is shown in Table 13, even a lighting level of 43 W/m^2 exceeds the annual energy budget of 132 kWh/year/m^2. At 43 W/m^2, the increase in energy usage for the code reference building is more than 33%.

The final parameter investigated in this study was the variation in annual energy usage as more efficient heating and cooling mechanical plants were used in the building design. Table 14 shows the code reference building energy consumption for the design heat pump [coefficient of

TABLE 11—*Fresh air ventilation.*

Ventilation, L/s/person	Annual Energy Usage, kWh/year/m^2	% Increase
2.4	112.5	. . .
3.1	113.4	0.84
4.7	115.0	2.24
8.0	118.1	5.04
10.9	121.3	7.8
15.6	126.3	12.3

TABLE 12—*Return air through fixtures.*

Installation Type	Annual Energy Usage, kWh/year/m^2
Vented and ducted	112.5
Vented only	111.2
Neither	109.6
Less than 2½% variation	

TABLE 13—*Annual energy usage as a function of lighting level.*

Lighting Load (Including Ballast), W/m^2	Annual Energy Usage, kWh/year/m^2	Percent Increase
29	112.5	. . .
32	121.2	7.8
37	132.6	17.9
43	149.6	33.1

TABLE 14—*Mechanical HVAC equipment annual energy usage.*

Reference System	Annual Energy Usage, kWh/year/m^2 (%)
COP[a] 2.6 to 2.79 EER[b] 7.5 to 7.99	112.5
COP 1.0 electric strip heat	115.3 (2.5)
EER 6.8 to 6.99	117.8 (4.8)
EER 8.55 to 8.99	107.1 (−4.8)
EER 9.5 to 9.99	102.7 (−8.7)
EER 10.5 to 10.99	98.6 (−12.3)
EER 12 and above	94.2 (−16.3)

[a]COP = coefficient of performance.
[b]EER = energy efficiency ratio.

performance (COP) = 2.6 to 2.79 and energy efficiency ratio (EER) = 7.5 to 7.99]. The next variation shows a 2½% increase if electric strip heat is to be used instead of an air-to-air heat pump. The reason for this is that heating is very seldom used in controlling the thermal environment of the office building. The recurring fact is that the lighting and occupancy load is used and is sufficient to provide heating for the building in the temperate winter months. This is substantiated by the fact that, for many years, commercial office and light manufacturing plants in central Florida did not even contain a heating plant. This practice was even more prevalent in the southern Florida district. The table also shows the reduced energy consumption as equipment energy efficiency ratios (EERs) are increased. This should not be considered a panacea since any particular package air conditioner can be made to show an increased EER if the saturated suction temperature is increased. This may not be desirable since, especially for high-humidity regions like Florida, occupants would then arbitrarily lower the thermostats, thus operating the building at a much lower dry bulb temperature. The increased running times under such conditions would cancel out any savings suggested by the higher EER machine.

Conclusions

This study adopted the Florida Model Energy Code manual calculations for determining the annual energy usage of a commercial building and demonstrated that the procedure can be used for rapid estimates of building energy usage as various design parameters are changed. The study showed, as expected, that significant changes in energy consumption occurred as the ventilation and lighting level were altered. Energy savings from extremely well-insulated walls and ceilings were not as large as expected because the occupancy heat gain and the lighting heat gain in semitropical climates usually provide more than sufficient heating for the building environmental thermal control. In effect, higher-transmission heat losses will provide some free cooling during the cooler winter months in semitropical climates.

References

[1] "Model Energy Efficiency Code for Building Construction," Florida Department of Community Affairs, Tallahassee, FL, September 1982.

[2] *ASHRAE Handbook of Fundamentals*, American Society of Heating, Refrigeration, and Air Conditioning Engineers, Atlanta, GA, 1981.

[3] Rudoy, W. and Cuba, J. F., *Cooling and Heating Load Calculation Manual*, ASHRAE GRP 158, American Society of Heating, Refrigeration, and Air Conditioning Engineers, Atlanta, GA.

T. J. Cardenas,[1] *Merle McBride,*[2] *George Barney,*[3] *Bruce Wilcox,*[4]
and Arthur Johnson[5]

Recommendations for the Development of a Residential Energy Standard

REFERENCE: Cardenas, T. J., McBride, M., Barney, G., Wilcox, B., and Johnson, A., **"Recommendations for the Development of a Residential Energy Standard,"** *Thermal Insulation: Materials and Systems, ASTM STP 922,* F. J. Powell and S. L. Matthews, Eds., American Society for Testing and Materials, Philadelphia, 1987, pp. 21–39.

ABSTRACT: In June 1983, the American Society of Heating, Refrigerating, and Air Conditioning Engineers (ASHRAE) Standards Committee appointed an *ad hoc* committee to revise Standard 90 for Residential Buildings. The charge to the *ad hoc* committee was to start the technical thinking process for the creation of a separate ASHRAE Standard 90 for residential buildings and to make recommendations to the ASHRAE Standards Committee as to how such a standard might best be produced.

This paper discusses the elements of the technical review process which included a quantitative assessment of the envelope requirements of American National Standards Institute (ANSI)/ASHRAE/Institute of Environmental Sciences (IES) Standard 90A-80 and comparisons between this standard and to the requirements of other standards and energy conservation guidelines.

During the review of Standard 90A-80, many requirements were identified that should be incorporated into the development of a new residential standard. In this regard, criteria for the technical and economic basis of a residential standard are outlined and discussed.

Finally, recommendations for the development of a methodology for producing a new data base for residential building envelopes, the means for updating the new data base in the future, new simpler compliance procedures, and more accurate trade-off compliance procedures are presented.

KEY WORDS: energy conservation standards, energy-efficient design, residential building envelopes, standards assessment, standards methodology, building performance optimization, code compliance procedures, thermal insulation

Within the past decade since Standard 90 for residential buildings was first introduced by the American Society of Heating, Refrigerating, and Air Conditioning Engineers (ASHRAE), advances in construction technology and building energy analysis have generated industrywide debate concerning the need for alternative design energy requirements for commercial and residential buildings. The debate has been fueled by several issues. On the technical side, comparative analysis of building energy load profiles supported the need for separate design strategies integrated with a broadened approach to varying climate parameters. Similarly, cost/benefit studies demonstrated the need for an economic basis for selecting and setting the optimum levels of building performance. Finally, within the regulatory domain, the appearance of alternative compliance approaches made evident the need for accurate, flexible, and simpler trade-off procedures.

[1]Senior research associate, Steven Winter Associates, Inc., New York, NY 10001.
[2]Senior engineer, Owens-Corning Fiberglas, Granville, OH 43023.
[3]Portland Cement Association, Skokie, IL 60077.
[4]Staff member, Berkeley Solar Group, Berkeley, CA 94703.
[5]Staff member, NAHB Research Foundation, Rockville, MD 20850.

To evaluate these broad industry concerns, the ASHRAE Standards Committee created the *Ad Hoc* Committee to Revise Standard 90 for Residential Buildings (*Ad Hoc* 90-R). The purpose of *Ad Hoc* 90-R was to provide ASHRAE with recommendations on the comprehensive, integrated, and coordinated development of a new residential building energy standard. The recommendations of *Ad Hoc* 90-R presented and discussed in this report include the following:

(*a*) technical and economic considerations to provide the basis for developing the standard,
(*b*) approaches and procedures for converting existing and new data base resources into a standard format, and
(*c*) engineering research directions and implementation strategies needed to accelerate the introduction and transfer of new technology and building construction practice into the energy standards development process.

Assessment of Standard 90A-80 on Residential Buildings

The current standard for residential buildings, Standard 90A-80, which was developed by the American National Standards Institute (ANSI), ASHRAE, and Institute of Environmental Sciences (IES), was reviewed by the *ad hoc* committee, which determined that there was a distinct need for ASHRAE to develop a new residential standard. Three major issues were considered in the review. First, how was the standard developed? In other words, the technology base used to develop the standard was reviewed. Second, this standard was compared with others for the level of energy conservation required. Based on this review, *Ad Hoc* 90-R formulated specific recommendations to upgrade and improve the development and implementation of a new residential standard.

Comparisons to Standard 90A-80

As a portion of the Standard 90A-80 review, a comparison was made with other standards and energy conservation guidelines, including the following:

(*a*) Housing and Urban Development Minimum Property Standards (HUD-MPS) [*1*],
(*b*) Farmers Home Administration (FmHA) [*2*],
(*c*) Title 24, California Energy Code (CEC) [*3*],
(*d*) Pennsylvania Energy Code (PENN),
(*e*) Housing Industry Dynamics Survey (HID) [*4*], and
(*f*) Thermal Performance Guidelines of NAHB (TPG) [*5*].

The basis for the comparisons was an investigation of the ceilings, walls, floors, and slabs for a typical 110-m^2 (1200-ft^2) ranch house with a 15% glazing area and a 4% door area. The results of the comparisons are shown in Figs. 1 through 4. The most immediate observation from these comparisons is that Standard 90A-80 requirements for ceilings, walls, and floors are the least stringent.

Requirements for a New Residential Energy Standard

Subsequent to the comparative analysis of Standard 90A-80, many requirements were identified that should be incorporated into the development and implementation of a new residential standard:

1. The level of energy conservation within Standard 90A-80 needs to be upgraded.
2. A basic methodology or framework on which to develop the new standard needs to be developed. The methodology must also be applicable for updating the standard in the future.

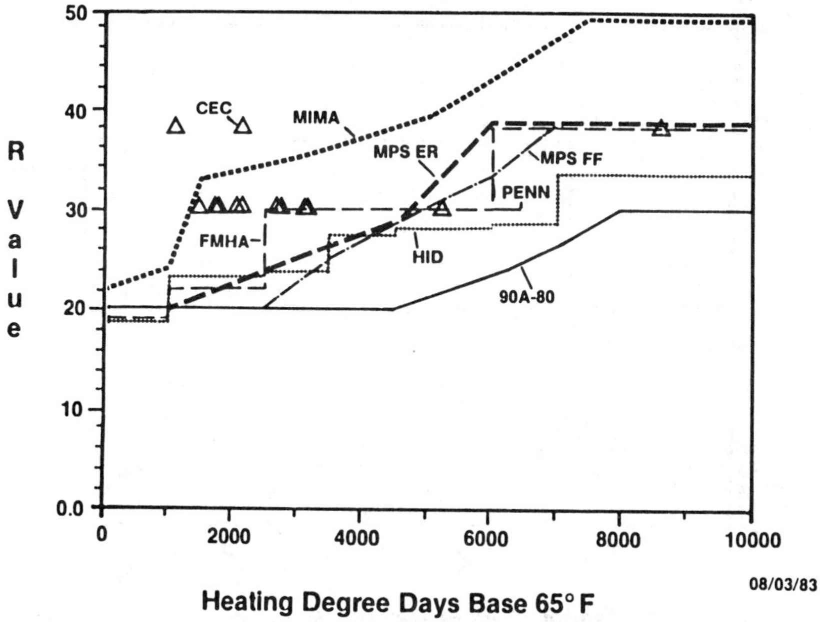

FIG. 1—*Comparative assessment of major code requirements for ceilings.*

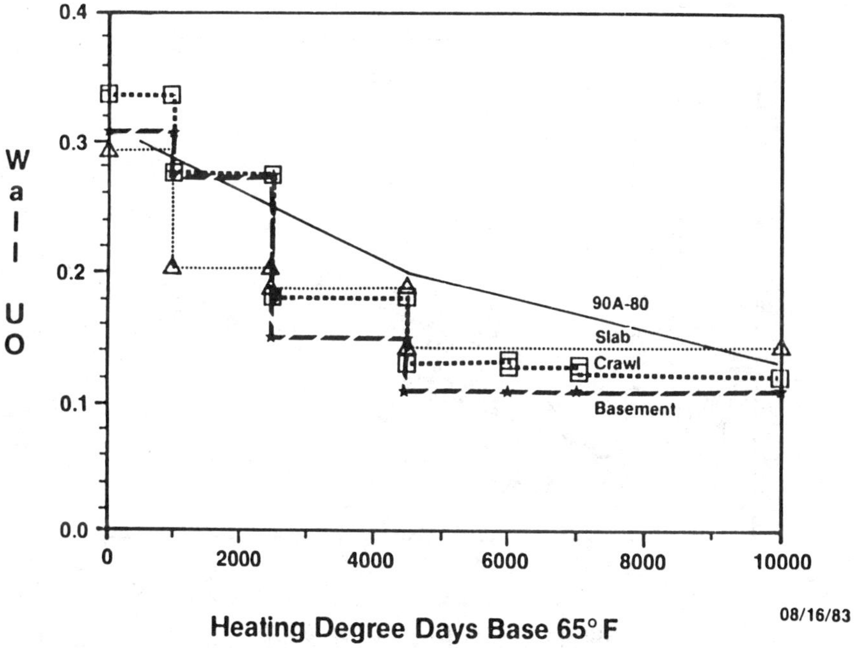

FIG. 2—*Comparative assessment of major code requirements for walls.*

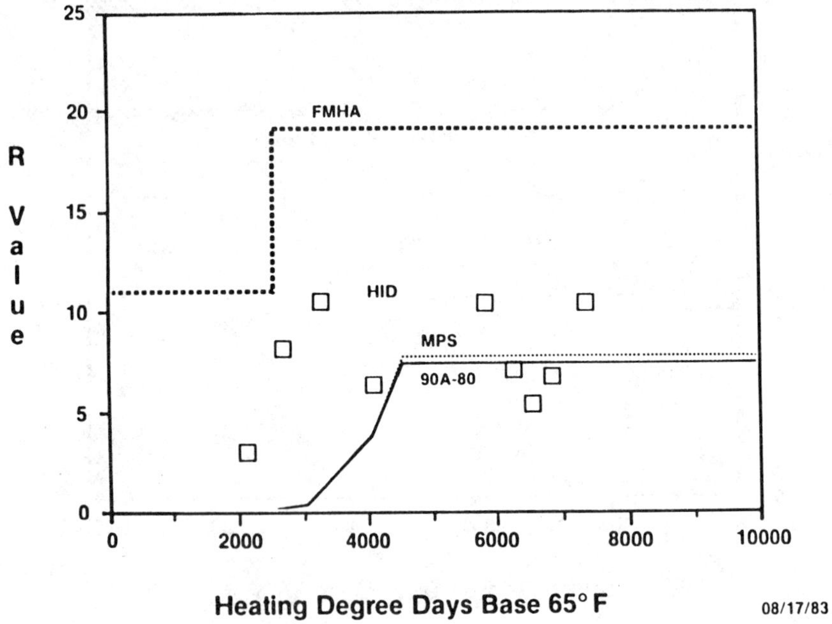

FIG. 3—*Comparative assessment of major code requirements for floors.*

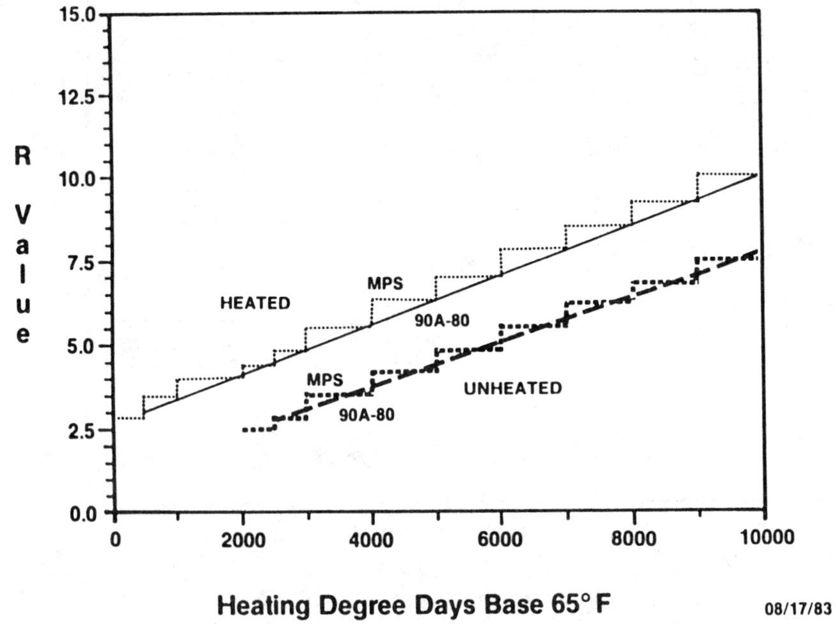

FIG. 4—*Comparative assessment of major code requirements for slabs.*

3. Simpler, flexible, and more accurate compliance and trade-off procedures need to be formulated.

Basis for a New Residential Standard

Based on the broad requirements identified during the review process, *Ad Hoc* 90-R developed recommendations focused on the comprehensive, integrated, and coordinated development of a new residential building energy standard. A purpose and scope were defined and outlined; and criteria for the selection of a data base and the development of a basic methodology were established. The following sections summarize the rationale and basis for the criteria recommended.

Technical Approach

The technical foundation of standards dealing with energy issues is typically a data base, developed by multiple simulations using an energy analysis procedure [*6*]. The selection of an energy analysis procedure raises two critical issues which are interrelated. The first issue concerns the modeling of energy transfer processes in a building; the second relates to modeling climate parameters that influence energy transfer. A variety of procedures have been employed in the past which generally utilize either steady-state calculation techniques or dynamic processors performing hour-by-hour simulations. In general, hour-by-hour dynamic simulations are preferable as a technical foundation because of their expected greater accuracy in projecting both absolute energy requirements and savings due to the use of energy conservation measures.

Energy Analysis Program Criteria—A number of candidate programs exist for use in developing a data base; the program selected should have the following characteristics:

(*a*) be in the public domain so as to be subject to industry review,
(*b*) be technically supported by validation efforts or using generally accepted algorithms for energy use projections,
(*c*) be able to provide data for a sufficient number of locations in the United States,
(*d*) be able to simulate a wide range of conservation techniques, and
(*e*) be able to analyze residential structures of various configurations and designs.

Data Base—A comprehensive effort to edit, updata, unify, and coordinate expansions of existing data base resources should be undertaken with input from cognizant technical committees; the data base should include the following:

(*a*) *definitions*—the current ASHRAE 90A-80 definitions will have to be modified or expanded in conjunction with the new standard;
(*b*) *equations*—provisions for simplified calculation procedures;
(*c*) *climatic data*—provisions for heating and cooling degree-days; and
(*d*) *performance data*—provisions for, or reference to, revised ASHRAE handbook data on components, materials, and building systems.

Economic Approach

Although rising energy costs have created incentives for consumers to capitalize on energy-conserving investments, the declining affordability of housing resulting from escalating material, labor, and interest costs generates an environment for improved standards that appears somewhat counterproductive. Undoubtedly, the upgrading of an energy standard redefines the thermal performance level of a given class of buildings; however, the variable impact on investment costs and benefits to consumers have not been fully appreciated in the current process of revision of ASHRAE Standard 90A-80.

An assessment of industry guidelines and state energy programs indicates the increasing role of economic analysis tools in the development of energy standards. While considerable analysis work has been performed to assess the economic impact of energy standards on both consumer and society, generalized solutions and specific economic parameters for setting residential standards are neither agreed upon nor deemed applicable for all residential construction types. The lack of consensus on a specific tool or parameters, however, does not preclude the necessity for a comprehensive economic approach to be applied in the development of cost-effective conservation levels for residential buildings. As aptly stated in the *National Program Plan,* the lack of consensus "does not diminish the value of, nor the need for, economic analysis in such problems. On the contrary, it points out the need for flexible calculation procedures, improved component and building energy performance data, credible forecasts of future regional energy price trends, and the recommendation of the most cost-effective methods for reducing energy requirements in new and existing buildings" [*7*].

Analytical Tool Selection—The choice of methodology involves a trade-off between accuracy and complexity. Life cycle cost analysis, cost/benefit ratio, and payback are the most common economic analysis tools used in conjunction with building energy analysis [*8,9*]. The analytical tool selected should have the following characteristics:

(*a*) the ability to measure the incremental costs and savings in energy of a conservation investment over selected time frames,

(*b*) the ability to identify the break-even point where costs equal benefits for the conservation options considered or the point where a positive cash flow can be achieved, and

(*c*) the ability to compare several energy-saving options competing for the same level of investment.

Economic Parameters—The authors recommend that a consumer versus a societal viewpoint be used in the development of the economic model. The implications of this recommendation are twofold. First, energy costs should be based on average costs to the consumer rather than incremental costs to the utilities. Second, the effect of investments in energy conservation should be judged with respect to their impact on the affordability of the home to the consumer. This means that the net increase in mortgage costs should be offset by energy savings early in the life of the investment. It is not in the best interests of the consumer or society to require investments that will adversely affect the consumer's mortgage qualifications or net cash flow during the early years of the mortgage [*9*].

1. *Economic life*—The impact of energy conservation investments on mortgage costs should be determined based on the length of a typical mortgage, currently 20 to 30 years. The authors recommend that a 25-year life be used.
2. *Time horizon*—Short time horizons that have proven acceptable to residential consumers have typically ranged from one to five years and in some instances as high as seven to eight years. The authors recommend that the time basis for achieving consumer positive cash flow be a short time span, preferably one year.
3. *Energy costs and energy cost inflation rates*—The authors suggest that these be determined on a regional basis and be used as a multiplier for the energy index discussed elsewhere. Fuel escalation rates can be established on either a regional or a national basis.
4. *Discount rates*—This rate should include the cost of capital for the consumer for his or her down payment and the tax-adjusted mortgage rate. This assumes a down payment percentage and a marginal tax rate. Alternatively, interest tax shields could be included as a benefit, and an unadjusted mortgage rate could be utilized.

5. *Incremental cost of an option*—The incremental cost of an option, including installation and maintenance, must be established. These could be estimated on a national basis and adjusted by a construction cost index on a regional basis if deemed necessary for accuracy.

6. *Taxes*—Taxes are based, directly or indirectly, on the market value of a home. Additional energy conservation features may increase the value of a home, and this should be accounted for in the analysis. Regional variations in tax rates should also be considered in the analysis.

7. *Salvage value*—Salvage value does not affect annual cash flow and should therefore be omitted in the analysis.

Optimization Approach

Integration of an energy analysis procedure with an economic methodology has been recommended as the basis for determining conservation levels for a new residential energy standard. Implicit with this approach are two economic issues that relate to optimization. The first issue concerns determining optimal conditions that produce a desired level of thermal performance in a given building. The second issue relates to the process of determining optimal design requirements for target performance levels in standards development. The determination of a prototype methodology for selecting optimal levels of energy-conserving practices is a departure from the current ASHRAE Standard 90A-80 approach, which is predicated on "providing minimum requirements for building envelope construction." The development of a new residential standard should preclude provision of minimum requirements; however, such provisions should be determined to be cost-effective. The approach is to select variables in the optimization that will result in cost-effective minimums (see Fig. 5).

Methodology—A target optimum level will either increase or decrease, depending on a particular parameter and how it enters the analysis. To ensure that the optimum represents a minimum, a consistent approach must be used to select the magnitude of the parameters used in the analysis:

1. Parameters that *increase* the optimum when they increase (fuel costs and present worth factors) should be selected at the *lower bound* of the range of acceptable values.

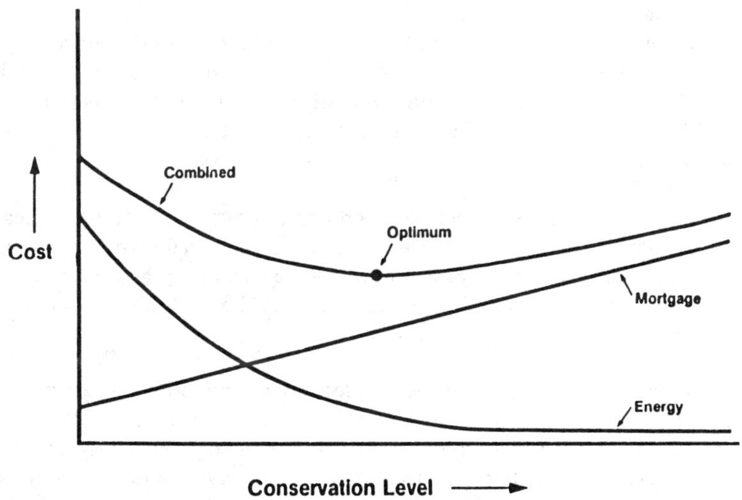

FIG. 5—*Optimization approach.*

2. Parameters that *decrease* the optimum when they increase (equipment efficiencies and labor plus material costs) should be selected at the *upper bound* of the range of acceptable values.

Parameter Selection and Analysis—Depending on which economic analysis procedure is selected, the optimum construction level may depend on nine parameters:

1. Cost of the heating fuel, C_H.
2. Cost of the cooling fuel, C_C.

As the cost of either or both fuels increases, the optimum construction level increases. Adoption of consumer fuel costs, which are lower than societal costs, generates an optimum level that is minimal. To minimize artificial boundaries in optimum construction levels by differentiating construction by regional fuel costs, the use of national cost averages for fuels would generate a desired level of uniformity in buildings located in similar weather zones.

3. Present worth factor for heating, PWF_H.
4. Present worth factor for cooling, PWF_C.

The other parameters that increase the optimum construction level when they increase are the present worth factors, PWFs. The critical variables in determining the PWFs are the time horizon, fuel escalation rates, and discount rate. The time horizon should be selected based on consumer preferences. Fuel escalation rates should be determined for each fuel type based on projections over the time horizon. The authors recommend that national averages be used. Discount rates should be selected based on current fixed mortgage rates, after typical tax benefits to the consumer are taken into account.

5. Cost of materials and labor, k.

The impact of material and labor costs, k, on mortgage costs must be considered in determining the economic optimum. Increases in the material and labor costs decrease the optimum construction level. Since k changes with construction, it introduces discontinuities into the analysis. For example, with ceilings, k changes when the truss system is modified to a raised truss for accommodating increased insulation levels. For walls, k changes if the cavity dimensions change (2 by 4 to 2 by 6) or if sheathing is used. Changes of this type create discontinuous functions but must be included in the analysis to properly identify optimums.

6. Heating equipment efficiencies, AFUE, SCOP.
7. Cooling equipment efficiencies, SEER.

These two parameters decrease the optimum construction level when they increase. Selecting the heating and cooling equipment efficiencies can be simplified by just using the minimums currently set forth in ASHRAE Standard 90A-80. Equipment with improved performance characteristics would lower the optimums but at an increased cost for the equipment.

8. Heating load, E_H.
9. Cooling load, E_C.

The last two variables that have impact on the optimum construction level—the heating and cooling loads—are a function of the heating and cooling climates. In determining the optimum, heating and cooling would remain independent variables. Thus, the optimum construction level varies with both heating and cooling climates, and this should be reflected in both the development and the presentation of the standard.

Analysis—Consistency must be considered in selecting the values for specific parameters so that discontinuities are avoided in the development of the standard. Values of parameters used in the analysis should be selected to minimize the sum of energy and mortgage costs to the consumer early in the period of home ownership. This will serve the dual purpose of saving energy and making housing affordable, both of which are in the best interests of the consumer and the nation.

To illustrate the consistency between optimization and minimums, as it relates to specific parameters, a simplified analysis was completed. See Appendix I for the complete derivation of the optimum insulation level in a residence.

Building Design Requirements and Trade-Off Approaches

Based on the requirements identified in the review process, *Ad Hoc* 90-R identified four alternative methods of developing building design requirements and trade-off procedures:

(*a*) a prescriptive approach, which lists component *R* values to be developed as a set of basic requirements;
(*b*) alternative packages, which contain prescriptive requirements for passive solar homes and alternative forms of construction;
(*c*) a point system, which accurately allows simplified trade-off calculations to be performed; and
(*d*) expansion of a systems analysis approach similar to ASHRAE Standard 90B, Section 10–11, since it provides opportunity for new and innovative design.

Prescriptive Approach

The purpose of prescriptive requirements is to provide a simple procedure to determine constructions that would demonstrate compliance with the desired level of building energy performance set by the standard. Development of prescriptive requirements is considered necessary to simplify the understanding, use, and compliance of the standard for typical builders/contractors and code officials. A prescriptive approach can be interpreted directly, based on construction, without requiring any elaborate calculations. Once developed, the prescriptive requirements also serve as the basis for other compliance procedures that must exhibit equal thermal performance.

Analysis and Methodology—The prescriptive requirements should be determined using a detailed energy analysis procedure in conjunction with an economic and optimization approach, as outlined previously.

Format/Data Base/Components—Prescriptive requirements should be developed for each of the items in the envelope, including, but not limited to, the following:

(*a*) ceilings—cathedral, vented attics;
(*b*) walls above grade—frame, masonry;
(*c*) walls below grade—basements and crawl spaces;
(*d*) glazing—number of panes, area, and sash;
(*e*) doors;
(*f*) slab—heated and unheated; and
(*g*) infiltration control.

The prescriptive requirements should be presented in a table using variables to characterize both the heating and cooling severity. Heating degree-days to a base of 18.3°C (65°F) are recommended because they have been used extensively. Cooling can be characterized by cooling degree-days to a base of 18.3°C (65°F) or cooling hours over 26.6°C (80°F). See Fig. 6 for an illustrative format.

Alternative Package Approach

The diverse selection and configuration of building construction types suitable for residential occupancy coupled with varying climate conditions, fuel sources, equipment efficiencies, and

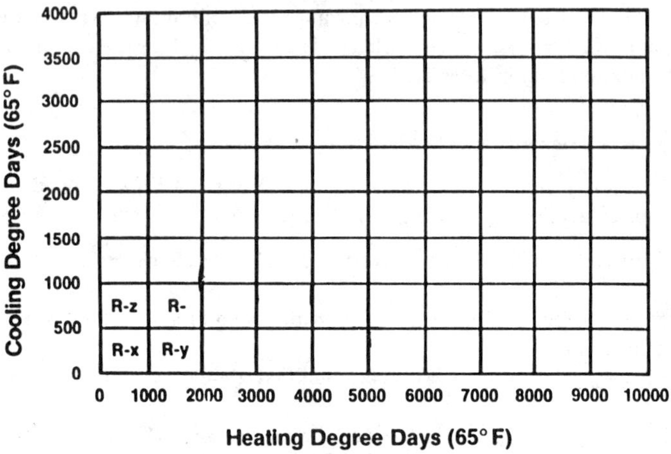

FIG. 6—*Illustrative prescriptive format.*

material and labor costs invariably limit the degree to which a single set of prescriptive energy standards can be applied equitably. Optimal standards developed for one type of residential building may incur significant cost and performance penalties when applied to several other generic types of residential construction. Historically, these types have included mobile, masonry, adobe, log, or solid timber construction. Hybrids of these, along with lightweight frame construction, are becoming increasingly common in light of passive design strategies for both heating and cooling.

Typically, efforts to broaden the applicability of prescriptive measures have resulted in the adoption of various methods at federal, state, and local levels. Generally, the alternative models are based on balanced envelope trade-offs, equivalency packages, or simple modifiers. More recently, attempts to standardize passive design strategies, including minimum requirements for thermal mass, maximum percentage south glazing, and shading coefficients have been pioneered.

Analysis and Methodology—Component packages should be developed in conjunction with the minimum requirements established for the prescriptive approach. Calculated annual energy consumption for the component packages should be no greater than that calculated for the basic prescriptive requirements. Components packages should be determined in consideration of climatic variations, building type, type of construction, and cost-effectiveness.

Format/Package Categories—For each building type covered by the standard, component packages should be offered in the following broad categories:

Common trade-offs—Provision of component packages allowing for common variations from basic prescriptive requirements in building envelope design, including allowable increases or decreases in glazing area and insulation level.

Alternative construction types—Provision of component packages specifying acceptable energy-conserving design alternatives for the construction types not specifically covered in the basic prescriptive requirements, including masonry and adobe homes, log or solid timber homes, and manufactured housing.

Passive solar construction—Provision of component packages for passive solar construction, including sun-tempered, direct gain, Trombe wall, and attached sun space construction. Component packages should specify requirements for several parameters critical to passive construction, including insulation levels above and below grade, the percentage of glazing at various orientations, night insulation and shading, and the envelope or interior thermal mass area.

These packages should be determined considering the impact of the proposed designs on both heating and cooling. Provisions for combinations of passive strategies should also be addressed. See Appendix III.

Point System Approach

In the regulatory domain, the development of energy codes has resulted in widespread use of simple energy analysis methods. Recently, performance codes have replaced prescriptive codes in some jurisdictions. Under this type of regulation, an allowed energy budget is stipulated, and the designer is required to demonstrate compliance with an energy analysis of some sort. Approved analysis methods are usually developed in parallel with the code, so that a standardized compliance scheme is available [*10–11*].

A point system is a simplified performance method that allows designers to determine whether a design will use more or less energy than a prescribed design without a computer simulation. To make any regulation as flexible as possible, a point system can be developed in addition to more rigid prescriptive standards. Point systems generally allow building component trade-offs, by denoting positive and negative points for building components, with the zero line representing the energy budget for a base prescriptive package, and allow for variation of energy-efficient design strategies. Point systems have been proposed and adopted by cities, states, and utilities. These point systems or rating systems vary in degree of accuracy and flexibility. Attempts at field validation of individual tools and rating systems have been extremely limited. Some of the states that have adopted points or rating systems are California, Pennsylvania, Florida, Wisconsin, and Minnesota, in addition to several large utilities. The development of a point system approach in conjunction with the residential standard maximizes the range of design solutions over that possible using a prescriptive or alternative package approach [*10*].

Analysis and Methodology—A point system can be derived from the results of hourly computer simulations of variations in conservation features of a prototypical "base" house in a given climate. To give point credit for any measure, a data base of single parameter variations needs to be developed. A curve of energy consumption versus parameter values can be established to show the change in energy consumption versus the change in parameter value. The energy budget level from which to calculate positive or negative point values can be set at any level, depending on climate, energy costs, and minimum acceptable conservation levels [*11*].

Format/Component Trade-Offs—The point system data base can be developed for different building types and take into account differences in internal loads and in surface-to-volume ratios. Roof and wall insulation, infiltration control, glazing type, area and orientation, thermal mass, and various heating, ventilating, and air-conditioning (HVAC) systems can all be considered variables in the system. Positive or negative point assignments can reflect net energy savings or losses, respectively, for each design variation. With a zero-value reference point as defined by the base building, variations are easily evaluated (see Appendix IV).

Systems Analysis Approach

The flexibility within ASHRAE Standard 90A-80, which allows a designer to "deviate from the specific design criteria" of 90A by demonstrating that the deviation "will result in annual energy consumption equal to or less than that resulting from compliance" with the intended level of conservation required by the standard, encourages innovative energy-conserving design and construction in residential building.

The development of the new residential building standard should provide for the incorporation and development of a systems analysis approach to residential building design.

Analysis Methodology—The analysis methodology contained within ASHRAE Standard 90B—75, Section 10, is appropriate for the new residential standard and is recommended for adoption.

Energy Analysis/Performance Criteria—Compliance with this section requires performance of a comparative annual energy analysis for the proposed alternative design with a design that meets the standard. Proposed designs may include alternative provisions for the envelope, air leakage control, HVAC equipment, double-hung windows (DHW) equipment, and active and passive solar, provided the preceding condition is statisfied.

Air Leakage

Current estimates indicate that infiltration can account for up to 40% of the energy use in residential buildings [*1*]. An assessment of current standards, proposed guidelines, and test data reveals a large potential for energy conservation through the specification of requirements that would effect the reduction and control of air leakage in residential buildings [*1-3, 7, 8*].[6] Target rates of air infiltration, as well as nomenclature regarding 'levels of tightness' vary; however, quality construction practices and products that can effectively reduce the air infiltration rate to 0.2 air changes per hour (ACH) [*7*] are becoming common in new residential construction (see Fig. 7).

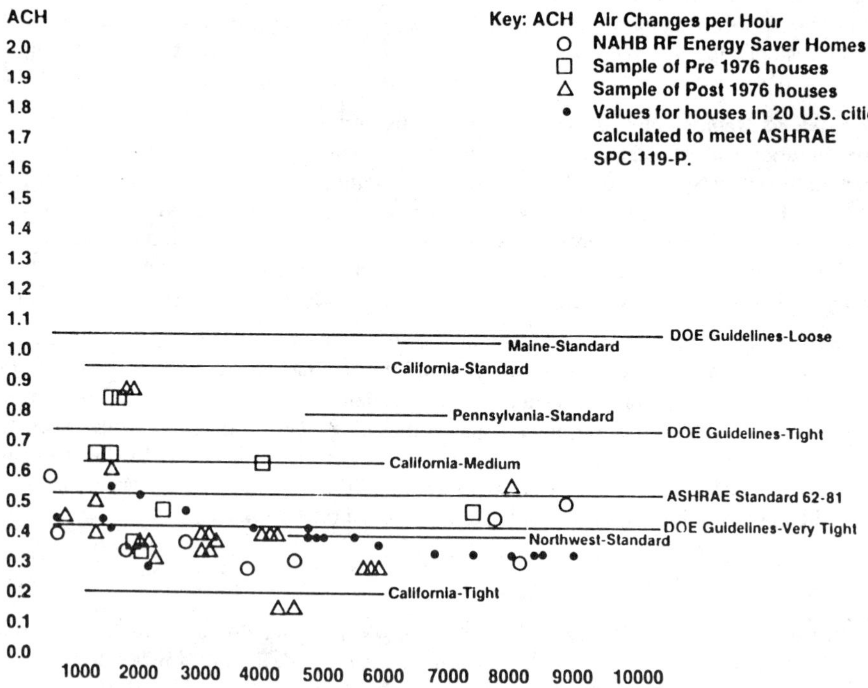

FIG. 7—*Comparative assessment of major code requirements for air leakage/infiltration control.*

[6]"Pennsylvania Energy Code," Pennsylvania.

Methodology—Minimum requirements and acceptable rates of infiltration in residential buildings for the prescriptive, alternative package, point system, or systems analysis approaches should be selected as follows:

Prescriptive approach—Minimum requirements for the reduction of infiltration should be specified for all residential buildings covered by the proposed standard. This is similar to the present ASHRAE Standard 90, Section 4.5, approach. However, the criteria used to determine the target rates of infiltration should also include consideration of the following: climate severity, including such variables as total degree days, wind speed, and stack effect.

Package approach—The package approach requires the selection of a base house and the testing (by calculation or by field verification) of various combinations in order to determine, in advance, compliance with the target prescriptive criteria of the standard. Various infiltration control packages and elements of such packages would be a part of the combinations so tested or simulated. Package elements targeting various levels of air leakage control should be developed and a simplified methodology adopted. Specific leakage area targets for buildings provide such a methodology.

Point system approach—By assigning a zero point value to the minimum criteria included in the prescriptive approach and relative points to the other elements of the package upgrades from the prescriptive approach, various trade-off provisions can be developed, as discussed previously.

Systems analysis—Depending on the analysis tool selected, the treatment of air leakage should be in the form of an air infiltration value that varies through the heating and cooling season depending on weather. Simpler analysis tools will use a long-term seasonal average infiltration rate, which should be different for the heating and cooling analysis. More complex analysis tools will use an equation that will vary the infiltration on an hourly, daily, or weekly basis, depending on the weather. The initial value for the infiltration function or the long-term seasonal average should be based on measured data.

Analysis—Figure 7 illustrates the variation in levels of airtightness as defined or calculated for current state codes and industry guidelines. What is regarded as a "standard" level of airtightness varies from 0.4 to 1.0 ACH; "tight" levels vary from 0.2 to 0.7 ACH.

In developing the standard, a general set of construction characteristics, as illustrated in Appendix II, should be empirically analyzed to establish a base case for standard infiltration rates. These correlations can be used to rank other construction characteristics that either increase or decrease the base rate in relation to energy savings achieved as a result of implementation of a given set of construction characteristics and can be utilized to establish minimum requirements for air leakage control that are optimal and cost-effective.

It should be noted that correlations between measurements of a small number of houses and predictions about the general behavior of all new "similar" houses, such as configuration of window area, microclimate, floor area, and volume, should be done with caution. An illustrative case in point is Pennsylvania's proposed ruling. A quantitative approach based on prescribed component leakage values was introduced into the proposed standard. Subsequent analysis and testing of the method by a large residential builder demonstrated that a calculated infiltration rate, based on a set of prescribed construction characteristics that were implemented, far exceeded the infiltration rate obtained by testing.

As proposed earlier, an energy analysis program coupled with an economic analysis should be utilized to establish minimum requirements. The methodology adopted to determine air leakage control measures should be consistent with the former recommendations. Particular attention should be given to the items enumerated here:

(*a*) variable load—differences in heating and cooling loads relative to high, medium, and low infiltration rates, climate severity, and wind speed;

(*b*) differential component leakage—window-related differences in air leakage due to sash types, air gaps, glazing area, orientation, movable insulation, and the use of reflective or absorptive glazing;

(*c*) differential volume/exposed wall area—volume-related differences due to building height (one-story, two-story, and so on); differences related to house geometry and configuration (middle town house versus end town house, multiple setbacks, and corners); and

(*d*) nomenclature—in conjunction with the work undertaken by ASHRAE SPC 119P, and the requirements of ASHRAE Standard 62-81, Ventilation for Acceptable Indoor Air Quality, inconsistencies should be avoided, or at least minimized, to facilitate analysis, understanding, and implementation.

Summary of Recommendations and Implementation Tasks

Rationale for a Special Project Committee

Development of a new residential standard, as outlined, is considered beyond the scope of the *Ad Hoc* 90-R committee. Coordinated comparative assessment of existing data base resources as well as expanded research are recommended. The authors strongly recommended that ASHRAE consider funding a special project committee actually to consolidate and develop the required data base necessary for drafting a new standard.

Some of the tasks that should be implemented by such a special project committee include, but are not limited to, the following:

(*a*) coordination and assessment of modeling parameters required for acceptable accuracy in the modeling of the energy transfer processes in building envelope subsystems (HVAC and DHW) and whole buildings;

(*b*) coordination and determination of parameters required for the economic analysis procedures used to optimize energy conservation measures for various types of construction and building configurations used in the residential industry;

(*c*) coordination of experimental field demonstration and modeling program data base resources to provide comprehensive assessment and comparison of the effects of building envelope mass on the seasonal and annual energy requirements and life cycle costs of residential buildings;

(*d*) coordination of data base resources on laboratory and field studies of below-grade envelope construction;

(*e*) coordination and comparative assessment of technical analysis and field studies on the energy transfer performance of windows and skylights; and

(*f*) data base coordination of current guidelines and standards for controlling air infiltration, moisture transfer, and indoor air quality in the design and construction of residential buildings.

Synopsis of Recommended Approaches

The new residential standard should be based on an energy analysis procedure integrated with a simplified life-cycle cost analysis (annualized cash flow analysis) to produce minimum levels of energy conservation that are cost justified.

Four alternative methods of compliance are recommended:

1. A prescriptive approach, which lists component *R* values, should be developed to set the basic requirements.
2. Alternative packages that contain prescriptive requirements for passive solar options should be developed.

3. A point system should be developed that accurately allows simplified trade-off calculations to be performed.
4. A system analysis approach (Standard 90B—Section 10) should remain an integral part of the new standard and should be expanded, since it provides opportunity for new and innovative designs.

Provisions to reduce air leakage should be expanded and developed in conjunction with the four levels of compliance.

Acknowledgments

The authors wish to acknowledge their gratitude to Frank Powell of the Center for Building Technology at the National Bureau of Standards, who served as chairman of the *Ad Hoc* 90-R committee. His leadership, participation, and support fostered and engendered a spirit of cooperative endeavor, which enabled diverse segments of the building industry to collaborate in the preparation of this technical document. In addition, many individuals and organizations too numerous to name are acknowledged for their industry participation and sponsorship of the activities that made this document possible, in particular the National Association of Home Builders, the Mineral Insulation Manufacturers Association, the National Forest Products Association, the National Concrete Masonry Association, the Log Homes Council, the Department of Housing and Urban Development, and the Council of American Building Officials.

APPENDIX I

Optimization of Costs Approach

The basis for optimization is to minimize the consumer's cash flow for mortgage and energy costs in the early period of home ownership. For simplicity, the authors recommend that the full costs of investments in energy conservation be amortized over the life of the mortgage.

Cash flow analysis is defined as

$$\mathrm{CF} = C_H + C_C + \mathrm{MC} \tag{1}$$

where

CF = cash flow over a selected time period,
C_H = cost of heating over the selected time period,
C_C = cost of cooling over the selected time period, and
MC = mortgage costs over the selected time period for the energy conservation option (assuming the option is 100% mortgage financed).

Heating costs are calculated from the equation

$$C_H = \frac{E_H C_h}{\mathrm{AFUE}}\,(\text{selected time period}) \tag{2}$$

where

E_H = annual heating load,
C_h = cost of heating energy per therm, and
AFUE = annual fuel utilization efficiency or seasonal coefficient of performance (SCOP) depending on the type of equipment.

A similar equation for cooling costs is

$$C_C = \frac{E_C C_c}{\text{SEER}} \text{ (selected time period)} \tag{3}$$

where

E_C = annual cooling load,
C_C = cost of cooling energy, and
SEER = seasonal energy-efficient ratio.

Mortgage costs can be calculated by

$$\text{MC} = \text{cost of conservation} \times \text{CRF} \times \text{(selected time period)}$$

where CRF = capital recovery factor, defined as

$$\text{CRF} = \frac{i(1 + i)^n}{(1 + i)^n - 1} \tag{4}$$

where

i = mortgage interest rate and
n = length of mortgage in years.

The optimization process consists of first using Eq 1 to determine the cash flow (CF) for a base case. The second step is to increase the construction option (that is, insulation, number of glazings, equipment efficiency, and so on) to the next level and determine the cash flow for this alternative from Eq 1. This process is repeated for additional increments until a minimum level of CF is found. This level of conservation is the optimum.

APPENDIX II

Requirements for Air Leakage Control

All residences subject to this standard must meet the following minimum requirements:

1. Doors and windows (including skylights) between conditioned and unconditioned spaces must be fully weather-stripped.
2. The following openings in the building envelope must be covered, caulked, sealed, or weather-stripped:
   (*a*) exterior joints around window and door frames, between wall soleplates and floors, and between exterior wall siding panels;
   (*b*) openings for plumbing, electricity, and gas lines in walls, ceilings, and floors;
   (*c*) openings in the attic floor (such as where ceiling panels meet interior and exterior walls and masonry fireplaces); and
   (*d*) all other such openings in the building envelope.
3. Windows and doors manfactured off-site must bear a label indicating that they meet appropriate industry standards. Site-built windows and doors are exempt from this requirement.

4. Exhaust fans and other exhaust systems must be equipped with back-draft or automatic dampers. Adequate supply air must be provided in the same space as the exhaust fan to reduce the use of the other parts of the residence for makeup air.
5. Masonry and factory-built fireplaces must have the following:
   (*a*) tight-fitting, closeable metal or glass doors covering the entire opening of the firebox. This requirement may be omitted if such doors would interfere with devices, such as heat tubes, permanently installed in the fireplace that are designed to increase the circulation of heat. The fireplace must be designed so that the doors can be closed when the fire is burning.
   (*b*) a combustion air intake to draw air from the outside of the building directly into the firebox, which is at least 38.7 cm^2 (6 $in.^2$) in area and is equipped with a readily accessible, operable, and tight-fitting damper (the damper may be behind the doors specified in point 5(*a*); and
   (*c*) a tight-fitting flue damper with a readily accessible control. This control may also be behind the doors.

These requirements do not apply to fireplace inserts or free-standing stoves.

Continuous-burning pilot lights and the use of indoor air for cooling a firebox jacket, when that indoor air is vented to the outside of the building, are prohibited. However, gas logs without pilot lights and gas log lighters are allowed.

6. Fossil-fuel-fired heating equipment must have a sealed combustion chamber that does not utilize indoor conditioned air for combustion purposes.

In addition to these minimum requirements, the following requirements for climate-related air leakage must be adhered to, depending on the climate in which the residence is located.

	Heating Degree Days		
	Less Than 2500	2500 to 4000	More Than 4500
Windows	0.5-ft^3/min/ft crack or less	0.34-ft^3/min/ft crack or less	0.25-ft^3/min/ft crack or less
Doors	0.5 ft^3/min/ft^2 or less	0.34 ft^3/min/ft^2 or less	0.25 ft^3/min/ft^2 or less
Walls	sheathing, panel siding, or masonry veneer, 0.75 ft^3/min/ft^2 of wall area or less	air infiltration barrier, 0.5 ft^3/min/ft^2 of wall area or less	air infiltration barrier, 0.25 ft^3/min/ft^2 of wall area or less

Note that these requirements are more illustrative of an approach using component leakage performance as a basis.

APPENDIX III

Package Approach

The following table provides an illustrative format of the package approach as proposed for the Northwest Power Planning Council grid, for 4000 to 6000 heating degree-days.

Component	Sun-Tempered	Passive Solar
	BUILDING ENVELOPE	
Insulation minimums, *R* value		
Ceiling	*R*-38	*R*-30
Wall	*R*-25	*R*-19
Slab floor perimeter	*R*-10	*R*-10
Floor over unconditioned space	*R*-19	*R*-19
Exterior doors	20	20
Glazing		
Maximum *U* value	0.47	0.47
Maximum total area, %	15	no requirement
Minimum effective solar glazing area, %	8	10
Thermal mass	not required	required
	SPACE CONDITIONING SYSTEM	
Heating system type	no special requirement	no special requirement
Infiltration control package	A	B

APPENDIX IV

Point System Approach

The following table provides an illustrative format of the point system approach as currently implemented in California under Title 24. The zero-level reference point represents the "base" point for the prescriptive standard. Positive and negative point values indicate energy saved or lost.

	South-Facing Glazing Points for Different Glazing Types		
Total Percentage of Floor Area	Single, $U =$ 0.66 to 1.10	Double, $U =$ 0.42 to 0.65	Triple, $U =$ 0.42 Down
Up to 1.1	+5	+6	+6
1.2 to 2.0	+2	+4	+4
2.1 to 2.7	0	2	3
2.8 to 3.7	−3	0	+1
3.8 to 4.9	−5	−2	−1
5.0 to 6.1	−8	−4	−3
6.2 to 7.3	−10	−6	−4
7.4 to 8.7	−13	−8	−6
8.8 to 10.4	−16	−10	−8
10.5 to 12.9	−19	−13	−10
13.0 to 15.0	−23	−16	−12

References

[1] "Minimum Property Standards, One and Two Family Dwellings," U.S. Department of Housing and Urban Development, Office of Housing, Washington, DC, 1982.
[2] "FmHA Construction Standards," U.S. Department of Agriculture, Farmers Home Administration, Washington, DC, 1978.
[3] "Residential Building Standards," California Energy Commission, Scramento, CA, 1983.
[4] "Coefficients for Forecasting Product Usage in New Construction," Housing Industry Dynamics Inc., Crofton, MD, 1982.
[5] "Thermal Performance Guidelines," National Association of Home Builders, Washington, DC, 10 Sept. 1982.
[6] "Affordable Housing Through Energy Conservation, Data Base for Simplified Energy Analysis," DOE/SF/00098-2, U.S. Department of Energy, Washington, DC, June 1983.
[7] "National Program Plan," ORNL/Sub-7973/1, Oak Ridge National Laboratory, Oak Ridge, TN, 1982.
[8] "Economics & Energy Conservation in the Design of New Single-Family Housing," NBSIR 81-2380, National Bureau of Standards, Washington, DC, 1980.
[9] "Simplified Energy Design Economics," NBS Special Publication 544, National Bureau of Standards, U.S. Department of Commerce, Washington, DC, January 1980.
[10] "Project Report No. 2 Base Case Buildings," P400-80-035, California Energy Commission, Sacramento, CA, June 1980, pp. 60–62.
[11] "Assumptions Used with Energy Performance Computer Programs," Report No. 7, P400-80-026, California Energy Commission, Sacramento, CA, June 1980.

Industrial Applications

Francis A. Govan[1] *and Nicholas M. Demetroulis*[2]

Design Criteria for Underground Insulated Piping Systems

REFERENCE: Govan, F. A. and Demetroulis, N. M., **"Design Criteria for Underground Insulated Piping Systems,"** *Thermal Insulation: Materials and Systems, ASTM STP 922,* F. J. Powell and S. L. Matthews, Eds., American Society for Testing and Materials, Philadelphia, 1987, pp. 43–51.

ABSTRACT: The U.S. Government encountered a series of major failures of underground heat distribution systems in the early 1950s. A task force of engineers from nine federal construction agencies was formed, under the auspices of the National Academy of Sciences, to determine the cause of the failures, to develop design criteria for new systems, and to recommend their implementation.

This paper details the results of those investigations and discusses the specific design criteria for each system included in the study.

KEY WORDS: thermal insulation, design criteria, underground insulated piping systems, soil condition criteria, expansion of piping, waterproof coatings

The United States Government, through various federal agencies, is the largest purchaser of underground insulated piping systems in the world.

Since World War II, many installations have been made to serve various federal facilities. The initial investment was considerable. During the 1950s, system failures became a matter of major concern. Therefore, a Federal Construction Council (FCC) task group was formed in 1957 to determine the reasons for the failures and to develop design and installation criteria that would produce more reliable systems. Nicholas M. Demetroulis was engineering representative of the U.S. Army Corps of Engineers and Francis A. Govan, was study manager for all of the studies.

The National Academy of Sciences/National Research Council published the findings and recommendations of the FCC in 1958 as Technical Report No. 30, "Underground Heat Distribution Systems." This report was revised and updated three times, the last time in 1975 [*1*]. FCC Technical Report No. 39, "Evaluation of Components for Underground Heat Distribution Systems," which outlined test procedures for prequalification of systems, was issued in 1960 and revised in 1964.

These reports provided the basis for the construction specifications of various federal agencies. In 1964, the first interagency specification based on the work of the FCC was published [*2*]. This is the Tri-Service Specification still used by the U.S. Army, Navy, and Air Force and converted to the Federal Construction Guide System (FCGS) specification. In 1963, FCC Technical Report No. 47, "Field Investigation of Underground Heat Distribution Systems," [*3*] was issued. This covered 121 field investigations of 15 different types of buried, insulated heat distribution systems, both prefabricated and field-fabricated. The age of these installations ranged from 2 to 46 years. In two of these field investigations, closed-cell cellular glass (Foamglas) was

[1]Executive vice president, ZBA, Inc., Cincinnati, OH 45202.
[2]President, NMD Assoc., Alexandria, VA 22308.

inspected. Foamglas, because of its frangibility and susceptibility to destruction where potential boiling action is present, is limited in its use by the Tri-Service group to applications below 122°C (250°F).

The specification criteria developed through the efforts of the FCC dramatically reversed the failure trend of the early post-World War II period [*4*]. To take advantage of new developments in materials technology, however, and to reduce costs where lower temperatures and pressures might be safely handled with materials other than steel, another FCC task group was formed in 1974 to reuse underground heat distribution system design and evaluation criteria based on current technology and the experience gained through use of the criteria developed previously. The recommendations of the task group may be found in TCC Technical Report No. 66, "Criteria for Underground Heat Distribution Systems," published in 1975.

An FCC guide specification, Section 15705, "Underground Heat Distribution Systems (Prefabricated or Pre-Engineered Type)," was prepared. This guide specification is used by members of the Federal Interagency Group (which has superseded the Tri-Service Committee), which currently consists of the three armed services, the General Services Administration, and the Veterans Administration. The criteria are based on the premise that water will get into any system sometime in its 20-year service life. The system must therefore be designed so that the insulation will withstand boiling and, after drying, be restored to within 10% of its original thermal value.

Preliminary Design Considerations

Many factors must be considered when selecting an underground insulated piping system [*5*]. Among the most important are knowledge of soil conditions along the proposed route with respect to soil type, groundwater conditions, corrosivity, soil stability, alkalinity, and drainage patterns. All these factors have a major bearing on what types of systems should be considered for a specific installation.

FCC Technical Report No. 66 [*4*] goes into considerable detail on soil types, their relation to groundwater conditions, and corrosiveness. Four groundwater classifications are set forth in the document: severe, bad, moderate, and mild. Briefly, these classifications are based on how frequently the water table will or will not be above the bottom of the insulated piping system and how long accumulated surface water will remain in the soil around the pipe. Table 1 [*6*] shows the relationship of these two criteria with the four classifications. The Tri-Service specification allows only drainable, dryable, pressure-testable systems to be used in the severe classification. These systems and a water-spread-limiting system are considered suitable for the bad and moderate classifications, while the mild classification could encompass many systems.

TABLE 1—*Underground water conditions classification.*[a]

Classification	Water Table Above Bottom of Systems	Duration that Surface Accumulation Remains in Soil
Severe	frequently or occasionally	long
Bad	occasionally or never	long
Moderate	never	short
Mild	never	not expected to remain

[a]This table is based on criteria set forth in Ref *6*. Additional information regarding soil types encountered with those classifications, precipitation, and irrigation practices may be found in this reference.

One of the key factors in the successful record of these systems is a mandatory requirement that the systems undergo a rigorous third-party testing and evaluation program to prequalify for U.S. Government use. The results are reviewed by a federal committee, and a letter of acceptability is provided. Although there has been much gnashing of teeth over this requirement, it has been successful in saving the government significant monies.

A soil survey should be made to determine the conditions along the pipeline route. It should include the water table and porosity of the soil and determination of the soil pH, since the soils' acidity or the alkalinity may adversely affect some materials. For example, asbestos cement may be harmed by soils having a pH less than 5.5.

Soil corrosivity is most commonly determined by the soil resistivity test. This measures the electrical resistance of the soil—the lower the resistance, the higher the corrosivity. Wet soils and soils having a high soluble salt content generally are corrosive. A coated metal system will require cathodic protection of either the anode or the impressed current design when the soil resistivity is 30 000 Ω/cm or less.

Pipe expansion can be handled with expansion loops, expansion joints, or ball joints. Prefabricated expansion loop assemblies are available for both pressure-testable and non-pressure-testable conduit systems. Expansion chambers can be designed into the encasement at corners and loops of insulating concrete installation. Expansion loops may be used with various types of loose-fill insulation, usually with some limitation on the amount of lateral movement that may be accommodated.

The groundwater level is not as great a concern with either pressure-testable, drainable and dryable conduits or water-spread-limiting conduits it is with some other types of systems. Some types of soil will retain surface water for extended periods. Also, the trench wherein the pipe is installed may disrupt normal drainage patterns and become a catch basin for surface water. Installation of drain tile along the route may therefore be necessary to ensure satisfactory performance of insulating envelopes and most types of field fabricated trench systems.

The proper insulation to withstand boiling action based on the National Bureau of Standards (NBS) tests should be specified for all pipes, valves, and other appurtenances in heat distribution manholes. The insulation should be covered with sheet metal jacketing to prevent mechanical damage. Manholes should also be vented and provided with mechanical pumping facilities. Excessive manhole temperatures have been found to be a major cause of inadequate inspection and maintenance of underground heat distribution systems.

Insulation thicknesses are based on the pioneering work of Dr. Kasuda of the National Bureau of Standards. His work takes into consideration the entire insulated piping system, soil type, moisture content, ground temperature, and other factors. It has been verified by field measurements and evaluation. His is the only method the authors deem valid and proven.

Drainable and dryable systems should be sloped so that water will drain from the entire length of conduits, concrete trenches, and so on. Provisions should be made for water removal from manholes if a major inflow occurs.

Installation and Backfill

All piping should be installed in accordance with applicable codes and industry standards. The conduit or pipe should be firmly supported along its entire length by virgin earth, compacted sand, or insulating fill so as to minimize the possibility of excessive strain due to ground subsidence and the possibility of washouts that could remove protective fill from around the pipe. Pipe laid in insulating backfill should not rest on bricks, timbers, or similar materials. These provide a moisture path to the pipe and may result in severe concentration cell corrosion.

All fill-type installations must be installed to the proper depth and properly compacted in accordance with the manufacturer's requirements. Care must be taken to avoid contaminating the fill with dirt or other construction debris.

The concrete trench system should be set on a firm foundation and underdrained and pitched to drain.

Insulating concrete should be permitted to dry before it is covered with a waterproof membrane and backfilling.

Any damage to protective coatings incurred during shipping or construction must be repaired in accordance with methods compatible to the original coating.

All welding should be done in accordance with the American Society of Mechanical Engineers (ASME) B31.1 Code for Pressure Piping. Manufacturer's requirements for joining coupled pipe should be followed to avoid leaks on low-pressure lines and damage to high-pressure systems.

All lines should be tested for tightness prior to complete backfilling. All joints and fittings should remain exposed until the test is completed. Pressure should not be put on lines before concrete anchors and thrust blocks have cured sufficiently to withstand the stress.

Care should be taken during backfilling to prevent damage to protective coatings, conduits, and vapor sealing membranes.

System Classification

Many types of insulated underground piping systems are in use [*6*]. They may be classified in various ways, such as prefabricated, preengineered, and field fabricated.

Thermal distribution systems may also be grouped by temperature. Three ranges are generally accepted: above 122°C (250°F), steam and high-temperature hot water; 84 to 122°C (180 to 250°F), steam and low temperature hot water; and 2 to 84°C (35 to 180°F), chilled and dual-temperature water.

There are a number of piping materials commonly in use that are restricted by allowable pressure rating to applications with the lowest or the two lower-temperature ranges.

Systems can also be classified by types as follows:

(*a*) pressure testable conduit systems,
(*b*) nonpressure testable conduit trench systems, and
(*c*) insulating envelope systems.

Because moisture in the form of groundwater or pipe leaks has been the main cause of insulation and pipe failure (by corrosion), the emphasis today for systems above 122°C (250°F) is on the development and use of systems that are either drainable and dryable or capable of confining water to a limited section.

Pressure-Testable Systems

Pressure-testable systems, sometimes referred to as air gap systems, are drainable and dryable when properly installed. They consist of a carrier pipe, pipe insulation, spacers, and an outer casing pipe or conduit. An annular air space around the insulation provides added resistance to heat flow and also provides the means of leak testing the conduit (with pressurized air). It also permits the system to be drained if the conduit or pipe develop leaks.

The outer casing may be of steel, galvanized steel, reinforced plastic, or cast iron. The steel and galvanized steel conduits are generally covered with either a glass-reinforced coal tar enamel having an outer wrap of glass-fiber-reinforced pipeline felt or a glass-fiber-reinforced epoxy resin. Cast-iron casings are not coated.

Pipe insulation is usually calcium silicate. The choice of calcium silicate for applications above 100°C (212°F) as the only acceptable preformed insulation was based on an extensive test program conducted by the National Bureau of Standards in 1960–1961. The results will be discussed in the insulation portion of this paper. The systems can be designed to handle fluid temperatures ranging from below freezing to 426°C (800°F) or higher.

The carrier pipes are joined by welding, and the joints are covered with split preformed insulation sections. Steel conduits are joined by welding also. Cast-iron conduits are connected by sleeves (plain-end conduits), bolted together (flanged-end conduits), or connected with mechanical joint fittings (mechanical joint conduit). Plastic pipes are generally joined with chemical welding or O-rings.

Steel conduit assemblies are available in 6.1 and 11.9-m (20 and 39-ft) lengths. Cast-iron assemblies come in 4 and 5.6-m (13 and 18½-ft) lengths.

Pressurized monitoring alarm systems are available that maintain the conduit at a pressure above atmospheric—typically, 0.15 to 0.56 kg/cm^2 (5 to 8 psig). The unit signals any loss in pressure if a leak develops or if the casing is accidentally ruptured by other construction. In the case of smaller leaks, the pressure source can restore pressure and prevent water from entering the conduit. Either an air compressor or compressed nitrogen gas may be used.

Nonpressure Testable Systems

These are available in a variety of material combinations to meet the requirements of many types of applications.

Some conduits are factory fabricated and consist of a carrier pipe surrounded by insulation, which, in turn, is enclosed in an outer nonmetallic casing. The area between the carrier pipe and outer casing is completely filled with insulation, and generally the ends of the insulation are sealed with a watertight enclosure, thus limiting the spread of moisture if the pipe or outer cover fails.

Other types of non-pressure-testable conduits are field fabricated. These may be a factory-engineered system consisting of components specifically manufactured for underground heat distribution components, or they may be concrete trenches or other nonproprietary designs.

Prefabricated water-limiting conduits can be obtained with a variety of components. Carrier pipes may be made of copper, carbon steel, stainless steel, polyvinyl chloride (PVC), fiberglass-reinforced plastic (FRP), or epoxy-lined asbestos-cement. Outer casings generally are PVC, FRP, or asbestos-cement. Insulations commonly used include calcium silicate for applications above 122°C (250°F), and both plastic foam and preformed foam glass for applications below 122°C (250°F).

Copper, FRP, and epoxy-lined cement carrier pipes are commonly used below 122°C (250°F). Copper may be used with fluid temperatures up to 122°C (250°F). FRP pipes generally are not used above 107°C (225°F), and some carry a maximum rating of 66°C (150°F). Asbestos-cement pipe has an allowable operating temperature range of 2 to 99°C (35 to 210°F). Steel pipe is used exclusively on hot water and steam installations operating above approximately 122°C (250°F) because of the pressures and temperatures encountered.

The insulation selected for prefabricated insulated conduits depends primarily on the operating temperature of the pipe.

Foamed polyurethane insulation may be used in applications between approximately 10 and 122°C (−50 and 250°F). Polyurethane resists water penetration. Nevertheless, it must be covered with a vapor barrier of some type. Groundwater under a sufficient head may rupture the cells. Damage during construction or pipe flexing due to ground settling after installation may cause cracks to develop in the foam. Also, if the pipe is used for chilled water service without a vapor barrier, the difference in vapor pressure resulting from a pipe temperature lower than the groundwater temperature may draw moisture to the pipe.

Composite insulations, consisting of an insulation suitable for higher-temperature service next to the pipe, which in turn is covered with a lower-temperature-rated insulation, are also used. Calcium silicate covers the pipe and is, in turn, surrounded by polyurethane or similar materials. Pipe systems having this insulation construction are rated for temperatures to 232°C (450°F) and a maximum operating pressure of 35.1 kg/cm^2 (500 psig).

Formed cellular glass insulation is also used with prefabricated pipe construction. The insulation itself will withstand temperatures between −268 and 427°C (−450 and 800°F). Prefabricated piping with sufficient insulation thickness to protect the outer jacket is available for this temperature range. Foam glass has not been widely used by the federal government for application above 122°C (250°F) because of its frangibility and susceptibility to damage due to boiling action.

Prefabricated insulated pipe lengths vary with the manufacturer, carrier pipe materials, and pipe size. Typical lengths are 6.1 m (20 ft) for copper, steel, PVC, and FRP carrier pipe sizes up to 0.3 m (12 in.). Asbestos-cement carrier pipe sections typically come in 3 and 4-m (10 and 13-ft) lengths with carrier pipe sizes ranging from 0.1 to 0.4 m (4 to 16 in.). One manufacturer offers 16.8-m (55-ft) lengths with carrier pipe diameters (steel) up to 0.9 m (36 in.).

Field-Fabricated Systems

Field-fabricated pipe conduits are not pressure testable and may or may not have an air space around the pipe. They may be preengineered in that the conduit consists of an assembly specifically designed to house underground insulated pipelines. One type consists of a poured or precast concrete structure formed so that it has a trough for water—either groundwater or pipe leakage—to collect and be drained away. Cast-iron pipe supports, rollers, and so on are usually mounted on the base. Vitrified clay side pieces and half-round covers cemented together complete the conduit. The cemented joints are covered with a waterproof mastic to seal the conduit.

Concrete trenches are also applied. These may be considered small tunnels, since they are similar in concept to walk-through tunnel systems. The bottom and sides are either precast or formed and poured on site. The top consists of concrete slabs. After the pipes are installed in the trench, the top is set in place. Tar or other sealing material is used to provide a barrier to groundwater or surface water entrance at the joint. The top of the trench is often at grade level and serves as a walkway. This facilitates access to the pipe, since the ground does not have to be excavated when checking for leaks, adding a line, replacing a line, and so forth. It is recommended that the bottom of the trench be set on a vapor barrier and the sides be coated with a mastic material to provide greater resistance to moisture penetration. Concrete trench systems must have positive mechanical drainage systems.

Vitrified clay tile and concrete sewer tile are also sometimes used to construct pipe conduits. These and other field-fabricated conduits may serve well where drainage is good and the system is installed above the water level.

Years ago, concrete trenches and field-fabricated conduits were sometimes completely filled with insulation. Examination of such systems, however, after they had been in operation for a period of time revealed that it was not unusual for the insulation to be wet and slumped and the pipes corroded. Therefore, today the pipe is insulated, and the insulation is covered with a moisture barrier. The air space around the pipe permits water to drain from the system, and the heat from the line can dry the insulation if it becomes wetted.

Insulating Envelope Systems

Insulating envelope systems are preengineered systems designed to surround the pipe with a poured-in-place insulating medium, which may be any of the following:

(*a*) an insulated concrete,
(*b*) a granular hydrocarbon fill—both uncured and heat-cured types are available,
(*c*) a treated calcium carbonate fill,

(*d*) an insulating granular perlite fill, and
(*e*) field-installed polyurethane foam.

Insulating concretes are available with two types of aggregates, vermiculite or expanded polystyrene beads. It is the aggregate that provides the insulating value to the concrete.

A structural concrete-base pad is poured to the desired grade. Precast insulating blocks and drain vents are set on a waterproof membrane that covers the structural slab and lines the trench or forms. The pipes rest on top of the insulating blocks, and the insulating concrete is poured over the slab, pipes, and internal drains. The waterproof membrane is wrapped around the top of the concrete and sealed.

If required, cathodic protection can be achieved by the installation of a continuous ribbon zinc anode parallel to the pipes inside the concrete.

Internal electrical sensors can be installed inside the internal drain channels to detect and locate leaks.

Added drainage can be had, if conditions warrant it, by installing an external drain tile.

In addition to providing thermal insulation, insulating concrete provides continuous support and alignment of the piping and resists heavy loads.

Granular hydrocarbon fills are installed in an open trench. A bed of the material is placed in the bottom of the trench. The pipe is installed, and more fill is placed around and over the pipe. The material is then compacted to the proper density. The material resists wetting; however, installation of drain tiles along the pipe path may be required, depending on the natural drainage characteristics of the earth and groundwater conditions, to avoid buildup of hydrostatic heads that would penetrate the fill.

Natural hydrocarbon mineral material does not have to be heat cured. It is suitable for temperatures between 2 and 260°C (35 and 500°F).

Heat-cured granular fills are derived from petroleum residuals. These are available for various temperature ranges between 66 and 271°C (150 and 520°F). An asphaltic binder in the fill melts and bonds to the pipe during the curing process.

Surrounding the pipe and binder is a sintered zone, and the outer layer of the fill is unaffected by the curing process.

Since these materials are used with metal piping systems, cathodic protection should be installed whenever material selection and soil conditions dictate.

Powdered calcium carbonate, which is treated to resist water penetration, is another insulating material. It is rated for temperatures up to 248°C (480°F). Side forms in the trench are all that is required. The material is installed around and below the pipe to the proper dimensions, compacted, and covered on the top with a plastic film. The manufacturer states that, because of the hydrophobic properties of the material and its high electrical resistivity, cathodic protection is not required.

Expanded perlite particles are also used as a buried insulation. Powdered perlite is heated to a high temperature, causing it to pop or burst into granular air-celled particles. These particles are mixed in the field with an asphalt binder. A base pad, consisting of the insulating material and binder, is poured in a formed trench. The pipes are laid on the base, tested as required, and then the pad and pipes are primed with a corrosion-resisting compound. The installation is then completed by pouring more material over the pipes to the depth of the forms.

Operating temperatures range from −5 to 426°C (15 to 800°F), and the material is said to have a high electrical resistivity.

Polyurethane foam may be poured or blown around pipes in the field. The systems are ready to be backfilled within an hour after the foam has been placed around the pipes. Forms are required for the froth-pour type of insulation, and a vapor barrier liner and covering are generally used to seal off ground moisture. The direct spray method does not require side forms. Again, vapor barriers are recommended. They may be used up to 122°C (250°F).

Buried Tunnel Systems

Buried walk-through tunnels generally are the most expensive means of housing underground piping. However, they can be used for other utilities, facilitate piping maintenance, and when properly drained and ventilated, minimize the development of corrosion on pipe and appurtenances.

Concrete tunnels constructed by forming and pouring are the most expensive.

To reduce the cost of tunnels, prefabricated concrete or steel tunnels have been developed. These are complete with pipe racks and hangers. The sections are dropped into the excavation and joined together. The steel is factory coated, and cathodic protection can be applied to provide additional protection against corrosion.

Large-diameter concrete sewer pipe has also been used successfully to construct tunnels. The joints must be sealed and caulked, and as with other tunnel systems; provisions for drainage should be made as job conditions require.

System Maintenance

Careful inspection and regular maintenance will reduce the likelihood of premature failure of buried lines. Procedures to follow should include the following:

1. Monitoring of steam pressures or water temperatures at points of send out and use—this will warn of excessive heat loss caused by damaged or wet insulation.
2. Periodic patrolling of the line—burned grass over the route, water or steam coming to the surface, and melted snow indicate problems.
3. Periodic inspection of manholes and vaults—water in the vault, steam plumes, signs of moisture in insulation or in the ends of conduits, and end cap corrosion are further signs of problems.
4. Maintaining a regular program of checking the performance of cathodic protection systems—this includes taking anode readings, monitoring rectifier installations, and checking all electrical insulating fittings.

Central heating/cooling plants offer many advantages. They can be more efficient in the use of resources than individual building heating/cooling plants. They provide a means of alleviating the solid waste disposal problem by disposing of combustible trash. And it is easier to control air pollution at a large central installation than at many small ones.

Underground insulated piping systems are an integral part of the central plant concept. In the past, they have sometimes been the weakest link. With a vast amount of past experience to draw from, however, and new technology, buried heating and cooling distribution systems can be installed that will give reliable, economical service for any application that is not beyond the limitations of the system itself.

Recent Government Work

Despite the significant improvement in performance of underground systems, the government is continuing its attempts to improve standards and criteria. A recent survey of a number of facilities indicates that some problems still exist. These problems apparently are mainly due to improper design, inadequate installation, and, particularly, lack of maintenance. Test installations of systems are being made, and the systems are being instrumented for long-term evaluation. These tests, together with new criteria for manhole design and maintenance, should prove highly beneficial in maintaining the efficiency and longevity of systems. This work is primarily being accomplished through the U.S. Army Construction Engineering Laboratory in Champagne, Illinois.

References

[1] "Underground Heat Distribution Systems," Federal Construction Council Technical Report No. 3OR-64, Publication 1186, National Academy of Sciences-National Research Council, Washington, DC, 1964.

[2] Irvin, Jr., L. V., "Federal Agency Specification for Underground Heat Distribution Systems," Underground Heat and Chilled Water Distribution Systems, NBS Building Science Series 66, U.S. Department of Commerce, National Bureau of Standards, Washington, DC, 1975, pp. 73–77.

[3] "Field Investigation of Underground Heat Distribution Systems," Federal Construction Council Technical Report No. 47, Publication 1144, National Academy of Sciences-National Research Council, Washington, DC, 1963.

[4] "Criteria for Underground Heat Distribution Systems," Federal Construction Council Technical Report No. 66, National Academy of Sciences, Washington, DC, 1975.

[5] *ASHRAE Handbook and Product Directory,* Systems Volume, American Society of Heating, Refrigerating, and Air Conditioning Engineers, New York, 1976, Chapter 14, p. 2.

[6] "Criteria for Underground Heat Distribution Systems," Federal Construction Council Technical Report No. 66, National Academy of Sciences, Washington, DC, 1975, pp. 4–5, 49–51.

Tamami Kusuda,[1] *Jin B. Fang,*[1] *and Walter M. Ellis*[1]

A Method for Measuring Heat Loss from Underground Heat Distribution Systems

REFERENCE: Kusuda, T., Fang, J. B., and Ellis, W. M., **"A Method for Measuring Heat Loss from Underground Heat Distribution Systems,"** *Thermal Insulation: Materials and Systems, ASTM STP 922,* F. J. Powell and S. L. Matthews, Eds., American Society for Testing and Materials, Philadelphia, 1987, pp. 52–68.

ABSTRACT: A simple procedure to obtain the *in situ* determination of heat loss from an underground heat distribution system is described. The proposed procedure for estimating the heat loss from underground heat distribution systems consists of making a series of temperature and thermal conductivity measurements on the ground grid in the vicinity of the pipe system and deducing the heat loss by statistically determining the parameters of a mathematical model using the gathered data. Verification of the procedure was tested in a simulated scale model of an underground system using an insulated sandbox. The scale model is a two-pipe system buried 0.3045 m (12 in.) below the top surface of a 0.66-m (26-in.)-wide, 0.71-m (28-in.)-high, and 1.22-m (48-in.)-long sandbox. The test pipes are two electrically heated steel pipes having an outside diameter of 3.37 cm (1.325 in.). The walls of the sandbox are insulated by 5.08-cm (2-in.)-thick expanded polystyrene board. The heat loss measurements from this simulated underground heat distribution system were obtained for pipe surface temperatures of 66°C (150.9°F), 93.8°C (200.8°F), and 149.2°C (300.7°F) for three different sand moisture contents (dry, 10%, and 13%). Relatively good agreement was obtained between the measured electrical power input to the test pipes and the system heat loss estimated by the sand temperature profile around the pipes and sand thermal conductivity measured by the thermal needle technique. The paper also describes an application of the technique to a field measurement conducted at the James Madison University, Harrisonburg, Virginia.

KEY WORDS: field measurement, *in situ* measurement, pipe heat loss, sandbox, scale model, thermal conductivity, thermal insulation

In situ measurement of heat loss from existing underground heat distribution systems is needed to assess the performance of old systems, for which thermal insulation may have been deteriorated because of undesirable ingress of groundwater, and also for new systems, which may have been damaged during installation. There are numerous unreported cases of excessive heat loss from heat-generating plants, which are evidenced by the dead grass in the summer and melted snow during the winter over the underground systems. In spite of many apparent problems associated with excessive heat loss from underground heat distribution systems, no data are currently available to estimate the extent of the increased energy loss from defective heat distribution systems. This is mainly due to the difficulty in quantifying the amount of heat loss.

Currently, several methods can be used to determine the heat loss from a section of the underground system. These entail measuring the heat flows at two adjacent manholes near the sections of interest. The two most common methods are the calorimetric technique and the heat flux meter technique.

[1]Chief, Building Physics Division, mechanical engineer and engineering technician, respectively, National Bureau of Standards, National Engineering Laboratory, Center for Building Technology, Gaithersburg, MD 20899.

The calorimetric technique is an absolute method in which the fluid mass flow rate and enthalpy change across the test section are measured. It is, however, impractical in many cases because the underground system must be cut into to install various transducers for measuring the fluid flow rate, temperature, pressure, and condensate flow in the case of a steam system.

The heat flux meter technique entails excavating the pipe and installing several heat flux transducers around the pipe circumference. If the heat flux meter is temperature sensitive (which is the case for most of the commercial heat flux transducers), the temperature of the heat flux meter should also be monitored. The excavation process inevitably disturbs the natural soil condition around the pipe. It is difficult to restore the surrounding soil to its original compaction and moisture content. Thus, the heat losses measured by this method are not necessarily those of the undisturbed pipe system. Also, the output signals of the heat flux meter exhibit rapid fluctuation because of minor temperature change of the pipe fluid. It is, therefore, important that the output be time averaged to smooth out these fluctuations.

The method described in this paper is called the thermal probe method. This is a nondestructive method based on a steady-state line heat source theory in which the temperature profile around the source is proportional to the strength of the line source or the heat output per unit length of the pipe (see Fig. 1). Strictly speaking, the theory is valid only for the source of infinitely small diameter in the heat conduction medium of homogeneous properties or thermal conductivity. It is also based on the assumption that the temperature field far away from the source is constant with respect to depth. T_0 in Fig. 1 is the earth temperature far away from the source, which is assumed to be constant throughout the region. In a realistic situation, the source has a finite size dictated by the outside diameter of the underground system that includes insulation, air space, and protective conduit. Moreover, the soil property as well as its temperature are seldom uniform in the region, reflecting the seasonal and depth-wise variation of moisture content and temperature. In addition, the equation shown in Fig. 1 was derived by ignoring the interaction between the conductive heat flux and the sensible heat due to moisture migration in the vicinity of a heated pipe system.

In this paper the thermal probe technique is investigated in a scale model test box filled with sand in order to assess its potential use for the field measurement of heat loss from actual underground heat distribution systems.

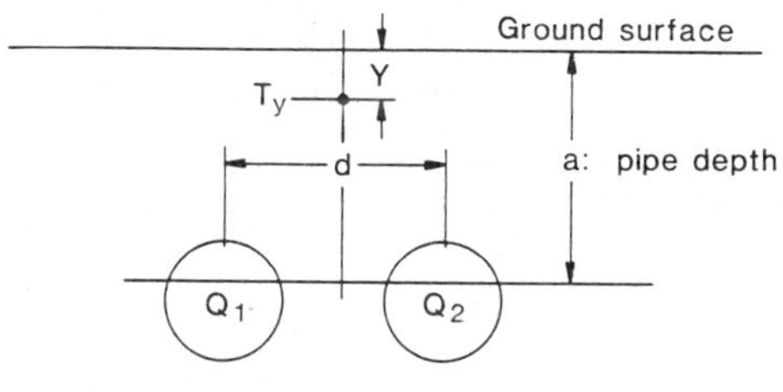

T_o: undisturbed soil temperature
k: soil thermal conductivity
Q_1: heat loss per unit length from pipe #1
Q_2: heat loss per unit length from pipe #2

$$T_y - T_o = \frac{(Q_1 - Q_2)}{2\pi k} \ln \frac{\sqrt{\left(\frac{d}{2}\right)^2 + (Y + a)^2}}{\sqrt{\left(\frac{d}{2}\right)^2 + (Y - a)^2}}$$

FIG. 1—*A schematic of a two-pipe underground heat distribution system.*

Temperature Field Around Buried Pipes

The soil temperature field caused by a buried heat source or sink, under steady-state conditions and negligible moisture migration, can be estimated using the method of images [*1*]. The underground temperature field around a two-pipe system can be obtained by the superposition of each pipe's contribution and is given by

$$T = \sum_{i=1}^{2} \frac{Q_i}{4\pi k} \ell n \left\{ \frac{(X - b_i)^2 + (Y + a_i)^2}{(X - b_i)^2 + (Y - a_i)^2} \right\} + T_0 \quad (1)$$

where T is the temperature of the soil at a given location, k is the thermal conductivity of the soil, Q_i is the strength of the ith heat source or sink per unit length, X and Y are cartesian coordinates, b_i and a_i are X and Y coordinates of the center of the ith pipe, and T_0 is the undisturbed soil temperature.

By using the temperature field in the vicinity of a pipe system, the heat fluxes, Q_i, can be estimated using statistical techniques. In an actual field test, one is also not certain of the exact locations of the pipes. If enough temperature measurements are made, Eq 1 can also be used to determine the location of the pipes.

The equation can be solved to give the locations, (b_1, b_2), depths, (a_1, a_2), and heat losses, (Q_1, Q_2), of pipes using the method of nonlinear least squares, as long as six or more earth temperature values, T, are available together with the values for soil thermal conductivity, k, and the separation distance of the pipes.

For two pipes buried at the same depth below the ground surface, the underground temperature (Eq 1) can be expressed by

$$T = \frac{Q_1}{4\pi k} \ell n \left\{ \frac{\left(X + \frac{d}{2}\right)^2 + (Y + a)^2}{\left(X + \frac{d}{2}\right)^2 + (Y - a)^2} \right\} + \frac{Q_2}{4\pi k} \ell n \left\{ \frac{\left(X - \frac{d}{2}\right)^2 + (Y + a)^2}{\left(X - \frac{d}{2}\right)^2 + (Y - a)^2} \right\} + T_0 \quad (2)$$

where d is the separation distance between the centers of the pipes, and a is the depth of the pipe.

Thermal Conductivity Measurement

Equations 1 and 2 indicate that the thermal probe method for determining the pipes' heat loss is directly affected by the accuracy of the soil thermal conductivity value, k. The thermal needle method [*2,3*] was used to determine the thermal conductivity or the thermal resistivity, which is the reciprocal of thermal conductivity, of the soil surrounding the buried pipes. This method is based on the measurement of the temperature rise along a line heat source dissipating heat into an infinite, homogeneous medium. In this method, the soil thermal conductivity is determined from the slope of a best-fit line based on a plot of temperature as a function of the logarithm of time.

The measurement system used includes the thermal conductivity probes and an automated microcomputer-controlled thermal property analyzer. The probe is a hollow stainless steel sheath with both ends closed and contains an electric heater and the measuring thermocouples located at the interior wall of the sheath. A 3.2-mm (1/8-in.)-diameter, single-thermocouple laboratory probe 0.1 m (3.94 in.) long was utilized for measuring the thermal conductivities of dry sand and wet sand at 10% and 13% moisture content. In field measurements, two types of the 12.7-mm (1/2-in.)-diameter probes used were 1 m (3.28 ft) in length with two thermocouples installed at 0.3 m (0.98 ft) and 0.6 m (1.97 ft) from the lower end of the probe, and 2 m (6.56 ft) long with three thermocouples positioned at 0.4-m (1.31-ft) intervals from the lower end. The thermal time constants for the laboratory and field probes are typically 100 and 600 s, respectively.

The thermal conductivity probe is simply inserted into the soil to be tested. The thermal property analyzer provides programmable electrical power to the thermal probes, reads thermocouples, probes electrical current and voltage, and computes thermal resistivity, in addition to automating the whole process of running a thermal conductivity test. During the test, the power inputs to each probe are shut down to protect against excess current to the heater contained in the probe if the temperature of any thermocouple exceeds 125°C (257°F). The measured thermal conductivity of dry sand and wet sands at 10% and 13% moisture content was found to be 0.64, 1.38, and 1.96 W/m · K (0.37, 0.80, and 1.13 Btu/h · ft · °F), respectively.

Small-Scale Test in Simulated Pipe System

A small-scale test was designed to simulate an underground pipe system. Figure 2 depicts the overall dimensions of the test box and relative locations of the test pipes. The box is 0.66 m (26 in.) wide, 0.71 m (28 in.) deep, and 1.22 m (48 in.) long and is made of 1.27-cm (1/2-in.)-thick plywood but coated with waterproof mastic cement over the inside surfaces, so that wet sand conditions can be maintained. The outsides of all four vertical walls were insulated by 5.08-cm (2-in.)-thick expanded polystyrene foam board. Two 2.54-cm (1-in.) nominal steel pipes of 3.34-cm (1.32-in.) outside diameter, 1.37 m (4 ft, 6 in.) in length, were installed 30.38 cm (12 in.) below the top surface of the box. The pipes were located at the center of the width of the box and were separated by a distance of 1.27 cm (0.5 in.). Both end sections of the sandbox are so constructed that the pipe depth can be varied at increments of 1.27 cm (0.5 in.).

Electrical resistance wires are embedded in the 10.6-cm (42-in.)-long section of the test pipe to generate the necessary heat to maintain a specified pipe surface temperature. The power input to these heaters was monitored by a watt-hour meter. Tests were conducted at three levels of pipe surface temperature—namely, 66°C (150.9°F), 93.8°C (200.8°F) and 147.3°C (300.7°F) and for sand moisture contents of 0, 10, and 13%. A thermal property analyzer [*2,3*] was used to measure the thermal conductivity of the sand. Thermocouple measurements of the sand temperature around the pipes were taken in a cross section of the test box normal to the pipes at 5.08-cm (2-in.) horizontal intervals. At each temperature measurement station, the temperature measurements were taken at 2.54-cm (1-in.)-depth intervals from the surface of the sand to a depth of 25.4 cm (10 in.). Thus, a grid of 120 temperature data points were monitored until the temperature distribution throughout the region became stationary—normally in a week.

Data from these tests are shown in Figs. 3 to 8. These data were used in two ways. First, using the measured values of sand, thermal conductivity and the pipes' locations and heat fluxes and the temperature fields depicted by Eq 2 were compared with the measured temperature field. Second, the measured temperature field was used in a least-squares technique to predict the parameters of Eq 1. Solid curves in Figs. 3 to 5, show isotherms over the test pipes based on the final and stationary 120 temperature data points recorded. The dashed curves superimposed in the same figures are profiles of the soil temperature calculated by Eq 2 using the measured values of Q_1 and Q_2 and the dimension of the pipe system.

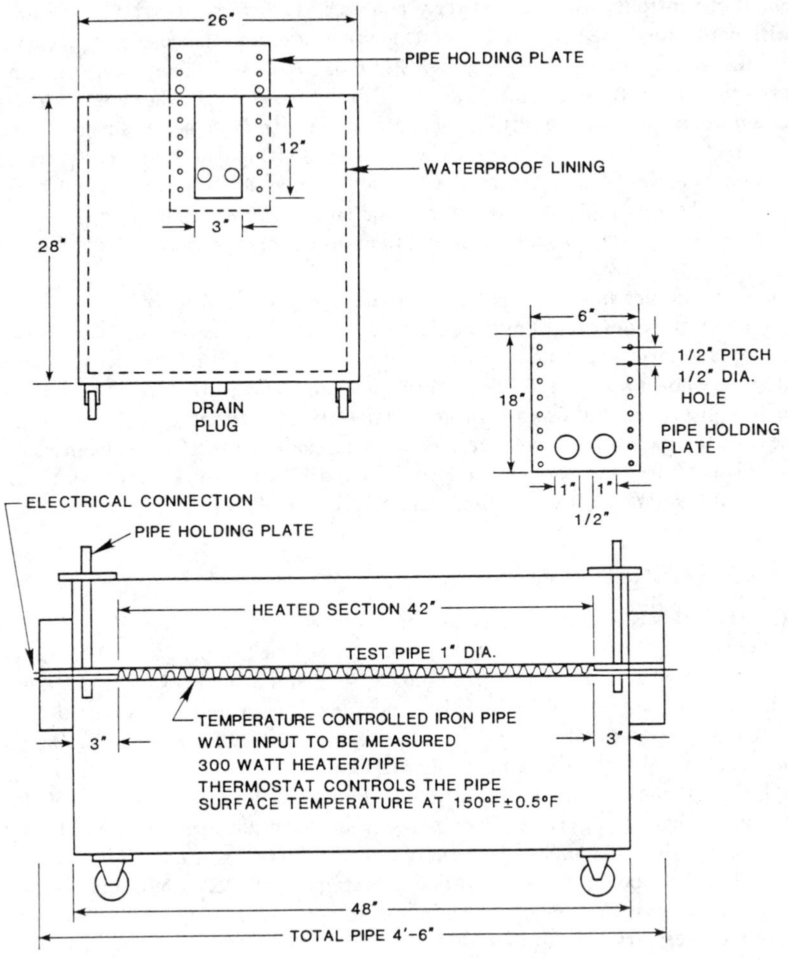

FIG. 2—*A schematic of the test scale-model sandbox for simulating an underground heat distribution system.*

Good agreement between the calculated and observed temperature profiles can be observed near the center of the box. The calculated sand temperatures deviate considerably at the sides of the box from the observed data. This is because the theoretical model is for an infinite medium and does not include the disturbance of the temperature field caused by the sides of the box.

To test the thermal probe technique, the temperature field in the scaled experiment was used to determine the parameters in Eq 1, and the results were compared with the known values. These parameters were determined by a nonlinear least-squares technique. Nonlinear least squares and curve fitting arising from data analysis in research activities are often solved by the method of Levenburg, Marquardt, and Morrison [*4,5*]. The computer algorithms using this method for nonlinear least-squares fitting based on a numerical method for an unconstrained optimization technique have been implemented in the computer program DATAPLOT [*6*] developed at the National Bureau of Standards (NBS) for statistical analysis.

Figures 6 to 8 show the measured and calculated temperature distribution within the sandbox filled with dry sand under an electrical power input to the pipes of 59 W/m (61.4 Btu/h · ft), wet

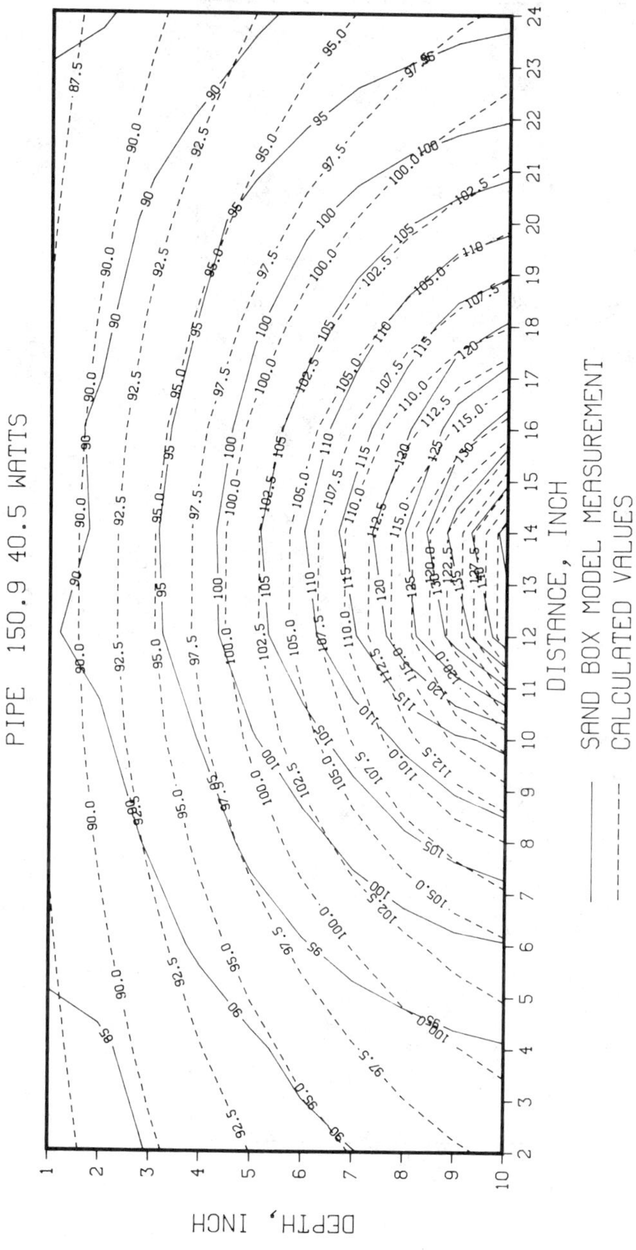

FIG. 3—Calculated and observed sand temperatures above two heated pipes at a temperature of 66°C (150.9°F).

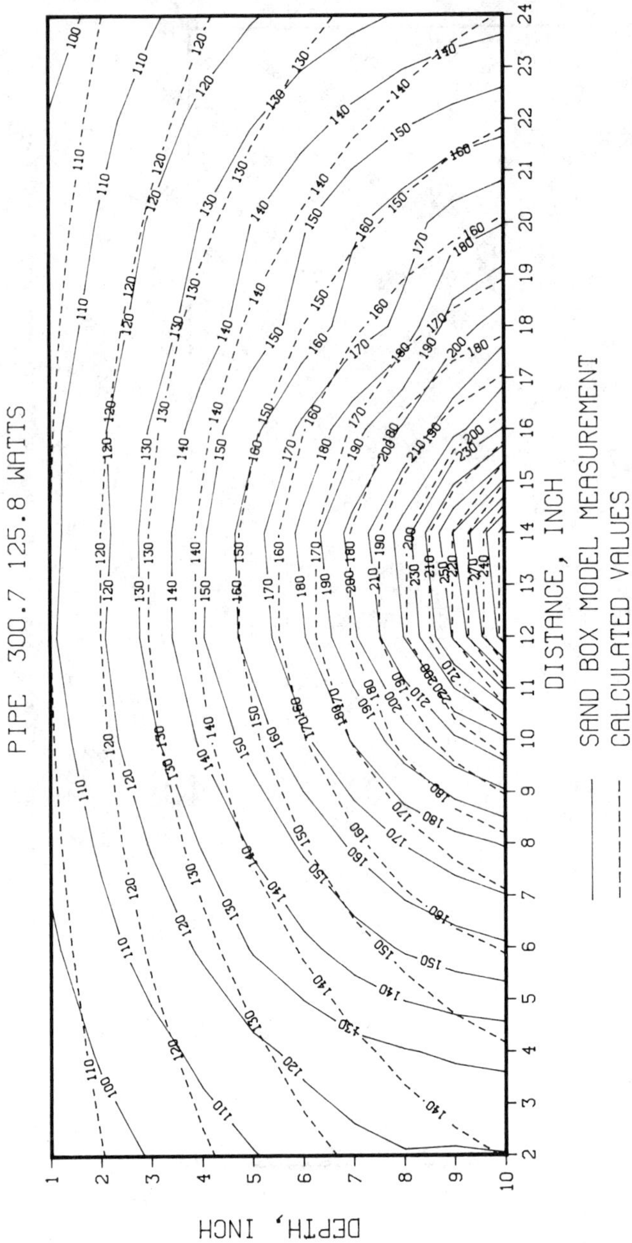

FIG. 4—Calculated and observed sand temperatures above two heated pipes at a temperature of 149°C (300.7°F).

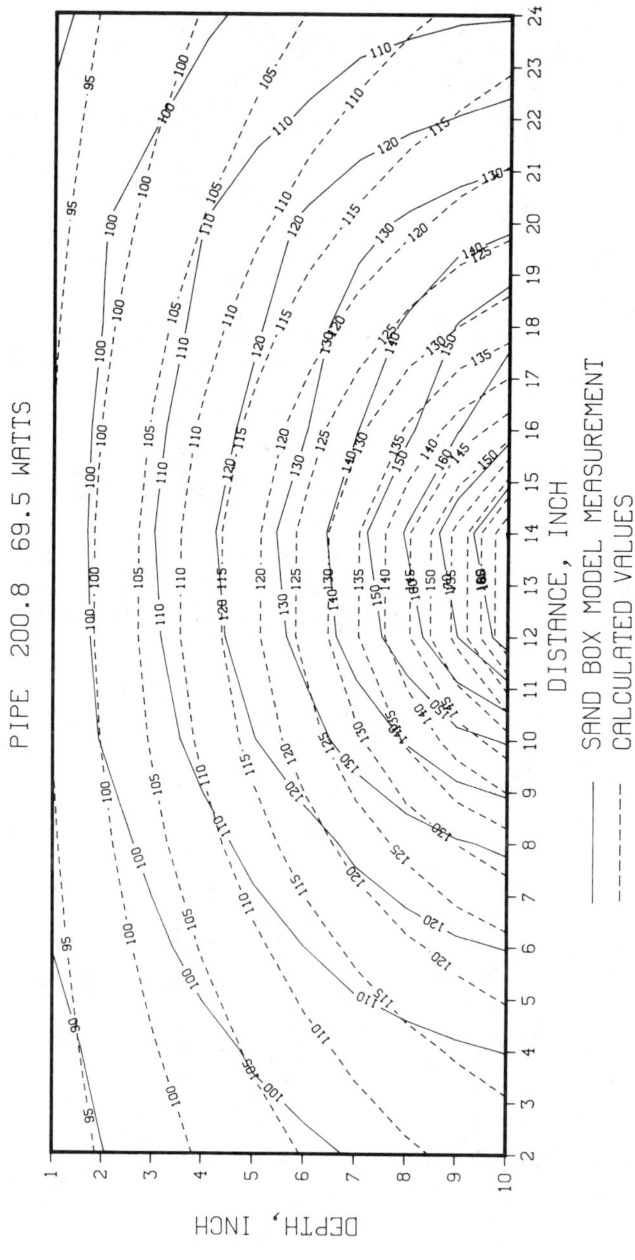

FIG. 5—*Calculated and observed sand temperatures above two heated pipes at a temperature of 94°C (200.8°F).*

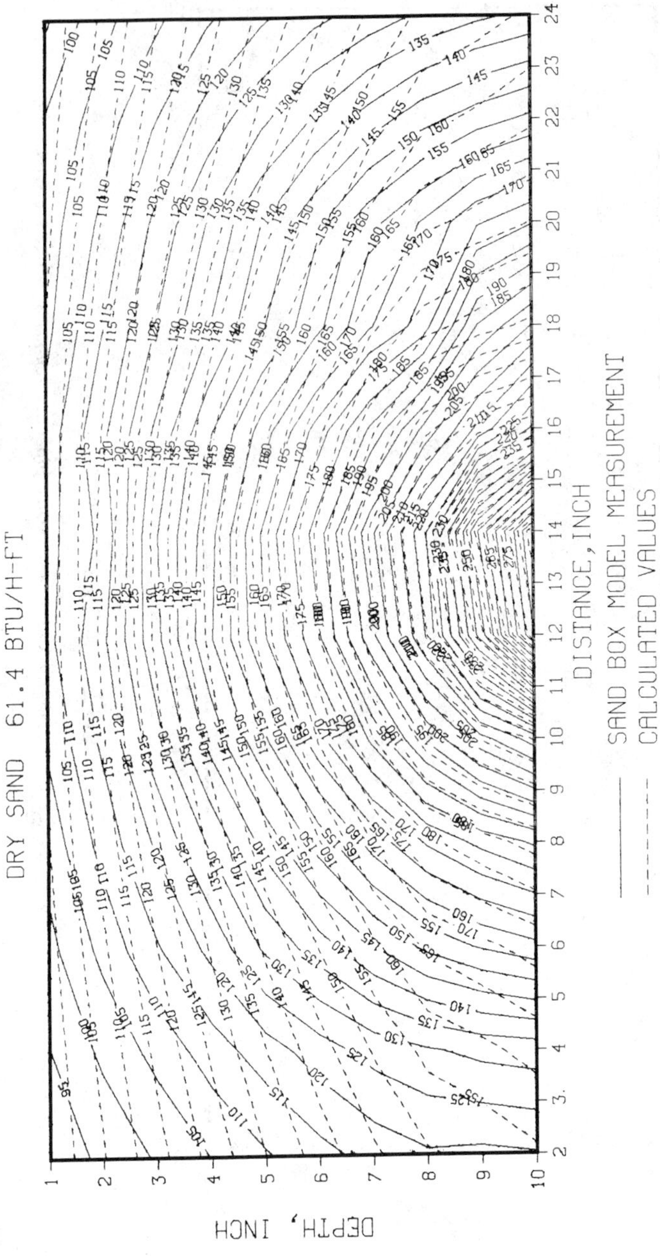

FIG. 6—*Calculated and observed sand temperatures above two heated pipes buried in dry sand.*

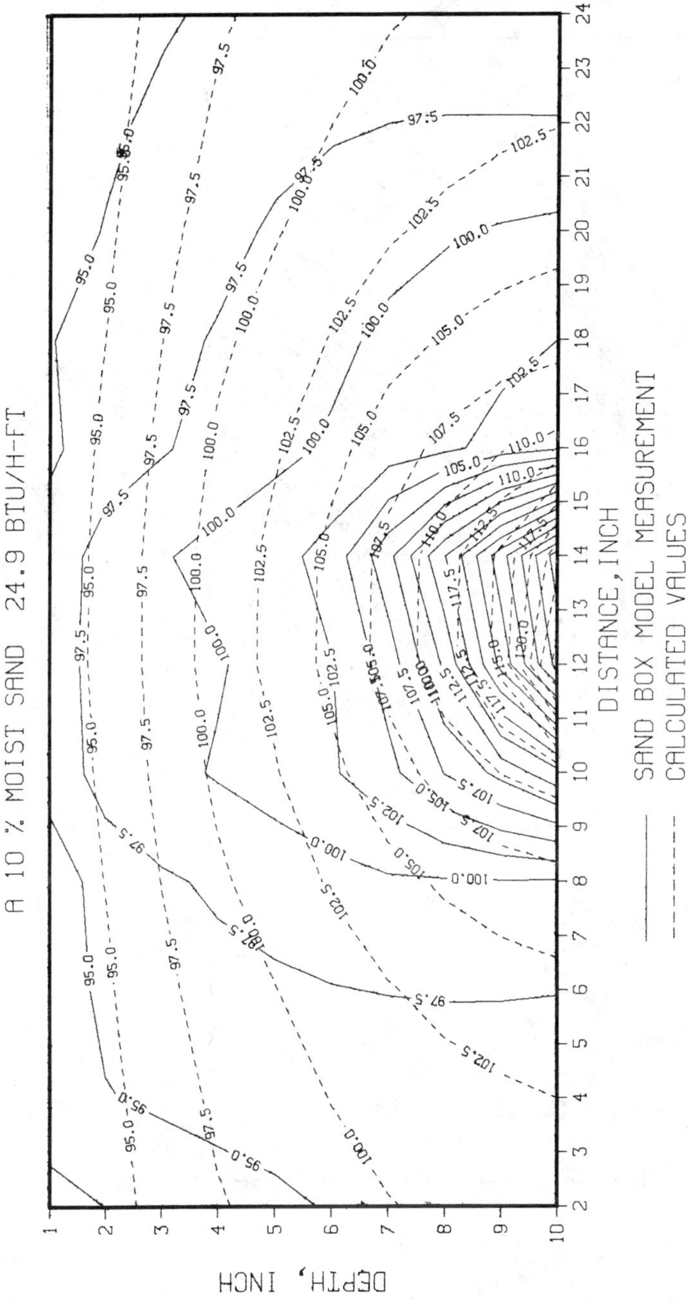

FIG. 7—*Calculated and observed sand temperatures above two heated pipes buried in a 10% moist sand.*

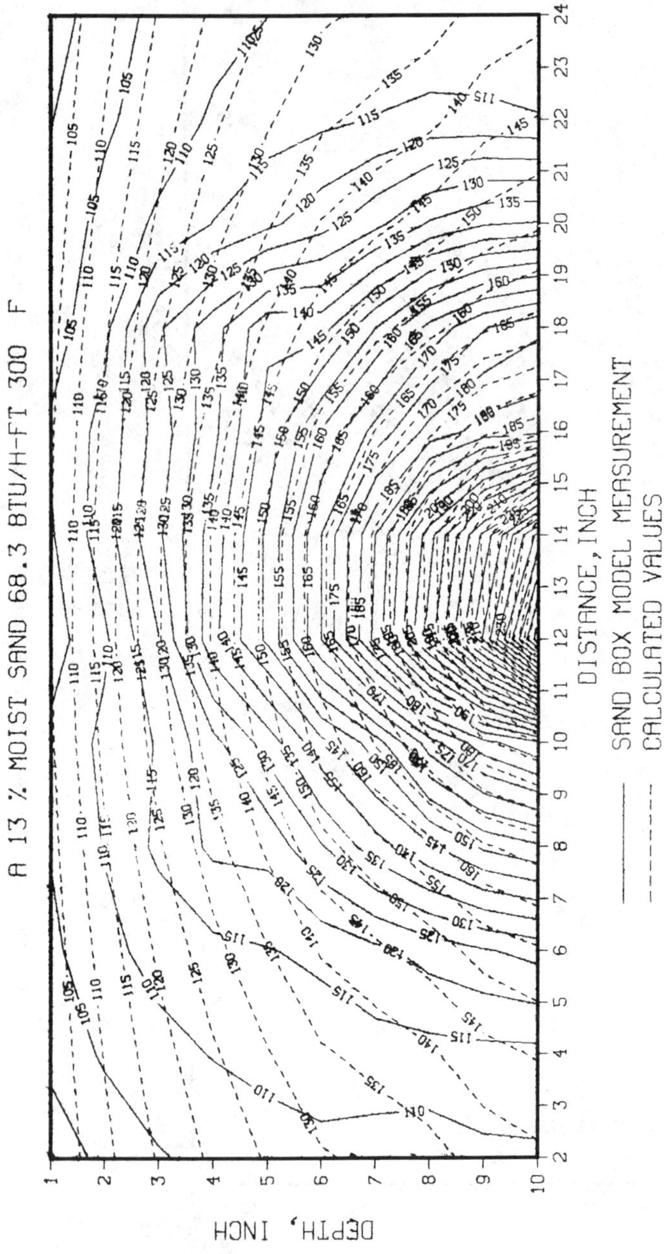

FIG. 8—Calculated and observed sand temperatures above two heated pipes buried in a 13% moist sand.

sand at 10% moisture content and a 23.9-W/m (24.9-Btu/h · ft) power input, and 13% moist sand with the pipes heated electrically at 65.6 W/m (68.3 Btu/h · ft). The solid curves in the figures represent the isotherms based on the observed sand temperature data. The dashed line curves are the profiles of the sand temperature calculated from Eq 1 with the values of coefficients determined by a nonlinear least-squares fit of Eq 1 to the temperature data using the computer program DATAPLOT. Because of excessive computation time, a total of 48 temperature data points were used for each scale-model test in the nonlinear least-squares fit. Excluding the data near the sides of the sandbox, the calculated temperature profiles agree reasonably well with those observed, as shown in the figures.

Method for Determining Total Heat Loss

In many instances, one is interested only in the total heat loss, ($Q_1 + Q_2$), from the underground pipe system. The equation shown in Fig. 1 can be used to determine ($Q_1 + Q_2$) by making use of sand temperature taken only at the center of the box. Table 1 shows the values of $Q_1 + Q_2$ as expressed by Q in the second column. These values are evaluated at each depth and show relatively good and encouraging agreement within ±20% with the power input to the test pipes except for the data near the surface, for all of the three test conditions with dry sand.

A comparison of the measured and the calculated values of the heat loss from the buried pipes, the horizontal location of Pipe 1, and the pipes' depths for scale-model experiments conducted is given in Table 2. For these tests, both heated pipes were buried 27.2 cm (10.7 in.) below the sand surface and 30.7 cm (12.1 in.) from the nearest side of the sandbox. The heat loss values determined by the thermal probe technique for the pipes embedded in dry sand are found to be approximately 8% less than the measured electrical power input to the simulated underground system. For the tests involving wet sand, the calculated and the measured heat loss values agree within ±25%. This decreased agreement may be caused by inhomogeneous settling of sand moisture found during the course of some tests. In general, there is a good correlation between the calculated and the actual locations and depths of the pipes.

Field Measurements

The *in situ* measurements of heat loss from the underground heat distribution system were carried out at James Madison University, Harrisonburg, Virginia, during the end of June 1984. The heat distribution system consists of a nominal 15-cm (6-in.) steam pipe and a 7.6-cm (3-in.) condensate return pipe buried side by side with a distance of 33 cm (13 in.) between the pipe centers. These pipes were installed at approximately 1.22 m (4 ft) below the ground surface, in a 0.91-m (36-in.)-wide by 0.76-m (30-in.)-high trench filled with protexulate insulation material and covered with earth. The earth is a clay type of soil.

The major equipment employed in field measurements include a mobile drilling rig, field thermal conductivity and temperature probes, thermal property analyzer, a thermocouple indicator, and an A-C generator to provide a constant power supply. The drilling rig is a trailer mounted and equipped with a motorized auger powered by a gasoline engine. The temperature probe was constructed from a nominal 12.7-mm (1/2-in.)-diameter by 1.83-m (6-ft)-long, flexible polyvinyl chloride (PVC) pipe and had five thermocouple junctions installed on the outer wall at 0.305-m (1-ft) intervals from the lower end of the probe.

Pilot holes using a drill bit, slightly smaller in diameter than the field probe, were drilled across the buried pipes at 0.305-m (1-ft) intervals for a horizontal distance extending from 1.22 m (4 ft) south to 1.83 m (6 ft) north of the steam pipe. The 12.7-mm (1/2-in.)-diameter field probes were inserted into the holes. Using a drill bit smaller than the diameter of the thermal conductivity probe ensured good probe contact with the earth, thus minimizing the contact resistance between the probe and the earth. A specially designed auger guide was used to keep

TABLE 1—*Scale sandbox test results.*[a,b]

Depth, in.	Calculated Pipe Heat Loss, W	Centerline Temperature, °F	Box Wall Temperature, °F
	PIPE HEATER OUTPUT = 40.5 W (138 Btu/h)		
1	59	87.2	82.8
2	47	90.8	83.7
3	41	94.4	85.1
4	39	98.4	86.2
5	42	104.2	87.5
6	43	110.0	88.9
7	46	117.4	89.9
8	47	125.2	90.8
9	51	137.3	91.5
10	50	146.5	91.8
	PIPE HEATER OUTPUT = 125.8 W (429.5 Btu/h)		
1	195	108.6	94.1
2	169	128.5	103.1
3	107	133.0	108.5
4	112	149.1	114.2
5	113	164.1	119.1
6	112	178.9	123.6
7	118	198.2	127.2
8	122	219.9	130.9
9	132	251.5	133.9
10	136	185.0	136.2
	PIPE HEATER OUTPUT = 69.5 W (237.3 Btu/h)		
1	99	95.8	88.4
2	61	101.9	92.8
3	62	109.8	95.7
4	63	118.2	98.8
5	62	126.3	101.8
6	63	135.2	104.3
7	68	147.2	106.3
8	73	161.2	107.6
9	75	175.2	108.9
10	74	190.7	109.3

[a]Thermal conductivity of dry sand = 0.64 W/m · K (0.37 Btu/h · ft · °F); pipe depth = 0.272 m (10.7 in.).

[b]Metric conversion factors: 1 in. = 0.0254 m; 1 W = 3.414425 Btu/h; °F = 9/5°C + 32.

the drill in line and minimize hole distortion caused by side-to-side movement of the drill. The thermal property analyzer was utilized in conjunction with the field probes to measure the soil thermal conductivity up to three vertical depths. The soil thermal resistivity for each thermocouple location was calculated from a least-squares fit to the time-temperature data collected at between 600 and 1500 s.

The holes were then enlarged with the use of a 15.9-mm (5/8-in.)-diameter drill bit for insertion of the temperature probe. The temperature probe was pushed carefully down into the hole to ensure good probe-soil thermal contact and minimal distortion of the hole. The earth temperatures above and on both sides of the buried pipes were measured using a thermocouple indicator and the temperature probe. The depth temperature measurements were also conducted on the undisturbed soils situated 7.92 m (26 ft) north of the steam pipe. A total of 60 temperature

TABLE 2—*A comparison of calculated and measured values with the scale model.* [a]

Sand Type	Average Heat Loss Rate, Btu/h · ft		Calculated Values, in.		
			Distance,	Depths	
	Measured	Calculated	b_1	a_1	a_2
Dry sand	19.8	17.6	12.3	10.0	9.6
	33.9	31.7	13.2	9.9	10.0
	61.4	56.6	12.2	9.9	9.7
Wet sand,	24.9	20.3	12.2	9.6	9.9
10% moisture	37.6	43.4	12.1	9.9	9.9
content	67.3	53.7	13.3	9.7	9.8
Wet sand,	27.3	20.4	11.9	10.0	10.0
13% moisture	39.0	51.7	11.9	10.0	9.9
content	68.3	100.5	12.1	10.0	9.8

[a]1 Btu/h · ft = 0.960876 W/m; 1 in. = 0.0254 m.

data points were taken around the underground pipes vertically at 0.305-m (1-ft) intervals between the depths of 0.305 and 1.52 m (1 and 5 ft) beneath the ground surface and horizontally at 0.305-m (1-ft) intervals for a distance of 3.35 m (11 ft).

For determining soil moisture content, a 76.2-mm (3-in.)-diameter helicoid bore auger was used to drill and excavate sample soils at various depths. Each soil sample taken with a soil sampler was sealed in a plastic wrap and placed in a glass container. The soil moisture content was determined in the laboratory by measuring the loss in weight of the sample after it was dried in an electric oven maintained at 100°C (212°F). The moisture contents of the soil samples taken above the steam pipe at 0.305, 0.61, and 0.91 m (1, 2, and 3 ft) below the ground surface were found to be 15.3, 19.4, and 9.2%, respectively. The undisturbed soil samples taken at 0.305-m (1-ft) intervals at depths of between 0.305 and 1.83 m (1 and 6 ft) had 15.5, 28.5, 27.7, 29.4, 31.9, and 36.2% moisture contents. The lower moisture levels found near the steam pipe were due to drying of the nearby soil by the heat of the pipe. The undisturbed soil has a relatively constant moisture content except for the soil in proximity to the ground surface, which is very susceptible to weather conditions.

The measured values of the thermal conductivity of soils surrounding the buried pipes are summarized in Table 3. The soil thermal conductivity values vary over a wide range from approximately 0.42 to 1.69 0.73 to 2.92 W/m · K (Btu/h · ft · °F), depending on their relative location with the heating pipes. The thermal conductivity of soil decreases with reduced moisture since the local heat transfer rate results in a redistribution of soil moisture, which affects the thermal conductivity.

Based on these 60 data points of earth temperatures and the average value of the thermal conductivity of soil at 0.61 to 0.91-m (2 to 3-ft) depths, nonlinear least-squares calculations were carried out to estimate the total heat loss from the underground system. Table 4 presents the calculated results, which include the heat loss from two-pipe systems and the horizontal locations and depths of the underground pipes. It is not possible to verify the calculated heat loss values since the actual values are unknown. However, the calculated depths and locations of the buried pipes agree reasonably well with the values found in the architectural drawings. The calculated isotherms in the vicinity of the underground pipes and experimental temperature values are shown in Fig. 9. The data indicate that a reasonably good correlation exists between the calculated and measured temperature profiles.

TABLE 3—*The thermal conductivity of soil around the underground pipes on the James Madison University campus.*[a]

Depth, ft	Soil Thermal Conductivity, Btu/h · ft · °F, at Various Distances, ft							
	2	4	6	7	8	10	12	22
1.31	1.693	0.732	0.737	1.062	0.843	...	...	...
2.30	1.116	0.632	0.364	0.417	0.416	...	...	...
2.62	...	...	...	...	...	0.637	0.676	0.529
3.94	...	...	...	...	...	0.578	0.718	0.599
5.25	...	...	...	...	...	0.906	0.695	0.777

[a]1 Btu/h · °F = 1.728 W/m · K; 1 ft = 0.3048 m.

TABLE 4—*The calculated results of field testing.*[a]

Total heat loss	93.7 Btu/h · ft
Heat loss from steam pipe	185.5 Btu/h · ft
Heat gain for return pipe	91.8 Btu/h · ft
Locations of the underground pipes	
Steam pipe	
Horizontal distance	6.04 ft
Depth	3.99 ft
Return pipe	
Horizontal distance	4.96 ft
Depth	4.44 ft

[a]The following values were used in the calculations:

the average value of soil thermal conductivity = 0.524 Btu/h · ft · °F; and

the undisturbed soil temperatues at 1-ft intervals between 1 and 5 ft below the ground surface are 76.5, 73.9, 70.7, 67.7, and 64.6°F, respectively.

1 Btu/h · ft = 0.960876 W/m; 1 ft = 0.3048 m; 1°F = 5/9°C + 32.

Conclusions and Recommendations

A scale sandbox or model was constructed to emulate a two-pipe underground heat distribution system. The heat loss determined by either the temperature profile along the centerline of the test box or the sand temperatures in the vicinity of the underground system, and measured thermal conductivity of sand showed good agreement with actual electrical power input to the two test pipes. This thermal probe technique was also applied to the protextulate insulation system installed on the James Madison University campus. The calculated depths and locations of the underground pipes agree reasonably well with the actual values found in architectural drawings.

Similar scale-model tests will have to be conducted for different types of soil and at various depths of the buried pipes before this thermal probe method can be safely field tested. The next phase of testing will be to conduct a similar test for wet soil to simulate more conductive soil.

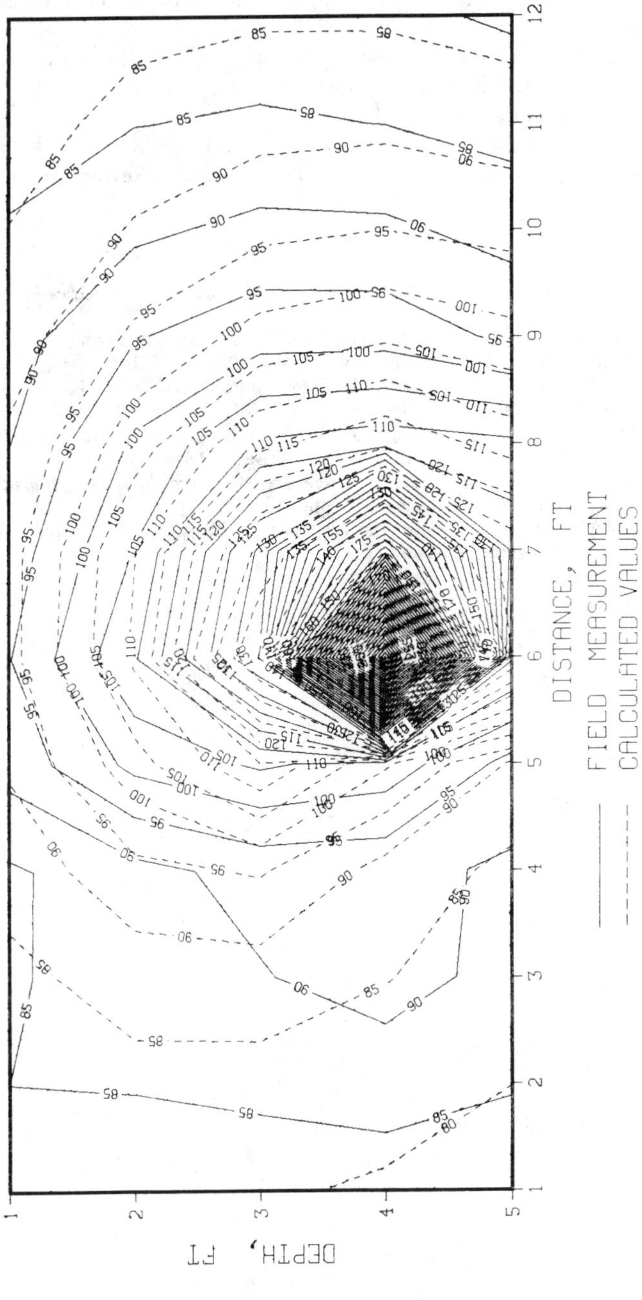

FIG. 9—*Calculated and measured temperatures of soil surrounding the underground pipes on the James Madison University campus, Harrisonburg, Virginia.*

Acknowledgments

The authors wish to thank Doug Pruitt and Erik Anderson of the National Bureau of Standards (NBS) for their help in obtaining the field data used in this paper. They are also grateful for the assistance and advice given by Dr. James Filliben of NBS on the use of the computer program DATAPLOT for nonlinear least-squares calculations. The authors also appreciate the cooperation provided by John Driver of Harrisonburg, Virginia, for granting permission to test the underground system of James Madison University.

This work was sponsored by the Tri-Services Committee on Underground Heat Distribution Systems, U.S. Department of Defense, under the NBS/Tri-Services Research Program.

References

[*1*] Eckert, E. R. G. and Drake, R. M., *Analysis of Heat and Mass Transfer,* McGraw-Hill, New York, 1972.

[*2*] Mason, V. V. and Kurtz, M., "Rapid Measurement of Thermal Resistivity of Soils," *Transactions of the American Institute of Electrical Engineers,* Vol. PAS-71, 1952, pp. 570–577.

[*3*] Boggs, S. A. and Chu, F. Y., "Measurements for Underground Thermal Transmission Design," EPRI Report, Electric Power Research Institute, Palo Alto, CA, 1982, pp. 134–146.

[*4*] Osborne, M. R., "Some Aspects of Nonlinear Least Squares Calculations" in *Numerical Methods for Nonlinear Optimization,* F. A. Lootsma, Ed., Academic Press, London, 1972.

[*5*] Osborne, M. R., "Nonlinear Least Squares—The Levenberg Algorithm Revisited," *Journal of Australian Mathematical Society,* Vol. 19, Series B, 1976, pp. 343–357.

[*6*] Filliben, J. J., "DATAPLOT—Introduction and Overview," NBS Special Publication 667, National Bureau of Standards, Washington, DC, June 1984.

Michael P. Duff[1]

Foamed-in-Place Polyurethane Foam Insulation System Design and Application for Low-Temperature Storage Tanks

REFERENCE: Duff, M. P., **"Foamed-in-Place Polyurethane Foam Insulation System Design and Application for Low-Temperature Storage Tanks,"** *Thermal Insulation: Materials and Systems, ASTM STP 922,* F. J. Powell and S. L. Matthews, Eds., American Society for Testing and Materials, Philadelphia, 1987, pp. 69-81.

ABSTRACT: Energy savings from reduced fuel consumption due to thermal insulation systems are frequently compared with the capital investment cost of the insulation system. Energy-saving thermal insulation programs for large storage tanks are not always implemented because the rate of return on the insulation investment may not be as attractive as the return on other investments competing for the same capital equipment expenditure dollars.

Seeking to develop a low-cost, thermally efficient, long-lived thermal insulation system that could increase energy savings and reduce first costs, the author selected foamed-in-place polyurethane foam as the system's basic insulation material. Insulation system details were selected to exploit the advantages and to overcome the disadvantages of polyurethane foam as a field-applied insulation material. A unique application method and application machinery were developed, specially suited to application to large, flat-bottomed, low-temperature storage tanks. The important features of the insulation system, Chicago Bridge and Iron Company's horizontal foamed-in-place (HFIP) insulation system, and the special application method are discussed in detail in the paper.

KEY WORDS: thermal insulation, insulation systems, low temperature, storage tanks, cellular polyurethane, polyurethane foam

Thermal insulation systems are not really installed to save energy, they are installed to save money. If energy were free, no good businessman would spend money on thermal insulation. Unfortunately, energy is not free, and in most areas it is not cheap either, so more than ever before industry is turning to thermal insulation as a means of reducing energy costs.

For large (15 m in diameter and up), low-temperature[2] storage tanks, the extensive surface areas that must be covered make large capital investments necessary to take advantage of the energy savings provided by thermal insulation. The expected energy savings can be calculated with relative ease and can be compared with the initial cost of the thermal insulation system to obtain an annual rate of return on the insulation system investment. The rate of return can then be compared with the expected rate of return on other capital improvements or investments that may be competing for the same capital investment dollars.

Thermal insulation systems, as competing investments, are somewhat handicapped in that

[1]Design engineer, Insulation System, Chicago Bridge & Iron Co., Oak Brook, IL 60521.

[2]Low temperature here refers to refrigerated tanks storing products at temperatures between 4.4 and −51.1°C (+40 and −60°F), which is the temperature range at which Appendix R of API Standard 620 is applicable [5].

they are not very exciting investments; they have no moving parts and you cannot check their value on the daily financial page. Insulation systems, generally speaking, are not very pretty. In addition, many single-wall insulation systems for large storage tanks, especially low-temperature tanks, have a poor reputation regarding longevity and maintenance requirements.

For these reasons, energy-saving thermal insulation programs for large storage tanks are not always implemented, even though they may be financially attractive. Often only "minimum-thickness" insulation options, which require the smallest capital expenditure, are purchased. Therefore, to be competitive, and to have any chance of being implemented, an insulation system must possess a very high thermal efficiency, be long lasting, require very little maintenance, and, of course, be low in cost.

Design Procedure

The functional design of an insulation system must include more than just selection of the thickness of insulation to be used. All of the materials, primers, sealants, fasteners, and application equipment necessary to ensure system performance must be selected, designed, and detailed before a meaningful thickness design can be made. Later, "economic thickness of insulation" analyses, which determine the thickness of insulation that will provide the greatest rate of return on investment, can be performed on a job-by-job basis to properly consider local variables such as labor rates, shipping costs, interest rates, and utility costs.

We can discuss the mechanical portion of an insulation system design using the design of the horizontal foamed-in-place (HFIP) system as an example. A list of thermal insulation system design objectives can be assembled from the desired insulation system properties:

1. *Low heat transmission*—This is the most obvious requirement for an insulation system, but unfortunately, low heat transmission is sometimes traded off for other features such as lower cost, simpler installation, or resistance to moisture. An initially low heat transmission is not enough, the system must maintain the low value.
2. *Tolerance for thermal and mechanical movements and thermal stresses*—Thermal and mechanical stress movements of large storage tanks can be quite large. Thermal movement of the insulation material itself will also occur because of variations in stored product temperature and ambient temperature. The insulation system must withstand these movements and stresses.
3. *Wind resistance*—Tank insulation systems or tank insulation coverings should not blow away in severe storms. Banded-on metal jackets especially can be a problem on large-diameter tanks.
4. *Long service life*—A tank insulation system should last at least ten years without requiring replacement or major repairs.
5. *Low maintenance requirement*—All insulation systems for single-wall tanks need periodic inspection and routine maintenance, but very few get them. Insulation systems should be easy to maintain and reasonably tolerant of lack of maintenance or delays in repairs in the event of mechanical damage.
6. *Fast application*—The costs often evaluated by the purchaser are the costs in lost storage or lost production from the tank's downtime. The insulation system design should allow for rapid application with a minimum of inconvenience and delay to the purchaser.
7. *Low cost*—Usually, only the lowest-cost, most financially attractive insulation programs get implemented. Purchasers should consider the risks for early rework or replacement of poorly designed insulation systems, but they often do not.

Selection of Primary Insulation Material

The most important step in designing an insulation system is the selection of the primary insulation material. All other features of the insulation system design are dependent on the primary material selection, as is the cost. No single insulation material will fulfill all design objectives, but some shortcomings can be "designed around."

Insulating efficiency is, of course, the most important factor in selecting an insulation material, followed closely by the cost. Selection of the primary insulation material can be limited to a group consisting of the six or seven most common commercially available insulation materials, listed in Fig. 1. The thermal conductivities of these materials are all well documented and easily obtainable from manufacturers' literature, technical papers, and other sources.

For comparison purposes, the thermal resistivities (*R* values) of the insulation materials and their approximate cost[3] per cubic metre or per board foot can be used to calculate a figure of merit (FM) for each insulation material. The FM is simply the *R* value divided by the basic material cost. Since both the *R* value and the cost are calculated for a unit area and thickness of insulation, the FM is a relative measure of the insulation value received per dollar spent. The *R* values and the FMs are depicted graphically in Figs. 1 and 2.

The merits of cellular polyurethane, also known as polyurethane foam,[4] are obvious from Figs. 1 and 2. Polyurethane foam has the highest thermal resistivity and the highest figure of merit of the insulation materials under consideration. It must clearly be the first-choice insulation material from the standpoint of both performance and cost. Polyurethane foam is available as either manufactured boards or slabs, or as liquid components that can be used to manufacture or foam the polyurethane in place on the job site. Foamed-in-place (FIP) polyurethane foam offers several advantages over foam boards. The liquid components occupy only about 1/30th of the volume of the finished insulation; therefore, shipping costs are drastically reduced. FIP polyurethane exactly fits the contours of the tank to be insulated, providing a better fit and eliminating cutting and waste. When foamed in place, the polyurethane's self-adhesive properties may be used to attach the insulation to the tank, eliminating the need for separate adhesives

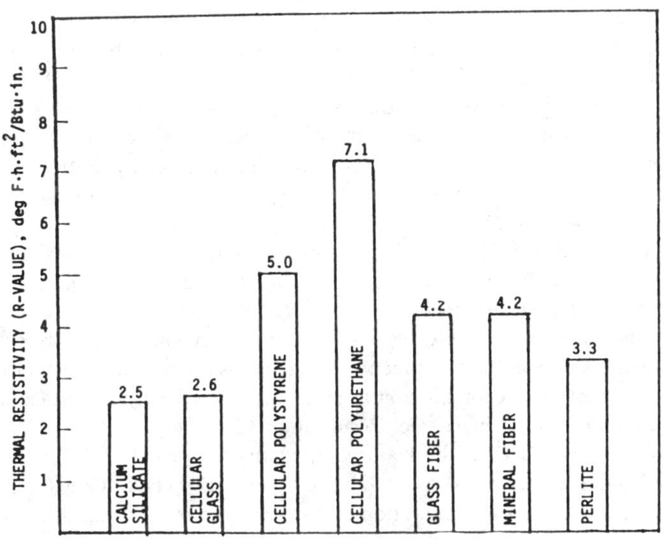

FIG. 1—*Comparison of thermal resistivity of several common types of insulation material.*

[3]The material costs used were average "contractor's costs" in the Midwestern United States, for summer 1984.

[4]Polyurethane foam here also includes polyisocyanurate foams and modified polyisocyanurate foams, because the thermal properties and application methods are essentially identical to those of polyurethane.

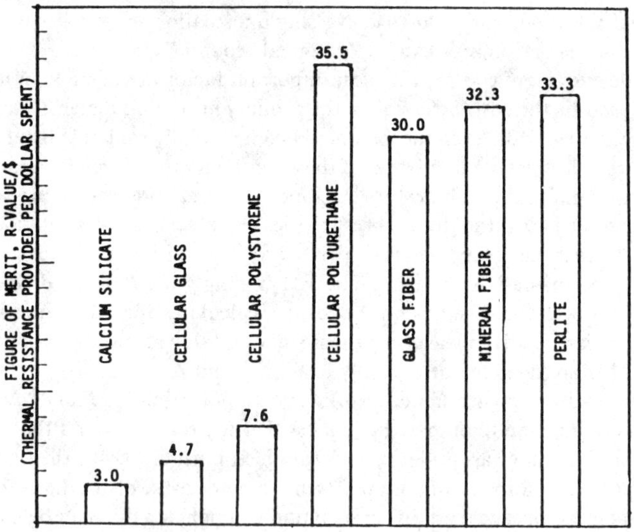

FIG. 2—*Comparison of the figure of merit of several common types of insulation materials.*

and fasteners. Foaming polyurethane in place also results in a single, monolithic piece of insulation material covering the entire shell of the tank. This monolithic structure eliminates discontinuities that often act as "thermal stress raisers"; thus, thermal movements and resulting thermal stresses are more uniformly distributed around the tank. The monolithic structure of FIP polyurethane also eliminates the internal voids associated with joints that most insulation systems have. Internal voids can increase the heat transfer, and when subjected to warming and cooling cycles, internal voids try to "breathe" and ingest air into the insulation system. This breathing of moist air is one of the mechanisms by which moisture enters a cold insulation system. Experience with other insulation systems that use composite FIP/board polyurethane foam insulations also shows that board polyurethane is much more susceptible to water penetration than FIP foam.

Polyurethane foam may be foamed in place by one of three general methods: pouring, spraying, or frothing. The main differences between these methods are the equipment used and the reaction and forming times of the foam.

In a polyurethane pour system, relatively slow-reacting foam components are mixed, and the mixed liquid is dispensed from the equipment by pouring. As the name implies, this method of foaming in place is best suited for horizontal applications and not for vertical applications, where the liquid mixture would be subject to running and sagging.

In a polyurethane spray system, the foam components are proportioned and mixed in the equipment under high pressure and dispensed by spraying. The polyurethane foam components are usually highly catalyzed for this operation in order to obtain a fast rise and set of the foam for application to vertical surfaces.

Froth systems may utilize either pour or spray equipment. In a froth system, Freon 12 blowing agent is added to the foam component liquids in addition to the normal Freon 11 blowing agent. Since Freon 12 boils at a temperature lower than normal ambient temperatures, it expands and foams up the component liquids immediately upon leaving the dispensing equipment to about one half of their final volume. The remainder of the expansion occurs after the heat of reaction raises the temperature sufficiently to boil the Freon 11 blowing agent. The advantage of a froth system is that the initial expansion allows for a much easier visual estimation of the amount of foam required to fill a given volume.

Because of the vertical position of the tank shell insulation system, and because a rapid rise and set were required to minimize foam application time and reduce cost, spray foam was selected for use in the HFIP tank insulation system. Spray foam provides an additional advantage in that it usually "wets out" the substrate it is sprayed on better than pour or froth foams, and a better foam-to-substrate bond is achieved. Spray foams are also available in a wider selection of densities, reaction speeds, and physical properties, from a wider selection of manufacturers, than are pour or froth foams.

Designing Around FIP Polyurethane's Shortcomings

FIP polyurethane foam unfortunately also has some disadvantages that must be compensated for in the insulation system design. Almost all the disadvantages can be overcome by the use of a barrier coating over the foam, which will isolate the foam from the environment. Selection of the barrier material should be aimed at elimination of all disadvantages with one material.

The main disadvantage is that polyurethane foam will absorb water. Water reduces the thermal resistance of polyurethane foam. This is especially a problem on low-temperature insulation systems because the difference in water vapor pressure between the outside air and the cold interior of the insulation system causes water vapor to diffuse into the insulation system. At a 26.6°C (80°F) air temperature and 80% relative humidity, and with a storage tank temperature of −33.3°C (−28°F), the vapor pressure difference is 2.75 kN/m^2 (0.4 psi). Other sources of water are leakage of air carrying water and leakage of surface water. To protect the polyurethane foam from water, a vapor barrier must be provided to completely cover the insulation system.

Polyurethane foam can also lose insulating efficiency because of "aging." The thermal conductivity of polyurethane foam, when it is new, is approximately 0.0187 W/m · K (0.13 Btu · in./h · ft^2 · °F). If polyurethane foam is left exposed to air, in time the fluorocarbon blowing agent in the cells of the foam is replaced by air, and since air has a higher thermal conductivity than the fluorocarbon blowing agent, the thermal conductivity of the polyurethane foam increases with time. If the foam sections are thick and are covered by an impermeable membrane, however, the exchange of cell gases and air is prevented and the initial thermal conductivity is maintained [*1*].

Another disadvantage of polyurethane foam is the fact that it burns. Low-flame-spread foams, with flame spreads of 25 or less, as measured in accordance with ASTM Test for Surface Burning Characteristics of Building Materials (E 84-84), are available, but they will still burn if exposed to a high-intensity heat source. This is of particular concern on tanks located in refineries, such as propane or butane tanks. When considering fire protection for an insulation system, consideration must be given to the type of fire from which the foam is to be protected. Several different kinds of coverings or coatings are available that afford different degrees of fire protection for the foam, but no ordinary coating or covering will protect polyurethane foam from a fire resulting from a major petroleum spill. Even nonburning insulation materials, and probably the storage tank itself, would be destroyed by a large petroleum product fire.

A more reasonable design approach is to protect the insulation system from small-spill fires and trash fires and to prevent the insulation from contributing to the fire in any substantial manner. The easiest way to provide such protection is by applying a barrier of some kind over the polyurethane to prevent flames from contacting the insulation in the event of a small fire.

Polyurethane foam is also sensitive to ultraviolet light and will degrade rapidly if exposed to sunlight. A barrier must be provided to protect the polyurethane from light.

Another disadvantage of polyurethane is that it is relatively soft and susceptible to mechanical damage. Mechanical protection must be provided.

Vapor Barrier Design

All of the preceding requirements for protection of polyurethane foam—vapor, fire, ultraviolet light, and mechanical barriers—should ideally be provided by a single material. After testing of various elastomeric coatings, cementitious coatings, and plastic and metal jacketing materials, corrugated aluminum jacketing was selected as the barrier material for the HFIP insulation system.

Aluminum jacketing has a water vapor permeability of essentially zero in the heavier thicknesses. (It must, of course, also be provided with suitable seal details at the seams.) It also provides a degree of fire protection from small fires by preventing flame and air from contacting the foam surface, at least at temperatures up to the melting point of aluminum—about 650°C (1200°F). It completely blocks ultraviolet light and, when corrugated, is still enough to afford a moderate amount of protection from mechanical damage.

To provide a degree of corrosion protection for the aluminum jacketing, the outside surface is coated with a factory-applied baked-on white polyester enamel. The inside surface is coated with Swift Fome-Bond 75, a primer that both protects the aluminum from corrosion and enhances the bond of the polyurethane foam to the aluminum jacket.

Jacket Seam Design and Testing

Because the aluminum jacket on the HFIP insulation system is corrugated only in the vertical direction, which provides for movement in the horizontal direction only, vertical movements in the insulation system must be provided for by maintaining flexibility in the horizontal seams. The sealant and fasteners must be selected to maintain a tight seal yet still allow for the necessary movement. A flexible butyl rubber tape sealant was selected for the horizontal seams because butyl rubber sealants can be formulated to remain flexible indefinitely and to possess an aggressive "tack" to make them stick to the aluminum jacket.

To confirm the suitability of the joint sealant and fasteners for cyclic joint movements, simulated insulation system segments, which included polyurethane foam, aluminum jacketing, several different types of sealants, and several different types of fasteners, were cycled in an Instron testing machine through movements that simulated a daily ambient temperature change from −34°C (−30°F) to 38°C (100°F), for over 36 500 cycles to simulate 100 years of daily thermal movements. In the tests, screws were found to be unsuitable because they sometimes loosened and "unscrewed" themselves from the seam under cyclic movement. Blind rivets, of the type that form "petals" on the inner side of the seam, were found to be the best fasteners. The blind rivet that was ultimately selected for use with the HFIP insulation system was actually found to clamp up the jacket seam too tightly because it did not allow movement of the jacket sheets; therefore, thermal movements caused repeated buckling of the aluminum jacket around the perimeter of the rivet heads and eventual fatigue failure of the aluminum jacketing. To prevent this, special rivets were obtained from the manufacturer with an enlargement or "shoulder" on the stem of the rivet that prevents clamp-up beyond a thickness which allows for approximately 50% compression of the butyl rubber sealant in the joints. This maintains the necessary flexibility in the seam to prevent buckling of the jacket without sacrificing the superior sealing properties and high pull-out strength of the rivet. The butyl sealant performed satisfactorily in all the joint tests, regardless of the type of fastener used.

Jacket Seam Permeability

To ensure that the jacket seams were sufficiently leak tight to allow the aluminum jacket to constitute a vapor barrier over the polyurethane foam, tests were performed to measure the water vapor transmission rate through the seams. The tests were performed in a manner very similar to the wet cup method of ASTM Test for Water Vapor Transmission of Materials (E 96-

80), except that a higher test temperature was used to simulate a Middle East location where the vapor pressure differential would be high. Both specimens of jacketing containing seams and control specimens containing solid pieces of aluminum jacketing without seams were tested, and the difference was assumed to be the water vapor transmission rate of the seam. The water vapor transmission rate was calculated in terms of grams per hour metre (lineal meter of seam) instead of grams per hour metre square.

The average measured rate of water vapor transmission was 0.0126 g/h · m^2 at 43°C (0.0591 grains/h · ft at 110°F) and 50% relative humidity. The calculated vapor pressure difference was 32.9 mm Hg (1.296 in. Hg). Since the seam spacing is 1.12 m (44 in.), there are 1.12 m^2 of insulation for every lineal metre of seam (3.67 ft^2 per foot of seam). For comparison purposes, an effective permeance can be calculated by assuming that the rate of water vapor transmission measured for 1 m of seam occurs over 1.12 m^2 of surface area. The "effective" permeance calculated in this manner is 0.718 metric perm (0.0125 grains/h · ft^2/in. Hg). This is one fourth of the maximum permeance recommended for pipe and vessels at −40 to 0°C (−40 to 32°F) in Table 2 of ASTM Recommended Practice for Selection of Vapor Barriers for Thermal Insulations [C 755-73 (1979)].

To demonstrate the efficiency of the vapor barrier, if the measured rate of water vapor transmission through the seams is assumed to occur for ten years and the average vapor pressure differential is assumed to be 3.39 kN/m^2 (1.0 in. Hg), the water gain of the insulation system is calculated to be 0.762 kg/m^2 (0.156 lb/ft^2). If the polyurethane foam insulation thickness is 102 mm (4 in.), the calculated moisture content of the insulation after ten years is 0.75% by volume.

The thermal conductivity of the polyurethane foam that has been exposed to moisture can be estimated by the following expression [*2*]

$$k_{\text{wet}} = k_{\text{dry}} + 0.008V$$

where

k_{wet} = thermal conductivity of polyurethane foam in moist condition, Btu · in./h · ft^2 · °F,[5]
k_{dry} = thermal conductivity of polyurethane foam in dry condition, Btu · in./h · ft^2 · °F, and
V = moisture content of the polyurethane foam, percentage by volume.

Thus, the initial thermal conductivity of 0.0187 W/m · K (0.13 Btu · in./h · ft^2 · °F) can be expected to increase to only 0.0196 W/m · K (0.136 Btu · in./h · ft^2 · °F) in ten years due to water vapor diffusion into the foam. This represents an increase of only 4.6% in ten years, which is an acceptable rate for a low-temperature tank insulation system.

Checking Thermal and Mechanical Stresses

Thermal stresses in the insulation system are the result of differential thermal expansion or contraction between the tank and the insulation system. Mechanical stresses are the result of tank movements due to hydrostatic loads and pressure. Both must be checked to ensure that cracking and damage to the insulation system will not occur.

Assuming the vertical and horizontal dimensions of the insulation system and the radius of the tank are large compared with the insulation thickness, and assuming that the tank wall imposes external constraint on the polyurethane foam, the temperature stress in the foam, near the steel tank, can be assumed to be equal to the following [*3*]

$$\text{thermal stress} = \frac{\Delta TE(\alpha_f - \alpha_s)}{1 - \nu}$$

[5] 1 Btu · in./h · ft^2 · °F = 0.144 W/m · K.

where

ΔT = temperature change in the foam and the steel tank,
E = modulus of elasticity of the foam,
α_f = coefficient of thermal expansion of the foam,
α_s = coefficient of thermal expansion of steel, and
ν = Poisson's ratio.

Mechanical stresses in the foam, if we assume the foam is constrained by the tank wall, are proportional to the mechanical stresses in the steel tank wall and can be calculated by the following (using any consistent units)

$$\text{mechanical stress} = \frac{E_f \sigma_s}{E_s}$$

where

E_f = modulus of elasticity of the foam,
E_s = modulus of elasticity of the steel, and
σ_s = mechanical stress in the steel.

Propane is usually the coldest product stored in single-wall, low-temperature tanks; therefore, propane tanks should experience the highest thermal stresses and the highest combination of thermal and mechanical stress. We will therefore use a propane tank as an example. Assuming a propane tank constructed at 25°C (77°F) with a service temperature of −45°C (−49°F), insulated with a 32-kg/m^3 (2-lb/ft^3) nominal-density polyurethane foam, the temperature change in the foam will be 70°C (126°F), and the stresses in the foam can be calculated as will be shown.

From Ref *4*, typical physical properties of a 32-kg/m^3 (2 lb/ft^3) polyurethane foam at −45°C are as follows:

E = 8275 kN/m^2 (1200 psi),
α_f = 5.5 × 10^{-5}/°C (3.05 × 10^{-5} in./in. · °F),
γ = 0.3, and
tensile strength = 310 kN/m^2 (45 psi).

If the propane tank shell is constructed of a typical low-temperature tank steel (ASTM A 516 Grade 70, using stresses allowable by American Petroleum Institute Standard 620) [5] and is stressed to its maximum allowable tension, we can assume the following steel properties:

α_s = 1.1 × 10^{-5}/°C,
E_s = 2.0 × 10^8 kN/m^2 (29 000 000 psi), and
σ_s = 1.45 × 10^5 kN/m^2 (21 000 psi).

Using these values, we get a thermal stress of 36.4 kN/m^2 (5.28 psi), a mechanical stress of 6.0 kN/m^2 (0.897 psi), and a combined (tensile) stress of 42.4 kN/m^2 (6.15 psi). The combined tensile stress in the foam is considerably less than the tensile strength of the foam (factor of safety of 7.3), therefore, the stress is acceptable, and cracking of the foam should not be a problem. Thermal and mechanical stresses should be checked similarly for each tank on an individual basis, but in most cases, stresses will be lower than given in the example.

Wind Resistance

Large storage tanks are typically designed to withstand a wind velocity of 44.7 m/s (100 mph) (higher velocities are specified for some coastal areas), and insulation system specifications usually require the insulation system to be designed for the same velocity as the storage tank. Because the insulation system has a considerably lower dollar value than the tank or the stored product, and because an insulation failure would not normally result in a life-threatening situation, it would seem reasonable to design the insulation system for a lower wind speed and accept the risk that it might be blown off in a heavy storm, but most purchasers seem unwilling to accept such a design. Multiple tank installations can also be a special problem because the funneling effect of the wind passing between tanks can increase the velocity substantially. For this reason, it is desirable to design a system that can tolerate the full design wind velocity.

The governing wind design consideration is usually the local negative (suction) pressure that occurs at a point approximately 90° from the wind direction (see Fig. 3). The local external wind pressure on the surface can be calculated using the following formula [*6*]

$$p = C_{pe}q = C_{pe}(0.002558)V^2$$

where

p = local pressure, lb/ft²,
q = dynamic wind pressure, lb/ft²,
C_{pe} = external pressure coefficient, and
V = wind velocity, mph.[6]

This formula is applicable to wind velocities in miles per hour and pressure in pounds (force) per square foot. Formulas for use with SI units can be found in the literature.

From Ref *6*, the external pressure coefficient, C_{pe}, for a point on the tank surface 90° from

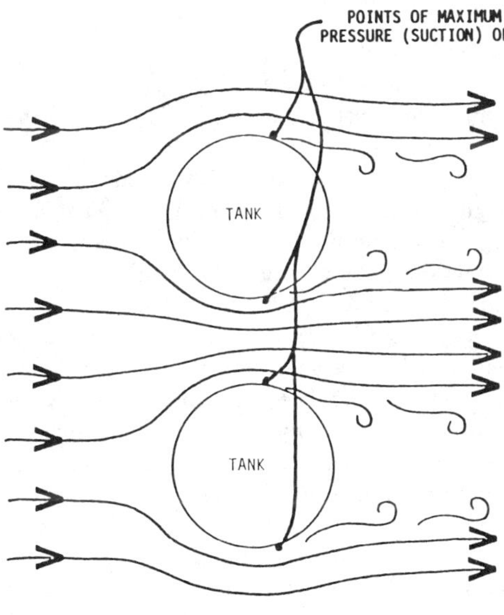

FIG. 3—*Pattern of wind flow around tanks.*

[6]1 mph = 1.6 km/h.

the incoming wind direction, is −1.7. The local negative (suction) pressure on the insulation surface for a 44.7-m/s (100-mph) wind velocity is

$$p = -1.7(0.002558)(100)^2 = -43.5 \text{ lb/ft}^2 \text{ (212 kg/m}^2\text{)}$$

If this pressure is applied to an insulation system that utilizes bands to hold the insulation on the tank, the negative pressure must be resisted by tension in the bands. The tension will be a function of the radius of the surface of the insulation and the spacing of the bands and can be calculated using the formula

$$T = pRs$$

where

T = tension in the band,
p = local pressure,
R = radius of insulation surface, and
s = band spacing (in any consistent system of units).

For example, for a 30-m-(98.4-ft) diameter tank (15 m in radius) with bands 1 m (3.28 ft) apart, the tension in the band would be 31.2 kN (7024 lb). Therefore, the bands, band splices, any spring tensioners, and all attachments should be designed to withstand a 3186-kN (7024-lb) load.

Compared to a banded-on insulation system, an insulation system that is continuously bonded to the tank can withstand a 45-m/s (100-mph) wind with relative ease. The local negative pressure is resisted by the bond of the insulation to the tank. For sprayed-in-place polyurethane foam insulation systems, foam-to-substrate bond strengths of 7030 kg/m² (1440 lb/ft² or 10 psi) are routinely achieved, which greatly exceed the 212-kg/m² (43.5 lb/ft²) strength required to resist the negative wind pressure.

Application Method

Application methods and equipment must, of course, be suitable for the insulation system details selected.

To apply the HFIP polyurethane insulation system and its bonded aluminum jacket, special application equipment was designed, constructed, and tested. In essence, the equipment manufactures, in place on the storage tanks, continuous insulation panels that extend all the way around the tank. The panel consists of the tank shell, the polyurethane foam, and the aluminum jacket, which are all bound together into a single unit by the adhesive property of the foam. The panel is manufactured by spraying polyurethane foam, with a reciprocating spray gun, onto the leading edge of a cavity formed between the tank and the aluminum jacket (as shown in Fig. 4). An extremely fast-reacting polyurethane foam formulation is used to enable each sprayed layer of foam to rise and partially set before the spray gun returns and sprays the next layer. The majority of the polyurethane foam in each layer is thus applied by spraying foam on foam, with only a relatively small amount being applied directly to the tank shell and aluminum jacket to provide the bond necessary to hold the system together. Because the output of the polyurethane spray gun is fixed, the horizontal speed of the application equipment is controlled to match the area of insulation installed, at a given thickness, to the spray gun output.

The aluminum jacketing is purchased 1.22 m (48 in.) wide; thus, with approximately a 102-mm (4-in.) overlap, each horizontal strip, or panel, covers an area 1.12 m (44 in.) high. As each successive panel is applied, the 76-mm-wide, 4.8-mm-thick (3-in.-wide, 3/16-in.-thick) butyl

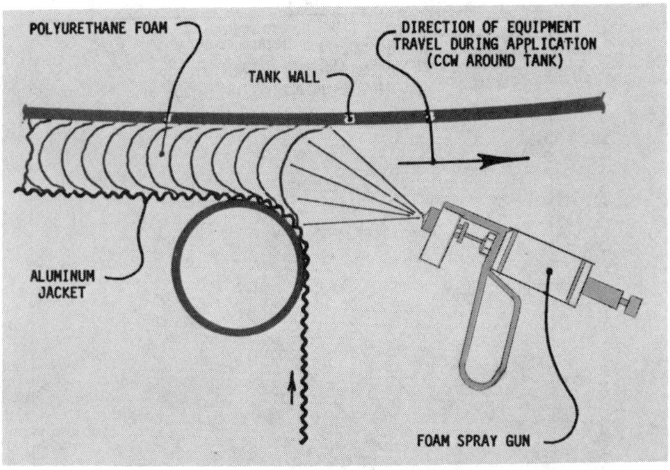

FIG. 4—*Horizontal cross section through tank wall during HFIP insulation system application.*

tape sealant is inserted into the overlap of the aluminum jacket, and force is applied to the lap to close the lap and embed the sealant into the corrugations in the jacket. The rivet holes are then automatically punched through both layers of the jacket and through the butyl tape sealant, and blind petal-type rivets are installed. The insulation application proceeds from the bottom of the tank upward until the entire tank is insulated. The insulation equipment is a completely self-contained unit that utilizes a four-man crew and requires no additional scaffolding. This minimizes field labor costs.

Summary—Getting the Most for the Money

For economic thickness of insulation analyses or designs, many tank owners are becoming more value conscious and are making a total systems evaluation of their low-temperature storage facility options. A typical evaluation will include the following:

(*a*) tank and insulation system present-day cost,
(*b*) refrigeration equipment cost,
(*c*) maintenance costs for Items *a* and *b*,
(*d*) allowance for replacement of Items *a* and *b*, and
(*e*) present-day utility costs for the life of the facility.

Many options can affect the costs for each item, but one reasonably predictable item is the cost of power, normally electricity, to operate the refrigeration system on larger low-temperature tanks such as ammonia tanks. Figure 5 can be used to assist in evaluating requirements for holding refrigeration and for utility costs on large refrigerated ammonia tanks with almost any FIP polyurethane thermal insulation system as long as the vapor barrier will maintain the system integrity. The figure is based on the following:

(*a*) typical dimensions for ringwall-supported tanks, 192 kN/m^2 (4000 lb/ft^2) allowable soil pressure,
(*b*) a design ambient temperature of 32°C (90°F) and a product temperature of −33°C (−28°F),

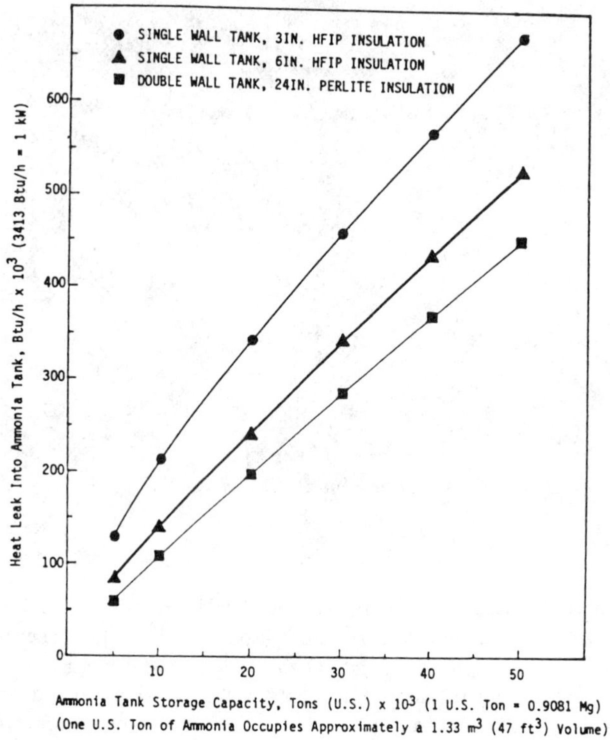

FIG. 5—*Refrigeration load versus storage capacity for typical ammonia storage tanks.*

(*c*) a 100-mm-(4-in.) thick cellular glass bottom insulation system and 150-mm-(6-in.) thick mineral fiber insulation on a suspended roof insulation deck,

(*d*) a design thermal conductivity of 0.020 W/m · K (0.14 Btu · in./h · ft^2 · °F) for polyurethane foam, and

(*e*) a 10% added contingency for heat leaking through piping and attachments.

FIP polyurethane foam thermal insulation systems that are both efficient and cost-effective can be designed for low-temperature storage tanks as long as the system vapor barrier provides protection for the polyurethane foam. The details of the insulation system must be designed, and the performance tested if necessary, to ensure that the integrity of the barrier will be maintained under both normal and severe service conditions. The HFIP insulation system that was used as a design example has, since 1976, competed successfully for 67 tank insulation systems, and all 67 are still in service.

References

[1] Backus, J. K. and Gemeinhardt, P. G. in *Plastic Foams*, Vol. II, K. C. Frisch and J. H. Sounders, Eds., Marcel-Dekker, New York, 1973, Chapter 9, p. 497.

[2] Dechow, F. J. and Epstein, K. A. in *Laboratory and Field Investigations of Moisture Absorption and Its Effect on Thermal Performance of Various Insulations, ASTM STP 660*, American Society for Testing and Materials, Philadelphia, 1978, pp. 234–262.

[3] Roark, R. J. and Young, W. C., *Formulas for Stress and Strain*, 5th ed., McGraw-Hill, New York, 1975, p. 583.

[4] "Properties of Rigid Urethane Foams," Du Pont Technical Bulletin E-08934, E.I. du Pont de Nemours and Co., Elastomer Chemicals Dept., Wilmington, DE, 1970.
[5] *Recommended Rules for Design and Construction of Large, Welded, Low Pressure Storage Tanks,* 7th ed., API Standard 620, American Petroleum Institute, Washington, DC, 1982.
[6] "Wind Forces on Structures," ASCE Paper No. 3269, *Transactions of the American Society of Civil Engineers,* Vol. 126, Part II, 1961, pp. 1149–1150.

Wartan J. Wartan[1]

Prevention of Common Pitfalls Through Engineered Specifications and On-Site Review

REFERENCE: Wartan, W. J., **"Prevention of Common Pitfalls Through Engineered Specifications and On-Site Review,"** *Thermal Insulation: Materials and Systems, ASTM STP 922,* F. J. Powell and S. L. Matthews, Eds., American Society for Testing and Materials, Philadelphia, 1987, pp. 82–88.

ABSTRACT: Poorly engineered specifications have a negative effect on thermal insulation systems. The availability, maintainability, and serviceability of piping and equipment are greatly affected by improper selection of thermal insulation materials and systems. System and equipment design criteria, that is, operating conditions including geographic, local, and environment, must be considered for each specific project.

The specification on file that was successfully used on a previous project at another time and at another location may not always be a wise choice. It is imperative that the manufacturer's literature be reviewed before insulation materials and accessories are specified, to preclude their use beyond the manufacturer's prescribed limits. Referencing of ASTM standards in the specification is a useful means of establishing a level of conformance and quality.

On-site reviews are a useful tool for review of the work and its conformance to the specification. The reviews must occur on a timely basis as the work progresses.

Provisions must be made to resolve problems that may arise because of construction restraints during the progress of the work. Close liaison with the insulation contractor will preclude guesswork and the resulting cost of rework.

KEY WORDS: thermal insulation, engineered specifications, on-site review

Insulation can make a major contribution to the conservation of energy. It is simple to use and cost-effective, with a quick return on the capital investment. Further, the reduction in heat loss associated with a fully engineered insulation system can reduce total energy requirements for a facility. However, when an insulation system is not properly engineered or installed, these benefits will not be fully realized.

The impact of engineered specifications and on-site review is great. An engineered insulation specification will optimize the capital expenditure and, when coupled with on-site review of the insulation, will improve the availability and maintainability of the project, thereby ensuring the intended service life.

Most specifications are composed of general conditions, project requirements, drawings, standards, a scope of work, schedules, pricing arrangements, and so on, including technical requirements with insulation thickness. But too often a passive attitude is taken toward the writing of insulation specifications, and more importance is given to the contractual language, resulting in a less-than-optimum insulation system. This sometimes results in technical require-

This paper has been previously published as "Conservation of Energy Through Engineered Specification and Onsite Review" in *Energy Progress,* Vol. 5, No. 3, 1985, pp. 161–164; it is reproduced with the permission of the American Institute of Chemical Engineers.

[1]Consultant, Sargent & Lundy, Chicago, IL 60603.

ments that are so vague they leave the contractor unsure of what the specification really requires.

Using a specification from a previous project may be expedient, but the existing specification should be reviewed to make sure it is suitable for the process or systems involved, including the geographic location. Many old specifications may contain obsolete insulation materials and methods of installation or inappropriate insulation thicknesses.

While insulation thickness is the primary means of energy conservation through the reduction of heat loss, this paper will show how ignoring other factors can cause an excessive loss of energy.

Factors That Affect Choice of Materials

The selection of insulation materials must be precise. The criteria for selecting insulation should, at a minimum, be influenced by the following questions:

1. Is the insulation designed to reduce energy loss from a hot surface or to reduce the absorption of heat and provide antisweat protection for cold surfaces?
2. Will the insulation be indoors or outdoors?
3. Will the insulation be aboveground or underground?
4. Must the insulation be protected from mechanical abuse?
5. What is the probable incidence of fire?

A wide variety of insulation materials and methods of application are employed to facilitate the application of insulation, thereby reducing the overall cost of the insulation contract. Up-to-date manufacturer's literature should be reviewed prior to specifying an insulation material to ensure that the material will perform as expected.

Types of Insulation Materials

The following is a list of some of the insulations currently available. Some are mass produced, and some are tailored to meet specific requirements that may be used by the industry:

(*a*) molded sectional and segmental pipe and block insulation made of calcium silicate, perlite, cellular glass, mineral fiber, and flexible elastomerics;
(*b*) factory-made preinsulated panels employing mineral fiber insulation;
(*c*) foam and sprayed insulation;
(*d*) sprayed antisweat cork-filled insulation;
(*e*) loose-fill insulation; and
(*f*) metallic reflective insulation.

Each of these products is governed by one or more ASTM specifications. When specifying materials, the use of these ASTM specifications will ensure a level of conformance and quality of the material used. In addition, the ASTM designation will assist the contractor in finding "equal" materials if the specification allows this.

Protective Finish

All installed insulation must be provided with some protective jacket or finish. Whether the jacket or finish is made of metal, cloth, mastic, plastic, or some other material, it must be applied in accordance with the manufacturer's recommendation. Where sheet or preformed jacketing is used outdoors, it must be installed in a manner to shed water with the longitudinal joint in the jacketing applied to horizontal pipes placed below the horizontal centerline, but not at a 6 o'clock angle. Circumferential joints in horizontal piping and all longitudinal joints must be lapped not less than 5.08 cm. Circumferential joints in vertical piping must be lapped not less than 7.62 cm.

Securement of Metal Jackets

All bands and screws used to secure metal jackets must be compatible with the jacketing material. Spacing of bands must be in accordance with the jacketing manufacturer's recommendations to maintain a continuous longitudinal tight lap.

Metallic Jacketing

Metallic jacketing has gained wide acceptance as protection for insulation surfaces. The selection of the proper metal must be determined on the basis of the atmosphere to which it will be exposed. The available jackets are stainless steel, galvanized steel, painted steel, and aluminum. One of the most widely used jacketing materials is aluminum.

Effect of Emissivity

Normally, aluminum and stainless steel jacketing materials are not painted, resulting in higher surface temperatures on the jacketing. This is due to the low emissivity of the shiny metal, which will cause the surface temperature to be 15% higher than surfaces covered with canvas, mastic, or painted metal. To reduce the higher surface temperature of an insulated surface jacketed with shiny metal would require a less cost-effective increase in insulation thickness.

To illustrate, a pipe surface insulated with a given thickness of insulation and covered with canvas may have a measured surface temperature of 65.6°C when the ambient temperature is 37.8°C, under theoretically still air conditions. Without altering the insulated surface in any way, an aluminum jacket is applied over this surface; the measured surface temperature can then reach 82.2°C. This temperature rise is caused by the low emissivity of the aluminum. Simply put, the outer surface temperature of any insulation is controlled by the resistance of the insulation and the heat lost to the surrounding air. Consequently, use of mastic, plastic, painted surfaces, or other surfaces that have higher emissivity than shiny metal surfaces will result in lower surface temperatures and higher heat loss from the system.

Application Procedures

In addition to specifying insulation thickness and the selection of insulation materials, an engineered specification must prescribe specific application procedures. These procedures should parallel the manufacturer's recommendations, with more stringent procedures included where the specifier feels they are necessary. Additional emphasis should be given to the following:

1. *Expansion and contraction*—Insulation should be installed with provision for expansion and contraction of the piping and equipment. Hot piping will expand while the insulation does not; insulation may even shrink. To accommodate linear expansion of piping, the insulation must be designed to preclude hot spots occurring in the insulation jacket. In single-layer applications of insulation, fibrous materials can be packed at suitable intervals at circumferential joints in the insulation. In two-layer applications, broken joint construction may be used. In either case, the vertical runs of piping must have insulation supports installed at frequent intervals to preclude the insulation from stacking and the pipe growing out of the insulation. At

371.1°C, the linear expansion of a carbon steel pipe will be approximately 50 cm per 100 m. To accommodate the contraction of cold piping, provision must be made in the insulation for the reduction in linear length of the piping. The insulation must be installed with circumferential gaps in the insulation, and the circumferential gaps must be packed with fibrous material. At −198.3°C, a stainless steel pipe will have linear shrinkage of approximately 33.3 cm per 100 m.

2. *Vapor Proofing*—Insulation installed on cold surfaces must be vapor proof. Where closed-cell insulation such as the flexible elastomeric type is used, the vapor barrier is built into the insulation. Most other forms of insulation must have the vapor barrier attached to the insulation or applied separately. The vapor barriers may be the sheet or mastic type. When mastics are used, they must have a 0.2-perm rating or better, depending on operating conditions. The vapor barriers must be monolithic to preclude vapor from entering the insulation to form condensation. If metal jacketing is specified for cold surfaces, it must be secured with bands. Screws and pop rivets must not be used to preclude puncturing the vapor barrier. All penetrations through the insulation must be caulked with vapor-proof caulking.

Removable Covers

Another means of energy conservation is to include in the engineered specifications removable and replaceable insulated covers for equipment, piping, and valves that require access for maintenance. These covers can be made of metal or fabric in a variety of shapes and sizes to fit the surface to be insulated and are provided with devices to hold them securely in place. The flexible covers may be specified with a variety of jackets from metal meshes to fabrics suitable for all temperature ranges and outdoor service. The use of removable covers precludes destroying the typical insulation applied to these surfaces, which often end up with no insulation after the maintenance procedures are performed.

Heat Tracing

Almost all power plants, be they chemical, industrial, or electric utility, require some form of heat tracing to maintain the specified fluid temperature in piping or equipment. The tracing medium may be electric resistance heating or steam. Additionally, in many areas of the sunny South, climatic changes have caused a reassessment of antifreeze protection. The duration of the cold spells is sufficient to cause freezing of existing heat-traced systems. The design of heat tracing must include sufficient heat input to overcome the effect of low ambient conditions and wind. All hangers, supports, valves, bonnets, and instrument taps connected to the piping that will act as heat sinks must be considered.

Antifreeze protection for piping penetrating grade must be provided with heat tracing down to the frost line. Special insulation procedures must be employed for such piping to preclude water from penetrating the insulation system. Insulation at the grade line must also be provided with a protective collar extending a minimum of 152.4 cm above grade to protect the insulation from mechanical abuse. Protective collars can be made of heavy plastic pipe or steel. The inside diameter of the collar must allow for the heat-traced pipe, including insulation.

Where electric tracing is used on equipment, it might be well to consider possible failure of the tracer. It might be prudent to make provision for two circuits so that the failure of one circuit will not cause the equipment to be out of service or require the immediate removal of the insulation to repair the tracer. The surface temperature of the equipment will determine whether or not the electrical tracer can be applied directly to the equipment surface.

At this point, it may be practical to consider the use of immersion-type heaters in caustic or outside water storage tanks. Since these heaters are installed inside the equipment, the insulation does not have to be removed to replace the heater. The resistance heater, however, must be suitable for the liquid to which it will be exposed.

The service temperature limits of the tracing medium may require the tracer to be installed over a layer of insulation and covered with an additional layer of insulation. The thickness of the outer layer must be determined by the design criteria and economic factors. Most tracing manufacturers will provide design criteria in the use of their product.

Underground Piping

At times, it becomes necessary to place piping in the ground. To use insulation to keep piping above the frost line from freezing is not possible without heat tracing. Insulation will only lengthen the time it will take for the freezable liquid to freeze below 0°C.

Insulated and heat-traced piping buried in the ground must be provided with a jacket that is impervious to groundwater. The jacket must also be protected during the backfilling procedure to prevent damage. Underground piping can also be protected using a prefabricated conduit system consisting of a pipe or pipes inside a casing. The conduit system can include insulation for piping where required. Also available are pourable insulation systems that, when properly installed, will protect the piping from groundwater and will retard heat loss. It is imperative that close adherence to the manufacturer's requirements be observed to preclude failures in the insulation.

Clearance for Insulation

For any insulation to perform as intended, sufficient clearances must be provided between the outer diameter of the insulation and other insulation or uninsulated pipes, structures, supports, wall sleeves, and equipment. These clearances must allow for the positioning of the piping in the hot or cold position due to expansion or contraction. Pipe supports, hangers, anchors, and other structures must all extend a sufficient distance from the pipe surface to allow for the application of the specific thickness of insulation. This is mandatory for outdoor piping and piping requiring a vapor-proof jacket to facilitate the caulking procedure. It is not uncommon to find these systems failing as a result of lack of foresight in the design stage. It is most practical to reduce to a minimum the amount of caulking required on an insulated surface.

Stress Corrosion

Insulated stainless steel piping and equipment may be subject to stress corrosion cracking, which can be induced by wet insulation. It is imperative that insulation systems be kept dry. Insulations with built-in inhibitors are not capable of protecting the insulated surface over the long term. Stress corrosion cracking can occur from external sources having nothing to do with the insulation, such as pollution and salt-air environments. The suitability of insulation for stainless steel surfaces must be determined from manufacturers' catalogs and possibly from laboratory certification.

Panel Insulation

Great advances have been made in the use of factory-made preengineered insulation panels wherein the insulation is attached to metal lagging and applied to tank and equipment surfaces. The installed panels have provisions for expansion and contraction. They are weathertight and can be designed to withstand the suction and wind pressure of the geographic region. Where these panels are installed away from the surface of the equipment, it is good practice to install draft barriers to preclude convection occurring between the insulation and the equipment surface. Draft barriers are usually installed at 3.05-m intervals. Panels installed on top surfaces must be sloped to shed water; the slope must be not less than 2.08 cm/m.

On-Site Review

Timely visits to the project site will establish a line of communication with the insulation applicator for resolving problems that may occur. The resolution of these problems may require an engineering approach and not just a decision to facilitate the application of insulation.

Infrared Thermography

After the insulation work has been completed and the plant or system is operating, a review of the insulation should be performed to ensure that the specification requirements have been met. Infrared-sensitive equipment is available for review of insulated surfaces. This equipment affords the user the capacity to locate discontinuities in the insulation without physical contact. The equipment can be hand held. It can also be provided with a camera to record the image.

Conclusion

Some of the more important design aspects of insulation systems made to meet industry requirements have been discussed in this paper. When preparing an insulation specification, the responsible engineer shall realize that only by providing an up-to-date specification and an open line of communication with the contractor can an optimum insulation system be achieved.

The materials and application procedure checklist in the Appendix may serve as a useful reference during an on-site review of thermal insulation work.

APPENDIX

Thermal Insulation Materials and Application Procedure Checklist

The following are to be used as a guide to achieve the minimum requirements of the thermal insulation contract:

1. Materials applied or stored must be protected from the weather at all times.
2. Piping insulation applied in two or more layers must be applied with staggered joint construction both ways.
3. A minimum of three tie-wires shall be used to hold each 9.14-m section of sectional pipe insulation in place. No fewer than three tie-wires shall be used on each preformed insulated fitting.
4. Insulation that is wet, broken, or damaged during installation or storage shall not be used.
5. Chipped corners and butt joints in insulation larger than 15.9 mm wide shall be filled with cement. This applies to scored insulation where the score is not applied against the insulated surface.
6. The dry thickness of insulating cement used for small fittings shall be equivalent to the specified thickness.
7. Longitudinal joints in metal jacketing shall be lapped no less than 5.08 cm. Circumferential joints in metal-jacketed vertical pipes shall be lapped no less than 7.62 cm. Circumferential joints in metal-jacketed horizontal pipes shall be lapped no less than 5.08 cm. Metal lagging for outdoor piping must be completely weathertight and installed to shed water. Longitudinal laps in horizontal pipe runs shall be located below the horizontal centerline of the pipe at approximately 4 or 8 o'clock, but not at 6 o'clock.
8. Spacing of screws and bands to hold metal jacketing in place must be sufficient to maintain a continuous tight longitudinal lap in the metal jacket.
9. Suitable means to accommodate expansion and contraction of the piping or equipment must be provided.

10. All penetrations through the insulation for cold surfaces must be completely caulked with vapor-proof caulking. The vapor barrier system must be continuously sealed and made vapor proof. If metal lagging is used, metal bands must be used to hold lagging in place to preclude penetrating the vapor barrier.

11. The piping penetrating grade must be provided with suitable collars to protect the heat tracing system.

12. Terminal ends of insulation such as manways, vent piping, and piping connecting to equipment must be sealed to make them weathertight.

13. Insulation and lagging applied to top surfaces of equipment and tanks must be sloped to drain such surfaces completely.

14. When panel system insulation is used, draft barriers must be installed between the insulation and the hot surface of the equipment.

Thermal Performance

Mark P. Modera, [1] *Max H. Sherman,* [1] *and Santiago G. de Vinuesa* [2]

In Situ Measurement of Wall Thermal Performance: Data Interpretation and Apparatus Design Recommendations

REFERENCE: Modera, M. P., Sherman, M. H., and de Vinuesa, S. G., **"*In Situ* Measurement of Wall Thermal Performance: Data Interpretation and Apparatus Design Recommendations,"** *Thermal Insulation: Materials and Systems, ASTM STP 922,* F. J. Powell and S. L. Matthews, Eds., American Society for Testing and Materials, Philadelphia, 1987, pp. 91–106.

ABSTRACT: Although the *U* values of many building materials have been determined by laboratory testing, the *in situ* thermal performance of walls, under either static or dynamic conditions, is not so well documented. This report examines the use of *field measurements* of heat flow and surface temperatures to determine the dynamic as well as static thermal performance of walls. The measurement strategies examined include both active devices, which generate their own heat fluxes on the wall surfaces, and passive devices, which rely on the weather to induce the required fluxes and temperature differences. Data obtained with both devices are analyzed with the simplified thermal parameter (STP) model, which was designed to characterize a wall from flux and temperature measurements rather than from assumed material characteristics. The active measurement data are also analyzed with a modified version of the STP model that takes into account lateral heat losses. Some possible sources of error for both active and passive measurement strategies are also examined, and recommendations for both measurement strategies are given.

KEY WORDS: walls, dynamic response, heat transfer, *in situ* measurement, *U* value, *R* value, time constant, periodic heat flux, Fourier transform, thermal insulation

A wall's thermal resistance (*R* value) and its thermal conductance (*U* value) are widely accepted parameters for characterizing its steady-state thermal performance. Techniques for determining the values of these parameters from *in situ* measurements of temperatures and fluxes are available in the literature, which also includes numerous reports describing the difficulties involved with making these measurements [*1–3*]. On the other hand, no widely accepted technique exists for measuring or interpreting the data necessary to determine the *dynamic* thermal performance of a building's walls *in situ*. Because the parameters conventionally used for characterizing dynamic performance (for example, response factors) [*4–6*] are not easily extracted from heat flux and temperature measurements, it is difficult to establish criteria for either making or interpreting these measurements.

As a means of extracting information about the dynamic thermal performance of a wall, two basic measurement strategies have been discussed in the literature—passive and active measurements [*7–10*]. The major difference between these strategies is that one uses time histories of naturally occurring heat fluxes and surface temperatures (passive measurement strategy) and the other generates fluxes on a wall surface and measures the resulting temperature response (active measurement strategy). The advantages of an active measurement strategy are that (1) the measurements are theoretically independent of the weather, not relying on naturally in-

[1]Staff scientists, Energy Performance of Buildings Group, Lawrence Berkeley Laboratory, University of California, Berkeley, CA 94720.
[2]Research scientist, Instituto Del Frio, Ciudad Universitaria, Madrid, 28003 Spain.

duced fluxes or temperature differences to provide measurable results, and (2) the desired flux/temperature frequencies and amplitudes can be specified directly. The major disadvantage of such a strategy is its complexity; it requires precise control of heat fluxes or temperatures, implying a specially designed apparatus for that purpose. Passive measurement strategies are usually much simpler, requiring only two temperature sensors and one or two heat flux meters. The major disadvantage of passive strategies is that they do rely on specific weather conditions to provide measurable temperature differences and fluxes; in other words, measurements can be made only during certain time periods. On the other hand, the dynamic temperatures and fluxes measured with a passive system are ostensibly the same as those that we expect to find when making wall performance predictions. Thus, the measurement period can be chosen based on the presence of dynamic temperature and flux effects characteristic of those expected over the course of the year (that is, the analysis automatically concentrates on naturally occurring dynamics).

The major issue in data interpretation is how to use heat flux and temperature data to generate a set of parameters that characterize the dynamic performance of a wall. Again, there are two basic aspects to this problem. The first is how to characterize the performance of a wall with a limited set of unique parameters. This problem is basically a modeling problem, the goal of which is to obtain the minimum number of independent parameters required to characterize the wall. The second aspect of the data interpretation problem is how to generate the parameter values from the measured flux and temperature histories. This part of the problem is mathematical, involving the selection of appropriate algorithms and statistical testing. The link between these two aspects of the problem is that the degree to which the model parameters are independent affects the effort required to separate them mathematically.

The purpose of this report is to explore the issues involved in characterizing the dynamic performance of a wall from *in situ* measurements, specifically (1) to examine data from passive and active measurement strategies, (2) to demonstrate the application of a particular data interpretation technique based on the simplified thermal parameter theory [*11*], and (3) to make recommendations for future dynamic characterization methodologies.

Measurements

As a means of comparing active and passive measurement strategies, measurements made with an active measurement prototype, the envelope thermal test unit [*12*], and passive measurements made in New Zealand[3] are examined.

Active Measurement Strategy

Active measurement strategies are little used for determining the thermal performance of walls. At Lawrence Berkeley Laboratory, however, we have made several laboratory and field tests with a prototype device designed and built by our technical staff. The device, called the envelope thermal test unit (ETTU), has been used to test sections of stud walls in the laboratory, a stud wall in a single-family residence, and a thick concrete wall in a university building [*12*].

ETTU is a microcomputer-controlled device that measures wall performance *in situ* by heating the wall surfaces and simultaneously measuring the heat fluxes and surface temperatures on both sides of the wall. It consists of two thermal insulation blankets (extruded polystyrene) fitted with temperature sensors and wafer-thin electric resistance heaters front and back. These blankets, which are pressed against opposite sides of the test wall with wooden support structures, serve as surface temperature probes and large-area heat flux meters (Fig. 1). They are

[3]Trethowen, H. A., Building Research Association of New Zealand, personal communication, 4 Oct. 1983.

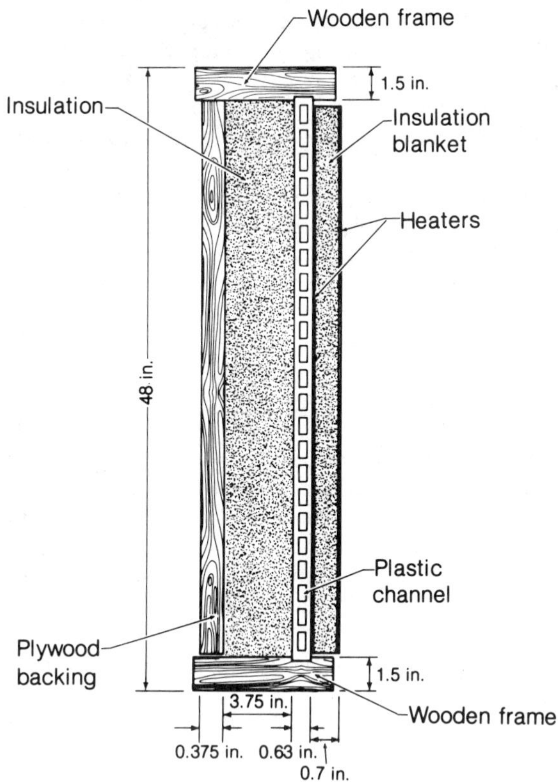

FIG. 1—*Cross-sectional view of the envelope thermal test unit blanket within its support structure.*

also used to specify the flux on the surface of the wall by controlling the power supplied to the electric-resistance heaters.

Using ETTU to evaluate the thermal performance of a wall involves *driving* the wall with prespecified fluxes on one side and measuring the resulting flux on the *receiving* side, as well as measuring the temperature responses on both sides. (The two ETTU blankets are functionally identical; the drive side is chosen by simply changing a parameter in the microcomputer program that controls the experiments and stores the data on a floppy disk.) On the drive side, the heater on the wall surface (primary heater) is supplied with the electric power required to provide the desired heat flux.[4] To ensure that the heat goes into the wall rather than being divided between the wall and the surroundings, the heater on the back side of the blanket (secondary heater) is also powered. The power to the secondary heater is controlled to minimize the temperature difference across the blanket, thus minimizing heat flow from the wall surface to the surroundings. On the receiving side of the wall, only the primary heater is powered, thus providing high-frequency, small-amplitude perturbations to the heat flux leaving the wall. The results of an ETTU test are time histories of surface temperatures and fluxes for a 0.6-m (2-ft) square wall section.

For the tests described in this report, ETTU was programmed to provide a pink-noise spectrum of heat fluxes on the wall surface. The pink-noise spectrum is similar to the better-known

[4]The primary heater is actually two separately controllable heaters, one for the central measurement section and one for the edge section.

white-noise spectrum, which contains all frequencies at equal amplitudes with random phase relationships. The pink-noise spectrum weights the amplitude at each frequency by the inverse of that frequency, thereby weighting lower frequencies more (that is, the lower frequencies have higher flux amplitudes). For the tests reported here, the fundamental frequency of the pink-noise spectrum was one cycle every 12 h, specifically chosen to be a harmonic of the diurnal flux/temperature cycle.

The ETTU tests described in this report include a laboratory test of an insulated stud wall specimen, a field test of a residential insulated stud wall, and a field test of an insulated concrete wall (Table 1). The wall specimen test was performed indoors on a built-up 1.8 by 1.8-m (6 by 6-ft) wall section. The residential stud-wall test was made on the wall between a house and its garage; both wall surfaces are gypsum board. The concrete wall test was on the thick exterior wall of the mechanical room of a university building.

Passive Measurement Strategy

Considerably more experience has been reported on the use of passive measurement strategies for determining the thermal performance of building walls [*2,8*]. As noted earlier, these tests require either one or two heat flux meters and a pair of temperature sensors. The data reported here, measurements made on the walls of two single-family residences, come from the Building Research Association of New Zealand. The data include indoor surface temperatures, outdoor surface temperatures, and indoor heat fluxes measured every 7.5 min for periods of approximately four days. In both cases, the fluxes were measured on a 0.6-m (2-ft) square section with sensors mechanically pressed against the interior surfaces (Fig. 2). Table 2 contains descriptions of the walls tested.

Data Interpretation

For both passive and active measurement strategies, the usual goal of data interpretation is to obtain a set of parameters that can be used either to characterize a wall or to make predictions of wall performance. Although numerous techniques exist for characterizing or predicting the dynamic thermal performance of a wall from the thermal properties of its components, few techniques exist for doing this from measurements of surface heat fluxes and temperatures. To interpret the data in this report, we use a simplified model of wall heat transfer, the simplified thermal parameter (STP) model developed at Lawrence Berkeley Laboratory [*11*], and a nonlinear search algorithm to obtain the parameter values.

The STP model characterizes a wall by its U value, its time constant, and several coefficients.

TABLE 1—*Descriptions of walls tested with active measurement strategy.*

Wall	Description (from Inside to Outside)
Insulated stud (laboratory)	1.3 cm (0.5 in.) gypsum board 5.1 cm (2.0 in.) polystyrene insulation with 5.1 by 5.1-cm (2.0 by 2.0-in.) studs spaced 41 cm (16 in.) on center 0.64 cm (0.25 in.) plywood
Insulated stud (*in situ*)	1.3 cm (0.5 in.) gypsum board 8.9 cm (3.5 in.) fiberglass insulation with 4.5 by 8.9-cm (1.75 by 3.5-in.) studs spaced 41 cm (16 in.) on center 1.3 cm (0.5 in.) gypsum board
Insulated concrete (*in situ*)	3.8 cm (1.5 in.) fiberglass insulation 20 cm (8 in.) reinforced concrete

FIG. 2—*Heat flux sensor in place in a New Zealand house.*

TABLE 2—*Descriptions of walls tested with the passive measurement strategy.*

Wall	Description (from Inside to Outside)
Brick-faced stud	1.3 cm (0.5 in.) gypsum 10 cm (4 in.) fiberglass insulation with 5 by 10-cm (2 by 4-in.) studs spaced 40 cm (16 in.) on center 3 cm (1.2 in.) air gap 10 cm (4 in.) face brick
Wood-faced stud	1.3 cm (0.5 in.) gypsum 10 cm (4 in.) fiberglass insulation with 5 by 10-cm (2 by 4-in.) studs spaced 40 cm (16 in.) on center 2 cm (0.8 in.) wooden sheathing

The U value and time constant have the conventional definitions, whereas the coefficients multiply the thermal filters derived in an analytical solution of the one-dimensional heat transfer equation for a homogeneous wall. These filter coefficients are what distinguish a multilayer wall from a homogeneous wall and can be interpreted as describing the distribution of thermal mass within the wall. Each wall surface has its own coefficient, a large coefficient implying that a large fraction of the wall's thermal mass is near that surface, and a small or negative coefficient implying very little thermal mass at that surface (the coefficients are exactly one for a homogeneous wall [*11*]).

To obtain the STPs of a wall from measured time histories of fluxes and temperatures, the time histories are first transformed into frequency representations using a fast Fourier transform algorithm. Using the Fourier transforms of the STP functions, the values of the parameters are determined by a nonlinear least-squares (χ^2) minimization routine. This routine adjusts the values of the parameters within the STP functions (that is, the U value, time constant, and filter coefficients for each side of the wall) to minimize the deviation between the measured fluxes and the fluxes predicted from temperatures with the STP model. If flux data are available for only one side of a wall, the analysis can determine the filter coefficients for that side of the wall only. In this case, the deviation of the wall from homogeneity is represented by a single parameter.

Active Measurement Analysis

The results of the STP analysis of the ETTU tests are compared with the results of one-dimensional-heat-flow computer simulations of the walls. The simulations were performed using handbook values for the thermal conductivities and specific heats of the wall materials. The exact materials in the laboratory specimens were known, whereas the materials in the field test walls were surmised from building plans and from observation. Table 3 compares the U values and time constants determined with the simulations with those obtained from measurements by STP analysis.

The filter coefficients from the STP analyses, a and b, are also presented in Table 3. The a coefficient refers to the drive side of the wall, and the b coefficient refers to the receiving side. For the laboratory test, the plywood sheathing was on the drive side, whereas the *in situ* stud wall was symmetric. The concrete wall was tested with the insulation on the drive side.

As is evident from Table 3, the results of the STP analysis are consistent with the computer simulations for the laboratory test but less so for the *in situ* tests. The results for the *in situ* test of the stud wall are acceptable,[5] but those for the concrete wall show large discrepancies between the simulation and the STP analysis. These discrepancies may be explained by an important difference between the laboratory and field tests that is not apparent in Table 3, namely,

TABLE 3—*Comparison of STP analyses and simulated wall performance (active measurement strategy).*

Wall	U Value, W/m^2 · K		Time Constant, h		Filter Coefficient, W/m^2 · K	
	STP Analysis	Simulated	STP Analysis	Simulated	a (STP)	b (STP)
Insulated stud (laboratory)	0.75	0.69	0.12	0.12	28.0	21.0
Insulated stud (*in situ*)	0.59	0.52	0.20	0.11	19.0	9.7
Insulated concrete (*in situ*)	0.47	0.92	2.4	7.2	−1.0	19.0

[5]Note that the percentage error in the time constant is rather large, but the absolute error is reasonably small.

that the average flux entering the wall was not equal to the average flux leaving the wall for the field tests. The average fluxes entering and exiting the walls tested are presented in Table 4.

Table 4 shows that, for both field tests, some fraction of the heat entering the wall on the drive side was evidently being removed from the measurement section by lateral conduction or convection. This effect was especially evident for the concrete wall, for which less than 50% of the heat entering on the drive side left on the receiving side. Lateral conduction was apparently minimal in the thin, carefully constructed laboratory wall, whereas the thicker, less controlled field walls had conductive or convective lateral heat flow paths. The *in situ* stud wall may have had internal air gaps that can convect away heat, whereas the concrete wall was four times the thickness of the laboratory wall and had highly conductive steel reinforcing rods. Because the STP analysis implicitly assumes that the heat leaving the wall is equal to that entering the wall, the lateral heat losses may well be the cause of the discrepancies between the simulations and the STP analyses for the *in situ* tests.

As described previously, the filter coefficients, *a*s and *b*s, in Table 3 are the parameters in the STP model that account for the nonuniform distribution of mass within most walls. This physical interpretation for the coefficients can be seen in Table 3. For example, in the laboratory test of the stud wall, *a* and *b* are very large, corresponding to the massive layers of plywood and gypsum board on the two surfaces. For the *in situ* test of the stud wall, both *a* and *b* are much larger than 1, indicating that the surfaces of the wall are more massive than the center, as is most surely the case for a stud wall with fiberglass insulation and a gypsum board surface. However, we would expect that *a* and *b* would be equal for the *in situ* stud-wall test, given that the wall was symmetrically constructed (gypsum board on both sides). These asymmetric results could be due to the lateral heat losses within the wall or to the difference between the heat flux spectrums on the two wall surfaces. For the insulated concrete wall, the trends are again correct: the insulated side of the wall had a small (negative) coefficient value, and the concrete side had a large coefficient value.

Modified Active Measurement Analysis

To account for the lateral heat flows induced by ETTU, we added a lateral heat-flow path into the STP analysis. The modification that we chose is an approximation, the correct solution being to rederive the simplified thermal parameter functions assuming multidimensional heat flow. Nevertheless, we were able to define physical limits for the new parameter introduced; that is, the average lateral heat flow is limited to the difference between the heat flows entering and

TABLE 4—*Average heat fluxes on test wall surfaces (active measurement strategy).*

Wall	Heat Flux Entering Wall, W/m^2	Heat Flux Exiting Wall, W/m^2	Exiting Flux/ Entering Flux Ratio
Insulated stud (laboratory)	9.72	9.67	0.995
Insulated stud (*in situ*)	11.7	8.06	0.689
Insulated concrete (*in situ*)	13.5	5.84	0.433

leaving the wall. This lateral heat flow path adds one additional adjustable parameter to the analysis—the lateral conductance. This lateral conductance is not meant to characterize the wall but rather to correct for heat flow anomalies created by the measurement apparatus.

Analyzing the field-test data sets using the modified simplified thermal parameter program did not significantly improve our comparisons with the computer simulations. The results of these comparisons, shown in Table 5, should be compared with the standard STP analysis results in Table 3.

Single-Sided Active Measurement Analysis

To determine whether or not single-sided flux measurements can provide satisfactory estimates of the thermal parameters describing a wall, we performed one additional test using the unmodified STP analysis on active measurement data. Taking the data from the field test of the insulated stud wall, we used only the flux measurements on the driven side of the wall. The parameter values thus obtained proved to be very similar to those obtained with two-sided analysis. The U value was calculated to be 0.58 W/m$^2 \cdot$ K, the time constant 0.23 h, and the filter coefficient, a, 17.0—all essentially equal to the values determined by two-sided analysis, as reported in Table 3 (the Filter Coefficient b cannot be determined with a single-sided analysis). It is encouraging that this analysis yields the same results as the two-sided analysis, although we do not have any information about the other side of the wall (Filter Coefficient b) or how similar our results would have been had the wall not been symmetric.

Passive Measurement Analysis

The simplified thermal parameter model was also used to analyze passive measurement data from the walls of two houses in the New Zealand studies. One wall was a standard fiberglass-insulated stud wall with a small air gap and brick facing on the exterior and gypsum on the interior. The second wall was also an insulated stud wall, only it had weatherboard rather than an air gap and brick on the exterior (see Table 2).

Because the New Zealand data include only the measured flux on the inside surface of the wall and the two surface temperatures, a one-sided STP analysis had to be performed. In Table 6, the results of the one-sided STP analysis on both walls are compared with the results of one-dimensional-heat-flow computer simulations and the results obtained by the Building Research Association in New Zealand. Because the exact material properties for the wall components were not available, we based the computer simulations on handbook properties for the de-

TABLE 5—*Comparison of modified STP analyses and simulated wall performance (active measurement strategy).*

	U Value, W/m$^2 \cdot$ K		Time Constant, h		Filer Coefficients, W/m$^2 \cdot$ K	
Wall	STP Analysis	Simulated	STP Analysis	Simulated	a (STP)	b (STP)
Insulated stud (*in situ*)	0.55	0.52	0.22	0.11	16.0	10.0
Insulated concrete (*in situ*)	0.41	0.92	3.6	7.2	−1.2	15.0

scribed construction materials. For each wall we performed the analysis for three different time periods to check for consistency in the results.

Looking at the *U* values in Table 6, we find that the STP analysis results are similar to those from the New Zealand analyses, but significantly different from the computer simulations. This outcome is not surprising, considering the uncertainty in the material properties used in the simulations. On the other hand, if we examine the STP analysis results for a given wall, we observe very little variation in the *U* value determined for the three separate time periods—standard deviations of 2 and 7% for the two walls. We can conclude from this consistency that the particular 24-h time period chosen does not have a strong effect on the *U* value determination.

Turning to the time constants and filter coefficients determined by the STP analysis (Table 6), we observe that these parameters are not as well determined as the *U* value. For both walls, an inverse correlation seems to exist between the values of the time constant and the filter coefficient (the larger the filter coefficient, the smaller the time constant). It appears that the search routine used in the STP analysis to determine the parameter values can arrive at different combinations that provide similar fits to the measured data, implying that the time constant and filter coefficient are not sufficiently independent. This inverse correlation suggests that either the single-sided analysis program or the single-sided analysis program combined with passive measurements are unable to provide unique parameters that describe the dynamic performance of a wall.

Another possible cause for the poor determination of the time constants and filter coefficients is the limited nature of the information available in the measurements. If the frequency spectrums of the passive heat fluxes and temperatures do not contain measurable amplitudes in the frequency range that invokes dynamic effects in the wall, any analysis program will have difficulties determining dynamic parameters that describe the wall. To determine a wall's dynamic characteristics, we must have information at low frequencies [the direct-current (D-C) component is sufficient] and at frequencies close to the inverse time constant of that wall. At frequencies much lower than the inverse time constant, the time constant of the wall has little effect on the heat transfer. For heat fluxes traversing the wall at these frequencies, the amplitude reduction and phase lag of the fluxes are very small and almost independent of the time constant. In the limiting case of D-C flux, the time constant of the wall has no effect on the heat transfer. At frequencies much higher than the inverse time constant, the wall's time constant once again has little effect on the heat transfer through the wall. At these frequencies, heat fluxes on one side of

TABLE 6—*Comparison of STP analyses, simulated wall performance, and New Zealand analyses (passive measurement strategy).*

		U Value, W/m² · K			Time Constant, h		Filter Coefficient, *a* (STP), W/m² · K
Wall	Test	STP Analysis	Simulated	New Zealand[a]	STP Analysis	Simulated	
Brick-faced stud	1	0.54	0.45	0.59	1.8	1.4	1.1
	2	0.53	0.45	0.59	1.8	1.4	1.6
	3	0.52	0.45	0.59	0.22	1.4	9.9
Wooden-faced stud	1	0.70	0.47	0.77	0.26	0.15	8.2
	2	0.72	0.47	0.77	0.44	0.15	7.4
	3	0.63	0.47	0.77	0.53	0.15	6.9

[a]Results obtained from steady-state analysis of the entire test period.

the wall do not traverse the wall to the other side. At frequencies close to the inverse time constant of the wall, the effects of the time constant of the wall are most clearly discernible.

To determine whether or not the New Zealand measurements contain measurable fluxes and temperatures at the appropriate frequencies, we performed fast Fourier transforms on the temperature and flux data. The resulting flux and temperature amplitudes are plotted in Figs. 3 through 6. If we assume that the simulation time constants are reasonably close to the true values, we find that the inverse time constant frequencies are approximately 0.7 and 7 rad/h (6.283 rad = 1 cycle), respectively, for the first and second walls. For both walls, the flux amplitudes are between 0.1 and 0.6 W/m^2 near the inverse time constant of the wall (see Figs. 3 and 4). However, the temperature magnitudes near the inverse time constant for the first wall (0.7 rad/h) are approximately 0.4 to 0.9 K (see Fig. 5), whereas near the inverse time constant for the second wall (7 rad/h) the temperature amplitudes are less than 0.1 K (see Fig. 6). Although the flux amplitudes for both walls are within the measurable range, it is clear that the temperature measurements near the inverse time constant of the second wall are pushing the limits of measurement accuracies. These results indicate that a large uncertainty is associated with the measurements of the time constant for the second wall.

Discussion

In our analysis of both active and passive measurement data with the simplified thermal parameter model, three important problem areas have been uncovered: (1) active measurements seem to be plagued by lateral heat losses; (2) passive measurements do not always contain enough information in the required frequency range; and (3) the parameters in the STP model do not seem to be mathematically independent.

With respect to the problem of lateral heat losses, we have established that active measurement systems are not, as originally conceived, independent of weather conditions. Rather, whenever an active system imposes an average flux that is very different from the weather-induced flux through the wall, the heat flux through the measurement section becomes nonuni-

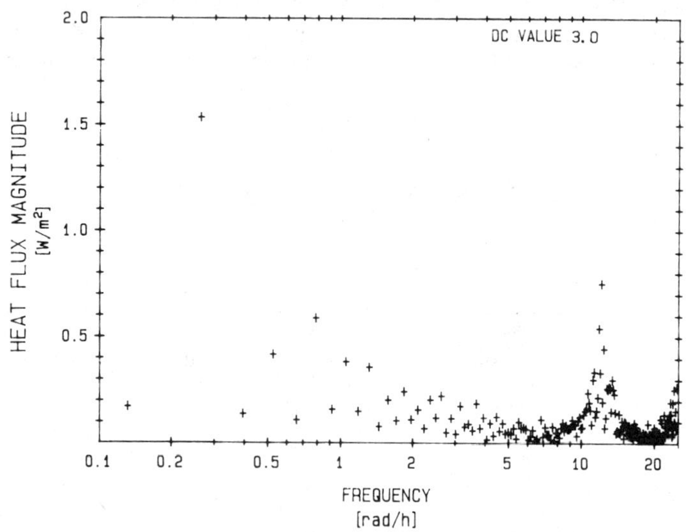

FIG. 3—*Fourier transform of New Zealand indoor heat fluxes for Wall 1 (D-C component of flux (average) = 3.0 W/m^2).*

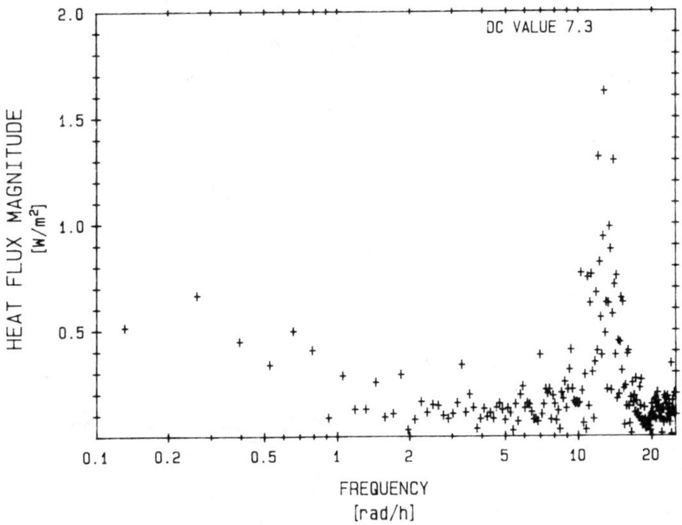

FIG. 4—*Fourier transform of New Zealand indoor heat fluxes for Wall 2 (D-C component of flux (average) = 7.3 W/m²).*

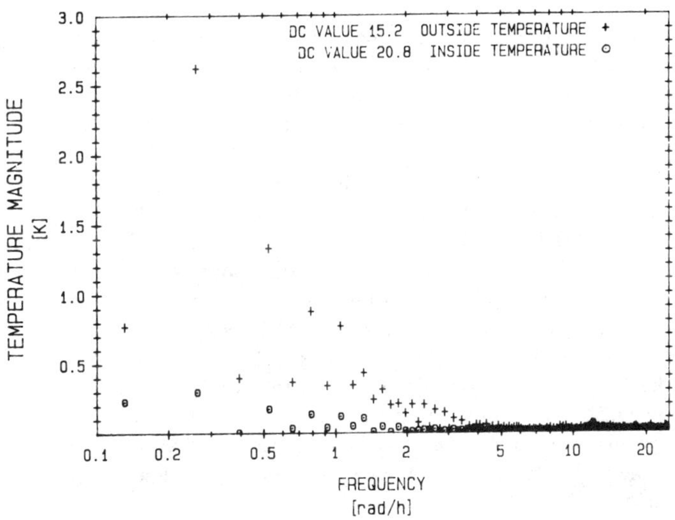

FIG. 5—*Fourier transform of New Zealand indoor and outdoor temperature data for Wall 1 (outdoor average = 15.2°C; indoor average = 20.8°C).*

form; as a consequence, data interpretation becomes quite difficult. Even when the STP analysis was modified to take lateral heat losses into account, the accuracy of the parameters determined was far from acceptable for the field test of a thick concrete wall.

The second problem area, that of obtaining measurable temperatures and fluxes in the required frequency range, can be further explored by means of Fourier transforms of the passive and active wall fluxes. Examining the passive flux amplitudes versus frequency for the two New

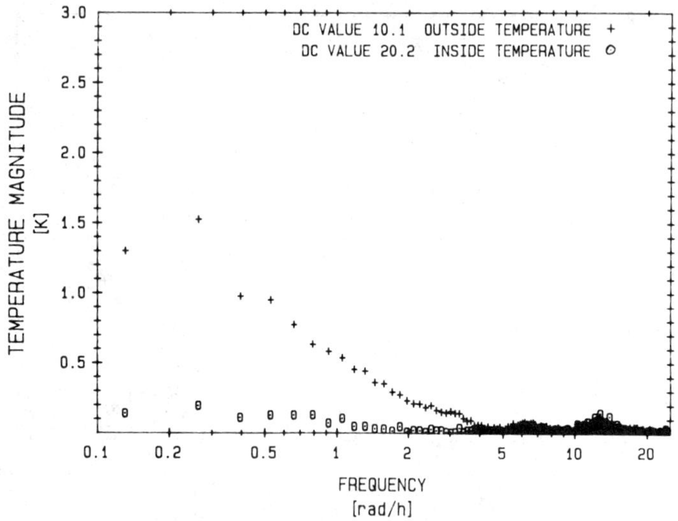

FIG. 6—*Fourier transform of New Zealand indoor and outdoor temperature data for Wall 2 (outdoor average = 10.1°C; indoor average = 20.2°C).*

Zealand tests (Figs. 3 and 4), we see that the majority of the dynamic flux is concentrated at approximately 12 rad/h (or two cycles per hour), and at less than 1 rad/h (corresponding to frequencies lower than 6 h/cycle). As described earlier, to make an accurate determination of a wall's dynamic properties, dynamic fluxes at frequencies close to the inverse time constant of the wall are necessary. Although these spectrums appear to be well suited to measurements of walls with inverse time constants near 12 rad/h or below 1 rad/h, they are clearly not optimal for measuring all walls. For the particular tests that we have analyzed, the concentration of dynamic fluxes near 12 rad/h corresponds to the cycling of the building's heating system, and the concentration of dynamic fluxes below 1 rad/h corresponds to the natural weather-induced dynamics. In general, the inside flux amplitudes at low frequencies will depend on the weather conditions, whereas the inside flux amplitudes at the heater cycling frequencies are affected by many different factors, including (1) the resistance and time constant of the wall, (2) the size of the building's heating system, (3) the dead band of the building's heating system, and (4) the severity of the weather conditions. From these two tests and from our general observations about the driving forces behind the flux spectrums, it appears that the accuracy or suitability of passive measurements is difficult to predict without having prior knowledge about the wall and the test conditions.

To compare naturally induced fluxes and temperatures with those generated by our active measurement system, we performed fast Fourier transforms of the pink-noise fluxes for the *in situ* test of a stud wall and plotted the flux amplitudes against frequency (see Fig. 7). Comparing Figs. 3, 4, and 7, we see that the frequency spectrum of the naturally induced fluxes is significantly different from that generated by our active pink-noise system. In general, the flux amplitudes at all frequencies are much higher for the active test (the passive flux amplitudes are higher at around 12 rad/h, which corresponds to the heater cycling frequency). These higher amplitudes for the active test spectrum suggest that the signal-to-noise ratio is higher for the active measurements. Assuming that the uncertainties of the temperature and flux sensors re-

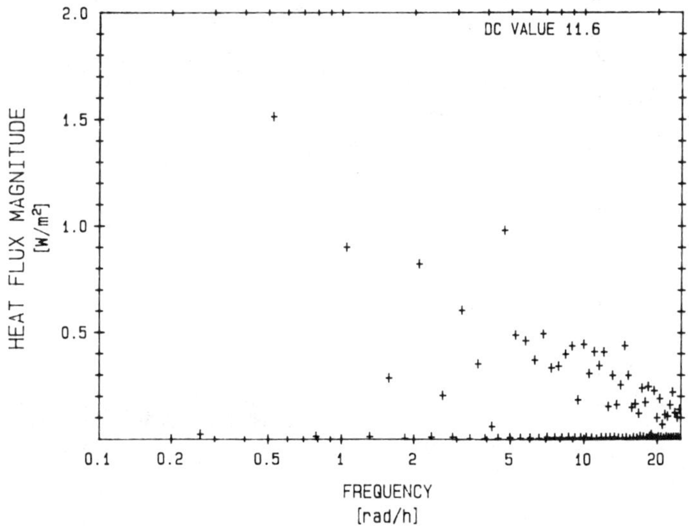

FIG. 7—*Fourier transform of pink-noise driveside heat fluxes for* in situ *test of insulated stud wall (D-C component of flux (average) = 11.3 W/m²).*

main constant, active measurements should thus provide more accurate determinations of dynamic properties.

The third problem area, the apparent interdependence of the dynamic parameters in the STP model, stems from the parameters chosen to characterize the distribution of thermal mass within a wall. Because the filter coefficients, *a*s and *b*s, in the STP model are not orthogonal to the time constant, at times the analysis will have difficulties separating the two parameters. These difficulties can arise when flux amplitudes near the inverse time constant are small, and they are amplified when using single-sided rather than two-sided analysis, both of which conditions were true for the passive measurements analyzed. Better measurement data would help to confirm the extent of this problem under normal circumstances, although theoretical work would be even more effective if it could develop an orthogonal parameter for quantifying the nonuniformity of thermal mass within walls.

One important issue that has not been addressed in this discussion is the effect of low-frequency (less than one cycle every 24 h) weather fluctuations on active or passive measurement analysis. Because the STP analysis does not take into account the total amount of heat stored in a wall at the beginning and end of a test, frequencies lower than one cycle every 24 h tend to bias the results. For example, if a low-frequency cycle causes the average temperature of the wall to be different at the beginning and end of the test, the heat stored in or removed from the thermal mass of the wall will not be accounted for by the STP analysis. One way to examine these effects is to use Fourier transforms once again, this time to analyze weather data from different climate regions [*13,14*]. As an example, a sol-air-temperature frequency spectrum for Madison, Wisconsin, in March is shown in Fig. 8. If the low-frequency weather fluctuations shown in Fig. 8 are typical, it appears that low-frequency cycles, on the order of one cycle every two to seven days, have rather large amplitudes. Although this study is not complete, it suggests that the STP analysis procedure (or any future analysis procedure) should be modified to take into account the effects of fractional cycles.

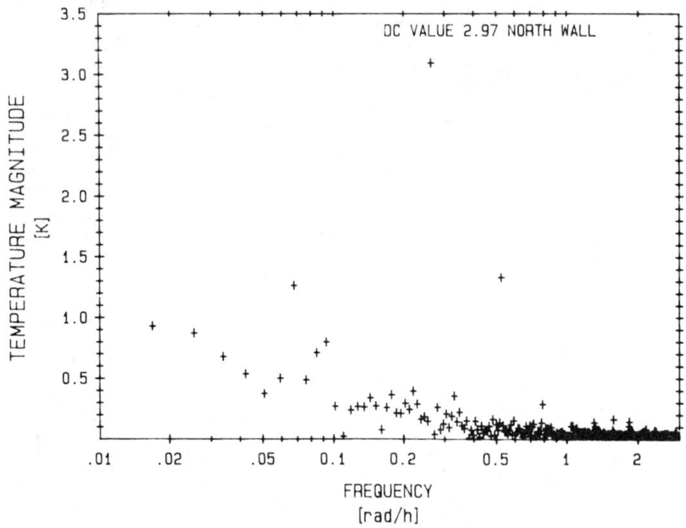

FIG. 8—*Fourier transform of sol-air temperature on a north-facing wall for March in Madison, Wisconsin (average temperature = 2.97°C).*

Conclusions and Recommendations

The most important conclusion to be drawn from the analyses described in this report is that the accuracy of both active and passive measurement strategies depends on weather conditions. Active measurement strategies depend on the weather to provide D-C (that is, average) heat fluxes through the wall large enough that the imposed active fluxes do not cause lateral heat fluxes in the wall. Passive measurement strategies depend on the weather to provide measurable D-C heat fluxes, as well as measurable dynamic heat fluxes at frequencies near the inverse time constant of the wall being tested. In addition, both strategies can be affected by low-frequency weather fluctuations.

If active measurement strategies are to be used in the future, it is clear that they must be designed to take weather conditions into account. This could be accomplished by devising a way to set the mean surface flux generated by an active measurement system equal to the mean weather-induced flux to be expected under the prevailing weather conditions. Two possible techniques for achieving this are (1) to use a passive heat flux sensor on the wall surface to control the D-C heat flux of the active system and (2) to use a wall classification scheme along with the weather to specify the D-C heat flux of the active system. This latter technique, wall classification, could also be used to optimize the frequency spectrum generated to analyze the wall. By concentrating heat fluxes at frequencies close to the inverse time constant of the wall being tested, a better signal-to-noise ratio could be ensured. This less general but potentially more accurate technique would use some easily available information about the wall to find the appropriate frequency range, no longer treating the wall as a black box. Walls could be put into several classes, such as light frame construction, light masonry, and heavy masonry. A sample classification scheme is shown in Table 7, where the quoted time constants were obtained by computer simulations with handbook material properties.

Our experience with passive measurement analysis can also provide some general recommendations for *in situ* dynamic performance determination. We saw that the inside surface flux spectrum was dominated by the cycling of the building's heating system at high frequencies and by weather fluctuations at low frequencies. Realizing that in general the cycling of a building's

TABLE 7—*Classification of walls by thermal time constant.*

Wall Type	Components	U Value, $W/m^2 \cdot K$	Time Constant, h
Light frame	gypsum insulation plywood	0.5	0.1
Light masonry	cinderblock	3	1
	gypsum insulation brick	0.5	2
Heavy masonry	gypsum concrete insulation	0.7	7
	gypsum cinderblock insulation plywood	0.4	6

heating system depends on building characteristics as well as the severity of the weather, we can conclude that passive heat flux measurements on the inside wall surface often will not provide the heat flux spectrum necessary to determine the dynamic performance of a wall. To expand our conclusions to outside surface flux measurements, we can once again use Fourier transforms of weather data to predict the flux spectrum to be expected during a passive test. The sol-air-temperature frequency spectrum for March in Madison, Wisconsin, in Fig. 8 shows the low-frequency amplitudes to be expected on the north wall of a building. However, the data sampling rate was only one point per hour. To perform this test for walls with short time constants, data at higher sampling frequencies would have to be used.

As we have already noted that passive measurements do not generally provide the information necessary for determining the dynamic thermal performance of a wall and that active measurement strategies must be modified to take into account the effects of weather, one final recommendation remains. This recommendation, based on the experiences behind this report, is that the modeling research required to improve or replace our present data interpretation strategy should proceed in parallel with any future measurement efforts. Based on the knowledge gained in the preceding efforts, such research should be able to provide a model with independent parameters and a simpler data analysis technique, which could then be used to better evaluate both active and passive measurement strategies, as well as to analyze laboratory hot box data.

Acknowledgment

This work was supported by the Assistant Secretary for Conservation and Renewable Energy, Office of Building Energy Research and Development, Building Systems Division of the U.S. Department of Energy under Contract No. DE-AC0376SF00098.

References

[*1*] Modera, M. P., Sherman, M. H., and Sonderegger, R. C., "Determining the *U* Value of a Wall from Field Measurements of Heat Flux and Surface Temperatures," *Building Applications of Heat Flux*

Transducers, ASTP STP 885, American Society for Testing and Materials, Philadelphia, 1985, pp. 203-220.

[2] Schwerdtfeger, P., "The Measurement of Heat Flow in the Ground and the Theory of Heat Flux Meters," Technical Report 232, U.S. Army Cold Regions Research and Engineering Laboratory, Hanover, NH, November 1970.

[3] Johannesson, G., "Heat Flow Measurements: Thermoelectrical Meters, Function Principles and Sources of Error," Report TVBH-3003, Division of Building Technology, Lund Institute of Technology, Lund, Sweden, 1979. (Draft Translation, U.S. Army Cold Regions Research and Engineering Laboratory, Hanover, NH, March 1982).

[4] Stephenson, D. G. and Mitalas, G. V., "Room Thermal Response Factors," *A.S.H.R.A.E. Transactions,* Vol. 73, Part 1, 1967, pp. III.2.1-III.2.10.

[5] Stephenson, D. G. and Mitalas, G. V., "Cooling Load Calculations by Thermal Response Factor Method," *A.S.H.R.A.E. Transactions,* Vol. 73, Part 1, 1967, pp. III.1.1-III.1.7.

[6] *ASHRAE Handbook of Fundamentals,* American Society of Heating, Refrigerating, and Air Conditioning Engineers, Atlanta, GA, 1981, Chapter 26, pp. 34-38.

[7] Raychaudhuri, B. C., "Simultaneous Determination of Overall Thermal Diffusivity and Conductivity of Composite Building Elements In-Situ," *Building Science,* Vol. 5, 1970, pp. 1-10.

[8] Condon, P. E. and Carroll, W. L., "Measurement and Analysis of In-Situ Dynamic Thermal Performance of Building Envelopes Using Heat Flow Meter Arrays," paper presented at the American Society of Heating, Refrigerating, and Air Conditioning Engineers/U.S. Department of Energy Conference on Thermal Performance of the Exterior Envelopes of Buildings, Orlando, FL, 3-5 Dec. 1979, Lawrence Berkeley Laboratory Report LBL-9821, Berkeley, CA, June 1980.

[9] Sonderegger, R. C., Sherman, M. H., and Adams, J. W., "The Envelope Thermal Test Unit (ETTU): Field Measurement of Wall Performance," paper presented at the Third International Symposium on Energy Conservation in the Built Environment, Dublin, Ireland, 30 March-1 April 1982, Lawrence Berkeley Laboratory Report LBL-13436, Berkeley, CA, October 1981.

[10] Sonderegger, R. C., Sherman, M. H., and Adams, J. W., "The Determination of the Dynamic Performance of Walls", *A.S.H.R.A.E. Transactions,* Vol. 88, No. 1, 1982, p. 689. (Lawrence Berkeley Laboratory Report LBL-13510).

[11] Sherman, M. H., Adams, J. W., and Sonderegger, R. C., "Simplified Thermal Parameters: A Model of the Dynamic Performance of Walls," paper presented at the Department of Energy/ASTM Conference on Thermal Insulations, Materials, and Systems for Energy Conservation in the 80s, Clearwater, FL, 8-10 Dec. 1981, Lawrence Berkeley Laboratory Report LBL-13503, Berkeley, CA, December 1981.

[12] Modera, M. P., "Technical Description: The Envelope Thermal Test Unit," *A.S.H.R.A.E. Transactions,* Vol. 91, Part 1B, January 1985, pp. 582-542.

[13] Cumali, Z. O., "Spectral Analysis of Coincident Weather Data for Application in Building Heating, Cooling Load and Energy Consumption Calculations," *A.S.H.R.A.E. Transactions,* No. 2159, Vol. 76, Part 2, 1970, pp. 240-257.

[14] Hittle, D. C., "A Comparison of Building Energy Use Calculated with Actual and Synthesized Weather Data," *A.S.H.R.A.E. Transactions,* Vol. 85, Part 2, 1979, pp. 167-189.

Jin B. Fang[1] *and Richard A. Grot*[1]

Field Measurement of the Thermal Resistance of Office Buildings

REFERENCE: Fang, J. B. and Grot, R. A., "**Field Measurement of the Thermal Resistance of Office Buildings,**" *Thermal Insulation: Materials and Systems, ASTM STP 922,* F. J. Powell and S. L. Matthews, Eds., American Society for Testing and Materials, Philadelphia, 1987, pp. 107-123.

ABSTRACT: The results of field tests conducted during the winter heating season on the thermal performance of the building envelopes of eight office buildings situated in various climatic regions are presented. The exterior envelopes of these buildings are masonry and metal-panel-faced construction of different designs. *In situ* measurements of heat flow through the envelopes and of air temperatures were made using heat flux transducers, portable calorimeters, and thermistors. The influences of the length of measurement period and the time lag between the air temperature difference across the envelope and the heat flow on the determination of the thermal resistance are described. The calorimeter measurement generally gives lower wall thermal resistance values than the heat flux transducer because of the additional heat flow through highly conductive framing members. Heat flow and air temperature measurements carried out over a period of 48 h and longer are found to provide satisfactory thermal resistance data. The measurements' accuracy can be improved through correction of the data for time lag. The masonry and metal-panel-faced exterior walls of these buildings tested have widely varying thermal resistances, ranging from approximately 0.5 to 5.6 $m^2 \cdot K/W$ (3 to 32 $ft^2 \cdot h \cdot °F/Btu$). The measured values deviated from the predicted values by an average of 22%, and in one case by as much as 71%.

KEY WORDS: exterior envelope, field test, heat flux, *in situ* measurements, office building, portable calorimeter, temperature, thermal insulation, thermal resistance, walls

Heat loss through the exterior envelope of a building can be estimated from the thermal resistances of the individual components constituting the envelope. These material thermal resistances are often determined in the laboratory under simulated steady-state test conditions. In actual situations, the rate of heat loss through the insulated envelopes may be greater than the estimated value because of thermal bridging, wind effects, and varying outdoor temperatures. Field testing of the thermal performance of a building component under actual use conditions can provide information about deterioration of the component with age, the construction quality, and the adequacy of the structural design. This information can be used as the basis for diagnostics and retrofit of the existing building envelopes. The data obtained from field measurements can be utilized to determine whether the laboratory test conditions, the variables employed for mathematical modeling, and the design criteria are representative of actual use. Field testing is also needed to determine whether construction practice is consistent with design objectives.

In situ measurement of the thermal performance of a building envelope is complicated by continuous variations of outdoor temperatures and surface wind conditions. The response of the building envelope also depends on the total amount of thermal mass and resistance contained in

[1]Mechanical engineer and leader, respectively, Indoor Air Quality and Ventilation Group, Center for Building Technology, National Bureau of Standards, Gaithersburg, MD 20899.

the envelope. A sufficient length of time is required for gathering the temperature and heat flux data for thermal resistance calculations and for interpreting the measurement data.

The work reported here resulted from a need to develop procedures to estimate the heat loss from highly conductive heat flow paths in buildings called thermal bridges. This paper presents the results of thermal resistance measurements of seven federal buildings and one state office building located in different geographical and various climatic regions: Ann Arbor, Michigan; Columbia, South Carolina; Springfield, Massachusetts; Huron, South Dakota; Norfolk, Virginia; Pittsfield, Massachusetts; Anchorage, Alaska; and Trenton, New Jersey. The exterior walls of these buildings are of masonry construction, with the exception of the Trenton building, which has metal panel facing and framings. This paper also describes the overall performance of a portable calorimeter used for quantification of the heat flow through building envelopes, and presents a comparison of the wall thermal resistances measured by this field apparatus with those measured by heat flux transducers.

Test Apparatus and Instrumentation

Rates of heat loss through major components of the building envelopes were determined using both heat flux transducers and portable calorimeters. The heat flux transducer was a 10.2-cm (4-in.)-diameter, thin, circular disk consisting of an embedded thermopile with its hot and cold junctions attached to the internal disk surfaces. The thermopile voltage outputs were directly proportional to the heat flux transferred through the transducer.

After a series of field tests were completed, calibration of heat flux transducers was accomplished using a standard guarded hot plate apparatus described in Refs *1* and *2* and the ASTM Test for Steady-State Thermal Transmission Properties by Means of the Guarded Hot Plate (C 177-76). The heat flux transducers were placed between the insulation boards installed next to the hot and cold plates of the apparatus, and exposed to a range of uniform heat fluxes at the temperature levels corresponding to those observed during the field measurements. The relationship between the voltage output from each heat flux transducer and the heat flux was determined. The accuracy of the transducer calibration was estimated to be within ±1%.

Portable calorimeters developed at the Division of Building Research, National Research Council of Canada, have been used extensively to investigate the influences of framing members on the overall thermal resistance and heat flow through wood frame walls of single-family houses [*3,4*]. In order to determine the effects of metal framing members on wall heat loss, two different sizes of portable calorimeters, similar to those employed by Brown and Schuyler [*3,4*] were designed, fabricated, and instrumented for *in situ* measurements of heat flow through the building envelopes of the selected office buildings.

The detailed construction of the small portable calorimeter is shown in Fig. 1. The calorimeter is a five-sided box fabricated from two 50-mm (2-in.)-thick layers of aluminum-foil-faced, semirigid glass fiber insulation boards, which were glued together with foiled side exposed, using construction contact adhesive. All the joints between the insulation boards were staggered, filled with small fragments of glass fiber insulation mixed with adhesive, and covered with aluminum foil duct tape. To further strengthen the complete calorimeter box, the four exterior sides of the calorimeter were covered by a 13-mm (1/2-in.)-thick plywood support frame. A narrow rubber foam strip attached to a 0.8-mm (1/32-in.)-thick aluminum angle frame was installed along the edges of the opening face of the calorimeter box to provide an airtight seal between the calorimeter and the wall surface being tested. The combined thermal resistance or the R value of the calorimeter wall was 3.06 $m^2 \cdot K/W$ (17.4 $ft^2 \cdot h \cdot °F/Btu$). The external dimensions of these two calorimeters were 0.86 by 0.97 by 0.2 m (34 by 38 by 8 in.) and 1.24 by 1.98 by 0.2 m (49 by 78 by 8 in.), respectively. These calorimeters had a metering area of 0.79 m (31 in.) wide by 0.89 m (35 in.) long, and inner dimensions of 1.17 by 1.91 m (46 by 75 in.).

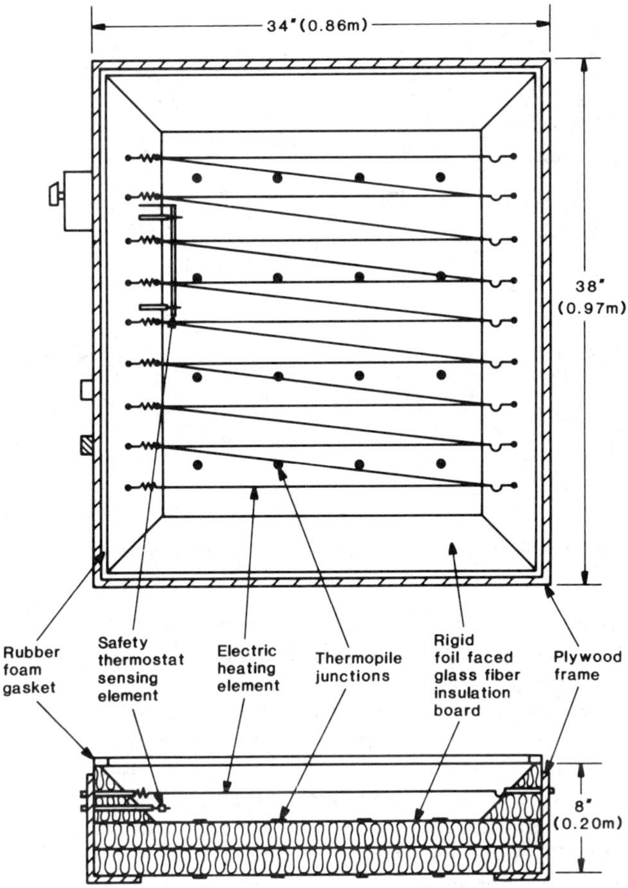

FIG. 1—*Construction details of a small portable calorimeter.*

Each portable calorimeter contained an electric resistance heater having an electrical power consumption rate of 90 or 140 W, depending on the size of the calorimeter, and consisting of a bare nickel chromium wire, with a B & S gage No. 30 or No. 27 wire diameter. Using steel springs and hooks, the heating wire, when heated electrically, remained tightly held to ceramic standoff insulators fastened to the plywood frame. A thermopile consisting of 16 or 18 thermocouple pairs, and having its thermocouple junctions placed on both the inner and outer faces of the calorimeter back wall, was used to monitor the temperature differential across the calorimeter wall. The thermocouples were constructed from B & S gage No. 28 copper and constantan wires with a 0.32-mm (0.0126-in.) diameter and solid nylon insulation.

The wiring schematic of the automatic measurement and control system for the calorimeter is shown in Fig. 2. A voltage controller, using the thermopile output as the feedback variable, was utilized for controlling the operation of the electric heater and maintaining a zero temperature difference across the calorimeter wall. A safety thermostat with a sensing element installed in the calorimeter provided control of the calorimeter air temperature within the safety limit by opening the electrical circuit at high temperatures. The total electrical energy consumed by the heater was monitored by a watt-hour meter equipped with an optoelectronic device consisting of

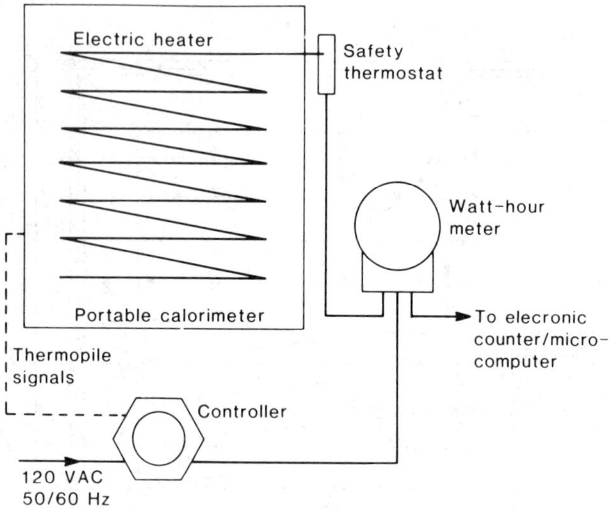

FIG. 2—*Schematic of the measurement/control system for the portable calorimeter.*

a light-emitting diode and a detector. The device generated an electric pulse for each period of time with an electrical energy consumption of 1.8 W · h. An electronic counter was used to register the total sum of these pulses.

Thermistors were used to measure the temperatures of the outdoor and indoor air in the vicinity of the heat flux transducers and of the air inside the portable calorimeters. Each temperature sensor consisted of a bead-shaped thermistor and an external network with fixed precision resistors. The thermistor under an applied voltage of 1.2 V direct current (DC) produced an output voltage proportional to the temperature over a temperature range between −30 and 50°C (−22 and 122°F). According to the technical data furnished by the manufacturer, the measurement error of the thermistors used was within ±0.3°C (0.5°F).

Test Procedure and Results

Field measurements of the heat flow through building envelopes were carried out between February and May of 1983 in the unoccupied rooms of selected floor levels of seven federal office buildings, and between March and April of 1984 in an occupied office and two conference rooms of a state judicial building. Construction details of the exterior walls and other major building components on which heat flow measurements were performed are summarized in the Appendix.

During the field measurements, the heat flux transducers were taped on the internal surfaces of the exterior walls, the structural columns and beams, and the roof deck and floor assemblies. A metal stud finder was used to ensure that the heat flux transducers were installed over thermal insulation filled cavities between wall framing members.

The open side of the calorimeter box was sealed with duct tape against the inner surface of the external wall to be measured. Each thermistor was installed with its sensing element placed 15.2 cm (6 in.) from the exposed measurement surface. The output signals from the heat flux transducers, the portable calorimeters and associated counters, and the thermistors were monitored by a microcomputer-based data acquisition system. The computer continuously scanned the channels of all transducers every 2 s, computed the time-averaged values, and recorded the test data as hourly averages on a floppy disk for further data reduction and analysis.

The hourly variations of the outdoor and indoor air temperatures, the wall surface heat fluxes measured by heat flux transducers and a large calorimeter, and the air-to-air temperature difference across the exterior wall at the locations of these heat flux sensors are shown in Fig. 3 for three daily cycles in the Huron building. The measured values for heat flux transducers are the average of the readings of eight transducers installed in the proximity of the calorimeter. As shown, the outdoor temperature was found to change in a periodic manner over a 24-h period, whereas the indoor air had a relatively constant temperature. By comparing the plots of the air temperature difference across the wall and the corresponding inside surface heat fluxes, one can see that the rate of heat flow into the wall section varied periodically with time and consistently lagged behind the temperature driving force because of the heat storage within the wall. From visual inspection of the curves, the thermal time lag of the wall was found to be approximately 6 h, which was estimated as the time required for the maximum heat flux to be reached after the peak temperature difference was attained.

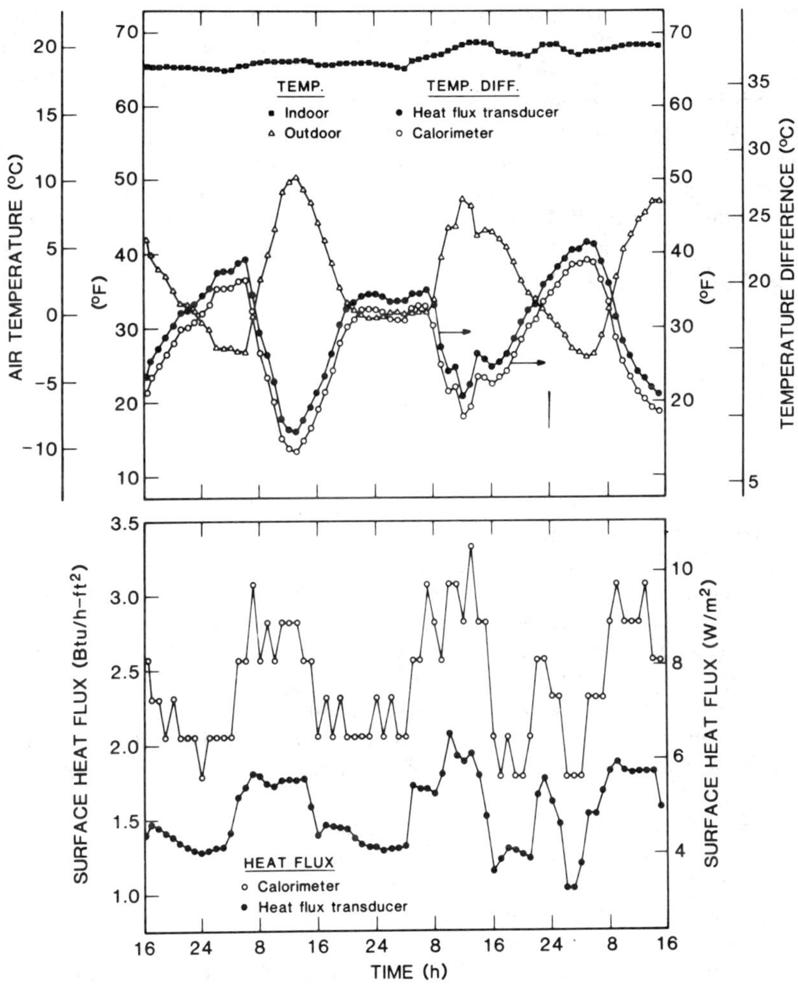

FIG. 3—*Indoor and outdoor temperatures, the temperature difference across the wall, and surface heat fluxes measured by heat flux transducers and the calorimeter.*

Also shown in Fig. 3, the rates of heat transmission through the wall determined by the calorimeter exhibited a periodic variation similar to those measured by the heat flux transducer, but the former was approximately 30% greater in magnitude than the latter. The calorimeter measured the amount of heat flow through the wall area backed by both highly conductive metallic framing members and insulation-filled cavities between the framings, while the heat flux transducer was positioned in the area without framing members. Calculations were performed to determine the influence of framing members on heat transmission through an exterior wall, using the series resistance and zone methods [5] and the published material thermal property data. The thermal resistances of air films at the inside and outside wall surfaces used in the calculations were 0.12 $m^2 \cdot K/W$ (0.68 $ft^2 \cdot h \cdot °F/Btu$) and 0.03 $m^2 \cdot K/W$ (0.17 $ft^2 \cdot h \cdot °F/Btu$), respectively. These values represent the conditions that the warm surface is exposed to still air and the cold surface is subjected to a 6.7-m/s (15 mile/h) wind. For the exterior wall of the Huron building, with construction details as described in the Appendix, and which has 38 by 89 by 1.5-mm (1.5 by 3.5 in. by 16-gage)-thick, light-gage steel studs spaced at 0.41-m (16-in.) intervals on center, the thermal transmittances or U values for the wall area backed by the studs and for the area over the insulation-filled cavities between the studs were calculated and found to be 1.28 and 0.34 $W/m^2 \cdot K$ (0.225 and 0.059 $Btu/h \cdot ft^2 \cdot °F$), respectively. The metallic framing members consequently increased the rate of heat flow through the wall section by 26%, which is in good agreement with the measured results.

The daily variations of the outdoor temperature, the temperature difference between the inside and outside air, and the wall thermal resistances determined by heat flux transducers and the calorimeter in the Huron building, are shown in Fig. 4 for a test duration of 15 days. The thermal resistance, R value, is defined as the ratio of the average temperature difference between the interior and exterior air across the building envelope to the average heat flow rate, and is calculated from the following relation

$$R = \frac{\int_0^\tau (\Delta T)dt}{\int_0^\tau q dt} \tag{1}$$

where T is the air-to-air temperature difference, q is the rate of heat flow through the exterior envelope, and τ is the length of the measurement period.

In order to minimize the transient effect due to the large thermal storage of the building envelope, the air temperature and surface heat flux data were averaged over a 24-h period. The wall thermal resistance values for the transducer, as shown in Fig. 4, were the average values of the resistance data obtained from eight heat flux transducers mounted on the interior wall surface below the suspended ceiling. Because of fluctuations in the outdoor temperatures, as depicted in the upper portion of Fig. 4, and in the speed and direction of the external wind, which affected the thermal resistance of the air film at the outer surface of the exterior wall, the overall wall thermal resistances determined by the heat flux transducers and the calorimeter were varied over a 24-h period.

Also plotted in Fig. 4 for comparison are the cumulative average of the daily wall thermal resistance values and the coefficient of variation of the daily thermal resistance data obtained by eight heat flux transducers. The cumulative average value was calculated by dividing the sum of the daily thermal resistance values by the number of days elapsed. In general, the cumulative average resistance was within 5% of its final value after two days of continual measurement. The coefficient of variation is defined as the ratio of the standard deviation to the mean value and is a dimensionless relative measure of dispersion with the mean value. As shown in Fig. 4, the coefficient of variation indicating the variability among the daily resistance values, deter-

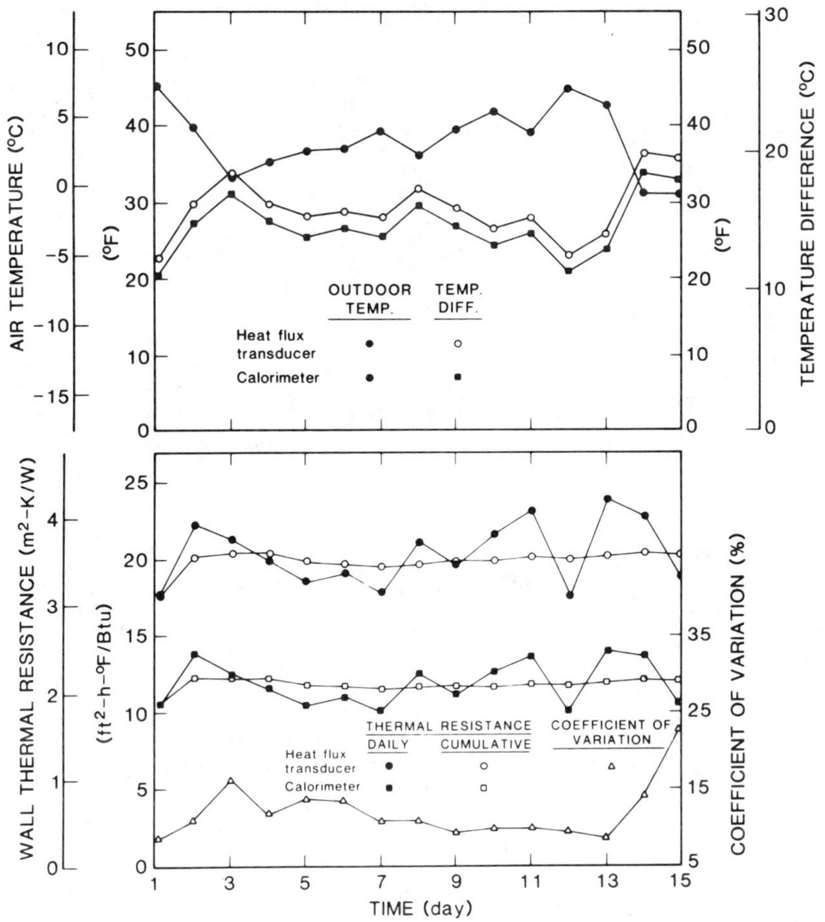

FIG. 4—*Daily variations of outdoor temperature, interior-to-exterior air temperature differences, wall thermal resistance, and coefficient of variation determined using heat flux transducers and the calorimeter.*

mined by eight heat flux transducers, was found to range from approximately 9 to 22%, with an average of 12%.

The thermal resistance values of various sections of the exterior walls in each building, as measured using heat flux transducers and portable calorimeters, are given in Table 1, along with the corresponding predicted steady-state values. Also presented in the table are the outdoor and indoor air temperatures, and heat flow rates determined by the heat flux transducers, including sections of the exterior envelope of the Trenton building, where measurements were made with the portable calorimeters. All the data listed in the table are the mean values of the data collected over a 24-h period for 2 to 16 consecutive days. The predicted steady-state thermal resistance values, which included the average *R* value for the wall section and the *R* value for areas between framing members, were calculated using the zone and the series resistance methods [5] and the published data on thermal properties of the building materials involved.

Inspection of Table 1 shows that the measured and predicted thermal resistance values are in general agreement, with the exception of the roof deck assembly in the Anchorage building and the exterior walls in the Trenton building. The discrepancy observed in the Anchorage building

TABLE 1—*Comparison of wall thermal resistance values measured with heat flux transducers (HFT) and portable calorimeters (PC) with the corresponding predicted values.*

Building Location	Building Component	Air Temperature, °F(°C)		Heat Flow Rate (HFT), Btu/h · ft² (w/m²)	Thermal Resistance, ft² · h · °F/Btu (m² · K/W)			
					Measured		Predicted	
		Hot Side	Cold Side		HFT	PC	Between Framings	Average
Huron, SD	wall	66.9 (19.4)	37.9 (3.3)	1.42 (4.48)	20.4 (3.59)	12.0 (2.11)	16.9 (2.97)	13.4 (2.35)
	upper wall	63.8 (17.7)	37.9 (3.3)	1.23 (3.88)	21.0 (3.70)		16.9 (2.97)	
	structural beam	63.8 (17.7)	37.9 (3.3)	2.49 (7.85)	10.4 (1.83)		15.6 (2.75)	
	structural column	66.6 (19.2)	37.9 (3.3)	3.02 (9.52)	9.5 (1.67)		9.7 (1.71)	
Ann Arbor, MI	wall	68.1 (20.1)	39.4 (4.1)	2.49 (7.85)	11.5 (2.03)	11.1 (1.95)	10.6 (1.87)	10.4 (1.83)
	structural beam	65.1 (18.4)	39.4 (4.1)	5.85 (18.44)	4.4 (0.77)		3.7 (0.66)	
	structural column	68.9 (20.5)	39.4 (4.1)	2.22 (7.00)	13.3 (2.34)		8.8 (1.55)	
Columbia, SC	wall	70.3 (21.3)	47.9 (8.8)	2.08 (6.56)	10.8 (1.89)	15.1 (2.66)	11.2 (1.97)	10.1 (1.78)
	upper wall	69.1 (20.6)	47.9 (8.8)	2.76 (8.70)	7.7 (1.35)		9.4 (1.66)	
	concrete wall	69.1 (20.6)	47.9 (8.8)	4.75 (15.0)	4.5 (0.79)		3.9 (0.68)	
	concrete beam	69.1 (20.6)	47.9 (8.8)	3.64 (11.5)	5.8 (1.03)		4.5 (0.78)	
Anchorage, AK	wall	73.2 (22.9)	50.6 (10.3)	0.70 (2.21)	32.0 (5.64)		29.9 (5.27)	

	upper wall	69.2 (20.7)	50.6 (10.3)	0.86 (2.71)	21.7 (3.81)		14.6 (2.56)	
	roof-deck webs lower	69.6 (20.9)	50.6 (10.3)	1.07 (3.37)	17.8 (3.13)		20.4 (3.59)	
	upper	69.6 (20.9)	50.6 (10.3)	0.64 (2.02)	29.6 (5.21)		17.1 (3.01)	
	structural column	69.2 (20.7)	50.6 (10.3)	1.59 (5.01)	11.7 (2.06)		15.1 (2.66)	
Springfield MA	wall	69.5 (20.8)	41.9 (5.5)	2.20 (6.92)	12.6 (2.21)	6.8 (1.20)	13.0 (2.29)	9.5 (1.68)
	upper wall	69.9 (21.1)	41.9 (5.5)	2.50 (7.88)	11.2 (1.97)		13.0 (2.29)	
	structural column	69.3 (20.7)	41.9 (5.5)	2.33 (7.35)	11.8 (2.07)		13.0 (2.29)	
Norfolk, VA	wall	77.2 (25.1)	48.4 (9.1)	3.36 (10.6)	8.6 (1.51)		15.7 (2.76)	
	structural column	77.7 (25.4)	48.4 (9.1)	1.98 (6.24)	14.8 (2.61)		15.7 (2.76)	
	structural beam	78.4 (25.8)	48.4 (9.1)	1.11 (3.48)	27.2 (4.78)		22.1 (3.88)	
Pittsfield, MA	wall	77.8 (25.4)	38.7 (3.7)	3.81 (12.0)	10.3 (1.81)		11.8 (2.07)	
	upper wall	78.0 (25.6)	38.7 (3.7)	4.02 (12.7)	9.8 (1.72)		10.9 (1.93)	
	structural column	77.5 (25.3)	38.7 (3.7)	6.62 (20.9)	5.9 (1.03)		4.6 (0.81)	
Trenton, NJ	wall	70.9 (21.6)	44.5 (6.9)	8.20 (25.9)	3.2 (0.56)		11.1 (1.95)	
	upper wall facings							
	gypsum board	71.9 (22.2)	44.5 (6.9)	8.33 (26.3)	3.3 (0.58)		11.1 (1.95)	
	aluminum panel	72.0 (22.2)	44.5 (6.9)	9.08 (28.6)	3.0 (0.53)		10.6 (1.87)	

TABLE 1—*Continued.*

Building Location	Building Component	Air Temperature, °F(°C) Hot Side	Air Temperature, °F(°C) Cold Side	Heat Flow Rate (HFT), Btu/h · ft² (w/m²)	Thermal Resistance, ft² · h · °F/Btu (m² · K/W) Measured HFT	Measured PC	Predicted Between Framings	Predicted Average
Trenton, NJ	wall at scaffold guide	73.4 (23.0)	44.5 (6.9)	22.6 (71.3)	1.3 (0.23)		4.1 (0.72)	
	window mullion	67.6 (19.8)	44.5 (6.9)	8.16 (25.7)	2.8 (0.49)		3.2 (0.56)	
	structural beam	64.4 (19.0)	44.5 (6.9)	8.09 (25.5)	2.5 (0.44)		10.6 (1.87)	
	walls							
	facing south	66.3 (19.1)	46.7 (8.2)	3.19[a] (10.1)		6.2 (1.1)		11.1 (1.95)
	facing north	70.3 (21.3)	46.7 (8.2)	13.8 (43.5)	1.7 (0.3)	6.8 (1.2)	11.1 (1.95)	11.1 (1.95)
	floor-assembly	59.1 (15.1)	46.7 (8.2)	1.45[a] (4.6)		8.5 (1.5)		5.9 (1.1)

[a]The heat flow rate values were measured by the portable calorimeter.

may be attributed to multidimensional heat flow conditions prevailing in the roof deck rather than the one-dimensional heat flow that is assumed to exist in the thermal resistance calculations. The considerable deviation of the measured values from the predicted results in the Trenton building—whose detailed constructions of the exterior wall and sections of the walls at window washing scaffold guide and at the window mullion are depicted in Fig. 5—was probably caused by excessive convection heat losses from the warmer to colder surfaces within the building envelopes. This convection process resulted from combined forced and buoyancy-driven flows of air in the vertical slots between the gypsum wallboard or aluminum back panel over metal studs and furring channels and the semirigid insulation board, as shown in Fig. 5. Both the leakage of air into the confined air spaces and the temperature difference between their side wall surfaces caused the motion of air along these parallel walls. Based on the wall thermal resistance data presented in Table 1, this thermal bypass due to a convection loop mechanism reduced the overall thermal resistance by approximately 70% in comparison with the same wall with closed air cavities.

The sections of the walls at the window washing scaffold guide and at the window mullion, as shown in Fig. 5, contained structural thermal bridges or highly conductive paths, such as the heat flow through aluminum cover and metallic guide components, which caused excessive heat transfer by means of the conduction process. These are indicated by the considerably lower values for measured resistance in comparison with the computed values in Table 1. It can be noted that the heat flux transducer installed on the north facing wall of the conference center in the Trenton building measured a much lower wall resistance than that measured by the portable calorimeter. This was probably because the heat flux transducer was placed on the area adjacent to the highly conductive metallic framing members. The measured value for the thermal resistance of the floor assembly is significantly greater than the calculated value because of the additional heat given up by the supply of hot air flowing inside the air space within the assembly, resulted in a reduced heat flow from the calorimeter.

As illustrated in Table 1, the wall thermal resistance determined by the portable calorimeters are generally lower than those measured with heat flux transducers because of the increased heat flow of the wall in the adjacent framing members. The metallic framing members and fasteners in insulated walls enhanced heat conduction through the wall segments and reduced the overall wall resistance by approximately 33%. It can also be noted that the significantly high wall resistance value obtained by the portable calorimeter for the Columbia building was probably caused by the additional heat conducted laterally into the metering area of the wall from a hot air duct in the vicinity.

As shown in Table 1, there is a wide variation in the measured thermal resistance values for the masonry and metal-panel-faced exterior walls, which ranged from approximately 0.5 to 5.6 $m^2 \cdot k/W$ (3 to 32 $ft^2 \cdot h \cdot {}^\circ F/Btu$) for the wall area between framing members. These wall resistance values departed from the predicted values by an average of 22% and in two cases by more than 45%.

The mean values of wall thermal resistances measured using eight heat flux transducers and a portable calorimeter for a test duration of 15 days at the Huron building are plotted as a function of the length of time of measurement in Fig. 6. Also plotted in the figure are the corresponding values of the standard deviation, which is an indication of the magnitude of the spread of the thermal resistance data about the mean. As stated earlier, the large thermal mass of the exterior wall exposed to a periodically varying air temperature at the outer surface induced the heat flow to lag behind the temperature driving force. A sufficiently long measurement period is required for the results of wall resistance calculations to converge to a constant value. As shown in Fig. 6, the standard deviation or the dispersion in the thermal resistance values generally decreases as the length of the measurement period increases. The coefficients of variation, which are a measure of the extent of variability of the resistance data, for both the heat flux

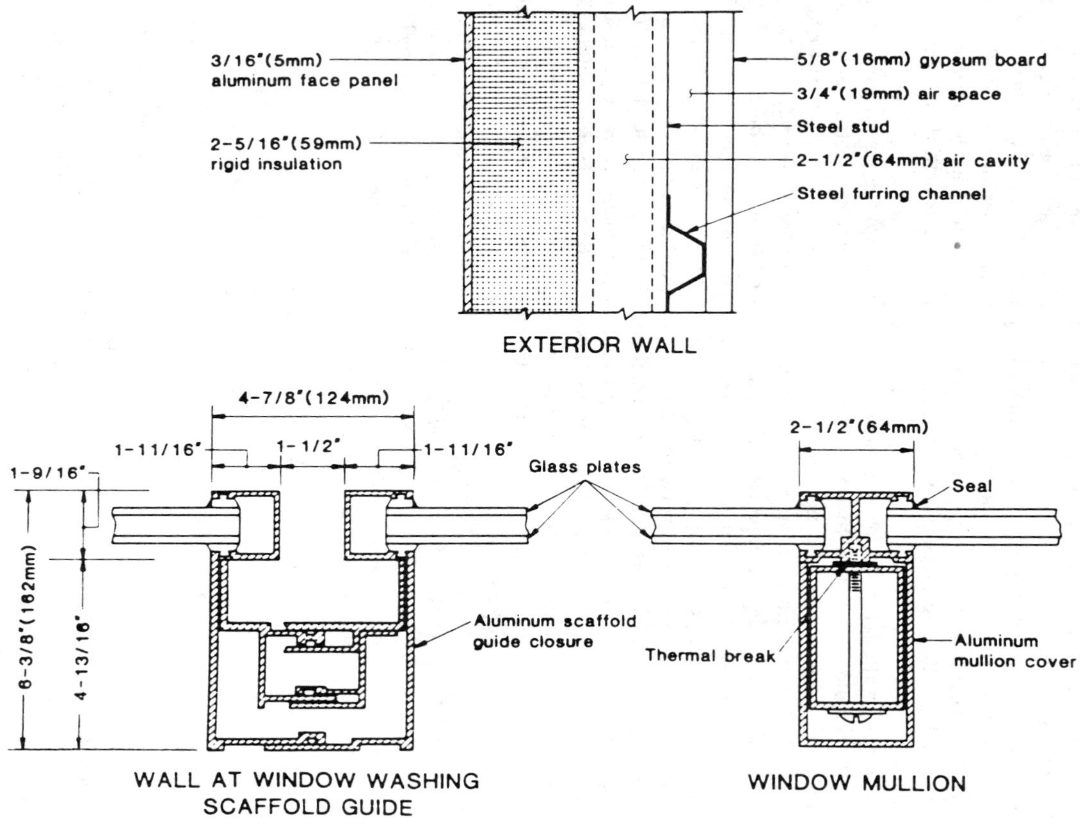

FIG. 5—*Construction details of the exterior wall and wall sections at the window washing scaffold guide and window mullion in the Trenton building.*

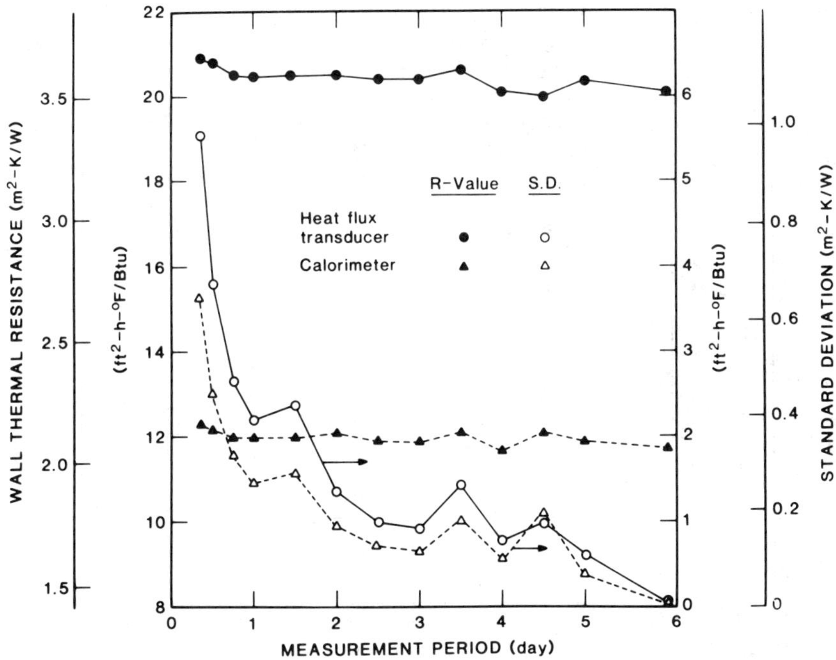

FIG. 6—*Effect of the length of measurement period on wall resistance values, determined by heat flux transducers and the calorimeter.*

transducers and the calorimeter were calculated and found to be within 8% for a measurement period greater than two days.

For the maximum measurement period of 15 days, the wall thermal resistances determined by heat flux transducers and the calorimeter approached the constant values of 3.58 and 2.10 $m^2 \cdot K/W$ (20.3 and 11.9 $ft^2 \cdot h \cdot °F$), respectively. The data in the figure show that the measurement period at a multiple of 24 h has significantly lower standard deviation values than the other remaining measurement period. This is attributed to the fact that the outdoor temperature, the air-to-air temperature difference across the wall, and the heat flow varied periodically over a 24-h cycle. Because of the effect of metallic framing members in insulated walls, the thermal resistance values determined from data obtained by the calorimeter were approximately 40% lower than the values derived from the heat flux transducers.

Figure 7 shows a comparison plot of the measurement period variation of the mean values of wall thermal resistance and their associated standard deviations, which were derived for conditions both with and without correction for thermal time lag between the air temperature difference and the heat flow. In order to correct the time lag, the indoor and outdoor air temperatures determined at the Huron building were treated in pairs, with the measured wall surface heat flux collected 6 h later using a heat flux transducer. This 6-h time lag was estimated from the time history plots of the temperature difference across the wall and the corresponding heat flux at the inside wall surface, using the data acquired during the first six days of the test. In spite of a slight reduction in the extent of variability in the wall resistance values—approximately 3% in the coefficient of variation, as shown in Fig. 7—the difference between the thermal resistance results of those with and without correction for time lag is less than 1.5%.

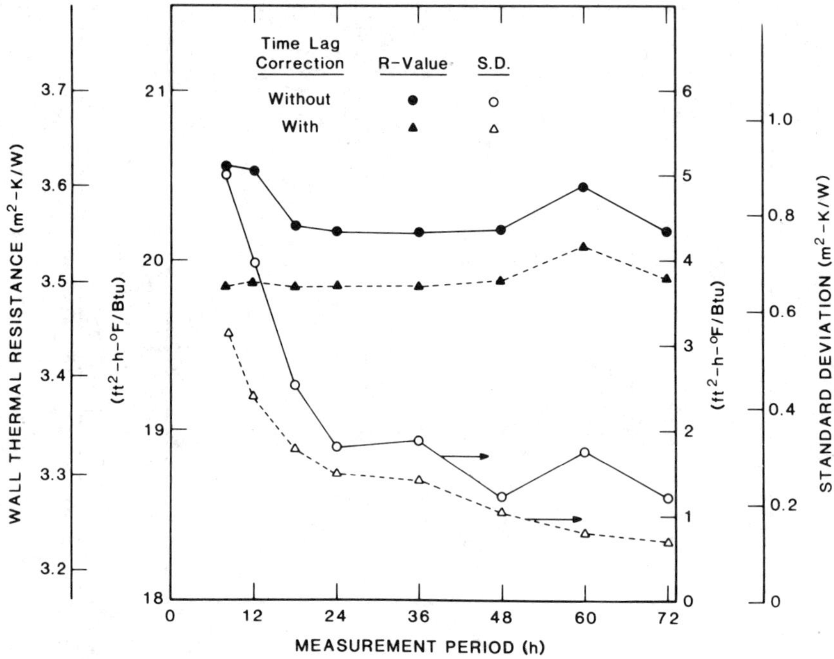

FIG. 7—*A comparison of wall thermal resistances determined with and without thermal time lag correction.*

Conclusions

The thermal resistances of the masonry and metal-faced exterior walls of eight office buildings were derived from the data obtained through continuous measurements of the indoor and outdoor air temperatures and the heat flow across the wall structure for test durations from 2 to 16 days. Because of the time variation of outdoor temperature conditions, a measurement period of two days or longer was found to provide satisfactory thermal resistance results. Both the heat flux transducer and the portable calorimeter were found to be useful tools for *in situ* measurement of heat flow through composite walls. The calorimeter was capable of being used to measure the average thermal resistance of an exterior wall containing highly conductive framing members in insulated wall construction.

The variability of wall thermal resistance values can be reduced by correcting for time lag between the indoor-to-outdoor air temperature difference and the heat flow. This thermal time lag can be estimated from visual inspection of the plots of temporal variations of the heat flow and the air temperature difference across the wall. The measured wall thermal resistances of the metal-faced building were found to be considerably lower than the predicted values. This excessive heat loss was probably caused by thermal bypass or the convection loop losses resulting from the combined forced and buoyancy-driven flows of air in the confined air spaces within the walls, especially cavity walls with air leakage. The thermal resistance values of exterior walls covering the wall area between framing members in eight masonry and metal-panel-faced office buildings have been found to vary widely, from approximately 0.5 to 5.6 $m^2 \cdot K/W$ (3 to 32 $ft^2 \cdot h \cdot °F/Btu$). These wall resistance values deviated from the predicted values by an average of 22%, with the most unfavorable circumstance being 71%.

Acknowledgments

The work reported here was sponsored by the Office of Building Energy Research and Development, U.S. Department of Energy, through an interagency agreement with the National Bureau of Standards (NBS). The authors would like to thank Doug Pruitt of NBS for his assistance in obtaining the field data used in this paper, and Dr. William C. Brown of the National Research Council of Canada for his help in providing valuable information during the construction of the portable calorimeter.

Appendix

Construction Details of Building Components in Test Buildings

The Huron Building

1. Wall located below acoustical tile ceiling: 4-in. (102-mm) face brick, 6-in. (152-mm) lightweight concrete masonry unit, 3-in. (76-mm) semirigid glass fiber insulation board, and 0.63-in. (16-mm) gypsum wallboard on steel studs spaced at 16-in. (406-mm) intervals on center.
2. Wall inside the ceiling plenum: the same as in Point 1.
3. Structural beam in the ceiling plenum: 4-in. (102-mm) face brick, 4-in. (102-mm) lightweight concrete masonry unit, 3-in. (76-mm) semirigid glass fiber insulation, and W24×84 steel beam[2] with a 2-in. (51-mm) minimum fireproofing.
4. Structural column: 4-in. (102-mm) face brick, 4-in. (102-mm) concrete masonry unit, structural column with 2-in. (51-mm) spray-on fireproofing, and 0.63-in. (16-mm) gypsum wallboard on metal studs as column enclosure.

The Ann Arbor Building

1. Wall below the suspended ceiling: 0.5-in. (13-mm) quarry tile, 1-in. (25-mm) metal lath and motor bed, 10-in. (254-mm) concrete masonry unit, 1.5-in. (38-mm) semirigid glass fiber insulation board, and 0.63-in. (16-mm) acoustic wall panel.
2. Beam section inside the ceiling plenum: 0.5-in. (13-mm) quarry tile, 1-in. (25-mm) metal lath and motor bed, 4-in. (102-mm) concrete masonry unit, and W21×44 steel beam with spray-on fireproofing.
3. Structural column: 0.5-in. (13-mm) quarry tile, 1-in. (25-mm) metal lath and motor bed, 10-in. (254-mm) concrete masonry unit, 1.5-in. (38-mm) glass fiber blanket insulation, structural column with spray-on fireproofing, and 0.63-in. (16-mm) plaster on 0.63-in. (16-mm) gypsum lath on 3.3-in. (84-mm) metal studs.

The Anchorage Building

1. Wall below the suspended ceiling: 5-in. (127-mm) precast concrete panel, 3-in. (76-mm) semirigid glass fiber insulation board, 4-in. (102-mm) glass fiber batt insulation, and 0.63-in. (16-mm) foil-backed gypsum wallboard on metal studs.
2. Wall inside the ceiling plenum: 5-in. (127-mm) precast concrete panel and 3-in. (76-mm) semirigid glass fiber insulation board with spray-on fireproofing.

[2]A wide flange shape of a nominal 24-in. depth, weighing 84 lb per linear foot.

[3]A wide flange shape of nominal 21-in. depth, weighing 44 lb per linear foot.

3. Roof deck in the ceiling plenum: 1-in. (25-mm) concrete paver, 3-in. (76-mm) semirigid glass fiber insulation board, and 6-in. (152-mm) lightweight aggregate concrete roof slab on a corrugated steel deck connecting to the wall structure.
4. Structural column: 5-in. (127-mm) precast concrete panel, 3-in. (76-mm) semirigid glass fiber insulation board, 20 by 20-in. (508 by 508-mm) steel column with spray-on fireproofing, and 0.63-in. (16-mm) gypsum board as column enclosure.

The Columbia Building

1. Wall below acoustic tile ceiling: 12-in. (305-mm) precast concrete panel, 2-in. (51-mm) glass fiber blanket insulation, and 0.63-in. (16-mm) gypsum wall board on 2-in. (51-mm) metal studs.
2. Wall above acoustic tile ceiling: 8-in. (203-mm) precast concrete panel, 24-in. (610-mm) air space inside concrete overhand, 3-in. (76-mm) precast concrete panel, 2-in. (51-mm) glass fiber blanket insulation, and 0.63-in. (16-mm) gypsum wallboard on 2-in. (51-mm) metal studs.
3. Concrete wall inside the ceiling plenum: 15-in. (381-mm) precast concrete panel.
4. Concrete beam inside the ceiling plenum: 10-in. (254-mm) precast concrete panel and 18-in. (457-mm) reinforced concrete beam.

The Springfield Building

1. Walls located below and above acoustical tile ceiling: 5.5-in. (140-mm) precast concrete panel, 2.5-in. (64-mm) semirigid glass fiber insulation board, and 0.5-in. (13-mm) gypsum wallboard.
2. Structural column: 5.5-in. (140-mm) precast concrete panel, 2.5-in. (64-mm) semirigid glass fiber insulation board, fireproofed steel beam, and 0.5-in. (13-mm) gypsum wallboard on 2.5-in. (64-mm) metal studs as the column enclosure.

The Norfolk Building

1. Wall below acoustical tile ceiling: 4-in. (102-mm) face brick, 0.75-in. (19-mm) air space, 0.5-in. (13-mm) gypsum board sheathing, 4-in. (102-mm) glass fiber blanket insulation, and 0.5-in. (13-mm) gypsum wallboard.
2. Structural column: 4-in. (102-mm) face brick, 0.75-in. (19-mm) air space, 0.5-in. (13-mm) gypsum board sheathing, 4-in. (102-mm) glass fiber blanket insulation, structural column with spray-on fireproofing, and 0.5-in. (13-mm) gypsum board enclosure.
3. Structural beam inside the ceiling plenum: 4-in. (102-mm) face brick, 0.75-in. (19-mm) air space, 0.5-in. (13-mm) gypsum wallboard, and structural beam with spray-on fireproofing.

The Pittsfield Building

1. Wall below the suspended ceiling: 4-in. (102-mm) face brick, 2-in. (51-mm) semirigid glass fiber insulation board, 6-in. (152-mm) solid masonry concrete block, and 4-in. (102-mm) face brick.
2. Wall inside the ceiling plenum: 4-in. (102-mm) face brick, 2-in. (51-mm) semirigid glass fiber insulation board, 4-in. (102-mm) face brick, 10-in. (2.54-mm) air space, and 4-in. (102-mm) masonry concrete block.
3. Structural column: 4-in. (102-mm) face brick, 0.5-in. (13-mm)-thick premolded expansion filler, 10 by 7-in. (254 by 178-mm) steel column, and 1-in. (25-mm) particle board with plastic laminated face.

The Trenton Building

1. Wall below windows: $^3/_{16}$-in. (5-mm) aluminum face panel, $2^5/_{16}$-in. (59-mm) semirigid glass fiber insulation, 2.5-in. (64-mm) air space between $2^1/_2$ by $1^5/_{16}$-in. (64 by 33-mm) steel studs, and $^5/_8$-in. (16-mm) moisture-resistant gypsum board on $^3/_4$ by $2^3/_4$-in. (19 by 70-mm) steel furring channels.
2. Gypsum board-faced wall above windows: the same as in Point 1.
3. Aluminum-faced wall above windows: $^3/_{16}$-in. (5-mm) aluminum face panel, $2^5/_{16}$-in. (59-mm) semirigid glass fiber insulation, 2.5-in. (64-mm) air space between $2^1/2$ by $^5/_{16}$-in. (64 by 33-mm) steel studs, and $^3/_{16}$-in. (5-mm) aluminum-back panel on $^3/_4$ by $2^3/_4$-in. (19 by 70-mm) steel furring channels.
4. Wall at window washing scaffold guide: $^3/_{16}$-in. (5-mm) aluminum-face panel and $^1/_8$-in. (3-mm) aluminum sheet enclosure measuring 5.5 by 5 in. (140 by 127 mm) in cross section, and containing a steel guide having a cavity with a slot opened to the outside atmosphere and $^3/_4$-in. (19-mm) semirigid glass fiber insulation.
5. Window mullion: $2^1/_2$-in. (64-mm)-wide by $6^3/_8$-in. (162-mm)-deep aluminum mullion with $^1/_8$-in. (3-mm) aluminum cover secured to a panel stiffener attached to the concrete floor deck.
6. Structural beam in the ceiling plenum: $^3/_{16}$-in. (5-mm) aluminum-face panel, $2^5/_{16}$-in. (59-mm) semirigid glass fiber insulation, 2.5-in. (64-mm) air space between steel studs, and steel brace beam installed over $^3/_{16}$-in. (5-mm) aluminum-back panel.
7. South and north facing walls in conference center: $^3/_{16}$-in. (5-mm) aluminum-face panel, $2^5/_{16}$-in. (59-mm) semirigid glass fiber insulation board, 2.5-in. (64-mm) air space between steel studs spaced at 16-in. (406-mm) intervals on center, and $^5/_8$-in. (16-mm) moisture-resistant gypsum board on steel furring channels.
8. Floor assembly: 5.5-in. (140-mm) concrete floor slab on corrugated steel deck enclosed in aluminum truss cover and closure, with a carpet over a rubber pad installed on the top of the floor.

References

[1] Powell, F. J. and Rennex, B. J., "NBS Line-Heat-Source Guarded Hot Plate for Thick Materials," *'Thermal Performance of the Exterior Envelopes of Buildings II,'* ASHRAE SP 38, *Proceedings*, American Society of Heating, Refrigerating, and Air Conditioning Engineers/Department of Energy Conference, Las Vegas, NV, 6–9 Dec. 1982.

[2] Rennex, B., "Error Analysis for the National Bureau of Standards 1016 mm Guarded Hot Plate," NBSIR 83-2674, National Bureau of Standards, Gaithersburg, MD, April 1983.

[3] Brown, W. C. and Schuyler, G. D., "A Calorimeter for Measuring Heat Flow Through Walls," *Proceedings*, American Society of Heating, Refrigerating, and Air Conditioning Engineers/Department of Energy Conference on Thermal Performance of the Exterior Envelopes of Buildings, Kissimmee, FL, ASHRAE SP 28, December 1979.

[4] Brown, W. C. and Schuyler, G. D., "In-Situ Measurements of Frame Wall Thermal Resistance," *A.S.H.R.A.E. Transactions*, Vol. 88, Part 1, 1982, pp. 667–676.

[5] *ASHRAE Handbook of 1985 Fundamentals*, American Society of Heating, Refrigerating, and Air-Conditioning Engineers, Atlanta, GA, 1985, Chapter 23, pp. 2–14.

Claes G. Bankvall[1]

Air Movements and the Thermal Performance of the Building Envelope

REFERENCE: Bankvall, C. G., "**Air Movements and the Thermal Performance of the Building Envelope,**" *Thermal Insulation: Materials and Systems, ASTM STP 922,* F. J. Powell and S. L. Matthews, Eds., American Society for Testing and Materials, Philadelphia, 1987, pp. 124–131.

ABSTRACT: Air movements will influence the thermal performance of the building envelope; to what degree will depend on the pressure situation in and around the structure, the permeability of the different materials, and the airtightness of joints between materials and building elements. In a multilayer structure with a high thermal resistance, different parts have different functions in order to protect against unwanted air flow and ensure airtightness.

Wind protection will give a degree of safety against air flow in the insulation, especially in situations involving large wind loads. Certain areas of the structure may be more sensitive and have a higher need for wind protection than others. In a similar way, the building envelope will be sensitive to air infiltration. The inside—that is, the vapor barrier, inside board, and joints—will greatly influence the air flow and the heat losses from the envelope.

Both types of increased heat losses from the building envelope (that due to air flow along the insulation and that due to air infiltration) will most influence envelopes with high thermal resistance. This type of envelope is typical of today's design in Scandinavia, and the results and discussions are related to this.

KEY WORDS: thermal insulation, air movements, permeability, wind protection, airtightness

Air movements will influence the thermal performance of the building envelope. The thermal insulation is sometimes protected by a special wind protection. This will give the structure a degree of safety against influence from wind. Wind and pressure situations around the envelope are of importance. Certain areas will be more sensitive and have a higher need for wind protection than others.

In a similar way, the building envelope will be sensitive to air infiltration. Infiltration through the envelope was previously a means of ventilating the building. In modern Scandinavian designs, this is not the case. The aim now is to reduce unwanted infiltration as much as possible. The inside—that is, the vapor barrier, board, and joints—will have the most influence on such infiltration.

Figure 1 shows an example of a building envelope. Behind the face wall and underneath the roofing there is an air flow that transports moisture from the structure. The different parts in the design intended to protect against wind (air flow along the insulation) and give the envelope its airtightness are indicated in the figure.

In Scandinavia there are requirements concerning airtightness of buildings and building components such as windows and doors. Wind protection is not quantified in the same manner, but its importance is underlined in the different building codes.

[1]Professor, Swedish National Testing Institute, S-501 15 Borås, Sweden.

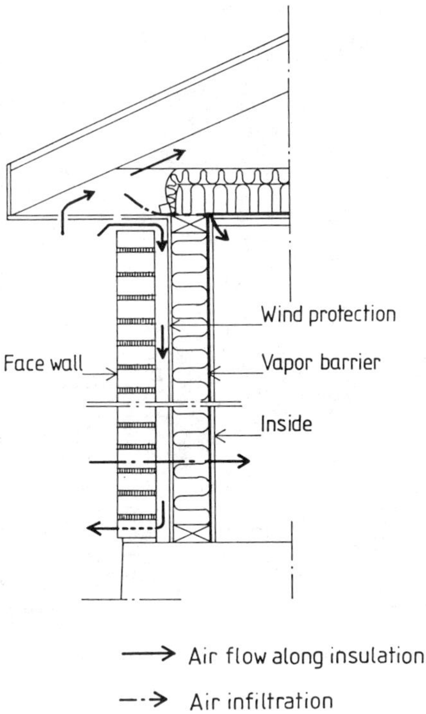

FIG. 1—*The building envelope, its ventilation, and the parts that provide wind protection and airtightness.*

Air Permeability

An important factor in the design of a modern structure is the airtightness of the building envelope and its joints. The air permeability of a material or a structural part, in combination with a pressure difference, leads to air movement in and through the structure, and consequently influences its thermal performance.

Figure 2 shows how a pressure difference will influence the air flow through different structural parts. The least permeable materials, such as plastic sheeting, are used for the vapor barrier, the main airtightness layer in the building envelope. Nailing a board joint against a crossbar can give fair airtightness, depending on the distance between the nails. A commonly used wind protection in Scandinavia is the asphalt-impregnated porous board. This has a lesser degree of airtightness but is still considered sufficient as wind protection. The face wall gives a certain degree of protection from direct wind. This will depend on the design of inlets and outlets to the air space behind the wall.

The examples in the figure have been based on practical measurements of materials and building elements, but also on theoretical analyses of air flow in different structures [*1–4*].

Wind on a building envelope will lead to pressure differences and air flow in and around the structure. The actual situation will depend on wind strength, building design, proximity to other buildings, and other factors.

It is therefore difficult to evaluate the true pressure situation for a building. Few field measurements have been done. Some investigations in wind tunnels are available. However, for a rough evaluation of the pressure changes at the outside of a one-family building, Fig. 3 has been

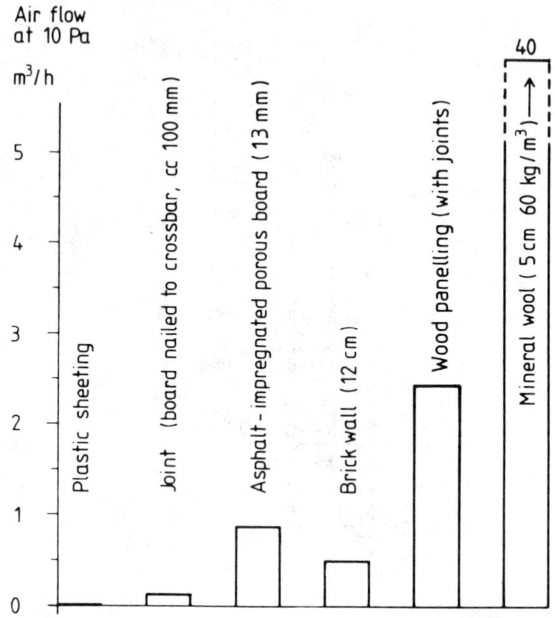

FIG. 2—*Air flow through different building components at a 10-Pa pressure difference.*

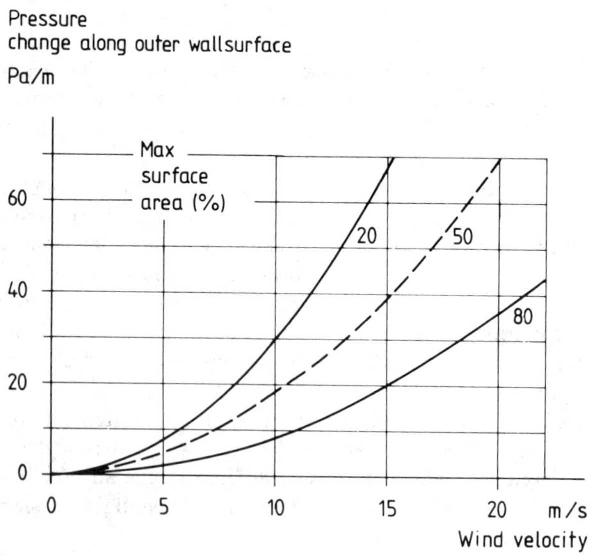

FIG. 3—*Estimated pressure changes along the wall surface at varying wind velocities at a one-family house.*

constructed. The figure shows that, for a given wind velocity, large pressure changes normally involve only minor parts of the total surface area. At a wind of 10 m/s, up to 50% of the surface area can experience a pressure change of about 20 Pa/m. This diagram can be used for rough evaluations of the importance of protecting thermal insulation from air flow along it in a given situation.

Estimations can also be made of pressure differences over the wall on the windward and leeward sides, based on building code requirements for wind loading in the Swedish building code (Fig. 4). The pressure differences may be somewhat overestimated. Airtightness and the consequences of air infiltration may be evaluated for a pressure difference from inside to outside of about 20 Pa. Experience from field measurements indicates that this is a reasonable value.

Face Wall

The pressure changes behind the face wall will depend on the wind, permeance of the wall, size of the ventilated airspace, and other factors. This is shown in Fig. 5. Two extreme situations are illustrated: one in which the air space has open inlets and outlets and the other in which these are closed. In the diagram, normal permeances for brick and wooden paneling, as measured on Swedish structures, are indicated.

Wind Protection

Air movements in the building envelope behind the face wall will increase the heat losses from the wall section. This may be described as an increase in the U value of the section (Fig. 6). The U value is indicated for thermal insulations of 10 and 30-cm thickness. The insulation is a mineral wool with λ value of 0.034 W/°C. Depending on the pressure change outside the insulated space and the wind protection of the insulation, the U value will increase as indicated in the figure. The unprotected insulation experiences a noticeable increase in U value, while the

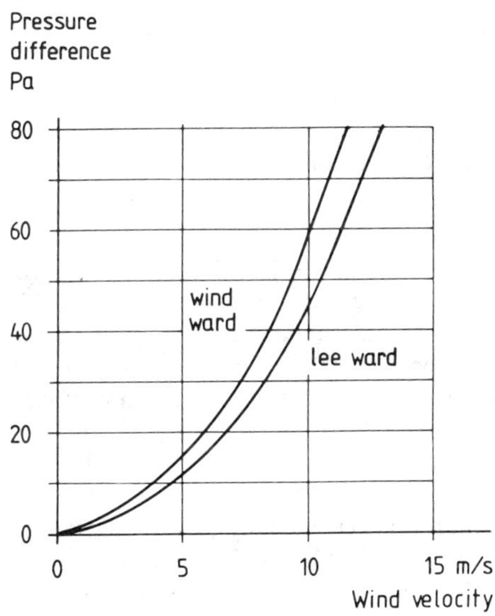

FIG. 4—*Estimated pressure difference over a wall at varying wind velocities at a one-family house.*

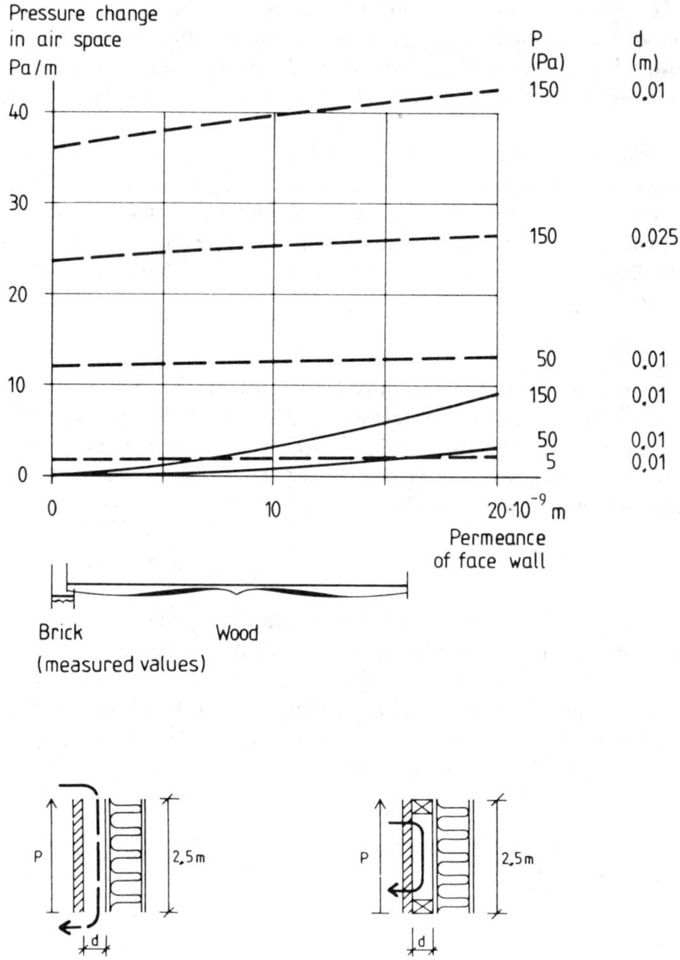

FIG. 5—*Pressure change behind the face wall at different outside wind pressures,* (P), *and varying wall permeances and airspace designs.*

increase for the protected one is very small. Relatively, the most sensitive situations are those with the highest thermal resistance. The basis for these evaluations is presented in Refs *1* and *3*.

Wind protection is especially important for those parts of a building in which pressure changes are normally large, such as at corners and eaves. Another situation in which the wind protection is of great importance occurs when the installation of insulation is deficient [*5*].

Airtightness

Air infiltration increases heat transfer. In many cases this is necessary for the normal ventilation of the building. In modern designs this air leakage means an unwanted increase in heat loss from the building. The airtightness of the building envelope is thus of importance [*6*].

Most of the air will pass through joints and openings that occur because of the design or installations in the envelope. The inside board and the vapor barrier give the building envelope its airtightness.

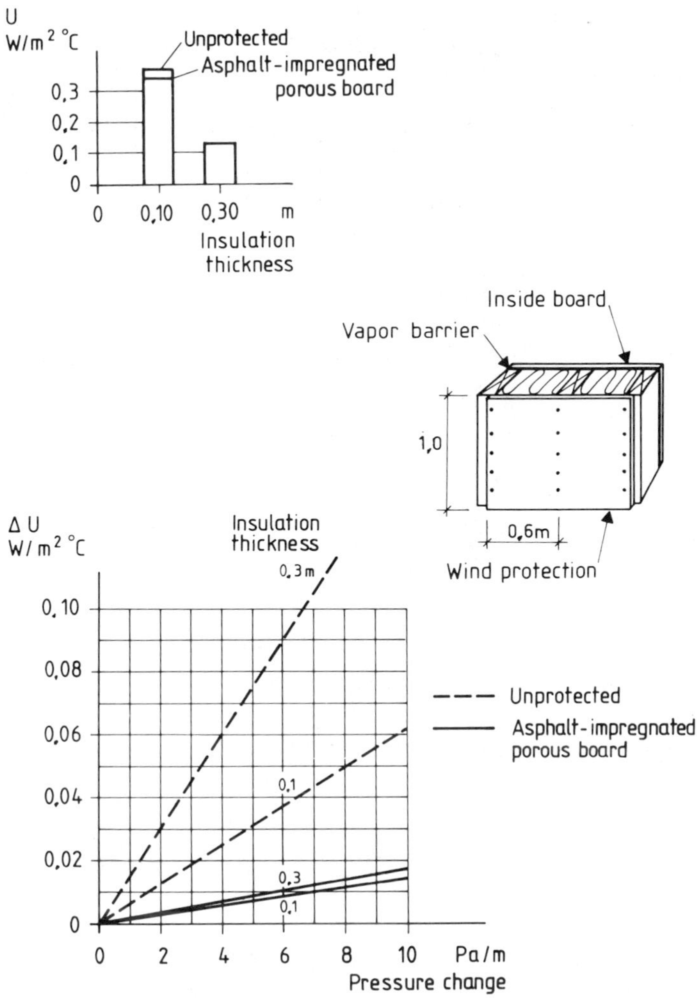

FIG. 6—*Increase in* U *value due to air flow along a wall section with varying insulation thicknesses and wind protection values.*

The pressure difference across a wall is exemplified in Fig. 7. It can be seen that the vapor barrier has the greatest effect on the pressure drop. When the vapor barrier is missing, the inside board is important. In the "normal" design, the wind protection is of less importance. The increase in heat transfer can be compared with the original U value of the wall. Additional heat transfer will depend on pressure drop, vapor barrier, and, to some extent, wind protection. Figure 8 shows the increased heat transfer as an increase in U value. The basis for this relationship will be found in Refs *1* and *3*. With a properly installed vapor barrier, flow through the structure is negligible and is not influenced by the choice of wind protection. Without the vapor barrier, the heat transfer increases markedly. This is even more the case when there are deficiencies in the inside board as well [*5*].

Air flow through the structure is generally independent of the thermal insulation. Heat transfer through the building envelope will therefore increase independent of the nominal U value. This will most influence structures with high thermal resistance.

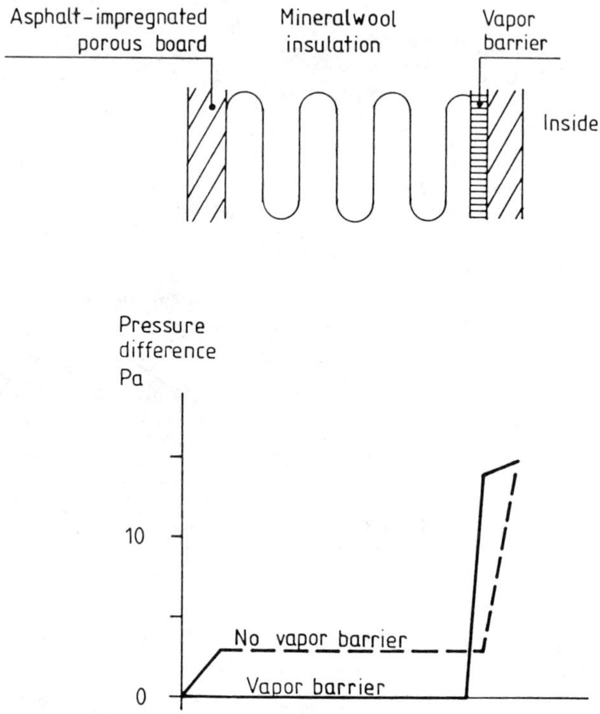

FIG. 7—*Pressure drop in a crossbar wall with and without a vapor barrier.*

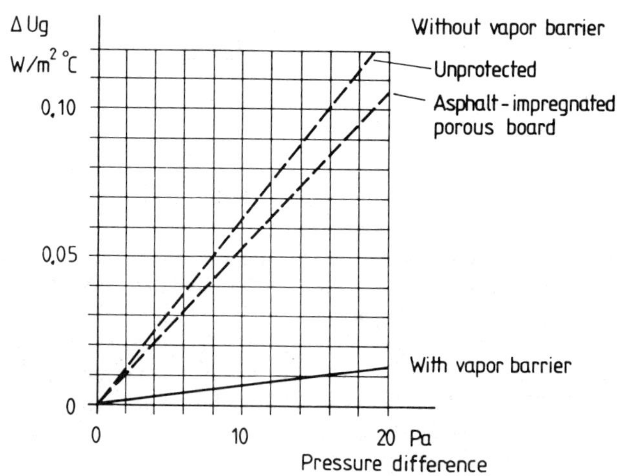

FIG. 8—*Increase in thermal transmittance due to pressure difference and air flow through the envelope.*

Conclusions

When air flows along the thermal insulated space, a special wind protection is useful. This is true for parts of a building where the pressure change is large, such as at corners, or for buildings frequently exposed to strong winds.

The airtightness of the structure depends on the vapor barrier and, to some extent, on the inside board. Air flow through the structure that markedly increases the heat loss is generally due to deficiencies in the vapor barrier and the inside board. Such deficiencies are difficult to compensate for by choice of wind protection.

Both types of increased heat loss from the building envelope, air flow along the insulation and air flow through the insulation, will have the most influence on envelopes with high thermal resistance.

References

[*1*] Bankvall, C. G., "Forced Convection. Practical Thermal Resistance Under the Influence of Wind and Workmanship," Swedish National Testing Institute, Borås, Sweden, 1979.

[*2*] Bankvall, C. G., "Forced Convection. Practical Thermal Conductivity in an Insulated Structure Under the Influence of Workmanship and Wind," in *Thermal Transmission Measurements of Insulations, ASTM STP 660,* American Society for Testing and Materials, Philadelphia, 1978, pp. 409–425.

[*3*] Bankvall, C. G., "Byggnadskonstruktioners värmeisoleringsförmåga," T18:1981, Swedish Council for Building Research, Stockholm (in Swedish, to be translated to English), 1981.

[*4*] Kronvall, J., "Air Flows in Building Components," *Building Technology,* TVBH-1002, Lund Institute of Technology, Lund, Sweden, 1980.

[*5*] Bankvall, C. G., "Thermal Performance of the Building Envelope as Influenced by Workmanship," this publication, pp. 679–684.

[*6*] Elmroth, A. and Levin, P., "Air Infiltration Control in Housing, a Guide to International Practice," D2:1983, International Energy Agency, Swedish Council for Building Research, Stockholm, Sweden.

Tamami Kusuda[1] *and John W. Bean*[1]

Design Heat Loss Factors for Basement and Slab Floors

REFERENCE: Kusuda, T. and Bean, J. W., "**Design Heat Loss Factors for Basement and Slab Floors,**" *Thermal Insulation: Materials and Systems, ASTM STP 922,* F. J. Powell and S. L. Matthews, Eds., American Society for Testing and Materials, Philadelphia, 1987, pp. 132–152.

ABSTRACT: The heat loss factors for selected types of earth contact surfaces, such as basement walls and floors and slab-on-grade floors, are presented to supplement data in the *ASHRAE Handbook of Fundamentals* for estimating the building design heat loss. The perimeter heat loss factors are derived from the procedure developed by Mitalas and from monthly normal temperature cycles of many localities in the United States. Except for those for cities in California, the heat loss factors are found to be a relatively well-defined function of degree-days and remain constant for degree-days beyond 3000 (engineering units). The heat loss factors show that the most effective way to reduce the earth contact surface heat loss is to insulate the inside corners of wall/floor joints. The insulation installed below the floor slab is found to be the least effective as long as the wall foundation is left uninsulated.

KEY WORDS: basement wall/floor heat loss, slab-on-grade floor heat loss, earth temperature, Mitalas method, ASHRAE handbook procedure, thermal insulation

Heat loss calculations from earth contact surfaces such as basement walls and floors have been considered one of the most uncertain areas in the building thermal analysis.

In spite of several noteworthy efforts of previous authors, such as Lachenbruch [*1*], Latta [*2*], Muncey [*3*], Wang [*4*], Kusuda [*5,6*], and Shipp [*7*], there have been no well-recognized design formulas applicable for the determination of earth contact heat loss. Previous work on this subject is divided into two different approaches.

The first is to develop analytical solutions using basic heat conduction equations and simple boundary conditions. One of the notable examples made in recent years is by Delsante et al [*8,9*] for the development of analytical solutions for slab-on-grade heat loss calculations. The Delsante solution, however, is limited to a square slab and is not applicable for floors with perimeter insulation. To the best of the authors' knowledge, no analytical solutions are available for insulated basement wall and floor heat loss analysis.

The second approach is to solve the basic heat conduction differential equation by numerical techniques such as finite element or finite difference methods. Several authors, such as Shipp [*7*], Wang [*4*], and Akasaka [*10*], published the application of numerical methods to selected types of earth contact heat transfer systems. The most comprehensive numerical calculations were made recently by Mitalas [*11,12*], who analyzed a total of 40 different earth contact systems involving shallow basements and deep basements. He later expanded the calculations to cover a total of 40 deep basements, 21 shallow basements, and 39 slab-on-grade floor configurations. Using the results of finite element analysis, Mitalas presented extremely simple empirical formulas to calculate the monthly heat loss from these earth contact systems.

[1]Chief and mathematician, respectively, Building Physics Division, National Bureau of Standards, Gaithersburg, MD 20899.

The purpose of this paper is to apply the Mitalas results for selected cities in the United States to generate design parameters suitable for estimating the design heat loss factors for different heat transfer systems.

Mitalas Equations

Mitalas formulated the results of his finite element analysis in the following simple manner: the basement heat loss calculations were treated in five area segments, A_1 through A_5, as follows (see Fig. 1):

(*a*) above-ground wall heat transfer, q_1,
(*b*) upper portion of the wall below grade, q_2,
(*c*) lower portion of the wall below grade, q_3,
(*d*) perimeter zone of the floor, 1 m wide, adjacent to the wall q_4, and
(*e*) central zone of the floor, q_5.

The monthly heat fluxes, q_n ($n = 1, 5$), computed by the finite element computer program were represented for each zone by the following equation

$$q_n(t) = S_n(\theta_B - \theta_G) + V_n \cdot \sigma_n \cdot \theta_v \sin \omega(t + \Delta t_n) \tag{1}$$

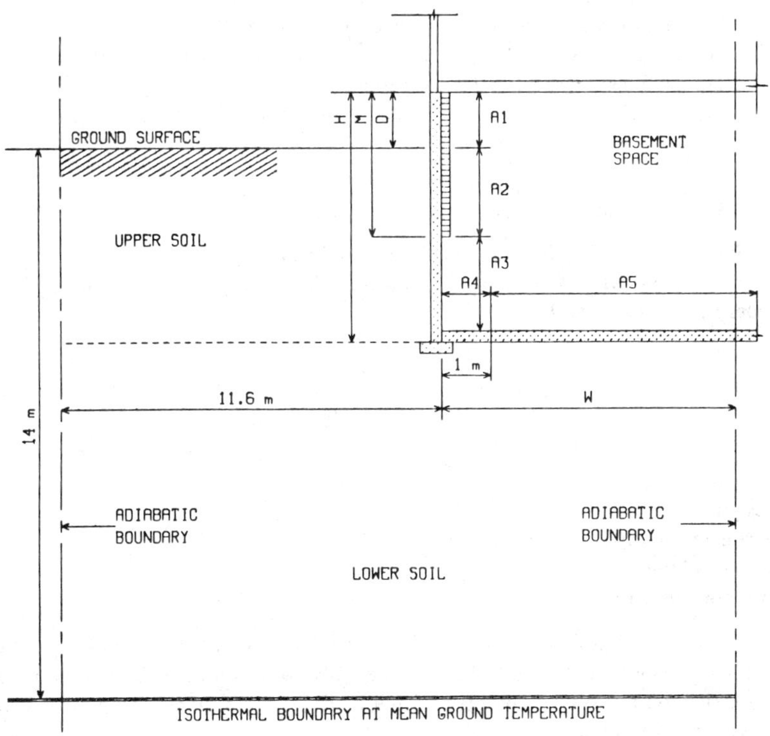

FIG. 1—*Basement model used by Mitalas.*

where

$q_n(t)$ = heat flux normal to the suface of Zone n,
S_n = shape factor for the steady-state heat loss through Segment n,
θ_B = room air temperature,
θ_G = annual average ground surface temperature,
Δt_n = time lag,
V_n = shape factor for periodic heat flow, and
σ_n = amplitude attenuation factor.

The total ground heat loss, $Q(t)$, was obtained by summing the products of this heat flux, q_n, multiplied by A_n, the area of the respective segments, as follows

$$Q(t) = \sum_{n=2}^{5} A_n \cdot q_n(t) \tag{2}$$

Slab-on-grade floor calculations were treated similarly to the basement floor problem, except that only Segments 4 and 5 were considered.

The Mitalas formula was developed for Canadian climates, which require monthly average ground surface temperature data and monthly average ambient temperature. Of course, the ground surface temperature is applicable for below-grade surface heat loss calculations, whereas the ambient temperature is applicable for above-grade basement wall heat loss analysis.

The Mitalas equation also includes the effect of the corners by adjusting the constants, S and V, for Segments 3, 4, and 5.

Mitalas varied the insulation thermal resistance, R, location of insulation, and earth thermal conductivity, resulting in a set of S, V, C, σ, and Δt for 40 different configurations of the basement system, as shown in Fig. 2. Also studied were 21 different configurations for a shallow basement, as shown in Fig. 3, and 39 different slab-on-grade configurations, as shown in Fig. 4.

ASHRAE Heat Loss Factor, F_2

The *ASHRAE Handbook of Fundamentals* [*13*], published by the American Society of Heating, Refrigerating, and Air-Conditioning Engineers (ASHRAE), defines the earth contact surface heat loss factor, F_2, as follows

$$F_2 = \frac{q}{P(\theta_i - \theta_0)} \tag{3}$$

where

q = design heat loss,
P = perimeter or exposed edge of floor,
θ_i = indoor temperature, and
θ_0 = outdoor temperature.

The same handbook lists the values of F_2 for four selected types of foundation construction with and without insulation. The usefulness of the ASHRAE handbook data are limited, however, because they do not reflect conditions such as the position of the insulation, insulation thickness, and soil condition.

Typical values for F_2 were generated for all of the Mitalas systems for the climatic conditions for approximately 300 weather stations in the United States. In this calculation, q is the maximum heat loss, θ_i is 21°C, and θ_0 is chosen to be the monthly normal temperature for February.

Indicated in Figs. 2, 3, and 4 is the loss factor, F_2 [*13*], expressed in British thermal units per hour per linear foot of perimeter per difference in degrees Fahrenheit between the indoor and outdoor temperature of February in Washington, DC. Each of the earth contact systems is identified by the same numbering system used by Mitalas [*12*].

The soil thermal conductivity data, for outside the wall and for below the floor, are indicated in these figures. Thermal conductivity values are given separately for the upper and lower regions of the earth contact systems for the range covering from 0.8 W/m · K (5.5 Btu · in./ft² · h · °F) to 2.0 W/m · K (13.9 Btu · in./ft² · h · °F).

These figures were obtained by using Mitalas Eqs 1 and 2 applied to the Washington, DC, weather conditions on the assumption that the ground surface temperature cycle follows very closely the monthly normal ambient air temperature cycle [*14*].

In these figures, thermal insulation is indicated by the dark solid bars and the concrete foundation walls and floors by the light bars. The heat loss factors indicated in these figures are all based on a fixed R value of insulation of 8.8 (1.55 RSI), except for the deep basement systems, 21 through 26. The Mitalas data are available to permit the evaluation of different R values of the thermal insulation for most of the earth contact systems in these figures, except for Systems 21 through 26. In these special cases, the solid bars identified by R_1 and R_2 were assumed to have R values of 10 (1.76 RSI) and 20 (3.52 RSI), respectively.

The insulated wood frame wall having the R value of 17 (3 RSI) is represented in the slab-on-grade systems by shaded bars in Fig. 4. For all these calculations, the overall heat transfer coefficient of the uninsulated concrete wall was assumed to be 0.45 W/m² · K (0.08 Btu/h · ft² · °F).

To show the sensitivity of thermal insulation on the earth contact walls and floors, the Appendix is provided to tabulate the results of F_2 calculations when R of the thermal insulation was assigned several typical values such as 4(0.70 RSI), 5.4(0.95 RSI), 8(1.4 RSI), and 8.8(1.55 RSI).

The column of the Appendix under RSI = 0.95 corresponds to the ASHRAE sample values and is used herein for the purpose of comparison with other R values used in Figs. 2 through 4. In the Appendix the heat loss factor, F_2, values are given in watts per metre kelvin, with the same data expressed in British thermal units per hour per linear foot per difference in degrees Fahrenheit in parentheses immediately below. It is obvious that the values of F_2 for slab-on-grade floor are less dependent on the earth contact insulation than those for the basement wall/floor. While the basement floor insulation is effective up to as much as an R of 8 (or RSI of 1.4), its advantage diminishes beyond an R value of 5.4 (or RSI of 0.95) for the slab-on-grade floor heat loss.

To test the climatic sensitivity of F_2, the calculations were repeated for Systems 1, 11, 41, and 53 for a total of 296 cities in the United States and results of these calculations are plotted in Fig. 5 against the heating degree-days [based on 18.3°C (65°F)]. In this calculation the slab insulation was assumed to have on R value of 8.8 (1.55 in RSI).

All the high spikes shown in Fig. 5 are for the California cities, and they are caused by a very small indoor/outdoor temperature difference in February. If one excludes these California cities, F_2 follows a relatively smooth relationship with respect to heating degree-days. Moreover, F_2 presented herein provides a useful index for evaluating the effectiveness of insulating earth contact surfaces.

The following observations can be made, for example, by observing Fig. 2 for the full basement heat transfer:

1. Comparison of Systems 3 and 11 shows that there is no advantage in insulating the outside of the basement wall over insulating the inside.

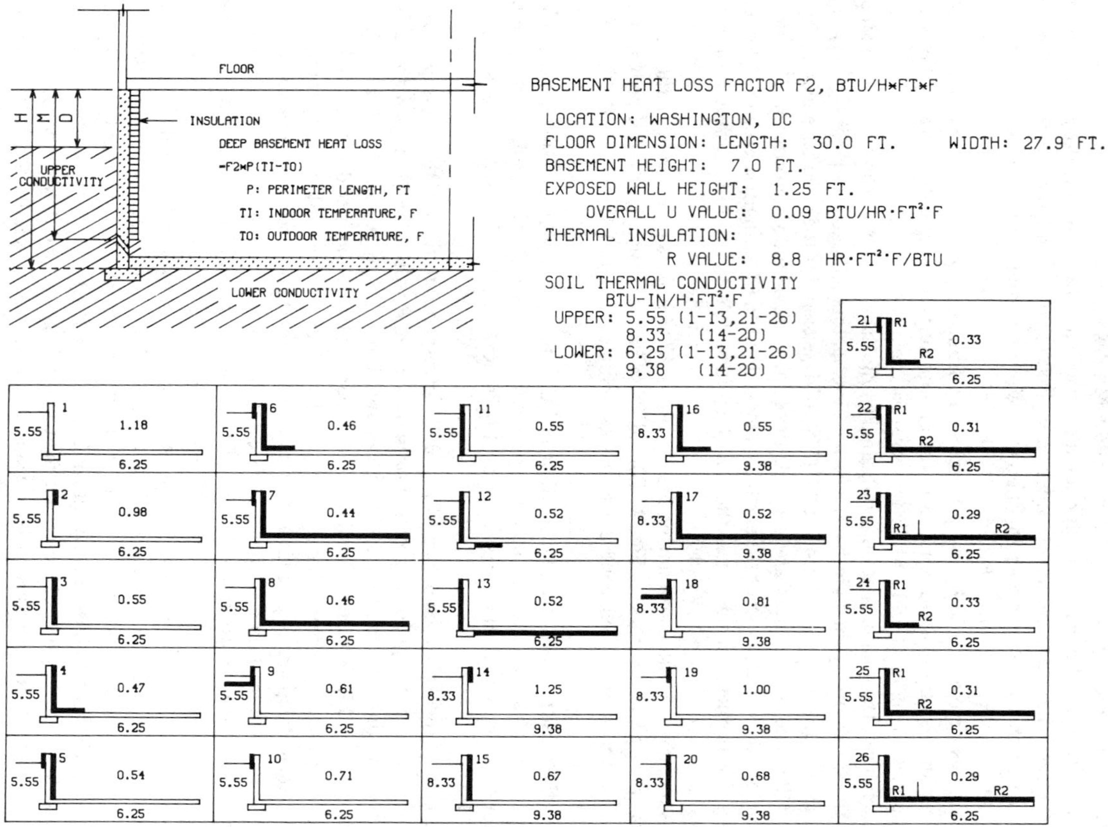

FIG. 2—*Basement heat loss factor in Washington, DC.*

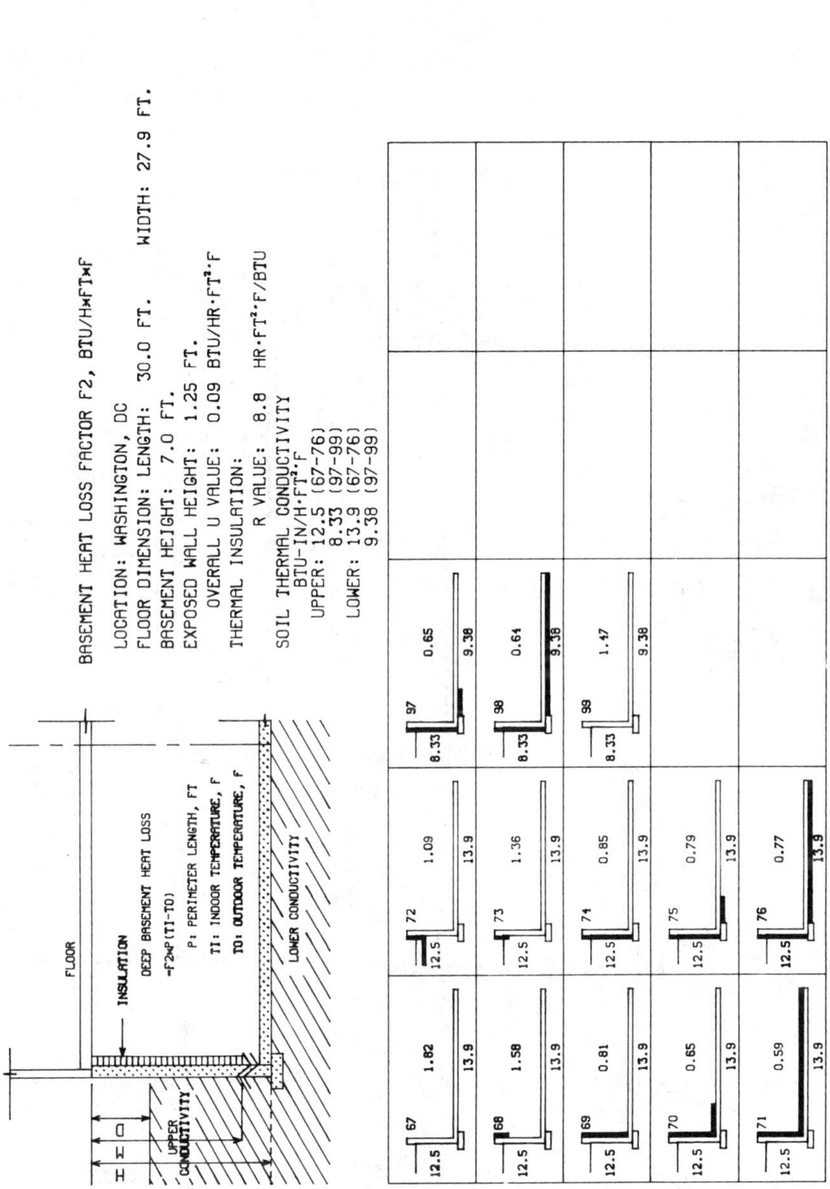

FIG. 2—*Continued.*

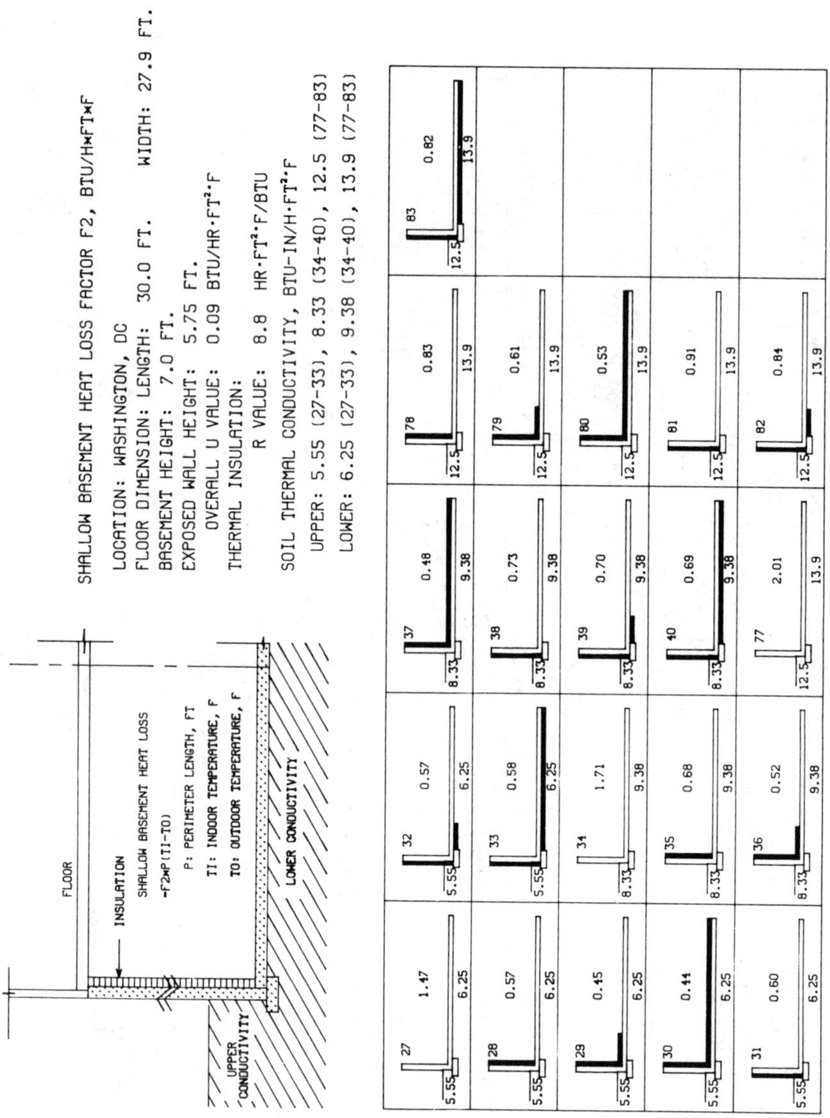

FIG. 3—*Shallow basement heat loss factor in Washington, DC.*

2. Comparison of Systems 4 and 12 shows that the basement floor should be insulated around its perimeter inside the floor. The outside insulation may be more effective if the foundation footing is completely covered with insulation.

3. Comparison of Systems 6 and 7 shows that insulating the basement floor beyond the perimeter area is not effective.

4. Comparison of Systems 2 and 14 shows that the earth thermal conductivity significantly changes the earth contact heat transfer.

5. Comparison of Systems 1 and 2 shows that insulating the upper half of the basement wall significantly reduces the basement heat loss.

6. Comparison of Systems 2 and 10 shows that if only the upper part of the basement wall is to be insulated, insulation should be placed outside the basement wall.

7. Systems 21 through 23, in comparison with the comparable Systems 6 through 8, show a significant reduction in basement heat loss by increasing the thermal resistance of insulation from an R of 8.8 to an R of 20.

8. Systems 24 through 26 show that no increase in heat loss results from eliminating the outside insulation along the top half of the basement wall, as far as the below-ground heat loss is concerned.

9. Comparing System 9 with System 10 and System 18 with System 19 shows the effectiveness of horizontal insulation extending outward near the ground surface.

10. Comparison of Figs. 2 and 3 shows that shallow basements definitely lose more heat than the deep basements.

For the shallow basement, Fig. 3 reveals the following:

11. Comparisons of Systems 28 and 31, 35 and 38, and 78 and 81 show that for a shallow basement, insulation should be placed inside the basement wall.

12. Comparisons of Systems 32 and 33, 39 and 40, and 82 and 83 also show that basement floors should not be insulated beyond the perimeter zones, especially when the insulation is placed outside.

13. Comparisons of Systems 28 and 39, 35 and 36, and 78 and 79 indicate a significant decrease in shallow basement heat loss by the use of floor perimeter insulation.

14. Comparisons of Systems 32 and 33, 39 and 40, and 82 and 83 clearly show that insulation below the floor slab is very ineffective as long as the foundation footing is uninsulated.

15. Comparisons of Systems 27 through 33, 34 through 40, and 77 through 83 clearly indicate the strong influence of soil thermal conductivity.

Figure 4 shows heat loss from slab-on-grade systems having three different kinds of foundation walls—namely, the concrete wall with insulation, insulated wall with a fixed R value of 17, and insulated wall with variable R values.

1. Comparisons of Systems 41 and 42 and 54 and 55 show the effectiveness of floor perimeter insulation as long as the insulation is inside.

2. Comparisons of Systems 41 and 44 and 54 and 57 show that perimeter insulation under the wall foundation is not as effective as inside insulation.

3. Comparisons of Systems 41 and 42, 54 and 55, and 84 and 85 indicate the importance of the insulation inside the floor slab near the wall foundation.

4. Although comparison of Systems 42 and 43 shows that extended floor slab insulation beyond the perimeter zone is practically useless over the dry soil, comparisons between Systems 55 and 56 and Systems 85 and 86 show its advantages over the more conductive soils.

5. Comparisons of Systems 41 through 45 with Systems 46 through 50, 54 through 58 with 59 through 63, and 84 through 88 with 89 through 93 show that the heat loss is significantly reduced as long as the concrete floor is isolated from the concrete foundation footing by insulated walls.

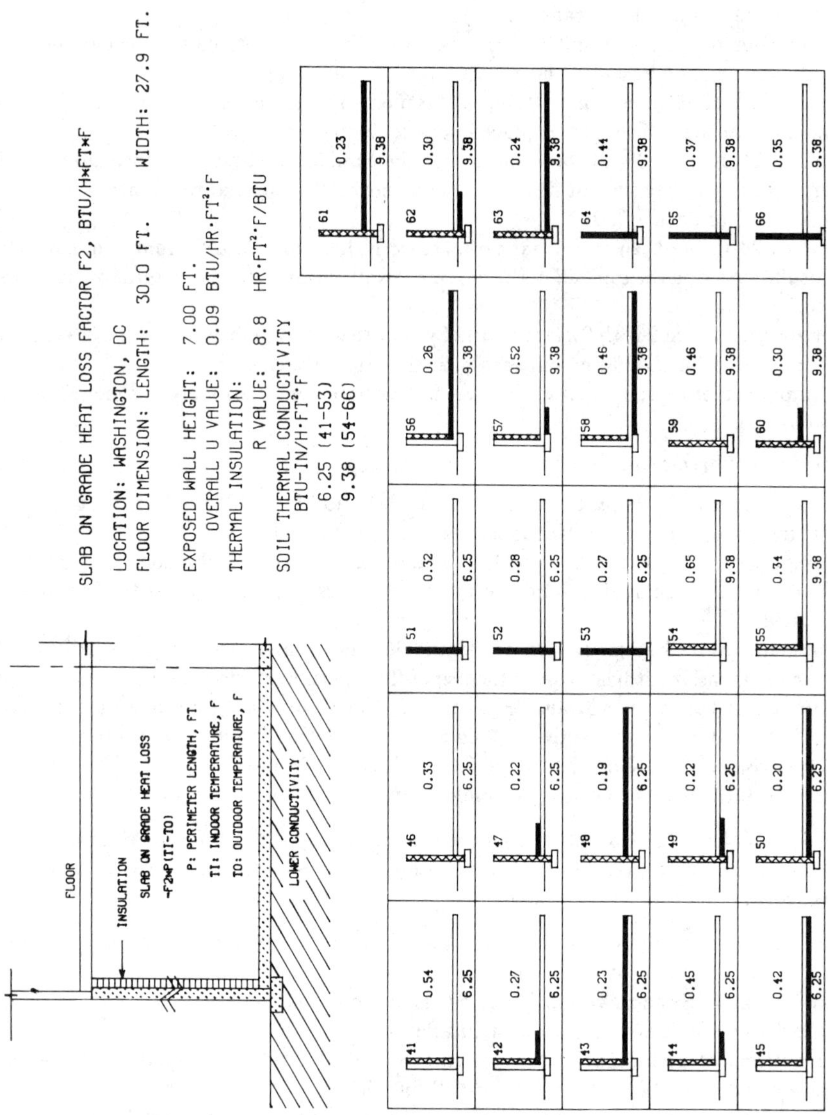

FIG. 4—*Slab-on-grade basement heat loss factor in Washington, DC.*

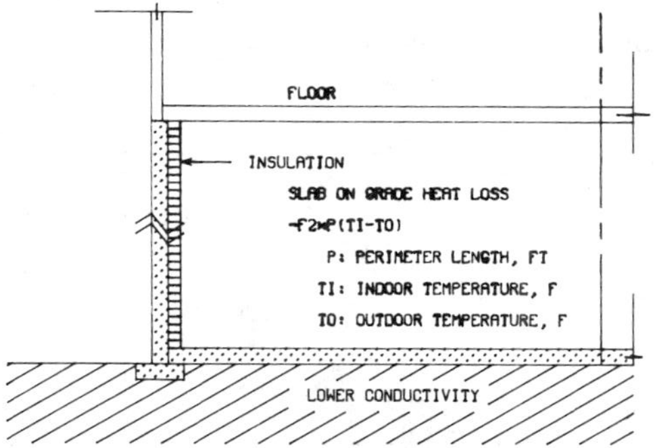

SLAB ON GRADE HEAT LOSS FACTOR F2, BTU/H*FT*F

LOCATION: WASHINGTON, DC
FLOOR DIMENSION: LENGTH: 30.0 FT. WIDTH: 27.9 FT.

EXPOSED WALL HEIGHT: 7.00 FT.
OVERALL U VALUE: 0.09 BTU/HR·FT²·F
THERMAL INSULATION:
R VALUE: 8.8 HR·FT²·F/BTU
SOIL THERMAL CONDUCTIVITY
BTU-IN/H·FT²·F
13.9

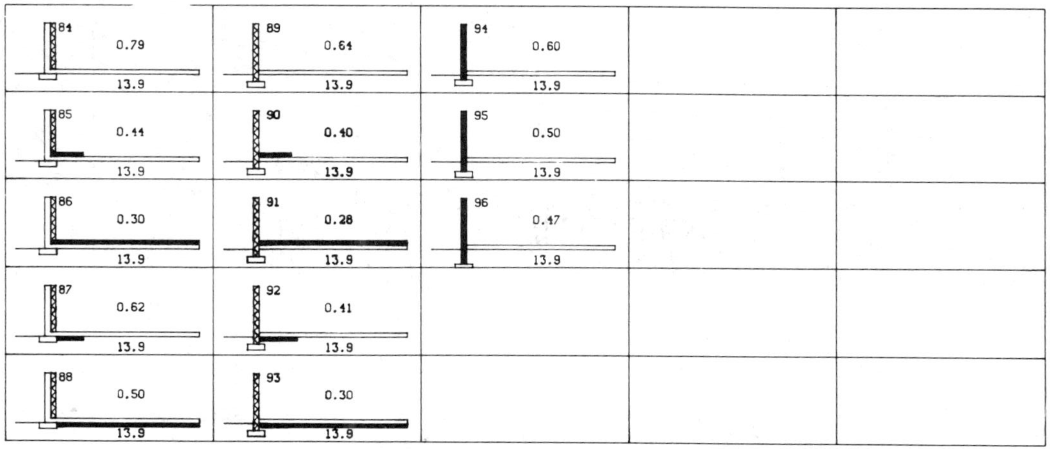

FIG. 4—*Continued.*

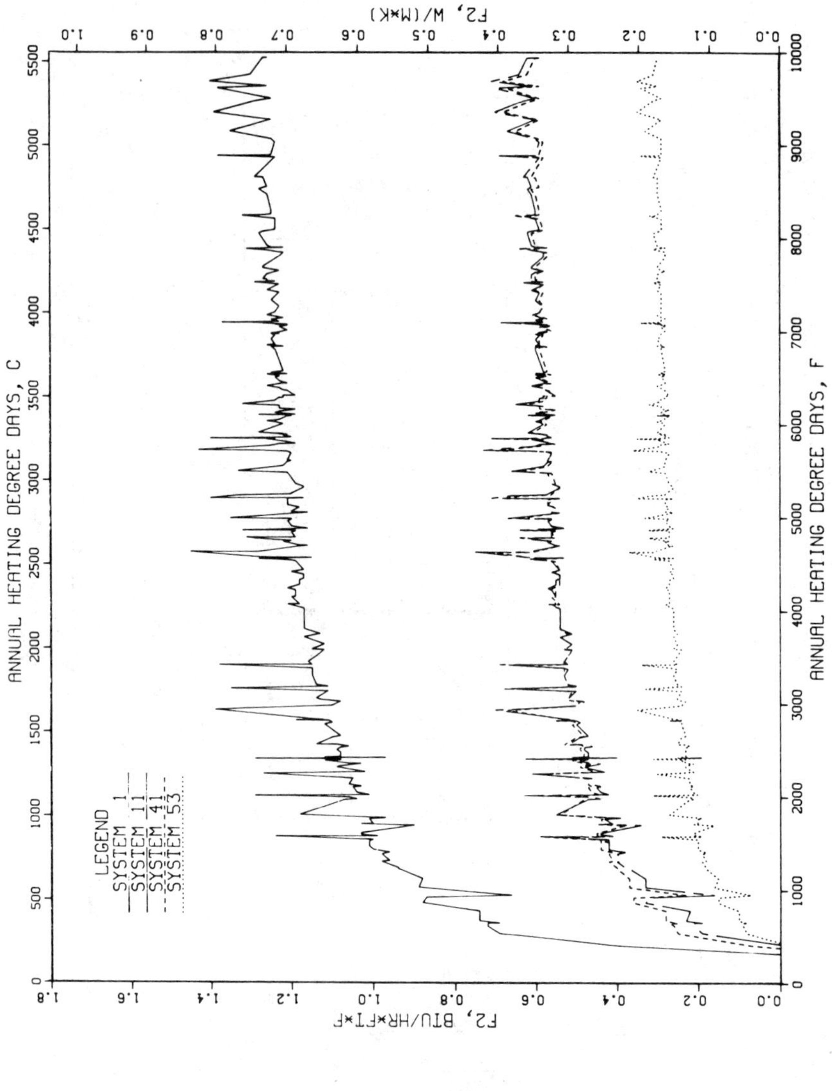

FIG. 5—*Climatic sensitivity of the basement heat loss factor,* F_2.

6. Systems 51 through 53, 64 through 66, and 94 through 96 represent the cases in which the insulated foundation wall was extended 0.5, 1.0, and 1.5 m below the grade for soil thermal conductivities of 0.94, 1.35, and 2.0 W/m · K, respectively. Extending the foundation footing more than 1 m below grade does not appear to be effective.

Summary and Conclusions

By using the earth contact surface heat loss formula developed by Mitalas and applying it to the climate in the United States, much important information was obtained with regard to the effectiveness of thermal insulation over earth contact surfaces.

1. Heat transfer factors, expressed in the maximum heat loss per hour per linear foot of perimeter, per maximum monthly temperature difference between indoors and outdoors, are practically independent of the climate as long as the degree-days are more than 3000.
2. A most effective way of reducing the earth contact heat loss is installation of floor perimeter insulation on the interior side of the slab floor. This is valid for the slab-on-grade floor as well as for shallow and deep basement floors. Soil side (or outside) insulation may be more effective than what is shown in this paper if the foundation footing is well insulated all around.

APPENDIX

Heat Loss Factors for a Full Basement, Shallow Basement, and Slab-on-Grade Floor in Washington, DC[a]

Mitalas System Identification	Soil Thermal Conductivity, W/m · K		Insulation Configuration[b]				Heat Loss Factors, W/m · K (Btu/h · ft · °F), for Various Thermal Insulation Resistances, RSI				
			Wall		Floor						
	Upper	Lower	Out	In	On	Under	0[c]	0.70	0.95	1.41	1.55
					Full Basement						
1	0.80	0.90					2.04 (1) (1.18)	. . .	. . .	. . .	. . .
2	0.80	0.90		P				1.70 (0.99)	1.70 (0.98)	1.69 (0.98)	1.69 (0.98
3	0.80	0.90		F				1.25 (0.72)	1.13 (0.65)	0.98 (0.57)	0.95 (0.55
4	0.80	0.90		F	P			1.13 (0.65)	1.00 (0.58)	0.84 (0.49)	0.81 (0.47
5	0.80	0.90	P	F				1.20 (0.69)	1.09 (0.63)	0.96 (0.55)	0.93 (0.54
6	0.80	0.90	P	F	P			1.08 (0.63)	0.97 (0.56)	0.82 (0.47)	0.79 (0.46
7	0.80	0.90	P	F	F			1.08 (0.62)	0.96 (0.56)	0.81 (0.47)	0.77 (0.44
8	0.80	0.90		F	F			1.12 (0.65)	1.00 (0.58)	0.83 (0.48)	0.79 (0.46
9	0.80	0.90	*					1.09 (0.63)	1.08 (0.62)	1.05 (0.61)	1.05 (0.61
10	0.80	0.90	P					1.27 (0.74)	1.27 (0.73)	1.26 (0.73)	1.26 (0.73

11	0.80	0.90	F					1.13 (0.65)	1.06 (0.61)	0.97 (0.56)	0.94 (0.55
12	0.80	0.90	F			P		1.09 (0.63)	1.02 (0.59)	0.93 (0.54)	0.91 (0.52
13	0.80	0.90	F			F		1.09 (0.63)	1.02 (0.59)	0.93 (0.54)	0.91 (0.52
14	1.20	1.35		P				2.17 (1.26)	2.16 (1.25)	2.16 (1.25)	2.16 (1.25
15	1.20	1.35		F				1.50 (0.87)	1.36 (0.78)	1.19 (0.69)	1.15 (0.67
16	1.20	1.35		F	P			1.32 (0.76)	1.17 (0.68)	0.99 (0.57)	0.95 (0.55
17	1.20	1.35		F	F			1.31 (0.76)	1.15 (0.67)	0.95 (0.55)	0.90 (0.52
18	1.20	1.35	*					1.45 (0.84)	1.44 (0.83)	1.41 (0.82)	1.41 (0.81
19	1.20	1.35	P					1.74 (1.01)	1.74 (1.01)	1.73 (1.00)	1.73 (1.00
20	1.20	1.35	F					1.40 (0.81)	1.32 (0.76)	1.21 (0.70)	1.18 (0.68
21	0.80	0.90	P	F	P		0.57 (2) (0.33)	...	...	...	...
22	0.80	0.90	P	F	F		0.53 (2) (0.31)	...	...	...	...
23	0.80	0.90	P	F	F		0.50 (2) (0.29)	...	...	...	...
24	0.80	0.90		F	P		0.57 (2) (0.33)	...	...	...	...
25	0.80	0.90		F	F		0.53 (2) (0.31)	...	...	...	...
26	0.80	0.90		F	F		0.50 (2) (0.29)	...	...	...	...
67	1.80	2.00					3.15 (1) (1.82)	...	...	...	...

Appendix—*Continued.*

Mitalas System Identification	*Soil Thermal* Conductivity, W/m · K		*Insulation Configuration*[b]				Heat Loss Factors, W/m · K (Btu/h · ft · °F), for Various Thermal Insulation Resistances, RSI				
			Wall		Floor						
	Upper	Lower	Out	In	On	Under	0[c]	0.70	0.95	1.41	1.55
						Full Basement					
68	1.80	2.00		P				2.76 (1.59)	2.75 (1.59)	2.74 (1.58)	2.74 (1.58
69	1.80	2.00		F				1.80 (1.04)	1.64 (0.95)	1.45 (0.84)	1.41 (0.81
70	1.80	2.00		F	P			1.55 (0.90)	1.38 (0.80)	1.17 (0.67)	1.12 (0.65
71	1.80	2.00		F	F			1.53 (0.88)	1.33 (0.77)	1.07 (0.62)	1.02 (0.59
72	1.80	2.00	*					1.93 (1.11)	1.91 (1.10)	1.88 (1.09)	1.88 (1.09
73	1.80	2.00	P					2.37 (1.37)	2.36 (1.37)	2.36 (1.36)	2.36 (1.36
74	1.80	2.00	F					1.71 (0.99)	1.62 (0.94)	1.50 (0.87)	1.47 (0.85
75	1.80	2.00	F			P		1.61 (0.93)	1.52 (0.88)	1.39 (0.81)	1.36 (0.79
76	1.80	2.00	F			F		1.59 (0.92)	1.49 (0.86)	1.36 (0.79)	1.33 (0.77
97	1.20	1.35	F			P		1.33 (0.77)	1.25 (0.72)	1.14 (0.66)	1.12 (0.65
98	1.20	1.35	F			F		1.33 (0.77)	1.25 (0.72)	1.14 (0.66)	1.11 (0.64
99	1.20	1.35					2.54 (1) (1.47)	...	...	...	...

Shallow Basement

27	0.80	0.90					2.54 (1) (1.47)	. . .	. . .	. . .	. . .
28	0.80	0.90		F				1.31 (0.76)	1.17 (0.68)	1.02 (0.59)	0.99 (0.57
29	0.80	0.90		F	P			1.14 (0.66)	1.00 (0.58)	0.82 (0.47)	0.78 (0.45
30	0.80	0.90		F	F			1.13 (0.66)	0.99 (0.57)	0.80 (0.46)	0.76 (0.44
31	0.80	0.90	F					1.14 (0.66)	1.11 (0.64)	1.05 (0.61)	1.03 (0.60
32	0.80	0.90	F			P		1.10 (0.64)	1.07 (0.62)	1.01 (0.58)	0.99 (0.57
33	0.80	0.90	F			F		1.10 (0.64)	1.07 (0.62)	1.01 (0.59)	1.00 (0.58
34	1.20	1.35					2.97 (1) (1.71)	. . .	. . .	. . .	. . .
35	1.20	1.35		F				1.52 (0.88)	1.37 (0.79)	1.22 (0.70)	1.18 (0.68
36	1.20	1.35		F	P			1.28 (0.74)	1.12 (0.65)	0.94 (0.54)	0.90 (0.52
37	1.20	1.35		F	F			1.28 (0.74)	1.10 (0.64)	0.89 (0.51)	0.84 (0.48
38	1.20	1.35	F					1.36 (0.79)	1.33 (0.77)	1.27 (0.74)	1.26 (0.73
39	1.20	1.35	F			P		1.32 (0.76)	1.29 (0.74)	1.23 (0.71)	1.21 (0.70
40	1.20	1.35	F			F		1.32 (0.76)	1.28 (0.74)	1.22 (0.70)	1.20 (0.69
77	1.80	2.00					3.48 (1) (2.01)	. . .	. . .	. . .	. . .
78	1.80	2.00		F				1.78 (1.03)	1.63 (0.94)	1.47 (0.85)	1.43 (0.83

Appendix—*Continued.*

Mitalas System Identification	*Soil Thermal* Conductivity, W/m · K		*Insulation Configuration*[b]				Heat Loss Factors, W/m · K (Btu/h · ft · °F), for Various Thermal Insulation Resistances, RSI				
			Wall		Floor						
	Upper	Lower	Out	In	On	Under	0[c]	0.70	0.95	1.41	1.55
					Shallow Basement						
79	1.80	2.00		F	P			1.46 (0.84)	1.29 (0.74)	1.09 (0.63)	1.05 (0.61
80	1.80	2.00		F	F			1.43 (0.83)	1.23 (0.71)	0.98 (0.56)	0.92 (0.53
81	1.80	2.00	F					1.69 (0.98)	1.65 (0.96)	1.59 (0.92)	1.58 (0.91
82	1.80	2.00	F			P		1.57 (0.91)	1.53 (0.89)	1.47 (0.85)	1.45 (0.84
83	1.80	2.00	F			F		1.56 (0.90)	1.51 (0.87)	1.44 (0.83)	1.42 (0.82
					Slab on Grade						
41		0.90		F			0.94 (1) (0.54)	. . .	. . .	. . .	. . .
42		0.90		F	P			0.52 (0.30)	0.50 (0.29)	0.48 (0.28)	0.47 (0.27
43		0.90		F	F			0.52 (0.30)	0.47 (0.27)	0.41 (0.24)	0.39 (0.23
44		0.90		F		P		0.80 (0.46)	0.79 (0.46)	0.78 (0.45)	0.78 (0.45
45		0.90		F		F		0.77 (0.45)	0.76 (0.44)	0.74 (0.43)	0.73 (0.42
46		0.90		F			0.57 (1) (0.33)	. . .	. . .	. . .	. . .
47		0.90		F	P			0.41 (0.24)	0.40 (0.23)	0.38 (0.22)	0.38 (0.22

48	0.90	F	F			0.40 (0.23)	0.38 (0.22)	0.33 (0.19)	0.32 (0.19
49	0.90	F		P		0.42 (0.24)	0.41 (0.23)	0.39 (0.22)	0.38 (0.22
50	0.90	F		F		0.41 (0.24)	0.39 (0.22)	0.35 (0.20)	0.34 (0.20
51	0.90	F				0.57 (0.33)	0.56 (0.32)	0.55 (0.32)	0.55 (0.32
52	0.90	F				0.51 (0.29)	0.50 (0.29)	0.48 (0.28)	0.48 (0.28
53	0.90	F				0.50 (0.29)	0.49 (0.28)	0.47 (0.27)	0.46 (0.27
54	1.35	F			1.13 (1) (0.65)	...	...	...	...
55	1.35	F	P			0.64 (0.37)	0.62 (0.36)	0.60 (0.35)	0.59 (0.34
56	1.35	F	F			0.61 (0.36)	0.56 (0.32)	0.47 (0.27)	0.45 (0.26
57	1.35	F		P		0.92 (0.53)	0.92 (0.53)	0.91 (0.52)	0.90 (0.52
58	1.35	F		F		0.87 (0.50)	0.85 (0.49)	0.81 (0.47)	0.80 (0.46
59	1.35	F			0.80 (1) (0.46)	...	...	...	...
60	1.35	F	P			0.55 (0.32)	0.54 (0.31)	0.52 (0.30)	0.52 (0.30
61	1.35	F	F			0.53 (0.30)	0.48 (0.28)	0.42 (0.24)	0.40 (0.23
62	1.35	F		P		0.55 (0.32)	0.54 (0.31)	0.52 (0.30)	0.52 (0.30
63	1.35	F		F		0.53 (0.31)	0.49 (0.28)	0.43 (0.25)	0.42 (0.24
64	1.35	F				0.77 (0.45)	0.77 (0.44)	0.76 (0.44)	0.76 (0.44

Appendix—*Continued.*

Mitalas System Identification	*Soil Thermal* Conductivity, W/m · K		*Insulation Configuration*[b]				Heat Loss Factors, W/m · K (Btu/h · ft · °F), for Various Thermal Insulation Resistances, RSI				
			Wall		Floor						
	Upper	Lower	Out	In	On	Under	0[c]	0.70	0.95	1.41	1.55
						Slab on Grade					
65		1.35		F				0.67 (0.39)	0.66 (0.38)	0.64 (0.37)	0.64 (0.37
66		1.35		F				0.65 (0.38)	0.64 (0.37)	0.62 (0.36)	0.61 (0.35
84		2.00		F			1.37 (1) (0.79)	. . .	. . .	. . .	. . .
85		2.00		F	P			0.80 (0.46)	0.79 (0.45)	0.76 (0.44)	0.75 (0.44
86		2.00		F	F			0.72 (0.42)	0.65 (0.37)	0.54 (0.31)	0.52 (0.30
87		2.00		F		P		1.08 (0.63)	1.08 (0.62)	1.07 (0.62)	1.06 (0.62
88		2.00		F		F		0.97 (0.56)	0.94 (0.54)	0.89 (0.51)	0.87 (0.50
89		2.00		F			1.11 (1) (0.64)	. . .	. . .	. . .	. . .
90		2.00		F	P			0.74 (0.43)	0.72 (0.42)	0.70 (0.41)	0.70 (0.40
91		2.00		F	F			0.67 (0.39)	0.60 (0.35)	0.51 (0.29)	0.49 (0.28
92		2.00		F		P		0.77 (0.45)	0.75 (0.43)	0.72 (0.42)	0.71 (0.41
93		2.00		F		F		0.72 (0.42)	0.65 (0.38)	0.55 (0.32)	0.52 (0.30

94	2.00	F		1.04 (0.60)	1.04 (0.60)	1.03 (0.60)	1.03 (0.60
95	2.00	F		0.89 (0.52)	0.88 (0.51)	0.87 (0.50)	0.86 (0.50
96	2.00	F		0.85 (0.49)	0.84 (0.48)	0.82 (0.47)	0.81 (0.47

[a]Floor length = 9.20 m; floor width = 8.50 m; wall height = 2.13 ; amplitude of ground surface temperature (monthly normal) = 11.42°C; ground surface temperature (annual mean) = 13.75°C; room temperature = 21°C.

[b]F = fully insulated; P = partially insulated; * = partially insulated and extending away from the wall.

[c](1) = uninsulated; (2) = special cases.

References

[*1*] Lachenbruch, A. H., "Three Dimensional Heat Conduction in Permafrost Beneath Heated Buildings," *Geological Survey Bulletin* 1052-B, U.S. Government Printing Office, Washington, DC, 1957.

[*2*] Latta, J. K. and Boileau, G. G., "Heat Loss from House Basement," *Canadian Building*, Vol. 19, No. 10, October 1964.

[*3*] Muncey, R. W. R. and Spencer, J. W., "Heat Flow into the Ground Under a House," *Energy Conservations of the Society of Heating, Cooling, and Ventilating Buildings*, Vol. 2, Hemisphere Publishing, Washington, DC, 1978, pp. 649-660.

[*4*] Wang, F. S., "Mathematical Modeling and Computer Simulation of Insulation Systems in Below Grade Application," *Proceedings*, American Society of Heating, Refrigerating, and Air Conditioning Engineers/Department of Energy-Oak Ridge National Laboratory Conference, Thermal Performance of the Exterior Envelopes of Buildings, Orlando, FL, December 1979, pp. 456-471.

[*5*] Kusuda, T., "An Approximate Method for Determining Monthly Heat Loss from Earth-Contact Floors," *Proceedings*, Fourth Symposium on the Use of Computer for Environmental Engineering Related to Buildings, Tokyo, Japan, 1983, pp. 110-118.

[*6*] Kusuda, T., Piet, O., and Bean, J. W., "Annual Variation of Temperature Field and Heat Transfer Under Heated Ground Surfaces: Slab-on-Grade Floor Heat Loss Calculations," *Proceedings*, American Society of Heating, Refrigerating, and Air Conditioning Engineers/Department of Energy Conference on Thermal Performance of the Exterior Envelopes of Buildings II, Las Vegas, NV, December 1982, pp. 67-94.

[*7*] Shipp, P. H., "Basement, Crawl Space and Slab-on-Grade Thermal Performance," *Proceedings*, ASHRAE SP Publication 38, American Society of Heating, Refrigerating, and Air Conditioning Engineers/Department of Energy-Oak Ridge National Laboratory Conference, Thermal Performance of the Exterior Envelopes of Buildings, Las Vegas, NV, December 1982, pp. 160-179.

[*8*] Delsante, A. E., "Modeling Heat Flow into the Ground in Program ZSTEP," *Proceedings*, Fourth International Symposium on the Use of Computers for Environmental Engineering Related to Buildings, Tokyo, Japan, 1983, pp. 104-109.

[*9*] Delsante, A. E., Stokes, A. N., and Walsh, P. J., "Application of Fourier Transform to Periodic Heat Flow into the Ground Under a Building," *International Journal of Heat and Mass Transfer*, Vol. 26, December 1982, pp. 121-132.

[*10*] Akasaka, H., "Calculation Methods of the Heat Loss Through a Floor and Basement Walls," *Transactions of the Society of Heating, Air-Conditioning, and Sanitary Engineers of Japan*, No. 7, June 1978, pp. 21-35.

[*11*] Mitalas, G. P., "Basement Heat Loss Studies at DBR/NRC," DBR Paper No. 1045, National Research Council of Canada, Ottawa, Canada, September 1983.

[*12*] Mitalas, G. P., "Calculation of Below Grade Heat Loss for Low Rise Residential Buildings," *Building Research Note*, National Research Council of Canada, Ottawa, Canada, October 1984.

[*13*] *ASHRAE Handbook of Fundamentals*, American Society of Heating, Refrigerating, and Air Conditioning Engineers, Atlanta, GA, 1981, Chapter 25, pp. 25.4-25.10.

[*14*] *Climatic Atlas of the United States*, U.S. Department of Commerce, Environmental Service Administration, Environmental Data Service, Washington, DC, June 1968.

Catherine Langlais[1] *and Eric Arquis*[2]

Thermal Efficiency of Counterflow Insulation Systems: Possible Applications

REFERENCE: Langlais, C. and Arquis, E., **"Thermal Efficiency of 'Counterflow' Insulation Systems: Possible Applications,"** *Thermal Insulation: Materials and Systems, ASTM STP 922,* F. J. Powell and S. L. Matthews, Eds., American Society for Testing and Materials, Philadelphia, 1987, pp. 153–172.

ABSTRACT: In the last few years, papers have been presented on so-called dynamic or counterflow insulation systems. These studies show significant reductions of thermal heat losses for such systems in comparison with "classic" insulation systems. Residential and industrial buildings using this technique already exist in Scandinavian countries and France. The aim of this study has been to obtain a better understanding of the actual efficiency of such systems.

After a brief reminder of the principle of this insulating technique, the authors present (1) a theoretical model by means of which values of thermal efficiency are derived, and (2) the experimental findings obtained under full-scale laboratory conditions. We then consider the limits and potential applications of dynamic systems.

KEY WORDS: air flow, building heat transfer, calibrated hot box, counterflow insulation, dynamic insulation, energy conservation, full-scale thermal testing, glass fibers, heat transmission, mineral fibers, thermal conductivity, thermal insulation, thermal resistance, roofs, ventilation, walls

Reducing the energy required for residential space heating is a growing concern throughout the world. This reduction can be achieved by different means, one of them being higher insulation levels.

Insulants of lower thermal conductivity or of larger thickness have been proposed: this theory is one of the basic principles of the so-called superinsulated house. A more original approach has been recently developed in France [*1*] with "dynamic" or "counterflow" insulations.

This technique was first introduced for cold chambers [*2*] and is applied in Scandinavian countries [*3*] mainly for industrial building roofs (60 000 m^2 in Sweden and 500 000 m^2 in Finland). In France, several residential houses [*4–6*] have recently been built to experiment with this unconventional insulating technique, and some authors [*7,8*] have already claimed very low levels of energy consumption.

The aim of the study presented here is to obtain a better understanding of the actual efficiency of such systems, which have often been a subject of controversy between specialists, in order to define more clearly their limits and their potential applications.

[1]Head, Thermal Laboratory, Isover Saint-Gobain, Centre de Recherches Industrielles, 60290 Rantigny, France.

[2]Assistant professor, Ensam, Laboratoire Energetique et Phenomenes de Transfert, Bordeaux-Talence, France.

Principle of Dynamic Insulations

The generic name "dynamic insulations" covers a large variety of configurations in which a fluid (generally air) is transported along or through a wall.[3] By limiting oneself to building applications, one can distinguish three main types:

Parietodynamic Insulation

Ventilation air needed for the building circulates along the wall; this air can be either fresh or used. In the first case, fresh air from the outside is preheated before it enters inside (Fig. 1*a*). In the second case, used air warms up the wall and consequently decreases conduction heat losses (Fig. 1*b*). This circulation requires the creation of one or two (Fig. 1*c*) air gaps. It is applicable to double- or triple-glazed windows.

Permeodynamic Insulation

Ventilation air needed for the building circulates through the walls, through two air gaps—more precisely, through a permeable porous medium, generally mineral fiber (Fig. 1*d-e*)—which acts as a heat exchanger. As in parietodynamic insulation, the circulation air can be either used or fresh; this last case, in which air flows in a direction opposite to the conduction heat flow, is sometimes referred to as counterflow or contraflux insulation (proflux for an inverse system). In both cases, the air flow changes the local temperature within the porous medium. In counterflow, the temperature will decrease in comparison with a situation in which there is no air flow. The slope of the temperature profile at the cold side is lower, and consequently heat losses are reduced. Similarly, the slope at the hot side is higher, but a certain amount of prewarmed air is brought inside. Our mathematical model will confirm this qualitative description of the phenomena.

Parietodynamic and permeodynamic systems are of interest only if ventilation air is needed. This is generally the case, and, in addition, the ventilation requirements correspond to an air velocity compatible with good working conditions of the system (about 1 to 3 m/h). However, the present trend of reducing the ventilation flow rates has led to the proposal of a third version of dynamic insulation, thermodynamic insulation.

Thermodynamic Insulation

This system is similar in principle to permeodynamic (counterflow configuration), but the air circulates in a closed circuit independent of the ventilation system. A heat exchanger is necessary to recover what is gained by the air as it flows through the porous medium. In practice, for appreciable gains, this heat exchanger is the evaporator of a heat pump (Fig. 1*f*).

Our study deals only with permeodynamic insulation. For simplicity, we will refer to it as dynamic as opposed to static insulation, corresponding to a classic configuration with no air flow.

The derivation of the theoretical model presented is general, but numerical applications and measurements are limited to the case of air flowing through a glass-fiber porous medium.

Theoretical Efficiency

To derive the theoretical efficiency of the permeodynamic systems, let us consider a house in which all the opaque surfaces are successively static and dynamic (Fig. 2) and compare the total heat losses in both cases, in a steady state.

[3] By wall, we mean a vertical or horizontal or tilted surface.

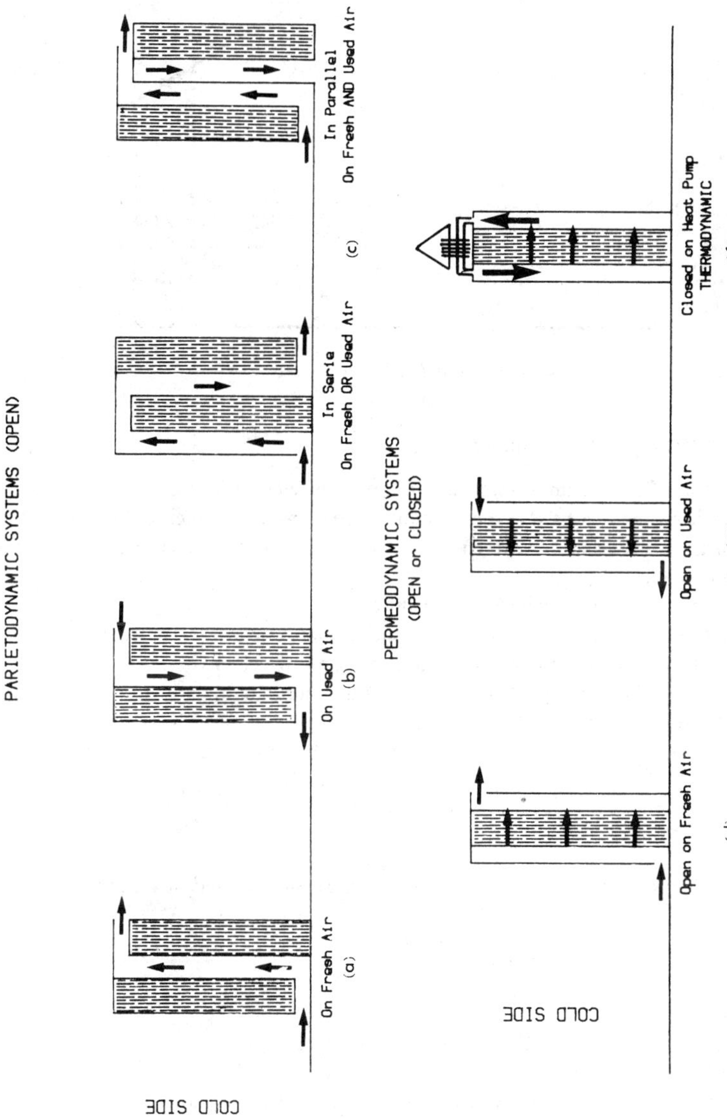

FIG. 1—*Different types of dynamic insulation.*

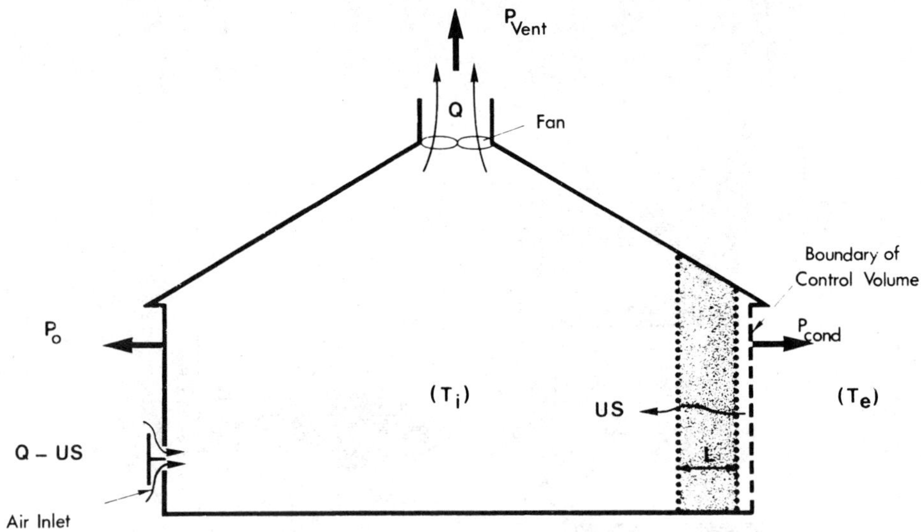

FIG. 2—*Model of heat exchanges.*

The following assumptions are made:

1. Heat and mass flow are one-dimensional along the x coordinate.
2. The dynamic surfaces consist of a permeable core between two facings (Fig. 1*d*) of negligible thermal resistance (both static and dynamic); the cores have all the same thermal conductivity, k^*, and thickness, L.
3. Surface heat transfer coefficients are infinite.
4. All the thermophysical properties (air and materials) are considered constant; particularly in dynamic insulations, k^* is considered independent of air velocity.
5. Dynamic air velocity is small enough to assume that the air and the material through which it flows have the same local temperature.
6. The air flow rate is uniformly distributed through the core, and possible convection effects are negligible.

Equations

Static Case—The total heat loss of the house (Fig. 2) is composed of three terms

$$P = P_{\text{cond}} + P_{\text{vent}} + P_0 \tag{1}$$

where

P_{cond} = conduction heat losses through opaque surfaces,
P_{vent} = ventilation losses, and
P_0 = all other types of heat losses (windows, thermal bridges, and so on).

We can define

$p = P_{\text{cond}}/P$ (part of conduction losses),
$r = P_{\text{vent}}/P$ (part of ventilation losses), and
$q = P_0/P \; (p + q + r = 1)$.

With our assumptions, Eq 1 can be written as

$$P = \frac{k^*S}{L}(T_i - T_e) + (\rho c)_f Q(T_i - T_e) + P_0 \tag{2}$$

where

S = total area of opaque surfaces,
ρ_f, c_f = density and specific heat of air,
Q = ventilation air flow rate,
T_i = inside temperature, and
T_e = outside temperature.

Dynamic Case—Equation 1 becomes

$$P' = P'_{\text{cond}} + P'_{\text{vent}} + P'_0 + P_{\text{fan}} \tag{3}$$

where P_{fan} is the power of the fan used to blow the dynamic air.

For the dynamic and static configurations to have comparable conditions, we have

$$P'_{\text{vent}} = P_{\text{vent}}$$

$$P'_0 = P_0$$

Q being the maximum allowable ventilation air flow rate (generally imposed by national building codes), the air velocity, U, through the dynamic surfaces is

$$U \leq \frac{Q}{S} \tag{4}$$

If $U < Q/S$, part of the ventilation air enters directly into the house (Fig. 2). From energy conservation and Darcy's law, we have the following equations

$$(\rho c)^* \frac{\partial T}{\partial t} + (\rho c)_f U \nabla T = k^* \nabla^2 T \tag{5}$$

$$-\nabla P_f - \frac{\mu U}{K} = 0 \tag{6}$$

where

ρ^*, c^* = density, specific heat of the dynamic core,
P_f, T = pressure and temperature,
μ = air viscosity, and
K = permeability of the dynamic core.

In steady state, with our assumptions and boundary conditions, we can derive from Eq 5

$$T = (T_i - T_e)\frac{\exp\left(-Pe^*\dfrac{x}{L}\right) - \exp(-Pe^*)}{1 - \exp(-Pe^*)} + T_e \tag{7}$$

where $Pe^* = (\rho c)_f UL/k^*$ = Peclet number.

With our notations and from Eq 4, it follows that

$$Pe^* \leq \frac{r}{p} \tag{8}$$

Equation 7 confirms the nonlinear shape of the temperature profile (see Permeodynamic insulation in the section on Principle of Dynamic Insulations) as opposed to the static case (Fig. 3). This equation gives

$$P'_{cond} = -k^*S\left(\frac{dT}{dx}\right)_{x=L} = k^*\frac{S}{L}\frac{Pe^*}{\exp(Pe^*) - 1}(T_i - T_e)$$

From Eq 6 we can write in first approximation

$$P_{fan} = \left(\frac{L\mu Sd}{K}\right)U$$

where d is a coefficient taking into account head losses. For most applications, one can show that P_{fan} is negligible compared with the other terms of Eq 3 [9]. We then finally obtain

$$P' = k^*\frac{S}{L}\frac{Pe^*}{\exp(Pe^*) - 1}(T_i - T_e) + (\rho c)_f US(T_i - T_e) + P_0 \tag{9}$$

Combining Eqs 2 and 9, we can write

$$P' = P_{cond}(1 - \eta) + P_{vent} + P_0 \tag{10}$$

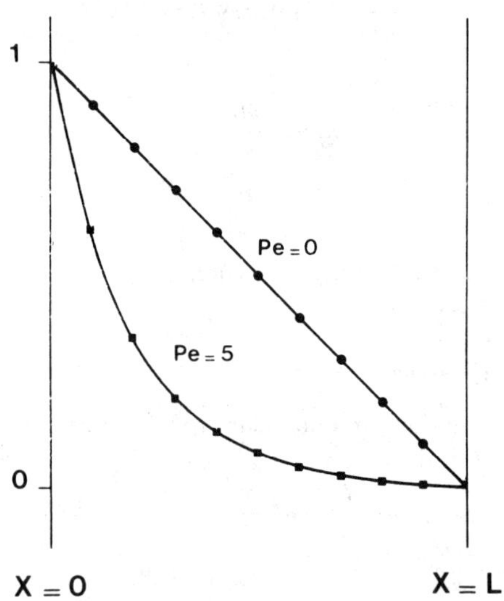

FIG. 3—*Temperature profiles within the permeable core, in steady state, with and without air flow.*

or

$$P' = P_{cond} + P_{vent}(1 - \eta') + P_0 \tag{11}$$

or

$$P' = P(1 - f) \tag{12}$$

with

$$\eta = 1 - \frac{Pe^*}{\exp(Pe^*) - 1} \tag{13}$$

$$\eta' = \frac{1}{Pe^*} - \frac{1}{\exp(Pe^*) - 1} \tag{14}$$

$$f = p\eta \tag{15}$$

It is therefore possible to define three efficiencies, η, η', and f:

1. η represents the relative gain on conduction losses. Introducing the notion of "dynamic thermal conductivity," k_D^*, we have $k_D^* = k^*(1 - \eta)$.
2. η' represents the relative gain on ventilation losses. It is directly comparable to the efficiency of a heat exchanger working on ventilation air.
3. f represents the overall efficiency of the system, that is, the relative gain on total heat losses when working in a dynamic compared to a static system.

The formulations of η, η', and f are general; it is of interest to compute and compare the values obtained in different practical cases.

Results and Discussion

Figure 4 shows the variation of η with air velocity for different static thermal resistances, ($R = L/k^*$), of the permeable porous medium. For a given velocity, η increases with the R value. At a velocity of 1 m/h (average value of the current applications considering the usual values of S and Q), we have

$$\eta = 30\%$$

for $R = 2\ \text{m}^2 \cdot \text{K/W}$. The dynamic resistance is then equal to

$$R_D = \frac{R}{1 - \eta} = 2.8\ \text{m}^2 \cdot \text{K/W}$$

For $R = 3\ \text{m}^2 \cdot \text{K/W}$

$$\eta = 42\%$$

and

$$R_D = 5.2\ \text{m}^2 \cdot \text{K/W}$$

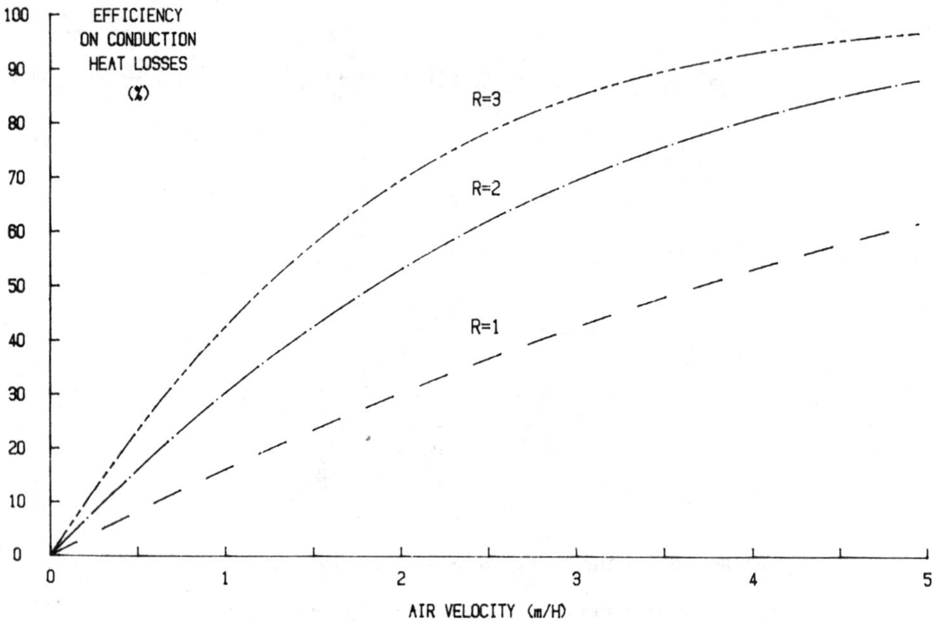

FIG. 4—*Variation of the efficiency of conduction heat losses with air velocity for different static thermal resistances (*R *in* $m^2 \cdot K/W$*).*

At first sight, permeodynamic insulation then appears as a means to increase thermal resistance when the limits (in terms of thermal conductivity and thickness) of what can be achieved in a static system are reached.

Figure 5 shows the variation of η' with air velocity. In all cases, η' is lower than 50%, which means that permeodynamic systems are less efficient than most, if not all, commercial heat exchangers working on ventilation air. Gains higher than 50% on ventilation losses are sometimes reported, but then conduction losses increase simultaneously.

In the limit case, we can, indeed, rewrite Eq 9 in suppressing the ventilation term and obtain

$$P' = P_{\text{cond}}(1 + \eta'') + P_0$$

with

$$\eta'' = \frac{Pe^* \exp(Pe^*)}{\exp(Pe^*) - 1} - 1$$

Since we can check that $\eta'' > 0$, this equation means that ventilation losses can be covered by dynamic systems only if conduction losses are higher than in static systems.

In Fig. 5, we can also notice that, contrary to η, η' decreases when R increases. Since ventilation air is warmed by the conduction heat flux, as this flux decreases (high R value), the gain becomes less. In fact, in permeodynamic systems, conduction and ventilation losses cannot be dissociated, and it can be misleading to consider only η or η'. The overall efficiency, f, that we introduced is, in our mind, more representative of the actual performances. The variation of f with the Peclet number is represented in Fig. 6. As expected, f varies with the values of p, q, and r, previously defined. For a given p, all the curves are bounded since the Peclet number

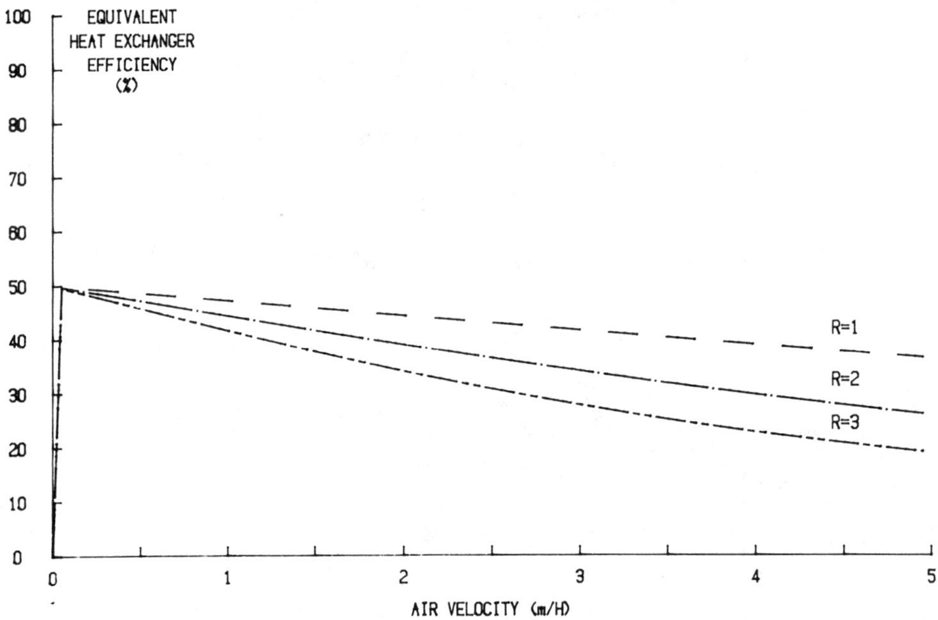

FIG. 5—*Variation of the equivalent heat exchanger efficiency with air velocity for different static thermal resistances (* R *in* $m^2 \cdot K/W$*).*

must satisfy conditions in Eq 8. It is possible to draw a curve of maxima (see Fig. 6) corresponding to the ideal case in which the building heat losses are limited to conduction and ventilation losses ($q = 0$; $p + r = 1$). This curve gives us a maximum theoretical efficiency of 23%, corresponding to $p = 35\%$ and $r = 65\%$. For the values of p, q, and r encountered in practice, it seems that the efficiency will lie somewhere between 10 and 20%. The total elimination of conduction losses ($\eta = 100\%$ and $f = p$) is never achieved, contrary to some claims. This elimination would be possible only for ventilation rates higher than in static systems, and computations show that this additional ventilation heat loss is not compensated by the gain made on conduction.

All our assumptions, in particular the stationary conditions (no influence of inertia), the infinite surface coefficients, the uniform air distribution, or the nonconsideration of solar gains can be considered oversimplified. Our aim was to derive a reasonable estimation of the overall efficiency and test some of our assumptions with full-scale experiments, knowing that more sophisticated computations could be made if the system turned out to be really attractive.

Full-Scale Experiments

Test Apparatus

A full-scale installation has been specially designed in the thermal laboratory of the Centre de Recherches Industrielles de Rantigny (CRIR) for testing vertical permeodynamic walls, in steady state, in both the static and dynamic configurations. The test wall (2.40 by 2.40 m^2) is placed between a hot chamber ($20 \leq T_C \leq 40°C$) and a cold chamber ($-20 \leq T_F \leq 10°C$).

In static configuration, the principle of measurement is similar to the calibrated hot box technique: heat flow through the test wall is estimated from the electrical power dissipated in

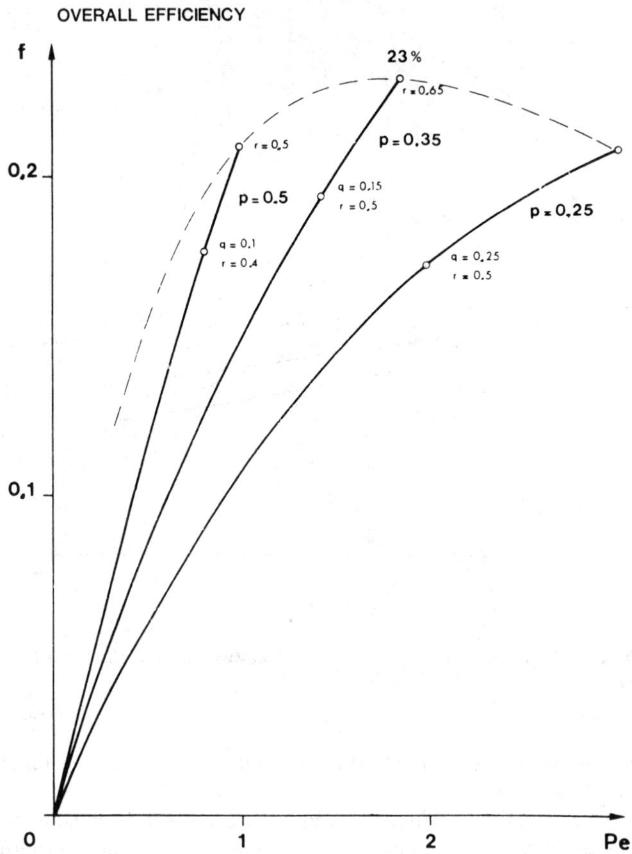

FIG. 6—*Variation of the overall efficiency with Peclet number (*p, q, *and* r *are defined in the text).*

the heating element of the hot chamber, while the heat losses of the chamber itself are controlled as indicated in Fig. 7.

For dynamic tests, an extraction duct is installed in the center of the hot chamber door with a fan that draws dry air from the top of the cold chamber through the test wall, at adjustable velocities.

Eighty thermocouples (Type K) measure air and wall temperatures; a hot wire anemometer is used for velocity measurements inside the duct. All the sensors are connected to a data logging system associated to a computer for presentation of the results. A complete description of this installation can be found in Ref *10*.

Test Wall

All the measurements presented in this paper have been performed on a glass-fiber board of the following nominal characteristics:

$L = 80$ mm,
$\rho = 70$ kg/m^3, and
$K = 10^{-9}$ m^2.

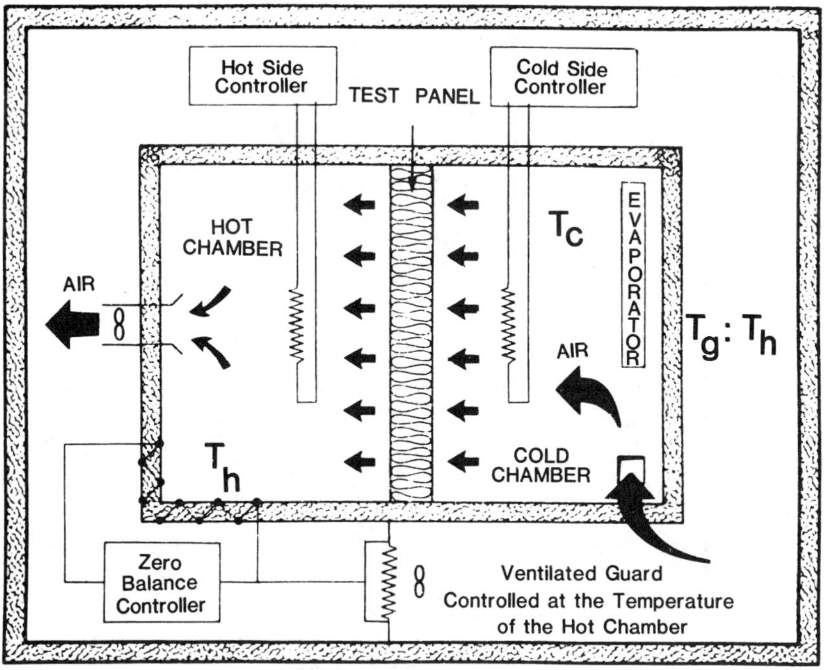

FIG. 7—*Full-scale installation.*

Although the installation allows one to test a complete permeodynamic system—that is, a permeable core (glass fiber) + two air gaps + two facings—we have tested only the core itself, for a simpler configuration for comparison with the theory. The test wall therefore consisted of three adjoining glass-fiber panels, with brown paper glued on both sides. Special care has been taken to avoid any air leak between the hot and the cold chamber, and the tightness of the whole system has been checked by a smoke test.

Measurements

Apart from calibration tests, measurements have been performed for a mean temperature of 10°C and a temperature difference of about 20 degrees Celsius.

Static Measurements—The aim of these measurements was to calibrate our box. The details of this calibration are given in Ref *10*.

Measurements performed for different temperature differences are in good agreement with the results obtained on a guarded hot plate. For our test conditions, the accuracy is estimated to lie within ±5%. Figures 8*a* and *b* and 9 show the good uniformity of hot and cold surface temperatures and the linearity of the temperature profiles within the insulant measured at different locations noted by the symbol # in Fig. 8.

Dynamic Measurements—Two series of tests have been performed. The second series followed a smoke test, which was a tentative visualization of the air flowing through the insulant (see section on Airflow visualization). The results will be referred to as "before" or "after smoke test."

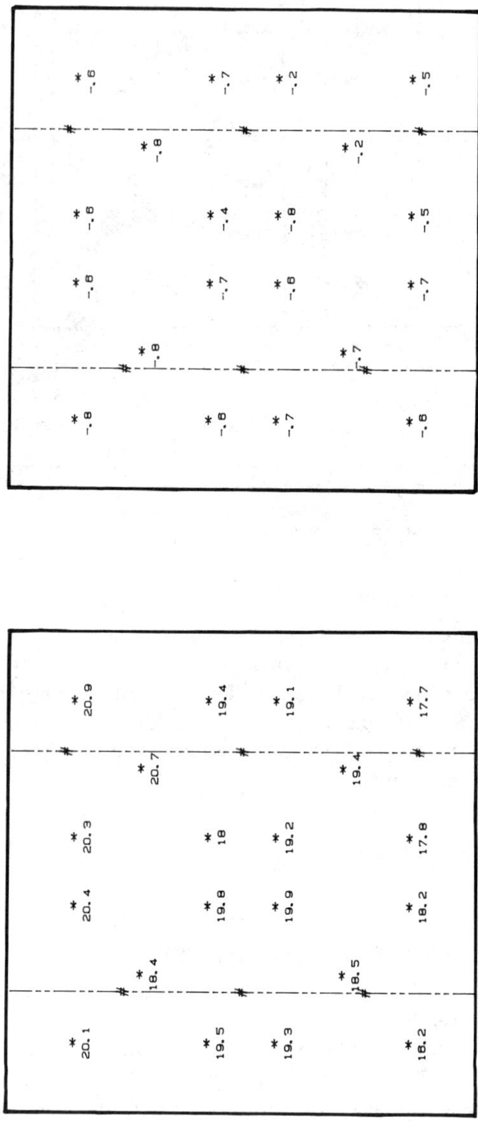

FIG. 8—*Surface temperatures in the static configuration: (a) hot side; (b) cold side.*

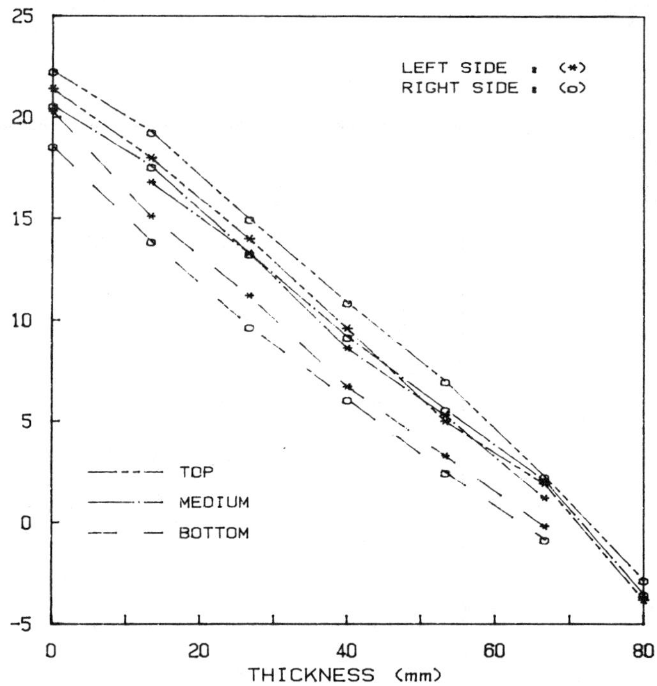

FIG. 9—*Temperature profiles within the insulant in static configuration.*

Temperature measurements—Temperature profiles within the insulant, recorded for different air velocities, are shown on Fig. 10*a* to *d*. Several points can be drawn from these figures:

1. The shape of the profiles is qualitatively in accordance with theory; the deformations increase with velocity.
2. Before the smoke test (Fig. 10*a* to *c*), the scattering between the different profiles is large compared with static conditions, but it is considerably reduced after the smoke test (Fig. 10*d*).
3. The hot side surface temperatures are lower than in static conditions (for similar air temperatures), which means that the assumption of an infinite surface heat transfer coefficient is not valid, contrary to what happens on the cold side. This point is illustrated on Fig. 11*a* and *b*.

The deformation of the temperature profile being a function of air velocity, the scatter observed on the data points is evidence of a bad air flow repartition through the test wall.

Using Eq 7, we can attribute a value of local velocity to each of the measured profiles. An example of this is given in Fig. 12. In this particular case, for a mean velocity of 2.8 m/h we obtained local velocities lying between 1.8 and 4.7 m/h. This bad repartition can be attributed to the local inhomogeneities of the wall, this hypothesis being supported by the results obtained after the smoke test. The observed deposit of smoke particles within the product (mainly on the cold side) could have partially eliminated this problem by creating a larger pressure drop through the wall for the same mean velocity. The examination of surface hot temperatures also showed lower temperatures at the junctions between the three constitutive panels of the wall, therefore indicating a preferential air flow at those two junctions.

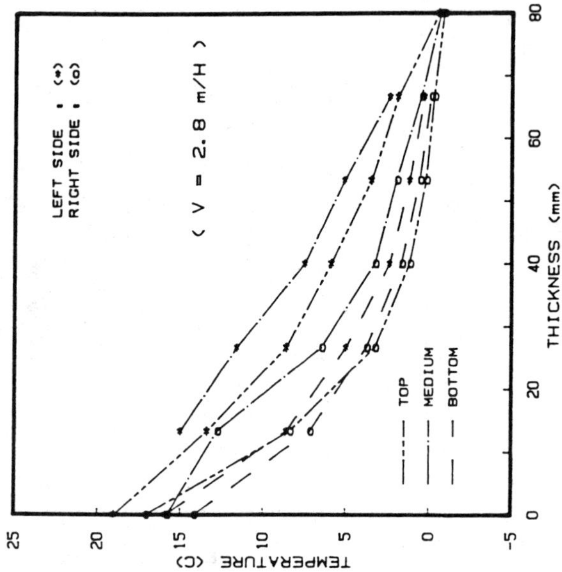
LEFT SIDE : (*)
RIGHT SIDE : (o)
(V = 2.8 m/H)
TOP
MEDIUM
BOTTOM
TEMPERATURE (C)
THICKNESS (mm)
25
20
15
10
5
0
-5
0
20
40
60
80
(b)

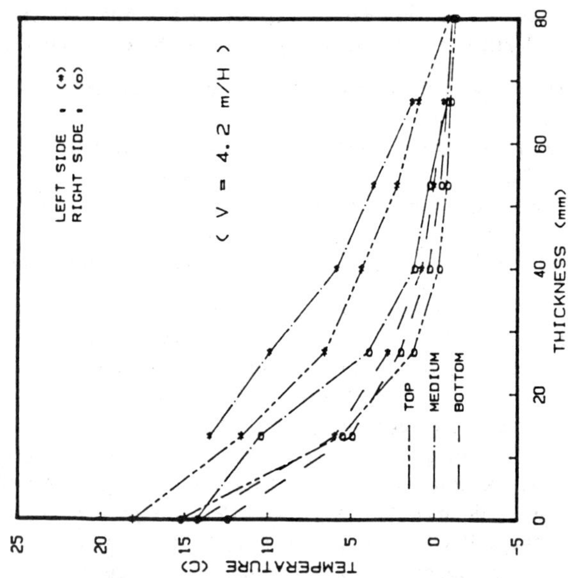
LEFT SIDE : (*)
RIGHT SIDE : (o)
(V = 4.2 m/H)
TOP
MEDIUM
BOTTOM
TEMPERATURE (C)
THICKNESS (mm)
25
20
15
10
5
0
-5
0
20
40
60
80
(a)

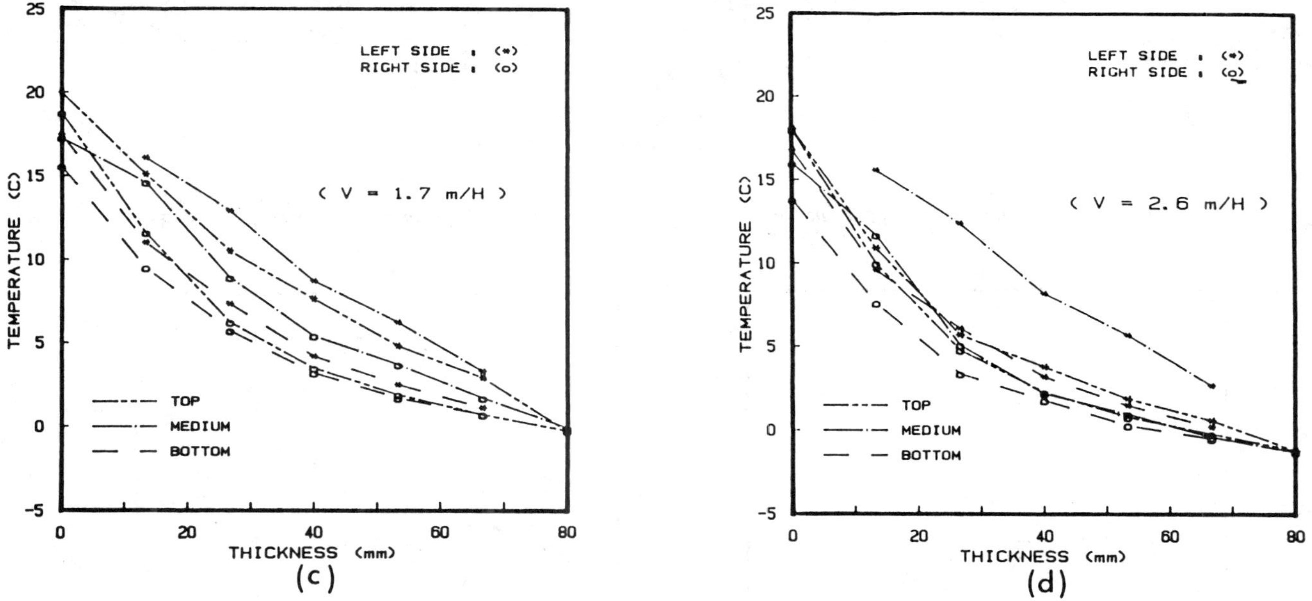

FIG. 10—*Temperature profiles within the insulant for different mean air velocities,* V *(*V *is equivalent to* U *in the text):* (a, b, c) *before the smoke test;* (d) *after the smoke test.*

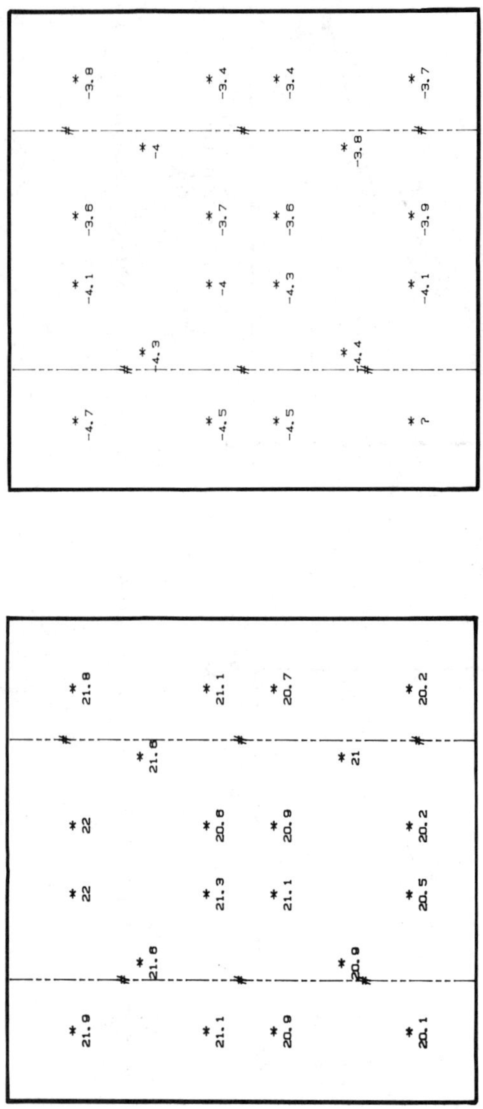

FIG. 11—*Surface temperatures in dynamic configuration for a mean velocity of 2.8 m/h: (a) hot side; (b) cold side.*

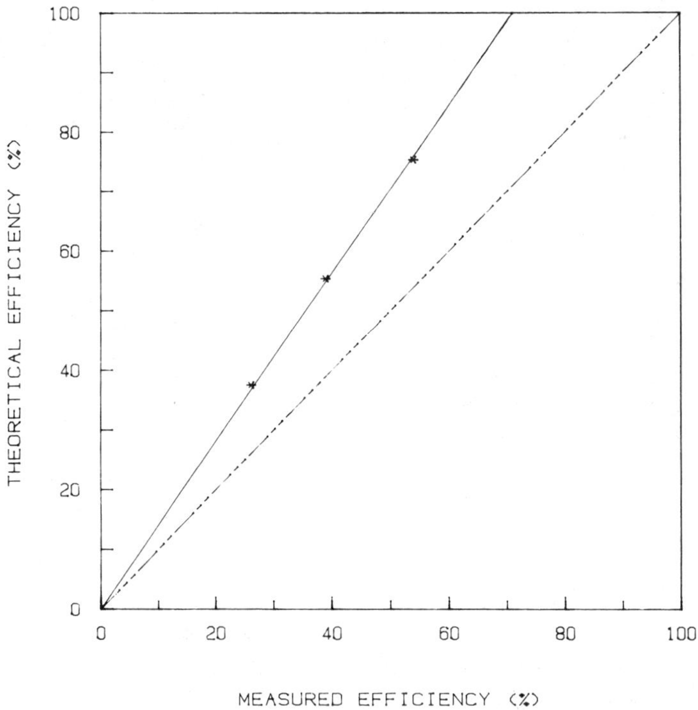

FIG. 12—*Computed local velocities for a mean velocity of 2.8 m/h.*

Efficiency measurements—The measured efficiency of conduction losses, as defined in the Equations section (Dynamic Case) can be written as

$$\eta_m = \frac{\phi_s - (\phi_D - \rho c U S \Delta Ta)}{\phi_s}$$

where

ϕ_D = heat flow measured in the dynamic test,
ϕ_S = heat flow measured in the static test for similar temperature conditions, and
ΔTa = air temperature difference between the hot and cold chambers.

Figure 13 compares the theoretical efficiency derived from Eq 13 and η_m obtained for the first series of measurements (before the smoke test). The value of η_m is always lower than the theoretical value, η_t. This result can possibly be attributed to the bad distribution of air flow. Simple calculations, illustrated in Fig. 14, indeed, show that for the same mean velocity, nonuniform air flow distributions always give lower efficiencies.

Several points might also explain these discrepancies:

(a) the influence of a noninfinite surface heat transfer coefficient on the hot side, although it seems of little influence on theoretical efficiency [*11*],

(b) the influence of air flow on the thermal conductivity, k^*, which was assumed negligible, and

(c) last, a small measurement error on the air velocity needed to evaluate both theoretical

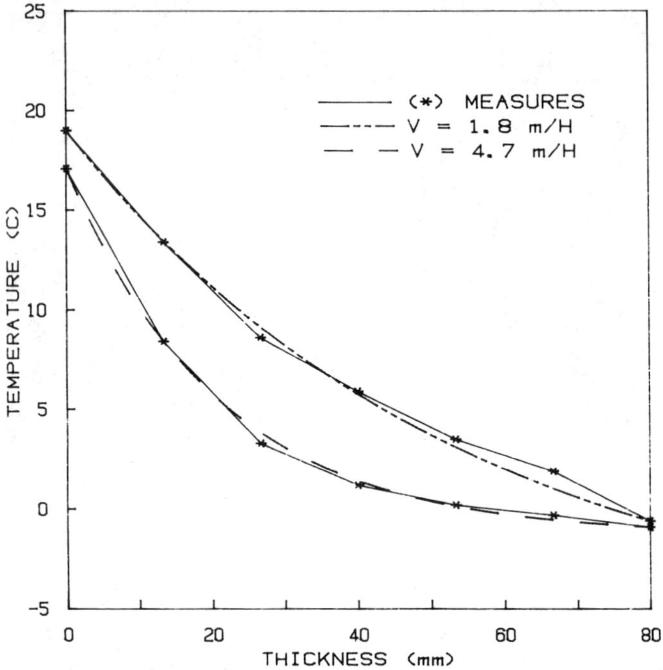

FIG. 13—*Comparison of theoretical and measured efficiencies.*

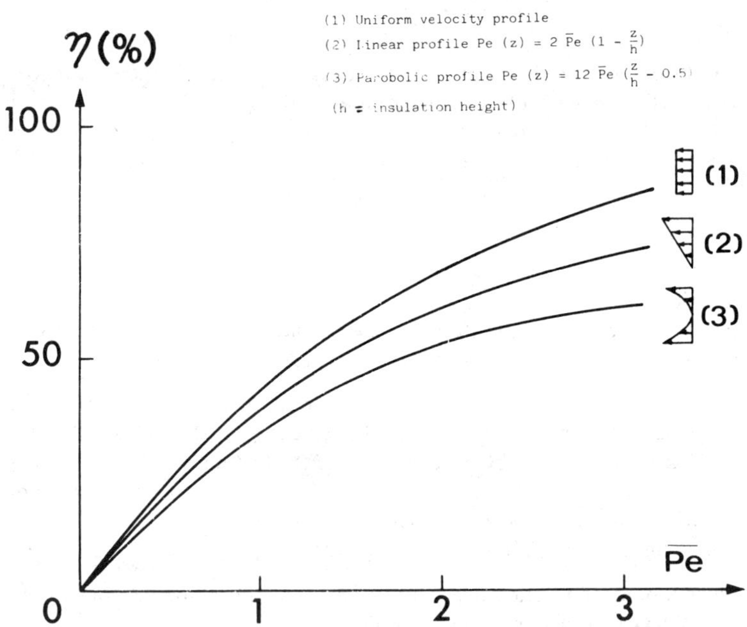

FIG. 14—*Variation of the efficiency,* η, *for different air flow distributions:* (1) *uniform velocity profile;* (2) *linear profile,* $Pe(z) = 2\overline{Pe}(1 - z/h)$; (3) *parabolic profile,* $Pe(z) = 12\overline{Pe}(z/h - 0.5)$; h = *insulation height.*

and measured efficiencies, which can significantly affect the results: it is possible to obtain $\eta_m = \eta_t$ by assuming a mean velocity 10% lower than the measured value.

Air flow visualization—As mentioned, a smoke test has been performed to visualize air flow; this technique was not suitable as the insulant filtered the smoke produced on the cold side, making its passage through impossible. Because temperature and air velocity are related, an infrared camera has also been used to evaluate surface temperature uniformity on the hot side. Cold spots corresponding to preferential air flow paths have been seen. Thermography appears as an interesting technique for checking dynamic systems.

Comments on the Results

Despite the limited number of measurements, our study allows us to make the following points:

1. The behavior of permeodynamic systems is in qualitative agreement with the theory. Experiments have confirmed the deformation of the temperature profile through the insulation as well as the simultaneous decrease of conduction losses. The efficiency increase with air velocity is also verified.
2. The critical point appears to be the realization of a uniform air flow through the vertical wall; moreover, in practice, in a residential house air leaks will be difficult to avoid and would reduce efficiency. In Sweden [*12*], a wooden prefabricated element, less than 2 cm thick, fixed directly against the mineral wool on the warm side and creating an overall pressure drop of about 50 Pa, has been reported as a means of regulating an even flow of air over the entire wall surface.

Measurement problems make the comparison of experimental results and theory difficult. The literature is sparse on the subject, especially concerning heat flow. Authors [*3*] have reported good agreement between theoretical and measured temperature profiles. Our feeling is that for a uniform air distribution, the simplified theory presented in the Theoretical Efficiency section is a good approximation of reality.

Conclusions

Permeodynamic systems have been proposed as a mean of reducing the energy required for space heating. The following are the main conclusions of our study on the actual behavior of such systems:

1. A simplified theory describing permeodynamic insulation has shown that the maximum relative gain on the total heat losses of a given building compared with that for conventional insulation is equal to 23%. This efficiency depends on the conduction and ventilation heat losses, and for residential houses, it appears to lie between 10 and 20%. Commercialized heat exchangers working on ventilation air have higher performance as long as their own efficiency is higher than 50%.
2. Full-scale laboratory experiments have indicated that in vertical configuration, a uniform air-flow distribution through the insulation is difficult to achieve. Additional work is needed to solve this problem, since it penalizes the actual efficiency of the system.
3. In the authors' opinion, considering the present state of knowledge, conventional insulation techniques remain competitive for residential houses. The constant improvement of the thermal conductivity of insulating materials already contributes to the reduction of conduction losses. Moreover, as permeodynamic systems require two air gaps, it seems safer in terms of actual gains, to fill them with insulation and work in the static configuration.

4. The situation is somewhat different for industrial buildings. In this case, one generally deals with large horizontal (or slightly tilted) areas of roofing and with high ventilation requirements. A permeodynamic roof can then turn out to be an interesting solution as air-flow distribution is easier to handle (horizontal surface). For high buildings, this system also has the advantage of minimizing the vertical gradient of temperature within the building, as air is forced inside from the top.

This type of roof has been installed in Scandinavian countries and gives full satisfaction to its users; no particular problem of condensation or pollution has been reported.

As a general conclusion, we can say that permeodynamic systems can indeed reduce the energy required for space heating. However, in most cases, they are too sophisticated to be really appropriate. They require perfect house airtightness and excellent workmanship, conditions which are not always fulfilled in practice even with prefabricated elements.

The problem is common to other types of dynamic systems such as thermodynamic or parietodynamic insulations. These may turn out to be attractive for some applications, such as industrial roofings or windows (parietodynamic), but additional work is still needed for their optimization. For instance, if counterflow can be considered a very safe technique from the standpoint of moisture, the risk of condensation in proflux constructions is obvious. Other aspects such as possible solar gains, thermal comfort, and reversed air flow for summer cooling purposes should also be investigated.

We hope this study will constitute a basis for future research in this area.

Acknowledgments

The authors wish to thank Thierry Picard for his assistance with the test wall preparation, facility operation, data collection, and other aspects of the experimental work.

The research described in this paper was sponsored by the French Plan Construction Ministère de l'Urbanisme et du Logement under Contract No. 82-61-208 with Isover Saint-Gobain.

References

[*1*] Olive, G., "L'espoir des isolations dynamiques," *Sòlaire,* Vol. 1, No. 15, June–July 1983, pp. 55–59.

[*2*] Golovkin, N. A., "New Cold Store Insulation System," *Bulletin de l'Institut International du Froid,* Annexe 1960-3 Commission V, Marseille Conference, 1960, pp. 443–447.

[*3*] Anderlind, G. and Johansson, B., "Dynamic Insulation," D8 Swedish Council for Building Research, Stockholm, Sweden, 1983.

[*4*] Lenat, Y., "A l'I.U.T. de Lorient, un bâtiment expérimental chauffé par l'isolation dynamique," *Chaud et Froid Plomberie,* April 1982, pp. 100–114.

[*5*] Michel, J. E., "Une première expérience d'isolation thermodynamique," *Promoclim Tome 13 E,* No. 4, October 1982, pp. 210–222.

[*6*] Michel, J. E., "Un second bâtiment à isolation thermodynamique à Dignes," *Promoclim Tome 13 E,* No. 4, October 1982, pp. 224-232.

[*7*] Gimberg, C. and van Migom, M., "Isolation Dynamique," doctoral dissertation, Option Génie Civil, Ecole Centrale de Lyon, Lyon, France, 1982.

[*8*] Programme H2 E 85, "Une première réalisation en Bretagne," *Le Moniteur,* No. 24, June 1982, pp. 71–74.

[*9*] Arquis, E., Langlais, C., Caltagirone, J. P., and Klarsfeld, S., "Quelques aspects fondamentaux sur l'isolation dynamique," *Revue Générale de Thermique*, April 1983, pp. 321-330.

[*10*] Langlais, C. and Picard, T., "Etude théorique et expérimentale d'une paroi perméodynamique," Centre de Recherches Industrielles de Rantigny Internal Report No. 53/84, 1984.

[*11*] Langlais, C. and Arquis, E., "Etude d'un système d'isolation dynamique," Institut International du Froid, Institut International de la Réfrigération, Commission B1, 16th International Congress on Cold, Paris, France, 1983.

[*12*] Boman, C. A., "Provhus med dynamisk isolering," R 142 Swedish Council for Building Research, Stockholm, Sweden, 1981.

Thermographic Measurements

Y. May Chang,[1] *Richard A. Grot,*[1] *and Lawrence S. Galowin*[1]

Infrared Inspection Techniques for Assessing the Exterior Envelopes of Office Buildings

REFERENCE: Chang, Y. M., Grot, R. A., and Galowin, L. S., **"Infrared Inspection Techniques for Assessing the Exterior Envelopes of Office Buildings,"** *Thermal Insulation: Materials and Systems, ASTM STP 922,* F. J. Powell and S. L. Matthews, Eds., American Society for Testing and Materials, Philadelphia, 1987, pp. 175–195.

ABSTRACT: In order to assess the thermal integrity of new and existing office buildings, the National Bureau of Standards (NBS) developed and performed a series of diagnostic tests on eight federal office buildings. These buildings have different sizes, ranging from 1800 m^2 to 48 000 m^2 of floor area, and are located in various climate zones. Thermographic inspection is one of the diagnostic procedures used to identify the locations of heat loss areas, to determine the extent of deterioration of building materials, and to establish a basis for recommendation of retrofit options. Other test methods applied to these buildings for envelope measurements include heat flow for the thermal resistances, tracer gas for air infiltration, and pressurization tests for building tightness. Three types of thermographic inspections were conducted on these buildings: ground-based infrared surveys, serial infrared surveys, and inspection with a spot radiometer. These buildings were designed to meet the new federal energy guidelines of less than 630 MJ/m^2 per year of on-site energy consumption and less than 1200 MJ/m^2 per year of off-site energy consumption. The thermographic inspections reveal the existence of several classes of thermal deficiencies in the buildings, which were due to both inadequate building design and poor construction quality. Comparisons of the results from several thermographic techniques are shown. Analysis of thermographic data with examples of thermal defects observed from all inspections is presented. Quantitative comparisons of the wall thermal resistance values measured by the spot radiometer with those from heat flow meter measurements are also included.

KEY WORDS: aerial thermography, building component thermal resistances, building diagnostics, building envelope heat loss, ground-based thermography, spot radiometer inspections, roof inspections, thermal insulation

A research program was undertaken by the National Bureau of Standards (NBS) for the General Services Administration (GSA) to develop measurement techniques and inspection procedures and to perform diagnostic tests on eight federal buildings [*1*]. The object of the project was to assess the thermal integrity of these buildings, which are located in various climatic zones across the nation, as shown on the map in Fig. 1. Other factors considered in selecting these buildings as test sites included the construction, size, design, and operating characteristics. These buildings were constructed according to the U.S. government's energy guidelines of less than 630 MJ/m^2 per year of on-site energy usage and 1200 MJ/m^2 per year of off-site energy usage in the last five to ten years.

Other tests performed by NBS on these buildings reported earlier included thermal resistance measurement on exterior walls by heat flow meter and calorimeter [*2*], pressurization tests for measuring the tightness of building envelopes [*3*], and air infiltration measurement by tracer

[1]General engineer, mechanical engineer, and senior research engineer, respectively, Center for Building Technology, National Bureau of Standards, Gaithersburg, MD 20899.

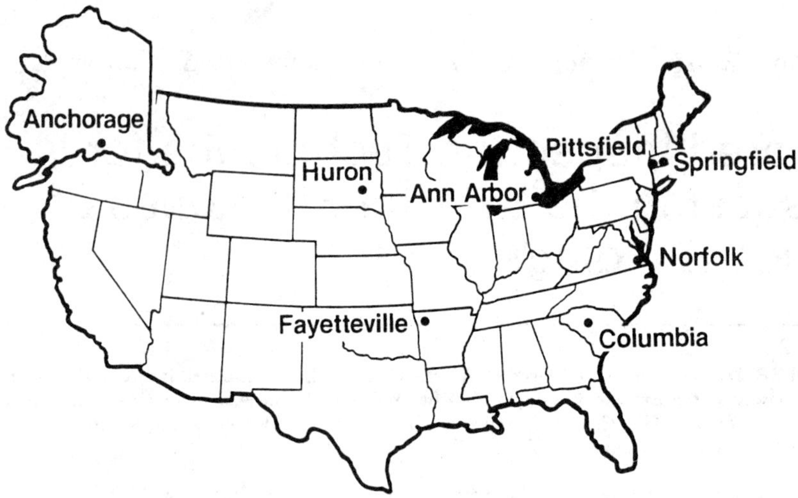

FIG. 1—*Building sites selected for diagnostic tests.*

gas [*4*]. These measurement methods are applicable to new buildings at different stages during their design and construction and to existing buildings to recommend remedial actions.

Three types of thermographic inspections were performed on these buildings; ground-based infrared surveys, aerial thermography, and thermal resistance measurement by infrared thermometer (spot radiometer). Ground-based thermography is useful for detecting thermal deficient areas caused by missing insulation, uneven insulation, air leakage paths, and thermal bridges in the exterior wall. Aerial thermography is effective in locating defects in roofs, which may be due to moisture condensation or other causes. Inspection by spot radiometers on building components can be used to estimate the thermal resistance (or *R* values) of such components.

This paper presents the analysis, comparison, and results of these three types of thermographic inspections performed on eight office buildings. Comparisons of the thermal resistances of exterior walls determined by spot radiometers with those from heat flow meter and calorimeter measurements are also included. Ground-based infrared inspections were performed by NBS on all eight buildings. Under the NBS guidelines, aerial infrared surveys and spot radiometer inspections were conducted by the contractors on three and two buildings, respectively.

Applications of Infrared Thermography

Diagnostic techniques for locating and identifying thermal defects in building envelopes, and for selection of cost-effective remedial or retrofit actions, are detailed in Ref *1*. There, the technical requirements are described for the measurement techniques and for the procedures for inspection and data analysis used to assess various aspects of the integrity of the building envelope. The applications of infrared technologies were demonstrated in all phases of test procedures for evaluations of existing buildings [*5*].

Ground-Based Infrared Thermography

The ground-based inspection of buildings using infrared scanning systems can produce information that is useful for locating and assessing heat loss paths from buildings, such as (1) areas of the exterior wall that were not insulated, (2) areas of the roof not properly insulated, (3) shrinkage or fissures in fully insulated wall areas, (4) air leakages around doors and windows, (5) excessive heat loss around door and window frames, (6) excessive heat loss in the joints between walls, ceilings, and floors, (7) excessive heat loss in unconditioned spaces, such as eaves, (8) air penetration into interior cavities, and (9) moisture damage to insulation.

Both exterior and interior thermographic surveys are applicable for building inspections. In general, the temperature variations existing on interior surfaces are greater than those on exterior surfaces, and the sensitivity of infrared equipment is greater under interior ambient conditions than under normally encountered exterior ambient temperatures. If it is possible to observe a class of defects from the interior, it is generally preferable to obtain data from the inside. However, certain practical considerations, such as the ease of performing the survey and accessibility, often dictate the type of survey used. Also, certain classes of defects such as heat bypasses into attics, excessive basement losses, and roof inspections for moisture damage are best observed from the exterior.

Although it is preferred that steady-state conditions exist at the time of inspection, thermographic inspections can be undertaken under other conditions if sufficient knowledge is used in taking and interpreting the thermographic data. Under different conditions, other classes of thermal patterns will exist. For example, a wall exposed to direct solar radiation will, after a sufficient time, experience a temperature reversal—the wall areas at the front of studs and voids will appear to be warm, and those at the insulated section will appear cold on interior inspection. An inspection performed about an hour after sunset will show that the studs are warmer than the insulated section on both inside and outside surfaces because of their greater thermal capacities. In southern climates, roofs suspected of having moisture damage can be inspected in the evenings after being exposed to solar radiation. In such cases, the interpretation of the thermograms is based not on differences in the thermal conductance between the damaged and undamaged sections but on the differences in their heat storage.

The interior thermographic inspection has the advantage of avoiding the problem of wind "scrubbing away" surface temperature differences, which will otherwise indicate thermal anomalies in the building envelope, and degrading of instrument sensitivity because of reduced object temperature. Surveys may be carried out during either the day or night. For day surveys it is important that solar loading not negate the outward heat conduction resulting from inside-to-outside temperature differences, thereby causing thermal anomalies to become indistinguishable. Night surveys are preferred because more stable heat transfer conditions will usually prevail; meaningful survey results will be obtained by eliminating interference from solar loading and reflections.

For exterior thermographic surveys, the infrared imaging system used for the survey may be portable, cart-mounted, ground-vehicle-mounted, or mounted on an aerial platform. The field of view of the infrared imaging system should be at least 18°. In carrying out the survey, an attempt should be made to include as much of the exterior surface as possible. The distance between the sensor and the surface can be chosen as long as the system's total field of view subtends at least 6 m^2 in object space.

The interpretation of thermograph data [*6, 7*] is a process that consists of comparing the thermal images of a building with previously observed and cataloged thermal images for various types of thermal defects. The following are the major steps in this process:

(*a*) determination of anticipated thermal images from given physical building data and the environmental conditions to which the building was subjected,

(*b*) acquisition of actual thermal images under environmental conditions that will produce thermal irregularities identifiable by the infrared system employed and for the defects one desires to observe,

(*c*) deduction of the temperature distribution on the inspected surface, correcting for emissivity changes, and eliminating apparent surface temperature distributions produced by the reflection of radiation from extraneous sources,

(*d*) comparison of the observed temperature distributions with the anticipated ones,

(*e*) deduction of the types of defects in the structure that could produce the observed irregularities in the thermal images, and

(*f*) verification of the conclusions, using auxiliary measurements, if reasonable uncertainty remains as to the cause of the thermal irregularities.

Aerial Infrared Survey

Aerial infrared thermography is an imagery process that produces an apparent radiance temperature of a partial map of the terrain. Infrared equipment for ground-based surveys or a line scanner are airborne to conduct a survey. The technique is effective in identifying defective areas that are either uninsulated or moisture damaged in individual built-up roofs. Variations in ceiling resistance cause differences in surface temperatures among the roofs displayed in an aerial infrared image. Field surveys have shown that under optimum climatic conditions (that is, clear sky and low wind), a highly sensitive infrared imaging system viewing a small cluster of residential roofs having the same roof total emittance and exposed to the same environmental condition will produce a thermal picture with a clear image. This thermogram can provide information for ranking the buildings according to their heat loss characteristics. Under nonoptimum conditions, the variations in roof emittance, local wind speed, and local outdoor temperature (throughout the macroclimate) may often cause differences in apparent surface temperature, which mask out those differences in the intensity of radiation emitted from roof surfaces as a result of variations in ceiling resistance.

Flights are most often conducted at altitudes between 305 and 610 m (1000 and 2000 ft). The instantaneous field of view of the infrared line scanner varies with the particular system but will generally range between 1 and 2.5 mrad.

For low-slope roofs, variations in roof thermal resistance cause differences in apparent surface temperature among the roofs displayed in an aerial infrared photograph, which are approximately a factor of three greater than those for pitched ventilated roofs. When aerial infrared surveys are performed under optimum conditions (clear sky and low wind), uninsulated roofs can be distinguished from insulated roofs provided that the roofs are exposed to the same microclimate and have the same emittance. For a large group of houses, typical variations in roof emittances and variations in local wind speed and local outdoor temperatures throughout the macroclimate cause differences in the apparent radiance temperature of roofs. Those variations may frequently mask out differences due to variations in ceiling thermal resistance.[2]

If all the built-up roofs displayed in an aerial infrared thermogram are well insulated, then variations in other parameters such as roof emittance, local wind speed, and local outdoor temperature may also cause particular roofs to appear warmer than others, which, without knowledge from ground inspections, may be incorrectly interpreted as the absence of roof insulation. A common problem with roofs is that the exterior membrane becomes ruptured, permitting water to penetrate the roof system and wet the insulation. Regions having wet insulation will conduct more heat and will appear warmer than regions having dry insulation. The merit of such a survey technique is that defective regions can be located, permitting local repairs to be carried out instead of replacing the whole roof system.

Spot Radiometer

The spot radiometer is a hand-held device used to measure radiometrically the equivalent blackbody temperature of a relatively small surface area of a target. Spot radiometers are gener-

[2]See the discussion in Ref *1*, pp. 39–41.

ally calibrated against a reference surface of known temperature. A hand-held infrared thermometer (spot radiometer) is used to measure the interior surface temperature of a building component and the indoor air temperature. Another measuring device is used to measure the coincident air temperature at the exterior surface. The three measured temperatures are applied in mathematical relations for estimating the apparent thermal resistance of the component.

A suitable application is to determine qualitatively whether a wall is insulated or if insulation voids or other thermal defects are present. The device is pointed at a surface area of interest and senses the total infrared radiation over a particular wavelength band emanating from the surface, including both the self-emitted surface radiation and reflected radiation through the surrounding surfaces. The device is calibrated to read out the apparent radiance temperature on either a digital or meter display. The apparent radiance temperature is defined as the temperature of a blackbody that would radiate the same amount of thermal radiation as a real surface at the same temperature and having a surface emittance different from unity. Typical specifications for spot radiometers include a temperature measurement resolution within ± 0.3 K (0.5°F). A spot radiometer with resolution of within ± 0.05 K (0.1°F) is also commercially available. Spot radiometers generally have their principal spectral region at 2 s.

Building Diagnostic Services

NBS staff performed ground-based thermographic studies and other diagnostic tests on eight office buildings. The projects also required establishment of contractor services to conduct the energy audits with spot radiometers and aerial infrared thermography studies. A two-step process was undertaken which determined the availability of specialized representative services from competitive organizations, followed by negotiations based on submitted proposals. Cross sections of contractor costs, equipment and usage, exceptions to requirements, diversity of geographical locations, and period for performance were sought. Compilations for contractor service responding organizations with interest, experience, and background specific to building investigations and staff capabilities were established from a "sources sought" solicitation in the *Commerce Business Daily.* Competitive proposals for building inspections were negotiated for contract awards.

The solicitation resulted in 25 responses for the aerial thermography contractor services. The respondees included companies with building experience, universities, and others with specialized infrared (IR) services for military or forest fire surveillance.

The responses to proposal bid requests for aerial thermographic surveys are indicated in the following tabular material (bids for more than one building are included). Contracts for three buildings were awarded.

Federal Building	Bids Received	Declined Bids[3]	Returned with No Response
Anchorage, AK	2	...	...
Springfield, MA	4	1	2
Columbia, SC	3	1	1

Comments from potential contractors expressed concern that aerial surveys may not be useful (particularly for a single building), and walk-on roof inspections with IR and other equipment

[3]Responses indicated that walk-on roof inspection with IR and other equipment is required.

(for example, core sampling) are essential. Core sampling was not allowed by the GSA because of concern for the long-term durability of the removed and repaired section.

Responses from twelve companies were received to conduct energy audit inspections with hand-held spot radiometers. A summary of responses to conduct energy audits and inspections with a spot radiometer is shown in the following table (with bids for more than one building). Contracts for three buildings were awarded. Because of the lack of cold weather conditions, the Pittsfield study was not performed.

Federal Building	Bids Received	Declined Bids[4]	Returned with No Response
Pittsfield, MA	3	1	2
Ann Arbor, MI	3	1	2
Huron, SD	3	1	2

Thermographic Inspections of Office Buildings

The occupiable floor spaces of the eight federal office buildings range from a 2-story building of about 1800 m^2 to a 15-story building of about 45 000 m^2. The construction materials used for the facing of their exterior walls include precast concrete panel, quarry tile, granite siding, and brick. Most of them have large window areas as well. Table 1 gives a general description of these buildings, including the detailed construction of their exterior walls.

Ground-Based Infrared Survey

The equipment used by NBS for thermographic inspection for all eight buildings included two scanning systems with spectral response regions ranging between 3 and 5.6 μm, and between 8 and 14 μm, respectively. Each building was inspected externally, and internal inspection was conducted at certain locations where thermal anomalies were detected or suspected during outside inspections. All inspections were conducted on these buildings during the winter heating season of 1982–83. Data were collected on videotapes for later analysis. Polaroid pictures of thermal images are also available for some buildings taken during inspections at the site. Table 2 is a summarized description of thermal anomalies observed in these buildings by ground-based thermographic inspections. Figures 2 to 7 are examples showing typical defects found in the buildings by ground-based thermographic inspections.

Air leakage at precast concrete panel joints as well as areas of thermal bridges existed in many buildings. Figure 2 exhibits such thermal deficiencies in the Anchorage building. Note that the vertical heat loss paths within each panel, which do not locate nearby seams, represent thermal bridges. Missing insulation in overhang ceilings appears to be the problem area, as shown in Fig. 3, where the heat loss locations above and beneath the mechanical room on the fifth floor of the Fayetteville building are given. Figure 4 also indicates that the second floor overhang ceiling of the Norfolk building lacks insulation. Figure 5 exhibits the defective insulation in wall areas as well as leakage through indentation of the Pittsfield building. An exterior infrared survey of the Springfield building revealed uniform air leakage paths, and interior inspection further identified that they leaked through both sides of the supporting columns, as shown in Fig. 6. In the Columbia building, the bright area above the window sections, as observed in Fig. 7, was discovered by inspection as missing insulation behind window panels.

[4]Responses indicate spot radiometers were not useful for the intended purpose.

TABLE 1—*The building size and exterior wall construction of each test building.*

Building Location	Floor Area, m^2	No. of Modules	No. of Stories	Typical Wall Construction
Anchorage, AK	48 470	6	2 to 6	precast concrete panel, semirigid glass-fiber insulation, gypsum wallboard on metal studs
Ann Arbor, MI	5 270	1	4	quarry tile, metal lath and mortar bed, semirigid glass-fiber insulation, acoustic wall panel
Columbia, SC	21 600	2	2, 15	granite siding, concrete, glass-fiber blanket insulation gypsum wallboard
Fayetteville, AR	3 660	1	5	face brick, concrete block
Huron, SD	6 910	1	4	face brick, lightweight concrete masonary unit, semirigid glass-fiber insulation board, gypsum wallboard
Norfolk, VA	18 570	1	8	face brick, air space, gypsum board sheathing, glass-fiber blanket insulation, gypsum wallboard
Pittsfield, MA	1 860	1	2	face brick, semirigid glass-fiber insulation board, brick
Springfield, MA	14 560	1	5	precast concrete panel, semirigid glass-fiber insulation board, gypsum wallboard on metal studs

Figure 8 is a sketch of the cross section of a window panel; the void area behind it will cause the heat stored in the space above the window to be lost to the outside through the panel air pocket.

Evaluation of Roof Systems by Aerial Thermography

In reference to the responses from the contractors and other investigations [*5–8*], it appears that the use of flyover infrared survey to identify heat loss locations of office building roof systems is not sufficient. In addition, walk-on ground-based thermographic inspection as well as visual investigation must be used to detect thermal anomalies accurately. Roof systems of three office buildings inspected by three different contractors were evaluated.

Figure 9 is a thermogram from an aerial thermographic survey performed on the Springfield building. After the walk-on infrared inspection, no major thermal anomalies or water-damaged insulation were found. The locations of heat loss regions shown in Fig. 9 were analyzed and explained by equipment under the area (Point *2*),[5] warm air vents and roof drains (Point *4*), metal hatch covers, and "corner effects" (Point *7*). The extended bright area at Point *7,* has vented warm air from the wall louvers of the penthouse (adjacent to the courthouse roof), which may obscure actual roof surface conditions. One unexplained area in Fig. 9 requiring further study is Area *1*.

[5]Numbers in parentheses refer to roof areas noted in Fig. 9.

TABLE 2—*Summary of thermal anomalies observed in each building.*

Building Location	Total Wall Area, m^2	Defective Wall Area, m^2	Percentage of Defective Wall Area, %	Major Defects Observed (Besides Leakage Around Windows and Seams)
Anchorage, AK	7 600	1 405	18	thermal bridges within panels, leakage at panel joints
Ann Arbor, MI	904	159	18	missing insulation, leakage at seams and joints
Columbia, SC	10 584	1 814	17	thermal bypass through convection loss from air spaces, missing insulation behind window panels
Fayetteville, AR	1 480	88	6	leakage at floor seams, heat loss underneath overhangs
Huron, SD	3 561	328	9	leakage through columns, thermal bridges between floors
Norfolk, VA	6 171	1 067	17	ceiling defects, thermal bridges between floors
Pittsfield, MA	1 013	187	18	insulation voids, leakage at indentations
Springfield, MA	4 388	303	7	leakage through columns, thermal bridges between floors

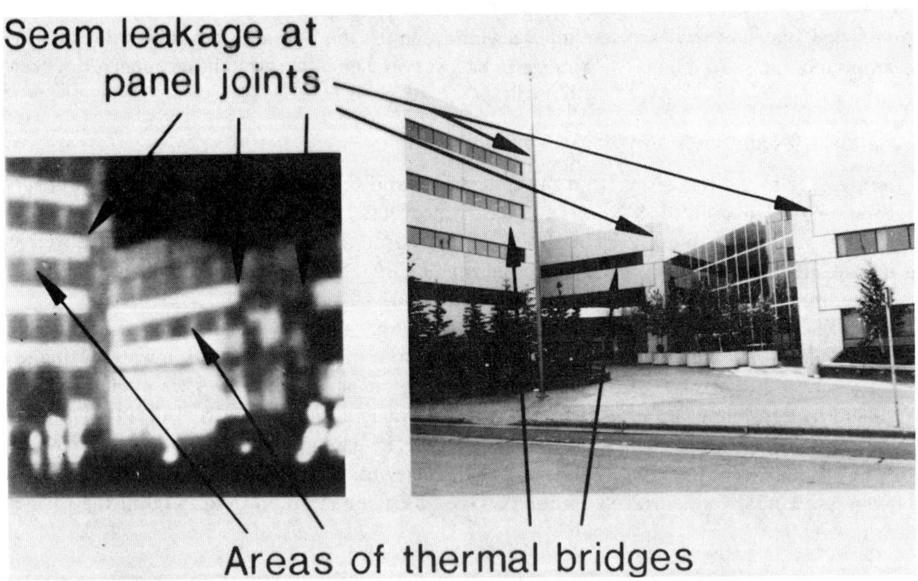

FIG. 2—*Thermal bridges observed within the precast concrete panels of the Anchorage building.*

FIG. 3—*Heat loss locations above and beneath the mechanical room of the Fayetteville building.*

FIG. 4—*Missing insulation and leakage observed in the overhang over garage of the Norfolk building.*

FIG. 5—*Defective insulation and leakage through indentation in the Pittsfield building.*

FIG. 6—*Leakage through supporting columns of the Springfield building.*

FIG. 7—*Heat loss locations above the window sections of the Columbia building.*

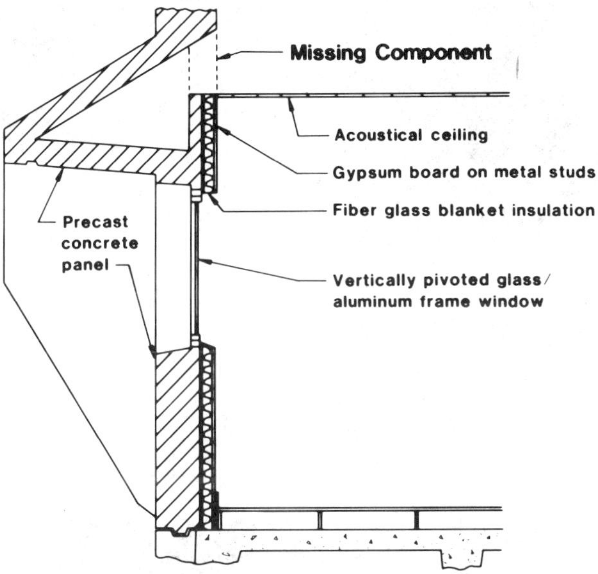

FIG. 8—*Missing components behind window panels of the Columbia building.*

After the flyover infrared survey was performed on the Columbia building, roof anomalies apparent from the IR images were explained by roof surface conditions, including vents, skylights, drains, and hatches. Two locations required further investigation. One was found to be a buildup of gravel and bitumen; the other produced an IR pattern common to divisions between wet and dry fiberous insulation board. The areas are shown in Figs. 10 (aerial) and 11 (walk-on). The cause of water intrusion was presumed to be channeling of precipitation behind the flashing (from grooves in the precast exterior wall panels). Since no core samples were allowed to be taken from the roof, no further investigation was made to determine the composition of roof materials, deterioration, and moisture content.

Th roof systems of both the Springfield and Columbia buildings consist of built-up roofs. The Anchorage building has inverted membrane roofs with blocks of polystyrene foam thermal insu-

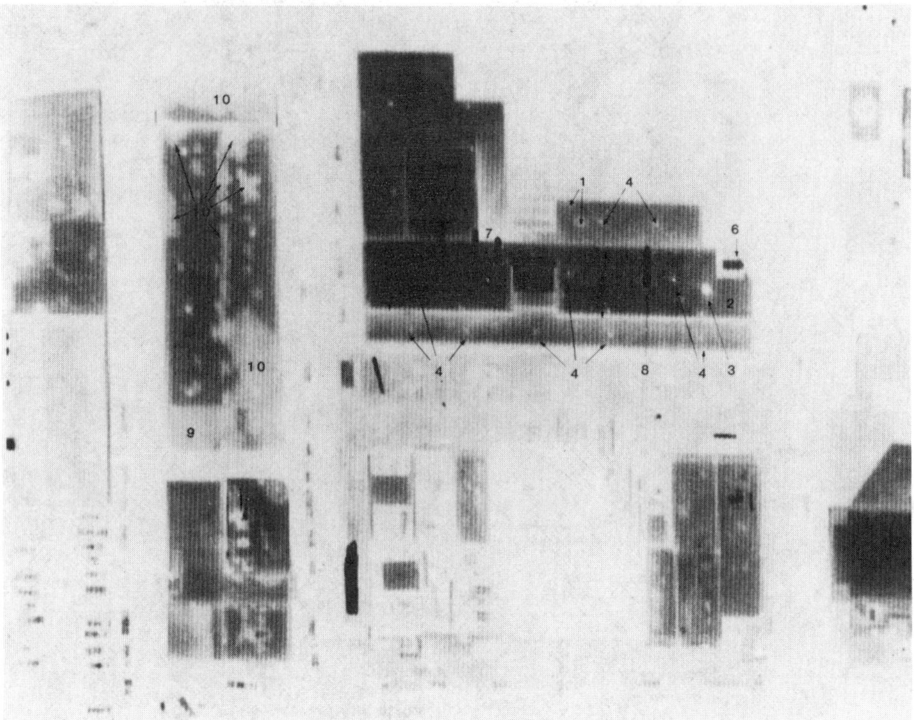

FIG. 9—*Heat loss areas of the Springfield roof system observed by aerial thermography.*

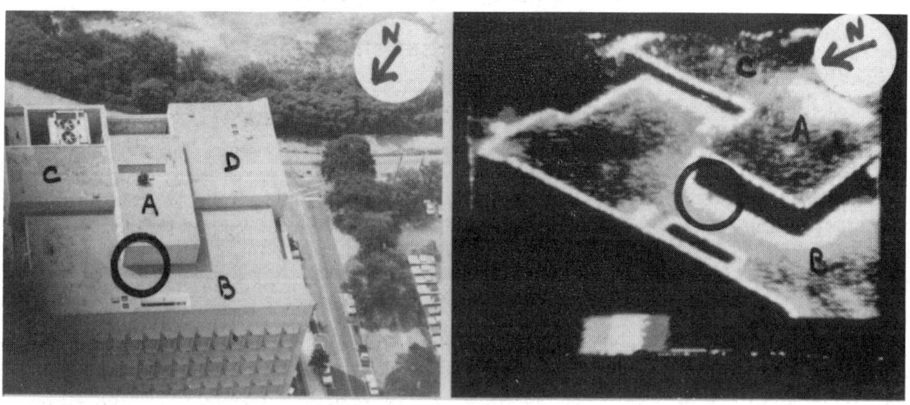

FIG. 10—*Aerial photograph* (left) *and aerial thermogram* (right) *of the Columbia building roof system circled area of anomaly.*

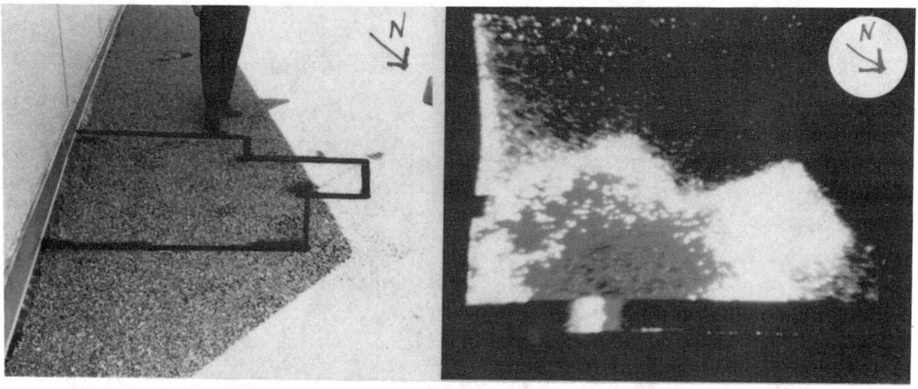

FIG. 11—*Photograph* (left) *and thermogram* (right) *from walk-on infrared survey of the Columbia building roof system—the lines enclose the area of concern.*

lation covered by cement blocks. Aerial thermographic inspection and walk-on infrared survey performed on the six modules of the building did not detect any areas with thermal deficiencies except some locations with explainable conditions, such as air vents. However, by walking on the rooftop, one could visualize that several problems were due to "sunken" surfaces. Removal of the cement blocks in a few areas for inspection showed that large gaps between the insulation blocks were heavily laden with water and deteriorated, as shown in Fig. 12. Therefore, thermography is not a useful diagnostic tool for the inverted membrane roof since the thermal problems are masked. The extent of damage to the entire roof surface was not determined, since the IR study was the only contracted service.

For detection of roof anomalies in a single building, the economic benefits of aerial thermography have not been determined. For the multiple-building complex or large built-up areas, aerial surveys have shown that many buildings can be evaluated by flyovers. Since it is necessary to followup with walk-on inspections, the cost factors for conducting the aerial survey may not be justifiable. On the other hand, the demonstration that all roof thermal problems on a building can be detected without an aerial survey should be undertaken. If it is proven, cost savings by elimination of the aerial study may then be shown.

Roof anomalies revealed through aerial infrared surveys are oftentimes recognized as explainable conditions (that is, not related to deficiencies resulting from the built-up roofing system) related to building system components or equipment. The analysis of such sources requires building and roof inspections by experienced or trained staff.

Roof-top level inspections with infrared equipment were required to define the identified potential deficiencies. Additional testing, such as core sampling or water content measurements in thermal insulation by destructive testing that required subsequent repair, was recommended by contractors for evaluation of the extent of deterioration.

Thermal Resistance Measurements by Spot Radiometer

A spot radiometer is a nonimaging infrared instrument designed to measure the emitted and reflected radiation from the surface. At steady-state conditions, the thermal resistance of an exterior surface of a building can be estimated from the following equation

$$R = \frac{T_i - T_0}{h_i(T_i - T_{is})} \tag{1}$$

FIG. 12—*Wet styrofoam insulation* (top) *and deterioration* (bottom) *of the Anchorage building roof system.*

where

R = the thermal resistance, $m^2 \cdot K/W$ ($h \cdot ft^2 \cdot °F/Btu$),
T_i = the interior air temperature, K (R),
T_0 = the exterior air temperature, K (R),
T_{is} = the inside surface temperature, K (R), and
h_i = the inside surface heat transfer coefficient, $W/m^2 \cdot K$ ($Btu/h \cdot ft^2 \cdot °F$).

The inside surface heat transfer coefficient, h_i, can be expressed in Eq 2, which includes thermal radiation and convection.

$$h_i = \frac{\sigma \cdot E(T_i^4 - T_{is}^4)}{T_i - T_{is} + 0.19(T_i - T_{is})^{0.33}} \tag{2}$$

where σ is the Stefan-Boltzmann constant, 5.67×10^{-8} $W/m^2 \cdot K^4$ (0.1714×10^{-8} $Btu/h \cdot ft^2 \cdot R^4$), and E is the emissivity factor, which is determined from

$$\frac{1}{E} = \frac{1}{E_1} + \frac{1}{E_2} - 1 \tag{3}$$

where E_1 and E_2 are the total emissivities of the subject surface and other room enclosure surfaces, respectively.

Therefore, the thermal resistance of an exterior surface can be calculated from Eqs 1 to 3 with temperature measurements and estimated emissivity.

According to the contractor's report [*9,10*], both the Huron and Ann Arbor buildings were inspected by a general purpose–dual range spot radiometer with an indoor scale of 283 K (50°F) to 311 K (100°F) and an outdoor scale of 244 K (−20°F) to 339 K (150°F). The instrument was equipped with a red filter of 8 to 14 μm to eliminate the effects of humidity, color, and other emissivity conditions that would affect the surface temperature readings. Calibration of the spot radiometer was made by adjusting its zero control to match its temperature to the attached surface thermometer of a factory-supplied reference block. The survey of each building, which took place in the spring of 1983, was conducted 3 h after sunset and ½ h after the heating system was turned off. It was found that the spot radiometer should be left in the "on" position throughout the inspection to avoid drifts from the reading at calibration. The temperatures of inside and outside air were recorded using a digital thermometer with a range of 238 K (−31°F) to 323 K (122°F). The accuracy of both the spot radiometer and the digital thermometer is within ±0.3 K (0.5°F). Data were collected on the temperature measurements at a certain region of a surface, and the thermal resistance (R value) of the region was calculated by assuming an emissivity factor of 0.9. During the inspection, several spots on each surface were inspected to determine their R values, and the average value of these were used to represent the thermal resistance of the surface.

From the architectural drawings of these two buildings, the R value of each building surface component was also calculated following the *ASHRAE Handbook of Fundamentals* [*14*].

Tables 3 and 4 are summaries of the measured R values compared with the design R values of the Huron and Ann Arbor buildings, respectively. In general, the measured thermal resistances of wall areas are lower than the design values, and R values of windows and doors are relatively close to the design values in these two buildings, as given in Tables 3 and 4. These discrepancies between the measured and design R values may be due to measurement errors, the method of calculation, or deficiency in construction or a combination of these factors. From the accuracies of both spot radiometer and digital thermometer, the variation of the measured air-to-surface

TABLE 3—*Comparison of thermal resistance measured by spot radiometer with design values in the Huron building.*

	Average Measured *R* Values, h · ft² · °F/Btu[a]							
	North Wing				South Wing			Design *R* values, h · ft² · °F/Btu[a]
	West	South	North	East	North	East	Lobby	
Office wall	6.18	6.35	5.44	5.86	4.54	6.10	...	15.75
Walls under windows	3.28	3.46	2.69	4.03	...	4.07	3.12	12.15
Stairway walls	...	...	...	4.03	4.72	...	...	7.47
Garage wall	...	1.73	...	...	...	...	...	2.35
Garage roof	...	1.65	...	...	...	...	...	5.47
Column walls	...	...	3.54	...	...	...	5.10	...
Doors	1.50	...	...	...	...	...	1.26	1.69
	...	...	...	...	...	...	...	0.90
Windows with curtains	3.40	3.19	3.18	...	...	3.65	2.14	...
Windows without curtains	...	...	...	...	...	1.79	...	2.04
Lobby wall	...	...	...	...	...	...	4.06	14.50

[a]Conversion factor for *R* values in SI units: 1 h · ft² · °F/Btu = 0.176 m² · K/W.

TABLE 4—*Comparison of thermal resistance measured by spot radiometer with design values in the Ann Arbor building.*

	Average Measured *R* Values, h · ft² · °F/Btu[a]					Design *R* Values, h · ft² · °F/Btu[a]
	West	South	East	North	Lobby	
Office walls						
Type 1	7.85	...	...	9.19	8.53	8.40
Type 2	5.71	6.71	...	...	...	8.77
Type 3	...	4.42	5.71	...	...	9.04
Type 4	...	4.42	3.15	6.08	5.65	4.54
Lobby walls						
Type 1	...	...	...	...	3.40	4.62
Type 2	...	...	...	...	2.67	2.12
Windows						
Vertical	...	...	1.22	1.53	...	2.04
Horizontal	...	...	1.27	1.38	...	1.69
Bay	...	1.09	...	0.94	...	0.9
Doors	...	...	...	1.52	...	1.69

[a]Conversion factor for *R* values in SI units: 1 h · ft² · °F/Btu = 0.176 m² · K/W.

temperature difference is within 0.6 degree kelvin (1 degree Fahrenheit). Therefore, the percentage variation of the measured *R* value by spot radiometers versus the temperature difference between inside air and surface can be illustrated in Fig. 13. According to Eq 1, the temperature measurements corresponding to building components having large *R* values will have a lower temperature difference between inside air and surface, and thus introduce a higher variation percentage in measured thermal resistance values, as shown in Fig. 13. On the other hand, for building components with low *R* values, such as windows and doors, the air-to-surface tem-

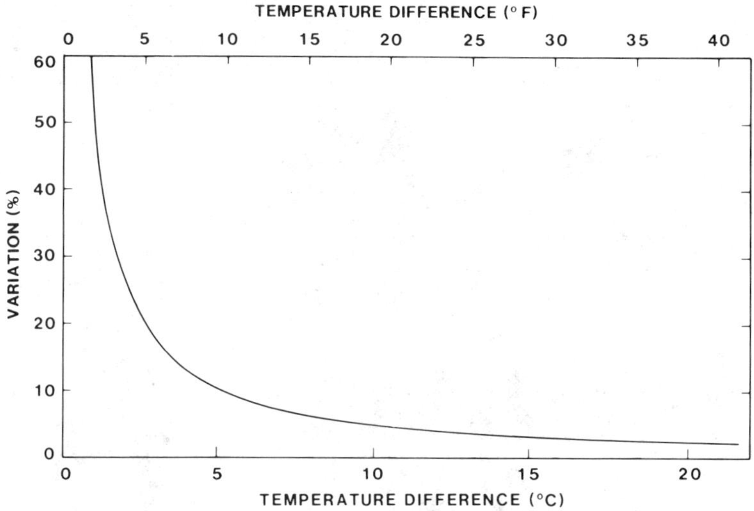

FIG. 13—*Percentage variation of measured* R *values from spot radiometer versus temperature difference of air and surface.*

perature differences are expected to be high, and the percentage variation in thermal resistance measurements are relatively low. Hence, the measured and design R values of these surfaces, also shown in Table 3, are much closer to each other. In addition, the equations used to calculate the thermal resistance from temperature measurements are based on a steady-state condition without considering the transient effects of thermal storage in construction materials, and they may cause errors in computing the rate of heat transfer across the building envelope as well as R values. As for the building construction, comparison of test results from various measurement methods should be used for qualification of thermal deficiencies.

Table 3 indicates that the measured R values of office and lobby walls at the Huron building are much lower than the design values. However, thermal conductance measurements on the north wall of the south wing made by heat flow meter and calorimeter give much higher R values of 3.59 and 2.11 $m^2 \cdot K/W$ (20.4 and 12.0 $h \cdot ft^2 \cdot °F/Btu$), respectively [*2*], compared with 0.80 $m^2 \cdot K/W$ (4.54 $h \cdot ft^2 \cdot °F/Btu$) from spot radiometer inspections and with 2.77 $m^2 \cdot K/W$ (15.75 $h \cdot ft^2 \cdot °F/Btu$) from design. Furthermore, the ground-based infrared survey performed on the Huron building identified about 9% of the wall area as defective (see Table 2), and results from air tightness measurements indicate that the Huron building has the lowest infiltration rate (0.10 to 0.20 vol/h) among the test buildings [*4*]. Therefore, the large discrepancy between the measured R values and the design values of wall areas are mainly due to the limitation of equipment sensitivities, which causes inaccuracy in measurements.

Figure 14 is a photograph of the Ann Arbor building, which has a particular design with various construction intervals for different wall sections. The measured thermal resistances of various building components in this building are relatively close to the design values, as given in Table 4. The thermal resistances measured by heat flow meter and calorimeter of the west wall are 2.02 and 1.95 $m^2 \cdot K/W$ (11.5 and 11.1 $h \cdot ft^2 \cdot °F/Btu$), respectively [*2*], compared with 1.38 $m^2 \cdot K/W$ (7.85 $h \cdot ft^2 \cdot °F/Btu$) from the design. However, ground-based thermographic inspection located about 18% of its wall areas with thermal anomalies (see Table 2), and the results from building tightness measurements indicated this building had an infiltration rate between 0.55 and 0.65 vol/h [*4*]. In addition, the Ann Arbor building is not considered to be energy efficient since the architectural design R values of all the wall areas are less than 17.6

FIG. 14—*Photograph of Ann Arbor federal building.*

$m^2 \cdot K/W$ ($10\ h \cdot ft^2 \cdot °F/Btu$). Therefore, the measured thermal resistances of surface components by spot radiometer survey are considered to be reasonably accurate.

A spot radiometer for thermal resistance measurements has the advantage of portability, speed, and simplicity. However, laboratory evaluation of using a low-resolution—for example, ±0.3 K (0.5°F)—spot radiometer to determine the thermal resistance of building walls showed that it does not provide the desirable accuracy [*15*]. The results from the measurements by spot radiometer for both the Huron and Ann Arbor buildings show discrepancies between the measured and the design *R* values. Figures 15 and 16 illustrate the comparisons of these *R* values as well as the least-squares regression lines fitting the entire data sets for the Huron and Ann Arbor buildings.

For the Huron building, the correlation coefficient, r, was determined to be 0.72 and r^2 (=0.52) indicated that about half (52%) of the total variation in measured *R* values was explainable by the regression of design values. For the Ann Arbor building, r was determined to be 0.876 and r^2 (=0.77) showed that about three quarters (77%) of the total variation in measured *R* values was explainable by the regression of the design values. The difference in r^2 values between the two buildings may be due to the higher thermal resistance of building materials in the Huron building, which introduced a smaller inside air-to-surface temperature difference, and thus a high percentage variation in calculated *R* values (see Fig. 13) and a lower r^2. In addition, the environmental conditions during inspection will also affect measurement results.

Another field investigation (previously reported [*16*]) using a high-resolution (±0.05 K or ±0.1°F) spot radiometer and radiant flow radiometer (with which the intensity of thermal radiation is measured directly without being converted to temperatures) to determine thermal resis-

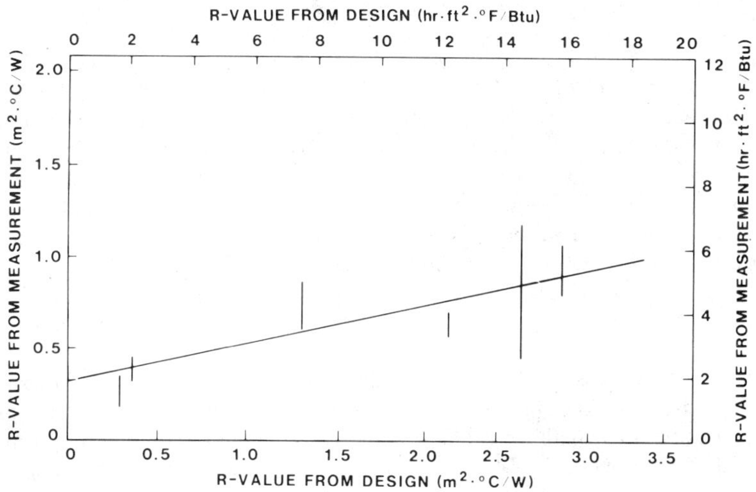

FIG. 15—*Thermal resistance (*R *value) from spot radiometer measurement versus design value for the Huron building.*

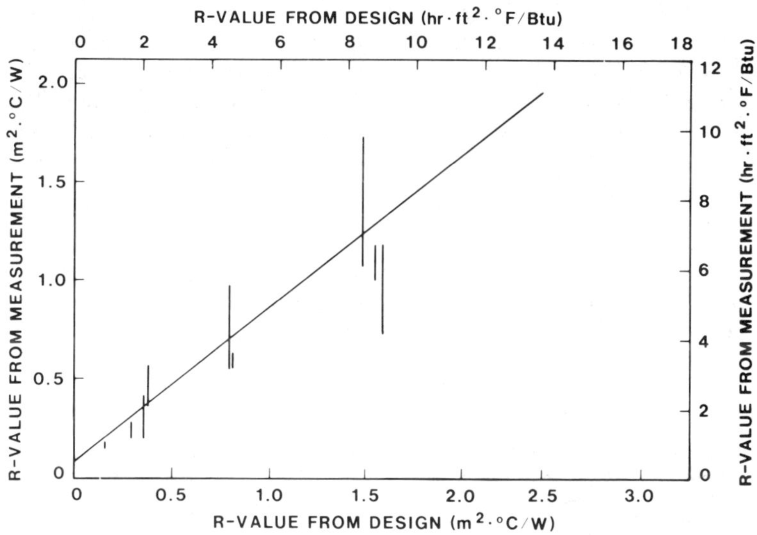

FIG. 16—*Thermal resistance (*R *value) from spot radiometer measurement versus design value for the Ann Arbor building.*

tance of the test house walls revealed that systematic errors were related to the heat capacity of wall materials. Furthermore, Burch and Krintz [*16*] suggested that spot radiometer inspections should be conducted with the furnace on, in contrast to the recommendation of American National Standards Institute/American Society of Heating, Refrigerating, and Air Conditioning Engineers (ANSI/ASHRAE) Standard 101-1981 to turn the furnace off.[6]

[6]"Application of Infrared Sensing Devices to the Assessment of Building Heat Loss Characteristics," ANSI/ASHRAE Standard 101-1981.

Conclusions

Various thermographic inspection techniques were used to assess the thermal integrity of federal office buildings. These buildings were built to meet the new energy guidelines; however, infrared thermographic surveys revealed that serious thermal defects exist in each building.

Ground-based thermography performed on all eight buildings observed 6 to 18% of their wall areas as exhibiting problems in insulation, in addition to common anomalies such as thermal bridges; leakages around windows, seams, and columns; defective ceilings; and heat loss through overhangs and indentations. The thermographic technique can be used in quality control by performing inspection on new buildings during the construction stage to locate defects for building owners. Also, after buildings are occupied, thermographic inspections are capable of identifying degradation of materials as well as the causes of poor thermal performance of building envelopes.

The locations of heat loss on three office building roof systems through aerial thermography were identified. However, the majority of them were recognized by walk-on infrared survey and visual inspections as explainable conditions related to building system equipment or components instead of deficiencies in the built-up roof systems. Furthermore, thermography was found not to be a useful diagnostic tool for the inverted membrane roof systems.

Results from spot radiometer inspection performed on two office buildings indicated there were larger variations in thermal resistance measurements of building components of high R value than those of low R values. This may be due to the sensitivity of the instrument as well as the equations used for calculations. However, the qualitative measurements of thermal resistance of building exterior surface were still valid, though the quantitative measurements were found to lack accuracy. The linear regression coefficients, r^2, of measured R values versus design R values were determined to be 0.52 and 0.77 for the Huron building (with high R values) and the Ann Arbor building (with low R values), respectively. Because of their simplicity and portability, spot radiometers can be used as a tool for energy audit purposes to check on suspected regions of building surface components.

Acknowledgment

This project was sponsored by the Public Building Service of the General Services Administration through an interagency agreement with the National Bureau of Standards.

References

[1] Grot, R. A., Burch, D. M., Silberstein, S., and Galowin, L. S., "Measurement Methods for Evaluation of Thermal Integrity of Building Envelopes," NBSIR 82-2605, National Bureau of Standards, Washington, DC, Sept. 1983.

[2] Fang, J. B. and Grot, R. A., "In-Situ Measurement of Thermal Resistance of Building Envelopes of Office Buildings," *A.S.H.R.A.E. Transactions*, Vol. 91, Part 1B, 1985, pp. 543–557.

[3] Persily, A. K. and Grot, R. A., "Pressurization Testing of Federal Buildings," Symposium on Measured Air Leakage Performance of Buildings, ASTM meeting, Philadelphia, PA, 1984.

[4] Grot, R. A., Chang, Y. M., Persily, A. K., and Fang, J. B., "Interim Report on NBS Thermal Integrity Diagnostic Tests of Eight Federal Office Buildings," NBSIR 83-2768, National Bureau of Standards, Washington, DC, Sept. 1983.

[5] Chang, Y. M. and Grot, R. A., "The Role of Thermography in the Assessment of the Thermal Integrity of Federal Office Buildings," *Proceedings of the Society of Photo-Optical Instrumentation Engineers*, Vol. 446, 1983, pp. 47–55.

[6] Grot, R. A., "Interpretation of Thermographic Data for the Identification of Building Heat Loss," *Proceedings of the Society of Photo-Optical Instrumentation Engineers*, Vol. 254, 1981, pp. 24–29.

[7] Mill, P. A. D., "Thermographic Interpretation of Building Enclosure Deficiencies in Canadian Public Buildings," *Proceedings of the Society of Photo-Optical Instrumentation Engineers*, Vol. 254, 1981, pp. 121–137.

[8] Ouzts, K. B., Vanier, J. G., and Brotherson, D. E., "A Comparison of Infrared Flyover, Infrared Walkover, and Visual Inspection Techniques for Detecting Roof Moisture Anomalies," *Proceedings of the Society of Photo-Optical Instrumentation Engineers,* Vol. 446, 1983, pp. 122–129.

[9] "Aerial Infrared Thermography of the Federal Building," Springfield, MA, Aeromarine Surveys, New London, CT, 7 May 1983.

[10] Wallace, J. R., "Thermographic Inspection Report," Wallace Thermographics Co., Richmond, VA, June 1983.

[11] Town, R. H., "Infrared Survey for Federal Building, C Street, Anchorage, AK," No. AP6546, Infrared Technology of Alaska, Anchorage, AK, April 1983.

[12] "Spot Radiometer Energy Audit, Federal Building in Huron, SD," Donahue and Associates, Madison, WI, 1983.

[13] "Spot Radiometer Energy Audit, Federal Building in Ann Arbor, MI," Donahue and Associates, Madison, WI, 1983.

[14] *ASHRAE Handbook of Fundamentals,* American Society of Heating, Refrigeration, and Air Conditioning Engineers, Atlanta, GA, 1981.

[15] Treado, S. J. and Burch, D. M., "Evaluation of Hand-Held Infrared Thermometers for Wall Thermal Resistance Determinations," NBSIR 79-1736, National Bureau of Standards, Washington, DC, July 1979.

[16] Burch, D. M. and Krintz, D. F., "Using High-Resolution Hand-Held Radiometers to Measure In-Situ Thermal Resistance," *Proceedings of the Photo-Optical Instrumentation Engineers,* Vol. 446, 1983, pp. 24–32.

George A. Tsongas[1] and Timothy J. Ball[2]

Thermographic Inspection of the Installation Quality of Retrofitted Wall Insulation

REFERENCE: Tsongas, G. A. and Ball, T. J., **"Thermographic Inspection of the Installation Quality of Retrofitted Wall Insulation,"** *Thermal Insulation: Materials and Systems, ASTM STP 922,* F. J. Powell and S. L. Matthews, Eds., American Society for Testing and Materials, Philadelphia, 1987, pp. 196–201.

ABSTRACT: Infrared thermographic scans of the walls of 79 homes in Spokane, Washington, were made during the winter of 1983 as a means of observing the installation quality of retrofitted wall insulation. A special attempt was made to determine quantitatively the amount of void area that is typical in insulated walls since it can have a significant impact on thermal performance. Valid results were obtained for 30 of the test homes in this study, and the average uninsulated or void area was 9%, with a range of 1 to 35%. This result is in excellent agreement with those from other studies. For a 9% uninsulated area, estimates of the increase in heat loss relative to a fully insulated wall vary from a low of about 9% to as high as about 40%.

As suggested by the wide range in results, some of the retrofit wall insulation jobs were excellent and some were extremely poor. The newer installations were generally poorer. The important point is that it is clear from these results that wall insulation can be properly installed with little or no void area. Improved quality control on the part of the installers is certainly possible and needed. The fact that the average installation quality is relatively poor and the range in results is very large indicates that some kind of inspection program is advisable.

During this thermographic study, an assessment was also made of blown-in wall insulation installation problems. A number were observed, most of which could be eliminated with proper care and attention on the part of the installers. In addition, an effort was also made to locate high heat loss areas. One notable finding is that often the exposed portions of heated basement walls were uninsulated and therefore exhibited relatively high heat loss.

KEY WORDS: thermal insulation, quality control, retrofit, thermography, weatherization

During the winter of 1982–83, a major field study was undertaken in Spokane, Washington [*1*]; the walls of 103 homes were opened and examined to determine whether moisture damage is caused by insulation being blown into a wall cavity without a vapor barrier. The walls were scheduled for opening in December 1982 and January and February 1983 at a rate of two to three homes per day (three holes per home).

As a special additional part of that field study, it was decided that infrared thermographic scans of the walls of the test homes would be made as a means of observing the insulation installation quality. The need for this infrared study arose because during a previous similar moisture field study in Portland, Oregon [*2*], the quality of installation was found to be generally poor with numerous instances of empty or incompletely filled cavities. However, the Portland observations were based mainly on opening the tops and bottoms of two stud cavities in the walls of each of 96 homes. Infrared thermography was used only in a few cases to check for incomplete filling. While an estimate was made of roughly a 10% void area in the walls, many of

[1]Professor of mechanical engineering, Portland State University, Portland, OR 97207.
[2]Energy consultant, Olympia, WA 98506.

the observations were qualitative rather than quantitative, and all the walls of a house were not examined. Furthermore, almost half of the homes were retrofitted with urea-formaldehyde foam, whereas only houses with blown-in cellulose, fiberglass, or rock wool were to be observed in the Spokane study.

Finally, it was recognized that poor installation quality could drastically reduce the effectiveness of wall insulation as a retrofit measure, and so it would be worthwhile to determine the percentage of uninsulated area in the walls of a large sample of homes. Thus, a decision was made to use infrared thermography to make a quantitative assessment of the void area in the insulated walls of the Spokane test houses and also to observe installation problems and high heat loss areas.

Data Collection and Analysis Procedure

During the time the walls were opened and closed, an infrared scan of the house walls was made from the outside using an AGA Thermovision unit borrowed from the U.S. Navy (it had no videotape capability). An attempt was made to locate all uninsulated areas, any installation problems, and high heat loss areas. A standard procedure was developed and refined using several homes as a proving ground. Because of the mild winter weather during the test period, and also because of solar heating of south and other walls, it was not possible to achieve satisfactory scans of all the houses or of some walls. Moreover, because the major effort of the Spokane study involved opening as many walls as possible, time and engineering manpower limitations prevented nighttime or indoor scans from being made except in a few cases, although daytime viewing from the outside was known to be complicated [*3*]. Because Bonneville Power Administration, the sponsoring agency, was aware of these limitations, they requested that the infrared scans provide sufficient documentation to determine the percentage of void area with reasonable confidence for only 25 of the 79 insulated homes; 30 valid cases were actually achieved.

One of the major outcomes of the infrared scans was a determination of the percentage of void area in an insulated wall relative to the fully insulated case. To be sure that voids were clearly distinguishable, a criterion was established that studs had to be clearly visible. Whenever conditions were appropriate, the void area in each wall of each house was estimated (to the nearest square foot) from the thermographic scans. Sketches were made of each wall using a standard form sheet with rulings corresponding to studs every 0.41 m (16 in.), similar to the approach used in the field by Allen and Allen [*3*]. The thermographer located voids on the wall sketch as they appeared on the monitor screen. The total void area was later calculated by scaling the sketch. The actual net wall area (gross wall area less window and door areas) was also determined by measurement (to the nearest square foot). Then the percentage of uninsulated area was calculated for each wall. Finally, an area-weighted average percentage of void area was calculated for each house. The valid results from all the houses were then simply averaged to give the overall percentage of void area. These results, along with notes pointing out where voids typically occur, as well as other areas with high heat leakage rates (for example, uninsulated foundation walls in homes with heated basements), were documented in a standard checklist for each house. When the homeowners were present they were shown what was seen on the screen. It is important to note that a representative of Oak Ridge National Laboratory came to Spokane for two days to do infrared (IR) thermography with their Inframetrics IR unit with videotape capability. This was done to compare IR system capabilities as well as to share experiences.

Results

All of the homes scanned had cavity wall insulation blown in by private contractors. The types of wall insulation installed were loose-fill cellulose, fiberglass, and rock wool; none of the homes tested in the moisture project had urea-formaldehyde foam. Most of the homes had wood shingles or siding, although some had metal or vinyl siding.

Void Area

While scans were made of almost all the test houses, valid results of the percentage of void area relative to the fully insulated wall (including framing) were obtained from only 30 of the 79 insulated homes. Voids could not be seen in 15 of the 79 homes because of the existence of metal-vinyl siding. In 16 of the 30 homes with valid results, voids were determined for all four walls and area-weighted as noted in the procedure section. For seven houses, the uninsulated area percentage was determined from three walls, and for seven other houses, valid results were obtained from only two walls (arbitrarily chosen as the minimum). Results were not included for walls where void areas were not clearly discernible, such as for walls heated by the sun. The results from each house were estimated to be accurate to within ±1 to 2%.

The average uninsulated or void area found in this study was 9%; the range was 1 to 35%, and the standard deviation was 8%. The percentage of void area for 22 homes with cellulose was 9%, whereas it was 5% for 3 rock wool homes and 14% for 5 fiberglass homes. The range of void area percentage results is broken down as follows: 0 to 5% for 33% of the homes, 6 to 10% for 40% of the homes, 11 to 15% for 10% of the homes, and 16 to 35% for 17% of the homes. The void area result as well as the insulation type and age for each of the 30 homes are presented in Fig. 1; the average of the measured void area results is shown as the dashed line.

It should be noted that the percentage of void area was also determined from only one wall in 6 additional houses. For those 6 houses, the average uninsulated wall area was 13%, with a

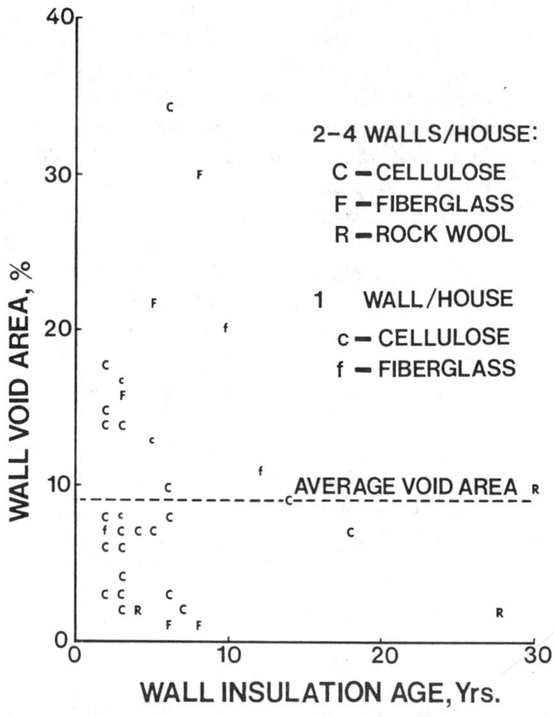

FIG. 1—*Wall insulation void area by insulation age and type.*

range of 7 to 20%. Had those 6 houses been included with the other 30, the overall average void area would have been 10%.

These measured results are in excellent agreement with the results from a number of other studies. Observations made during the Portland study [*2*] are in general agreement with the Spokane results. In an optimal weatherization demonstration project, 148 low-income houses across the nation were weatherized. Of those, many had wall insulation retrofitted. The insulation quality was subsequently assessed in detail using infrared thermography. Overall, an average void area of about 10% was observed; moreover, the range of results observed was essentially identical [*4*]. Hall and Jacobs [*5*] found the same void area result when using IR methods to investigate 35 homes in Madison, Wisconsin, although many of the homes they investigated had urea-formaldehyde foam insulation. The percentage of uninsulated area that results from shrinkage and incomplete filling of urea-formaldehyde foam was found from actual measurement in a number of homes during the Portland study [*2*] to be about twice that found in this study. Thus, there does appear to be a large difference between the results for urea-formaldehyde foam and those for blown-in loose-fill insulations.

For a 9% uninsulated area, estimates of the increase in heat loss relative to a fully insulated wall vary from a low of about 9% to as high as about 40% [*6*]. Although any wall insulation is expected to decrease heat loss, and it provides many advantages that are not usually accounted for [*7*], the fact that wall cavities are not all filled reduces the expected performance, possibly to a considerable degree. Yet that reduction is seldom accounted for and should be when considering any wall insulation conservation program. Derating the insulation's R value on the basis of the existence of typical void areas has been suggested as one possible means. However, that approach does have some drawbacks [*6*]. Whether or not that is a feasible approach was not addressed in this study.

Installation Problems

As noted, there was a wide range in the percentage of void area observed. From those results, as well as from scan observations of the other houses in the study, there were both excellent installation jobs and extremely poor jobs. The old rock wool jobs done in the 1950s were done very well; typically even all the very narrow stud spaces, such as around windows, were insulated. Generally speaking, the quality of the older installations is better than that of the newer ones, and the poorest jobs have been done within the last ten years. However, some of the newer jobs have been done very well. The important point is that insulation can be properly installed so that there is little or no void area; one third of the homes tested in this study had less than a 4% void area with an average of about 2%. While some areas are unquestionably hard to insulate well, the amount of the wall cavity that is typically left uninsulated can and should be reduced. Improved quality control on the part of installers is needed and certainly possible. The fact that such a large range of void area results exists indicates that some kind of better on-site checking or inspection is advisable.

Providing improved quality control is obviously in the best interest of insulation installers, and many are in favor of on-site inspection [*8*]. If contractors knew there was a good chance of being inspected, they would be more careful because it would cost them more to fix a bad job than to do it right in the first place. Those contractors with repeated problems could be dropped from a utility's weatherization program. Since one of the major problems with inspecting is that it is usually impossible to visually locate voids inside the wall cavity, selective testing using infrared thermography to locate any voids would seem prudent. In fact, such an inspection program has been successful in Rhode Island [*8*]. During any infrared thermography inspection it would also be important to check to see if sufficient clearance exists between insulation and any heat sources. While thermographic inspection does involve an additional cost (a full scan might cost $40 or so per house [*8*]), it is cheap in comparison with the cost of installing the insulation or in comparison with the full life-cycle dollar value of the energy savings from installing wall insulation or even from reducing voids.

The quality of insulation installation clearly depends on the applicators. It appeared from the results of this study that corners were sometimes cut in an effort to achieve a low bid. Often, whole cavities were left uninsulated, and at other times many cavities were only partially filled. Typically, the cavities next to those that were empty or partially filled were properly filled right to the top, suggesting that the voids observed were due to incomplete filling rather than to settling. On the other hand, sometimes the only voids were in narrow stud cavities that were hard to access and thus were missed. However, missing only those cavities can still result in a relatively large void area. A good applicator can find and fill almost all of those cavities.

The location and size of the fill holes also strongly influenced the quality of installation. Often all the holes were drilled below the window height. When the holes were at low height or midheight, the insulation generally did not fill the cavity. The hose apparently clogs when pressurizing in those cases. In addition, when two holes were drilled in each cavity, the insulation was usually filled more completely than when only one fill hole per cavity was used. Moreover, jobs with 5-cm (2-in.) holes were generally better than those using 2.5-cm (1-in.) holes. The larger hole allows for use of a tape to check for fire stops or other obstacles. In some homes with balloon construction, insulation was blown down into wall cavities from the attic, resulting in none being installed below the windows.

High Heat Loss Areas

In addition to using infrared thermography to determine void areas and assess installation problems, an effort was made to locate and qualitatively assess high heat loss areas. Voids at the tops of walls were, of course, prevalent. Compared with insulated frame walls, windows and doors (and to a lesser extent their frames) exhibited high heat loss, as expected. Infiltration heat loss around windows and doors was hard to detect because most were weather stripped. In addition, the exposed portions of basement walls were often noted to exhibit as much heat loss as windows, especially when the basements were heated. Typically, those exposed basement walls were about 0.6 to 0.9 m (2 to 3 ft) high and were not insulated. Since the thermographic results suggested that the heat loss from those exposed uninsulated basement walls was relatively large, it would seem worthwhile to check if retrofitting insulation in such cases would be an economically attractive conservation measure.

Conclusions

Infrared thermography has been used to assess qualitatively and quantitatively the quality of installation of retrofitted wall insulation. Generally speaking, the quality of installation is poorer than it need be, and that has an adverse impact on the thermal performance of the insulated wall.

To protect consumers and improve quality control and thermal performance, an on-site inspection program is needed. One that provides a check for uninsulated areas using infrared thermography scans appears cost-effective and in the best interest of consumers, installers, and utilities; it is recommended.

Acknowledgments

Special thanks must go to the Bonneville Power Administration (BPA) for recognizing the need for this study and for providing the necessary funding. Appreciation is extended especially to Phil Thor, BPA's contract monitor, for his assistance and able coordination of this project.

A note of appreciation is also extended to John King of the Navy Civil Engineering Laboratory for loan of the infrared thermography equipment and to George Courville and Phil Childs of Oak Ridge National Laboratory for their assistance with the thermography.

Finally, special thanks must go to the owners and families in the test homes; their interest, support, and patience made this study possible.

References

[*1*] Tsongas, G. A., "The Spokane Wall Insulation Project: A Field Study of Moisture Damage in Walls Insulated Without a Vapor Barrier," DOE/BP-541, U.S. Department of Energy/Bonneville Power Administration, Portland, OR, September 1985; also American Society of Heating, Refrigerating, and Air Conditioning Engineers/Department of Energy/Building Thermal Envelope Coordinating Council Conference on Thermal Performance of the Exterior Envelopes of Buildings III, Clearwater Beach, FL, 2-5 Dec. 1985.

[*2*] Tsongas, G. A., "A Field Study of Moisture Damage in Walls Insulated Without a Vapor Barrier," ORNL/SUB-78/97726/1, Oak Ridge National Laboratory, Oak Ridge, TN, 1980; also *Proceedings*, American Society of Heating, Refrigerating, and Air Conditioning Engineers/Department of Energy/Oak Ridge National Laboratory Conference on the Thermal Performance of the Exterior Envelopes of Buildings, *ASHRAE SP 28*, 1980, pp. 801-815.

[*3*] Allen, L. R. and Allen, S., "Development of Economic Techniques for Residential Thermography," *Proceedings of the Society of Photo-Optical Instrumentation Engineers, Thermosense V*, Vol. 371, 1982, pp. 145-149.

[*4*] Grot, R. A. and Chary, Y. L., "Thermographic Inspection of Cavity Wall Insulation Retrofits," *Thermal Insulations, Materials, and Systems for Energy Conservation in the '80s, ASTM STP 789*, American Society for Testing and Materials, Philadelphia, 1983, pp. 321-336.

[*5*] Hall, L. F. and Jacobs, T., "A Thermographic Study of Thirty-Five Residences in Madison, Wisconsin" *(Vanlen Views)*, Report for City of Madison Energy Conservation Program, Madison, WI, March 1978.

[*6*] Vinieratos, E. R. and Verschoor, J. D., "Influence of Insulation Deficiencies on Heat Loss in Walls and Ceilings," in *Thermal Insulation Performance*, D. L. McElroy and R. P. Tye, Eds., *ASTM STP 718*, American Society for Testing and Materials, Philadelphia, 1980, pp. 142-159.

[*7*] Tsongas, G. A., "A State of the Art Review of Retrofitted Wall Insulation," *Proceedings, ASHRAE SP 38*, 1982 American Society of Heating, Refrigerating, and Air Conditioning Engineers/Department of Energy Conference on the Thermal Performance of the Exterior Envelopes of Buildings II, 1983, pp. 548-470; also, DOE/BP-30575-1, U.S. Department of Energy/Bonneville Power Administration, Portland, OR, Sept. 1985.

[*8*] Ward, R. J., "Thermographic Inspections and the Residential Conservation Service Program (RCS)," *Proceedings of the Society of Photo-Optical Instrumentation Engineers, Thermosense V*, Vol. 371, 1982, pp. 129-132.

Stephen N. Flanders[1]

Measured Insulation Improvement Potential for Ten U.S. Army Buildings

REFERENCE: Flanders, S. N., "**Measured Insulation Improvement Potential for Ten U.S. Army Buildings,**" *Thermal Insulation: Materials and Systems, ASTM STP 922,* F. J. Powell and S. L. Matthews, Eds., American Society for Testing and Materials, Philadelphia, 1987, pp. 202–220.

ABSTRACT: As-built drawings and handbook calculations of *R* values are often inadequate bases for investment decisions regarding improved insulation of U.S. Army buildings. Reported field and laboratory experience indicates that a technique employing surface-mounted heat flux sensors (HFSs) in conjunction with infrared thermography (IRT) can yield reliable estimates of *R* values. This technique employs IRT to position HFSs and thermocouples at representative locations on walls and roofs or attics to acquire heat flow and temperature data for estimating *R* values. This paper reports on the application of this technique at Ft. Carson, Colorado, and Ft. Richardson, Alaska, to eight family housing units, a temporary office building, and a barracks.

Infrared thermography of these buildings detected few thermal anomalies, but measurement of several walls with HFSs and thermocouples (typically at six locations spaced vertically on each wall) revealed significant variation in estimated *R* values; this variation is attributable to convection, even within fully insulated walls. This is significant for proper placement of sensors and indicates that installed fibrous insulation can lack the ability to quell convection. The insulating ability of walls containing poorly installed mineral fiber batt insulation was much worse than would be indicated by the design handbook values. Some attic insulation performed exactly as expected; some was at least 40% worse than expected.

Plotted curves using the life-cycle cost assumptions required by the U.S. Army's Energy Conservation Investment Program show the budget necessary for each level of insulation improvement to break even over a 15-year economic life. At Ft. Carson, filling empty stud cavities in brick veneer construction to an overall *R* value of 13 would warrant an investment of about $17.6 per square metre ($1.63 per square foot). A contractor estimates that the actual cost would be only $4.5 per square metre ($0.42 per square foot). At Ft. Richardson an *R*-20 level of insulation would justify an expenditure of between $12.9 and $30.1 per square metre ($1.20 and $2.80 per square foot) of wall on the wood frame buildings that were measured. Attic spaces at both installations were heavily insulated and had much less potential for improvement than walls.

The IRT/HFS measurement technique is most useful on poorly insulated buildings, where the inaccuracy of information about the initial insulating level has the greatest impact on the allowable budget for additional insulation.

KEY WORDS: heat flux sensors, *in situ R* value measurement, heat flux sensor calibration, measurement error analysis, building thermal diagnosis, natural convection, radiation heat transfer, conductive heat transfer, insulation retrofit, energy conservation, economic analysis, thermal insulation

The U.S. Army has stringent goals for reducing building energy consumption. The Energy Conservation Investment Program (ECIP) is a major source of funding for improving existing buildings that are responsible for much of the Army's energy consumption. ECIP employs life-cycle cost (LCC) principles to determine the order of priority for investing in energy conservation improvements. In the past, improvements in energy controls and in combustion and energy delivery systems have received the most attention on Army bases. Recently, more has been spent

[1]Research civil engineer, U.S. Army Cold Regions Research and Engineering Laboratory, Hanover, NH 03755.

to diminish heat loads and gains attributable to building envelopes. The Army wants to know how well building walls, attics, and roofs are insulated, so the best investment possible for energy conservation can be determined.

As-built drawings and handbook calculations do not necessarily reflect the material properties of the construction and cannot represent flaws in workmanship or deterioration of thermal performance due to moisture or other causes. Direct measurement of building *R* values could diminish these uncertainties. However, thermal measurement introduces uncertainties of its own.

Previous work [*1*] has described techniques for obtaining temperature and heat flux data for building walls, roofs, and attics using infrared thermography to place heat flux sensors (HFSs) and thermocouples at appropriate locations for monitoring for several days. With proper care, this technique has a standard deviation of less than 10% of the mean [*2*]. Earlier experience [*3*] with nine buildings at Ft. Devens, Massachusetts, demonstrated that similarly constructed Army buildings would render similar measurements. This makes sampling of buildings appear feasible.

The purpose of the field measurements was to determine how frequently the estimated building *R* values differ significantly from those that would be expected from handbook calculations, according to Ref *4*. Are the differences due to the measurement procedure, or do they reflect actual performance? We can be reasonably confident that the measurements are realistic [*2*]. Where differences are found between expectation and measurement, the investigator wishes to determine the causes (faulty installation, convection within the insulation, moisture, and so on).

Although the technique employed has potential as an inspection tool for new construction or newly performed retrofit construction, the purpose of the present study was to determine the economic potential for investing in insulation improvements in Army buildings. This requires not only reasonably accurate measurements but also knowledge of the policy guiding investments, the heating costs at the installation, the climatic variables, and the potential retrofit options. Therefore, the specific economic conclusions obtained for buildings at the Army bases studied do not apply to Army buildings at different locations or to civilian buildings in the same locale. Both the measurement technique and the analysis method can be generalized, however.

Field Investigation

The investigation of wall, attic, and roof thermal performance took place at Ft. Carson, Colorado, from 14 to 25 March 1983 and at Ft. Richardson, Alaska, from 3 to 15 April 1983. The effort concentrated on family housing units, visitors' quarters, a barracks dining hall, and a temporary office structure. Typically, the quarters were unoccupied.

Climates and Building Types

Ft. Devens [3597 heating degree-days Celsius (6475 heating degree-days Fahrenheit)] offered a moderately cold climate in the northern tier of the contiguous United States and a mixture of frame buildings, brick veneer and frame buildings, and masonry buildings. This study expands the experience to Ft. Carson, Colorado, which has about the same number of heating degree-days [3541 Celsius (6373 Fahrenheit)] as Ft. Devens but has a more southerly, sunnier climate; more masonry was used in the construction at Ft. Carson. Ft. Richardson, Alaska, was also a subject of this study; it has almost twice the heating degree-days [5957 Celsius (10 722 Fahrenheit)] of the other two locations, less sun in the winter, and mostly frame construction.

Two basic building types were studied at Ft. Carson: frame and frame with brick veneer, both with and without insulation. Figure 1 shows examples of the buildings, and Table 1 summarizes the constructions and their expected *R* values. The brick on much of the housing at Ft. Carson provides some storage of solar energy.

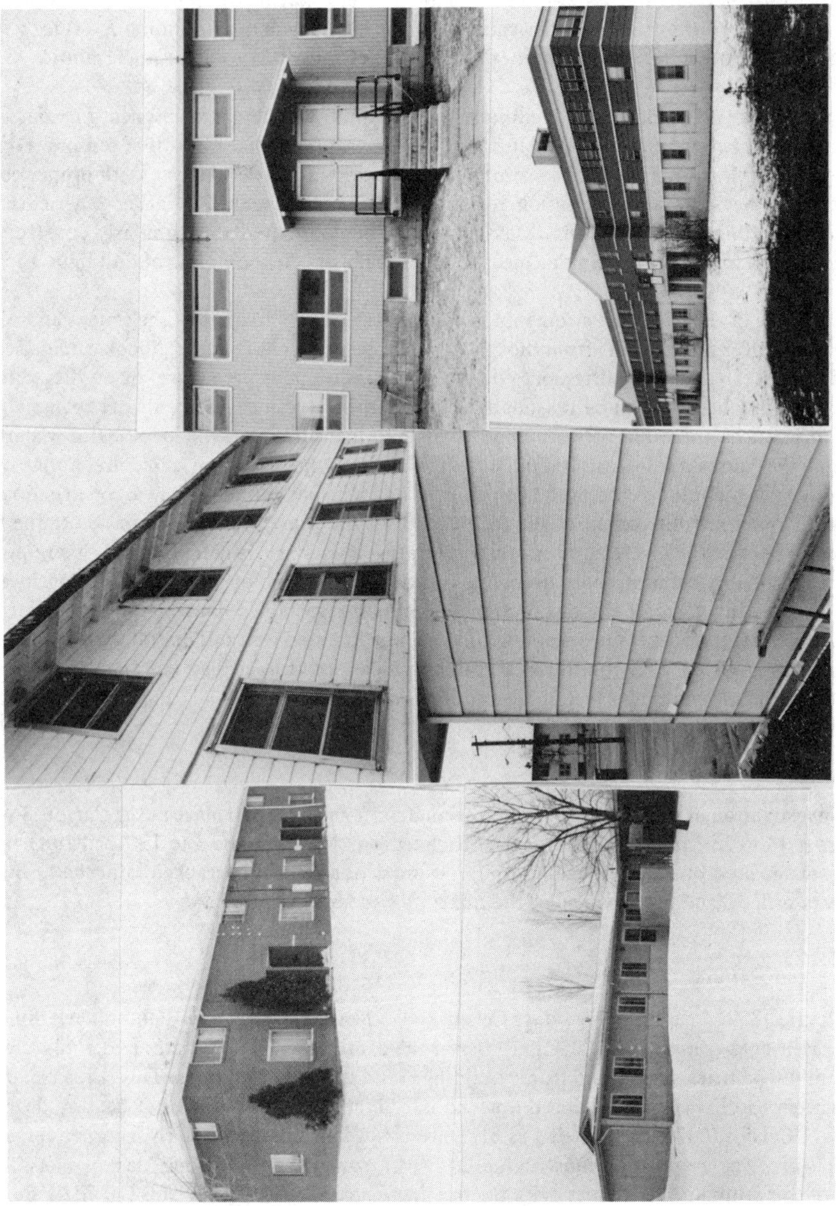

FIG. 1—*Typical construction of buildings studied at Ft. Carson and Ft. Richardson.*

Four types of construction were present at Ft. Richardson: 2-by-4 frame, 2-by-4 frame with exterior insulation and steel siding, 2-by-6 frame, and masonry with 50.8 mm (2-in.) rigid insulation between wythes of concrete masonry units (CMUs). Figure 1 shows examples of the building types studied; Table 2 summarizes their characteristics. All five buildings had nominally fully insulated walls.

Measurement Procedure

The measurement procedure entailed four steps: (1) establishing sensor sites with infrared thermography, (2) emplacing thermal sensors, (3) recording the sensor responses to temperature and heat flow changes in the construction, and (4) mapping the region of measurement thermographically. The last step was omitted if Step 1 located no anomalies. Predictable thermal discontinuities, such as studs, also did not warrant carrying out Step 4.

Establishing Sensor Sites—Infrared thermography determined the strategy for locating sensors. In frame walls there was seldom evidence of random thermal anomalies. Vertical studs and horizontal fire blocking at midheight were almost always evident. Often there was evidence of convection in the stud space (Fig. 2), visible as lighter (warmer) regions at the top of a space than at the bottom. Therefore, the sensors were placed along a vertical line midway between joists, typically three above the fire blocking and three below. This practice is consistent with the laboratory observations of previous researchers.

TABLE 1—*Summary of construction types of Ft. Carson, Colorado.*[a]

Building No.	Construction	Wall Insulation, in. (mm)	Wall/Attic *R* Values	Wall/Attic, RSI
4822	frame	1.5 (75)	12/29	2.1/5.2
340	frame	none	6.2/. . .	1.2/. . .
5635B	brick/frame	none	3.9/16	0.68/2.9
7273B	brick/frame	none	3.9/. . .	0.68/. . .
7305	brick/frame	3.5 (89)	14/. . .	2.4/. . .

[a] *R* values are calculated from test borings, as-built drawings, and handbook thermal properties [*4*]. *R* values are computed from interior to exterior surface, in $ft^2 \cdot h \cdot °F/Btu$ and $m^2 \cdot K/W$. Note that *R* values were calculated for only two building attics, one with an average of 216 mm (8.5 in.) of fiberglass insulation topped with an inch of cellulose, the other with 89 mm (3.5 in.) of fiberglass insulation topped with 51 mm (2 in.) of cellulose.

TABLE 2—*Summary of construction types at Ft. Richardson, Alaska.*[a]

Building No.	Construction	Ceiling Insulation, in. (mm)	Wall/Roof *R* Values	Wall/Roof, RSI
533F	2 by 4	attic 10 (254)	12/37	2.1/6.5
386C	2 by 4	attic 5.2 (132)	12/29	2.1/5.1
505E	2 by 4 and siding	no attic measured	12/. . .	2.1/. . .
1107-11	2 by 6 and siding	roof 7.5 (191)	19/9.4	3.4/1.7
600	insulated masonry	roof 2 (51)	10/11	1.8/1.9

[a] *R* values in customary and SI units are calculated from test borings, as-built drawings, and handbook thermal properties [*4*].

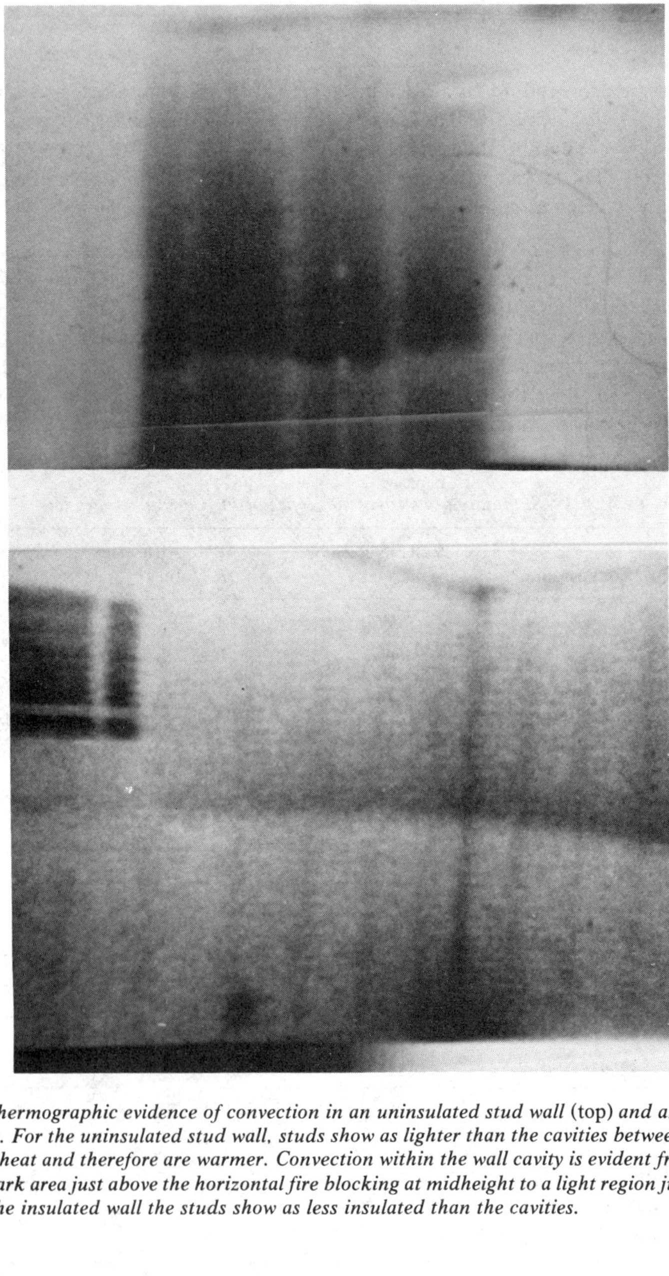

FIG. 2—*Thermographic evidence of convection in an uninsulated stud wall* (top) *and an insulated stud wall* (bottom). *For the uninsulated stud wall, studs show as lighter than the cavities between studs because they lose less heat and therefore are warmer. Convection within the wall cavity is evident from the progression from a dark area just above the horizontal fire blocking at midheight to a light region just below ceiling height. For the insulated wall the studs show as less insulated than the cavities.*

The objective is to characterize the thermal performance of all like-insulated surfaces from a small sample. Thermography revealed the consistency among stud bays. A vertical array of sensor sites on the northern and southern exposures of buildings, where possible, provided maximum potential variation in the thermal performance. Previous experience [3] with CMU construction demonstrated that there is little difference in heat flow between a site over a web in the block and a site over a core.

The investigation included four attic spaces and two low-slope roofs. In each case, thermography detected the location of rafters or joists, and the sensors were evenly spaced along a line between those structural members.

Emplacing Thermal Sensors—Each site hosted two temperature sensors (thermocouples, in this case), one each on the inside and outside surfaces, and a heat flow sensor (HFS) on the inside surface (Fig. 3). Indoor thermocouples were integral to the 2 by 20 by 30-mm HFSs.

The HFSs were stuck to the wall with a gel toothpaste and covered with masking tape. The outside thermocouple was also affixed with masking tape along 300 mm of the wire leading to

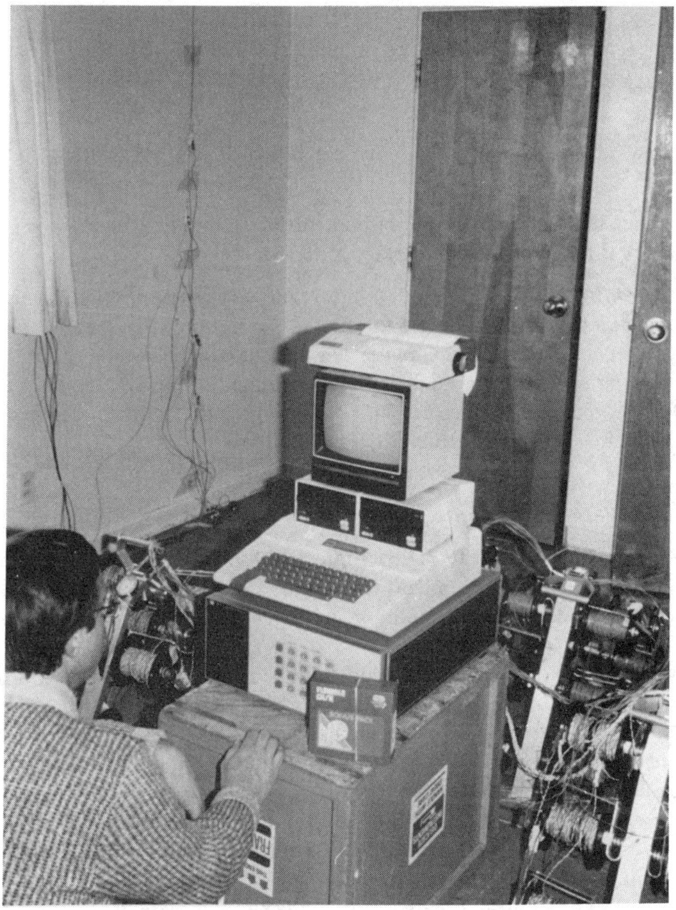

FIG. 3—*Sensors and data acquisition system. Heat flux sensors under tape with built-in thermocouples together with outdoor thermocouples at corresponding heights feed data to the computer for averaging and storing on floppy disk.*

the sensing end. Masking tape over the HFS serves two functions: it matches the absorptance of the sensor with its surroundings, and it fairs the HFS onto the wall smoothly, so it will not produce turbulent convective air flow over the sensor that is not also occurring adjacent to the HFS. Both factors help the HFS "feel" like the adjacent wall or ceiling.

Earlier experience [2] indicates that the HFS may affect the precision of the measurement with a standard deviation that is 10% of the mean value obtained by independent measurements.

Data Acquisition—Twenty sensor combinations fed data on heat flow and inside and outside surface temperatures to a data acquisition system (DAS), consisting of a personal computer with floppy disk storage. The computer controlled a multiplexing digital voltmeter with temperature compensation on the thermocouple channels. The DAS collected data from each sensor site every 40 s, averaged it, and wrote it to disk every 20 min to record a time series. At the same time the program calculated an estimate of R value according to Eq 1 and determined whether that estimate had remained stable for the previous 6 h.

$$R_e = \frac{\sum_{i=1}^{n} \Delta T_i}{\sum_{i=1}^{n} Q_i} \tag{1}$$

where

ΔT_i = difference between indoor and outdoor surface temperatures at the ith reading and
Q_i = heat flux through the inside surface at the ith reading.

Independent Confirmation—A 9.5-mm (3/8-in.) hole was drilled through the indoor surface under a sensor location at midheight in either the top or bottom half of each wall to give independent verification of the construction to the depth of the outside sheathing. Frame walls without insulation may draw a significant portion of their insulating value from the sheathing, especially if it is fibrous. Therefore, the drill hole has limitations. In attics, depth measurements with a wire probe offered independent verification.

Determining Heat Flux Sensor Conversion Factors

Conversion factors to translate HFS output into heat flux were obtained by simulating under known heat flow the thermal environment the sensors would find in the field: a gypsum wallboard surface, natural convection, similar mounting methods, similar surface temperature, and so forth.

Forty HFSs, associated with two data acquisition systems, were placed on a test wall in a calibrated hot box at the University of Massachusetts Thermal Measurements Laboratory. The specimen wall consisted of 51-mm (2-in.) beadboard polystyrene at a 29-kg/m^3 (1.8-lb/ft^3) density with 16-mm (5/8-in.) gypsum wallboard adhered to it. The experimental R value agreed well with handbook calculations [4].

The HFSs were arrayed on a grid according to a randomized plan for three separate test runs. The HFSs were mounted with gel toothpaste and covered with masking tape as in the field and exposed only to natural convection on the warm side of the testing chamber. The average R_e values from three runs and known R value of the wall specimen were the basis for establishing a factor for converting the sensor millivolt signal to a value for Q. The average standard deviation of the three runs was 6% of the mean R_e obtained from each sensor.

Field Study Results

The purpose of making thermal measurements in the field is to determine whether actual buildings perform significantly differently than expected. If they do, we want to understand what thermal mechanism is involved and whether the discrepancy is due to error in measurement.

Expectations Versus Measurement

At Ft. Carson or Ft. Richardson only one building was found to be significantly worse than expected, based on much more consistent independent verification of the wall construction. This was a 2-by-6 frame building with added steel siding. Boring confirmed that the insulation was inconsistently thick and did not touch both sides of the stud cavity. Measurements of the four brick veneer walls with empty stud bays gave at least 24% greater R_e than expected. This diminished the case for adding insulation to building walls. On one of these buildings the measurement took place upstairs; on the other buildings they were downstairs. The Appendix tabulates the percentage difference between R_e and calculated R by building number and by orientation.

Measured depths of attic insulation were the basis for calculating expected insulation value in four instances, two each at Ft. Carson and Ft. Richardson. In all cases the attic had fiberglass roll or batt insulation with an additional layer of poured or blown cellulose (Ft. Carson) or fiberglass (Ft. Richardson). Table 3 gives the details and results from measuring attic insulation. No conclusive evidence was apparent to explain why some measurements revealed R values close to expectation and some did not. Convection within the attic insulation is a possible explanation.

Convection in Walls

Convection can have a significant impact on a measurement employing heat flux sensors both on a wall and on a ceiling or roof. A wall is prone to convection currents within when there is incomplete or porous insulation and a substantial difference in temperature across the wall. There might be a continuous loop of rising and falling air, with a warm pocket at the top of the space and a cool pocket at the bottom, apparent in Fig. 2. Clearly, with heat moving within the plane of the wall, there must be enough appropriately placed HFSs to account for the average effect, as outlined under Measurement Procedure.

The measurement results bear out the expectation of convection in walls. The estimated R values (Fig. 4) of two uninsulated brick veneer buildings at Ft. Carson, normalized to the aver-

TABLE 3—*Average difference between estimated* R *value and expected* R *value as a percentage of expected* R *value (%D) for attics of housing units at Ft. Carson and Ft. Richardson.*

Building No.	Attic Insulation, mm (in.)	%D
	FT. CARSON	
5635B	89 (3.5) fiberglass batt with 51 (2) blown cellulose fiber	+04
4822	89 (3.5) fiberglass batt with 25 (1) blown cellulose fiber	−38
	FT. RICHARDSON	
533F	127 (5) fiberglass batt with 1137 (5.4) blown fiberglass	−01
386C	89 (3.5) fiberglass batt with 132 (5.2) blown fiberglass	−37

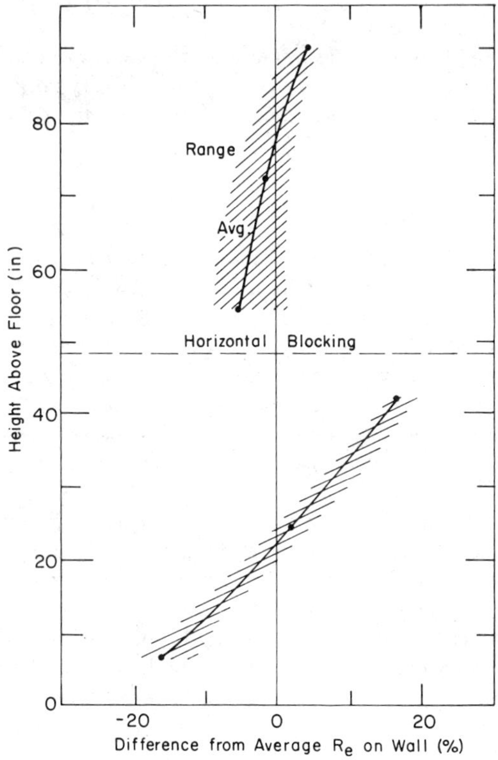

FIG. 4—*Percentage difference of* R_e *from the average* R_e *for the wall for two uninsulated brick veneer frame buildings at Ft. Carson, Colorado, at different vertical locations on the wall.*

age for a vertical run of measurements, show a larger *R* value at the top than at the bottom of each half of the stud space divided by horizontal fire blocking. The average *R* value for each half is approximately the same, but the influence of convection appears much stronger in the lower half. Contrast this with Fig. 5, which depicts normalized *R* values up an insulated 2-by-4 frame wall. In this case, conduction near the plate, fire blocking, and shoe may be the dominant cause for the variation. In an insulated 2-by-4 (with brick veneer or siding) or an insulated 2-by-6 wall, an area with framing underneath would be expected to have a 50 to 65% smaller *R* value than one with fibrous insulation. For an uninsulated 2-by-4 frame wall with or without a brick veneer, the framing would improve the *R* value by 50 to 90% over the value for an air space alone.

Insulated frame buildings at Ft. Richardson showed evidence of variation in measured *R* value attributable to both convection and conduction through nearby framing (Fig. 6). In both the 2-by-4 and 2-by-6 constructions, the variation was greater than 30%. In these cases the upper half of the wall showed the greatest variation attributable to convection.

Convection in Attics

Another convection process can occur in attic insulation, which is heated from below and cooled from above, if there are sufficient defects to permit vertical and lateral air flow. Figure 7 is a thermograph of the ceiling beneath the attic represented in the upper left example of Fig. 8. There is strong evidence of intrusion of cold air beneath the attic insulation. Natural convection is a probable cause.

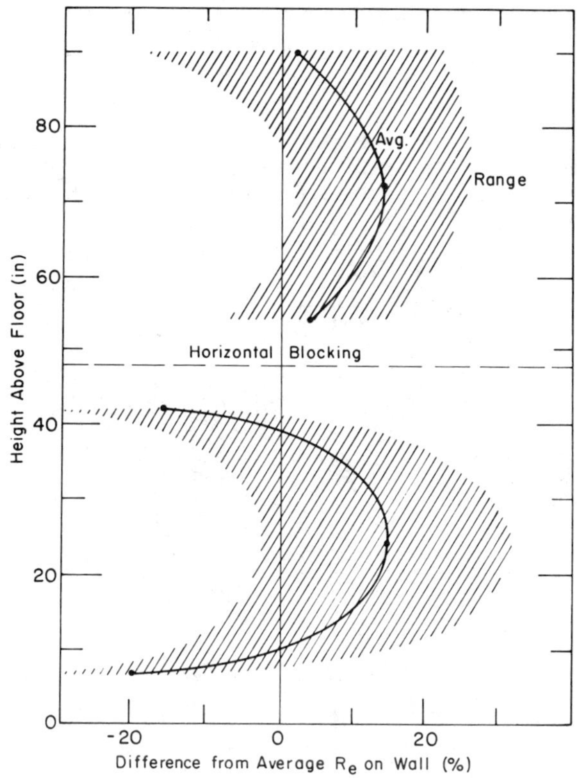

FIG. 5—*Percentage difference of* R_e *from the average* R_e *for the wall for a fully insulated 2-by-4 frame building at Ft. Carson at different vertical locations on the wall.*

Convection flow is not as likely to organize within attic insulation as it is through defects in the insulation. However, the work of Besant and Miller [5] indicates that convection can be significant when attic insulation is thick or temperature differences are large.[2]

Convection in attic insulation also requires an average reading from several sensors to formulate an average R_e.

If we employ the measured depths of attic insulation to determine an expected R value at each point and then compare the percentage difference (%D) between measurement and expectation, we see variation that does not appear to be related to the depth of insulation (Fig. 8). The sensors were 203-mm (8-in.) apart; %D appears to vary on about double that scale in three of the four attics looked at. In a different three of the four, the %D is lower toward the outside wall than toward the center of the attic. This suggests that cool air from the roof may be invading the insulation at the perimeter of the building. In one building (top right in Fig. 8) the temperature sensors were underneath a sheet of cardboard that covered much of the width of the insulation between trusses. This would have had a substantial effect on quelling convection that penetrates the top of the insulation.

[2]Public Works, Canada, has an unpublished thermogram of the underside of a thickly insulated ceiling at a large difference between inside and outside temperature that shows strong evidence of polygonal Benard convection cells within the insulation, so we cannot rule out the possibility of significant convection within insulation. If convection creates sideways heat flux paths between inside and outside sensors, then the measurement at any given location will not properly account for heat flux through the ceiling.

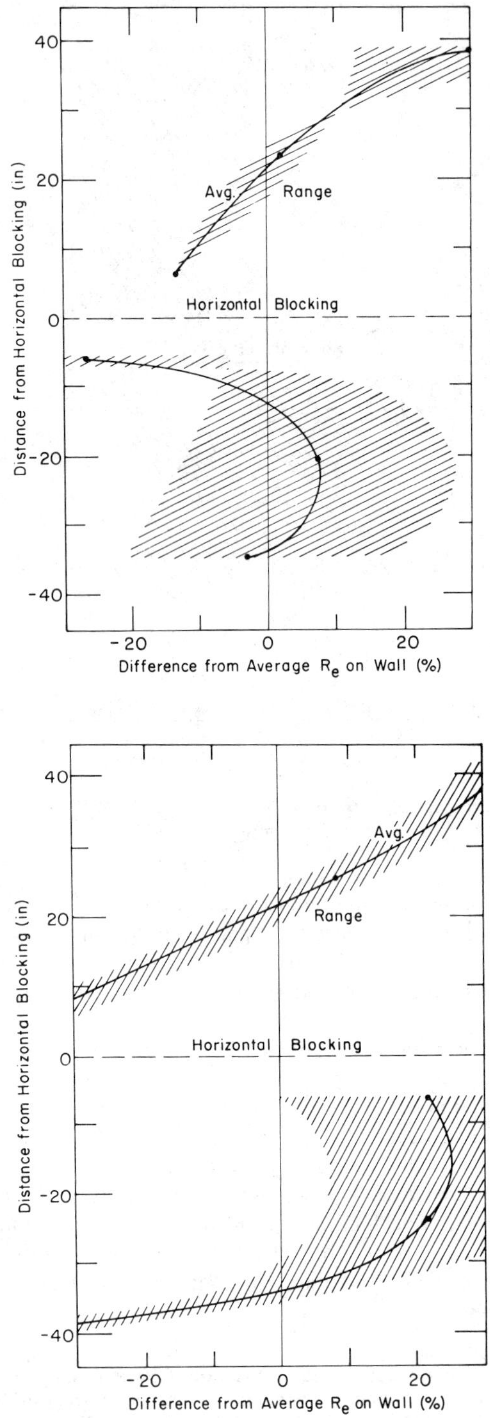

FIG. 6—*Percentage difference of* R_e *from the average* R_e *for three frame buildings at Ft. Richardson at different vertical locations on the wall:* (a) *2 by 4.* (b) *2 by 6 with siding and exterior insulation.*

FIG. 7—*Thermogram of ceiling beneath insulated attic of Ft. Carson Building 4822A.*

This evidence from measuring attics suggests possible patterns for the occurrence of convection but also contradicts each possibility at least once:

1. Convection within the full attic space may influence *R* value measurements with lower values toward the perimeter of the attic, except for that shown in the top left of Fig. 8.
2. Convection cells are about (16 in.) across, except for that shown in the bottom right of Fig. 8.

Possible Inaccuracy of Estimate Due to Air Leakage in Walls

Previous work [*2*] discussed the accuracy limitations of employing surface-mounted heat flow sensors (HFSs). It did not address fully the problems of using the technique described here to find *R* values of walls, ceilings, and insulated roofs with HFSs and thermocouples. It mentioned convection within walls but not air leakage through walls nor convection within attics or roofs.

Air leakage through walls does not show up statistically as a problem with measurements made at Ft. Devens, Ft. Carson, or Ft. Richardson. We could expect leakage of outdoor air into a wall downstairs and indoor air out through a wall upstairs in a two-story building because of the stack effect. This would cool down the interior of the downstairs wall, cause higher heat flux through the inside wall surface, and result in a lower estimated *R* value from heat flux and temperature measurements. Upstairs warm indoor air invading the wall would have the opposite effect—a higher estimated *R* value than expected. However, a compilation of differences between estimated and expected *R* values reveals just as many occurrences of higher-than-expected *R* values as lower-than-expected *R* values whether we consider an upstairs (four higher versus five lower), downstairs (three versus three), or single-story (two versus two) building wall.

Summary of Field Measurements

These results demonstrate that average estimates of *R* values from measurements and handbook calculations can differ significantly. The variation among *R* values at different locations

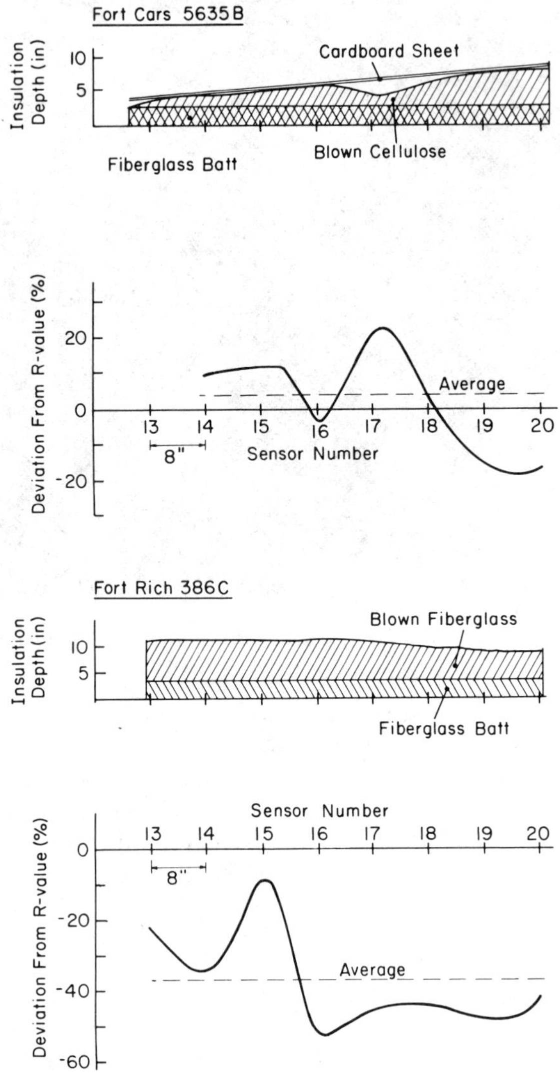

FIG. 8—*Attic insulation depth and percentage difference between expected* R *value and measured* R *value as a function of location across the attic.*

underscores the need to take several measurements to obtain a valid average estimate of *R* values both for walls and for attics.

Potential for Energy Conservation Investment

The Department of Defense (DOD) Energy Conservation Investment Program (ECIP) funds improvements of existing DOD buildings and energy systems to improve their efficiency and economy.

Life-cycle costing is the core principle of the program. Projected future savings for the programmed lifetime of the project are discounted against present construction costs to determine

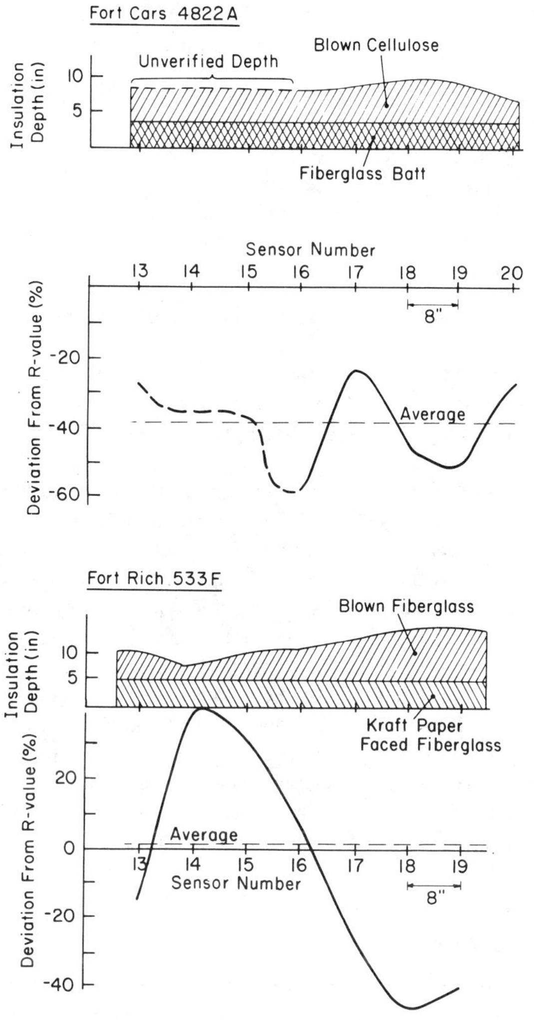

FIG. 8—*Continued.*

whether the savings-to-investment ratio (SIR) is greater than one with a discount rate of 7% and project life of 15 years.

ECIP criteria include present worth factors for a variety of fuels in a variety of regions to account for their anticipated rises in costs at a rate faster than general inflation.

Generic Investment Curves

ECIP accepts the heating degree-day (HDD) method of estimating annual energy consumption, despite its imperfections with respect to heat storage effects and to internal and solar gains. With this assumption, it is easy to develop a climate-heating cost parameter (CHC) specific to each facility, requiring only a description of the change in U value or other HDD-dependent variable to determine the change in energy costs over the project lifetime in terms of present worth

$$CHC = \frac{24(HDD)(PWF)(FC)}{(EFF)} \qquad (2)$$

where

24 = number of hours per day,
HDD = heating degree-days, Fahrenheit or Celsius,
PWF = present worth factor for fuel used (dimensionless),
FC = fuel cost, dollars/Btu or dollars/W · h, and
EFF = efficiency of energy delivery (fraction).

Table 4 outlines the values assumed for computing the CHC parameter for Ft. Carson and Ft. Richardson.

Ft. Richardson has a central plant that distributes heat through steam lines to the buildings on post. Accepted estimates of the efficiency of this system indicate that 70% of the energy leaving the plant makes it to the heated space in the building. This study assumes plant efficiency of 80% for both Ft. Carson and Ft. Richardson.

The CHC determines the economic return for each increment of additional insulation based on each starting level of insulation (Fig. 9). The steeper lines from the same starting point reflect the higher worth of each increment of investment at Ft. Richardson than at Ft. Carson. The curves demonstrate that the lower the initial insulating value, the greater the acceptable construction budget for improving the R value. (Actually, the curves are all the same shape but are shifted up and down to intersect the horizontal axis at the appropriate point.)

The curves demonstrate that the first increments of improvement are the most cost-effective. In fact, when a building insulated against the cold reaches the higher R values shown, it becomes "superinsulated." This means that indoor heat gain and solar gain become more significant in offsetting the heat lost through the building exterior. Consequently, the temperature at which it is necessary to turn on the heating plant is lower than that assumed in standard heating degree-day data. This would cause the incremental value of increasing insulation to be higher than depicted, but this effect is so specific to individual buildings that the analysis does not include it.

Investment Potential at Ft. Carson

The uninsulated frame buildings with brick veneer, Buildings 5635A and 7273B, had the greatest ECIP potential at Ft. Carson. The measurements indicated that these overperformed the expected insulation value by 25 and 37%. Nevertheless, Fig. 10 demonstrates that if the stud cavity were filled with cellulose to bring the overall insulating level up to R-14, this would justify a budget somewhere between \$16.1 and \$18.8 per square metre (\$1.50 and \$1.75 per square foot) of wall. A Colorado Springs contractor said that cellulose insulation blown from the in-

TABLE 4—*Parameters to determine CHC according to Eq 2 for Ft. Carson and Ft. Richardson.*[a]

Fort	Fuel	HDD, °C-day (°F-day)	PWF	Fuel Cost, $\$/10^6$W · h ($\$/10^6$ Btu)	Efficiency	CHC, \$/°C/W (\$/h · °F/Btu)
Carson	natural gas	3541 (6373)	12.46	18.7 (5.48)	0.80	29.64 (13.00)
Richardson	natural gas	5957 (10 722)	10.84	13.8 (4.03)	0.56	37.91 (20.00)

[a]The data were obtained from the energy office at each base and from Refs *6* and *7*.

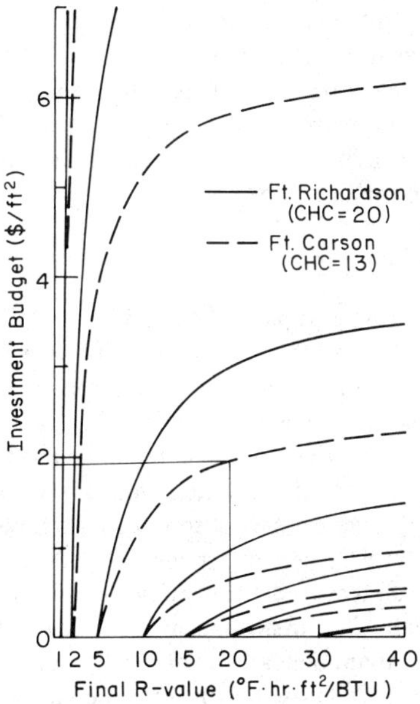

FIG. 9—*Generic insulation investment budget curves for Ft. Richardson and Ft. Carson. If the construction is initially* R-5 *and is improved to* R-20 *at Ft. Carson, we follow the vertical line from 20 on the horizontal axis up to the curve that originates at* R = 5 *and then horizontally to the vertical axis to determine that such an improvement would have to cost less than $20.5 per square metre ($1.90 per square foot) to be economically justified.*

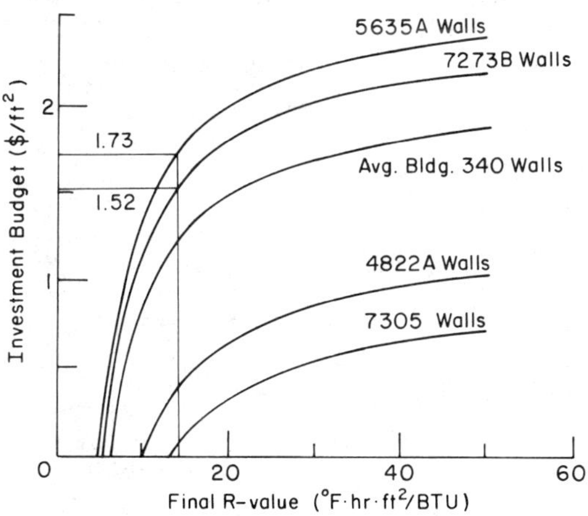

FIG. 10—*Investment potential for insulating 2-by-4 walls at Ft. Carson. The two uppermost curves represent uninsulated frame brick veneer buildings.*

doors into a 89-mm (3.5-in.) wall cavity would cost about \$4.5 per square metre (\$0.42 per square foot) in 1984. This offers an SIR of about four.

Attics at Ft. Carson have already been retrofitted. It is possible that a contractor could add another *R*-10 for [\$3.2 per square metre (\$0.30 per square foot)] (Fig. 10). However, this would have been easiest when the current insulation was added. This illustrates the necessity of adding the right amount of insulation the first time.

Investment Potential at Ft. Richardson

Ft. Richardson's climate-heating cost parameter of \$37.9 °C/W (\$20.0/h · °F/Btu) suggests greater investment prospects than at Ft. Carson, which has a CHC of \$24.6 °C/W (\$13.0/h · °F/Btu). However, all the buildings on post currently have insulation in the walls, which allows the possibility of having an inexpensive means of installation. Consequently, Building 533 (shown in Fig. 11) would warrant about the same budget as the empty frame/brick veneer buildings at Ft. Carson, but additional insulation would have to be applied to the inside or outside of the existing wall. If this were a part of some other remodeling effort, an additional *R*-5 would warrant spending about \$3.2 per square metre (\$0.30 per square foot) through ECIP.

Building 1107 (Fig. 11) had a measured performance much worse than expected for a nominally insulated 2-by-6 frame wall. The outside has metal siding with a reported 12.7-mm (1/2-in.) layer of beadboard polystyrene foam beneath. It has the potential to become about *R*-20 from insulation within the wall cavity. Cellulose blown from the inside would be justified at costs up to \$10.8 per square metre (\$1.00 per square foot).

At most Army installations in the 48 contiguous states, masonry buildings are among the poorest insulated. At Ft. Richardson the measurement of Building 600 indicated much better insulation than expected. This is contrary to the expectation that high room temperatures may have decreased the thermal resistance of the insulation. Since the material was sandwiched between two layers of masonry, verification of its thickness was impractical.

Attics at Ft. Richardson were insulated with a measured *R*-20 to *R*-30 of batt and loose fiberglass insulation.

Conclusions

The data suggest that sampling a few representative buildings of a given type will adequately represent the thermal performance of the entire class of similar buildings. This makes it unnecessary to measure each building and develop individual assessments.

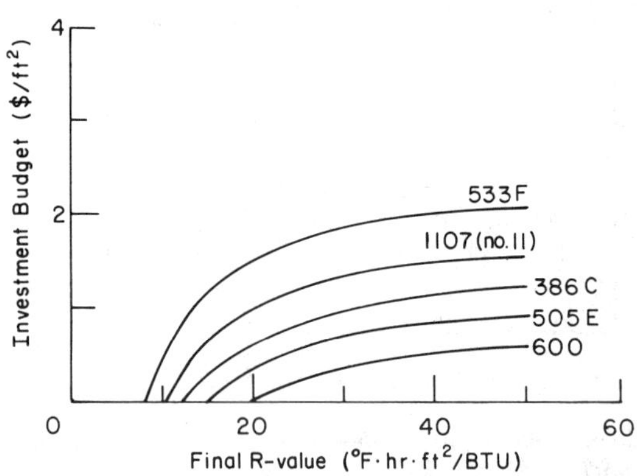

FIG. 11—*Investment potential for insulating walls at Ft. Richardson.*

Independent verification of building wall construction agrees fairly well with the measured results. In cases with discrepancies, faulty installation of insulation, unverified sheathing in low-R-value walls, and unverified insulation thickness in masonry walls are among the plausible explanations for differences between measurements and expectation in walls.

In attics, the reason for discrepancies between expectation and measurement is not evident in the four instances surveyed. Figure 7 indicates that natural convection within the attic and its insulation may be a factor.

Convection was an apparent mechanism of heat flow in both insulated and uninsulated buildings in the study. The literature [*6*] suggests that fibrous insulation is not likely to support convection within the body of the insulation. However, if the insulation does not press completely against the warm side of the cavity, it can promote convection within walls. The survey uncovered some instances in which the wall R value was much worse than predicted, because of compressed or dislocated insulation.

Experience at Ft. Devens [*3*] had indicated that uninsulated frame buildings performed better than the calculations performed according to the *ASHRAE Handbook—1981 Fundamentals* [*4*] would indicate. This was also true of the uninsulated brick veneer frame buildings at Ft. Carson. Possibly the unverified sheathing material was a significantly better insulator than assumed, or perhaps the cavity was more capable of slowing convection currents than assumed.

One goal of making measurements is to find a worse performance than handbook assumptions would indicate for uninsulated buildings to improve drastically the economic justification for adding insulation. This was not the case for any of the uninsulated or masonry buildings at Ft. Carson or Ft. Richardson that were surveyed. However, the survey demonstrated the economic potential for reinsulating a large class of buildings at Ft. Carson and demonstrated that fibrous insulation performance is very much dependent on the quality of installation.

APPENDIX

Percentage Differences in Wall *R* Value Between Those Estimated by Field Measurement and Those Calculated Based on Information from Drilled Holes and As-Built Drawings

Building Number and Type	Percentage Difference (Number)				
	North	East	South	West	Average
		FT. CARSON WALLS			
340 Frame uninsulated	−5 (20)	. . . (. . .)	. . . (. . .)	. . . (. . .)	−5 (20)
4822A Frame insulated	+38[a] (6)	. . . (. . .)	+8 (6)	. . . (. . .)	+23 (12)
7305 Brick veneer insulated	−5 (8)	−4 (12)	. . . (. . .)	. . . (. . .)	−4 (20)
7273 Brick veneer uninsulated	. . . (. . .)	+32 (6)	+39 (14)	. . . (. . .)	+37 (20)
5635 Brick veneer uninsulated	+24 (6)	. . . (. . .)	+30 (6)	. . . (. . .)	+27 (12)
Total	+6 (40)	+8 (18)	+30 (26)	. . . (. . .)	+14 (84)

Building Number and Type	Percentage Difference (Number)				
	North	East	South	West	Average
			FT. RICHARDSON WALLS		
1107 Frame 2 by 6	. . . (. . .)	. . . (. . .)	. . . (. . .)	−48 (12)	−48 (12)
505[b] Frame 2 by 4	−1 (8)	. . . (. . .)	+15 (12)	. . . (. . .)	+7 (20)
386 Frame 2 by 4	. . . (. . .)	+7 (6)	. . . (. . .)	−5 (6)	+1 (12)
533[c] Frame 2 by 4	. . . (. . .)	−5 (6)	. . . (. . .)	+2 (6)	−3 (12)
600[d] Masonry insulated	. . . (. . .)	. . . (. . .)	. . . (. . .)	+76 (12)	+76 (12)
Total	−1 (8)	+1 (12)	+15 (12)	+9 (36)	+7 (68)

[a]Inconsistent insulation thickness or placement.

[b]Has 12.7 mm (0.5-in.) expanded polystyrene insulation over the sheathing with steel siding.

[c]Contains 38.1 mm (1.5-in.) fiberglass batt.

[d]Has 50.8 mm (2 in.) of unspecified "rigid insulation" between two wythes of 203 mm (8-in.) concrete block, according to as-built drawings.

References

[1] Flanders, S. N. and Marshall, S. J., "Toward In-Situ Building *R* Value Measurements," CRREL Report 84-1, U.S. Army Cold Regions Research and Engineering Laboratory, Hanover, NH, 1984.

[2] Flanders, S. N., "Heat Flow Sensors on Walls—What Can We Learn?" *Building Applications of Heat Flux Transducers, ASTM STP 885,* American Society for Testing and Materials, Philadelphia, 1985.

[3] Flanders, S. N., "Measuring Thermal Performance of Building Envelopes—Nine Case Studies," CRREL Report 85-7, U.S. Army Cold Regions Research and Engineering Laboratory, Hanover, NH, March 1985.

[4] *ASHRAE Handbook—1981 Fundamentals,* American Society of Heating, Refrigerating, and Air Conditioning Engineers, Atlanta, GA, 1981.

[5] Besant, R. W. and Miller, E., "Thermal Resistance of Loose-Fill Fiberglass Insulation in Spaces Heated from Below," in *Thermal Performance of the Exterior Envelopes of Buildings II,* B. S. Clevenger, Ed., American Society of Heating, Refrigerating, and Air Conditioning Engineers, Atlanta, GA, 1983.

[6] Holman, J. P., *Heat Transfer,* McGraw-Hill, New York, 1976.

Materials

Thomas G. Matthews[1] *and Rita R. Westley*[2]

Formaldehyde Emissions from Selected Fibrous Glass Insulation Products

REFERENCE: Matthews, T. G. and Westley, R. R., **"Formaldehyde Emissions from Selected Fibrous Glass Insulation Products,"** *Thermal Insulation: Materials and Systems, ASTM STP 922,* F. J. Powell and S. L. Matthews, Eds., American Society for Testing and Materials, Philadelphia, 1987, pp. 223–237.

ABSTRACT: The formaldehyde (CH_2O) emission potential of fibrous glass insulation acquired from the three largest manufacturers in the United States has been investigated. Screening measurements of unfaced and kraft-faced R-19, foil-faced R-11, and 25.4-mm (1-in.) unfaced insulation made with a formaldehyde surface emission monitor (FSEM) demonstrate low interproduct and intermanufacturer variation in CH_2O emission rates. The FSEM screening data and environmental chamber measurements of unfaced R-19 insulation taken at 23°C and 50% relative humidity both indicate CH_2O emission rates of 0.02 ± 0.01 mg/m^2 h after one to three months of product conditioning in atmospheres of low CH_2O concentration. Threefold to fourfold increases in CH_2O emissions were measured in environmental chamber tests taken at 38°C and 68% relative humidity, simulating potential seasonal exposures in attic and wall cavities adjacent to indoor compartments. However, a two-compartment, diffusion model describing the potential impact of attic insulation at 38°C and 68% relative humidity on indoor CH_2O concentrations indicates a maximum increment of 0.03 mg/m^3 (0.02 ppm).

KEY WORDS: formaldehyde, fibrous glass insulation, formaldehyde surface emission monitor, environmental chamber, indoor concentrations, thermal insulation

Formaldehyde (CH_2O)-based resins are incorporated in a wide variety of consumer and construction materials. Generic product lines include pressed-wood products, urea-formaldehyde foam insulation (UFFI), fibrous glass insulation, ceiling panels, and textiles [*1*]. The CH_2O-based resins incorporated in these products slowly degrade in the presence of water vapor at normal indoor temperatures to release CH_2O vapor. The potential impact of these emissions on indoor air quality (IAQ) has been the subject of increased research, especially with the current emphasis on energy-efficient homes with reduced air exchange rates [*2*]. Pressed-wood products containing urea-formaldehyde resins and UFFI are generally expected to have the strongest impact on indoor CH_2O concentrations [*3*]. Recent surveys of particleboard, hardwood plywood paneling, and medium-density fiberboard products, and UFFI encased in simulated wall panels have shown highly variable CH_2O emission rates ranging from 0.1 to 0.8, 0.03 to 0.6, 0.6 to 2.3, and 0.05 to 0.8 mg/m^2 h, respectively. The impact of formaldehyde-emitting pressed-wood and insulation products on indoor air quality in domestic environments is enhanced because of their widespread use in homes and frequent exposure to indoor air through ineffective CH_2O permeation barriers such as gypsum board or carpeting. The strong temperature and humidity dependence of CH_2O emissions from these products [*4*] is an additional complicating factor, particularly for insulation products exposed to harsh environmental conditions in wall and attic cavities during summer periods.

[1]Research scientist, Measurement Applications Group, Health and Safety Research Division, Oak Ridge National Laboratory, Oak Ridge, TN 37831.

[2]Graduate student, University of Tennessee, Knoxville, TN.

Fibrous glass insulation products have received little attention in comparison with UFFI and urea-formaldehyde resin-containing pressed-wood products. This is because fibrous glass products incorporate predominantly phenol-formaldehyde resins, which are more stable under normal environmental exposures and result in typically much weaker CH_2O emissions [*3*]. An extensive literature search yielded very limited CH_2O emissions data for fibrous glass products. Pickrell et al report CH_2O emission rates of 0.02 mg/m^2 h from a static desiccator test at approximately 24°C and >90% relative humidity [5] and 0.13 mg/m^2 h from a dynamic chamber test at 25°C and 90% relative humidity [*6*]. Humidity cabinet tests have been performed by Prusaczyk to measure CH_2O emissions from freshly manufactured, unfaced R-19 insulation at 38°C and 90% relative humidity [*7*]; the CH_2O emission rates decayed from about 0.1 to 0.01 mg/m^2 h during the first 1.5 months of testing.

To quantify better the CH_2O emission rates of fibrous glass insulation products and their potential impact on indoor CH_2O concentrations, the U.S. Consumer Product Safety Commission (CPSC) sponsored a survey of products manufactured in the United States. In the spring of 1983, representatives of CPSC selected a total of nine products from three of the largest manufacturers in the United States and sent these samples directly to Oak Ridge National Laboratory for testing. Screening measurements were performed using a formaldehyde surface emission monitor (FSEM) [*8*] to evaluate the interproduct and intermanufacturer variation and the temporal decay in CH_2O emission rates. Environmental chamber tests were performed to quantitate the CH_2O emission rates of selected samples with the strongest emissions under the potential temperature and relative humidity conditions of attic and wall cavities adjacent to indoor compartments. The potential impact of fibrous glass insulation on indoor CH_2O concentrations was modeled taking into account the influence of environmental parameters and the presence of CH_2O permeation barriers in conventional home construction.

Screening Measurements Using the FSEM

The FSEM has been used to measure CH_2O emission rates from a variety of pressed-wood products and urea-formaldehyde foam insulations [*8*]. An approximate one-to-one correlation has been observed between the measured CH_2O emission rates from these products using the FSEM and environmental chamber tests. Since an analogous intermethod correlation has not been established for fibrous glass insulation, the FSEM was used only for intersample and intermanufacturer comparative measurements in this study.

The screening measurements consisted of three phases: specimen selection, specimen conditioning, and FSEM measurements. Specimens of fibrous glass insulation were cut from random interior positions on each roll of product. A minimum of two specimens was cut from each of three rolls received from each of three manufacturers (that is, a total of ≥18 test specimens). All of the specimens were conditioned at 24 ± 2°C, 50 ± 15% relative humidity, and <0.06 mg/m^3 (<0.05 ppm) CH_2O. Uniform product conditioning before and between measurements was used to minimize the variation in CH_2O emissions due to product storage and handling. The low CH_2O concentration ensured that potentially low-CH_2O-emitting materials would sorb minimal quantities of CH_2O from the conditioning atmosphere prior to FSEM testing. FSEM measurements were taken after conditioning periods ranging from 1 to 100 days. Only the fibrous glass side of specimens with "facing" was measured using the FSEM. Preliminary FSEM measurements of the kraft paper and foil facing yielded CH_2O emission results comparable to those of the analytical blank. Near-zero CH_2O transport rates are expected directly through these facing materials in FSEM measurements (see the discussion of permeation barriers, later in this paper).

The FSEM consists of a 20-cm inside-diameter mechanical sieve, a top cover, and a 3-cm-wide circular flange attached concentrically to the bottom of the mechanical sieve [*8*]. Formaldehyde emissions (sampled selectively from the test surface beneath the monitor) are sorbed by 10 g of × 13 molecular sieve, supported in a uniform planar distribution on the mesh screen inside the FSEM. Sorbed CH_2O is analyzed using a water rinse desorption and pararosaniline colorimetric analysis procedure. The FSEM protocol for the measurement of fibrous glass insulation was slightly modified from the published protocol for pressed-wood products [*9*]. The typical 2-h sampling period was increased to 3 h to achieve a lower detection limit of about 0.015 mg/m^2 h. A 4-kg container of water was placed on top of each FSEM to seal the monitor on the test specimen. With fibrous glass insulation, this was necessary to compress the specimen to minimize air movement through the interface between the FSEM flange and the insulation under the monitor.

The results of the FSEM screening measurements of the fibrous glass insulation are shown in Table 1. For each product sample (that is, Samples A, B, and C in Table 1) the average CH_2O emission rate is listed for FSEM measurements of typically two test specimens cut from the product sample. The number of FSEM measurements and the conditioning periods before the measurement (that is, 0 to 10, 11 to 30, and 31 to 100 days) are specified for each product sample. An approximate twofold decline in the average CH_2O emission rate from about 0.04 ± 0.02 to 0.02 ± 0.01 mg/m^2 h is observed after 0 to 10 and 31 to 100 days of conditioning, respectively. There is no explanation for the small increase in measured CH_2O emission rates from FI2 specimens between the 0 to 10 and the 10 to 30-day conditioning periods. The intersample and intermanufacturer variation in CH_2O emission rates are low. For data taken after 31 to 100 days of product conditioning, the average standard deviation, σ, for intersample variation and σ for intermanufacturer variation are 0.003 and 0.006 mg/m^2 h, respectively. These variations include both inherent variation in product emission rates and measurement uncertainty. However, a minimum σ of about 0.005 mg/m^2 h, anticipated for the FSEM using a 3-h sampling period, may account for most of the observed variation in CH_2O emission rates.

Environmental Chamber Testing

The experimental procedures and test apparatus used for small-scale environmental chamber tests of the CH_2O emission rates of fibrous glass insulation were patterned after those used for extensive studies of pressed-wood products [*4*]. The 0.206-m^3 test chamber was Teflon lined to minimize losses of CH_2O on the chamber walls. This chamber design with a Teflon lining had previously been used in the generation of CH_2O atmospheres of as low as 0.02 mg/m^3 (0.015 ppm) [*10*] with negligible losses of CH_2O on the chamber walls [*4*]. Temperature control (that is, ±0.2 to 0.4°C) was achieved by placing the test chamber inside a Hotpack temperature-humidity chamber. The relative humidity was adjusted using a temperature-controlled water bubbler in the air input line to the chamber and monitored inside the chamber using a hygrometer, calibrated to ±1 to 3% relative humidity against wet, dry bulb measurement devices. The air exchange rate was controlled using Matheson mass-flow valves calibrated to ±1 to 2% against a wet test meter. The CH_2O concentration inside the test chamber was measured by sampling air from an exhaust port with a CEA Instruments Inc. Model 555 analyzer, having a 3 σ detection limit of approximately 0.01 mg/m^3 [*10*].

Test specimens of fibrous glass insulation were cut from random interior positions of the insulation rolls that were received from the manufacturers. The specimens were conditioned for a minimum of one to two months at 24 ± 2°C, 50 ± 15% relative humidity and <0.06 mg/m^3 (0.05 ppm) CH_2O to minimize potential CH_2O sorption by the insulation. (Conditioning at low CH_2O concentrations is an important precaution in the study of weak CH_2O emitters.) An additional two to three-day conditioning period at the precise temperature and relative humidity conditions of the environmental chamber tests was subsequently performed. During chamber measurements, a single 34 by 47-cm test specimen was held inside a rectangular wooden box

TABLE 1—*FSEM screening data summary for fibrous glass insulation (FI).*

			Average CH_2O Emission Rate, mg/m^2 h, as a Function of Conditioning Period					
Manufacturer	Product	Sample	0 to 10 Days	(Number of Measurements)	11 to 30 Days	(Number of Measurements)	31 to 100 Days	(Number of Measurements)
FI1	R-19	A	...	...	0.029 ± 0.005	(2)	0.020 ± 0.005	(12)
	unfaced	B	...	...	0.039 ± 0.005[a]	(1)	0.018 ± 0.005	(12)
		C	...	...	0.034 ± 0.005	(2)	0.019 ± 0.006	(11)
manufacturer average			...	...	0.033 ± 0.005	(5)	0.019 ± 0.005	(35)
FI2	R-19	A	0.030 ± 0.013	(3)	0.049 ± 0.011	(2)	0.028 ± 0.011	(12)
	unfaced	B	0.034 ± 0.013	(2)	0.048 ± 0.005	(2)	0.021 ± 0.006	(11)
		C	0.053 ± 0.016	(2)	0.042 ± 0.005	(2)	0.024 ± 0.009	(11)
manufacturer average			0.038 ± 0.015	(7)	0.046 ± 0.006	(6)	0.025 ± 0.009	(34)
FI3	1-in. unfaced	...	0.026 ± 0.011	(3)	...		0.011 ± 0.005	(12)
	R-19 kraft[b]	...	0.043 ± 0.015	(5)	...		0.017 ± 0.005	(8)
	R-11 foil[b]	...	0.030 ± 0.008	(4)	...		0.013 ± 0.005	(8)
manufacturer average			0.034 ± 0.013	(12)	...		0.014 ± 0.005	(28)

[a]A minimum σ of 0.005 mg/m^2 h is anticipated for 3-h FSEM measurements.
[b]FSEM measurements were performed only on the fibrous glass side of the insulation.

with open sides to reduce CH_2O emissions from the edges of the 16-cm-thick specimen (see Fig. 1). This was done to simulate the studwork in attics and walls of homes. The box was completely covered with Teflon to minimize the sorption of CH_2O on the walls of the holder.

The environmental chamber measurements were limited to two specimens of unfaced fibrous glass insulation at two different temperature and relative humidity conditions because of resource limitations. A temperature of 23°C and relative humidity of 50% was selected for one test condition because of its common use as a standard test condition for CH_2O emissions from consumer and construction products [3]. As a consequence, the fibrous glass emissions data taken in these chamber tests could be compared with those of other CH_2O resin-containing products. In addition, these chamber data could be qualitatively compared with the FSEM results for fibrous glass insulation that were taken at similar temperature and relative humidity conditions. Environmental chamber tests were also performed at 38°C and 68% relative humidity to simulate the potentially harsh environmental conditions that can occur in walls and attics adjacent to indoor compartments, particularly during warmer seasons. This specific temperature and relative humidity condition was used in prior studies of the environmental dependence of CH_2O emissions from pressed-wood products [4]. Thus, the measured variation in CH_2O emissions from fibrous glass insulation at conditions between 23°C and 50% relative humidity, and 38°C and 68% relative humidity, could be compared directly with the experimental results of the more extensive pressed-wood product studies. At both temperature and relative humidity test conditions, each fibrous glass insulation specimen was tested typically at three different air flow rates through the environmental chamber. This study design was chosen to vary the CH_2O vapor concentration generated by the insulation inside the test chamber. The CH_2O vapor concentration dependence of the CH_2O emission rate from the insulation could then be evaluated. A negative linear dependence of CH_2O emission rates on the CH_2O concentrations in the atmosphere surrounding the test specimens has been observed for a variety of pressed-wood products [3].

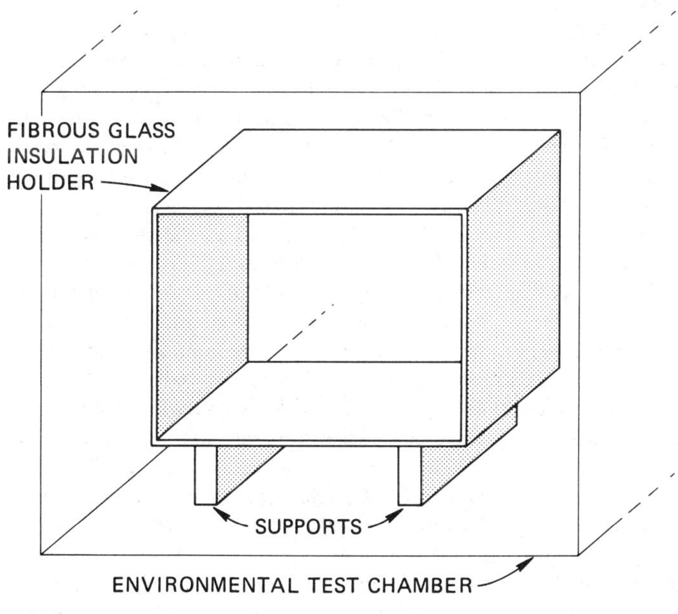

FIG. 1—*Holder for fibrous glass insulation inside the environmental chamber.*

The results of the environmental chamber tests taken at 23°C and 50% relative humidity and those taken at 38°C and 68% relative humidity are summarized in Table 2. For most test conditions, the measured CH_2O concentrations from separate tests of the specimen holder and the holder plus insulation are listed. The contribution of the fibrous glass insulation (FI) to the measured CH_2O concentration of the insulation plus specimen holder is estimated as

$$CH_2O_{FI} = CH_2O_{FI+holder} - CH_2O_{holder} \tag{1}$$

This simple model assumes that the Teflon-covered holder is a weak CH_2O emission source that does not positively or negatively influence the measured CH_2O emissions from the insulation specimen because of (1) sorption of CH_2O on the Teflon-covered holder or (2) any CH_2O concentration dependence of CH_2O emissions from either the insulation or the holder. However, if the CH_2O emissions for the holder (that is, measured separately from the insulation) are suppressed at the elevated CH_2O concentrations caused by the combination of the insulation plus holder, then the modeled CH_2O concentration for the insulation may be somewhat low.

The CH_2O emission rates of the insulation specimens (CH_2O_{ER}) are calculated as

$$CH_2O_{ER}\ (mg/m^2\ h) = CH_2O\ (mg/m^3) \cdot N/L\ (m/h) \tag{2}$$

where N/L is the ratio of the air exchange rate, $N(h^{-1})$, and the product loading, $L(m^2/m^3)$, inside the chamber. A flat surface area of 0.32 m^2, equal to the sum of the front and back openings of the insulation holder, is assumed for the front and back sides of each insulation specimen.

The environmental chamber results for fibrous glass insulation taken at 23°C and 50% relative humidity (see Table 2) indicate very low CH_2O emission potential. The estimated CH_2O concentrations and CH_2O_{ER}s for the insulation range from 0.009 to 0.020 mg/m^3 (0.007 to 0.016 ppm) and 0.009 to 0.025 mg/m^2 h, respectively. The mean CH_2O_{ER} measured in the environmental chamber tests (0.019 ± 0.006 mg/m^2 h) correlates well with the mean CH_2O_{ER} determined with the FSEM (that is, 0.02 ± 0.01 mg/m^2 h), using similar product conditioning periods. A negative dependence of CH_2O_{ER} on CH_2O vapor concentration is clearly indicated only in the results for manufacturer FI1. At CH_2O concentrations of 0.026, 0.020, and 0.012 mg/m^3 (0.021, 0.016, and 0.010 ppm for the insulation plus the holder), the calculated CH_2O_{ER}s for the insulation are 0.009, 0.021, and 0.025 mg/m^2 h, respectively.

The environmental chamber results taken at 38°C and 68% relative humidity typically demonstrate threefold to fourfold increases in the estimated CH_2O concentration for the insulation over corresponding tests at the same N/L at 23°C and 50% relative humidity. This indicates that the CH_2O emission potential of fibrous glass insulation is somewhat less sensitive to changes in temperature and relative humidity than that of pressed-wood products fabricated with urea-formaldehyde resins. A fourfold to sevenfold increase in CH_2O concentration has been observed at conditions between 23°C and 50% relative humidity, and 38°C and 68% relative humidity at constant N/L for particleboard underlayment and hardwood plywood paneling [*11*]. A negative dependence of the CH_2O_{ER} of fibrous glass insulation on CH_2O concentration is clearly indicated in the environmental chamber results for both test specimens taken at 38°C and 68% relative humidity. The results are illustrated graphically in Fig. 2, which shows a linear least-squares fit of the calculated CH_2O_{ER} data for the insulation against the measured CH_2O concentration for the insulation and holder; this concentration is assumed to be the CH_2O level to which the surface of the insulation is exposed.

Extrapolating from the environmental chamber data to a zero CH_2O concentration, a maximum CH_2O_{ER} of about 0.09 to 0.12 mg/m^2 h is predicted. Extrapolating the experimental data to a CH_2O_{ER} of zero, a maximum CH_2O concentration of about 0.13 to 0.15 mg/m^3 (0.10 to

TABLE 2—*Environmental chamber data summary.*

Manufacturer	Product	Air Flow, L/min	N/L, m/h	Formaldehyde Concentration, mg/m^3: FI + Holder	Holder	FI[a]	CH_2O Emission Rate, mg/m^2 h
			DATA TAKEN AT 23°C, 50% RH				
FI1	R-19	5	0.94	0.026 ± 0.001	0.016 ± 0.004	0.010 ± 0.004	0.009 ± 0.004
	unfaced	10	1.88	0.020 ± 0.004	0.009 ± 0.005[b]	0.011 ± 0.006	0.021 ± 0.011
	insulation	15	2.81	0.012 ± 0.001	0.004 ± 0.002	0.009 ± 0.002	0.025 ± 0.006
FI2	R-19	5	0.94	0.036 ± 0.006	0.016 ± 0.004	0.020 ± 0.007	0.019 ± 0.007
	unfaced	10	1.88	0.020 ± 0.005	0.009 ± 0.005[b]	0.011 ± 0.007	0.021 ± 0.013
	insulation						
			DATA TAKEN AT 38°C, 68% RH				
FI1	R-19	5	0.94	0.068 ± 0.007	0.022 ± 0.002	0.046 ± 0.007	0.043 ± 0.007
	unfaced	10	1.88	0.047 ± 0.004	0.014 ± 0.002	0.032 ± 0.005	0.060 ± 0.009
	insulation	15	2.81	0.032 ± 0.002	0.007 ± 0.002[c]	0.025 ± 0.004	0.070 ± 0.011
FI2	R-19	5	0.94	0.082 ± 0.021	0.022 ± 0.002	0.060 ± 0.021	0.056 ± 0.020
	unfaced	10	1.88	0.053 ± 0.001	0.014 ± 0.002	0.040 ± 0.002	0.075 ± 0.004
	insulation	15	2.81	0.041 ± 0.002	0.007 ± 0.002[c]	0.033 ± 0.004	0.093 ± 0.011

[a] $CH_2O_{FI} = CH_2O_{FI+holder} - CH_2O_{holder}$.
[b] Estimated by linear interpolation between data at 5 and 15 L/min.
[c] Estimated by linear extrapolation of data taken at 5 and 10 L/min.

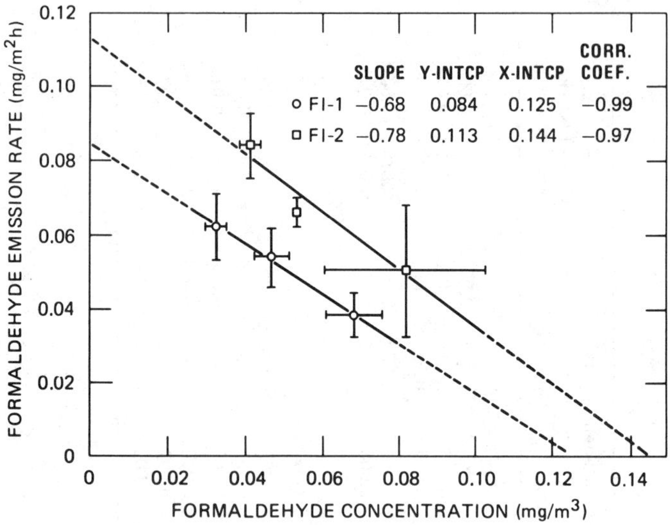

FIG. 2—*Negative linear dependence of the CH_2O emission rate of fibrous glass insulation on the CH_2O concentration at 38°C and 68% relative humidity.*

0.12 ppm) is predicted. This indicates that, under the environmental conditions of the chamber tests, the test specimens are not anticipated to contribute CH_2O to any atmosphere containing greater than about 0.15 mg/m^3 CH_2O.

Potential Impact of Fibrous Glass Insulation on Indoor Formaldehyde Concentrations Using a Diffusion Model

To estimate the potential impact of fibrous glass insulation on indoor CH_2O concentrations, the effects of basic product-use parameters must be considered. Fibrous glass insulation is generally mounted outside the indoor compartment under the floors, in the exterior walls, or in the attics of homes (Fig. 3). In each case, there is a different diffusion barrier between the insulation and the indoor air volume. Under the floor, CH_2O emitted by insulation must typically diffuse through softwood plywood subflooring, particleboard underlayment, and a decorative floor covering. In exterior walls, emitted CH_2O must generally diffuse through kraft paper, foil facing on the insulation, or a continuous plastic (for example, polyethylene) barrier. Above the ceiling, the emitted CH_2O must diffuse through gypsum board, which has been shown in previous work to be highly permeable to CH_2O [*12*]. Potential convective transport of CH_2O from the insulation through the floor, wall, or ceiling because of air infiltration is strongly complicated by environmental, building construction, and occupant activity dependent air flow and mixing rates through insulation-containing compartments into and out of the building. Such transport processes are not considered in the modeling. However, any potential CH_2O contribution from the insulation to the air that is convectively transported indoors will be limited by (1) the weak CH_2O emission potential of the insulation and (2) the negative, CH_2O-concentration dependence of the CH_2O emission rates from the insulation, illustrated in Fig. 2 for test conditions of 38°C and 68% relative humidity.

Fick's law can be used to estimate the diffusion rate of CH_2O through various physical barriers from a region of high CH_2O concentration to a region of low concentration.

$$J = \frac{D}{L}(CH_2O_{high} - CH_2O_{low}) \tag{3}$$

where

D = CH_2O diffusion coefficient, m^2/h,
L = thickness of barrier, m,
CH_2O = CH_2O concentration, mg/m^3, and
J = flux of CH_2O passing through the barrier (that is, CH_2O_{ER}, mg/m^2 h).

The CH_2O diffusion coefficient for softwood plywood, kraft paper, foil, plastic sheathing, gypsum board, and fibrous glass insulation are estimated from permeability data for water vapor (Table 3) [*13*]. It is recognized, however, that the relationship between H_2O and CH_2O diffusion coefficients has been validated only for selected materials, such as gypsum board at high CH_2O concentrations (for example, 300 ppm) [*12*], and may vary at low CH_2O levels. The maximum CH_2O concentration differential between the fibrous glass insulation and indoor air [approximately 0.15 mg/m^3 at 38°C and 68% relative humidity (Fig. 2)] is then substituted into Eq 3, resulting in an order-of-magnitude estimate of the worst-case CH_2O_{ER}s through the barriers. For kraft paper, foil, softwood plywood, and polyethylene, the CH_2O_{ER} for diffusion is typically less than 0.01 mg/m^2 h. For 13-mm (0.5-in.) gypsum board, the worst-case CH_2O_{ER} for diffusion is about 0.1 mg/m^2 h. Therefore, only the application of fibrous glass insulation above gypsum board ceilings of indoor compartments is considered further.

To estimate the potential impact of attic insulation on indoor CH_2O concentrations, a simple CH_2O transport model based on adjoining indoor and attic compartments separated by 13-mm (0.5-in.) gypsum board and R-19 unfaced fibrous glass insulation is considered (Fig. 4). The CH_2O transport model is used to describe the diffusion of CH_2O from the interface between the fibrous glass insulation and the gypsum board into the attic and indoor compartments. Convec-

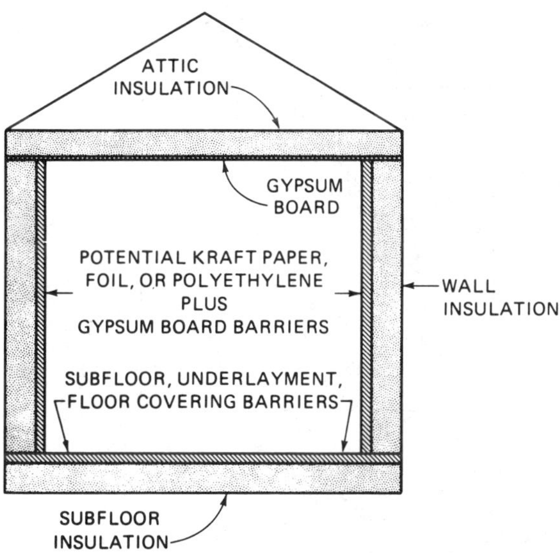

FIG. 3—*Potential applications of fibrous glass insulation in homes and potential physical barriers between the insulation and the indoor compartment.*

TABLE 3—*Summary of water vapor and estimated formaldehyde vapor permeability data.*

Product	Water Permeability Data[a]				Estimated D/L, m/h, for Formaldehyde[b]
	Perms	Thickness, (in.)[c]	Perm · in.[d]	D/L, m/h	
Fibrous glass	116	1.0	116[e]	3.2	2.4
	33	3.5	116	0.91	0.68
	12	9.5	116	0.34	0.26
Gypsum board	50[e]	0.38	19	1.4	1.1
	38	0.50	19	1.1	0.84
Softwood plywood	0.7[e]	0.25	0.18	0.020	0.015
	0.35	0.50	0.18	0.010	0.008
Kraft paper	0.6 to 4.2[e]	. . .	. . .	0.017 to 0.12	0.013 to 0.089
Foil	0.05[e]	3.5×10^{-3}	1.8×10^{-4}	0.001	0.001
Polyethylene	0.06[e]	0.006	3.6×10^{-4}	0.002	0.001

[a] 1 perm = 0.028 m/h for water vapor [*12*].
[b] $D_{CH_2O}/D_{H_2O} = 0.76$ [*12*].
[c] 1 in. = 25.4 mm or 0.025 m.
[d] 1 perm · in. = 0.025 perm · m or 25.4 perm · m.
[e] Data are taken from ASHRAE tables [*13*].

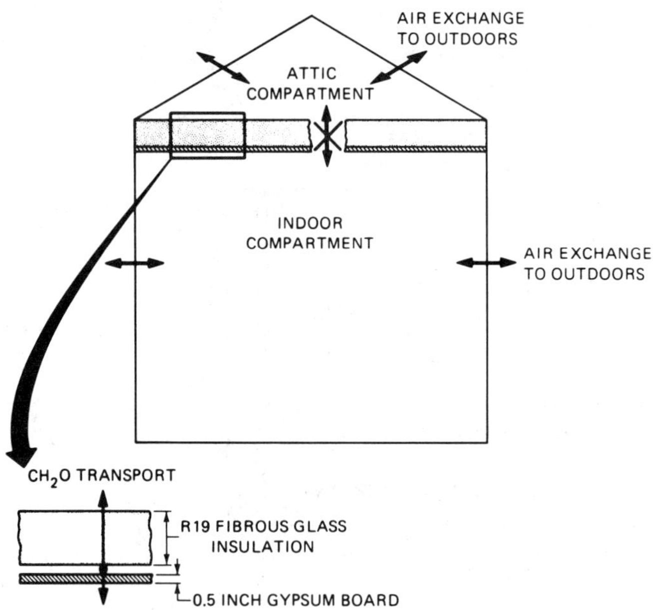

FIG. 4—*Two-compartment model of CH_2O emissions from attic insulation into an indoor compartment.*

tive transport of CH_2O resulting from direct air leakage between attic and indoor compartments is not considered in the modeling. The temperature and humidity conditions in the attic are assumed to be 38°C and 68% relative humidity to make direct use of the CH_2O emissions data taken in the environmental chamber experiments. These environmental conditions are potentially representative of those found in attic insulation adjacent to indoor compartments during the summer months in much of the United States. Several additional assumptions are made to simplify the mathematical treatment of CH_2O transport in the two-compartment model:

1. A steady-state CH_2O concentration exists in both the attic and indoor compartments.
2. Uniform mixing of air exists in both compartments.
3. The contribution of CH_2O emission sources other than the fibrous glass insulation to indoor and attic CH_2O concentrations does not change with the addition of the insulation.
4. Permanent CH_2O loss mechanisms resulting from mechanical or natural air filtration systems or CH_2O sinks are absent in both attic and indoor compartments.
5. The temperature and relative humidity throughout the insulation and attic compartments are uniformly 38°C and 68% relative humidity. Temperature gradients across the insulation are not considered.
6. The negative linear dependence of the CH_2O_{ER} of fibrous glass insulation on the CH_2O concentration measured in the environmental chamber experiments is assumed to describe the CH_2O concentration dependence of the CH_2O emission rates of the insulation at the interface between the gypsum board and the insulation.

The following is a list of terms to be used in the development of the model:

A = attic compartment,
I = indoor compartment,
ITF = interface between the fibrous glass insulation and the gypsum board,
CH_2O_0 = initial CH_2O concentration before the addition of the insulation, mg/m^3,
CH_2O = final CH_2O concentration after the addition of the insulation, mg/m^3,
area = area of ceiling, insulation, m^2,
N = air exchange rate, h^{-1}, and
CH_2O_{TR} = CH_2O transport rate into a compartment from the interface between the fibrous glass insulation and the gypsum board, mg/m^2 h.

Fick's law (Eq 3) is used to describe CH_2O transport from the insulation-gypsum board interface into adjacent indoor and attic compartments. Substituting D/L values of 0.84 m/h and 0.26 m/h for gypsum board and fibrous glass insulation, respectively, then

$$CH_2O_{TR_I} = 0.84(CH_2O_{ITF} - CH_2O_I) \tag{4}$$

and

$$CH_2O_{TR_A} = 0.26(CH_2O_{ITF} - CH_2O_A) \tag{5}$$

Separate, steady-state, single-compartment models [3] are used to describe the CH_2O concentration in the indoor and attic compartments.

$$CH_2O_I = CH_2O_{I0} + \frac{CH_2O_{TR_I} \cdot \text{area}}{V_I \cdot N_I} \tag{6}$$

and

$$CH_2O_A = CH_2O_{A0} + \frac{CH_2O_{TR_A} \cdot \text{area}}{V_A \cdot N_A} \tag{7}$$

Substituting Eq 4 into Eq 6

$$CH_2O_I = \frac{CH_2O_{I0} + a \cdot CH_2O_{ITF}}{1 + a} \tag{8}$$

where

$$a = \frac{0.84 \cdot \text{area}}{V_I \cdot N_I} \tag{9}$$

Substituting Eq 5 into Eq 7

$$CH_2O_A = \frac{CH_2O_{A0} + b \cdot CH_2O_{ITF}}{1 + b} \tag{10}$$

where

$$b = \frac{0.26 \cdot \text{area}}{V_A \cdot N_A} \tag{11}$$

At this point in the development of the model, CH_2O_A and CH_2O_I are both expressed in terms of one variable, CH_2O_{ITF}, and several constants concerning the initial CH_2O concentrations, air exchange rates, and dimensions of the compartments.

Substituting Eq 8 back into Eq 4

$$CH_2O_{TR_I} = \frac{0.84(CH_2O_{ITF} - CH_2O_{I0})}{1 + a} \tag{12}$$

Substituting Eq 10 back into Eq 5

$$CH_2O_{TR_A} = \frac{0.26(CH_2O_{ITF} - CH_2O_{A0})}{1 + b} \tag{13}$$

Noting that

$$CH_2O_{TR\text{total}} = CH_2O_{TR_I} + CH_2O_{TR_A} \tag{14}$$

then

$$CH_2O_{TR\text{total}} = \left(\frac{0.84}{1 + a} + \frac{0.26}{1 + b}\right) CH_2O_{ITF} + K \tag{15}$$

where

$$K = \frac{-0.84CH_2O_{I_0}}{1 + a} - \frac{0.26CH_2O_{A_0}}{1 + b} \tag{16}$$

Equation 15 is a simple linear equation that relates the CH_2O concentration at the interface of the insulation and gypsum board to the total CH_2O_{TR} into both the attic and indoor compartments.

An analogous equation can be developed from the experimental data illustrated in Fig. 2. Combining the CH_2O_{ER} data taken in the environmental chamber at 38°C and 68% relative humidity for Manufacturers FG1 and FG2, by simply averaging the slope and intercept values for FI1 and FI2

$$CH_2O_{ER} = -0.81CH_2O_C + 0.11 \tag{17}$$

where

$$CH_2O_C = CH_2O \text{ concentration in the environmental chamber} \tag{18}$$

Assuming that CH_2O_{ITF} is represented by CH_2O_C and CH_2O_{TR} by CH_2O_{ER} (that is, Assumption 6 listed previously), then Eqs 15 and 17 are two simultaneous equations with two unknown variables. The solution to both equations represents a solution to the two-compartment CH_2O transport model that is consistent with the experimental CH_2O_{ER} data for the insulation. This is illustrated graphically in Fig. 5. The intersection of the lines for the model and the experimental data represents a discrete solution to Eqs 15 and 17. However, this solution is dependent on the initial CH_2O concentrations, air exchange rates, and dimensions that are selected for the attic and indoor compartments.

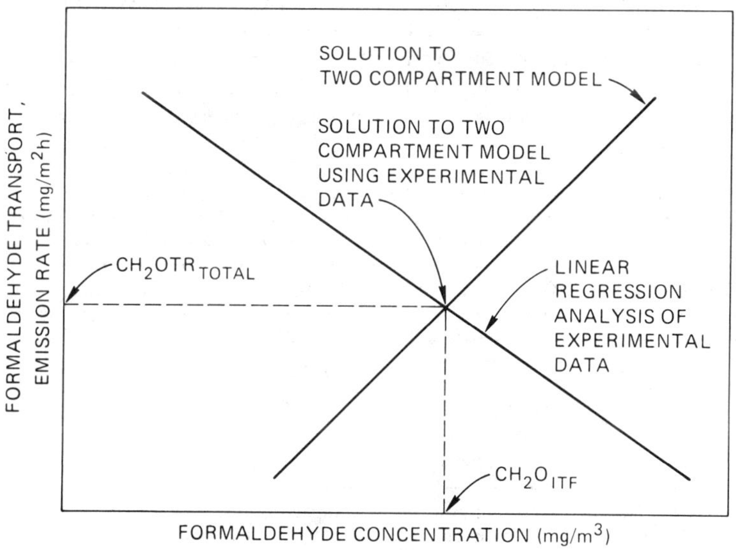

FIG. 5—*Graphical illustration of a solution to a two-compartment attic insulation model using experimental CH_2O emission rate data taken in environmental chamber experiments.*

To illustrate the two-compartment model, numerical values have been chosen for the following parameters.

$$
\begin{aligned}
V_A &= 10 \text{ m}^3, \\
V_I &= 40 \text{ m}^3, \\
N_A &= 5.0 \text{ h}^{-1}, \\
N_I &= 0.5 \text{ h}^{-1}, \\
\text{area} &= 16 \text{ m}^2, \\
CH_2O_{A0} &= 0.006 \text{ mg/m}^3, \text{ and} \\
CH_2O_{I0} &= 0.030 \text{ mg/m}^3.
\end{aligned}
$$

The values of parameters V_I, N_I, and area were previously chosen to represent an average indoor compartment [3]. The values of parameters V_A and N_A were estimated using guidelines of the American Society of Heating, Refrigerating, and Air Conditioning Engineers (ASHRAE) for attic structures and ventilation [13]. The CH_2O_{I0} of 0.030 mg/m^3 (0.025 ppm) is a potential baseline concentration in homes with minimal CH_2O emitters. The CH_2O_{A0} of 0.006 mg/m^3 (0.005 ppm) was chosen to be much lower than CH_2O_{I0} because of the increased air exchange rate in the attic and the generally low outdoor CH_2O concentration.

Solving Eqs 15 and 17, $CH_2O_{ITF} = 0.082$ mg/m^3. Substituting this result into Eq 8, $CH_2O_I =$ 0.051 mg/m^3, corresponding to an increase of 0.021 mg/m^3 (0.017 ppm) CH_2O because of the insulation. The results of analogous calculations using different values of CH_2O_{I0} are summarized in Table 4. The maximum increase in the indoor CH_2O concentration, according to the two-compartment model and experimental data taken at 38°C and 68% relative humidity, is approximately 0.03 mg/m^3 (0.02 ppm).

Summary

Formaldehyde emission measurements have been performed on selected fibrous glass insulation products acquired from the three largest manufacturers in the United States. All of the insulation specimens were conditioned in atmospheres of low CH_2O concentration before being tested. Screening measurements of a total of 18 specimens from nine test products, using a formaldehyde surface emission monitor, indicate consistently weak CH_2O emissions. The measured standard deviation for intersample and intermanufacturer variation was less than 0.006 mg/m^2 h. Environmental chamber tests of two specimens of R-19 unfaced insulation at 23°C and 50% relative humidity are consistent with the results of the surface monitor, indicating an

TABLE 4—*Predicted impact of fibrous glass ceiling insulation on indoor formaldehyde concentrations using a two-compartment, diffusion-controlled model and environmental chamber data taken at 38°C, 68% relative humidity.*

CH_2O_{I0}, mg/m^3 [a]	CH_2O_{ITF}, mg/m^3 [b]	CH_2O_I, mg/m^3 [c]	ΔCH_2O_I, mg/m^3 [d]
0	0.068	0.027	0.027
0.013	0.073	0.037	0.024
0.03	0.078	0.049	0.019
0.063	0.088	0.073	0.010
0.094	0.098	0.096	0.002

[a] Initial CH_2O concentration in the indoor compartment.
[b] CH_2O concentration at the interface between the fibrous glass insulation and gypsum board.
[c] Final CH_2O concentration in the indoor compartment.
[d] Increase in CH_2O concentration because of the addition of the fibrous glass.

average CH_2O emission rate of approximately 0.02 mg/m^2 h for all test products after one to three months of conditioning. Threefold to fourfold increases in CH_2O concentration are typically observed in chamber tests at 38°C and 68% relative humidity over results taken at 23°C and 50% relative humidity, using constant air exchange and product loading values. A negative CH_2O concentration dependence of the CH_2O emission rates of fibrous glass insulation is clearly indicated in CH_2O emission data taken for one test specimen at 23°C and 50% relative humidity, and for two test specimens at 38°C and 68% relative humidity. Equilibrium CH_2O concentrations (that is, with no air exchange) of approximately 0.04 and 0.14 mg/m^3 are indicated at 23°C and 50% relative humidity, and at 38°C and 68% relative humidity, respectively. However, a simple, two-compartment, diffusion-controlled model describing the impact of CH_2O emissions from attic insulation at 38°C and 68% relative humidity on steady-state CH_2O concentrations indoors indicates an increase of approximately 0.002 to 0.03 mg/m^3 CH_2O for background levels of 0.09 to 0.00 mg/m^3, respectively, assuming an air exchange rate of 0.5 h^{-1}.

Acknowledgment

This research was sponsored by the U.S. Consumer Product Safety Commission under Interagency Agreement CPSC-IAG-82-1181 and was conducted under Martin Marietta Energy Systems, Inc., Contract No. DE-AC05-840R21400, with the U.S. Department of Energy.

References

[1] Committee on Aldehydes, Board of Toxicology and Environmental Health Hazards, Assembly of Life Sciences, *Formaldehyde and Other Aldehydes,* National Academy Press, Washington, DC, 1981.

[2] Spengler, J. D., Committee on Indoor Pollutants, *Indoor Pollutants,* National Academy Press, Washington, DC, 1981.

[3] Matthews, T. G., Reed, T. J., Tromberg, B. J., Daffron, C. R., and Hawthorne, A. R., *Formaldehyde: Analytical Chemistry and Toxicology,* Advances in Chemistry Series No. 210, American Chemical Society, Washington, DC, 1985, pp. 131-150.

[4] Matthews, T. G., Reed, R. J., Daffron, C. R., and Hawthorne, A. R., *Proceedings,* 18th International Particleboard/Composite Materials Symposium, Washington State University, Pullman, WA, 1984, pp. 41-70.

[5] Pickrell, J. A., Mokler, B. V., Griffis, L. C., and Hobbs, C. H., *Environmental Science and Technology,* Vol. 17, 1983, pp. 753-757.

[6] Pickrell, J. A., Griffis, L. C., Mokler, B. V., Kanapilly, G. M., and Hobbs, C. H., *Environmental Science and Technology,* Vol. 18, 1984, pp. 682-686.

[7] Prusaczyk, J. E., *Test Information on Formaldehyde Release from Fibrous Glass Insulation Products,* Insulation Technology Laboratory, Owens Corning Corp., Granville, OH, 1983.

[8] Matthews, T. G., Hawthorne, A. R., Corey, M. D., Daffron, C. R., and Schrimsher, J. M., *Journal of Analytical Chemistry,* Vol. 56, 1983, pp. 448-454.

[9] Matthews, T. G., Daffron, C. R., and Corey, M. D., "Formaldehyde Surface Emission Monitor Protocol I: Pressed-Wood Products," Technical Report ORNL/TM-8656, Oak Ridge National Laboratory, Oak Ridge, TN, 1983.

[10] Matthews, T. G., *American Industrial Hygiene Association Journal,* Vol. 43, 1982, pp. 547-552.

[11] Matthews, T. G., Reed, T. J., Tromberg, B. J., Fung, K. W., Thompson, C. V., Simpson, J. O., and Hawthorne, A. R., "Modeling and Testing of Formaldehyde Emission Characteristics of Pressed-Wood Products," ORNL/TM-9867, Report XVIII to the U.S. Consumer Product Safety Commission, Oak Ridge National Laboratory, Oak Ridge, TN, 1985.

[12] Hawthorne, A. R., Gammage, R. B., Matthews, T. G., Blackman, G. S., Howell, T. C., and Allen, R. J., "An Evaluation of Formaldehyde Emission Potential from Urea-Formaldehyde Foam Insulation: Panel Measurements and Modeling," Technical Report ORNL/TM-7959, Oak Ridge National Laboratory, Oak Ridge, TN, 1981.

[13] *ASHRAE Handbook of 1981 Fundamentals,* American Society of Heating, Refrigeration and Air-Conditioning Engineers, Atlanta, GA, 1981.

T. J. Cardenas[1] *and G. Thomas Bible*[1]

The Thermal Properties of Wood—Data Base

REFERENCE: Cardenas, T. J. and Bible, G. T., **"The Thermal Properties of Wood—Data Base,"** *Thermal Insulation: Materials and Systems, ASTM STP 922,* F. J. Powell and S. L. Matthews, Eds., American Society for Testing and Materials, Philadelphia, 1987, pp. 238-282.

ABSTRACT: A data base of steady-state thermal properties for 15 wood species representative of those used commercially in building construction has been developed from available test data. Selected mean values of conductivity, density, moisture content, and thermal resistance are tabulated for representative groups of softwoods and hardwoods. The methodology for organizing the test data incorporating ASTM standards on wood is presented, and the statistical methods utilized for the empirical analysis are discussed. A proposed table of thermal design coefficients for softwoods doubles the reliability of predicted values currently listed in the *ASHRAE Handbook of Fundamentals.*

All test data are documented in a format suitable to a broad range of user groups. Design professionals undertaking computer modeling of composite assemblies and systems for building energy analysis will find the data base a useful reference tool. Recommendations on the need to verify the steady-state thermal properties for wood and wood-based building materials are presented and discussed.

KEY WORDS: wood, thermophysical properties, thermal conductivity, density, moisture content, mean values, thermal insulation

In the building community, the present capability of the design professional to review and evaluate original test data from a variety of sources for a particular building design project is limited and not cost-effective. The vast array of building components, assemblies, and systems coupled with a myriad of code regulations magnifies the scale of performing such a task. In short, state-of-the-art building envelope design relies heavily on handbook data and voluntary or consensus standards.

Among the research objectives on materials and components summarized in the National Program Plan [*1*], primary emphasis is assigned to the experimental evaluation and verification of material properties. This report concludes and documents an initial phase of a much larger industrywide research effort directed toward the National Program Plan goals of establishing and verifying the thermal properties of wood under steady-state and dynamic thermal performance conditions: "Even though dynamic test procedures for building envelope subsystems are urgently needed to provide a better basis for determining energy requirements are equipment sizing, there will be a continuing need for steady-state thermal conductance and resistance data for a number of years until the dynamic test procedures are developed and standardized" [*1*].

The experimental work on the effects of thermal mass being pioneered largely by the National Bureau of Standards (NBS) [2] and the New Mexico Energy Research and Development Insti-

[1]Senior research associate and associate engineer, respectively, Steven Winter Associates, Inc., New York, NY 10001.

tute (NMERDI) [*3*], and the analytical work of the Oak Ridge National Laboratory (ORNL) [*4*] and others strongly indicate innovative energy-conserving design opportunities using insulated envelope construction in conjunction with envelope mass. The thermal performance of solid wood construction, in particular, as demonstrated in the NBS and NMERDI field studies, highlights the potential for broadening the use of solid wood as an envelope mass material. Consistent with the goals embodied in the National Program Plan, this inventory of the fundamental steady-state thermal properties of wood is clearly necessary as an initial consistency check to facilitate subsequent correlation studies of dynamic and steady-state building performance data.

It is the intent of this report to present a table of thermal design values that can be utilized by the broadest spectrum of design professionals. Fortunately, the commercial development of wood is presently governed by a number of recognized industry standards, which will be discussed later. Every effort was utilized to organize the available experimental data within the framework of these standards. Hence, this report is concerned with the organizational aspects of the mechanical properties of wood insofar as they provide a uniform and consistent frame of reference for the development of thermal design values of maximum use and reliability to the design engineer.

Methodology

In any analysis in which data are compiled by a number of researchers, there is cause for circumspection in making generalities based on that data. A wide number of variables resulting from differing testing procedures, reporting procedures, and original intentions in gathering the data could lead to systematic discrepancies in the data, which would obviously bias the results.

In compiling the data, a number of assumptions were made. The intent of this portion of the discussion is to outline those assumptions and to attempt to apply statistical procedures as a way of justifying the assumptions.

Compilation of the Experimental Data

The first assumption was that all the data were "good" and compatible. This assumption was linked with the notion that using the largest amount of data available would tend to damp out random errors. We were able, in most cases, to read the original researcher's experimental procedures to determine the compatability of the processes. Data were rejected for tests in which all the basic variables—mean temperature, density, moisture content, and conductivity—were not recorded. A complete list of referenced test data can be found in the section on references. All the test data, statistical analysis, and notes are organized in tables, as follows:

Summary tables (Tables 1 through 9)—contain the results of the statistical analysis, which will be discussed later.

Data base tables (Tables 10 through 24)—contain the results of the data tabulations from the references obtained. Further identification of the test data is given under the column heading Specimen Description/Remarks. Where required, conversions to either customary or SI units were performed in accordance with the chapter on units and conversion factors contained in the *ASHRAE Handbook of Fundamentals* [5].

Table 25 abstracts the individual test procedures used by the original researchers referenced in the data base tables.

TABLE 1—*Mean value of thermal conductivity for selected wood species, MC < 15%.*

Table Reference	Wood Species	Data, No.	Mean Temperature		Density		Moisture Content, %OD Weight	Conductivity		Resistance	
			K	°F	kg/m^3	lb/ft^3		W/m · K	Btu · in./h · ft^2 · °F	m · K/W	h · ft^2 · °F/Btu · in.
					SOFTWOODS						
10	cedar	25	298	77	353.0	22.04	8.83	0.0932	0.646	10.73	1.55
11	spruce	33	300	80	383.8	23.96	8.27	0.0983	0.682	10.17	1.47
12	cypress	86	300	81	402.4	25.12	8.51	0.1128	0.782	8.87	1.28
13	redwood	40	300	81	399.7	24.95	6.98	0.1100	0.763	9.09	1.31
14	pine	110	299	79	432.6	27.01	7.57	0.1093	0.758	9.15	1.32
15	fir	27	300	80	435.5	27.20	8.47	0.1033	0.716	9.68	1.40
16	hemlock	39	297	76	458.6	28.63	6.52	0.1026	0.711	9.75	1.41
17	Douglas fir	27	300	80	496.6	31.00	6.50	0.1103	0.765	9.07	1.31
18	larch	23	301	82	523.3	32.67	10.26	0.1392	0.965	7.18	1.04
19	southern pine	76	298	77	559.8	34.95	7.18	0.1364	0.946	7.33	1.06
					HARDWOODS						
20	elm	9	299	78	554.2	34.60	7.08	0.1350	0.936	7.41	1.07
21	birch	45	300	79	634.8	39.63	9.06	0.1465	1.016	6.83	0.98
22	maple	26	299	79	630.0	39.33	6.18	0.1500	1.040	6.67	0.96
23	ash	17	298	77	629.9	39.32	9.60	0.1456	1.010	6.87	0.99
24	oak	84	298	78	686.6	42.86	7.88	0.1638	1.136	6.11	0.88

TABLE 2—*Mean value of thermal conductivity for selected wood species, S-DRY (MC < 19%).*

Table Reference	Wood Species	Data, No.	Mean Temperature		Density		Moisture Content, %OD Weight	Conductivity		Resistance	
			K	°F	kg/m³	lb/ft³		W/m · K	Btu · in./h · ft² · °F	m · K/W	h · ft² · °F/Btu · in.
					SOFTWOODS						
10	cedar	41	300	81	340.2	21.24	10.58	0.0956	0.663	10.46	1.51
11	spruce	37	299	79	387.5	24.19	9.17	0.0999	0.693	10.01	1.44
12	cypress	91	300	81	400.8	25.02	8.94	0.1129	0.783	8.86	1.28
13	redwood	...[a]	...	...	...	...	...	...	...	...	...
14	pine	136	300	80	423.0	26.41	8.33	0.1102	0.764	9.07	1.31
15	fir	...	...	...	...	...	...	...	...	...	...
16	hemlock	42	297	76	461.8	28.83	7.29	0.1031	0.715	9.70	1.40
17	Douglas fir	58	301	82	477.4	29.80	12.86	0.1261	0.875	7.93	1.14
18	larch	...	...	...	...	...	...	...	...	...	...
19	southern pine	138	302	81	548.9	34.27	10.39	0.1474	1.022	6.78	0.98
					HARDWOODS						
20	elm	16	301	81	596.5	37.24	11.82	0.1605	1.113	6.23	0.90
21	birch	50	299	78	638.2	39.84	9.75	0.1467	1.017	6.82	0.98
22	maple	52	301	82	645.3	40.29	8.94	0.1759	1.220	5.69	0.82
23	ash	28	299	78	628.4	39.23	12.03	0.1587	1.101	6.30	0.91
24	oak	114	300	80	669.3	41.78	8.86	0.1723	1.195	5.80	0.84

[a] Indicates insufficient test data available for statistical correlation.

TABLE 3—*Standard deviation of the mean thermal conductivity.*

Table Reference	Wood Species	MC 15				S-DRY				S-GRN			
		No.	k	SD	SD/k	No.	k	SD	SD/k	No.	k	SD	SD/k
10	cedar	25	0.646	0.0815	0.1261	41	0.663	0.0671	0.1012	...	...	...	...
11	spruce	33	0.682	0.0632	0.0927	37	0.693	0.0788	0.1136	...	...	...	...
12	cypress	86	0.782	0.0744	0.0951	91	0.783	0.0731	0.0935	5	0.793	0.0194	0.0244
13	redwood	40	0.763	0.0486	0.0637	...[a]	...	...	...	...	...	...	...
14	pine	110	0.758	0.0700	0.0924	136	0.764	0.0684	0.0895	5	0.989	0.0860	0.0870
15	fir	27	0.716	0.0809	0.1129	...	...	...	...	...	...	...	...
16	hemlock	39	0.711	0.0743	0.1045	42	0.715	0.0731	0.1023	8	0.904	0.0670	0.0741
17	Douglas fir	27	0.765	0.0410	0.0536	58	0.875	0.1060	0.1212	...	...	...	...
18	larch	23	0.965	0.0440	0.0455	...	...	...	...	...	...	...	...
19	southern pine	76	0.946	0.1055	0.1114	138	1.022	0.1176	0.1150	3	1.124	0.0925	0.0823
20	elm	9	0.936	0.1013	0.1083	16	1.113	0.2144	0.1927	...	...	...	...
21	birch	45	1.016	0.1161	0.1143	50	1.017	0.1132	0.1114	2	1.211	...	...
22	maple	26	1.040	0.0815	0.0842	52	1.220	0.1906	0.1562	...	...	...	...
23	ash	17	1.010	0.0918	0.0909	28	1.101	0.1386	0.1260	...	...	...	...
24	oak	84	1.136	0.0796	0.0701	114	1.195	0.1208	0.1011	...	...	...	...

TABLE 4—*Standard deviation of the mean density.*

Table Reference	Wood Species	MC 15				S-DRY				S-GRN			
		No.	d	SD	SD/d	No.	d	SD	SD/d	No.	d	SD	SD/d
10	cedar	25	22.04	4.088	0.1855	41	21.24	3.344	0.1574	...	...	...	...
11	spruce	33	23.957	2.749	0.1147	37	24.192	2.687	0.1111	...	...	...	...
12	cypress	86	25.121	2.651	0.1055	91	25.02	2.7487	0.1098	5	28.454	0.652	0.0229
13	redwood	40	24.95	2.026	0.0812	...[a]	...	...	...	...	...	...	...
14	pine	110	27.008	2.905	0.1076	136	26.408	3.1884	0.1207	5	29.562	2.2028	0.0745
15	fir	27	27.195	3.5013	0.1287	...	...	...	...	...	...	...	...
16	hemlock	39	28.634	3.4332	0.1199	42	28.829	3.3981	0.1179	8	29.741	3.0524	0.1026
17	Douglas fir	27	31.004	2.1656	0.0699	58	29.799	1.9006	0.0638	...	...	...	...
18	larch	23	32.667	5.1075	0.1563	...	...	...	...	...	...	...	...
19	southern pine	76	34.948	4.2186	0.1207	138	34.271	3.3342	0.0973	3	34.636	1.0169	0.0294
20	elm	9	34.60	5.0774	0.1467	16	37.239	4.842	0.1300	...	...	...	...
21	birch	45	39.626	3.2585	0.0822	50	39.843	3.2601	0.0818	2	47.38	...	...
22	maple	26	39.325	4.641	0.1180	52	40.288	3.420	0.0849	...	...	...	...
23	ash	17	39.321	5.361	0.1363	28	39.230	5.382	0.1372	...	...	...	...
24	oak	84	42.863	3.879	0.0905	114	41.781	3.790	0.0907	...	...	...	...

TABLE 5—*Standard deviation of the mean moisture-content.*

Table Reference	Wood Species	MC 15				S-DRY				S-GRN			
		No.	MC	SD	SD/MC	No.	MC	SD	SD/MC	No.	MC	SD	SD/MC
10	cedar	25	8.83	3.982	0.4508	41	10.58	3.797	0.3590	...	...	...	...
11	spruce	33	8.271	4.251	0.5139	37	9.169	4.772	0.5205	...	...	...	...
12	cypress	86	8.51	4.134	0.4854	91	8.940	4.387	0.4907	5	21.43	0.9806	0.0458
13	redwood	40	6.98	4.305	0.6166	...[a]	...	...	...	...	...	...	...
14	white pine	110	7.572	3.469	0.4581	136	8.329	3.764	0.4519	5	25.682	3.3729	0.1313
15	fir	27	8.470	3.524	0.4161	...	...	...	...	...	...	...	...
16	hemlock	39	6.524	2.951	0.4523	42	7.289	3.992	0.5476	8	22.544	0.9133	0.0405
17	Douglas fir	27	6.498	4.619	0.7310	58	12.859	6.722	0.5227	...	...	...	...
18	larch	23	10.258	4.107	0.4003	...	...	...	...	...	...	...	...
19	southern pine	76	7.182	4.0493	0.5638	138	10.370	4.7269	0.4550	3	24.39	3.5594	0.1459
20	elm	9	7.084	2.6212	0.3700	16	11.816	5.714	0.4836	...	...	...	...
21	birch	45	9.061	2.6525	0.2927	50	9.751	3.2657	0.3349	2	24.04	...	...
22	maple	26	6.181	3.6101	0.5841	52	8.941	3.7591	0.4205	...	...	...	...
23	ash	17	9.595	2.0327	0.2119	28	12.033	3.425	0.2846	...	...	...	...
24	oak	84	7.876	3.1895	0.4050	114	8.861	3.213	0.3626	...	...	...	...

[a] Insufficient test data available for statistical correlation.

TABLE 6—*Analysis of selected groupings of softwoods (S-DRY, MC < 19%).*

GRP	Table Reference	Wood Species	Test Runs, No.	Density			Thermal Conductivity		
				d	SD	SD/d	k	SD	SD/k
A	10,11	cedar, spruce	78	22.64	3.387	0.1496	0.677	0.0744	0.1099
B	11,14,15	spruce, pine, fir	200	26.10	3.287	0.1259	0.744	0.0779	0.1047
C	11,12,13	spruce, cypress, redwood	168	24.82	2.602	0.1048	0.758	0.0781	0.1030
D	12,13,14	cypress, redwood, pine	267	25.72	2.977	0.1158	0.770	0.0681	0.0884
E	12–16	cypress, redwood, pine, hem-fir	336	26.22	3.256	0.1242	0.759	0.0733	0.0966
F	15,16	hemlock, fir	69	28.19	3.530	0.1252	0.715	0.0762	0.1066
G	17,18	Douglas fir, larch	81	30.61	3.416	0.1116	0.901	0.1012	0.1124
H	17,19	Douglas fir, southern pine	196	32.95	3.614	0.1097	0.979	0.1325	0.1354
I	17,18,19	Douglas fir, larch, southern pine	219	32.92	3.780	0.1154	0.977	0.1263	0.1292
J	18,19	larch, southern pine	161	34.04	3.684	0.1082	1.014	0.1119	0.1104
K	10,19	cedar, yellow pine	179	31.28	6.412	0.2044	0.939	0.1856	0.1975
L	10,14,19	cedar, pine, southern pine	315	29.18	5.796	0.1986	0.864	0.1708	0.1977
M	11–19	all softwoods[a]	633	28.10	5.060	0.1800	0.824	0.1485	0.1801

[a]The k value of 0.824 demonstrates agreement with the k value currently listed as the average for all softwoods in the *ASHRAE Handbook of Fundamentals.*

TABLE 7—*Comparison of softwood data with standard estimates of precision (S-DRY, MC < 19%).*

Table Reference	Wood Species	Test Data, No.	Mean Conductivity, k	Standard Deviation (SD)[a]	Points Within Range, %	
					$k \pm 1$ SD[b]	$k \pm 2$ SD[c]
10	cedar	41	0.663	0.0671	70.7	92.7
11	spruce	37	0.693	0.0788	91.9	97.3
12	cypress	91	0.783	0.0731	74.7	100
13	redwood	40	0.763	0.0486	40.0	97.5
14	pine	136	0.764	0.0684	66.6	100
15	fir	27	0.716	0.0809	69.8	100
16	hemlock	42	0.715	0.0731	66.6	100
17	Douglas fir	58	0.875	0.106	81.0	100
18	larch	23	0.965	0.044	82.6	98.3
19	southern pine	138	1.022	0.1176	75.4	98.6

[a] $SD = [(\Sigma x - (\Sigma x)^2/N)/N]^{1/2}$.
[b] Gaussian distribution would result in 68.2% of the data within the range of one standard deviation each side of the mean.
[c] Gaussian distribution would result in 95% of the data within the range of two standard deviations each side of the mean.

TABLE 8—*Comparison of mean density data with mechanical property data.*

Table Reference	Wood Species	Data, No.	Mean MC, %	Mean Density, lb/ft^3 [a]	Specific Gravity, g	Mechanical Property Data		
						g^b	g^c	g^d
10	cedar	41	10.58	21.24	0.34	0.31 to 0.35	0.30 to 0.31	0.29 to 0.47
11	spruce	37	9.17	24.19	0.39	0.36 to 0.43	0.35 to 0.41	0.33 to 0.41
12	cypress	91	8.94	25.02	0.40	0.48	0.42 to 0.43	0.42 to 0.46
13	redwood	40	6.98	24.95	0.40	0.37 to 0.42	0.34 to 0.39	0.34 to 0.40
14	pine	136	8.33	26.41	0.42	0.38 to 0.49	0.36 to 0.44	0.34 to 0.46
15	fir	27	8.47	27.20	0.44	0.38 to 0.44	0.33 to 0.36	0.31 to 0.43
16	hemlock	42	7.29	28.83	0.46	0.43 to 0.48	0.39 to 0.42	0.38 to 0.45
17	Douglas fir	58	12.86	29.80	0.48	0.48 to 0.51	0.45	0.43 to 0.50
18	larch	23	10.26	32.67	0.52	0.51	0.55	0.48 to 0.52
19	southern pine	138	10.39	34.27	0.55	0.55	0.46 to 0.51	0.47 to 0.59
20	elm	16	11.82	37.24	0.59	...	0.46 to 0.57	0.46 to 0.57
21	birch	50	9.75	39.84	0.64	0.66	0.48 to 0.60	0.48 to 0.65
22	maple	52	8.94	40.29	0.65	0.66	0.44 to 0.57	0.44 to 0.63
23	ash	28	12.03	39.23	0.63	0.62	0.45 to 0.53	0.45 to 0.60
24	oak	114	8.66	41.78	0.67	0.67	0.53 to 0.81	0.52 to 0.88

[a] 1 lb/ft^3 = 16.018 46 kg/m^3.

[b] Specific gravity values based on weight and volume when oven-dry, refer to the Table 8.1A Grouping of Species for Fastening Design, National Design Specification for Wood.

[c] Specific gravity values at 12% moisture content, refer to Tables 2 and 3, Clear Wood Strength Values, ANSI/ASTM Method D 2555-781.

[d] Specific gravity values based on weight when oven-dry and volume when green or at 12% moisture content, refer to Table 4-2, Mechanical Properties of some commercially important woods grown in the United States, *Wood Handbook*, No. 72, Forest Products, Laboratory, 1974.

TABLE 9—*Comparison of mean data with average computed conductivity.*

Table Reference	Wood Species	Data, No.	Mean Temperature, °F	Mean Density, lb/ft³ [a]	Specific Gravity, g	Mean MC, %	Mean Conductivity, k	Average Computed Conductivity			
								k^b	k^c	k^d	k^e
10	cedar	41	81	21.24	0.34	10.58	0.663	0.638	0.751	0.673	0.665
11	spruce	37	79	24.19	0.39	9.17	0.693	0.703	0.769	0.75	0.733
12	cypress	91	81	25.02	0.40	8.94	0.783	0.722	0.855	0.763	0.752
13	redwood	40	81	24.95	0.40	6.98	0.763	0.720	0.855	0.763	0.750
14	pine	136	80	26.41	0.42	8.33	0.764	0.753	0.889	0.794	0.784
15	fir	27	80	27.20	0.44	8.47	0.716	0.771	0.924	0.809	0.802
16	hemlock	42	76	28.83	0.46	7.29	0.715	0.806	0.959	0.841	0.839
17	Douglas fir	58	82	29.80	0.48	12.86	0.875	0.829	0.993	0.827	0.861
18	larch	23	82	32.67	0.52	10.26	0.965	0.892	1.062	0.920	0.927
19	southern pine	138	81	34.27	0.55	10.39	1.022	0.929	1.11	0.969	0.964
20	elm	16	81	37.24	0.59	11.82	1.113	0.995	1.18	1.035	1.032
21	birch	50	78	39.84	0.64	9.75	1.017	1.053	1.27	1.104	1.091
22	maple	52	82	40.29	0.65	8.94	1.220	1.063	1.28	1.122	1.10
23	ash	28	78	39.23	0.63	12.03	1.101	1.040	1.25	1.087	1.07
24	oak	114	80	41.78	0.67	8.86	1.195	1.096	1.32	1.157	1.14

[a] 1 lb/ft³ = 16.018 46 kg/m³.

[b] Values computed using the empirical model developed by MacLean [*11,13*] with mean values for g and M

$$k = g(1.39 + 0.028M) + 0.165 \tag{1}$$

[c] Values computed using Eq 1, where M is assumed to be 12%.

[d] Values computed using the empirical model developed by Siau [*14*], where G and M are normalized to 12%.

$$k = 3.54 - 3.28[1 - g(0.667 + 0.01M)]^{1/2} \tag{2}$$

[e] Values computed using the empirical model developed by Wilkes [*12*], where $k_{air} = 0.1791$, d is the density in pounds per cubic foot, M is assumed to be 12%; and regression constants: $a_0 = 1.874 \times 10^{-2}$, $a_1 = 5.753 \times 10^{-4}$

$$k_{T=75°F} = \frac{k_{air} + (a_0 + a_1 M)d}{(1 + 0.01M)} \tag{3}$$

TABLE 10—*Thermal conductivity of cedar.*

Data Reference	Test Runs	Mean Temperature K	Mean Temperature °F	Density kg/m³	Density lb/ft³	Moisture Content, %OD Weight	Conductivity W/m · K	Conductivity Btu · in./h · ft² · °F	Specimen Description/Remarks
(G)23	1	293	68	490.1	30.60	13.4	0.1214	0.842	cedar;
	1	293	68	480.6	30.00	12.7	0.1055	0.732	see Note 1, Table 25
	1	293	68	480.6	30.00	10.6	0.1159	0.804	
(R)33	1 (*)[a]	297	75	422.9	26.40	11.00	0.1116	0.774	
	1 (*)	297	75	401.9	25.09	11.58	0.1010	0.700	
	1 (*)	297	75	360.4	22.50	10.15	0.0890	0.617	
	1 (*)	297	75	362.3	22.62	9.97	0.0888	0.616	
	1 (*)	297	75	365.2	22.80	9.70	0.0883	0.612	
	1 (*)	297	75	349.5	21.82	8.25	0.0870	0.603	
	1 (*)	297	75	350.2	21.86	8.14	0.0808	0.560	
	1 (*)	297	75	357.4	22.31	8.46	0.0871	0.604	
	1	297	75	453.2	28.29	5.05	0.1178	0.817	
	1	297	75	331.6	20.70	6.43	0.0844	0.585	
(W)40	1	301	82	289.9	18.10	9.4	0.0834	0.578	western red cedar;
	1	301	82	289.9	18.10	15.1	0.0901	0.625	see Note 3, Table 25
	1	301	82	289.9	18.10	13.3	0.0914	0.634	
	1	301	82	280.3	17.50	8.9	0.0851	0.590	
	1	301	82	280.3	17.50	9.3	0.0825	0.572	
	1	301	82	289.9	18.10	8.9	0.0838	0.581	
	1	301	82	289.9	18.10	9.2	0.0813	0.564	
	1	301	82	299.5	18.70	7.5	0.0838	0.581	
(W)40	1	301	82	289.9	18.10	13.8	0.0930	0.645	southern white cedar; see Note 3, Table 25
(M)41	3	303	85	340.4	21.25	0.00	0.0923	0.640	western red cedar
	16[b]	303	85	320.4	20.00	13.3	0.0995	0.690	
(L)42	1	297	75	352.4	22.00	[c]	0.0839	0.582	red cedar; 13/16-in. thick; see Note 5, Table 25
Total data	25	7467	1947	9147.18	551.04	220.83	2.3301	16.156	MC < 15% data
Mean value									
MC 15	25	298	77	353.0	22.04	8.83	0.0932	0.646	
S-DRY	41	300	81	340.2	21.24	10.58	0.0956	0.663	
S-GRN	...[d]	...	...	...	...	...	...	...	

[a]See Note 2, Table 25.
[b]MC 10.2 to 20.4%; S-Dry; See Note 4, Table 25.
[c]Data are not reported. The mean is adjusted for the runs given.
[d]Insufficient data.

TABLE 11—*Thermal conductivity of spruce.*

Data Reference	Test Runs	Mean Temperature K	Mean Temperature °F	Density kg/m³	Density lb/ft³	Moisture Content, %OD Weight	Conductivity W/m · K	Conductivity Btu · in./h · ft² · °F	Specimen Description/Remarks
(G)23	1	**[a]	**	**	**	3.40	0.1222	0.8476	spruce
	1	**	**	**	**	5.80	0.1264	0.8767	
	1	**	**	**	**	7.70	0.1289	0.8941	
	1	**	**	**	**	9.95	0.1331	0.9231	
	1	**	**	**	**	17.00	0.1424	0.9870	
	1	293	68	410.6	26.63	17.0	0.1428	0.9899	
	1	293	68	410.6	26.63	16.9	0.1105	0.766	
	[1][b]	[293]	68	[410.6]	[26.63]	[15.9]	[0.2302]	[1.596]	
	1	293	68	410.6	25.63	16.2	0.1014	0.703	
	1	293	68	410.6	25.63	16.2	0.0992	0.688	
	[1][b]	[293]	[68]	[410.6]	[25.63]	[15.4]	[0.2130]	[1.477]	
(R)33	2	297	75	472.5	29.50	6.26	0.0976	0.677	Sitka spruce; see Note 2, Table 25
	2	297	75	450.1	28.10	6.26	0.0909	0.630	
	1	297	75	466.0	29.09	6.36	0.0958	0.664	
	1	297	75	423.7	26.45	6.55	0.0940	0.652	
	3	297	75	413.0	25.78	6.01	0.0959	0.665	
	1	297	75	409.6	25.57	6.35	0.0920	0.638	
	2	297	75	370.0	23.10	6.58	0.0835	0.579	
	2	297	75	351.3	21.93	6.93	0.0886	0.614	
(W)40	1	301	82	419.7	26.2	9.2	0.0939	0.651	Engelmann spruce; see Note 3, Table 25
	1	301	82	419.7	26.2	9.7	0.0975	0.676	
	1	301	82	429.3	26.8	8.7	0.0908	0.630	
(M)41	4	303	85	340.4	21.25	0.00	0.0894	0.620	Engelmann spruce; heartwood and sapwood
	12[c]	303	85	350.6	21.89	13.00	0.1096	0.760	
Total data	33	9909	2656	12664.0	790.59	272.95	3.2459	22.506	MC < 15% data
Mean value									
MC 15	33	300	80	383.8	23.96	8.27	0.0983	0.682	
S-DRY	37	299	79	387.5	24.19	9.17	0.0999	0.693	
S-GRN	...[d]	...	...	...	...	...	...	...	

[a]Double asterisk indicates that data are not reported. The mean is adjusted for the runs given.
[b]For numbers in brackets, see Note 1, Table 25.
[c]MC 12.6 to 13.2%; MC 15; See Note 4, Table 25.
[d]Insufficient data.

Generic Identification

The second working assumption was that industry standards should be followed where possible. This assumption was applied as follows:

1. The standard nomenclature applied in the classification, grading, testing, and quality control of wood and wood products is used to identify the generic wood species or species groups associated under a single commercial name. The standard nomenclature and commercial names identified are in accordance with the provisions of the following standards:

(*a*) Product Standard (PS) 20-70, American Softwood Lumber Standard,
(*b*) National Lumber Grading Authority (NLGA) Grading Rules (Standard Grading Rules for Canadian Lumber),
(*c*) ANSI/ASTM Definitions of Terms Relating to Wood (D 9-81),
(*d*) ASTM Standard for Establishing Structural Grades and Related Allowable Properties for Visually Graded Lumber (D 245-81),
(*e*) ANSI/ASTM Nomenclature of Domestic Hardwoods and Softwoods (D 1165-80),
(*f*) ASTM Standard for Moisture Content of Wood [D 2016-74 (1983)],
(*g*) ANSI/ASTM Test for Specific Gravity of Wood and Wood-Base Materials (D 2395-83),
(*h*) ASTM Test for Establishing Stress Grades for Structural Members Used in Log Buildings (D 3957-80), and
(*i*) ANSI/ASTM Standard for Establishing Clear-Wood Strength Values (D 2555-81).

In determining the initial categories of species, for example, the grouping of ponderosa pine, northern white pine, lodgepole pine, and western white pine into one group called white pine (see Table 14), and of southern yellow, shortleaf, loblolly, and longleaf yellow into another group called yellow pine (see Table 19) is justified based on the way these species are grouped for mechanical properties [*6*]. Similarly, in the groupings of several species, spruce, pine, and fir are typically grouped and designated as S-P-F; or hem-fir, for hemlock and fir; or D fir-larch, for Douglas fir and larch [*7*] (Fig. 1).

Typically, wood is visually graded by designated quality control inspectors. Grading stamps denoting mill, grade, species, and moisture content, as further explained later, are affixed to individual batches of wood on inspection NLGA Grading Rules; ASTM Method D 245-81). Hence, data for 15 groups of wood species commonly used for domestic residential construction are tabulated in this manner in Tables 10 through 24.

2. The test data on individual groups of species were further tabulated according to the percentage of moisture content recorded at the time of testing. In accordance with the industry standards referenced earlier, the following designations of moisture content applied in the commercial grading of wood were used:

(*a*) MC 15 (moisture content 15%)—standard-size lumber surfaced at a specific moisture content of 15% or less,
(*b*) S-DRY (surface dry)—standard-size lumber surfaced at a moisture content of 19% or less, and
(*c*) S-GRN (surface green)—standard-size lumber surfaced at a moisture content of 20% or more.

Independent Variables

The third basic assumption dealt with the significance of the independent variables. Previous researchers have determined the sensitivity of conductivity in relation to density, moisture content, and mean temperature [*8–11*]. Since density (specific gravity) is a function of species, the major effort in the analysis was to correlate the conductivity with the density and, as a result,

TABLE 12—*Thermal conductivity of cypress.*

Data Reference	Test Runs	Mean Temperature		Density		Moisture Content, %OD Weight	Conductivity		Specimen Description/Remarks
		K	°F	kg/m³	lb/ft³		W/m · K	Btu · in./h · ft² · °F	
(R)33	1 (*)[a]	297	75	431.1	26.91	7.35	0.1080	0.749	red cypress
	1 (*)	297	75	419.7	26.20	7.08	0.1040	0.721	
	1 (*)	297	75	445.5	27.81	6.79	0.1132	0.785	
	1 (*)	297	75	528.8	33.01	6.97	0.1255	0.870	
	1 (*)	297	75	545.7	34.07	6.57	0.1222	0.847	
	1 (*)	297	75	549.3	34.29	7.27	0.1266	0.878	
	1 (*)	297	75	356.7	22.27	6.95	0.1059	0.734	
	1 (*)	297	75	365.1	22.79	6.84	0.1098	0.761	
	1 (*)	297	75	361.1	22.54	6.98	0.1070	0.742	
	1	297	75	368.6	23.01	5.28	0.1005	0.697	
	1	297	75	413.1	25.79	5.58	0.1023	0.709	
	1	297	75	462.6	28.88	4.60	0.1086	0.753	
	1	297	75	529.4	33.05	4.38	0.1248	0.865	
(W)40	1	301	82	410.0	25.6	14.4	0.1124	0.779	southern cypress; see Note 3, Table 25
	1	301	82	410.0	25.6	14.4	0.1060	0.735	
	1	301	82	410.0	25.6	8.9	0.1056	0.732	
	1	301	82	410.0	25.6	8.5	0.1047	0.726	
	1	301	82	419.7	26.2	7.4	0.1014	0.703	
	1	301	82	419.7	26.2	6.3	0.1089	0.755	
	1	301	82	419.7	26.2	6.5	0.1106	0.767	
	1	301	82	419.7	26.2	6.7	0.1115	0.773	
	1	301	82	419.7	26.2	5.8	0.0926	0.642	
	1	301	82	419.7	26.2	5.9	0.1089	0.755	
	1	301	82	419.7	26.2	5.6	0.1056	0.732	
	1	301	82	419.7	26.2	9.2	0.1126	0.781	
	1	301	82	430.9	26.9	9.6	0.1076	0.746	
	1	301	82	430.9	26.9	10.4	0.1060	0.735	
	1	301	82	430.9	26.9	8.0	0.1051	0.729	
	1	301	82	430.9	26.9	9.6	0.1027	0.712	
	1	301	82	430.9	26.9	7.8	0.0927	0.643	
(M)41	5	303	85	390.5	24.38	0.00	0.1082	0.750	bald cypress; mostly heartwood; MC 7.1 to 15.0%
	30[b]	303	85	380.5	23.75	11.7	0.1240	0.860	

(R)33	1	297	75	431.7	26.95	0.29	0.0974	0.675	red cypress; see Note 2, Table 25
	1	297	75	436.7	27.26	1.95	0.1007	0.698	
	1	297	75	452.5	28.25	7.08	0.1067	0.740	
	1	297	75	459.7	28.47	15.75	0.1154	0.800	
	⟨1⟩[c]	⟨297⟩	⟨75⟩	⟨471.6⟩	⟨29.44⟩	⟨21.33⟩	⟨0.1177⟩	⟨0.816⟩	
	1	297	75	426.4	26.62	0.51	0.0968	0.671	
	1	297	75	427.7	26.70	2.36	0.1009	0.700	
	1	297	75	444.4	27.74	6.67	0.1057	0.733	
	1	297	75	447.5	27.92	14.22	0.1129	0.783	
	⟨1⟩	⟨297⟩	⟨75⟩	⟨463.6⟩	⟨28.94⟩	⟨20.02⟩	⟨0.1151⟩	⟨0.798⟩	
	1	297	75	405.4	25.31	0.71	0.0952	0.660	
	1	297	75	410.4	25.62	2.78	0.0979	0.679	
	1	297	75	422.9	26.40	7.29	0.1033	0.716	
	1	297	75	435.6	27.20	16.38	0.1118	0.775	
	⟨1⟩	⟨297⟩	⟨75⟩	⟨449.2⟩	⟨28.04⟩	⟨22.16⟩	⟨0.1157⟩	⟨0.802⟩	
	1	297	75	398.9	24.90	0.60	0.0922	0.639	
	1	297	75	409.8	25.58	2.04	0.0930	0.645	
	1	297	75	424.0	26.47	7.08	0.0968	0.671	
	⟨1⟩	⟨297⟩	⟨75⟩	⟨442.4⟩	⟨27.62⟩	⟨20.82⟩	⟨0.1093⟩	⟨0.758⟩	
	⟨1⟩	⟨297⟩	⟨75⟩	⟨452.2⟩	⟨28.23⟩	⟨22.81⟩	⟨0.1138⟩	⟨0.789⟩	
(W)40	1	301	82	430.9	26.9	8.6	0.1018	0.706	southern cypress; see Note 3; Table 25
	1	301	82	310.8	19.4	15.3	0.1248	0.865	
	1	301	82	310.8	19.4	15.9	0.1252	0.868	
	1	301	82	310.8	19.4	14.6	0.1173	0.813	
	1	301	82	320.4	20.0	13.5	0.1240	0.860	
	1	301	82	329.9	20.6	16.6	0.1124	0.779	
	1	301	82	339.6	21.2	16.5	0.1042	0.723	
	1	301	82	339.6	21.2	14.2	0.1187	0.823	
	1	301	82	350.8	21.9	11.8	0.1044	0.724	
	1	301	82	350.8	21.9	11.2	0.0988	0.685	
	1	301	82	360.4	22.5	11.0	0.0972	0.674	
Total data	86	25551	6975	34296.2	2160.44	732.42	9.7016	67.266	MC < 15% data
Mean value									
MC 15	86	300	81	402.4	25.12	8.51	0.1128	0.782	
S-DRY	91	300	81	400.8	25.02	8.94	0.1129	0.783	
S-GRN	5	297	75	455.7	28.45	21.43	0.1144	0.793	

[a] See Note 2, Table 25.
[b] See Note 4, Table 25.
[c] Numbers in ⟨ ⟩ signify S-GRN; MC > 20%.

TABLE 13—*Thermal conductivity of redwood.*

Data Reference	Test Runs	Mean Temperature K	Mean Temperature °F	Density kg/m³	Density lb/ft³	Moisture Content, %OD Weight	Conductivity W/m · K	Conductivity Btu · in./h · ft² · °F	Specimen Description/Remarks
(R)33	1	297	75	331.3	20.67	4.92	0.0953	0.661	California redwood, see Note 2, Table 25
	2	297	75	352.4	22.00	6.30	0.0995	0.690	
	1	297	75	365.2	22.80	5.48	0.1023	0.709	
	2	297	75	383.0	23.91	6.09	0.1017	0.705	
	1	297	75	392.6	24.51	7.06	0.1028	0.713	
	1	297	75	399.7	24.95	6.63	0.1028	0.713	
	1	297	75	417.4	26.06	6.67	0.1028	0.713	
	5	297	75	433.3	27.05	5.75	0.1064	0.738	
	3	297	75	487.8	30.45	6.50	0.1108	0.768	
(M)41	8	303	85	390.5	24.38	0.00	0.1067	0.740	redwood; heartwood
	15[a]	303	85	390.5	24.38	11.7	0.1183	0.820	
Total data	40	12018	3230	15988.8	998.15	279.29	4.4008	30.513	MC < 15% data
Mean value									
MC 15	40	300	81	399.7	24.95	6.98	0.1100	0.763	
S-DRY	...[b]	...	...	...	...	...	...	...	
S-GRN	...	...	...	...	...	...	...	...	

[a]MC 6.8 to 12.7%; see Note 4, Table 25.
[b]Insufficient data.

TABLE 14—*Thermal conductivity of white pine.*

Data Reference	Test Runs	Mean Temperature K	Mean Temperature °F	Density kg/m³	Density lb/ft³	Moisture Content, %OD Weight	Conductivity W/m · K	Conductivity Btu · in./h · ft² · °F	Specimen Description/Remarks
(G)23	1	293	68	301.1	18.8	. . .[a]	0.0760	0.527	Kingia pine;
	1	293	68	600.7	37.5	. . .[a]	0.1057	0.733	see Note 1, Table 25
(R)33	1	297	75	531.5	33.18	6.75	0.1187	0.823	Norway pine (red pine);
	1	297	75	517.4	32.30	6.35	0.1160	0.804	see Note 2, Table 25
	1	297	75	515.6	32.00	7.05	0.1219	0.845	
	1	297	75	505.5	31.56	5.87	0.1183	0.820	
	1	297	75	493.5	30.81	6.49	0.1170	0.811	
	1	297	75	486.6	30.38	6.20	0.1092	0.757	
	1	297	75	441.0	27.53	6.39	0.1041	0.722	
	1	297	75	415.4	25.93	6.38	0.1038	0.720	
	1	297	75	391.5	24.44	6.08	0.1066	0.739	
(W)40	1	301	82	410.1	25.6	14.9	0.1218	0.845	Norway pine (red pine);
	1	301	82	410.1	25.6	14.5	0.1194	0.828	see Note 3, Table 25
	1	301	82	419.7	26.2	8.2	0.1103	0.765	
	1	301	82	419.7	26.2	8.1	0.1168	0.810	
	1	301	82	429.3	26.8	7.6	0.1086	0.753	
	1	301	82	429.3	26.8	7.0	0.1124	0.779	
	1	301	82	429.3	26.8	7.2	0.1131	0.784	
(R)33	1	297	75	555.4	34.67	8.83	0.1263	0.876	ponderosa pine
	1	297	75	525.2	32.79	8.90	0.1204	0.835	(western yellow pine);
	1	297	75	502.2	31.35	8.19	0.1154	0.800	see Note 2, Table 25
	1	297	75	470.9	29.40	6.32	0.1080	0.749	
	1	297	75	441.9	27.59	8.05	0.1124	0.779	
	1	297	75	422.1	26.35	8.63	0.1067	0.740	
	1	297	75	422.1	26.35	9.11	0.1073	0.744	
	1	297	75	397.7	24.83	8.73	0.1036	0.718	
	1	297	75	505.1	31.52	6.02	0.1140	0.792	
	1	297	75	503.3	31.42	5.68	0.1168	0.810	
	1	297	75	425.1	26.54	10.75	0.1175	0.815	
	1	297	75	475.4	29.68	9.88	0.1193	0.827	

TABLE 14—*Continued.*

Data Reference	Test Runs	Mean Temperature K	Mean Temperature °F	Density kg/m³	Density lb/ft³	Moisture Content, %OD Weight	Conductivity W/m · K	Conductivity Btu · in./h · ft² · °F	Specimen Description/Remarks
(W)40	1[b]	301	82	429.3	26.8	18.1	0.1278	0.886	ponderosa pine
	1	301	82	450.1	28.1	10.6	0.1174	0.814	(western yellow pine);
	1	301	82	450.1	28.1	10.2	0.1156	0.802	see Note 3, Table 25
	1	301	82	469.3	29.3	9.8	0.1207	0.837	
	1	301	82	469.3	29.3	7.2	0.1157	0.802	
	1	301	82	479.0	29.9	9.1	0.1224	0.849	
	1	301	82	479.0	29.9	8.4	0.1212	0.840	
	1	301	82	479.0	29.9	8.5	0.1243	0.862	
(R)33	1	297	75	519.0	32.40	6.25	0.1269	0.880	northern white pine
	1	297	75	501.1	31.28	6.78	0.1314	0.911	(eastern white pine);
	1	297	75	447.7	27.95	6.24	0.1093	0.758	see Note 2, Table 25
	1	297	75	429.9	26.84	6.93	0.1147	0.795	
	1	297	75	429.1	26.79	6.65	0.1145	0.794	
	1	297	75	392.0	24.47	6.55	0.1015	0.704	
	1	297	75	382.4	23.87	6.14	0.0948	0.657	
	1	297	75	368.6	23.01	6.54	0.0955	0.662	
	1	297	75	336.7	21.02	6.32	0.0956	0.663	
(W)40	1	301	82	360.4	22.5	7.6	0.0997	0.691	northern white pine
	1	301	82	370.0	23.1	7.8	0.0954	0.662	(eastern white pine);
	1	301	82	400.5	25.0	14.0	0.1155	0.801	See Note 3, Table 25
	1	301	82	410.1	25.6	7.6	0.1122	0.778	
	1	301	82	410.1	25.6	5.9	0.1034	0.717	
	1	301	82	410.1	25.6	6.3	0.1025	0.711	
(W)40	1	301	82	410.1	25.6	7.4	0.1080	0.749	northern white pine;
	1	301	82	410.1	25.6	6.7	0.1017	0.705	see Note 3, Table 25
	1	301	82	419.7	26.2	13.6	0.1080	0.749	
	1	301	82	430.9	26.9	8.1	0.1004	0.696	
	1	301	82	430.9	26.9	8.0	0.0992	0.688	
	1	301	82	430.9	26.9	6.7	0.1004	0.696	
	1	301	82	430.9	26.9	9.0	0.1089	0.755	
	1	301	82	430.9	26.9	6.8	0.1030	0.714	
	1	301	82	430.9	26.9	8.9	0.1051	0.729	

	1	301	82	430.9	26.9	14.3	0.1173	0.813	
	1	301	82	438.9	27.4	5.5	0.1047	0.726	
	1	301	82	438.9	27.4	7.7	0.1147	0.795	
(M)41	10	303	85	400.5	25.00	0.00	0.1012	0.720	white pine; heartwood
	20[c]	303	85	360.4	22.50	9.80	0.1110	0.770	
(R)33	2	297	75	433.9	27.09	8.44	0.1040	0.721	sugar pine;
	1	297	75	418.4	26.12	8.43	0.0935	0.648	see Note 2, Table 25
	2	297	75	388.0	24.22	8.60	0.0896	0.621	
	2	297	75	383.8	23.96	8.27	0.0926	0.642	
	2	297	75	374.5	23.38	8.34	0.0917	0.636	
	1	297	75	369.7	23.08	8.86	0.0914	0.634	
(W)40	1	301	82	459.7	28.7	11.6	0.1165	0.808	lodgepole pine;
	1	301	82	480.6	30.0	14.8	0.1312	0.910	see Note 3, Table 25
	1	301	82	490.2	30.6	9.1	0.1287	0.892	
	1	301	82	490.2	30.6	9.7	0.1260	0.875	
	1	301	82	490.2	30.6	9.2	0.1212	0.840	
(B)82	1	294	70	431.2	26.92	8.0	0.1092	0.757	see Note 6, Table 25
(W)40	1	301	82	370.0	23.1	14.7	0.1038	0.720	western white pine;
	1	301	82	379.6	23.7	7.7	0.0879	0.610	see Note 3, Table 25
	1	301	82	390.9	24.4	13.5	0.1190	0.825	
	1	301	82	400.5	25.0	7.0	0.1089	0.755	
(L)42	1	297	75	390.9	24.34	[a]	0.1014	0.703	soft pine; 3/8 in.
	1	297	75	388.9	24.28	[a]	0.1036	0.791	soft pine; 1 1/4 in.
	1	297	75	408.5	25.5	[a]	0.0883	0.612	Idaho white pine; 13/16 in. see Note 5, Table 25
(R)33	1	297	75	382.2	23.86	1.85	0.0959	0.665	ponderosa pine;
	1	297	75	386.0	24.10	3.47	0.0979	0.679	(western yellow pine)
	1	297	75	395.7	24.70	7.54	0.1028	0.713	see Note 4, Table 25
	1	297	75	397.7	24.83	8.73	0.1036	0.718	
	1[b]	297	75	407.9	25.47	15.83	0.1116	0.774	
	⟨1⟩[d]	⟨297⟩	⟨75⟩	⟨421.1⟩	⟨26.29⟩	⟨22.65⟩	⟨0.1206⟩	⟨0.836⟩	
	1	297	75	506.8	31.64	3.62	0.1142	0.792	
	1	297	75	508.1	31.72	4.41	0.1145	0.794	
	1	297	75	514.8	32.14	7.24	0.1175	0.815	
	1	297	75	523.3	32.67	11.30	0.1241	0.861	
	1[b]	297	75	531.7	33.19	15.48	0.1299	0.901	
	1[b]	297	75	539.9	33.71	18.85	0.1363	0.945	
	1	297	75	398.9	24.90	1.99	0.1049	0.728	

TABLE 14—*Continued.*

Data Reference	Test Runs	Mean Temperature K	Mean Temperature °F	Density kg/m^3	Density lb/ft^3	Moisture Content, %OD Weight	Conductivity W/m · K	Conductivity Btu · in./h · ft^2 · °F	Specimen Description/Remarks
	1	297	75	412.8	25.77	6.68	0.1135	0.787	
	1	297	75	412.3	25.74	7.01	0.1128	0.782	
	1[b]	297	75	442.9	27.65	18.40	[a]	[a]	
	1[b]	297	75	442.6	27.63	19.37	0.1337	0.927	
	⟨1⟩	⟨297⟩	⟨75⟩	⟨452.2⟩	⟨28.23⟩	⟨23.65⟩	⟨0.1419⟩	⟨0.984⟩	
	⟨1⟩	⟨297⟩	⟨75⟩	⟨473.2⟩	⟨29.54⟩	⟨29.69⟩	⟨0.1526⟩	⟨1.058⟩	
	1	297	75	[a]	[a]	5.78	0.1158	0.803	
	1	297	75	474.3	29.61	12.55	0.1248	0.865	
	1	297	75	469.9	29.34	13.67	0.1242	0.861	
	1	297	75	476.4	29.74	13.94	0.1230	0.853	
	⟨1⟩	⟨297⟩	⟨75⟩	⟨498.8⟩	⟨31.14⟩	⟨22.54⟩	⟨0.1422⟩	⟨0.986⟩	
	⟨1⟩	⟨297⟩	⟨75⟩	⟨522.4⟩	⟨32.61⟩	⟨29.88⟩	⟨0.1562⟩	⟨1.083⟩	
(W)40	1	301	82	379.6	23.7	9.1	0.1022	0.709	limber pine; see Note 3, Table 25
	1	301	82	378.6	23.7	11.1	0.1027	0.712	
	1	301	82	379.6	23.7	10.1	0.1018	0.706	
	1[b]	301	82	379.6	23.7	16.2	0.1086	0.753	
	1	301	82	389.2	24.3	10.5	0.1034	0.717	
Total data	110	32907	8660	47589.7	2970.93	838.72	12.0237	83.366	MC < 15% data
Mean value									
MC 15	110	299	79	432.6	27.01	7.57	0.1093	0.758	
S-DRY	136	300	80	423.0	26.41	8.33	0.1102	0.764	
S-GRN	5	297	75	473.5	29.56	25.68	0.1426	0.989	

[a] Data are not reported. The mean is adjusted for the runs given.
[b] S-DRY; MC > 15.
[c] MC 4.8 to 17.9%; S-DRY; see Note 4, Table 25.
[d] Figures in ⟨ ⟩ signify S-GRN; MC > 20%.

TABLE 15—*Thermal conductivity of fir.*

Data Reference	Test Runs	Mean Temperature K	Mean Temperature °F	Density kg/m³	Density lb/ft³	Moisture Content, %OD Weight	Conductivity W/m · K	Conductivity Btu · in./h · ft² · °F	Specimen Description/Remarks
(G)23	1	293	68	600.7	37.5	15.0	0.1182	0.820	fir; see Note 1, Table 25
(R)33	1	297	75	351.1	21.92	8.80	0.0815	0.565	white fir; see Note 2, Table 25
	1	297	75	358.3	22.37	8.28	0.0826	0.573	
	1	297	75	360.1	22.48	8.42	0.0848	0.588	
	1	297	75	374.5	23.38	8.50	0.0857	0.594	
	1	297	75	386.8	24.15	8.37	0.0858	0.595	
	1	297	75	405.6	25.32	8.49	0.0903	0.626	
	1	297	75	424.5	26.50	8.49	0.0955	0.662	
	1	297	75	402.1	25.10	8.60	0.0946	0.656	
	1	297	75	435.7	27.20	8.35	0.0999	0.693	
(W)40	1	301	82	440.5	27.5	8.6	0.1093	0.758	lowland white fir (western balsam); (grand fir); see Note 3, Table 25
	1	301	82	440.5	27.5	8.5	0.1096	0.760	
	1	301	82	450.1	28.1	5.0	0.0988	0.685	
	1	301	82	450.1	28.1	5.0	0.1060	0.735	
	1	301	82	450.1	28.1	5.0	0.1149	0.773	
	1	301	82	450.1	28.1	5.4	0.1064	0.738	
	1	301	82	450.1	28.1	14.4	0.1124	0.779	
	1	301	82	459.7	28.7	8.6	0.1101	0.764	
	1	301	82	469.3	29.3	11.6	0.1218	0.845	
	1	301	82	478.9	29.9	11.5	0.1161	0.805	
	1	301	82	499.8	31.2	13.8	0.1168	0.810	
	1	301	82	520.6	32.5	8.3	0.1103	0.765	
	1	301	82	520.6	32.5	8.3	0.1103	0.765	
(M)41	2	303	85	410.6	25.63	0.00	0.1024	0.710	white fir; heartwood
	2[a]	303	85	380.4	23.75	11.70	0.1125	0.780	
Total data	27	8091	2149	11762.04	734.28	228.7	2.7885	19.334	MC < 15% data
Mean value									
MC 15	27	300	80	435.5	27.19	8.47	0.1033	0.716	
S-DRY	...[b]	...	...	...	...	...	...	...	
S-GRN	...	...	...	...	...	...	...	...	

[a]MC 10.8 to 12.0%; see Note 4, Table 25.
[b]Insufficient data.

TABLE 16—*Thermal conductivity of hemlock.*

Data Reference	Test Runs	Mean Temperature		Density		Moisture Content, %OD Weight	Conductivity		Specimen Description/Remarks
		K	°F	kg/m³	lb/ft³		W/m · K	Btu · in./h · ft² · °F	
(R)33	2	297	75	531.5	33.18	5.58	0.1089	0.755	West Coast hemlock; see Note 2, Table 25
	1	297	75	497.7	31.07	6.42	0.1056	0.732	
	3	297	75	464.2	28.98	6.32	0.0962	0.667	
	3	297	75	433.5	27.06	5.84	0.1084	0.752	
	2	297	75	382.8	23.90	6.81	0.0844	0.585	
	2	297	75	370.7	23.14	6.19	0.0831	0.576	
	1	297	75	390.2	24.36	5.38	0.0948	0.657	
	1	297	75	357.5	22.32	5.30	0.0965	0.669	
	1	297	75	453.0	28.28	6.18	0.1098	0.761	
	1	297	75	425.0	26.53	9.04	0.1124	0.779	
(M)41	2	303	85	460.5	28.75	0.00	0.1139	0.790	West Coast hemlock; heartwood
	⟨5⟩[a]	⟨303⟩	⟨85⟩	⟨440.5⟩	⟨27.5⟩	⟨23.0⟩	⟨0.1370⟩	⟨0.95⟩	
(R)33	1	297	75	520.1	32.47	7.06	0.1227	0.851	eastern hemlock; see Note 2, Table 25
	1	297	75	488.4	30.49	7.17	0.1125	0.780	
	1	297	75	452.0	28.22	7.63	0.1087	0.754	
	1	297	75	439.5	27.44	6.76	0.1080	0.749	
	1	297	75	431.2	26.92	7.18	0.1079	0.748	
	1	297	75	428.0	26.72	6.92	0.0999	0.693	
	1	297	75	407.0	25.41	7.37	0.1004	0.696	
	1	297	75	375.9	23.47	6.89	0.1018	0.706	

(R)33	1	297	75	458.1	28.60	1.72	0.0879	0.610	West Coast hemlock; see Note 2, Table 25
	1	297	75	467.7	29.20	5.38	0.0947	0.657	
	1	297	75	486.9	30.30	16.14	0.1053	0.730	
	⟨1⟩	⟨297⟩	⟨75⟩	⟨503.5⟩	⟨31.43⟩	⟨23.15⟩	⟨0.1116⟩	⟨0.774⟩	
	1	297	75	442.1	27.60	0.93	0.8437	0.585	
	1	297	75	455.6	28.44	5.30	0.0909	0.630	
	1	297	75	470.9	29.40	12.13	0.1010	0.700	
	1	297	75	491.3	30.67	19.79	0.1113	0.772	
	1	297	75	543.8	33.95	6.18	0.0879	0.761	
	1	297	75	552.9	34.52	11.28	0.1139	0.790	
	1	297	75	551.7	34.44	11.93	0.1197	0.830	
	1	297	75	562.2	35.10	14.76	0.1188	0.824	
	⟨1⟩	⟨297⟩	⟨75⟩	⟨565.5⟩	⟨35.30⟩	⟨20.36⟩	⟨0.1288⟩	⟨0.893⟩	
	1	297	75	499.8	31.20	2.98	0.0919	0.637	
	1	297	75	516.9	32.27	9.35	0.1060	0.735	
	1	297	75	516.6	32.25	9.55	0.1069	0.741	
	1	297	75	530.2	33.10	15.78	0.1134	0.786	
	⟨1⟩	⟨297⟩	⟨75⟩	⟨539.8⟩	⟨33.70⟩	⟨21.84⟩	⟨0.1173⟩	⟨0.813⟩	
Total data	39	11595	2945	19817.2	1116.73	254.43	4.4461	27.744	MC < 15% data
Mean value									
MC 15	39	297	76	458.6	28.63	6.52	0.1026	0.711	
S-DRY	42	297	76	461.8	28.83	7.29	0.1031	0.715	
S-GRN	8	301	81	476.4	29.74	22.54	0.1304	0.904	

[a]Numbers in ⟨ ⟩ signify S-GRN; MC > 20%.

TABLE 17—*Thermal conductivity of Douglas fir.*

Data Reference	Test Runs	Mean Temperature K	Mean Temperature °F	Density kg/m³	Density lb/ft³	Moisture Content, %OD Weight	Conductivity W/m · K	Conductivity Btu · in./h · ft² · °F	Specimen Description/Remarks
(R)33	1 (*)[a]	297	75	507.3	31.67	10.75	0.1057	0.733	Douglas fir
	1 (*)	297	75	501.5	31.31	10.75	0.1093	0.758	
	1 (*)	297	75	564.2	35.22	10.75	0.1121	0.777	
	1 (*)	297	75	527.2	32.91	10.14	0.1098	0.761	
	1 (*)	297	75	496.2	30.98	10.52	0.1038	0.720	
	1 (*)	297	75	544.8	34.01	10.28	0.1102	0.764	
	1 (*)	297	75	577.5	36.05	10.87	0.1139	0.790	
	1 (*)	297	75	542.4	33.86	10.87	0.1102	0.764	
	1 (*)	297	75	521.6	32.56	10.87	0.1064	0.738	
	1	297	75	467.6	29.19	5.62	0.0988	0.685	
	1	297	75	510.8	31.89	5.93	0.1033	0.716	
	1	297	75	555.5	34.68	5.56	0.1076	0.746	
	1	297	75	449.8	28.08	6.53	0.1020	0.707	
(W)40	1	301	82	478.9	29.9	9.6	0.1152	0.799	Douglas fir; see Note 3, Table 25
	1	301	82	490.2	30.6	7.4	0.1148	0.796	
	1	301	82	490.2	30.6	10.0	0.1190	0.825	
	1	301	82	490.2	30.6	14.0	0.1311	0.909	
	1	301	82	499.8	31.2	7.7	0.1143	0.793	
	1	301	82	509.4	31.8	7.3	0.1147	0.795	
(M)41	8	303	85	460.5	28.75	0.00	0.1096	0.760	Douglas fir; heartwood
	31[b]	303	85	460.5	28.75	18.4	0.1399	0.970	
	⟨6⟩[c,d]	⟨303⟩	⟨85⟩	⟨**⟩[e]	⟨**⟩	⟨**⟩	⟨**⟩	⟨**⟩	
	⟨15⟩	⟨303⟩	⟨85⟩	⟨510.5⟩	⟨31.8⟩	⟨33.2⟩	⟨**⟩	⟨**⟩	
Total data	27	8091	2147	13409.2	837.11	175.44	2.9789	20.654	MC < 15% data
Mean value									
MC 15	27	300	80	496.6	31.00	6.50	0.1103	0.765	
S-DRY	58	301	82	477.4	29.80	12.86	0.1261	0.875	
S-GRN	...[f]	...	...	...	...	...	...	...	

[a] See Note 2, Table 25.
[b] MC 7.2 to 18.4%; S-DRY.
[c] MC 20 to 25%.
[d] Figures in ⟨ ⟩ signify S-GRN; MC > 20%; see Note 4, Table 25.
[e] The double asterisk indicates that data are not reported. The mean is adjusted for the runs given.
[f] Insufficient data.

TABLE 18—*Thermal conductivity of larch.*

Data Reference	Test Runs	Mean Temperature K	Mean Temperature °F	Density kg/m³	Density lb/ft³	Moisture Content, %OD Weight	Conductivity W/m · K	Conductivity Btu · in./h · ft² · °F	Specimen Description/Remarks
(R)33	1	297	75	714.7	44.62	10.00	0.1527	1.059	western larch;
	1	297	75	670.5	41.86	9.66	0.1527	1.059	see Note 2, Table 25
	1	297	75	661.2	41.28	10.46	0.1421	0.985	
	1	297	75	635.6	39.68	10.30	0.1422	0.986	
	1	297	75	600.7	37.50	10.30	0.1387	0.962	
	1	297	75	527.8	32.95	10.81	0.1282	0.889	
	1	297	75	525.6	32.81	10.61	0.1199	0.831	
(M)41	3	303	85	570.7	35.63	0.00	0.1356	0.940	western larch;
	13[a]	303	85	460.5	28.75	12.6	0.1399	0.970	heartwood
Total data	23	6927	1885	12035.3	751.34	235.94	3.201	22.201	MC < 15% data
Mean value									
MC 15	23	301	82	523.3	32.67	10.26	0.1392	0.965	
S-DRY	...[b]	...	...	...	...	...	...	...	
S-GRN	...	...	...	...	...	...	...	...	

[a]MC 11.2 to 13.4%; see Note 4, Table 25.
[b]Insufficient data.

TABLE 19—*Thermal conductivity of southern pine.*

Data Reference	Test Runs	Mean Temperature K	Mean Temperature °F	Density kg/m³	Density lb/ft³	Moisture Content, %OD Weight	Conductivity W/m · K	Conductivity Btu · in./h · ft² · °F	Specimen Description/Remarks
(G)23	1	293	68	[a]	[a]	15.0	0.1395	0.967	pitch pine; see Note 1, Table 25.
(R)33	2	297	75	646.3	40.35	7.24	0.1566	1.086	shortleaf yellow pine (pitch pine); see Note 2, Table 25
	2	297	75	627.6	39.18	6.13	0.1400	0.971	
	1	297	75	600.1	37.46	7.72	0.1390	0.964	
	2	297	75	547.2	34.16	6.33	0.1351	0.937	
	1	297	75	485.4	30.30	6.72	0.1274	0.883	
	2	297	75	552.8	34.51	5.88	0.1385	0.960	
	2	297	75	410.7	25.64	5.90	0.1147	0.795	
	1	297	75	412.0	25.72	4.55	0.1158	0.803	
	2	297	75	418.4	26.12	6.04	0.1134	0.786	
	1	297	75	508.7	31.76	6.85	0.1227	0.851	
	1	297	75	535.3	33.42	6.64	0.1331	0.923	
	1	297	75	499.9	31.21	6.83	0.1232	0.854	
	1	297	75	513.1	32.03	6.75	0.1265	0.877	
(W)40	1	301	82	530.2	33.1	13.1	0.1471	1.020	shortleaf pine (pitch pine); see Note 3, Table 25
	1	301	82	530.2	33.1	16.0	0.1486	1.030	
	1	301	82	539.8	33.7	11.3	0.1391	0.965	
	1	301	82	560.6	35.0	13.2	0.1558	0.080	
	1	301	82	629.5	39.3	10.9	0.1499	1.040	
	1	301	82	650.3	40.6	12.9	0.1774	1.230	
	1	301	82	490.2	30.6	14.4	0.1442	1.000	
	1	301	82	499.8	31.2	8.9	0.1287	0.892	
	1	301	82	518.9	32.4	9.0	0.1304	0.904	
(R)33	4	297	75	690.6	43.11	6.58	0.1441	0.999	longleaf yellow pine; (pitch pine); see Note 2, Table 25
	1	297	75	657.2	41.03	7.66	0.1354	0.939	
	2	297	75	640.1	39.96	7.14	0.1327	0.920	
	3	297	75	570.3	35.60	6.57	0.1279	0.887	
	2	297	75	552.2	34.47	6.65	0.1196	0.829	
	2	297	75	522.0	32.59	6.31	0.1178	0.817	
	1	297	75	493.0	30.78	7.39	0.1226	0.850	
	1	297	75	488.4	30.49	7.47	0.1248	0.865	

(W)40	1	301	82	610.3	38.1	17.2	0.1702	1.180	longleaf pine
	1	301	82	619.9	38.7	14.6	0.1673	1.160	(pitch pine);
	1	301	82	619.9	38.7	17.1	0.1788	1.240	see Note 3, Table 25
	1	301	82	629.5	39.3	13.3	0.1644	1.140	
	1	301	82	629.5	39.3	13.0	0.1601	1.110	
	1	301	82	639.1	39.9	12.2	0.1644	1.140	
	1	301	82	639.1	39.9	16.6	0.1745	1.210	
	1	301	82	650.3	40.6	14.4	0.1846	1.280	
	1	301	82	659.9	41.2	12.2	0.1644	1.140	
(M)41	7	303	85	560.6	35.00	0.00	0.1356	0.940	southern yellow pine;
	52[b]	303	85	530.7	33.13	13.8	0.1616	1.120	heartwood and sapwood
(W)40	1	301	82	520.6	32.5	16.2	0.1558	1.08	loblolly pine
	1	301	82	539.8	33.7	11.9	0.1572	1.09	(pitch pine);
	1	301	82	549.4	34.3	11.6	0.1499	1.04	see Note 3, Table 25
	1	301	82	549.4	34.3	17.7	0.1486	1.03	
	⟨1⟩[c]	⟨301⟩	⟨82⟩	⟨560.6⟩	⟨35.0⟩	⟨21.0⟩	⟨0.1760⟩	⟨1.22⟩	
	1	301	82	570.3	35.6	15.7	0.1630	1.13	
	1	301	82	579.9	36.2	12.3	0.1471	1.02	
(R)33	1	297	75	584.9	36.52	0.06	0.1243	0.862	shortleaf pine
	1	297	75	595.4	37.17	3.89	0.1308	0.907	(pitch pine)
	1	297	75	609.6	38.06	10.05	0.1398	0.969	see Note 2, Table 25
	1	297	75	617.2	38.53	13.54	0.1439	0.998	
	1	297	75	528.8	33.01	0.47	0.1278	0.886	
	1	297	75	536.9	33.52	3.13	0.1309	0.908	
	1	297	75	545.9	34.08	5.45	0.1356	0.940	
	1	297	75	544.8	34.01	6.95	0.1405	0.974	
	1	297	75	559.8	34.95	12.34	0.1467	1.017	
	1	297	75	562.9	35.14	13.91	0.1473	1.021	
(R)33	1	297	75	487.8	30.45	1.23	0.1111	0.770	shortleaf pine
	1	297	75	507.8	31.70	6.65	0.1230	0.853	(pitch pine)
	1	297	75	508.7	31.76	6.85	0.1227	0.851	see Note 2, Table 25
	1	297	75	530.2	33.10	15.31	[a]	[a]	
	⟨1⟩	⟨297⟩	⟨75⟩	⟨[a]⟩	⟨[a]⟩	⟨24.11⟩	⟨0.1543⟩	⟨1.070⟩	
	⟨1⟩	⟨297⟩	⟨75⟩	⟨571.2⟩	⟨35.66⟩	⟨29.31⟩	⟨0.1663⟩	⟨1.153⟩	
	1	297	75	489.8	30.58	0.21	0.1188	0.824	
	1	297	75	501.1	31.28	3.33	0.1219	0.845	
	1	297	75	513.1	32.03	6.75	0.1265	0.877	
	1	297	75	509.2	31.79	11.99	0.1344	0.932	

TABLE 19—*Continued.*

Data Reference	Test Runs	Mean Temperature K	Mean Temperature °F	Density kg/m³	Density lb/ft³	Moisture Content, %OD Weight	Conductivity W/m · K	Conductivity Btu · in./h · ft² · °F	Specimen Description/Remarks
	1	297	75	522.2	32.60	16.90	0.1393	0.966	
	1	297	75	527.8	32.95	17.78	0.1408	0.976	
	1	297	75	525.1	32.78	19.15	0.1431	0.992	
	⟨1⟩	⟨297⟩	⟨75⟩	⟨532.6⟩	⟨33.25⟩	⟨22.87⟩	⟨0.1441⟩	⟨0.999⟩	
Total data	76	22682	5889	42546.31	2656.08	545.85	10.3747	71.933	MC $<$ 15% data
Mean value									
MC 15	76	298	77	559.8	34.95	7.18	0.1364	0.946	
S-DRY	138	302	81	548.9	34.27	10.39	0.1474	1.022	
S-GRN	3	298	77	554.9	34.64	24.39	0.1621	1.124	

[a]Data are not reported. The mean is adjusted for the runs given.
[b]MC 5.2 to 23.8%; S-DRY; see Note 4, Table 25.
[c]Numbers in ⟨ ⟩ signify S-GRN; MC $>$ 20%.

TABLE 20—*Thermal conductivity of elm.*

Data Reference	Test Runs	Mean Temperature K	Mean Temperature °F	Density kg/m³	Density lb/ft³	Moisture Content, %OD Weight	Conductivity W/m · K	Conductivity Btu · in./h · ft² · °F	Specimen Description/Remarks
(R)33	1	297	75	564.8	35.26	7.67	0.1408	0.976	soft elm; see Note 2, Table 25
	1	297	75	552.9	34.52	7.29	0.1327	0.920	
	1	297	75	527.2	32.91	7.58	0.1298	0.900	
	1	297	75	516.9	32.27	7.83	0.1232	0.854	
	1	297	75	486.8	30.39	7.51	0.1240	0.860	
	1	297	75	458.3	28.61	7.18	0.1135	0.787	
(W)40	1	301	82	579.9	36.20	9.30	0.1420	0.985	American elm; see Note 3, Table 25
(M)41	1	303	85	760.9	47.50	0.00	0.1673	1.160	rock elm; see Note 4, Table 25
(M)41	1	303	85	540.62	33.75	9.4	0.1413	0.980	American elm; heartwood; see Note 4, Table 25
(M)41	7[a]	303	85	650.8	40.63	17.9	0.1933	1.34	rock elm; heart and sapwood
	⟨2⟩[b]	⟨303⟩	⟨85⟩	⟨690.8⟩	⟨43.13⟩	⟨31.2⟩	⟨[c]⟩	⟨[c]⟩	
Total data	9	2689	702	4988.3	311.41	63.76	1.2147	8.422	MC < 15% data
Mean value									
MC 15	9	299	78	554.23	34.60	7.08	0.1350	0.936	
S-DRY	16	301	81	596.53	37.24	11.82	0.1605	1.113	
S-GRN	...[d]	...	...	...	...	...	...	...	

[a]MC 12.1 to 17.9%; S-DRY.
[b]Figures in ⟨ ⟩ signify S-GRN; MC < 20%; see Note 4, Table 25.
[c]Data are not reported. The mean is adjusted for the runs given.
[d]Insufficient data.

TABLE 21—*Thermal conductivity of birch.*

Data Reference	Test Runs	Mean Temperature K	Mean Temperature °F	Density kg/m³	Density lb/ft³	Moisture Content, %OD Weight	Conductivity W/m · K	Conductivity Btu · in./h · ft² · °F	Specimen Description/Remarks
(R)33	1	297	75	726.3	45.34	9.71	0.1474	1.022	yellow birch; see Note 2, Table 25
	1	297	75	703.9	43.94	9.14	0.1449	1.005	
	1	297	75	665.4	41.54	9.20	0.1385	0.960	
	1	297	75	637.7	39.81	9.28	0.1421	0.985	
	1	297	75	621.8	38.82	9.49	0.1234	0.856	
	1	297	75	604.4	37.73	9.84	0.1336	0.926	
	1	297	75	654.4	40.85	6.03	0.1281	0.888	
	1	297	75	736.8	46.00	6.63	0.1491	1.034	
(W)40	1	301	82	539.8	33.7	14.6	0.1383	0.959	paper birch; see Note 3, Table 25
	1	301	82	560.6	35.0	9.0	0.1217	0.913	
	1	301	82	560.6	35.0	9.4	0.1348	0.935	
	1	301	82	560.6	35.0	8.5	0.1252	0.868	
	1	301	82	560.6	35.0	8.2	0.1364	0.946	
(W)40	1	301	82	579.9	36.2	14.7	0.1457	1.010	yellow birch; see Note 3, Table 25
	1	301	82	589.5	36.8	9.6	0.1307	0.906	
	1	301	82	589.5	36.8	7.1	0.1364	0.946	
	1	301	82	589.5	36.8	7.0	0.1421	0.985	
	1	301	82	589.5	36.8	7.0	0.1408	0.976	
	1	301	82	589.5	36.8	6.9	0.1393	0.966	
	1	301	82	600.7	37.5	8.8	0.1305	0.905	
	1	301	82	600.7	37.5	10.5	0.1305	0.905	
(M)41	11[a]	303	85	640.7	40.0	10.8	0.1716	1.19	yellow birch; heartwood
(R)33	1	297	75	632.1	39.46	5.69	0.1382	0.958	yellow birch; see Note 2, Table 25
	1	297	75	636.1	39.71	7.75	0.1403	0.973	
	1	297	75	637.7	39.81	9.28	0.1421	0.985	
	1	297	75	647.1	40.40	15.73	0.1504	1.043	
	1	297	75	591.1	36.90	5.60	0.1263	0.876	
	1	297	75	604.4	37.73	9.84	0.1336	0.926	
	1	297	75	606.8	37.88	11.78	0.1356	0.940	
	1	297	75	614.8	38.38	16.90	0.1409	0.977	

(R)33	1	297	75	**[b]	**	0.0	0.1377	0.955	yellow birch; see Note 2; Table 25
	1	297	75	**	**	3.28	0.1431	0.992	
	1	297	75	**	**	6.43	0.1499	1.039	
	1	297	75	**	**	13.24	0.1558	1.080	
	1	297	75	**	**	13.64	0.1587	1.100	
	⟨1⟩	⟨297⟩	⟨75⟩	⟨**⟩[c]	⟨**⟩	⟨34.83⟩	⟨0.1866⟩	⟨1.294⟩	
	1	297	75	631.3	39.41	0.87	**	**	
	1	297	75	641.4	40.04	3.51	0.1227	0.851	
	1	297	75	654.4	40.85	6.03	0.1280	0.888	
	1	297	75	672.5	41.98	15.48	0.1385	0.960	
	1	297	75	673.9	42.07	16.57	0.1396	0.968	
	⟨1⟩	⟨297⟩	⟨75⟩	⟨697.1⟩	⟨43.52⟩	⟨27.36⟩	⟨**⟩	⟨**⟩	
	⟨1⟩	⟨297⟩	⟨75⟩	⟨706.7⟩	⟨44.12⟩	⟨30.88⟩	⟨**⟩	⟨**⟩	
	1	297	75	720.5	44.98	1.69	0.1385	0.960	
	1	297	75	725.3	45.28	4.69	0.1478	1.025	
	1	297	75	736.8	46.00	6.63	0.1491	1.034	
	1	297	75	730.4	45.60	12.02	0.1664	1.154	
	1	297	75	737.0	46.01	13.82	0.1664	1.154	
	1	297	75	738.9	46.13	15.14	0.1702	1.180	
	⟨1⟩	⟨297⟩	⟨75⟩	⟨758.9⟩	⟨47.38⟩	⟨22.81⟩	⟨0.1765⟩	⟨1.224⟩	
	⟨1⟩	⟨297⟩	⟨75⟩	⟨758.8⟩	⟨47.37⟩	⟨25.27⟩	⟨0.1726⟩	⟨1.197⟩	
	1	297	75	721.3	45.03	0.60	**	**	
	1	297	75	727.2	45.40	3.30	**	**	
	1	297	75	745.0	46.51	13.20	**	**	
	⟨1⟩	⟨297⟩	⟨75⟩	⟨770.8⟩	⟨48.12⟩	⟨27.19⟩	⟨**⟩	⟨**⟩	
Total data	45	13483	3575	34735.9	1783.18	407.75	8.0779	45.710	MC < 15% data
Mean value									
MC 15	45	300	79	634.8	39.63	9.061	0.1465	1.016	
S-DRY	50	299	78	638.2	39.84	9.75	0.1467	1.017	
S-GRN	2	297	75	758.9	47.38	24.04	0.1747	1.211	

[a] MC 6.8 to 11.7%; see Note 4, Table 25.
[b] Data are not reported. The mean value is adjusted for the runs given.
[c] Numbers in ⟨ ⟩ signify S-GRN; MC > 20%.

TABLE 22—*Thermal conductivity of maple.*

Data Reference	Test Runs	Mean Temperature K	Mean Temperature °F	Density kg/m³	Density lb/ft³	Moisture Content, %OD Weight	Conductivity W/m · K	Conductivity Btu · in./h · ft² · °F	Specimen Description/Remarks
(R)33	1	297	75	625.2	39.03	6.03	0.1387	0.962	soft maple;
	1	297	75	582.0	36.33	6.54	0.1408	0.976	see Note 2, Table 25
	1	297	75	606.0	37.83	6.07	0.1350	0.936	
	1	297	75	605.0	37.77	6.04	0.1373	0.952	
	1	297	75	624.4	38.98	6.33	0.1441	0.999	
	1	297	75	633.3	41.41	6.12	0.1425	0.988	
(R)33	1	297	75	673.3	42.03	5.63	0.1561	1.082	hard maple;
	1	297	75	755.6	47.17	6.08	0.1601	1.110	see Note 2, Table 25
	1	297	75	748.4	46.72	6.10	0.1653	1.146	
	1	297	75	707.1	44.14	6.11	0.1608	1.115	
	1	297	75	679.8	42.44	6.03	0.1598	1.108	
	1	297	75	659.3	41.16	6.57	0.1592	1.104	
	1	297	75	664.0	41.45	5.87	0.1512	1.048	
	1	297	75	687.0	42.89	5.99	0.1553	1.077	
(W)40	1	301	82	551.0	34.4	10.2	0.1370	0.950	silver maple;
	1	301	82	560.6	35.0	10.6	0.1586	1.10	see Note 3, Table 25
(W)40	1	301	82	551.0	34.40	10.20	0.1370	0.950	striped and bigleaf
	1	301	82	560.6	35.00	10.60	0.1587	1.100	maple;
	1	301	82	490.2	30.60	11.10	0.1244	0.863	see Note 3, Table 25
	1	301	82	490.2	30.60	11.10	0.1276	0.885	
	1	301	82	490.2	30.60	11.40	0.1336	0.926	
(M)41	5	303	85	680.8	42.50	0.00	0.1629	1.130	sugar maple;
	26[a]	303	85	660.8	41.25	11.70	0.2019	1.40	heartwood and sapwood
Total data	26	15658	2049	33557.9	1022.45	160.71	9.1479	27.027	MC < 15% data
Mean value									
MC 15	26	299	79	630.0	39.33	6.18	0.1500	1.040	
S-DRY	52	301	82	645.3	40.29	8.94	0.1759	1.220	
S-GRN	...[b]	...	...	...	...	...	...	...	

[a] MC 5.2 to 27.8%; S-DRY; see Note 4, Table 25.
[b] Insufficient data.

TABLE 23—*Thermal conductivity of ash.*

Data Reference	Test Runs	Mean Temperature K	Mean Temperature °F	Density kg/m³	Density lb/ft³	Moisture Content, %OD Weight	Conductivity W/m · K	Conductivity Btu · in./h · ft² · °F	Specimen Description/Remarks
(G)23	1	293	68	740.9	46.25	15.0	0.1558	1.080	ash;
	1	293	68	740.9	46.25	16.0	0.1961	1.360	see Note 1, Table 25
	1	293	68	740.9	46.25	16.6	0.1759	1.220	
	1	298	77	740.9	46.25	15.0	0.1572	1.090	
	1	298	77	740.9	46.25	16.0	0.1975	1.370	
	1	298	77	740.9	46.25	16.0	0.1774	1.230	
(R)33	1	297	75	736.5	45.98	8.59	0.1677	1.163	white ash;
	1	297	75	727.6	45.42	9.73	0.1620	1.123	see Note 2, Table 25
	2	297	75	679.2	42.40	8.81	0.1470	1.019	
	2	297	75	669.9	41.82	8.46	0.1522	1.055	
	1	297	75	619.4	38.67	8.66	0.1354	0.939	
	1	297	75	611.9	38.20	8.56	0.1473	1.021	
	1	297	75	514.8	32.14	8.13	0.1310	0.908	
(W)40	1	301	82	499.7	31.20	9.7	0.1261	0.874	red ash;
	1	301	82	509.4	31.80	9.8	0.1315	0.912	see Note 3, Table 25
	1	301	82	509.4	31.80	8.3	0.1243	0.862	
	1	301	82	520.6	32.50	8.9	0.1278	0.886	
(W)40	1	301	82	639.1	39.90	9.4	0.1587	1.100	white ash;
	1	301	82	639.1	39.90	8.8	0.1529	1.060	see Note 3, Table 25
(M)41	7[a]	303	85	560.7	35.0	15.6	0.1745	1.21	white ash
Total data	17	5070	1312	10707.54	668.45	163.11	2.4758	17.166	MC < 15% data
Mean value									
MC 15	17	298	77	629.9	39.32	9.59	0.1456	1.010	
S-DRY	28	299	78	628.4	39.23	12.03	0.1587	1.101	
S-GRN	...[b]	...	...	...	...	...	...	...	

[a]MC 12.1 to 17.2%; S-DRY; see Note 4, Table 25.
[b]Insufficient data.

TABLE 24—*Thermal conductivity of oak.*

Data Reference	Test Runs	Mean Temperature K	Mean Temperature °F	Density kg/m^3	Density lb/ft^3	Moisture Content, %OD Weight	Conductivity $W/m \cdot K$	Conductivity $Btu \cdot in./h \cdot ft^2 \cdot °F$	Specimen Description/Remarks
(R)33	1	297	75	794.5	49.60	5.43	0.1846	1.280	red oak;
	2	297	75	776.1	48.45	5.81	0.1751	1.214	see Note 2, Table 25
	2	297	75	704.2	43.96	6.37	0.1640	1.137	
	1	297	75	687.0	42.89	7.07	0.1647	1.142	
	1	297	75	685.9	42.82	6.71	0.1576	1.093	
	2	297	75	661.2	41.28	5.58	0.1552	1.076	
	1	297	75	632.9	39.51	5.76	0.1507	1.045	
	4	297	75	615.6	38.43	5.87	0.1500	1.040	
(W)40	1	301	82	559.1	37.4	10.2	0.1644	1.140	red oak;
	1	301	82	559.1	37.4	14.8	0.1644	1.140	see Note 3, Table 25
	1	301	82	610.3	38.1	8.8	0.1687	1.170	
	1	301	82	610.3	38.1	14.5	0.1731	1.200	
	1	301	82	619.9	38.7	8.5	0.1514	1.050	
	1	301	82	619.9	38.7	8.3	0.1644	1.140	
	1	301	82	629.5	39.3	8.4	0.1630	1.130	
(W)40	1	301	82	619.9	38.7	11.7	0.1471	1.020	southern red oak;
	1	301	82	629.5	39.3	8.1	0.1356	0.940	see Note 3, Table 25
	1	301	82	679.2	42.4	13.7	0.1836	1.270	
	1	301	82	690.4	43.1	6.2	0.1572	1.090	
	1	301	82	690.4	43.1	5.9	0.1659	1.150	
	1	301	82	700.0	43.7	11.2	0.1745	1.210	
	1	301	82	720.8	45.0	7.6	0.1644	1.140	
	1	301	82	669.6	41.8	10.8	0.1601	1.110	
	1	301	82	679.2	42.4	10.3	0.1514	1.050	
	1	301	82	679.2	42.4	10.1	0.1442	1.000	
	1	301	82	679.2	42.4	11.0	0.1514	1.050	
	1	301	82	679.2	42.4	10.3	0.1514	1.050	
	1	301	82	690.4	43.1	10.1	0.1514	1.050	
	1	301	82	619.9	38.7	13.4	0.1529	1.060	
	1	301	82	639.1	39.9	8.4	0.1730	1.120	
	1	301	82	639.1	39.9	9.7	0.1702	1.180	
	1	301	82	639.1	39.9	9.1	0.1687	1.170	
	1	301	82	650.3	40.6	8.1	0.1587	1.100	
(M)41	5	303	85	670.9	41.88	0.00	0.1716	1.190	red oak
	12[a]	303	85	620.7	38.75	12.40	0.1947	1.350	
	18[b]	303	85	620.7	38.75	11.1	0.1976	1.370	heartwood and sapwood; see Note 4, Table 25

(G)23	1	293	68	600.7	37.5	13.7	0.1172	0.8128	oak; see Note 1, Table 25
(R)33	2	297	75	822.5	51.35	6.75	0.1794	1.244	white oak; see Note 2, Table 25
	1	297	75	793.6	49.54	7.18	0.1761	1.221	
	2	297	75	778.7	48.61	7.00	0.1696	1.176	
	2	297	75	705.3	44.03	7.26	0.1702	1.180	
	2	297	75	685.4	42.79	6.74	0.1638	1.136	
	2	297	75	640.6	39.99	6.92	0.1565	1.085	
	1	297	75	787.8	49.18	5.13	0.1726	1.197	
	1	297	75	793.6	49.54	7.18	0.1761	1.221	
	1	297	75	790.0	49.32	8.67	0.1781	1.235	
	1	297	75	796.9	49.75	10.61	0.1810	1.255	
	1	297	75	638.3	39.85	6.46	0.1563	1.084	
	1	297	75	640.1	39.96	7.38	0.1566	1.086	
	1	297	75	639.8	39.94	9.35	0.1598	1.108	
	1	297	75	643.3	40.16	11.43	0.1618	1.122	
(R)33	1	297	75	766.5	47.85	4.83	0.1702	1.180	red oak; see Note 2, Table 25
	1	297	75	773.1	48.26	5.54	0.1734	1.202	
	1	297	75	779.1	48.64	6.07	0.1768	1.226	
	1	297	75	778.0	48.57	7.69	0.1761	1.221	
	1	297	75	776.4	48.47	8.76	0.1755	1.231	
	1	297	75	777.5	48.54	10.09	0.1789	1.241	
	1	297	75	694.4	43.35	5.46	0.1615	1.120	
	1	297	75	704.0	43.95	5.54	0.1639	1.137	
	1	297	75	703.2	43.90	5.98	0.1634	1.133	
	1	297	75	704.9	44.01	6.75	0.1644	1.140	
	1	297	75	703.4	43.91	9.13	0.1667	1.156	
	1	297	75	708.5	44.23	11.93	0.1690	1.172	
(W)40	1	301	82	595.8	37.2	15.0	0.1615	1.12	white oak; see Note 3, Table 25
	1	301	82	599.1	37.4	9.3	0.1442	1.00	
	1	301	82	610.3	38.1	8.9	0.1499	1.04	
	1	301	82	640.7	40.0	15.0	0.1817	1.26	
	1	301	82	669.6	41.8	7.1	0.1716	1.19	
	1	301	82	759.3	47.4	9.2	0.1587	1.10	
	1	301	82	679.2	42.4	10.8	0.1644	1.14	
	1	301	82	679.2	42.4	8.9	0.1687	1.17	
Total data	84	25106	6574	42474.5	3600.48	661.57	10.2599	95.417	MC < 15% data
Mean value									
MC 15	84	298	78	686.6	42.86	7.88	0.1638	1.136	
S-DRY	114	300	80	669.3	41.78	8.86	0.1723	1.195	
S-GRN	...[c]	...	...	...	...	...	...	...	

[a]MC 6.5 to 19.4%, red oak.
[b]MC 5.4 to 20.1%, white.
[c]Insufficient data.

TABLE 25—*Specimen preparation and selection/test conditions.*[a]

Note No.	Data Reference	Description of Conditioning	Kiln Temperature, °F	Test Facility	Test Apparatus	Test Temperature, °F	
						Hot Plate	Cold Plate
1.	(G)23	Specimens selected were gradually oven dried at 100°C, measurements of the conductivity being taken with progressively reduced moisture contents. Longitudinal and transverse (radial and tangential) conductivities were measured using a hot plate apparatus. (Transverse conductivities are listed as reported; longitudinal conductivities are indicated in other tables by brackets [ ], and are excluded from the mean value.	212 (100°C)	National Physical Laboratory, Great Britain	electric hot plate	110	64
2.	(R)33	Specimens were conditioned to approximately 8% moisture content at 175°F. Pairs of specimens, 12 by 12 by 1 in. were used in each run. Thickness of the specimens was taken as the distance between the test plates during the test. Conductivity was measured with a 12-in. guarded hot plate. Readings at constant conditions were continued for average periods of 5 h. An asterisk in parenthesis (*) indicates that test specimens were 24 by 24 in., and were measured with a 24-in. hot plate.	175	Experimental Engineering Laboratories, University of Minnesota	guarded hot plate	100	50
3.	(W)40	Specimens selected were subjected to different moisture conditioning treatments to reach and 8 and 14% moisture content. Pairs of specimens 5 by 5 by 1 in. were used in each run. Conductivity was measured with a guarded hot plate. Specimens were kept under test for 2 to 7 h with instrument readings for computation of conductivity taken upon attainment of steady state; the volume of each	221	New York State College of Forestry, Syracuse, NY	U.S. Bureau of Standards guarded hot plate	82	average

		specimen was computed from a measurement of its dimensions. Upon test conclusion, each specimen was dried to constant weight at 221°F to obtain a direct measurement of its moisture content at the time of testing as well as to permit a direct calculation of its specific gravity on the basis of oven-dry weight and volume at the time the conductivity determination was made.				
4.	(M)41	Specimens selected were conditioned to 0% moisture content (oven dry). Twin specimens, 13 1/2 by 13 1/2 by 1/2 in. to 3/4 in. were used in each run. Conductivity was measured with a guarded hot plate. Specimens were kept under test for 24 h with instrument readings for computation of conductivity taken at frequent intervals for a 1-h period. Results were averaged.	Forest Products Laboratory, Forest Service, USDA, Madison, WI	guarded hot-plate	58	average
5.	(L)42	Specimens of dressed lumber were tested as received from the lumber yard. Specimen finish and thicknesses vary as noted.	Engineering Experiment Station, University of Minnesota	guarded hot plate	average	NR
6.	(B)82	A test panel consisting of a 6 by 6-ft wall constructed of lodge pole pine logs, 7 by 7 in. square, butt jointed and caulked. The wall specimen was sandwiched between a hot box maintained at approximate 70°F (21°C) and a cold box maintained at approximate 35°F (1.7°C) and permitted to reach thermal equilibrium. The resulting heat transfer rate through a 48 by 32-in. (1.2 by 0.81-m) metering section was measured. The measurement precision was determined to be better than 5%.	Dynatech R/D Company, Cambridge, MA	guarded hot box	70	35

[a]Metric conversion factors: 1 in. = 25.4 mm.
°C = (°F − 32)/1.8.

Grade Stamp

The Western Wood Products Association grade stamp contains five elements identifying the manufacturer, grade, species, moisture content and certification.

Certification—This symbol indicates that a piece bearing this mark has been graded under supervision of Western Wood Products Association.

Manufacturer—The originating mill is identified by an assigned number or by the firm's name or brand.

Grade—The grade is shown by the grade name or abbreviation. Dimension grade names include:

CONST	= Construction	SEL STR	= Select Structural
STAND	= Standard	1	= Number 1
UTIL	= Utility	2	= Number 2
STUD	= Stud	3	= Number 3

Species Mark—Species or species grouping is indicated by appropriate symbol:

DOUG. FIR-L = Douglas Fir-Larch WW = White Woods

HEM FIR = Hem-Fir = Ponderosa Pine

Moisture Content—Moisture content at time of surfacing is shown by appropriate abbreviation:

S-GRN = (Surfaced Green) Standard size unseasoned lumber with moisture content of 20% or more.

S-DRY = (Surface Dry) Standard size lumber dried to moisture content of 19% or less.

MC 15 = (Moisture Content 15%) Standard size lumber dried to a specific moisture content of 15% or less.

FIG. 1—*Western Wood Products Association grade stamp.*

with the individual species. Thermal conductivity, finally, was reported for major groupings of species.

As documented in the research cited [*8,12*], wood thermal conductivity will vary anisotropically. The data base contained in Tables 10 through 24 includes test data reported in the technical literature. The test measurements utilized in determining a mean value for a given species or grouping of species included only the values reported for transverse conductivity, that is, conductivity measured across the grain of the test specimen. Test data reported for radial or longitudinal conductivity were tabulated but not included in the mean value. The authors recognize that misreporting in the technical literature might result in the inclusion of values measured radially or longitudinally; however, the result would be a higher mean value of transverse conductivity for a given species or group. Finally, the authors established that more than 95% of the data reported in the literature and tabulated in Tables 10 through 24 are test measurements across the grain; hence, anisotropy as an independent variable in this study was ignored.

The influence of moisture content as a variable was maintained as a function of the classifications MC 15, S-DRY, and S-GRN, as defined and discussed previously. Thus, in the analysis and tabulation of each species (Tables 10 through 24) density and k values with a moisture content of 15% or less reported at the time of testing were averaged, and the mean k and mean density values at MC 15 were listed as shown. The identical procedure was used for the S-DRY and S-GRN categories. It seemed reasonable to assume that this procedure would minimize the effect of the variation in reported moisture contents.

The mean temperatures at the time of testing reported by individual researchers ranged from 20 to 29.4°C (68 to 85°F). It was initially assumed that the mean range corresponded to 23.8°C (75°F) and, therefore, was not considered to be a significant variable. However, as will be discussed later, the variation in mean temperature at the time of testing does contribute to discrepancies in reporting design values.

Consistency of Data for Individual Species

Two questions were of concern in analyzing the data within individual species: to what extent were the various test results consitent, and how close was the distribution of data?

In reviewing the data, it occasionally appeared that some researchers' values for conductivity seemed inconsistent with the rest of the data. A statistical analysis was undertaken on what appeared to be the worst species. The analysis was conducted by computing a least-squares linear approximation for the density versus conductivity for the species. Through this procedure the correlation coefficient, R, was calculated as a measure of the consistency of the data.

These calculations were then performed for subsets of the data with each researcher's results removed successively. The correlation coefficients and the precision in predicting the excluded value were then compared. It was assumed that, if the correlation coefficient improved significantly or the prediction precision dropped significantly, the excluded set of data would not be compatible with the total set. In the cases analyzed, the correlation coefficient varied slightly (from 0.888 to 0.808 in white pine), as did the precision of prediction (94 to 98%). Based on these values, the authors decided that none of the data was significantly less precise than the rest; consequently, no researcher's set of data was rejected on a statistical basis.

The compatability of the form of the data posed another problem. Whereas most researchers had published the results of each test, one set of data [*11*] listed the range and distribution of densities and moisture content at the time of testing but not the range of conductivity. This had no effect on calculating the average values, but it obviously affected the standard deviation. A number of procedures to estimate the effect on the standard deviation were considered but were rejected as being either too cumbersome or arbitrary, and unjustifiable. In the calculations, the reported number of test runs from each researcher was used as a weighting factor in arriving at the mean value and the standard deviation.

For all the softwood species, the data were checked to find their conformance to a normal distribution (Table 7). In almost all cases the actual number of points corresponded with the predicted confidence level at one standard deviation and two standard deviations from the mean. The implication from this result is that the data correspond very well with the normal distribution and that all statistical predictions based on this distribution hold for the data.

Grouping of the Data

One of the most crucial steps in this report was grouping the species together into five groups of softwoods and two groups of hardwoods. Three criteria were used to arrive at these groups:

(*a*) standard industry practice, as previously discussed;
(*b*) density and conductivity relations, in which moisture content is minimized as a variable, as previously discussed; and
(*c*) statistical analysis.

When moisture content can be minimized as an independent variable, the thermal conductivity of wood was found to be directly proportional to the density. On this basis, it seemed reasonable to group species by densities, generally following other groupings established by the lumber grading associations. A number of trial groupings were developed and tested statistically for accuracy (Table 6). All groupings taken at a moisture content level of S-DRY, composed of two or three species with similar densities, resulted in the same range of accuracy in predicting the conductivity. The measure of accuracy used was the ratio of the standard deviation to the mean value of conductivity; this value ranged from 9.6% for the proposed group of cypress, redwood, pine, fir, and hemlock to 11.2% for the group of Douglas fir and larch. It should be noted that the current American Society of Heating, Refrigerating, and Air Conditioning Engineers (ASHRAE) [5] listing of all softwoods has a ratio of standard deviation to mean conductivity of 18%. The proposed groups, therefore, almost double the reliability of predicted values.

The statistical analysis justified the concept of presenting the data in the form of individual species as well as smaller groups of two to five species. The precision within the small groups was commensurate with the precision of the individual species and significantly better than that of the ten species taken together. However, the difference in precision between trial groups was not great enough to warrant selection of the one group over another based only on the statistical criterion. The similarity of precision did, however, allow us to base the grouping decision on the other criteria—industry practice and density—without sacrificing any precision.

Empirical Correlations

As a consistency check, the conductivity values derived by mean tabulation as discussed previously, were compared with computed values using the empirical models developed by MacLean [*11*] and Wilkes [*12*]. The results of the comparison are shown in Table 9 and graphically represented in Table 26.

Equation 1 outlines the form of MacLean's empirical correlation, and Eq 2 outlines the form of Wilkes's regression model. Two fundamental differences are observed in the comparison of mean values and the computed values obtained from these equations. First, the MacLean model (Eq 1) was developed using test data obtained at a mean temperature of 29.4°C (85°F), whereas the Wilkes model (Eq 2) was developed using a larger base of test data in which the results are normalized to a mean temperature of 23.8°C (75°F), consistent with ASHRAE handbook reporting criteria. Second, Table 26 shows that the MacLean model provides a conservative estimate of conductivity, where the computed values in some cases are in excess of two

TABLE 26—*Density/conductivity relationship of softwoods at a nominal service moisture content of 12%.*[a]

	Density, lb/ft^3 [b]														
Conductivity, k	21	22	23	24	25	26	27	28	29	30	31	32	33	34	35
1.10														x	
1.05													x		
1.00															
														• southern	
										x				+ pine	
0.95									x				+	*	
0.90						x	x			*	• Douglas fir, larch				
0.85					x					+			*		
0.80					+	*	*		*						
0.75	x			x	*	+	• cypress, redwood, hem-fir, pine								
0.70	*			* • spruce			+		+						
	• cedar			+											
	+														
0.65															
0.60															

[a]Key to symbols:
- • Proposed values for Table 3A in Chapter 23 of *ASHRAE Handbook of Fundamentals.* See density and conductivity values in Table 27.
- \+ Mean values based on measured test data. See mean density and mean conductivity values in Table 9.
- x Conductivity values derived using the empirical model of MacLean. See mean density and Footnote *b* in Table 9.
- * Conductivity values derived using the empirical model of Wilkes. See mean density and Footnote *d* in Table 9.

[b]1 lb/ft^3 = 16.018 46 kg/m^3.

standard deviations from the tabulated means. In comparison, the Wilkes model produces results that are within one standard deviation of the mean or coincident with the mean values.

$$k = g(1.39 + 0.028M) + 0.165 \tag{1}$$

where

k = conductivity,
g = specific gravity, and
M = moisture content.

$$k_{T=75°F} = \frac{k_{air} + (a_0 + a_1 M)d}{(1 + 0.01M)} \tag{2}$$

where

$k_{T=75°F}$ = conductivity at 23.8°C (75°F) mean temperature,
k_{air} = 0.1791,
a_0 = 1.87×10^{-2},
a_1 = 5.753×10^{-4},
M = moisture content, and
d = density, lb/ft^3.

Conclusions

The data tabulated in this paper, to our knowledge, incorporate all the major research done on the conductivity of wood in North America. The grouping of species corresponds very well with similar groupings used by the timber industry in recording other mechanical properties of wood; this grouping, as shown, allows a much greater accuracy in estimating values of conductivity or resistance and consequently affords a more refined procedure for the thermal designer. Within individual species, and within the proposed groupings, the data base shows a close correspondence to a normal Gaussian distribution. This result implies that the values derived in Table 27 can be used with the usual levels of confidence. Further, a consistency check utilizing the empirical model developed by Wilkes demonstrates that mean conductivity values calculated at a mean temperature of 23.9°C (75°F) show close agreement to the subject study. These reasons form the basis for the following conclusions:

1. A statistical analysis of published test data on the thermal conductivity of wood reveals that the use of mean values for representative wood species groups offers increased design reliability over current handbook values.
2. Calculated values for wood thermal conductivity based on the regression model developed by Wilkes in 1979 [*12*] show close agreement with the mean values for representative wood species groups.
3. Calculated values for wood thermal conductivity based on the empirical model developed by MacLean in 1941 must be used with caution because of technical inconsistencies related to thermal transmission test methodology and specimen calibration.

Recommendations

1. The mean values of thermal conductivity for 15 representative wood species groups should be validated experimentally (see Table 27).

TABLE 27—*Summary of the steady-state heat transmission design coefficients for wood (revised 2/84).*[a]

Description of Woods	Density, lb/ft³	Conductivity, k, Btu · in./h · ft² · °F	Conductance (C)	Customary Unit Resistance (R): Per Inch Thickness ($1/k$)	Customary Unit Resistance (R): Per Thickness Listed ($1/C$)	Specific Heat, Btu/lb · °F	SI Unit Resistance (R): m · K/W	SI Unit Resistance (R): m² · K/W
HARDWOODS								
	...	...	...	...	...	0.39	...	...
Oak	46.8	1.06	...	0.94	...		...	...
Maple, birch	46.1	1.05	...	0.95	...		...	...
Ash	43.3	0.99	...	1.01	...		...	...
SOFTWOODS								
	...	...	...	...	...	0.39	...	...
Southern pine	38.4	0.90	...	1.11	...		...	...
Douglas fir-larch (DF-L)	35.6	0.87	...	1.14	...		...	...
Southern cypress	33.5	0.81	...	1.24	...		...	...
Hem-fir (H-F), spruce-pine-fir (S-P-F), redwood (close grain), eastern softwoods	29.4	0.73	...	1.37	...		...	...
White woods (western woods), western cedars	24.5	0.64	...	1.56	...		...	...
Northern white cedar	21.7	0.59	...	1.70	...		...	...

[a]The thermal design values are recommended for incorporation into Table 3A, Chapter 23, 1985 Edition, *ASHRAE Handbook of Fundamentals.* Densities and thermal conductivities are average values for wood in service at 12% moisture content. (Sources: *National Design Specification for Wood Construction,* National Forest Products Association, 1982, Table 8.1A; *Wood Handbook,* U.S. Department of Agriculture, 1974, Tables 3-7 and 4-2, Figs. 3-4 and 3-5; and Wilkes, K. E., "Thermophysical Properties Data Base," *Proceedings,* American Society of Heating, Refrigerating, and Air Conditioning Engineers/Department of Energy/Oak Ridge National Laboratory Conference, ASHRAE SP 28, 1979, Table 1.)

2. Given the relationship of density and moisture content to wood thermal conductivity, standard calibration specimens need to be developed.

3. Test measurement and reporting of other thermophysical properties of wood including density and specific heat must be performed.

4. To maintain data base consistency, test values for thermal conductivity, density, and specific heat should be reported at 4.4, 23.8, and 43.3°C (40, 75, and 100°F) mean temperatures and at a service moisture content of 12%.

Acknowledgments

The authors wish to express their gratitude to the Log Homes Council of the National Association of Home Builders for their sponsorship of this research. The interest and efforts of members of the LHC Technical Committee, particularly Ed Perazone, chairman, were invaluable to the completion of the project.

Sincere appreciation is expressed to the reviewers from industry, educational, and governmental organizations who submitted editorial and technical comments during the revision of this report. These reviewers were selected for their knowledge and experience in wood technology, testing procedures, and material thermophysical property data. The reviewers' names and their parent organizations are as follows: E. King, National Forest Products Association; B. Bohannan and G. Hans, Forest Products Laboratory; N. Pinson, Western Wood Products Association; W. A. Baker, American Plywood Association; G. Kyanka and J. Siau, State University of New York College of Environmental Science and Forestry; D. R. Fairman, Construction Standards Division, Housing and Urban Development; T. Kusuda, National Bureau of Standards; R. Tye, Dynatech R/D Company; R. Godfrey, W. Goss, and K. Wilkes, Owens-Corning Fiberglas; and S. L. Matthews, Rockwool Industries.

References

[*1*] "The National Program Plan for the Thermal Performance of Building Envelope Systems and Materials," ORNL, Sub-7973/1, Oak Ridge National Laboratory, Oak Ridge, TN, March 1982.

[*2*] Burch, D., "A Field Study of the Effect of Wall Mass on the Heating and Cooling Loads of Residential Buildings," National Bureau of Standards, Thermal Mass Effects in Buildings Seminar, Knoxville, TN, 2–3 June 1982.

[*3*] Gustinis, J. and Robertson, D., "Southwest Thermal Mass Study for the First Research Phase," New Mexico Energy Research and Development Institute, Los Alamos, NM, 17 Sept. 1982, draft.

[*4*] "Thermal Mass Assessment," Thermal Mass Effects in Buildings Seminar, Knoxville, TN, 2–3 June 1982, Oak Ridge National Laboratory, Oak Ridge, TN.

[*5*] *ASHRAE Handbook of Fundamentals,* American Society of Heating, Refrigerating, and Air Conditioning Engineers, Atlanta, GA, 1981.

[*6*] "National Design Specification for Wood Construction," National Forest Products Association, Washington, DC, 1982.

[*7*] "Design Values for Wood Construction," National Forest Products Association, Washington, DC, March 1982.

[*8*] Griffiths, E. and Kaye, G. W. C., "The Measurement of Thermal Conductivity," *Proceedings of the Royal Society* (London), Vol. A 104, 1923, pp. 71–98.

[*9*] Rowley, F. B., "The Heat Conductivity of Wood at Climatic Temperature Differences," *Transactions of the American Society of Heating and Ventilating Engineers,* Vol. 39, 1933, p. 329.

[*10*] Wangaard, F. F., "Transverse Heat Conductivity of Wood," *Heating Piping and Air Conditioning,* Vol. 12, July 1940, pp. 459–464.

[*11*] MacLean, J. D., "The Thermal Conductivity of Wood," *Transactions of the American Society of Heating and Ventilating Engineers,* Vol. 47, 1941, p. 323.

[*12*] Wilkes, K. E., "Thermophysical Properties Data Base at Owens-Corning Fiberglas," *Proceedings,* American Society of Heating, Refrigerating, and Air Conditioning Engineers/Department of Energy/Oak Ridge National Laboratory Conference (ASHRAE Special Publication 28), Orlando, FL, 1979.

Jean-Laurent Jaouen[1] *and Sorïn Klarsfeld*[1]

Heat Transfer Through a Still Air Layer

REFERENCE: Jaouen, J. L. and Klarsfeld, S., **"Heat Transfer Through a Still Air Layer,"** *Thermal Insulation: Materials and Systems, ASTM STP 922,* F. J. Powell and S. L. Matthews, Eds., American Society for Testing and Materials, Philadelphia, 1987, pp. 283–294.

ABSTRACT: One question is at the origin of this study—Could a still air layer be used as a reference material in interlaboratory measurements by the guarded hot plate or heat flow meter method? Such a "material," which is easy to get, has indeed been recommended by the International Organization for Standardization (ISO) Technical Committee 163 in a round-robin test, but very few measurement results have been reported from the participating laboratories.

The authors have investigated this configuration of particular interest using a one-specimen guarded hot plate apparatus with a thermal gradient directed downwards. By studying the variation of thermal conductance as a function of air layer thickness, we have been able to determine the thermal conductivity of still air, λ_g, and the total hemispherical emissivity, ϵ, of the measurement surfaces and to compare the experimental data with the numerical results derived from a simple model, taking into account simultaneous radiative and conductive heat transfers and giving the values of λ_g and ϵ.

This study allows us to answer the initial question positively and to show the possible sources of misunderstanding in comparing and analyzing the thermal conductivity measurements from different laboratories.

KEY WORDS: air layer, conductance, conductive heat transfer, emissivity, guarded hot plate, heat transfer, radiative heat transfer, reference material, round-robin test, thermal conductivity, thermal insulation

In 1981, The National Bureau of Standards, on behalf of Technical Committee 163/SC 1/WG 6 of the International Organization for Standardization (ISO), organized a round-robin test to measure with guarded hot plate and fluxmeter apparatus the thermal conductance of a reference glass fiber board and of a 2.54-cm-(1-in.) thick air layer as a second reference material [*1*].[2] The preliminary information sent to the participants included the following instructions:

> The round-robin which you have agreed to participate in begins with guarded hot-plate and heat flow meter measurements on glass fiber board (density approximately 160 kg/m^3), and a measurement of an enclosed air gap 2.54-cm thick for each apparatus. . . .
>
> You are also asked to seal and enclose a 25.4-mm (1-in.)-thick air gap(s) between your hot plate and cold plate(s) and to measure and report the thermal conductance of this air gap at a mean temperature of 23.9°C and a temperature difference of 28°C. This can be accomplished by using an empty thin wood frame at the edges of the plates or by use of other suitable nonconducting spacers as a means to enclose the air gap and seal it airtight.

But is it possible to use an air layer as a reference material for thermal conductivity measurements? This study yielded information additional to that obtainable in ordinary inter-laboratory comparative measurement. As a matter of fact, the authors, by using an elementary model

[1]Physicist and research engineer, respectively, Isover Saint-Gobain, Centre de Recherches Industrielles de Rantigny, Rantigny 60290, France.

[2]ISO/DP 8301, "Thermal Insulation: Determination of Steady-State Specific Thermal Resistance and Related Properties—Heat Flow Aparatus," and ISO/DP 8302, "Thermal Insulation: Determination of Steady-State Specific Thermal Resistance and Related Properties—Guarded Hot Plate Apparatus."

that takes into account combined conductive and radiative heat transfer, could determine the following:

(*a*) the thermal conductivity of still air, comparing it to calculated or measured values found in scientific literature; and

(*b*) the total hemispherical emissivity of the plates of apparatus used by the Centre de Recherches Industrielles de Rantigny (CRIR). This parameter is essential for combined conductive and radiative heat transfer measurements.

Of course, such an analysis can be carried out only by considering heat transfer restricted to conduction and radiation, that is, free of convective movements. To accomplish this, a single-sided guarded hot plate with the hot side on top (downward heat flux) was used. This configuration ensures air stillness. The measurements have not been restricted to the round-robin test goals in that three air layer thicknesses (8, 17, and 26 mm) and three mean temperatures (24, 40, and 60°C) were used.

The reader will find the following information in this paper:

(*a*) a description of the elementary radiative and conductive model used in this study,
(*b*) a description of the method and the facilities used, and
(*c*) an analysis of the results.

Heat Transfer in a Still Air Layer

Consider two infinite, parallel planes at temperatures hot, T_H, and cold, T_C, with a layer of air between them. Assume that the total heat flux, Φ_T, may be written as the sum of conduction, Φ_g, and radiation, Φ_r, heat fluxes, thus

$$\Phi_T = \Phi_g + \Phi_r \tag{1}$$

The conductive heat flux in gas may be expressed by Fourier's law as

$$\Phi_g = \lambda_g \frac{\Delta T}{L} \tag{2}$$

where

$\Delta T = T_H - T_C$,
λ_g = gas thermal conductivity (still air), and
L = the air layer thickness.

Finally, the radiative heat flux through the transparent air layer is as follows

$$\Phi_r = \frac{\sigma(T_H^4 - T_C^4)}{\dfrac{1}{\epsilon_H} + \dfrac{1}{\epsilon_C} - 1} \tag{3}$$

where

σ = the Boltzmann constant,
ϵ_H = the hot plate emissivity, and
ϵ_C = the cold plate emissivity.

Assuming a small temperature gradient, the total heat flux, Φ_T, is given by Fourier's law as

$$\Phi_T = \lambda_T \frac{\Delta T}{L} \tag{4}$$

where λ_T is the total apparent or equivalent thermal conductivity, which will now be called total conductivity.

In the same way, we may express the radiative apparent or equivalent conductivity, λ_r, as follows

$$\Phi_r = \lambda_r \frac{\Delta T}{L} \tag{5}$$

λ_r is now called radiative thermal conductivity.

If we replace $(T_H^4 - T_C^4)$ in Eq 3 by its approximation, $4\bar{T}^3\Delta T^3$, and combine it with Eq 5, we get

$$\lambda_r = \frac{4\sigma\bar{T}^3}{\dfrac{1}{\epsilon_H} + \dfrac{1}{\epsilon_C} - 1} \cdot L \tag{6}$$

When both plates have the same emissivity, $(\epsilon_H = \epsilon_C = \epsilon)$

$$\lambda_r = \frac{4\sigma\bar{T}^3}{\dfrac{2}{\epsilon} - 1} \cdot L \tag{7}$$

Combining Eqs 1, 3, 4, 5, and 7, we get

$$\lambda_T = \lambda_g + \frac{4\sigma\bar{T}^3}{\dfrac{2}{\epsilon} - 1} \cdot L \tag{8}$$

Thus, in the absence of any convective heat transfer and for a given mean temperature, there is a linear relation between the air layer total conductivity, λ_T, and its thickness, L.

The measurement of the total conductivity, λ_T, for different thicknesses and mean temperatures permits the calculation of λ_g, λ_r, and $\epsilon(\bar{T})$, using Eq 8.

Experimental Results

Method

By measuring the total conductivity, λ_T, versus the air layer thickness, L, one can fit $\lambda_T(L)$ by linear regression using the expression

$$\lambda_T = A + B \cdot L \tag{9}$$

If the experimental results represent a good fit of the linear equation, we may evaluate the parameters.

[3]Where $\bar{T} = (T_H + T_C)/2$. This approximation is better than 1%. See the Appendix.

$$\lambda_g = A$$

is the ordinate intercept of Function 9 and

$$\frac{\lambda_r}{L} = B = \frac{4\sigma\bar{T}^3}{\frac{2}{\epsilon} - 1}$$

is the slope of Function 9.

We then plot the total still air layer thermal conductivity versus the layer thickness at different mean temperatures and plate emissivities. From the ordinate intercepts, A, at different temperatures, we plot the still air thermal conductivity versus mean temperatures and, finally, calculate the thermal emissivity values for the three mean temperatures from the slope values, B.

Description of Facilities

Table 1 and Figs. 1 through 3 describe the biguarded hot plate apparatus [*2*] in terms of its characteristics, operating principle, and air-layer specimen frames. Before the biguarded hot plate was operated in this study, its operation was checked. Using a glass fiber reference specimen ($\rho = 80$ kg/m^3), the authors plotted Φ versus ΔT (Fig. 4) for $7 < \Delta T < 20$°C. According to the ordinate intercept of this linear fitted curve, one adjusts the setting point of the rear guard and ring guard temperature differential controllers to set it to zero.

Analysis of the Results

Figure 5 shows the linear dependence of the total apparent thermal conductivity on the air layer thickness for three mean temperatures, and this confirms the validity of the model.

TABLE 1—*CRIR biguarded hot plate apparatus technical performances and characteristics.*

Performance and Characteristics	Values
Specimen:	
Surface, mm^2	400 × 400
Thickness range, mm	5 to 75
Hot plate + cold plate:	
Total surface area: core + guarding, mm^2	400 × 400
Measurement surface area (core), mm^2	230 × 230
Temperature range, °C	0 to 80
Temperature heterogeneity, °C	±0.1
Temperature stability, °C	±0.05
Flatness, mm	0.1
Parallelism, mm	0.1
Thickness measurement, mm	±0.01
Measurement instrument	200.000 points, 1-μV resolution
Time required to obtain a steady state (fibrous insulating material), h	1 to 2
Repeatability, %	±0.5
Accuracy, %	±1

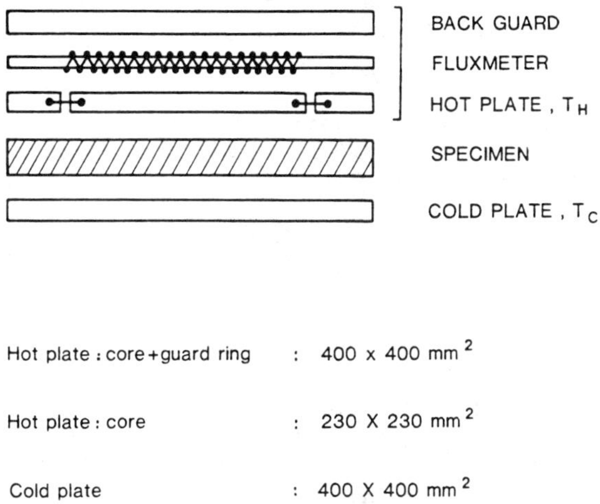

FIG. 1—*Biguarded hot plate apparatus (schematic diagram).*

Numerical regression gives the following values for the still air thermal conductivity and plate surface emissivity:

T, °C	ϵ	λ_g, mW/m · K
24	0.894	26.42
40	0.897	27.68
60	0.904	29.03

Plate Surface Emissivities—A slight dependence of ϵ on temperature was found. The variation obtained cannot be considered significant—~1% between 24 and 60°C—the random error of the apparatus being ±0.5%. The conclusion is that the plate surface emissivity of the CRIR apparatus is equal to about 0.90 between 24 and 60°C.

Air Thermal Conductivity—The authors compared the "measured" still air thermal conductivity with the theoretical values of Kreith [*3*], White [*4*], and Sacadura [*5*] (Fig. 6). The measured data are the ordinate intercept of $\Phi_T = f(L)$ curves. The values are very close to the data of Kreith and Sacadura and 2 to 2.5% greater than White's data obtained by the following expression

$$\lambda_g = 0.2572 \times T^{0.81}$$

with λ_g expressed in watts per metre kelvin, and T in degrees kelvin.

This data agreement may be considered a quantitative check of the elementary heat transfer model. Assuming infinite and isotherm surfaces, the CRIR biguarded hot plate apparatus results agree with those of the model. Thus, for thicknesses up to 26 mm, lateral leakage errors are negligible.

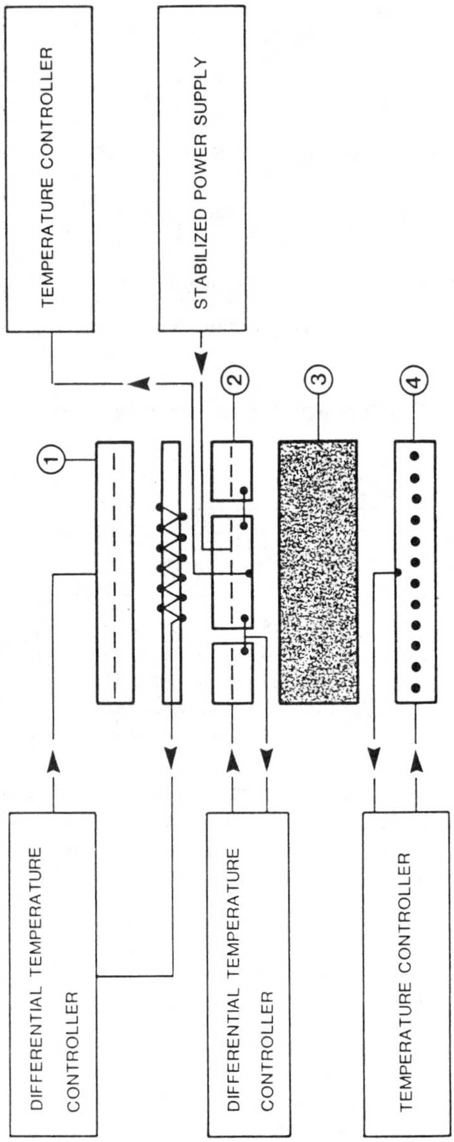

FIG. 2—*Biguarded hot plate apparatus. Temperature and zero-balance controllers.*

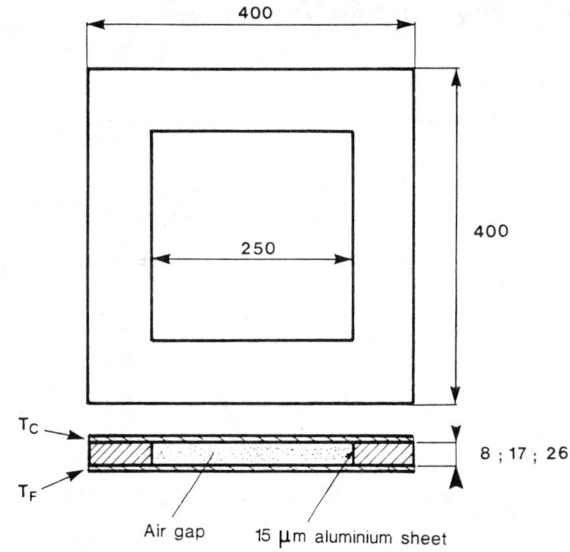

FIG. 3—*Polystyrene frame limiting the air layer.*

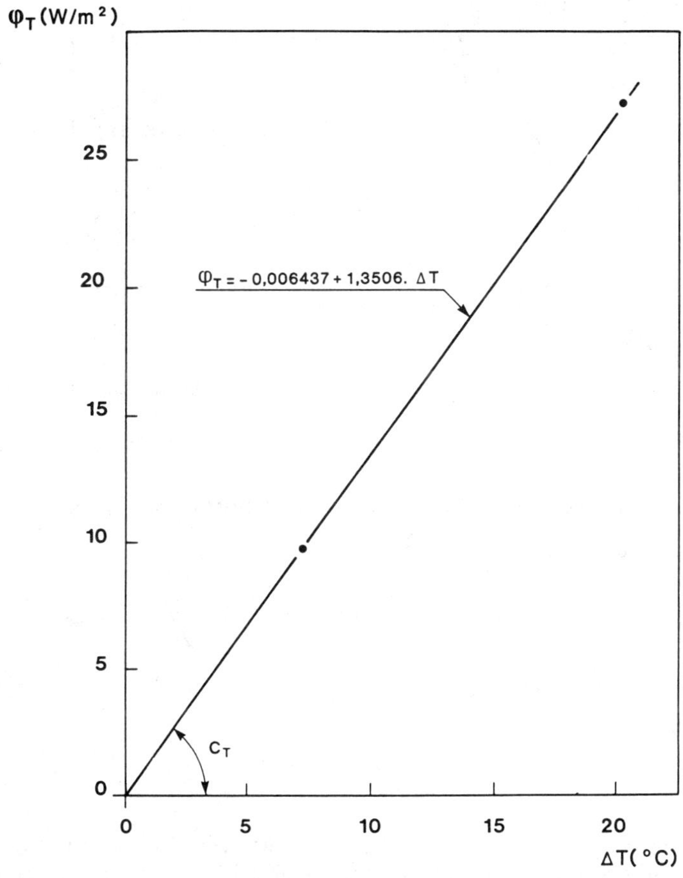

FIG. 4—*Heat flux versus temperature difference at a constant mean temperature ($\bar{T}$ = 24°C).*

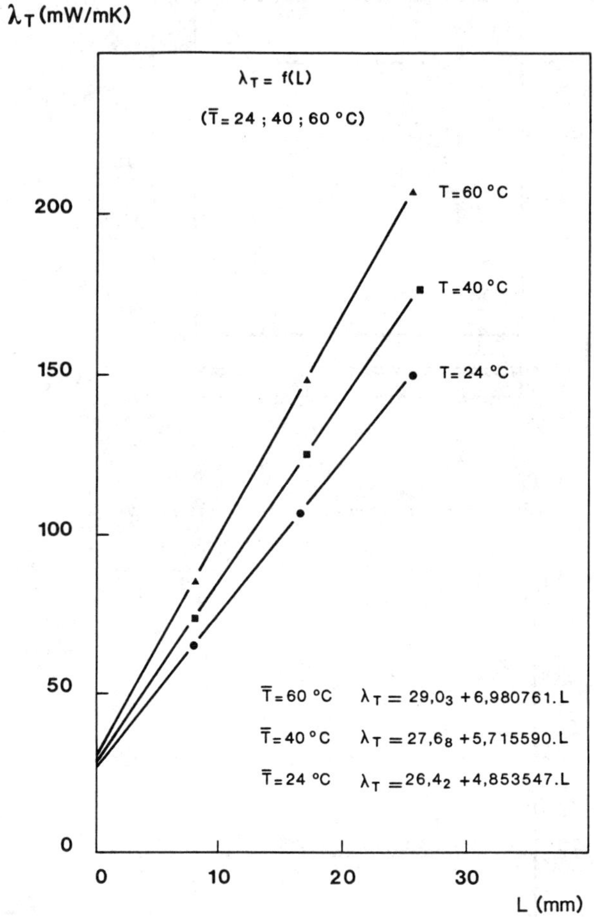

FIG. 5—*Total "apparent" air layer thermal conductivity versus thickness at three different mean temperatures.*

ISO Air Layer Reference—According to these results, the thermal conductance of a 25.4-mm-thick air layer at 24°C would be

$$C_{24} = \frac{\lambda_{g24}}{L} = 1.040 \text{ W/m} \cdot \text{K}$$

and the thermal resistance would be

$$R_{24} = \frac{1}{C_{24}} = 0.962 \text{ m} \cdot \text{K/W}$$

provided that only the gaseous conductive heat transfer separated from the radiant component is considered. This result is independent of surface emissivity and layer thickness [4]. If both

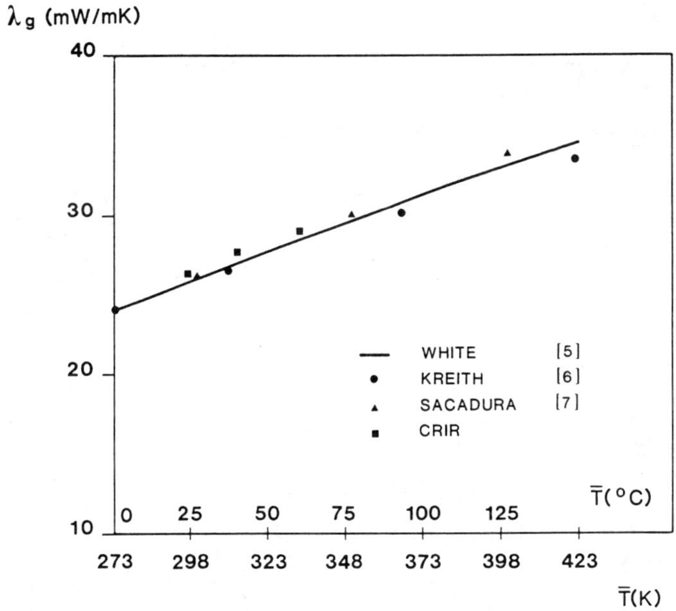

FIG. 6—*Air thermal conductivity versus temperature.*

conductive and radiative heat transfer are considered (with a plate emissivity value of $\epsilon = 0.90$), the total conductance would be

$$C'_{24} = \frac{\lambda_{g_{24}}}{L} + \frac{4\sigma\bar{T}^3}{\frac{2}{\epsilon} - 1} = 5.906 \text{ W/m}^2 \cdot \text{K}$$

or

$$R'_{24} = 0.169 \text{ m}^2 \cdot \text{K/W}$$

In this case, the result depends strongly on surface emissivity. The great difference between C_{24} and C'_{24} is due to the emissivity values.

Figure 7 shows the total thermal conductivity versus specimen thickness, for three emissivity values—0.9, 0.5, and 1.0—at 24°C. The plate emissivities influence the measurement results to a large extent. For $L \to 0$, we have $\lambda_T \to \lambda_g$. Thus, the plate emissivities should be a significant factor for the results of the ISO round-robin test.

Discussion and Conclusions

The total conductivity of an air layer is not an intrinsic property of the layer gas but depends also on experimental conditions, such as the layer thickness and the boundary surface emissivities.

Such comparative interlaboratory measurements make sense if we take into account the respective emissivity of each apparatus before comparing thermal conductivity differences on a

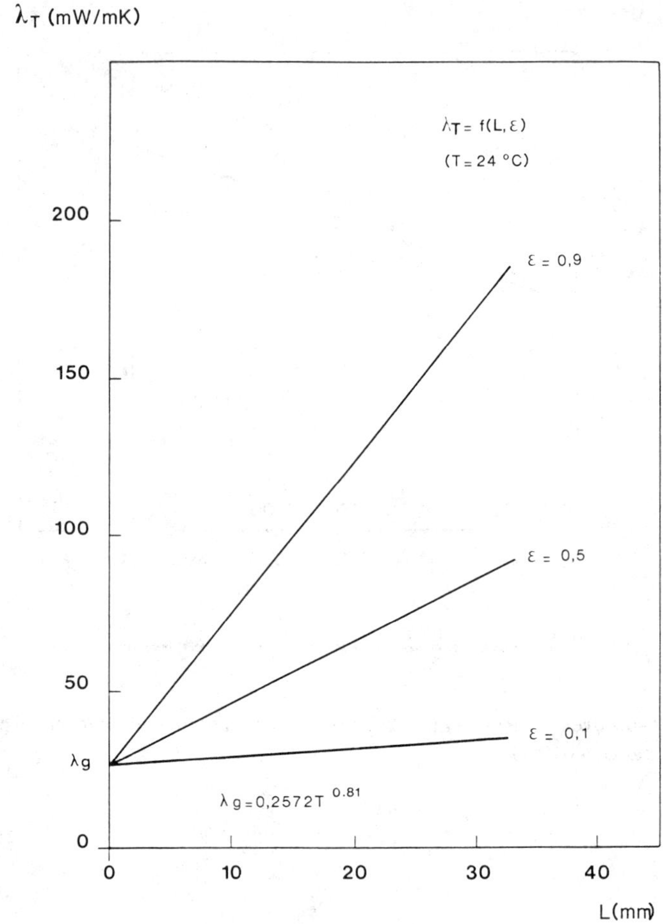

FIG. 7—*Total "apparent" air layer thermal conductivity versus thickness for three (hot and cold) surfaces emissivities.*

given material placed under the same conditions. In addition, if air stillness is ensured, then an air layer could be used as a heat transfer standard reference.

All the measurement data are within a scattering range of $\pm 1\%$, which is in agreement with the accuracy range of the thermal conductivity (or conductance) measurements around the ambient temperature. At the present time, both the model and the experimental results have provided the following data:

(*a*) a better estimate of heat leakages in guarded hot plate apparatus, which are negligible, at least up to a thickness of 26 mm with CRIR 400-mm square plate apparatus;

(*b*) a new approach to measuring phenomena caused by radiation within a range of temperature where optical measurements are difficult to get; and

(*c*) an ability to extend the model to a similar case—combined heat transfer by radiation and gaseous-solid conduction in a semitransparent porous medium. In this case, the porous medium optical parameters will be taken into account.

APPENDIX

Differences between the two expressions, $T_C^4 - T_F^4$ and $4\bar{T}^3\Delta T$, for different ΔT and two mean temperatures.

$\bar{T}$, K	ΔT, K	T_C, K	T_F, K	$T_C^4 - T_F^4$, K^4	$4\bar{T}^3\Delta T$, K^4	$\frac{(T_C^4 - T_F^4) - 4\bar{T}^3\Delta T}{T_C^4 - T_F^4}$, %
297 (24°C)	10	302	292	1.0482×10^9	1.0479×10^9	0.03
	20	307	287	2.0982×10^9	2.0958×10^9	0.11
	30	312	282	3.1517×10^9	3.1437×10^9	0.25
	50	322	272	5.2767×10^9	5.2396×10^9	0.70
	100	347	247	10.776×10^9	10.479×10^9	2.76
	200	397	197	23.334×10^9	20.958×10^9	10.18
333 (60°C)	10	338	328	1.4773×10^9	1.4770×10^9	0.02
	20	343	323	2.9567×10^9	2.9540×10^9	0.09
	30	348	318	4.4401×10^9	4.4311×10^9	0.20
	50	358	308	7.4268×10^9	7.3852×10^9	0.56
	100	383	283	15.103×10^9	14.770×10^9	2.20
	200	433	233	32.204×10^9	29.540×10^9	8.27

References

[1] Powell, F. J., Calibration activities at the National Bureau of Standards, Forum on the Guarded Hot Plate and Heat Flow Meter, ASTM Committee C16 on Thermal Insulation Meeting, 7-8 Oct. 1982, Quebec City, Canada.

[2] Fournier, D., and Klarsfeld, S., "Mesures de conductivité thermique des matériaux isolants par un appareil orientable à plaque chaude bi-gardée," Institut International du Froid, Commissions 2 and 6, Liège, 1969, *Bulletin de L'Institut International du Froid*, Annexe 1969-7, pp. 321-331.

[3] Kreith, F., *Principles of Heat Transfer*, 3rd ed., IEP-A Dun-Donnelet Publisher, New York, 1976.

[4] White, F. M., *Viscous Fluid Flow*, McGraw-Hill, New York, 1974.

[5] Sacadura, J. F., *Initiation aux Transferts Thermiques*, Technique et Documentation, Lavoisier, Paris, 1978.

Hot Boxes

Alan G. Guy[1] *and Jeffrey A. Nixon*[1]

A Detailed Verification Procedure for a Guarded Hot Box

REFERENCE: Guy, A. G. and Nixon, J. A., **"A Detailed Verification Procedure for a Guarded Hot Box,"** *Thermal Insulation: Materials and Systems, ASTM STP 922,* F. J. Powell and S. L. Matthews, Eds., American Society for Testing and Materials, Philadelphia, 1987, pp. 297–309.

ABSTRACT: A verification procedure carried out for a particular guarded hot box facility is described. This relied on measurement of a uniform test specimen of known well-defined properties, these properties having been measured on an accredited guarded hot plate. An analysis of the various energy flows in the box is given, together with procedures for application of various correction measures required. These included a common zeroing of any out-of-balance flux through the metering box walls with that in the specimen across the metering box periphery. The effects of specimen mean temperature were included, and specimen thermal conductances and metering box thermal properties were determined using a linear regression approach. Resulting conductances were in good agreement with those obtained from the hot plate measurements.

KEY WORDS: guarded hot box, verification procedures, metering box, heat exchange coefficients, surface heat transfer coefficients, thermal insulation

A hot box facility was installed at Pilkington Brothers, Research and Development Laboratories, United Kingdom, in 1982 for the measurement of thermal performance levels of building elements containing structural insulation products, glass and glazed assemblies, and other components intended for practical use. The hot box is a "dual" facility (that is, it can operate in either calibrated or guarded mode) and was built by Wiss, Janney, Elstner, and Associates (Illinois). The bulk of the experience accumulated to date has resulted from operating in the guarded hot box (GHB) mode.

Principles of GHB operation are now well founded, and standard test methods are available [for example, the ASTM Test for Steady-State Thermal Performance of Building Assemblies by Means of a Guarded Hot Box (C 236-80)]. However, although in principle GHBs provide an absolute method of thermal property measurement, it would be imprudent to apply any GHB facility without prior verification procedures, as most standard methods emphasize. Such a procedure was carried out for the preceding installation and is described in the following sections. It was aimed at providing detailed evaluation of the various factors influencing operation of this particular facility and assessing likely accuracy levels for eventual practical measurement situations.

Summary Details of the Facility

The facility at Pilkington is completely self-contained, with the box situated in a temperature-controlled room (for calibrated mode measurement purposes), as schematically shown in Fig. 1. The box itself is fairly conventional and was designed according to the guidelines given in an

[1]Senior technologist and technologist, respectively, Pilkington Brothers, Research and Development Laboratories, Lathom, Lancashire, United Kingdom 695 73801.

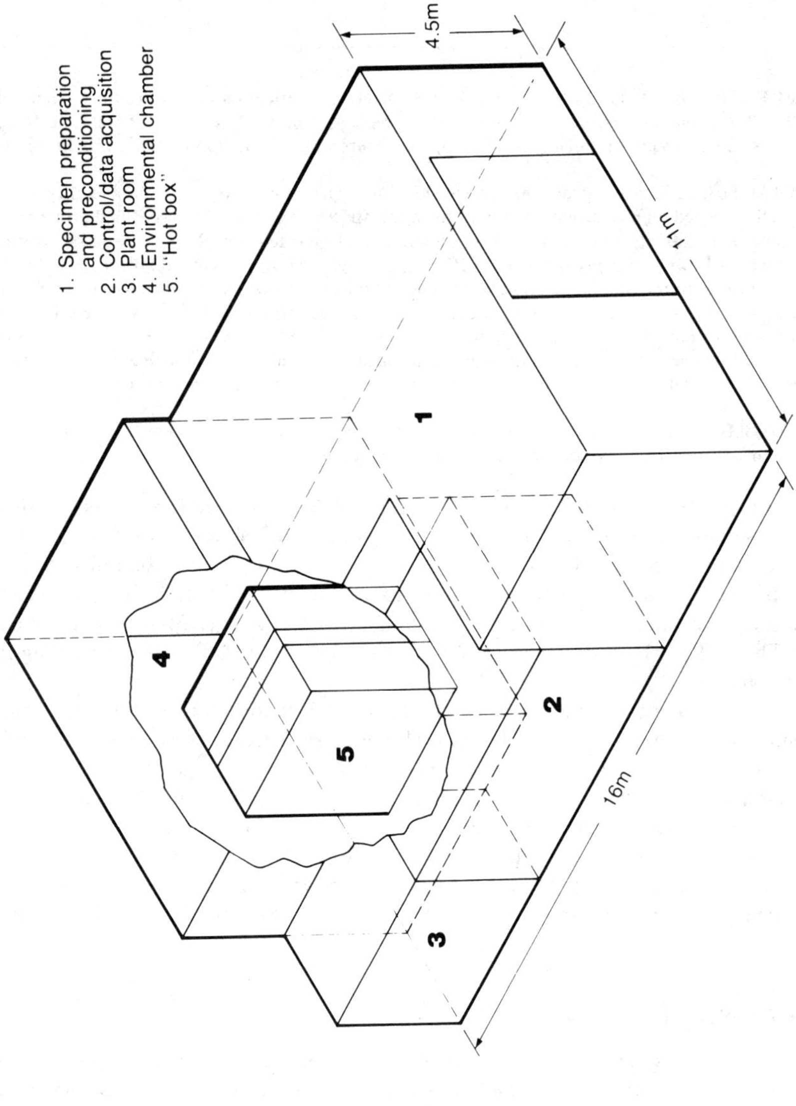

FIG. 1—*Layout of Pilkington thermal transfer test facility.*

earlier version of ASTM Test C 236-80. The main details of the facility are summarized in Table 1 (for the GHB mode).

Principles of Operation of the Guarded Hot Box

The principles of operation of a GHB are fairly simple. Controlled known steady environments (temperatures and air flows, usually appropriate to those in practical use) are set up on either side of a test specimen. The resulting energy flux density (in watts per square metre) through the specimen is then measured through a central test area of the specimen, so that edge effects are small (see Fig. 2 for an illustration).

Two thermal properties of the specimen are defined as follows:

1. Thermal transmittance, or U value, in watts per square metre Kelvin, is given by

$$U = \frac{Q}{T_{HE} - T_{CE}} \tag{1}$$

where Q = heat flux density through specimen, in watts per square metre, and $T_{HE}(T_{CE})$ is an environmental temperature on the warm (cold) side of the specimen, in degrees Celsius or Kelvin. (Environmental temperature, in this instance, is the temperature of a hypothetical uniform environment, with surroundings and air at equal temperature, which gives the same rate of heat transfer through the test element as occurs under the prevailing conditions of surface temperature and heat transfer coefficients.)

2. Thermal conductance, C (in watts per square metre Kelvin), is given by

$$C = \frac{Q}{T_{HS} - T_{CS}} \tag{2}$$

where $T_{HS}(T_{CS})$ is a constant (or appropriately averaged) surface temperature on the warm (cold) side of the specimen (in degrees Celsius or Kelvin). This property is clearly independent of environmental effects, although it may be difficult to measure for specimens that are significantly nonuniform.

TABLE 1—*Main details of Pilkington GHB facility.*

Characteristic	Values
Specimen size, m	2.8 by 2.8, overall (and calibrated) 1.9 by 1.9, guarded area
Design temperatures, °C	−10 to 40, cold side 10 to 40, warm side and laboratory
Nominal wall thermal resistances, m^2K/W	1.25, metering box 7.5, guard box, cold box, and specimen frames
Design air velocities, m/s	0 to 4, cold side 0 to 0.25, warm side
Design specimen transmittance range, W/m^2K	0.1 to 6

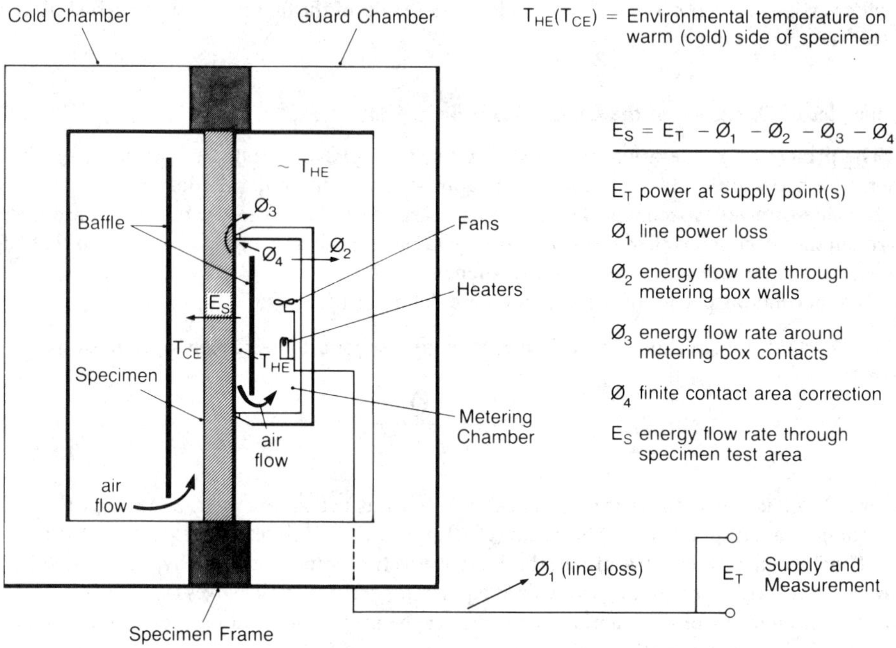

FIG. 2—*Schematic of the guarded hot box and its energy flows.*

Figure 2 also illustrates the various energy flows that can be present and may need to be accounted for over a test measurement. These are as follows:

E_T = total power to the box (heaters plus fans) metered at supply point or points,
ϕ_1 = power dissipated in leads from supply point or points to the metering box,
ϕ_2 = energy flow rate through metering box walls due to (small) temperature differences across the metering box,
ϕ_3 = energy flow rate through the specimen around the contact edges of the metering box due to (small) specimen surface temperature differences in metering and guard areas,
ϕ_4 = a correction factor to account for the finite contact area of the metering box periphery, and
E_S = energy flux through the test specimen.

In relation to the definitions of transmittance and conductance, E_S is the eventual quantity required and

$$Q = \frac{E_S}{A} \tag{3}$$

where A is the metering or measurement area, in square metres, and by above

$$E_S = E_T - \phi_1 - \phi_2 - \phi_3 - \phi_4 \tag{4}$$

Lead losses, ϕ_1, should be negligible if power supply voltages are sensed at the metering box. If sensed more remotely (as for the box here), they will usually form a (small) almost constant proportion of the total supply power (heaters and fans). Of course, if heater/fan configurations are changed, this proportion may also change.

Ideally, a GHB should be operated so that ϕ_2 and ϕ_3 are as small as possible, and preferably both zero. This latter may not always be achievable in practice, as it usually demands very close matching of metering and guard space environments.

The factor ϕ_4 is usually small. It can be allowed for up to fairly large contact widths (100 to 200 mm or so) by suitable definition of an effective test area. However, for large metering boxes with narrow contact widths (as in the box here) it can be shown to be negligible. It will not be considered further here.

Energy Measurement, Corrections, and Accuracies

In principle, a GHB provides an absolute method of thermal conductance/transmittance measurement, and so calibration procedures should be minimal, if not unnecessary. However, some calibration measures are advisable, particularly for determining the ϕ_2 and ϕ_3 corrections in Eq 4. While insulation and construction properties of the metering box are known, so that its heat loss coefficient should be calculable, it would be surprising if this were very accurate (because of uncertainties due to its tapered edge/neoprene gasket, cables passing through, and so on, these factors being difficult to account for fully). It appears preferable to determine the metering box heat transmission properties by direct measurement (for example, ASTM Test C 236-80). This procedure is described later.

For the box here, the line losses, ϕ_1, were fairly straightforward to obtain. Oscillator bridge measurements showed nonohmic reactance effects for the alternating-current (A-C) heater circuits to be negligible [the metering side fans are all direct current (DC)]. The heater line ohmic losses were determined simply by measuring resistances with heaters in and (shorted) out of circuit. Combinations of these line measurements agreed well with each other when various series/parallel combinations of heaters and associated power line connections were switched in and out. As would be expected, low-current (series) heater operation is most beneficial for reducing line losses (that is, high heater resistance). For the Pilkington GHB, this gives line losses of 0.3% (of the total to heaters) compared with losses of 1.0% for high-current (parallel) heater operation, and it would seem that low-current operation is favored. However, this turns out to be incorrect.

The energy measurement devices are watt meters with pulsed outputs (0.1 W · h pulses for the heaters, 0.01 W · h for the fans—Ohio Semitronics Inc. instruments). The characteristics of the watt meter used for measuring heater power (here supplied by pulsed "burst firing" at constant voltage) are such that pulsed high-current heater operation is advantageous, since accuracy levels (based on manufacturer's figures and allowing for up to a 4-degrees-Celsius control room temperature swing) are ±0.4% and ±0.7% of reading, respectively, for high and low-current heater operation. Line losses can probably be corrected for to a confidence level of ±0.1% at high current and ±0.03% at low current (that is, 10% relative). Thus, high-current heater operation succeeds as the favored mode of operation, with final likely accuracy to ±0.5%, compared with ±0.73% or thereabouts for low current.

Fan energy measurement accuracies, though ostensibly relatively coarse at low power inputs (typically ±5%), have much less influence on the overall accuracy levels, since air circulation conditions within the metering box will normally be maintained close to natural convection, requiring little power input. For the test measurements described later, metering box fan energies represented roughly about 3% of the total energy input, so the uncertainty contribution to the total was about ±0.15%. This yields an overall uncertainty level in measured total energy of about ±0.7% for high-current heater operation.

To this must be added uncertainty contributions arising from the following:

(*a*) heater energies only being measurable in pulses of 0.1 W · h, and
(*b*) energy variations due to (small) temperature fluctuations coupled with associated thermal inertia effects (of the specimen and metering box).

Both of these factors can be made small by adopting suitable measurement procedures, which are, first, extending the time intervals between energy measurements (so that 0.1 W · h becomes less significant) and, second, repeated sampling to reduce standard errors.

Temperature Measurements and Accuracies

The main method of temperature measurement uses Type T (copper-constantan) thermocouples. All are made from the same wire batch. Each is ice referenced, and there is facility for up to 256 of these. The thermocouples measuring guard box air temperatures, and all the cold box temperatures, are hard-wired through the boxes to their ice-reference points to avoid junction effects. Each of the remaining couples has a single Type-*T* plug/socket connector either in the guard box (metering box thermocouples) or outside the box (test specimen couples). Although junction effects should be small for these, since the connectors are situated within controlled environments, all the connectors are additionally enclosed within thermally insulated boxes to reduce junction effects even further.

All the thermocouples supplied had welded junctions, and the level of measurement agreement between them (checked by the suppliers by "bathing up" 100-plus couples chosen at random) was quoted as "to within ±0.06 degrees Celsius." (This was largely borne out by the subsequent test measurements.)

Thermocouple voltages, and all other required outputs are monitored by a digital volt meter (DVM) scan system (Microconsultants Group) with automatic minicomputer (PDP 11/23, Digital Equipment Corp.) initiation. At each scan, measurements are also taken of a standard short circuit and a standard cell voltage for zero referencing and scaling purposes. (This is usually unnecessary, but it allows for possible drift of scanner components and compensates for consistent apparatus measurement errors.) Measurement uncertainties associated with the scan equipment (manufacturer's data) are equivalent to ±0.08 degrees Celsius. However, any remaining consistent attributes of this should be removed by the eventual temperature differencing process required for thermal property determination.

The zero reference junction compensation equipment is designed to maintain reference temperatures to "better than ±0.05 degrees Celsius" (with stability characteristics of ±0.002 degrees Celsius). This could give measurement uncertainty in temperature difference calculations if mean temperatures are referenced to different units. This does occur in the Pilkington box for thermal transmittance measurements (but not thermal conductance, in which case temperatures on both surfaces of a test specimen are referenced by the same unit). However, resultant uncertainty levels associated with these effects are quite satisfactory; for a mean temperature difference of 20 degrees Kelvin, they would amount to, at most, ±0.5%

In practice, other factors associated with the box itself may influence accuracy levels (for example, lateral variations in air flows, inconsistencies in surface fixing of thermocouples, variations in surface emittances, and so on). These are not easy to account for, and it is probably wise to allow an additional ±0.5% or so uncertainty for these.

In light of the observations of this and the previous section, the minimum limits that can be set for overall uncertainty levels under typical conditions of measurements will be ±1.2% for thermal conductance measurements, and some 0.5% higher for thermal transmittance measurements. By measuring at many points, random measurement uncertainties can be made small. Typically the number of measurements taken for any one regimen in the Pilkington GHB is about 20 (for each of the specimen's surface, air, and radiating baffles). In the test measurements described in the following sections, random (standard) errors in measured total energies and mean temperature differences across the test specimen could readily be kept at the ±0.2% level. Hence, estimated overall accuracy figures were about 1.5 to 2% for the eventual conductance measurements. This is very satisfactory, as current guarded hot plate test methods[2] regard ±5% as acceptable, and ±3% as "correct."

Measurement and Verification Procedures

An obvious means of evaluating GHB performance is to measure a "standard" test specimen of known, well-defined thermal properties. This was the method adopted here, and for this purpose about three dozen high-density glass-fiber boards were used, having the following nominal characteristics:

thickness	50 mm
size	1.4 by 1.2 m
density	110 kg/m^3
facing	unreinforced aluminum foil
backing	plain paper

Each board was cut down to 0.93 by 0.70 m (to remove edge variations and also to tessellate it into an overall specimen size of 2.8 by 2.8 m). These were individually weighed, and boards of extreme density and thickness were excluded. A 2.8-m by 2.8-m by 99-mm (double thickness) test specimen of an average measured density of 105 kg/m^3 was made up from 24 of the remaining boards and set in a specimen frame. This was then instrumented with thermocouples (approximately 40 per side, 20 of these being uniformly distributed in the metering area). The specimen surfaces were then painted matt black to enable radiation effects to be determined more accurately.

Two boards of lowest and highest densities from those originally excluded were measured internally for thermal conductivity (single thickness) on a heat flow meter hot plate. In addition, two specimens (both of normal density) were submitted externally for similar measurements on a guarded hot plate having British Calibration Society (BCS) accreditation.

With the test specimen assembled into the GHB, test conditions were set up as follows:

warm side (metering and guard)	20°C, almost natural convection (0.05 to 0.1 m/s air speed),
cold side	0°C, approximately 1 m/s air speed (specimen mean temperature, 10°C).

The first phase of the measurements was geared to produce operating conditions that allowed accurate determination of the energy flux through the specimen test area. In essence, this meant determining the behavior of the energy correction terms, ϕ_2 and ϕ_3. Both of these are dependent on temperature differences between metering and guard areas, ϕ_2 being explicitly related to metering box average surface temperature differences, ΔT_m, and ϕ_3 to specimen average surface temperature differences, ΔT_s, between these areas (that is, "edge" effects).

It was anticipated that there would be little difficulty in operating the box, so that ΔT_m (and thus ϕ_2) and ΔT_s (and thus ϕ_3) could, with suitable guard space air flow adjustment, be made zero at the same time, and this was confirmed. Figure 3 shows the measured relative behavior induced by varying guard space temperatures of the two temperature differences involved at eventual preferred guard space air flow conditions. Regression procedures indicated that the

[2]BSI 874, Methods for Thermal Conductivity Determination, 1973, currently under revision.

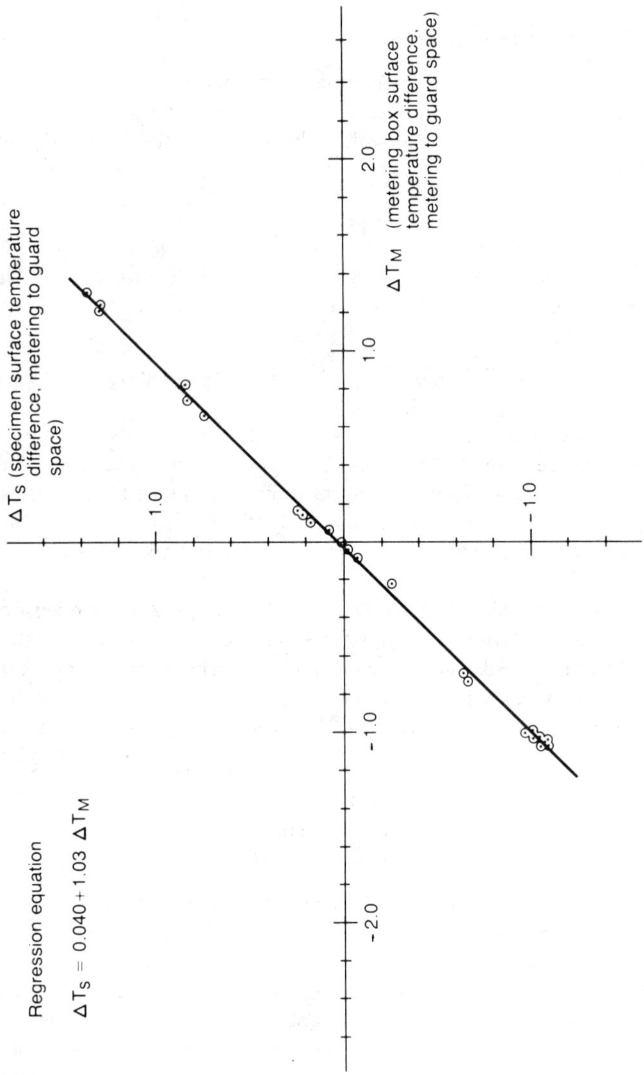

FIG. 3—*Metering to guard space temperature difference behavior.*

temperature difference behavior was predominantly linear (linear equation given on Fig. 3) with only a 0.04-degree-Kelvin difference if either ΔT_m or ΔT_s was controlled to zero. (Other less-preferred guard space air flow conditions produced about a 0.1 to 0.2-degree-Kelvin difference). Finer tuning could probably have further reduced this, but on grounds of accuracy it was not worth the effort.

The advantage of the preferred mode of operation is clear. Temperature difference across the metering box surfaces, ΔT_m, now becomes the sole determining factor for correcting energy inputs (after correcting for line losses). Equation 4 may be written (assuming ϕ_2 versus ΔT_m behavior is predominantly linear)

$$E_S = E_T' - \alpha \Delta T_m \tag{5}$$

where E_T' is the energy supplied, corrected for any line losses, and α is a total heat exchange coefficient accounting for metering box and edge effects (in watts per degree Kelvin).

Under these conditions, if the GHB can now be operated so that ΔT_m is controlled to zero, the only corrections necessary are those for line losses. Such control was available here, though it was not used at the time (in order to determine heat loss characteristics of the metering box system for future test purposes).

The remainder of the test measurements were directed at obtaining both metering box heat transfer properties and sample thermal conductances for the following three nominal temperature ranges.

Warm Side, °C	Cold Side, °C	Mean, °C
35	15	25
27.5	7.5	17.5
20	0	10

Because during a sequence of test measurements it was not always possible to control temperatures on either side of the specimen to give exact matching for each measurement within that sequence, it proved convenient to normalize Eq 5. There are a variety of ways of doing this, and the one chosen was as follows.

Letting ΔT be the measured average surface temperature difference across the test specimen for the (equilibrium) period of measurement, Eq 5 may be written

$$E_T' = AC\Delta T + \alpha \Delta T_m \tag{6}$$

where A is the metering area, in square metres, and C is the required specimen conductance, in watts per square metre Kelvin (see Eq 2).

It follows that

$$\left[\frac{E_T'}{A\Delta T}\right] = C + \alpha \left[\frac{\Delta T_m}{A\Delta T}\right] \tag{7}$$

and plotting variations between the two bracketed terms (induced by altering guard space temperatures) for a series of measurements should produce approximately straight-line behavior of Intercept C and Slope α.

Results and Comparison with Other Measurements

Thermal Conductances

Figure 4 shows the measured behavior resulting from Eq 7 for all the measurements made at the different mean temperatures of the specimen. Over the range of ΔT_m variations investigated (approximately $\pm 1.5°C$), the expected linearity of Eq 7 was confirmed, with appropriate (linear) regression equations as given in Fig. 4.

The thermal conductance properties of the specimen (in watts per square metre Kelvin) given by the regression equations are shown in Table 2, together with those obtained from the hot plate measurements normalized to double thickness.

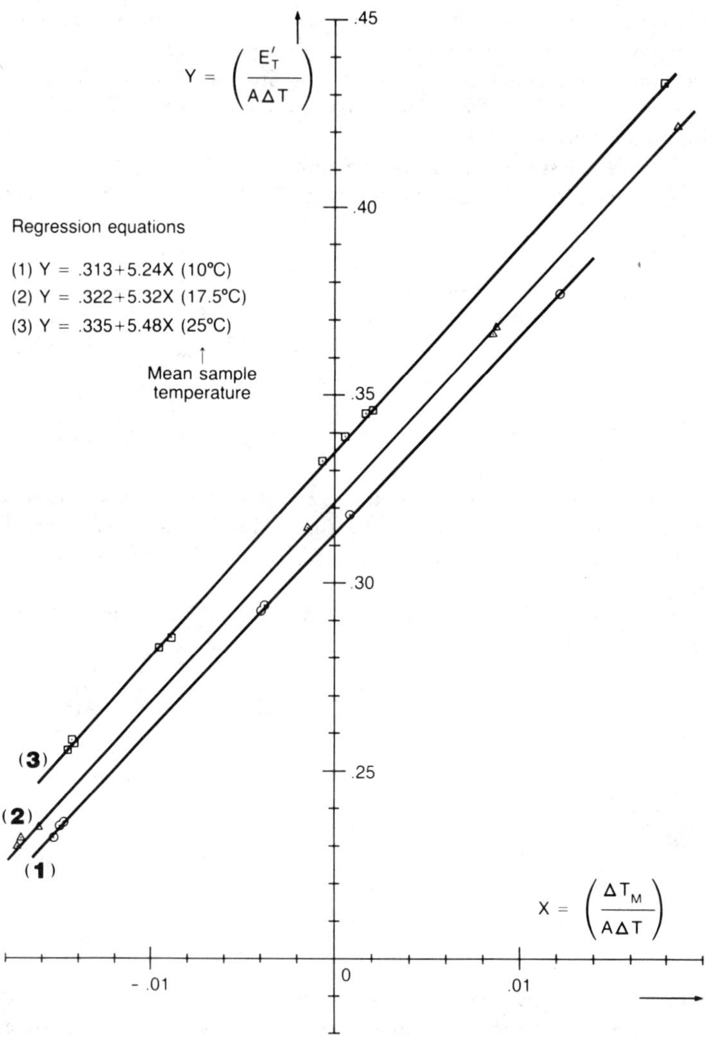

FIG. 4—*Normalized energy behavior.*

TABLE 2—*Conductance comparisons.*

Test Apparatus	Specimen		Measured Thermal Conductances, $W/m^2 \cdot K$ at Three Mean Temperatures, °C		
	Thickness, mm	Density, kg/m^3	10	17.5	25
Heat flow meter hot plate	50 by 2	67[a] to 112	0.308 to 0.315	...	...
BCS accredited guarded hot plate	49.4 by 2	102	0.312	0.322	0.333
Pilkington GHB	99	105	0.313	0.322	0.335

[a]This low density resulted from a "layer" deficiency for the original pressing of that particular glass-fiber board.

Agreement of the Pilkington GHB results with those from the accepted hot plates is clearly very satisfactory. In the longer term, it is preferable to repeat verification checks of this nature (for example, annually). The standard test specimen measured in Table 2 is being retained for this purpose along with other similar ones.

Metering Box Heat Transfer Coefficients

The heat transfer coefficients, α (in watts per degree Kelvin), given by the regression equations are shown in Table 3.

Although these figures represent a combined heat exchange coefficient for metering box and specimen edges in the test area, the probable contribution due to the latter, while not explicitly required, can be estimated theoretically, for example, by numerical means. It depends mainly on specimen thermal conductivity, thickness, and metering area perimeter length. Here, a difference ΔT_s in specimen surface temperature between points situated 50 mm from each side of the metering box periphery (the measured position) was predicted to give a contributory heat exchange coefficient of about 0.08 W/K. For the nominal test conditions, it meant that heat flow through the metering box walls accounted for about 98.5% of any energy correction required to compensate for temperature imbalance between metering and guard boxes. Thus, actual metering box heat exchange coefficients should be about 0.985 of the measured α values. For example, at a mean metering box temperature of 20°C, the metering box heat exchange

TABLE 3—*Measured heat exchange coefficients,* α.

	Mean Temperature of Metering Box, °C		
	20	27.5	35
α, W/K	5.24	5.32	5.48

coefficient would then be 5.16 W/K. The corresponding value based on known properties and dimensions (including estimation of tapered edge and gasket effects, but not wiring passing through the box) is 4.92 W/K, so agreement is reasonable. Future GHB measurements on different types of test specimens will provide further information on this.

Surface Heat Transfer Coefficients

From the measured temperatures and energy flows, estimates of surface heat transfer coefficients on hot and cold sides of the test specimen can be obtained. For example, the energy flux density, Q, through the specimen is given by

$$Q = h_r \Delta T_r + h_a \Delta T_a \tag{8}$$

where h_r is a radiation coefficient of heat transfer for the surface and its radiating surroundings (temperature difference ΔT_r), and h_a is a convection coefficient (natural or nearly so on the warm side, forced on the cold side, appropriate temperature difference ΔT_a).

The radiation coefficient h_r is given by $4\sigma T_r^3 f$, where T_r is an appropriate "mean radiating temperature" (in degrees Kelvin), and f is the view factor for the specimen surface to its radiating surroundings, which here can be taken to be that for two parallel planes of high emittance (for this GHB, with blackened surfaces, emittances should be >0.9, and probably at the 0.95 level). These may readily be estimated for hot and cold sides, so that surface convection heat transfer coefficients may then be determined from measured values of Q and ΔT_a. These are given in Table 4 (averaged), together with corresponding values as calculated from standard relations (see, for example, Ref *1*) applied for the experimental conditions in the GHB (assuming surface emittances of 0.95 within the GHB). Because of the now relatively low determining temperature differences, ΔT_r and ΔT_a (typically <1 K), uncertainties in the calculated convection coefficients are much greater than those in conductance or transmittance measurements. In this instance, accuracy levels were about ± 10 to 20%.

Agreement was, for the most part, within the estimated levels of experimental uncertainty. The slightly high nature of the measured warm side coefficients (in relation to the standard, natural convection, values) may possibly have been due to the slight air movement (0.05 to 0.1 m/s) used in the metering box for aiding control purposes. Future tests will undoubtedly provide more information in this area.

Conclusions

The verification procedure used here was found to provide a very satisfactory means of checking the performance of a particular guarded hot box facility. Good agreement was obtained (within 1% over a range of temperatures) between the results from the hot box and those from an accredited guarded hot plate for the thermal conductances of a measured test specimen. In addition, the procedures have also provided valuable data (for example, metering box heat exchange coefficients, surface heat transfer behavior) for the subsequent operation of the facility.

TABLE 4—*Average convection coefficients,* h_a, *W/m² · K.*

Warm side	1.9 to 2.5	(approximately 1.4 to 1.9 [*1*])
Cold side	9.5 to 11.5	(9.1 to 10.3 [*1*])

Acknowledgments

The authors wish to thank A. Simpson of Salford University and D. Brown of Fibreglass PLC for carrying out the hot plate measurements.

This paper is published with the permission of the directors of Pilkington Brothers and A. S. Robinson, director of Group Research and Development.

Reference

[*1*] *CIBS Guide,* Section C3, Curwen Press, London, 1980.

David J. McCaa,[1] Edward D. Pentz,[1] John Carre,[1] and Lester J. Infante[1]

Experiences in Identification of Thermal Bridging and Elimination of the Thermal Short

REFERENCE: McCaa, D. J., Pentz, E. D., Carre, J., and Infante, L. J., **"Experiences on the Identification of Thermal Bridging and Elimination of the Thermal Short,"** *Thermal Insulation: Materials and Systems, ASTM STP 922,* F. J. Powell and S. L. Matthews, Eds., American Society for Testing and Materials, Philadelphia, 1987, pp. 310–317.

ABSTRACT: Our experiences encompass basic laboratory studies using the guarded hot box, field identification of the building envelope defect, and correction of the problem based on proven techniques and products.

In the laboratory, a total of 24 wall configurations were tested using ASTM Test for Steady-State Thermal Performance of Building Assemblies by Means of a Guarded Hot Box (C 236-80). Comparing theoretical *R* versus actual *R* values enabled us to identify several causes of thermal shorts along with techniques to correct the defects.

This work was related to the field as we observed:

(*a*) thermal bridging due to construction design,
(*b*) insulation gaps due to workmanship defects, and
(*c*) inability of certain construction systems to recover from high humidity and condensation conditions.

Considerable attention was given to these factors during our introduction of two new fiberglass products. In both the retrofit and new construction field, examples illustrate how the insulation was used to correct the envelope defects as well as how to avoid the defects.

KEY WORDS: thermal insulation, heat transmission, guarded hot box, thermal resistance, thermal short circuit, wall systems

Our experiences encompass basic laboratory studies using the guarded hot box, field identification of the building envelope defects, and correction of the problem based on proven techniques and products.

Extensive field tests have been conducted in connection with two new insulation products, an unbonded loose-fill fiberglass insulation and a spray-applied fiberglass insulation system. Two areas of excessive energy losses were clearly revealed.

(*a*) thermal bridging due to construction design, and
(*b*) insulation gaps due to workmanship defects.

[1]Group leader, senior project leader, project engineer, and manager, Thermal Testing Laboratory, Residential Projects Development, respectively, Insulation Group, Research and Development, CertainTeed Corp., Blue Bell, PA 19422.

Thermal Bridging

These problems were reproduced in the thermal laboratory for further analysis with the guarded hot box (GHB). This testing confirmed field identification of thermal bridges and shorts and enabled the recommendation of various uses of insulation products to eliminate the thermal short.

The test wall sections had overall dimensions of 3 by 3 m (9 by 9 ft) and were tested in the guarded hot box with a 1.17 by 1.17-m (4 by 4-ft) metering area. All the wall sections were tested in accordance with the ASTM Test for Steady-State Thermal Performance of Building Assemblies by Means of a Guarded Hot Box (C 236-80). As background information, a series of 23 wall configurations were tested, which included conventional walls and a few new wall concepts. The basic components of the wall, from interior to exterior skin, were gypsum wallboard, metal or wood studs, fiberglass insulation, sheathing, and vinyl siding. Figure 1 shows the wall configurations constructed. Figure 2 is a sketch of the guarded hot box test panels.

For all the tests discussed here, the guarded hot box was run with the wall sections in their normal vertical position, resulting in horizontal heat flow from the gypsum side of the wall to the vinyl siding. Nominal hot and cold side air temperatures [75 mm (3 in.) from the surfaces] were 37.7 and 10.0°C (100 and 50°F), respectively. In addition, fans were used to set up air flows on the hot and cold surfaces with average velocities of 9.12 m/min (30 ft/min) on the hot surface and 11.5 m/min (38 ft/min) on the cold surface.

In the metering area, nine surface and nine air thermocouples are used to obtain a good average value on both the hot and cold sides of the wall.

Of interest in this testing was a comparison of calculated results using the American Society of Heating, Refrigerating, and Air Conditioning Engineers (ASHRAE) standards parallel path, including studs to the actual results (ASTM Method C 236-80). Our past experience has shown the calculated and experimental data to be within a 10% tolerance, unless a thermal short or construction defect is present. Since some of the sheathing material was insulated sheathing, it was necessary to establish the *R* value variability (measured versus advertised). This was completed using a heat flow meter apparatus as in the ASTM Test for Steady-State Heat Flux Measurements and Thermal Transmission Properties by Means of the Heat Flow Meter Apparatus (C 518-80). Without stating trade names, the comparisons of *R* values are given in Table 1.

With the data shown, all but one sheathing material was within a 10% tolerance (Foam D at 75% of advertised). A decision was made to use the actual value rather than the claimed value in the calculated *R* value.

Table 2 is a summary of several wall systems tested.

Of interest are the results with metal studs. In most of the walls tested, the experimental *R* value was higher than calculated, but with the metal studs, the measured *R* value was *24.5%* lower. This difference was caused by the thermal short of the metal studs, as similar panels with wood studs did not show this trend. This energy loss is significant and should be taken into consideration in wall designs.

Included in the overall laboratory test program was a panel made by a commercial manufacturer. The panel consisted of corrugated metal, one full-size metal purlin and 25.4-mm (1-in.) vinyl-faced fiberglass board. The panel tested as received had an *R* value of 15.78. After correcting shorts caused by metal clips holding the board to the purlin, and correcting a 3% void area, the experimental *R* was increased 6.4% to an *R* of 16.79.

Tests were also conducted on measuring the effect of twelve metal pins inserted through a sheet of 25.4-mm (1-in.) foil-faced polyisocyanurate foam sheathing 1.22 by 1.22 m (48 by 48 in.). These were typical fastening pins and clips. The 2.54-mm-(0.10-in.) diameter pins, 50.8 mm (2 in.) in length, were welded to 50.8-mm-(2-in.) square sheet metal tabs, and the clips were 38.1 mm (1.5 in.) in diameter. Each pin and clip formed a thermal short, which reduced the effective *R* by more than *50%*.

R value ($ft^2 \cdot h \cdot °F/Btu$)—1-in. foam only = 6.94
R value ($ft^2 \cdot h \cdot °F/Btu$)—1-in. foam and clips = 3.26
R value ($ft^2 \cdot h \cdot °F/Btu$)—1-in. foam and clips with 1-in. sprayed fiberglass = 6.94

As the data show, applying 25.4 mm (1 in.) of sprayed fiberglass over the foam with the metal clips still present brought the R value back to the original value of the foam alone. Based on a resistivity of 4.0/25.4 mm (4.0/in.), 45 mm (1.75 in.) of the sprayed fiberglass would equal the performance of the combined system (foam and spray), with obvious reductions in material and labor costs.

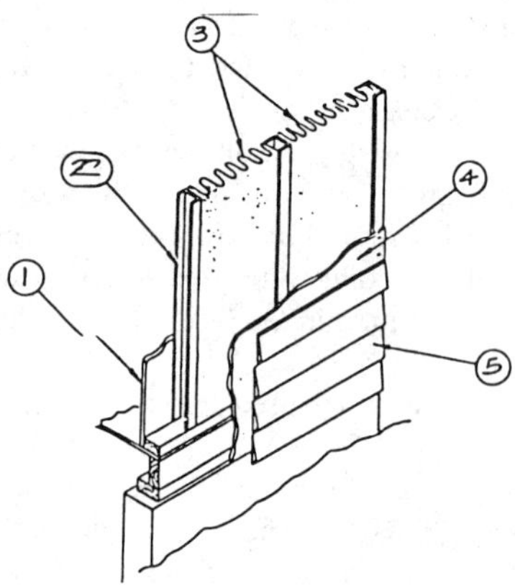

1 - Gypsum Wallboard
2 - Studs, Metal
3 - Insulation
4 - Sheathing, 1"
5 - Vinyl Siding

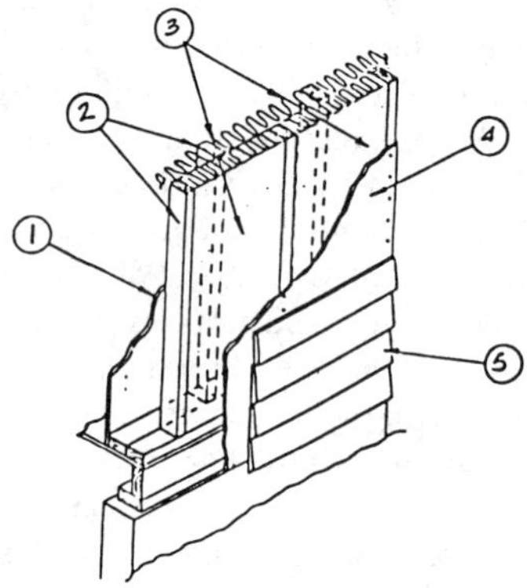

1 - Gypsum Wallboard, 1/2"
2 - Studs, 2" x 4"-16" O.C.
3 - Insulation
4 - Sheathing, 1/2"
5 - Vinyl Siding

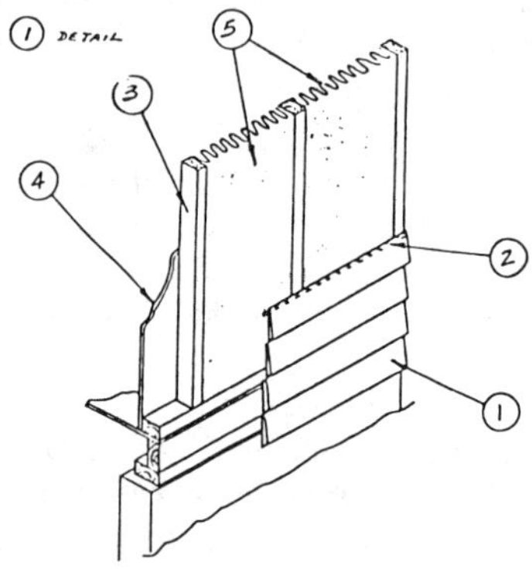

1 - Vinyl Siding
2 - 1" Furring Strip
3 - Studs, 2" x 4"-16" O.C.
4 - Gypsum Wallboard, 1/2" thick
5 - Insulation

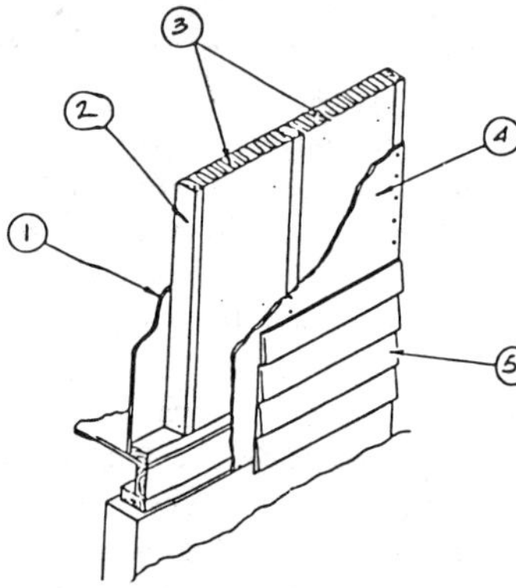

1 - Gypsum Wallboard
2 - Studs, 2" x 4" 16" O.C. or 2"x6" - 24" O.C.
3 - Insulation
4 - Sheathing
5 - Vinyl Siding

FIG. 1—*Typical test wall constructions.*

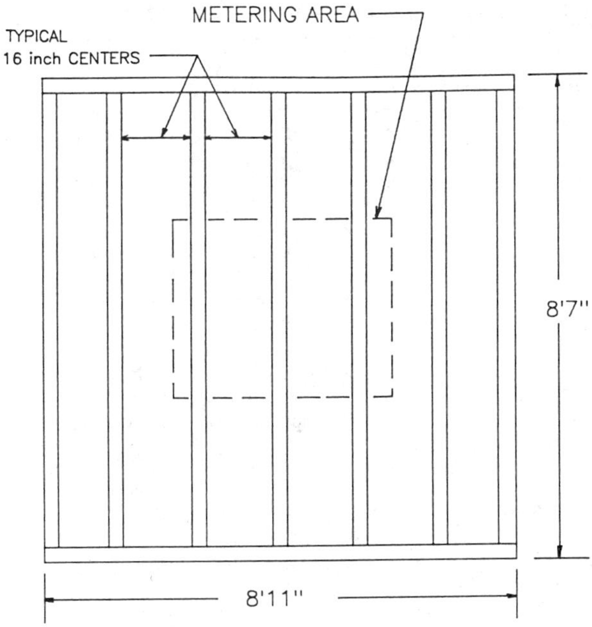

FIG. 2—*Typical frame wall construction.*

TABLE 1—*Comparisons of* R *values.*

Product	*R* Values at 75°F Mean Temperature, $ft^2 \cdot h \cdot °F/Btu$[a]	
	Advertised	Measured
Foam A—1-in. foil-faced polyisocyanurate	7.20	8.07
Foam B—1-in. polystyrene	5.00	4.79
Foam C—1-in. polystyrene	4.00	3.71
Foam D—½-in. foil-faced polyisocyanurate	4.00	2.94
Foam E—1-in. foil-faced polyisocyanurate	7.20	7.32
Foam F—½-in. foil-faced polyisocyanurate	3.60	3.49
1-in. fiberglass sheathing—vinyl faced	4.40	4.44
½-in. particle board	1.32	1.22
⅛-in. cardboard—foil faced	0.20	...

[a] $1\ ft^2 \cdot h \cdot °F/Btu = 0.1762\ K \cdot m^2/W$.

The GHB testing presented to this point clearly shows the need to be aware of metal shorts in construction and how a formation of an insulation blanket without shorts can improve thermal performance.

Insulation Gaps

On the subject of insulation gaps or workmanship defect, Fig. 3 summarizes the impact of voids in wall insulations. The basic wall panel was constructed of 50-by-200-mm (2-by-4) studs

TABLE 2—*Wall systems tested as recommended in ASTM Method C 236-80.*

Wall Number	Interior Skin (Wallboard), in.[a]	Frame	Insulation	Sheathing	R_T[b]	R_E[c]
1	1/2	2 by 4[d]	*R*-13[h]	1/2-in. particle board	15.62	17.55
2	1/2	2 by 4	*R*-13	1-in. Foam A	22.80	24.95
3	1/2	2 by 4	*R*-13	1-in. Foam B	19.40	21.98
4	1/2	2 by 4	*R*-13	1-in. Foam C	18.27	19.18
5	1/2	2 by 4	*R*-13	1/2-in. Foam D	17.46	21.49
6	1/2	2 by 4	*R*-13	1-in. sheathing FG	19.03	21.60
7	1/2	2 by 4	sprayed[i]	1-in. furring strip	17.48	20.07
8	1/2	2 by 4	sprayed[i]	1/2-in. Foam F	18.77	21.57
9	3/8	2 1/2 metal	*R*-11	1-in. Foam E	14.35	10.83
10	1/2	2 by 4[e]	*R*-13	1/2-in. Foam D	33.01	36.51
11	1/2	2 by 6[f]	*R*-19	1-in. Foam E	26.92	30.20
12	1/2	2 by 6	loose fill	1/8-in. cardboard	20.33	22.37
13	1/2	2 by 6[g]	sprayed[j]	infiltration barrier[k]	23.23	21.15[l]

[a]Interior skin was gypsum wallboard in the thickness shown. All the test panels had an exterior skin of vinyl siding.
[b]R_T = theory (ft^2 · h · °F/Btu); 1 ft^2 · h · °F/Btu = 0.1762 K · m^2/W.
[c]R_E = *R* experimental — guarded hot box (ft^2 · h · °F/Btu).
[d]Test walls numbered 1 through 8 are wood studs on 0.4-m (16-in.) centers.
[e]Wood studs, double offset.
[f]Walls 11 and 12, wood studs on 0.4-m (16-in.) centers.
[g]Wood studs on 0.6-m (24-in.) centers.
[h]Test wall numbered 1 through 11 are insulated with fiberglass blanket.
[i]Spray-applied fiberglass insulation.
[j]Spray-applied fiberglass insulation with poly vapor barrier.
[k]Air infiltration barrier of high-density polyethylene fibers.
[l]Lower R_E due to moisture present; the sprayed product was not completely dried.

on 406-mm (16-in.) centers with the wall cavities filled to different initial densities, as listed in the figure.

The data were established with the GHB with an unbonded fiberglass loose fill with voids created in the insulation. A 2% total void area (full cavity depth), reduced overall *R* values by approximately 13 to 21%, depending on the density. The objective of this test was to relate poor application procedures (poor workmanship) to actual thermal performance.

New Wall Systems

As a final phase of the laboratory testing, new wall designs were built and evaluated. Referring to the 13 wall panels tested (see Table 2), Item 13 is of interest. This test wall has the sheathing replaced by a thin, air-infiltration barrier that has a moisture vapor transmission rate of 51. With a measured *R* value of 21 (versus a theoretical value of 23), this wall is equal in performance to many of the walls tested with foam sheathing. Since the foam sheathings tested do not offer structural support to the framing, the air-infiltration barrier is not a disadvantage in this respect. Structural support still needs to be achieved with both systems by other means (for example, cross bracing).

In addition to the cost considerations, this new wall system offers a lower percentage of non-combustible materials. Comparing the 25.4-mm (1-in.) foam sheathing Walls 2, 3, and 4 with the experimental Wall 13, you find the following:

Wall	Percentage by Volume of Combustible Material
Wall 13	6
Walls 2, 3, and 4	27

The gypsum wallboard and fiberglass were considered noncombustible materials, and the siding material was left out of the calculation since this component is common to both systems. The experimental wall (No. 13) had 94% noncombustible material.

In the experimental designs of new approaches for insulation of walls, the elimination of sheathing material appears to be an appealing construction design of the future.

In the field, the use of foam-insulated sheathing may not cover 100% of the exterior walls of a residential construction. Quite often, we see combinations of foam and plywood or particle board sheathing, the latter two being used on corners and around doors and windows. Based on the laboratory data shown, the foam sheathing wall may have an average *R* of 16 to 22 when corrected for the other sheathing used. Considering this, the 50-by-150-mm (2-by-6) wall construction shown without insulated foam sheathing can yield an average *R* for the structure of 20 to 24. The laboratory data also showed that some wall design systems are capable of a 17 to 20 *R* value, using 50-by-200-mm sections (2 by 4s) and no foam sheathing. Such systems can increase

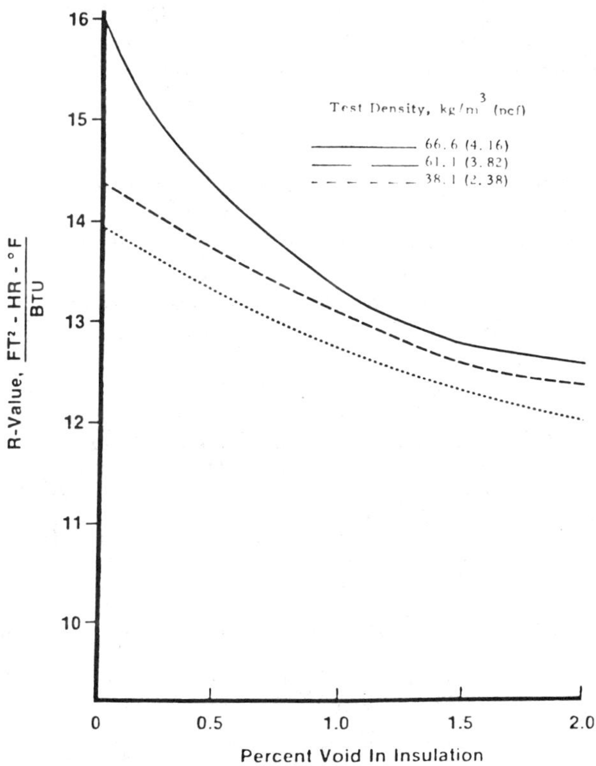

FIG. 3—*The effect of voids in a fiberglass-filled wall.*

the total noncombustible material used and maintain the proper use of a vapor barrier on the interior of the wall's system. A total lower cost for an R-19 wall system appears to be possible.

Thermal Shorts

In field introduction of the new insulation products, the laboratory testing was useful in identifying construction faults that could be overcome with the new products, in both new and retrofit work. Many examples can be stated of the obvious shorts shown from metal purlins and beams. However, some shorts are not that obvious. In the case of a 16-story, new construction high-rise building, the exterior skin called for 50.8-mm (2-in.) granite walls with a spray-applied foam insulation. The insulation was to be covered with a fireproofing material as required by the codes, but the specifications did not call for insulating the metal supports and beams for the granite walls. This entire system, insulation plus fireproof overcoat, was replaced by a 88.9-mm (3½-in.) layer of sprayed fiberglass. This approach not only was lower in cost but covered up the extensive metal components to yield a continuous fiberglass barrier with no thermal shorts. Since the interior wall was offset with gypsum board, the thermal seal was maintained. In this example, to the delight of the architects who had designed a maximum energy use limit of 189 300 $W/m^2 \cdot$ year (60 000 $Btu/ft^2 \cdot$ year), the actual energy usage has been closer to 141 975 $W/m^2 \cdot$ year (45 000 Btu/ft^2). That is about one third less than the company's maximum energy use limit and half that of similar-sized buildings of a decade ago.

An example of reinsulating a structure that was insulated but had many thermal shorts (metal bracing, columns, doors, and so on) was the case of an adhesive storage room in an unheated warehouse. The existing 12.7-mm (½-in.) board insulation was inadequate and was removed. The entire storage room was sprayed with fiberglass [50.8 mm (2 in.) on the ceiling; 76.2 mm (3 in.) on the walls], and new seals were installed on the sliding doors. The British-thermal-unit-per-hour heat loss was reduced from 102 400 to 22 300 (50% of the reduction was from improved door seals).

The return on investment (ROI) for this job was one year. Table 3 shows the breakdown of heat losses for each section of the adhesive room. Note that the calculated infiltration heat loss is dependent on ambient conditions and will not be a constant loss.

Buildings that have insulation should not be overlooked as retrofit possibilities. The retrofit of the insulated structure could improve the overall performance as well as show an acceptable ROI. In many cases, it is more practical to take out the old insulation, for example, removing 19 to 25.4-mm (¾ to 1-in.) metal building insulation and replacing it with 76.2 to 127 mm (3 to 5

TABLE 3—*Adhesive room heat losses (calculated).*

Location	Before Spraying, Btu/h[a]	% of Total	After Spraying, Btu/h[a]	% of Total	After Sealing Doors (Theoretical)[b]	% of Total
Ceiling	17 800	17.4	4 400	4.3	3 100	13.9
East wall	11 700	11.4	2 000	2.0	1 400	6.3
North wall	5 300	5.2	1 500	1.5	1 000	4.5
South wall	5 300	5.2	1 500	1.5	1 000	4.5
West wall	5 300	5.2	3 200	3.1	2 000	9.0
Walls and ceiling	45 400	44.4	12 600	1.24	8 500	38.2
Infiltration (doors)	57 000	55.6	89 800	87.6	13 800	61.8
Total	102 400	00.0	102 400	100.0	22 300	100.0

[a] 1 Btu/h = 0.2931 W.
[b] Assuming a room temperature of 70°F (21.1°C).

in.) of a sprayed fiberglass system that covers all the thermal shorts. This is especially effective in buildings in which the insulation standards have changed over the years.

Regarding void areas in insulation, better understanding of energy conservation by all trades can reduce this problem. Electricians should not run electric wiring or even telephone wire down the center of framing construction, otherwise the insulation contractor may compress the insulation either in front or behind the wire, reducing the effective R. If the wire is placed closer to the interior surface, then the insulating contractor can obtain the full thickness of the rated insulation.

Installers should extend sheathing above the top plate, especially when blown-in loose fill is applied. Our field testing revealed many examples of lack of insulation on the top plate of the framing. Installation of cut blanket or boards on the top plate appears to yield better results than the use of thin air baffles.

In summary, construction faults that contribute to heat losses in building envelopes can be duplicated in the laboratory, can be detected in the field, and in most cases *can be eliminated.*

Martha G. Van Geem[1]

Heat Transfer Characteristics of a Masonry Cavity Wall

REFERENCE: Van Geem, M. G., **"Heat Transfer Characteristics of a Masonry Cavity Wall,"** *Thermal Insulation: Materials and Systems, ASTM STP 922,* F. J. Powell and S. L. Matthews, Eds., American Society for Testing and Materials, Philadelphia, 1987, pp. 318–342.

ABSTRACT: Heat transfer characteristics of building elements must be known to evaluate energy losses through a building envelope. Laboratory tests of walls under dynamic temperature conditions provide data that can be used to determine thermal properties. Dynamic testing is particularly important for massive envelope components that store as well as transmit heat.

A block-brick cavity wall was tested in the calibrated hot box facility at the Construction Technology Laboratories, a division of the Portland Cement Association. The wall consisted of 100-mm (4-in.) clay brick on the exterior face, a 70-mm (2³/₄-in.) cavity, 150-mm (6-in.) concrete block, and 3 mm (¹/₈ in.) of plaster on the interior face. Metal rectangular ties were used for tying brick and block wythes.

Laboratory tests were performed on the wall with and without expanded perlite fill in the cavity. The wall was subjected to steady-state, transient, and periodically varying temperature conditions. Steady-state results are used to define heat transmission coefficients, such as *U* and *R* values. Data obtained during transient and periodic temperature variations are used to define dynamic thermal response of the wall. Dynamic response includes heat storage capacity as well as heat transmission characteristics of the wall assembly. Test data are also used to determine the effects of perlite insulation in the cavity.

The cavity wall construction is similar to that of a test building monitored by the National Bureau of Standards (NBS) in Gaithersburg, Maryland. Two of the dynamic temperature cycles applied to the wall during laboratory tests were derived from actual wall surface temperatures at the NBS test building. Heat flow meter data from the test building are compared with calibrated hot box test results.

Laboratory test results provide a data base for evaluation of building envelope performance where cavity walls are used. The results also provide information on the effectiveness of expanded perlite as a cavity-fill material.

KEY WORDS: buildings, calibrated hot box, cavity walls, concrete, energy, heat transmission, masonry, thermal resistance, thermal inertia, thermal insulation

Nomenclature

q_{ss} = Heat flux predicted from steady-state analysis
q_w = Heat flux measured by calibrated hot box
t_i = Indoor chamber air temperature
t_1 = Wall surface temperature, indoor side
t_3 = Block temperature on surface facing cavity
t_4 = Brick temperature on surface facing cavity
t_2 = Wall surface temperature, outdoor side
t_o = Outdoor chamber air temperature

[1] Senior research engineer, Construction Technology Laboratories, Portland Cement Association, Skokie, IL 60077.

Tests were conducted to evaluate thermal performance of a masonry cavity wall with and without expanded perlite loose-fill insulation. Steady-state test results were used to obtain average heat transmission coefficients including total thermal resistance, R_T, and thermal transmittance, U. Data obtained during periodic temperature variations were used to define dynamic thermal response under selected temperature ranges.

The wall with expanded perlite insulation is similar to walls of a test building monitored by the National Bureau of Standards (NBS) in Gaithersburg, Maryland. Two of the dynamic temperature cycles applied to the wall during laboratory tests were derived from actual wall surface temperatures of the NBS test building.

A third temperature cycle, a simulated sol-air cycle, was selected to permit comparison of results with walls previously tested [*1–6*].

The objectives of the experimental investigation were to evaluate and compare the performance of walls with and without expanded perlite insulation and to compare laboratory results with NBS field test results. The walls were tested in the calibrated hot box facility of the Portland Cement Association's Construction Technology Laboratories (CTL).

Test Specimens

A 300-mm (12-in.) block-brick cavity wall without insulation and the same wall with expanded perlite loose-fill insulation were tested in the calibrated hot box. The walls were built using techniques representative of field construction practices. The overall nominal wall dimensions were 2.62 by 2.62 m (103 by 103 in.).

Wall Construction

The 300-mm (12-in.) uninsulated, reinforced block-brick cavity wall, designated Wall M9, was constructed of 143-mm (5⅝-in.) concrete block and 86-mm (3⅜-in.) clay brick separated by a 70-mm (2¾-in.) air space. The measured unit weight of the wall was 395 kg/m^2 (81 lb/ft^2). The average measured wall thickness was 305 mm (12 in.). The measured wall area perpendicular to heat flow was 6.84 m^2 (73.7 ft^2).

Two-core concrete block was used to construct the wall. Measured unit weights of the oven-dry block and brick were 1980 kg/m^3 (124 lb/ft^3) and 2160 kg/m^3 (135 lb/ft^3), respectively. The measured solid volume percentages of the block and brick were 56 and 82%, respectively. The measured moisture contents of the block and brick after the calibrated hot box tests had been completed were 1.2 and 0.3% of the oven-dry weight, respectively.

The mortar consisted of one part Type S masonry cement to three parts masonry sand by volume. The block and brick were laid in a running bond pattern. The joints were tooled. Metal rectangular ties were placed across the cavity at every other block joint. The tie dimensions are given in Ref *7*.

A 3-mm (⅛-in.)-thick smooth plaster coating was applied to the inside block surface. The plaster consisted of one part Type S masonry cement to 2½ parts masonry sand by volume. Two coats of off-white flat latex paint were applied to the plaster prior to testing.

Instrumentation

Thermocouples corresponding to the ASTM Temperature Electromotive Force (EMF) Tables for Standardized Thermocouples (E 230-83), Type T, were used to measure temperatures. The thermocouples were located on the indoor and outdoor wall faces, on the interior faces of the cavity, and in the air space of each chamber.

Sixteen thermocouples were taped to the indoor plaster surface. On the outdoor wall face, eight thermocouples were taped to vertical mortar joints, and eight were taped to brick. The thermocouples were uniformly distributed on a 500-mm (20-in.)-square grid over the wall area.

Thermocouples taped to the cavity side of the block face were also uniformly distributed on a 500-mm (20-in.)-square grid. The 16 thermocouples were wired so that electrical averages of four thermocouple junctions, located along a horizontal line across the grid, were obtained.

Thermocouples on the cavity side of the brick face were embedded in the horizontal mortar joints closest to the grid pattern of the other thermocouples. The thermocouple junctions were located 6 to 13 mm (1/4 to 1/2 in.) into the back of the plane of the brick surface facing the cavity. The thermocouples on this surface were also wired to obtain one average reading of four thermocouples in a horizontal row. Average temperatures reported for this surface were an average reading of twelve thermocouples because one row of thermocouples gave erroneous readings.

Surface thermocouples were securely attached to the wall over a length of approximately 75 mm (3 in.). The tape covering the sensors on the indoor side was painted the same color as the plaster surface. Gray tape used on other surfaces was not painted.

Sixteen thermocouples were used to measure temperatures in the air space of each chamber. The thermocouples were uniformly distributed on the same 500-mm (20-in.) grid over the wall area as the other thermocouples and were located approximately 75 mm (3 in.) from the face of the test wall.

Addition of Expanded Perlite Insulation

After Wall M9 was tested, the cavity was filled with expanded perlite loose-fill insulation. The wall was redesignated Wall M10 and tested in the calibrated hot box. Figures 1 and 2 show the outside and inside surfaces of Wall M10 after tests were completed.

The silicone-treated expanded perlite, as received, had a unit weight of 69 kg/m^3 (4.3 lb/ft^3). The measured unit weight of Wall M10 was 400 kg/m^2 (82 lb/ft^2).

The top portion of the test frame was removed after calibrated hot box tests were completed to determine whether the expanded perlite insulation had settled. At the end of the calibrated hot box tests, 3% of the cavity area perpendicular to heat flow did not contain expanded perlite insulation. Locations of expanded perlite voids are illustrated in Fig. 3.

Insulation settlement was probably accelerated by vibrations from dynamic load testing at a site located approximately 9 to 18 m (30 to 60 ft) from the calibrated hot box. Dynamic load tests occurred most of the time the cavity wall was being tested.

NBS Test Building No. 6

The National Bureau of Standards (NBS) constructed six one-room test buildings in Gaithersburg, Maryland, to determine the effect of wall mass on heating and cooling loads of residential buildings. The 6.1 by 6.1-m (20 by 20-ft) buildings have the same floor plan and orientation. The buildings are identical except for materials used for wall construction. The six buildings were extensively instrumented for measuring heating loads, cooling loads, wall heat transmission, indoor temperature, and humidity [*8*].

Test Building 6 has walls similar to the insulated cavity wall tested in the calibrated hot box. The walls were constructed of hollow core concrete block and clay brick separated by a 89-mm (3$^1/_2$-in.) cavity containing expanded perlite loose-fill insulation. The nominal dimensions of the block and brick were 200 mm (8 in.) and 100 mm (4 in.), respectively. The block weighed 1680 kg/m^3 (105 lb/ft^3) [*8*].

Calibrated Hot Box Test Facility

The block-brick cavity wall with and without expanded perlite loose-fill insulation was tested in the calibrated hot box facility shown in Figs. 4 and 5. Tests were performed in accordance

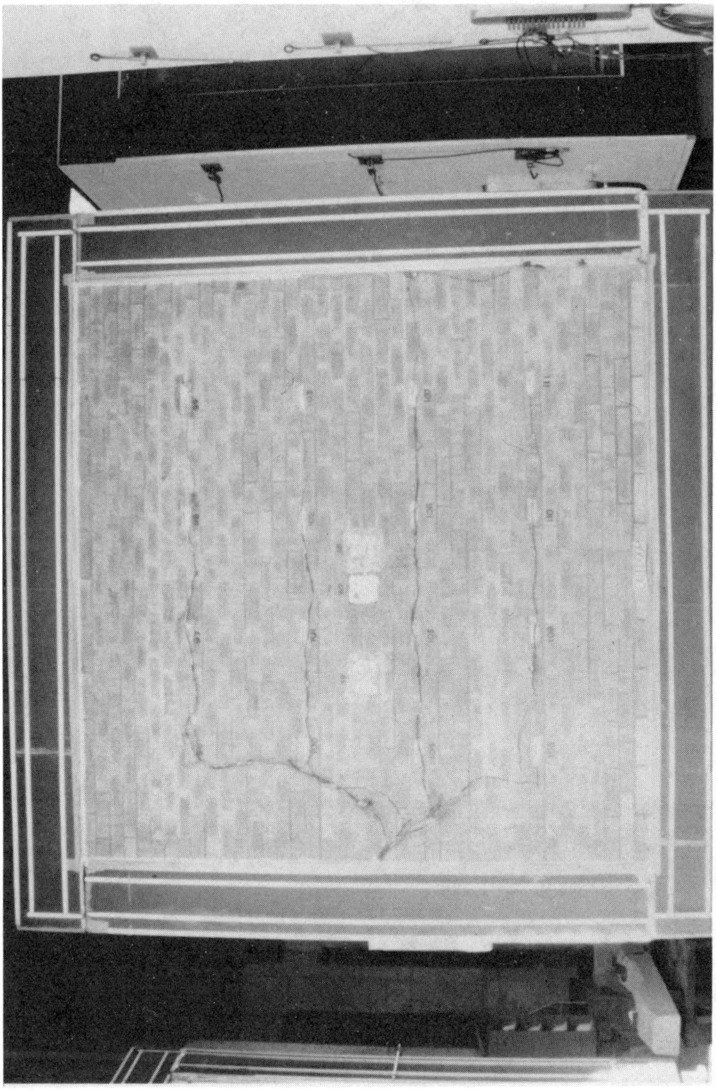

FIG. 1—*Wall M10 outdoor surface after testing.*

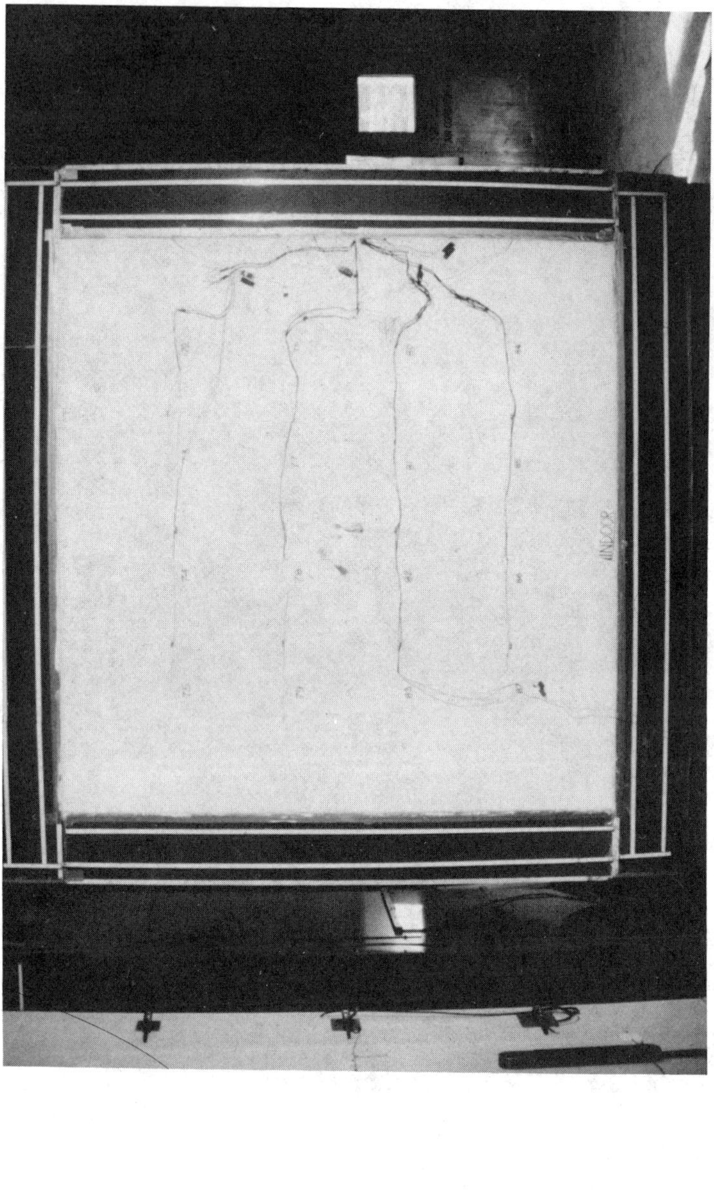

FIG. 2—*Wall M10 indoor surface after testing.*

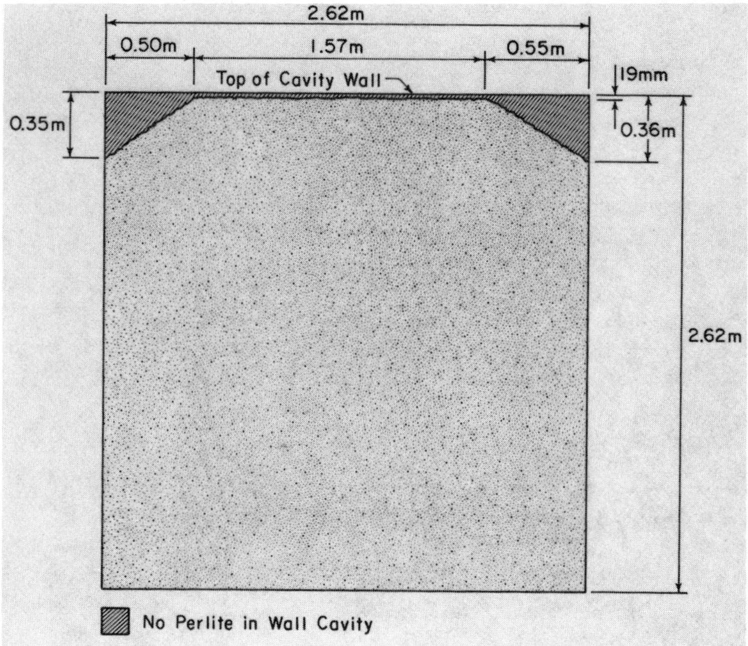

FIG. 3—*Locations of expanded perlite voids after the calibrated hot box tests.*

with the ASTM Test for Thermal Performance of Building Assemblies by Means of a Calibrated Hot Box (C 976-82).

The following is a brief description of the calibrated hot box. Instrumentation and calibration details are described in Refs *7* and *9.* The facility consists of two highly insulated chambers, as shown in Fig. 5. The walls, ceiling, and floors of each chamber are insulated with foamed urethane sheets to obtain a nominal thickness of 300 mm (12 in.). During tests, the chambers are clamped tightly against an insulated frame that surrounds the test wall. Air in each chamber is conditioned by heating and cooling equipment to obtain desired temperatures on each side of the test wall.

The outdoor (climatic) chamber can be held at a constant temperature or cycled between −26 and 54°C (−15 and 130°F). Temperature cycles can be programmed to obtain the desired time-temperature relationship. The indoor (metering) chamber, which simulates an indoor environment, can be maintained at a constant room temperature between 18 and 27°C (65 and 80°F).

The facility was designed to accommodate walls with thermal resistance values ranging from 0.26 to 3.52 $m^2 \cdot K/W$ (1.5 to 20 $h \cdot ft^2 \cdot °F/Btu$).

Heat flow through a test specimen is determined from measurements of the amount of energy input to the indoor chamber to maintain a constant temperature. The measured energy input must be adjusted for heat losses. Since the net energy into the indoor chamber equals zero, heat transfer through the test wall can be expressed by the following energy balance equation.

$$Q_w = Q_c - Q_h - Q_{fan} - Q_\ell - Q_f \qquad (1)$$

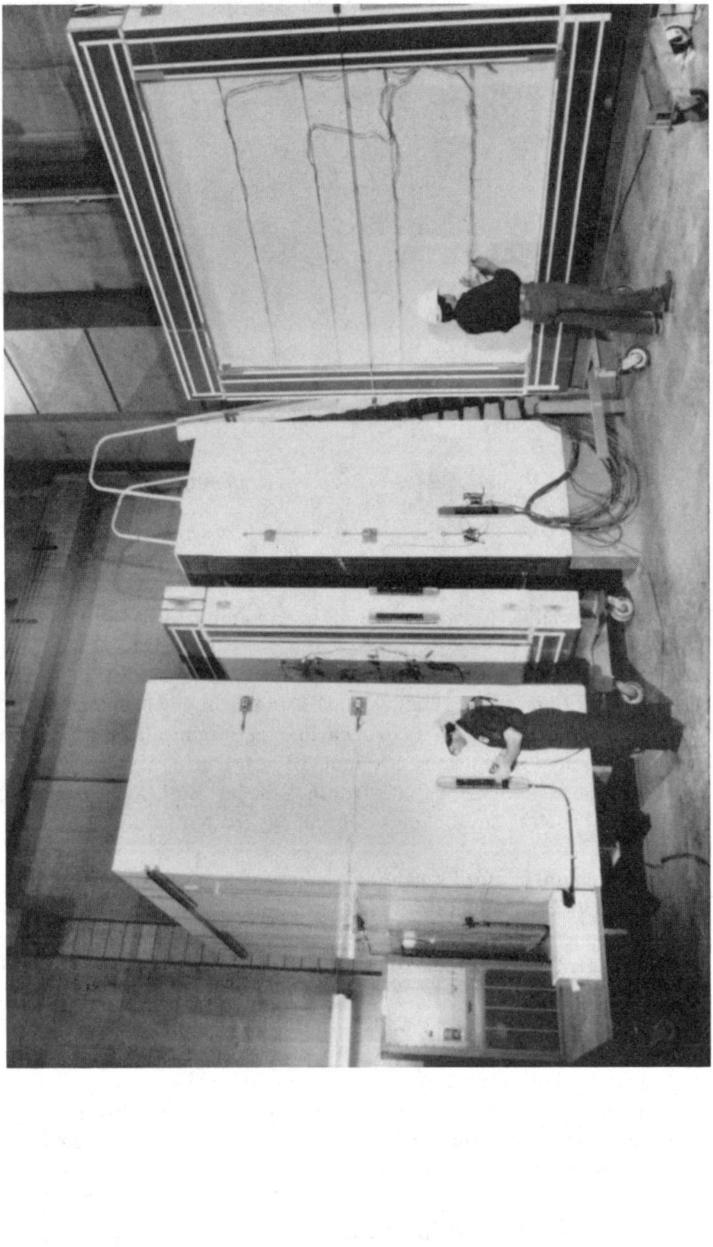

FIG. 4—*Calibrated hot box test facility.*

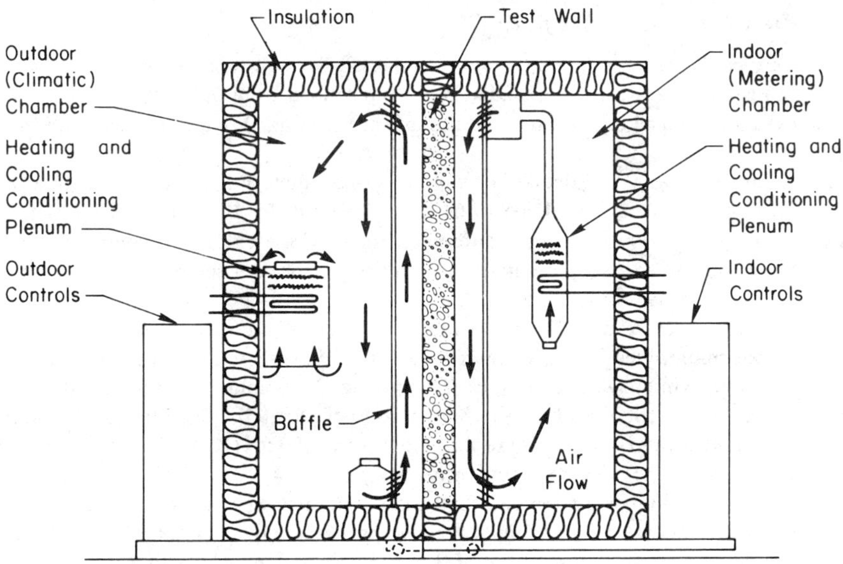

FIG. 5—*Schematic of a calibrated hot box.*

where

Q_w = heat transfer through the wall from the outdoor chamber to the indoor chamber,
Q_c = heat removed by indoor chamber cooling,
Q_h = heat supplied by indoor electrical resistance heaters,
Q_{fan} = heat supplied by the indoor circulation fan,
Q_ℓ = heat gain (loss) from the laboratory, and
Q_f = heat gain (loss) to the indoor chamber from the flanking path around the specimen.

The units for the terms of Eq 1 are watt-hours per hour (British thermal units per hour).

A watt-hour transducer is used to measure Q_h and Q_{fan}. The value of Q_ℓ is calculated from measured laboratory and indoor chamber temperatures. Heat flux transducers are used to check calculations of Q_ℓ. Heat removed by indoor chamber cooling, Q_c, is calculated from the refrigerant enthalpy and mass flow rate, assuming an ideal basic vapor compression refrigeration cycle. Steady-state calibrated hot box tests of two "standard" calibration specimens were used to adjust for inefficiencies in the actual refrigeration cycle and to determine Q_f. In addition to flanking losses, other miscellaneous losses from the indoor chamber are included in Q_f.

Steady-State Tests

Two steady-state calibrated hot box tests were performed on the walls with and without expanded perlite insulation. Energy and temperature measurements were used to calculate the average thermal properties, including total thermal resistance, R_T, and transmittance, U.

Design heat transmission coefficients were calculated for the walls and compared with measured values. The results were compared with values for NBS Test Building 6 walls.

Measured Heat Transmission Coefficients

Steady-state calibrated hot box tests were conducted by maintaining constant indoor and outdoor chamber temperatures. Resistances were determined from data collected when specimen temperatures reached equilibrium and the rate of heat flow through the test wall was constant.

Total resistance values were calculated using measured values of heat flux and standard surface resistances equal to 0.12 $m^2 \cdot K/W$ (0.68 $h \cdot ft^2 \cdot °F/Btu$) for the indoor surface and 0.03 $m^2 \cdot K/W$ (0.17 $h \cdot ft^2 \cdot °F/Btu$) for the outdoor surface. These values are commonly used in design and are considered to represent still air on the indoor wall surface and an air flow of 24 km/h (15 mph) on the outdoor wall surface.

Steady-state tests were run at two temperature differentials. For the first case, indoor air temperature was maintained at approximately 23°C (73°F) while outdoor air temperature was maintained at approximately 52°C (125°F). This provided a nominal temperature differential of approximately 29°C (53°F) and a mean temperature of 38°C (100°F). Measured total resistances for Walls M9 and M10 under these conditions were 0.59 $m^2 \cdot K/W$ (3.37 $h \cdot ft^2 \cdot °F/Btu$) and 1.67 $m^2 \cdot K/W$ (9.47 $h \cdot ft^2 \cdot °F/Btu$), respectively.

In the second case, indoor air temperature was maintained at approximately 21°C (70°F) while outdoor air temperature was maintained at approximately −23°C (−10°F). This provided a nominal temperature differential of 44°C (80°F) and a mean temperature of −1°C (30°F). Measured total resistances for Walls M9 and M10 under these conditions were 0.64 $m^2 \cdot K/W$ (3.64 $h \cdot ft^2 \cdot °F/Btu$) and 1.64 $m^2 \cdot K/W$ (9.32 $h \cdot ft^2 \cdot °F/Btu$), respectively.

Design Heat Transmission Coefficients

Design values of total resistance for Walls M9 and M10 and NBS Test Building 6 walls were calculated in accordance with procedures established by the American Society of Heating, Refrigerating, and Air-Conditioning Engineers [*10*]. Resistances for construction materials were taken from the *ASHRAE Handbook—1981 Fundamentals* [*10*] and the *Concrete Masonry Handbook* [*11*]. Standard surface resistances given in the Measured Heat Transmission Coefficients portion of this section were used.

The isothermal planes (series-parallel) method of calculating resistances was used to determine reduced wall resistances due to metal ties spanning the cavity. Reduced resistance due to insulation voids in the cavity of Wall M10 was also calculated using the isothermal planes method. As mentioned in the Test Specimens section, 3% of the cavity area perpendicular to heat flow was void of insulation.

Design values of total resistance for Wall M9, Wall M10, and NBS Test Building 6 walls were 0.61, 1.56, and 2.37 $m^2 \cdot K/W$ (3.46, 8.83, and 13.48 $h \cdot ft^2 \cdot °F/Btu$), respectively. Reference 7 describes design resistance calculations more thoroughly.

Thermal Resistance Comparisons

The total thermal resistance values, R_T, for Walls M9 and M10 are summarized in Table 1. The design values are within 6% of the calibrated hot box test results for Walls M9 and M10. The total thermal resistance of the 300-mm (12-in.) cavity wall was increased 170% by the addition of perlite insulation.

Table 1 also lists thermal resistances of a wall similar to NBS Test Building 6 walls measured by Dynatech R/D Company [*12*]. The wall tested at Dynatech R/D was built by the same contractor who built the NBS test building. Tests were performed in accordance with the ASTM Test for Steady-State Thermal Performance of Building Assemblies by Means of a Guarded Hot Box (C 236-80). The design resistances calculated by Dynatech R/D Company [*12*] are also listed in Table 1.

TABLE 1—*Total thermal resistance,* R_T.

Wall Designation	Wall Description	Calculated (Design) R_T, $m^2 \cdot K/W$ ($h \cdot ft^2 \cdot °F/Btu$)	Measured R_T, $m^2 \cdot K/W$ ($h \cdot ft^2 \cdot °F/Btu$)	Mean Wall Temperature, °C (°F)	Calculated R_T/ Measured R_T Ratio
M9	300-mm (12-in.) block-brick cavity wall	0.61[a] (3.46)	0.61[b] (3.48)	22 (72)	0.99
M10	300-mm (12-in.) block-brick cavity wall with perlite	1.56[c] (8.83)	1.66[b] (9.41)	22 (72)	0.94
NBS	400-mm (16-in.) cavity wall with perlite; NBS Test Building 6 walls	2.69[d][*12*] (15.29)	2.21[e][*12*] (12.53)	2 to 3 (35 to 37)	1.22
		2.61[d][*12*] (14.83)	2.15[e][*12*] (12.19)	10 to 11 (50 to 53)	1.22
		2.37[a] (13.48)	. . .	22 (72)	. . .

[a] Considering the effect of metal ties.
[b] Interpolated from calibrated hot box test results.
[c] Considering the effects of metal ties and expanded perlite settlement.
[d] Ignoring the effect of metal ties and expanded perlite settlement.
[e] Measured values were determined by Dynatech on a wall similar to those used in NBS Test Building 6.

The design values from Ref *12* of NBS Test Building 6 walls do not consider the effects of metal ties or insulation settlement. The calculated R_T values would be lower, and therefore closer to the measured values, if these effects were considered.

Dynamic Tests

Exterior building walls are seldom in a steady-state condition. Outdoor air temperatures and solar effects cause cyclic changes in outdoor surface temperatures. Generally, indoor surface temperatures are relatively constant compared with outdoor surface temperatures.

Dynamic tests are a means of evaluating thermal response under controlled conditions that simulate temperature changes actually encountered in building envelopes. The response of walls to temperature changes is a function of both thermal resistance and heat storage capacity.

Test Procedures

Since outdoor temperature conditions are frequently similar for several consecutive days, they may be assumed to follow a 24-h periodic cycle. Dynamic tests were conducted by maintaining a constant calibrated hot box indoor air temperature while outdoor air temperatures were cycled over a predetermined time-versus-temperature relationship. The rate of heat flow through a test specimen was determined from hourly averages of data.

One 24-h (diurnal) temperature cycle, denoted the NBS test cycle, has been applied to every wall tested in the calibrated hot box. This cycle is based on a simulated sol-air[2] cycle used by the National Bureau of Standards in their evaluation of dynamic thermal performance of an experimental masonry building [*13*]. It represents a large variation in outdoor temperature over a 24-h period. The mean outdoor temperature of the cycle is approximately equal to the mean indoor temperature.

Dynamic performance is dependent on the temperature cycle applied to the test specimen. Use of the NBS test cycle permits dynamic performance comparisons of all wall assemblies tested in the calibrated hot box [*1–6*].

Two additional sol-air diurnal temperature cycles were applied to the 300-mm (12-in.) cavity walls with and without expanded perlite fill. Outdoor surface temperatures of the west wall of NBS Test Building 6 were used to create cycles that produced similar outdoor surface temperatures on the test specimens. Dynamic test cycles, denoting the Gaithersburg April and Gaithersburg May cycles, respectively, were derived from data collected at the NBS test building from 10:00 A.M. 23 April 1982 through 10:00 A.M. 24 April 1982, and from 9:00 A.M. 31 May 1982 through 9:00 A.M. 1 June 1982. These 24-h cycles represent the third day of relatively repetitive daily temperature conditions at the NBS test site.

For all tests, the dynamic cycles were repeated until conditions of equilibrium were obtained. Equilibrium conditions were evaluated by consistency of applied temperatures and measured energy response. Each test required approximately six to eight days for completion. After equilibrium conditions were reached, the tests were generally continued for a period of three days. The results are based on average readings for at least three consecutive 24-h cycles, unless otherwise noted.

[2] Sol-air temperature is the temperature of outdoor air that, in the absence of all radiation exchanges, would give the same rate of heat entry into the surface as would exist with the actual combination of incident solar radiation, radiant energy exchange, and convective heat exchange with outdoor air [*10*].

Dynamic Test Results

Measured Temperatures and Heat Flux

Measured temperatures for dynamic temperature cycles applied to Walls M9 and M10 are illustrated in Figs. 6, 7, and 8. For Walls M9 and M10, the outdoor air, t_o, outdoor surface, t_2, indoor air, t_i, and indoor surface, t_1, temperatures are average readings of 16 thermocouples placed as described in the Test Specimens section of this paper. The internal block surface temperatures, t_3, and internal brick surface temperatures, t_4 are average-readings of thermocouples placed on the block and brick surfaces, respectively, facing the cavity.

The measured and calculated heat flux values are illustrated in Figs. 9, 10, and 11. Heat flux determined from calibrated hot box tests is denoted q_w. Heat flux is positive when heat flows from the calibrated hot box outdoor chamber to the indoor chamber.

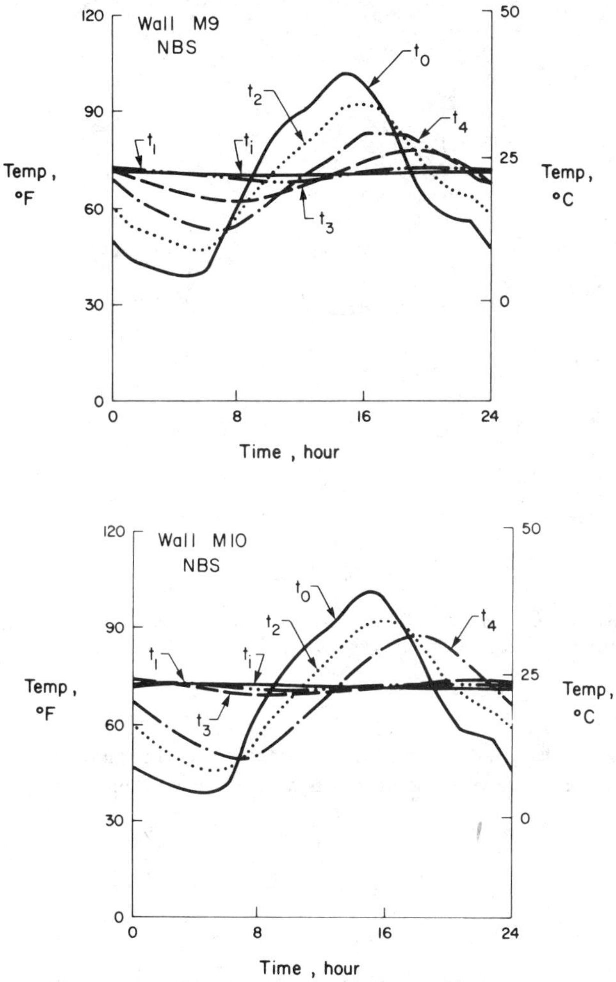

FIG. 6—*Measured temperatures for the NBS test cycle:* (top) *Wall M9 (no insulation in the cavity);* (bottom) *Wall M10 (insulation in the cavity).*

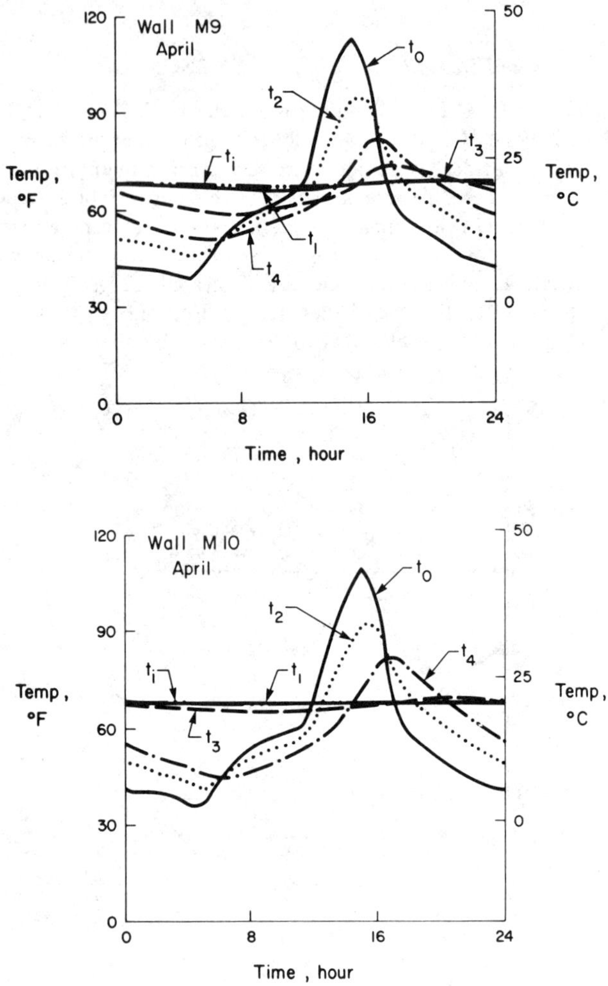

FIG. 7—*Measured temperatures for the Gaithersburg April test cycle:* (top) *Wall M9 (no insulation in the cavity);* (bottom) *Wall M10 (insulation in the cavity).*

Heat flux predicted by steady-state analysis is denoted q_{ss}. The values were calculated on an hourly basis from wall surface temperatures using the following equation

$$q_{ss} = \frac{(t_2 - t_1)}{R} \tag{2}$$

where

q_{ss} = heat flux through the wall predicted by steady-state analysis, W/m^2 (Btu/h · ft^2),
R = average thermal resistance of the wall, m^2 · K/W (h · ft^2 · °F/Btu),
t_2 = average temperature of the outdoor wall surface, °C (°F), and
t_1 = average temperature of the indoor wall surface, °C (°F).

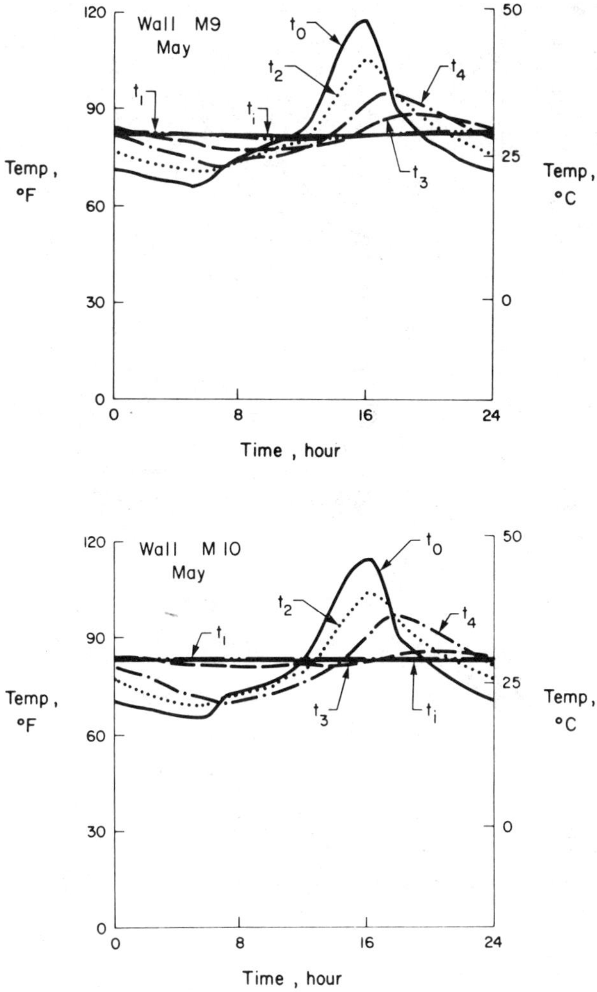

FIG. 8—*Measured temperatures for the Gaithersburg May test cycle:* (top) *Wall M9 (no insulation in the cavity);* (bottom) *Wall M10 (insulation in the cavity).*

Resistances for Walls M9 and M10 were derived from steady-state calibrated hot box tests. The resistance for the west wall of NBS Test Building 6 was determined from Dynatech R/D Company guarded hot box test results [*12*].

Descriptions of the symbols used in the figures appear in the Nomenclature section.

Heat Flux Comparisons

Figures 9, 10, and 11 illustrate heat flux measured by the calibrated hot box, q_w, and calculated using steady-state theory, q_{ss}, for Walls M9 and M10. The measured heat flux is less than the calculated heat flux for both the wall without perlite insulation (Wall M9) and the wall with perlite insulation (Wall M10). Both the measured heat flux, q_w, and calculated heat flux, q_{ss},

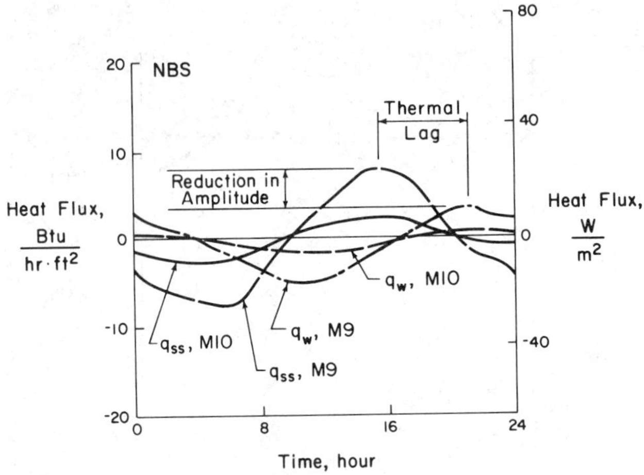

FIG. 9—*Heat flux for the NBS cycle applied to Walls M9 and M10.*

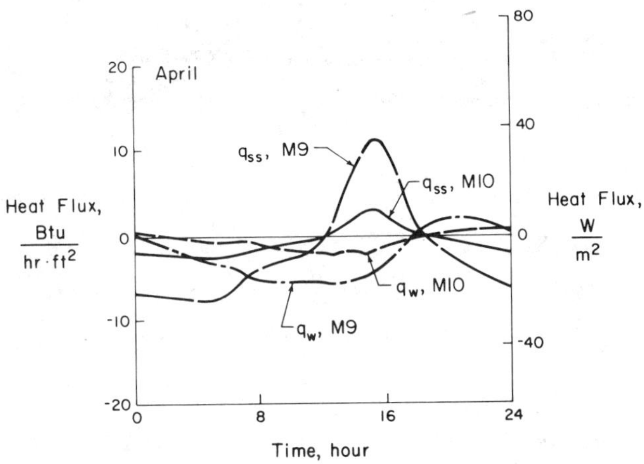

FIG. 10—*Heat flux for the Gaithersburg April cycle applied to Walls M9 and M10.*

are reduced for Wall M10 because of the addition of perlite insulation. The figures also show the shift in peaks, or thermal lag, from Wall M9 to Wall M10.

Figures 12 and 13 show calibrated hot box measurements for the Gaithersburg April and May test cycles, respectively, and measurements at NBS Test Building 6 for the same time period. The heat flux values measured by a heat flux transducer located on the west wall of NBS Test Building 6 are denoted q_{NBS} and were supplied by D. Burch of NBS.

For the Gaithersburg April test cycle (Fig. 12), calibrated hot box results for Wall M10 show less heat loss from the indoor environment than results from the NBS test building wall. The difference may be due to the test building wall having a cooler temperature history than the wall tested in the calibrated hot box (Wall M10).

Wall M10 was in the calibrated hot box approximately three days before dynamic equilib-

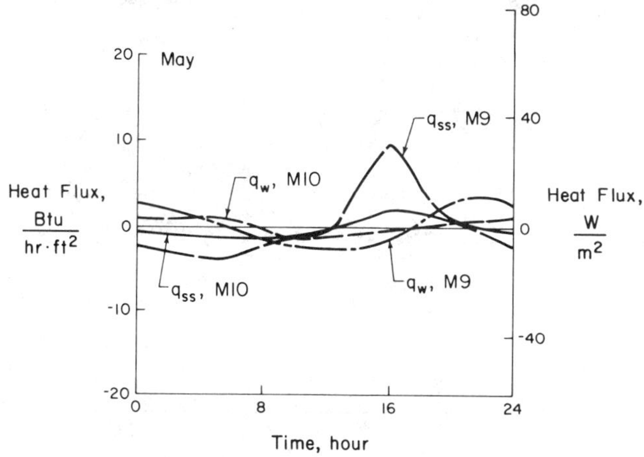

FIG. 11—*Heat flux for the Gaithersburg May cycle applied to Walls M9 and M10.*

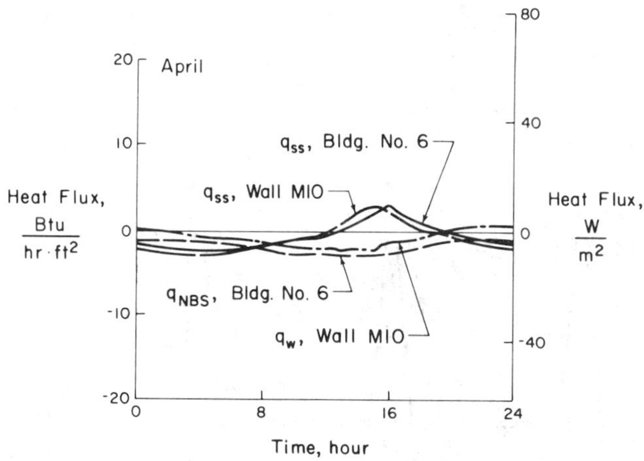

FIG. 12—*Heat flux for Wall M10 and NBS Test Building 6 for the Gaithersburg April test cycle.*

rium was reached. Dynamic equilibrium occurs when the measured temperatures and heat flux for the wall are relatively constant from day to day at any particular hour of the test cycle. The three-day temperature history for the NBS test building wall is shown in Fig. 14. The measured outdoor surface and indoor air temperatures are illustrated. Temperatures measured during dynamic equilibrium in the calibrated hot box are also shown in Fig. 14.

The average temperatures of the test building wall, on 22 and 23 April were 1.1°C (1.7°F) and 1.7°C (2.7°F) cooler, respectively, than those applied to Wall M10 during the Gaithersburg April test cycle. Temperatures for the third day preceding the comparative data may also have been cooler than temperatures for Wall M10. A cooler time history results in more heat loss from the indoor environment than predicted because temperatures within the wall are cooler than predicted.

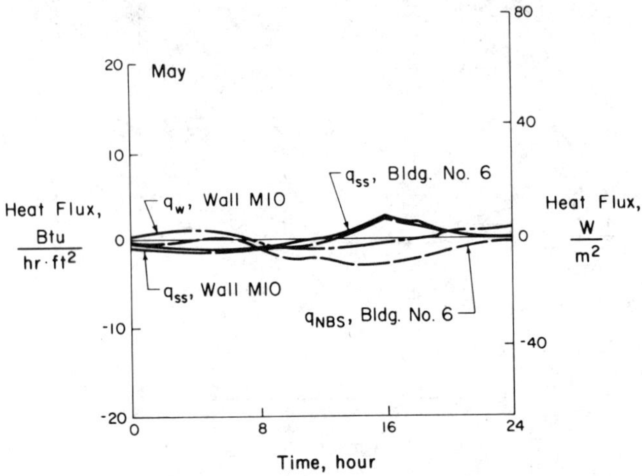

FIG. 13—*Heat flux for Wall M10 and NBS Test Building 6 for the Gaithersburg May test cycle.*

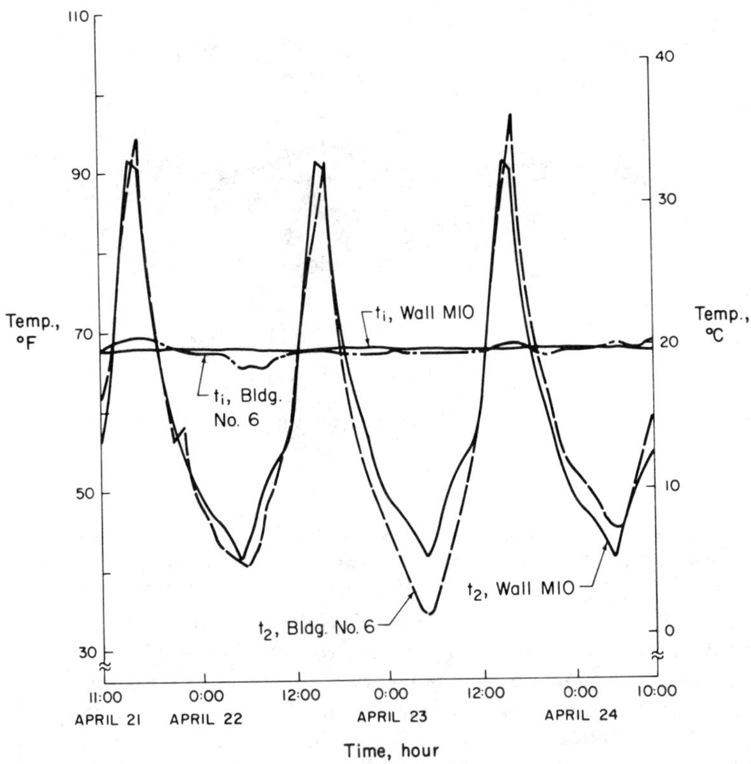

FIG. 14—*April temperatures for NBS Test Building 6 and Wall M10.*

Figure 13 shows that the measured heat loss for the Gaithersburg May cycle is also greater for the NBS test building wall than for Wall M10. The three-day temperature history for the NBS test building wall is shown in Fig. 15. Indoor air temperatures steadily rose during the three-day period; therefore, temperatures within the NBS test building wall might have been cooler than those within Wall M10.

Heat flux transducer measurements from the NBS test building wall may be lower than predicted for three reasons. First, the data comparisons were for sunny days during months when the average daily temperatures are generally increasing. Cooler temperatures within the wall cause more heat loss from the indoor environment than predicted. Heat energy is required to raise the average daily temperature of massive walls.

Second, the heat flux transducer readings were taken over a 50-mm (2-in.)-square area, and were instantaneous hourly readings. Therefore, the readings might not accurately reflect the integrated heat flux for the entire wall.

Third, the 400-mm (16-in.) wall was thick compared with the overall test building, which was 6.1 m (20 ft) square. Heat flow through the wall might have been two- or three-dimensional, rather than one-dimensional, as assumed. Heat distribution and transmission through the building slab and roof might have influenced heat flux through walls.

Thermal Lag and Reduction in Amplitude

Thermal lag and reduction in amplitude are used to describe wall dynamic thermal performance for a particular dynamic temperature cycle. Thermal lag, as illustrated in Fig. 9, is the difference in time between the actual heat flow, q_w or q_{NBS}, and the heat flow based on steady-state predictions, q_{ss}. Thermal lag is of interest because the time of occurrence of peak heat flows will have an effect on the overall response of the building envelope. If the envelope can be

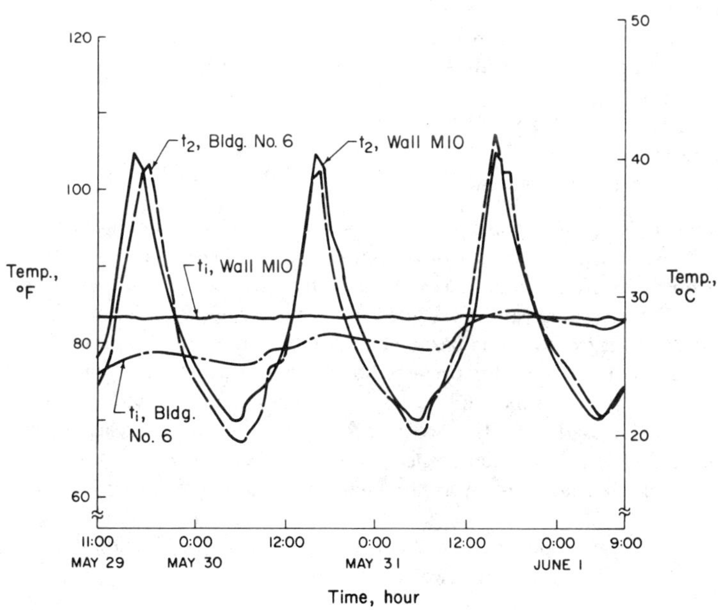

FIG. 15—*May temperatures for NBS Test Building 6 and Wall M10.*

used effectively to delay the occurrence of peak loads, it may be possible to improve overall energy efficiency. The "lag effect" is also of interest for passive solar applications.

The reduction in amplitude is the percentage reduction in actual peak heat flow compared with peak heat flow calculated using steady-state theory. The actual maximum heat flow through a wall is important in determining the peak energy load for a building envelope. Using actual peak heat flow rather than heat flow based on steady-state theory may reduce peak energy demands.

Thermal lag and reduction in amplitude are dependent on both the thermal resistance, R, and heat storage capacity

$$\rho c L$$

where

ρ = wall density, kg/m^3 (lb/ft^3),
c = wall specific heat, J/kg · K (Btu/lb · °F), and
L = wall thickness, m (ft).

Mass, ρL, is the predominant factor in determining the heat storage capacity of most building materials.

For homogeneous walls, thermal lag and reduction in amplitude have been shown to increase with an increase in the dimensionless parameter, M [*14*]

$$M = \left(\frac{L^2/\alpha}{P}\right)^{1/2} = \left(\frac{(\rho c L) \cdot (R)}{P}\right)^{1/2} \tag{3}$$

where

L = wall thickness, m (ft),
α = thermal diffusivity, $k/\rho c$, m^2/s (ft^2/h),
k = thermal conductivity of wall, W/m · K (Btu/h · ft · °F),
ρ = wall density, kg/m^3 (lb/ft^3),
c = wall specific heat, J/kg · K (Btu/lb · °F),
R = wall resistance, m^2 · K/W (h · ft^2 · °F/Btu), and
P = period of dynamic cycle, h.

The parameter M is equal to the square root of the ratio of two measurements of time. The numerator, L^2/α, is a measurement of time required for the wall to thermally respond to a change in temperature. The denominator, P, is the time required to complete a cycle. Dynamic temperature cycles generally can be assumed to have a period, P, of 24 h.

Changes in R affect the dynamic parameters of thermal lag and reduction in amplitude, as well as alter the maximum heat flux predicted by steady-state analysis. Changes in heat storage capacity affect only the dynamic parameters of thermal lag and reduction in amplitude.

The principles discussed in the last two paragraphs are valid for multilayered wall assemblies even though Eq 3 is derived for homogeneous walls only. Childs [*14*] suggests using the sum of M values for each wall layer as an approximate method of predicting lag and reduction in amplitude.

Table 2 lists material properties used to calculate M' values for individual layers of Wall M9, Wall M10, and NBS Test Building 6 walls.

$$M' = (L^2/\alpha)^{1/2} = (\rho c L \cdot R)^{1/2} \tag{4}$$

TABLE 2—*Calculation of* M′ *values for individual wall layers.*

Wall Designation	Wall Layer	R, Thermal Resistance, m² · K/W (h · ft² · °F/Btu)	ρ, Unit Weight, kg/m³ (lb/ft³)	c,[a] Specific Heat, J/kg · K (Btu/lb · °F)	L,[b] Equivalent Thickness, m (ft)	ρcL, Storage Capacity, W · h/K · m² (Btu/°F · ft²)	M' $(\rho cL \cdot R)^{1/2}$, (h)$^{1/2}$
M9	brick	0.08 (0.44)	2160[c] (135)	800 (0.19)	0.070 (0.23)	33.5 (5.90)	1.61
	air	0.17 (0.97)	0	. . .	. . .	0	0
	block	0.21 (1.18)	2000[c] (125)	840 (0.20)	0.079 (0.26)	36.9 (6.50)	2.77
	plaster	0.01 (0.03)	1860 (116)	840 (0.20)	0.003 (0.01)	1.3 (0.24)	0.08
M10	brick	0.08 (0.44)	2160[c] (135)	800 (0.19)	0.070 (0.23)	33.5 (5.90)	1.61
	perlite	1.51 (8.59)	69 (4.3)	1090 (0.26)	0.070 (0.23)	1.5 (0.26)	1.49
	block	0.21 (1.18)	2000[c] (125)	840 (0.20)	0.079 (0.26)	36.9 (6.50)	2.77
	plaster	0.01 (0.03)	1860 (116)	840 (0.20)	0.003 (0.01)	1.3 (0.24)	0.08
NBS	brick	0.08 (0.44)	2160 (135)	800 (0.19)	0.070 (0.23)	33.5 (5.90)	1.61
	perlite	1.93 (10.94)	69 (4.3)	1090 (0.26)	0.089 (0.29)	1.82 (0.32)	1.87
	block	0.30 (1.68)	1680 (105)	840 (0.20)	0.11 (0.35)	41.7 (7.35)	3.51
	plaster	0.02 (0.10)	1860 (116)	840 (0.20)	0.012 (0.04)	5.31 (0.93)	0.31

[a] Values of specific heat are determined from Refs *10* and *15*.
[b] Equivalent thicknesses for block and brick are equal to the product of the layer width and percentage of solid volume.
[c] Unit weight is equal to the sum of the oven-dry unit weight of the material and estimated moisture content.

The term P is omitted from the equation for M' because all walls are tested using cycles with a constant P, equal to 24 h. The thermal resistances are from Refs *10* and *11* and do not include surface resistances. Unit weights of block and brick for Walls M9 and M10 are determined from measured oven-dry material unit weights and estimated moisture contents. The unit weight of perlite used in Wall M10 is measured. Other unit weights are estimated using properties listed in Ref *10*. Values of specific heat are determined from Refs *10* and *15*. The equivalent thicknesses for block and brick are equal to the product of the layer width and percentage of solid volume.

The last two columns of Table 2 list heat storage capacity, ρcL, and M' values, $(\rho cL \cdot R)^{1/2}$, for each layer of the walls. The expanded perlite layers have M' values within the same range as the brick and block layers. Expanded perlite has relatively high thermal resistance and low storage capacity, while brick and block have low resistances and high storage capacities.

Table 3 lists M' values, thermal lag, and percentage of reduction in amplitude for Wall M9, Wall M10, and NBS Test Building 6 walls, denoted NBS. Values of M' for individual wall layers are summed to determine total wall M' values. The thermal lag and percentage of reduction in amplitude for Walls M9 and M10 are measured using the calibrated hot box. The values for NBS Test Building 6 are measured field data from a heat flux transducer mounted on the west wall.

Table 3 shows that measured thermal lag and calculated M' increase with the addition of expanded perlite insulation to the 300-mm (12-in.) cavity wall. The addition of expanded perlite insulation increases M' from 4.5, for Wall M9, to 6.0 $(h)^{1/2}$, for Wall M10.

Table 3 also shows that thermal lags increase with an increase in M' for walls with insulation, Walls M10 and NBS. The 400-mm (16-in.) insulated cavity wall, denoted NBS, has greater thermal lags and M' than the 300-mm (12-in.) insulated cavity wall, denoted M10.

Reduction in amplitude remains relatively constant when expanded perlite insulation is added to the 300-mm (12-in.) cavity wall. Table 3 shows that reduction in amplitude values range from 43 to 51% for all cycles on Walls M9 and M10.

The thermal lags and reductions in amplitude for the April and May data from the NBS test building wall differ more than those determined from calibrated hot box tests. The differences may be due to variability in heat flux transducer data or any of the phenomena described in the Heat Flux Comparisons portion of this section.

Total Energy Requirements

The results of dynamic tests for Walls M9 and M10 were also compared using measurements of energy expended in maintaining constant indoor temperature while outdoor temperatures were varied. The energy expended is a measurement of heat flow through the test wall.

TABLE 3—*Dynamic parameters of walls.*

		Wall Properties					
		Measured Thermal Lag, h, for Test Cycle			Measured Reduction in Amplitude, %, for Test Cycle		
Wall Designation	Calculated M', $h^{1/2}$	NBS	April	May	NBS	April	May
M9	4.46	5.5	5	6.5	43	51	49
M10	5.95	7	7.5	7.5	50	44.5	47
NBS	7.30	...	8	12	...	51	5

The measured and calculated total energy requirements for the NBS Gaithersburg April and Gaithersburg May test cycles applied to Walls M9 and M10 are summarized in Table 4. The curves marked q_w in Figs. 9 through 13 are measurements of heat flow through the test wall. The sum of the areas between each q_w curve and the zero heat flow rate line is taken to represent the total energy over a 24-h period. This value is denoted q_w^T in Table 4. A similar procedure is used to calculate the total energy over a 24-h period for predictions based on steady-state analysis. These values are denoted q_{ss}^T in Table 4.

The addition of expanded perlite fill to the masonry cavity of Wall M9 to form Wall M10 reduces the total energy requirements by 66 to 68%, as shown in the columns labeled Measured Total Energy Comparisons and Calculated Total Energy Comparisons. The results are consistent for both measured and calculated values. This reduction in energy requirements is due to the increased thermal resistance of the expanded perlite-filled walls.

Table 4 also shows that the measured total energy, q_w^T, for each cycle, is less than the total energy predicted by steady-state analysis, q_{ss}^T. The measured total energy ranges from 57 to 62% of the calculated total energy for all three test cycles on both walls. This is due to the wall storage capacity and reversals of heat flow through each wall. When the NBS Gaithersburg April and Gaithersburg May test cycle are applied to walls during the calibrated hot box tests, the outdoor wall surface temperatures fluctuate above and below the indoor wall surface temperatures, which causes reversals of heat flow through the walls. For example, during the NBS test cycle, the outdoor surface temperatures vary from approximately 10 to 38°C (50 to 100°F) while indoor surface temperatures vary in the range of 21 to 24°C (70 to 75°F).

For walls subjected to reversals in heat flow, total heat flow entering and leaving the indoor surface is reduced compared to steady-state predictions. The amount of heat flow entering or leaving the indoor surface is not as great as that predicted by steady-state analysis because steady-state equilibrium is never achieved within the wall. Heat storage within the wall reduces temperature fluctuations on the indoor surface.

It should be noted that comparison of the measured energy values for test walls is limited to specimens and dynamic cycles evaluated in this program. The results are for diurnal test cycles and should not be arbitrarily assumed to represent annual heating and cooling loads. In addition, the results are for individual opaque wall assemblies; as such, they are representative of only one component of the building envelope.

Summary and Conclusions

This report presents results of an experimental investigation of heat transmission characteristics for two block-brick cavity walls. Wall M9 contained no insulation. Wall M10 contained expanded perlite loose-fill insulation in the cavity. Tests were conducted under steady-state and dynamic temperature conditions. The results are compared with measured values from a test building with cavity walls monitored by the National Bureau of Standards in Gaithersburg, Maryland.

The following conclusions are based on results obtained in this investigation:

1. Measured overall thermal resistances, R_T, for Walls M9 and M10, respectively, were 0.61 and 1.66 $m^2 \cdot K/W$ (3.48 and 9.41 $h \cdot ft^2 \cdot °F/Btu$).
2. Design overall thermal resistances, R_T, for Walls M9 and M10 were within 6% of calibrated hot box test results.
3. Calibrated hot box results for the Gaithersburg April and Gaithersburg May test cycles on the wall with expanded perlite insulation (Wall M10), showed less heat loss from the indoor environment than heat flux transducer measurements from NBS Test Building 6.
4. As indicated by thermal lag, the heat storage capacities of the block-brick cavity walls

TABLE 4—*Measured and calculated total energy requirements.*

Test Cycle	Measured Total Energy, q_w^T, W · h/m² (Btu/ft²)		Measured Total Energy Comparisons, With Perlite/Without Perlite Ratio	Calculated Total Energy, q_{ss}^T, W · h/m² (Btu/ft²)		Calculated Total Energy Comparisons, With Perlite/Without Perlite Ratio	Total Energy Comparisons, $\frac{q_w^T}{q_{ss}^T}$	
	Wall M9, Without Perlite	Wall M10, With Perlite		Wall M9, Without Perlite	Wall M10, With Perlite		Wall M9	Wall M10
NBS	219.1 (69.5)	70.0 (22.2)	0.32	382.4 (121.2)	122.8 (38.9)	0.32	0.57	0.57
Gaithersburg April	247.4 (78.4)	83.5 (26.5)	0.34	396.9 (125.8)	136.4 (43.2)	0.34	0.62	0.61
Gaithersburg May	151.5 (48.0)	50.2 (15.9)	0.33	243.3 (77.1)	80.7 (25.6)	0.33	0.62	0.62

delayed heat flow through the specimens. The thermal lag values range from 5 to 6.5 h for Wall M9, from 7 to 7.5 h for Wall M10, and from 8 to 12 h for the west wall of NBS Test Building 6.

5. Measured thermal lag increases for each dynamic cycle when expanded perlite is added to the cavity.

6. As indicated by the damping effect, the heat storage capacities of Walls M9 and M10 reduced peak heat flows through the specimens. Reduction in the amplitude values ranges from 43 to 51% for Wall M9 and from 45 to 50% for Wall M10.

7. For the three diurnal temperature cycles applied to Walls M9 and M10, energy requirements for a 24-h period were less than would be predicted by steady-state analysis. Total measured heat flows for the 24-h cycles ranged from 57 to 62% of those predicted by steady-state analysis, for both test specimens. These reductions in total heat flow are attributed to wall storage capacity and reversals in heat flow.

8. The addition of expanded perlite fill to the cavity wall reduces total heat flow by 66 to 68% for the three diurnal cycles. This reduction is due to the increased thermal resistance of the expanded-perlite-filled wall.

Results described in this paper provide data on the thermal response of masonry cavity walls subjected to steady-state and diurnal sol-air temperature cycles. A complete analysis of building energy requirements must include consideration of the entire building envelope, building orientation, building operation, and yearly weather conditions. Data developed in this experimental program provide a quantitative basis for modeling the building envelope, which is part of the overall energy analysis process.

Acknowledgments

This paper was prepared as part of a project sponsored jointly by the Portland Cement Association, the Brick Institute of America, the National Concrete Masonry Association, and the Perlite Institute, Inc. The work was performed in the Engineering and Resource Development Division of the Construction Technology Laboratories (CTL), a Division of the Portland Cement Association, under the direction of Dr. W. G. Corley, executive director, and D. W. Musser, formerly director of the Construction Methods Department.

The brick for the cavity wall construction was provided by the Brick Institute of America. Silbrico Corp. supplied expanded perlite loose-fill insulation for the cavity. Metal ties were provided by Dur-o-Wall, Inc.

Data from NBS Test Building 6 for comparison with calibrated hot box test results were supplied by D. Burch, National Bureau of Standards. Information on NBS Test Building 6 wall construction was provided by A. Desjarlais, Dynatech R/D Company. S. C. Larson, associate research engineer, Fire Research Section, assisted with data reduction, reviewed the manuscript, and provided helpful comments and suggestions.

References

[1] Fiorato, A. E. and Cruz, C. R., "Thermal Performance of Masonry Walls," Research and Development Bulletin RD071, Portland Cement Association, Skokie, IL, 1980.

[2] Fiorato, A. E., "Heat Transfer Characteristics of Walls Under Dynamic Temperature Conditions," Research and Development Bulletin RD075, Portland Cement Association, Skokie, IL, 1981.

[3] Fiorato, A. E. and Bravinsky, E., "Heat Transfer Characteristics of Walls Under Arizona Temperature Conditions," Construction Technology Laboratories, Portland Cement Association, Skokie, IL, 1981.

[4] Van Geem, M. G., Fiorato, A. E., and Julien, J. T., "Heat Transfer Characteristics of a Normal Weight Concrete Wall," Construction Technology Laboratories, Portland Cement Association, Skokie, IL, 1983.

[5] Van Geem, M. G. and Fiorato, A. E., "Heat Transfer Characteristics of a Structural Lightweight

Concrete Wall," Construction Technology Laboratories, Portland Cement Association, Skokie, IL, 1983.

[*6*] Van Geem, M. G. and Fiorato, A. E., "Heat Transfer Characteristics of a Low Density Concrete Wall," Construction Technology Laboratories, Portland Cement Association, Skokie, IL, 1983.

[*7*] Van Geem, M. G. and Larson, S. C., "Heat Transfer Characteristics of a Masonry Cavity Wall With and Without Expanded Perlite Insulation," Construction Technology Laboratories, Portland Cement Association, Skokie, IL, 1984.

[*8*] Burch, D. M., Remmert, W. E., Krintz, D. F., and Barnes, C. S., "A Field Study of the Effect of Wall Mass on the Heating and Cooling Loads of Residential Buildings," *Proceedings,* Thermal Mass Effects in Buildings Seminar, Knoxville, TN, June 1982, Oak Ridge National Laboratory, Oak Ridge, TN, 1983, pp. 265-312.

[*9*] Fiorato, A. E., "Laboratory Tests of Thermal Performance of Exterior Walls," *Proceedings,* ASHRAE SP28, American Society of Heating, Refrigerating, and Air Conditioning Engineers/Department of Energy-Oak Ridge National Laboratory Conference on Thermal Performance of the Exterior Envelopes of Buildings, Orlando, FL, Dec. 1979, Atlanta, GA, 1981, pp. 221-236.

[*10*] *ASHRAE Handbook—1981 Fundamentals,* American Society of Heating, Refrigerating, and Air Conditioning Engineers, Atlanta, GA, 1981.

[*11*] Randall, F. A., Jr. and Panarese, W. C., *Concrete Masonry Handbook,* Bulletin EB008.04M, Portland Cement Association, Skokie, IL, 1980.

[*12*] Desjarlais, A., "Thermal Conductance and Thermal Transmittance of Six Residential Wall Structures," Report No. NBS-3, Dynatech R/D Company, Cambridge, MA, 1982.

[*13*] Peavy, B. A., Powell, F. J., and Burch, D. M., "Dynamic Thermal Performance of an Experimental Masonry Building," Building Science Series 45, U.S. Department of Commerce, National Bureau of Standards, Washington, DC, 1973.

[*14*] Childs, K. W., Courville, G. E., and Bales, E. L., "Thermal Mass Assessment," Oak Ridge National Laboratory for the U.S. Department of Energy, Oak Ridge, TN, 1983.

[*15*] Van Geem, M. G. and Fiorato, A. E., "Thermal Properties of Masonry Materials for Passive Solar Design—A State-of-the-Art Review," Construction Technology Laboratories, Portland Cement Association, Skokie, IL, 1983.

Moisture

Francesco De Ponte[1]

Design Considerations on Guarded and Calibrated Hot Box Apparatus

REFERENCE: De Ponte, F., **"Design Considerations on Guarded and Calibrated Hot Box Apparatus,"** *Thermal Insulation: Materials and Systems, ASTM STP 922,* F. J. Powell and S. L. Matthews, Eds., American Society for Testing and Materials, Philadelphia, 1987, pp. 345–359.

ABSTRACT: The precision and accuracy of guarded and calibrated hot box apparatus have seldom been evaluated, either during the design process or at the performance check stage. Even round robins can supply little information on errors if no theory of the apparatus is available. The purpose of this paper is to consider heat transfer in the apparatus, to supply some rough practical tips to aid in understanding the apparatus performance, even though much more has to be done to reach the same level of knowledge achieved for the guarded hot plate apparatus.

Convection heat transfer is considered first. Theoretical relationships are recalled to predict local surface coefficients in natural convection. These change along specimen surfaces, inducing temperature nonuniformities along the border of the metering box. Transitions from laminar to turbulent flow are also considered, which add hardly predictable temperature fluctuations. The problem of the position of the sensors that detect air temperatures in forced convection is then analyzed.

The analysis of coupled radiation and convection heat transfer allows a definition of the maximum expected sensitivity of the apparatus to imbalance conditions.

From the theoretical description of some sources of error, guidelines may be obtained for an apparatus performance check.

KEY WORDS: measuring techniques, hot box method, error analysis, imbalance detection, convection, radiation, thermal insulation

Nomenclature

A_1, A_2	Area of isothermal surfaces, m^2
c_p	Specific heat, J/kg · K
h, h_c, h_r	Surface coefficient of heat transfer; that due to convection; that due to radiation, W/m^2 · K
F_{12}	Shape factor
Gr	Grashof number
g	Gravity acceleration, m/s^2
Nu_x, Nu_δ	Nusselt number at distance x; Nusselt number for layer thickness δ
Pr	Prandtl number
q_r	Heat flow rate by radiation, W
Re_x, Re_δ, Re_c	Reynolds number at x; Reynolds number at layer thickness δ; critical Reynolds number
R	Resistance per unit area, m^2 · K/W
R_B, R_{BN}	Metering box resistances: surface to surface; longitudinal, K/W
R_{E1}, R_{E2}, R_G	Edge resistances; specimen resistance at the guard region, K/W
R_I, R_M	Imbalance resistance; specimen resistance at the metering area, K/W

[1]Professor, Istituto di Fisica Tecnica, Università di Padova, 35131 Padua, Italy.

R_{SC}, $R_{SC'}$	Cold-side surface resistance at the metering area; at the guard region, K/W
R_{SG}, R_{SM}	Hot-side surface resistance at the guard region; at the metering area, K/W
R_{SMB}, R_{SMG}	Surface resistance at the inner side of the metering box; at the outer side, K/W
T, T_a, T_s, T_{so}, T_{sx}	Temperature; of the air; of a surface; of a surface at a distance zero or x, K
T_{aa}, T_{aao}, T_{aax}	Bulk temperature; at a distance zero or x, K
T_1, T_2, T_m	Temperatures; mean temperature (in radiation), K
T_{AC}, $T_{AC'}$	Cold-side air temperature at the metering area; at the guard region, K
T_{AG}, T_{AM}	Hot-side air temperature at the guard region; at the metering area, K
T_E	Specimen edge temperature, K
T_{SG}, T_{SM}	Specimen temperature on the hot surface at the guard region; at the metering area, K
u, u_m, u_{max}, u_s	Air velocity; mean; maximum; out of the boundary layer, m/s
U	Thermal transmittance, $W/m^2 \cdot K$
x, x_c	Distance along the flow; critical distance, m
y	Distance from the surface in the flow, m
β	Coefficient of cubic expansion, K^{-1}
δ, δ_c	Boundary layer thickness; at critical distance, m
ϵ, ϵ_1, ϵ_2	Surface emittances
λ	Thermal conductivity (of the air), $W/m \cdot K$
μ	Dynamic viscosity, $kg/m \cdot s$
ν	Kinematic viscosity, m^2/s
ρ	Density, kg/m^3
σ_n	Stefan's constant, 5.67×10^{-8} $W/m^2 \cdot K^4$

The guarded hot box (GHB) and, more recently, calibrated hot box (CHB) have became popular equipment for testing large pieces of insulation of homogeneous materials or, in the majority of cases, composites. The method is standardized in many countries, for example, in France (NF X 10-022), Sweden (SS 02 42 12), West Germany (DIN 52611), the United Kingdom (BS 874), and the United States [ASTM Test for Steady-State Thermal Performance of Building Assemblies by Means of a Guarded Hot Box (C 236-80) and ASTM Test for Thermal Performance of Building Assemblies by means of a calibrated hot box (C 976-82)]. Standardization within the International Organization for Standardization (ISO) is in progress in Committee 163 on Thermal Insulation. Also available in the literature are descriptions of some equipment, for example, Mumaw [*1*], Orlandi et al [*2*], and Goss [*3*]. The variety of sizes and the large dimensions of the specimens makes it rather difficult to conduct round robins (one is in progress in the United States, under the sponsorship of ASTM), so that there is limited evidence on the accuracy of such equipment. Calibration and verification procedures have been developed by Lavine et al [*4*] and by Guy and Nixon [*5*], and theoretical analysis of flanking losses have been performed by Onega and Burns [*6*]; nevertheless, a comprehensive hot box error analysis is far more complex than a guarded hot plate (GHP) analysis, because of the combined heat transfer by convection and radiation from the specimen surface to the apparatus.

The intention of this paper is to draw some design and error evaluation guidelines through an elementary analysis of heat transfer in the apparatus based on the expressions that can be found in any textbook on heat transfer, for example, Eckert and Drake [*7,8*].

Convection

In many hot boxes the specimen is mounted vertically, and this analysis will be limited to this testing configuration.

Heat is transferred through the specimen surface, both by radiation and by convection. One important difference is that radiation heat transfer depends on the temperature of the surfaces "seen" by the specimen, while convection heat transfer depends on the temperature of the air. Even considering average values, the air and apparatus surface temperatures are never exactly equal, although they are usually assumed to be equal to define an overall surface coefficient of heat transfer.

Convection heat transfer may be either natural or forced. Natural convection is sometimes preferred on the hot side of the specimen to evaluate inside heat transfer in buildings. Forced convection is normally preferred on the cold side of the specimen because it gives a better approximation of outdoor conditions.

Natural and forced convection can raise completely different problems; consequently, they will be considered separately, starting with natural convection.

Natural Convection

Let us consider still air at temperature T_a and a flat vertical surface at a temperature $T_s < T_a$ (see Fig. 1). Air molecules above the top of the surface, those just in contact with the surface, and all those far away from it are still. Starting from the top air molecules close to the surface (but not those just in contact with it) move downwards because of the bouyancy forces. While moving downwards, their velocity increases and, because of friction, other molecules are pushed into motion. The region where air is in movement is called the boundary layer; up to a certain distance from the top the molecule movement is unidirectionally downwards (that is, the flow is laminar). Then there is a region where movement is unstable, with velocity fluctuations, which precedes the region of fully turbulent flow.

Two considerations come from this qualitative description. The air temperature to be referenced in heat transfer calculations should be that of still air, that is, air outside the boundary layer. Consequently, temperature sensors must be placed at appropriate distances from the specimen. In defining such a distance it must be kept in mind that the boundary layer thickness increases from top to bottom and that it increases considerably while evolving from laminar to turbulent flow.

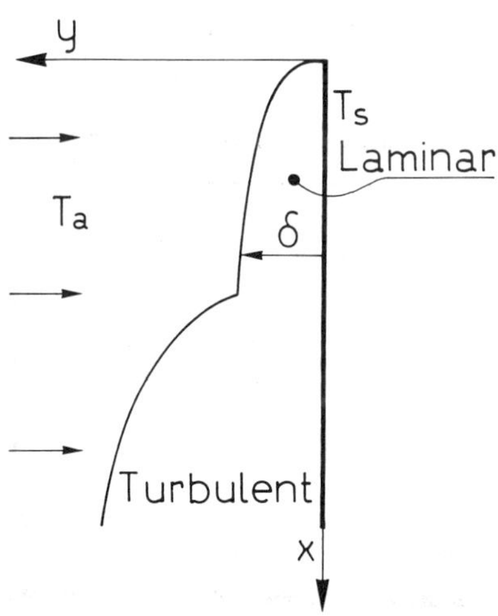

FIG. 1—*The boundary layer in natural convection when the surface is colder than the air.*

The second consideration is that, to allow the growth of the boundary layer, new molecules have to replace those that started moving; to allow this replacement, there must be a very slow air motion perpendicular to the boundary layer and hence to the surface, as shown by the arrows of Fig. 1. As a practical consequence, the presence of a baffle at a distance from the vertical surface comparable to the boundary layer thickness will prevent the described air motion, thus completely changing the flow pattern. The configuration of Fig. 1 allows an easy quantitative description, at least in laminar flow, yet the heat transfer inside buildings and in the hot box apparatus is somewhat different. The air moving downwards is impinging on the floor of the room or at the bottom of the apparatus, changing the direction of the motion. In the meantime, if the room walls are at the temperature T_a of still air, cold air at the bottom is heated and moves upwards. This creates air circulation, as shown in Fig. 2. From this sketch, it is evident that the temperature T_a is that of still air only in the central region of the room. These considerations may help in understanding the difficulties in correctly designing a hot box apparatus when natural convection takes place and in correctly measuring the temperature T_a.

Even though the ideal configuration of Fig. 1 does not exactly describe convection on a vertical wall in a room or in a hot box apparatus, we can still supply some relationships that will help in understanding the phenomenon. The type of flow of the air (laminar or turbulent) is essentially defined by the value of the Grashof number, Gr

$$\mathrm{Gr} = \frac{x^3 \beta g (T_a - T_s)}{\nu^2}$$

which depends on the temperature difference, the distance, x, from the top, and the nature of the fluid (βg is the coefficient of cubic expansion multiplied by the gravity acceleration, and ν is the kinematic viscosity). When $\mathrm{Gr} < 10^8$, the flow is definitely laminar; when $\mathrm{Gr} > 10^{10}$, the flow is definitely turbulent; the transition region is located between these values. These limits can also supply an idea of the possible length of the transition region.

In laminar flow, the velocity and temperature profiles are as shown in Fig. 3. In studying convection, it is common practice to introduce the dimensionless Nusselt number, Nu_x, defined as

$$\mathrm{Nu}_x = \frac{h_c x}{\lambda}$$

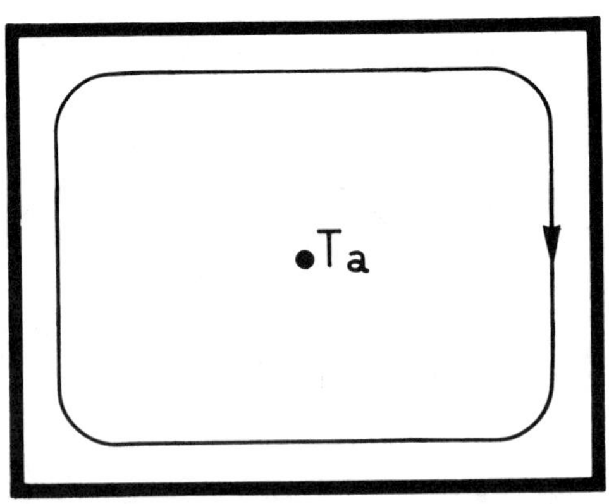

FIG. 2—*Air motion in a room or on the hot side of the hot box appratus.* T_a *is the temperature of still air.*

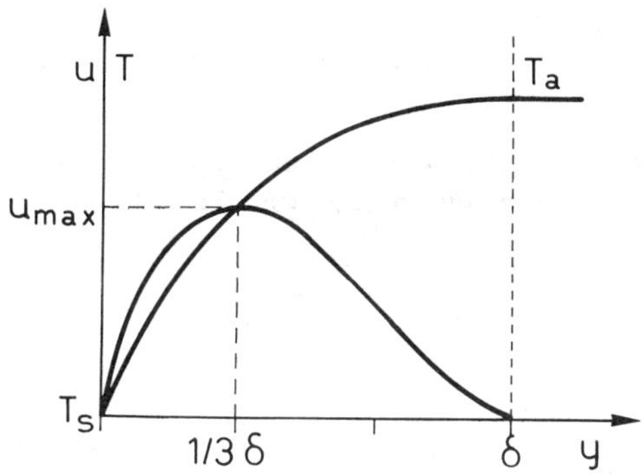

FIG. 3—*Temperature and velocity in natural convection (with no baffle installed).*

where h_c is the convection coefficient and λ is the thermal conductivity of the fluid. For natural convection in laminar flow, the following relationship is usually accepted

$$\text{Nu}_x = 0.509\,\text{Pr}^{1/2}(0.952 + \text{Pr})^{-1/4}\text{Gr}^{1/4}$$

Pr being the Prandtl number (specific heat, c_p, multiplied by the dynamic viscosity, μ, and divided by the conductivity, λ); for the air around room temperature Pr = 0.71, hence the previous expression becomes simply

$$\text{Nu}_x = 0.38\,\text{Gr}^{1/4}$$

From Nu_x the local convection coefficient at a distance x can be drawn immediately: it decreases while x increases. The mean convection coefficient of the portion of surface from the top ($x = 0$) to the distance x is 4/3 of the value at the distance x. Also, the boundary layer thickness, δ, can be derived from Nu_x; this value is

$$\delta = \frac{2x}{\text{Nu}_x}$$

The boundary layer thickness, δ, increases with x. The mean air velocity, u_m, in a cross section at a given distance, x, is

$$u_m = 0.431\left(\frac{\nu}{x}\right)(0.952 + \text{Pr})^{-1/2}\text{Gr}^{1/2} \qquad \text{m/s}$$

and for air it is

$$u_m = \frac{(5 \times 10^{-6}\,\text{Gr}^{1/2})}{x} \qquad \text{m/s}$$

The maximum velocity, u_{max}, is $u_{max} = 16/9\,u_m$.

Calling y the distance from the surface in the direction normal to the flow, the velocity distribution in a cross section at a distance, x, from the top of the surface, so that δ is the boundary layer thickness, is

$$u = 12u_m \frac{y}{\delta}\left(1 - \frac{y}{\delta}\right)^2$$

The temperature, T, in a cross section of the boundary layer and at a distance, x, so that δ is the boundary layer thickness at a given distance, y, from the surface is

$$T = T_a - (T_a - T_s)\left(1 - \frac{y}{\delta}\right)^2$$

For air around room temperature (Pr = 0.71) when a turbulent boundary layer is fully developed (typically for Gr > 10^9, but in some cases only when Gr > 10^{10}), the convection coefficient becomes independent of the distance, x. For 10^9 < Gr < 10^{12} the following expression may be used in turbulent flow.

$$\mathrm{Nu}_x = 0.12\,\mathrm{Gr}^{1/3}$$

Table 1 gives some values for the parameters just described when $T_a = 295$ K and $T_a - T_s = 2$ K. Under those conditions (which can exist in a hot box apparatus), it can be seen that the transition from laminar to turbulent flow will occur at approximately $x = 1.5$ m, although, in some cases, the transition can occur at x as low as 0.75 m. When $x = 3$ m, the flow is quite definitely turbulent and fully developed. At this distance Nu_x and h_c were evaluated with the expressions of turbulent flow.

Forced Convection

The heat transfer in forced convection requires far different considerations. To understand the problem, we can consider a flat surface immersed in an air flow of uniform velocity, u_s, running parallel to it. Again, a boundary layer exists. Fluid molecules in contact with the surface are still, but all those far from the surface are now moving at the velocity u_s. The shape of the boundary layer looks similar to that in Fig. 1; now the type of flow depends on the Reynolds number

$$\mathrm{Re}_x = \frac{u_s x}{\nu}$$

TABLE 1—*Natural convection data.*

x, m	Gr	u_m, m/s	Nu_x	δ, m	h_c, W/m^2 · K
0.05	3.72×10^4	0.019	5.24	0.019	2.73
0.125	5.81×10^5	0.031	10.4	0.024	2.17
0.25	4.65×10^6	0.043	17.5	0.029	1.82
0.50	3.72×10^7	0.061	29.5	0.034	1.53
0.75	1.25×10^8	0.074	40.0	0.038	1.39
1.00	2.97×10^8	0.086	49.6	0.040	1.29
1.25	5.81×10^8	0.096	58.6	0.043	1.22
1.50	1.00×10^9	0.106	67.2	0.047	1.17
2.00	2.38×10^9	...	...	...	...
3.00	8.03×10^9	...	240.0	...	2.08

where x has the same meaning as for natural convection. The transition from laminar to turbulent flow takes place at a distance x so that $Re_x = 500\ 000$. In laminar flow, the boundary layer thickness, δ, is expressed as

$$\delta = 4.64x\, Re_x^{-1/2} \qquad m$$

Hence, the thickness δ_c of the boundary layer at the transition from laminar to turbulent flow in air ($\nu \simeq 1.5\ 10^{-5}$ m²/s) is $\delta_c = 0.049/u_s$ m/s, and the corresponding distance, x_c, is

$$x_c = \frac{500\ 000\nu}{u_s}$$

Now let us consider the flow between two flat parallel surfaces at a distance 2 δ, as shown in Fig. 4. On each surface a boundary layer starts to develop, growing in thickness. If the thickness is $2\ \delta = 2\ \delta_c$, from the corresponding distance, x_c, on, the flow occupies all the space between the surfaces (fully developed flow) and is turbulent. Beyond the critical distance, x_c, the flow pattern in the boundary layer will change no more. This also implies that the convection coefficient will remain constant. Replacing the distance x_c in the Reynolds number, Re_x, it is possible to derive the new expression of the Reynolds number, Re_δ, as a function of the thickness

$$Re_\delta = \frac{u_s\delta}{\nu}$$

The condition $2\ \delta = 2\ \delta_c$ thus corresponds to $Re_c = 3280$. The velocity pattern in a section normal to the flow as a function of the distance, y, from the nearest surface ($0 \leq y \leq \delta$) is given as

$$u = u_s\left(\frac{y}{\delta}\right)^{1/7}$$

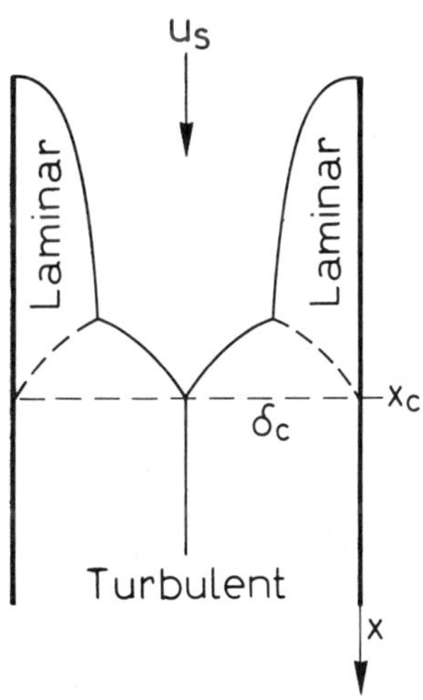

FIG. 4—*Air flow between two parallel surfaces.*

while the mean velocity, u_m, by integration of the last equation, is $u_m = {}^7/_8 u_s$. If we assume a distance beyond x_c, the flow pattern will not vary further, and therefore the convection coefficient will remain the same, making it possible to evaluate it for air at room temperature through the expression

$$\mathrm{Nu}_\delta = 0.014\,\mathrm{Re}^{3/4}$$

where $\mathrm{Nu}_\delta = h_c\delta/\lambda$ and Re_δ is evaluated with the half thickness, δ.

In the guarded and calibrated hot box, we may assume the specimen surface and the baffle to be flat parallel surfaces, as considered here. Their distance is 2 δ, and the velocities both on the specimen surface and on the baffle surface are zero, while in the middle the velocity is maximum when the condition $x > x_c$ is met. As regards heat transfer, the baffle may be considered adiabatic, hence the thermal boundary layer, when the flow is fully developed, has a thickness of 2 δ, and the distribution of the temperature, T, within the fully developed flow may be expressed in a section normal to the flow itself as

$$T = T_a + (T_s - T_a)\left(1 - \left(\frac{y}{2\delta}\right)^{1/7}\right)$$

T_s being the temperature of the surface and T_a that of the air just in contact with the adiabatic baffle. Figure 5 can give a qualitative idea of temperature and velocity distribution in a turbulent, fully developed flow between flat parallel surfaces when the air is colder than the specimen, as on the cold side of the hot box apparatus.

Here, again, some practical conclusions may be drawn. When a fluid flows in a closed loop of constant cross section at a sufficient velocity to develop a turbulent flow, the convection coefficient is uniform (it is equivalent to $x \to \infty$). When designing guarded and calibrated hot boxes, it is therefore mandatory to avoid large changes in cross sections to avoid local stagnations or transitions from turbulent to laminar flow, and vice versa. A transition from laminar to turbulent flow between the baffle and the specimen implies a change in the convection coefficient, and hence local temperature nonuniformities.

Considering the space behind the baffle, where heaters, fans, and coolers are located (see Fig. 6), an air velocity too low may result in a laminar flow. If heaters or coolers are immersed in a laminar flow, there is no mixing beyond them between hot and cold portions of the flow. If such a flow comes in contact with the specimen, unpredictable temperature nonuniformities will result. The ideal situation would be that of turbulent flow with a constant cross section both between the specimen and the baffle and behind the baffle, where the heaters and coolers lie.

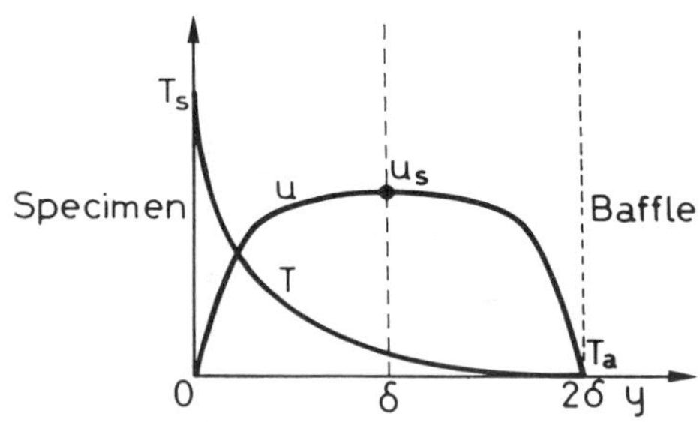

FIG. 5—*Temperature and velocity in forced convection (with a baffle installed).*

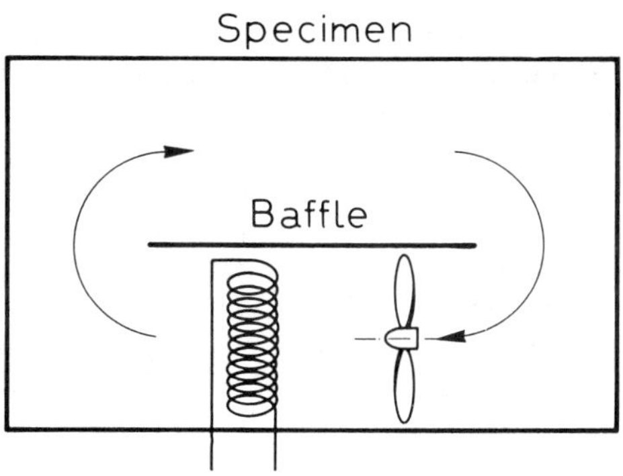

FIG. 6—*Baffle, fans, and heaters in a hot box.*

For the problem of temperature uniformity on the surface of the specimen, the conditions are completely different from those of natural convection. In natural convection a still-air temperature can be defined, but the convection coefficient is not uniform over the surface; in forced convection, on the contrary, the convection coefficient is uniform and independent of the temperature difference when the flow is turbulent and fully developed. Yet the air flow exchanges heat, so that its enthalpy changes, and consequently its temperature. But it has also been shown that in a cross section the temperature differs at each point (see Fig. 5), so that some criterion is needed to define a meaningful cross section temperature. Whenever the amount of heat transferred to or from the flow is considered, the only correct way of evaluating the temperature to be assigned to each cross section is the so-called bulk temperature, T_{aa}, which corresponds to the ideal assumption of mixing adiabatically all the flow portions that pass from a section at a distance x to a section at a distance $x + dx$ in a given time interval, Δt, that is

$$T_{aa} = \frac{\int_{2\delta} T c_p \rho u dy}{\int_{2\delta} c_p \rho u dy}$$

T_{aa} does not correspond to the surface temperature T_s or to the mean temperature from the integration of the preceding equation. Because the temperature profile from the preceding equation is rather flat for large distances, y, there will be a distance, y, not much larger than δ, where T_{aa} is approximated. It must be kept in mind, however, that this point is at a distance proportional to the distance of the baffle from the specimen. There is therefore no ground for substantiating the requirement in some standards for positioning the temperature sensors at a fixed distance from the specimen surface. The only way of measuring T_{aa} is to check both velocity and temperature distribution under some relevant testing conditions.

If the temperature T_s is assumed to be uniform, the bulk temperature, T_{aax}, at a given distance, x, from a section where the flow is already turbulent and fully developed, with a bulk temperature T_{aao}, can be computed as follows

$$T_{aax} + T_s - (T_s - T_{aao}) \exp\left(-\frac{h_c x}{2 c_p \rho \delta u_m} \right)$$

When the heat flow density on the surface is assumed to be uniform, that is, it is constant in each cross section, the difference between the bulk temperature, T_{aa}, and the surface temperature, T_s, is

$$(T_{aax} - T_{aao})\, c_p 2\delta\rho u_m = (T_{sx} - T_{so}) h_c x$$

where T_{so} and T_{sx} are the surface temperatures, respectively, at the distances 0 and x. Actual situations lie between those two extremes. These extremes may also be used to evaluate the temperature uniformity on the surface of the specimen.

Table 2 gives some values of the convection coefficient assuming air to be the fluid and the thickness to be $\delta = 0.05$ m.

Radiation

When two gray surfaces exchange heat by radiation, when each surface temperature is uniform, and when both the energy emitted and that reflected are uniform at each point and perfectly diffused, the heat flow rate, q_r, exchanged by radiation can be written as

$$q_r = \frac{\sigma_n (T_1^4 - T_2^4)}{\dfrac{1-\epsilon_1}{\epsilon_1 A_1} + \dfrac{1-\epsilon_2}{\epsilon_2 A_2} + \dfrac{1}{A_1 F_{12}}}$$

where σ_n is the Stephan-Boltzmann constant, ϵ_1 and ϵ_2 are the total hemispherical emittances, and F_{12} is the shape factor of Surface 1 versus Surface 2. The term $(T_1^4 - T_2^4)$, when $T_1 - T_2$ is small, may be approximated by $4T_m^3(T_1 - T_2)$, in which $T_m = (T_1 + T_2)/2$.

Replacing this expression in the preceding equation, we can write it as

$$q_r = h_r(T_1 - T_2)A_1$$

where h_r is a function of the emittances, of the shape factor F_{12}, and of T_m. Only when $T_1 = T_a$ or $T_1 = T_{aa}$ and $T_2 = T_s$ can a total surface coefficient $h = h_c + h_r$ be used. This discussion helps in understanding the difficulties in meeting such a condition and the need to define an environmental temperature that will take into account not only the air temperature, but also the effect of radiation from the apparatus surfaces "seen" by the specimen.

Temperature Uniformity and Imbalance

When thermal transmittance, U, for a wall is established, the following assumptions are made:

1. A temperature can be assigned to the air at each side of the wall.
2. Surface coefficients at each surface of the wall are uniform.

TABLE 2—*Forced convection data.*

u_s, m/s	Re	Nu	h_c, W/m$^2 \cdot$ K
0.5	1 666	...	...
1.0	3 333	6.14	3.19
3.0	10 000	14.00	7.28
5.0	16 667	20.50	10.70
10.0	33 333	34.50	20.00

Under these conditions, wall surface temperatures are uniform. In the section on convection it has been shown that, with natural convection, local surface coefficients usually change, while with forced convection the fluid temperature changes. Nonuniformities create problems in the definition of the U value to be reported as well as in the definition of imbalance in the guarded hot box. To understand the situation, let us imagine a test specimen having a thermal resistance per unit of area of $R = 3\ m^2 \cdot K/W$ mounted in a 1.5 by 1.5-m guarded hot box with natural convection on the hot side ($T_a = 295$ K) and forced convection on the cold side ($u_s = 5$ m/s; $T_{aao} = 270$ K; $\delta = 0.05$ m).

The apparatus surfaces are first assumed to be perfectly reflective—that is, no radiation heat transfer is implied. Surface temperatures can be derived only step by step. From Table 1 a mean tentative heat transfer coefficient, $h_c = 1.5\ W/m^2 \cdot K$, may be assumed on the hot side, while Table 2 gives $h_c = 10.7\ W/m^2 \cdot K$ for the cold side. These values give a temperature difference of $T_a - T_s \simeq 4$ K for the hot side and $T_s - T_a = 0.6$ K for the cold side.

The evaluation of T_{aax} shows that the temperature remains uniform within a few hundredths of a kelvin, so that it may be assumed to be uniform. From Table 1, tentative local surface coefficients also may be derived for the hot side and, assuming that longitudinal conduction along the specimen will not seriously influence surface temperatures, local surface temperatures may be obtained from those local surface coefficients. If local temperature differences differ from the starting temperatures, the convection coefficients will be computed again with the new temperature differences and the process will be repeated until the process converges.

In a guarded hot box we can imagine such a computation along a vertical line crossing the metering box as Path B in Fig. 7 and along a line lying in the guard area, as Path A of Fig. 7. For Path A, starting from the top and assuming the validity of the expression for evaluating convection coefficients on a vertical surface, given in the section on natural convection, h_c is maximum, then it decreases while in laminar flow. When it reaches the height of the bottom of the metering box, h_c increases again because of the transition to turbulent flow.

Using the iterative procedure described, $T_s - T_a$ lies between -2 and -4 K (see Fig. 8, Plot A on the right). Considering Path B, and starting again from the top, we have the same profile as for Plot A, but when inside the metering box, we start again with a new boundary layer, just as at the top of the guard area. Just out of the metering box, we start again with a new boundary layer. It is like shifting Profile A first to the top of the metering box and then to the bottom. The resulting profile could be like Path B on the right of Fig. 8.

If now we consider imbalance along the perimeter of the metering area of the specimen, we see that on the two vertical sides imbalance is due to the difference between Plots A and B, and it has a similar profile, even though it is not the same at each point.

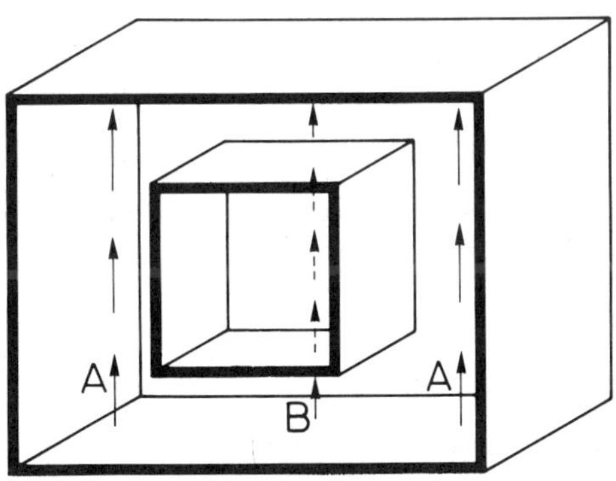

FIG. 7—*View of the hot side of a guarded hot box with the specimen removed.*

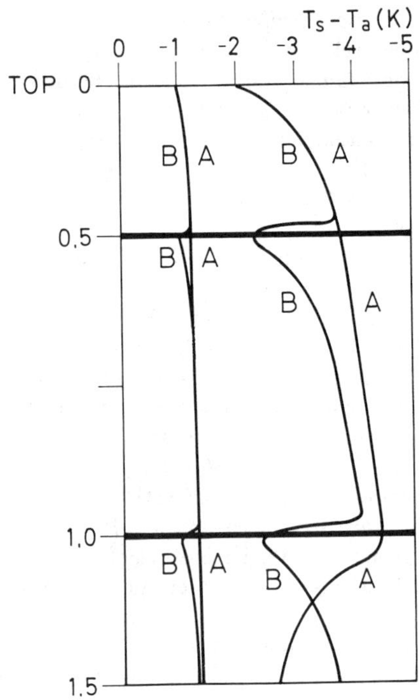

FIG. 8—*Surface temperatures of the specimen along vertical sections:* A *is in the guard area;* B *is in the center, through the metering area. Plots* A *and* B *on the right pertain to reflective surfaces; Plots* A *and* B *on the left pertain to emitting surfaces.*

On the top, the imbalance does not change from point to point (see the difference between Profiles A and B). On the bottom it is also uniform, but different from that of the top.

The imbalance is thus nonuniform through the perimeter of the metering area. Usually a mean value is used to operate the apparatus. Here a very important remark has to be made: the accuracy of the measurements is affected by the degree of imbalance, and in any measurement the amount of imbalance that can be allowed if a stated accuracy is not to be exceeded can be either computed or experimentally determined by detecting the effect of a known imbalance. If the needed level of imbalance is far smaller than the local imbalances, as described in Fig. 8, this level cannot be attained in practice, and hence the accuracy aimed at cannot be reached. This is therefore a testing limit for each testing configuration that can be called the "imbalance resolution." On the left in Fig. 8 another couple of plots are shown through Paths A and B; these were obtained assuming that the specimen surfaces and the apparatus surfaces had a total hemispherical emittance of $\epsilon = 0.9$. The benefit in temperature uniformity is evident, yet the problem of imbalance resolution still remains.

Balancing Criteria

The previous analysis pointed out the difficulties in reaching the balance on the surface of the specimen because of local temperature nonuniformities. A more comprehensive approach is possible, considering just a few relevant temperatures and resistances. This means that mean

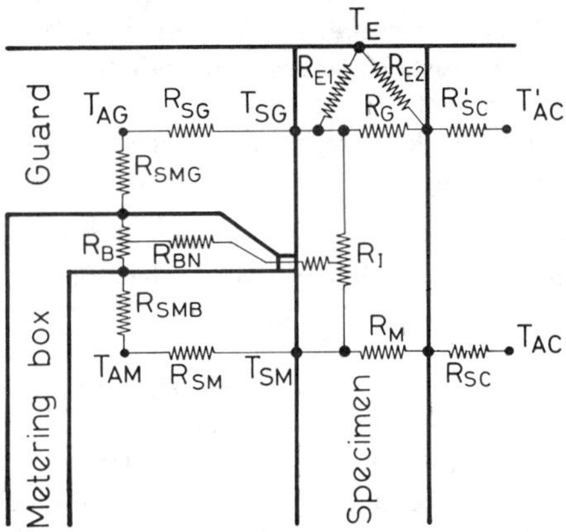

FIG. 9—*Imbalance analysis in a guarded hot box.*

values will be considered. Figure 9 can help in understanding the problem. Terms used in this figure are defined as follows:

(*a*) T_{AG} is the mean air temperature in the guard region;
(*b*) T_{AM} is the mean air temperature in the metering box;
(*c*) T_{SG} is the mean surface temperature of the portion of specimen facing the guard;
(*d*) T_{SM} is the mean surface temperature of the portion of specimen facing the metering box;
(*e*) T_{AC} is the mean temperature of the air on the cold side in a region corresponding to the metering box;
(*f*) $T_{AC'}$ is the mean temperature of the air on the cold side in a region corresponding to the guard area; and
(*g*) T_E is the mean temperature at the periphery of the specimen.

Let us also consider the following thermal resistances:

(*a*) R_{SMG} is the resistance between the surface of the outer side of the metering box and surrounding air;
(*b*) R_{SMB} is the resistance between the surface of the inner side of the metering box and the air enclosed there;
(*c*) R_B is the surface-to-surface resistance of the metering box;
(*d*) R_{BN} is an equivalent resistance globally representing the conduction from the box walls to the specimen through the nose of the metering box;
(*e*) R_{SM} is the resistance between the air inside the metering box and the corresponding surface of the specimen;
(*f*) R_{SG} is the resistance between the air outside the metering box and the corresponding surface of the specimen;
(*g*) R_{SC} is the resistance between the air on the cold side and the surface of the specimen corresponding to the metering box;
(*h*) $R_{SC'}$ is the resistance between the air on the cold side and the surface of the specimen corresponding to the guard region (when flow is fully developed and turbulent all over the specimen surface, and local convection coefficients are uniform);
(*i*) R_M is the resistance (surface to surface) of the portion of specimen facing the metering box;

(*j*) R_G is the resistance (surface to surface) of the portion of specimen facing the guard;

(*k*) R_I is an equivalent global resistance representing the effects of temperature imbalance on the hot surface of the specimen; and

(*l*) R_{E1} and R_{E2} are equivalent global resistances representing the effect of heat losses or gains on the periphery of the specimen.

The purpose of the test is to find the specimen resistance, R_M and possibly the surface resistances, R_{SM} and R_{SC}. To measure them without errors, no heat flow can cross both R_I and R_{SMB}. This is impossible, because it would imply $T_{AG} = T_{AM}$, $T_{SG} = T_{SM}$, and R_{BN} approaching infinity.

Those considerations open two possible approaches. The first is to balance the temperatures T_{SG} and T_{SM} on the surface of the specimen (this involves the problems of imbalance resolution mentioned in the previous paragraph), then to measure the temperature difference $T_{AM} - T_{AG}$ and, having calibrated the box, to measure the heat that flows through R_{SMB}. When specimen-surface temperatures are very uniform, this could be the most accurate testing solution. The alternative approach is to balance T_{AG} and T_{AM}, but this certainly produces an imbalance between T_{SG} and T_{SM}. This technique is the only one applicable when specimens are nonhomogeneous and, consequently, surface temperatures cannot be uniform.

From Fig. 9 it is evident that the choice of one of these two techniques depends on the relative relevance of R_I over R_G, R_M, R_{BN}, R_{SG}, R_{SM}, R_{SC}, and $R_{SC'}$.

As a general rule, when R_I is small balance may be hard or impossible to achieve because the temperature difference, $T_{SG} - T_{SM}$, may fall below the imbalance resolution defined in the previous paragraph.

The last point to be considered is the effect of the environment through R_{E1} and R_{E2}. Usually the minimum error can be obtained when the environment is at the mean specimen temperature, yet the effect can be reduced if the guarding periphery is wide enough. There is a strong belief that most guarded hot boxes are not guarded enough (a comparison with recommended sizes for GHB apparatus can help).

In the case of calibrated hot boxes, the effect of the environment depends, not only on the temperature, T_E, but also on the interaction between the frame and the specimen size and specimen resistance. Research on calibration procedures is still in progress.

Conclusions

The considerations of this paper do not provide any definite solution to the design and operation of calibrated and guarded hot boxes, yet they suggest trying a more conscious design, taking into account the flow patterns and consequent local distribution of surface coefficients and surface temperatures. After new equipment is built, temperatures and velocities must be checked. Finally, a performance check is required that will evaluate the sources of testing errors, in particular, those affecting imbalance and edge heat loss errors. Most of the performance check can be done by making known changes in T_{AG}, T_{AM}, T_E, and other temperatures with different specimens and then checking the effect.

Such efforts will also increase awareness of the difficulty of the problem and a better understanding of the whole testing technique.

References

[*1*] Mumaw, J. R. in *Heat Transmission Measurements in Thermal Insulations, ASTM STP 544*, American Society for Testing and Materials, Philadelphia, 1974, pp. 193–211.

[*2*] Orlandi, R. D., Howanski, J. W., Derderian, G. D., and Shu, L. S. in *Thermal Insulation, Materials, and Systems for Energy Conservation in the '80s, ASTM STP 789*, American Society for Testing and Materials, Philadelphia, 1983, pp. 205–214.

[3] Goss, W. P. and Olpak, A. in *Thermal Insulation, Materials, and Systems for Energy Conservation in the '80s, ASTM STP 789,* American Society for Testing and Materials, Philadelphia, 1983, pp. 215–233.
[4] Lavine, A. G., Rucker, J. L., and Wilkes, K. E. in *Thermal Insulation, Materials, and Systems for Energy Conservation in the '80s, ASTM STP 789,* American Society for Testing and Materials, Philadelphia, 1983, pp. 234–247.
[5] Guy, A. G. and Nixon, J. A. "A Detailed Verification Procedure for a Guarded Hot Box," this volume, pp. 297–309.
[6] Onega, R. J. and Burns, P. J. "Thermal Flanking Losses," NBSIR-83-2804, National Bureau of Standards, Gaithersburg, MD, 1983.
[7] Eckert, E. R. G. *Introduction to the Transfer of Heat and Mass,* McGraw-Hill, New York, 1950.
[8] Eckert, E. R. G. and Drake, R. M., Jr. *Analysis of Heat and Mass Transfer,* McGraw-Hill Kogakusha, Tokyo, 1972.

Steven M. Benner,[1] Debendra Modi,[2] and Donald C. Larson[1]

Effects of Moisture on the Thermal Performance of Spray-Applied Insulation Systems

REFERENCE: Benner, S. M., Modi, D., and Larson, D. C., **"Effects of Moisture on the Thermal Performance of Spray-Applied Insulation Systems,"** *Thermal Insulation: Materials and Systems, ASTM STP 922,* F. J. Powell and S. L. Matthews, Eds., American Society for Testing and Materials, Philadelphia, 1987, pp. 360-370.

ABSTRACT: The absorption of moisture in insulation systems can produce a degradation of the thermal performance of these systems. These effects have been studied with both laboratory and field testing methods in fiberglass and cellulose spray-applied insulation systems. In the laboratory, transient thermal resistance and moisture gain of horizontal, 25-cm-square, 5-cm-thick specimens have been measured. The upper metal substrate surfaces of the specimens were maintained at 10°C, and the bottom free surfaces were exposed to a flow of air at 30°C and 10 to 50% relative humidities. Moisture gain was measured by periodic weighing of the specimens, and the thermal resistance was calculated from thermocouple and heat-flow sensor readings. The fiberglass and cellulose materials gained as much as 5 and 8% moisture by volume, respectively, when exposed to air at 50% relative humidity; much lower moisture gains were observed at lower air humidity levels. The process of vapor flow toward the substrate surfaces contributed significantly to the apparent thermal conductivity. Field measurements of thermal resistances were made on horizontal 1.2-m-square, 5-cm-thick specimens. The corrugated metal substrates were exposed to outdoor ambient conditions while the bottom free insulation surfaces were exposed to a steady flow of indoor air. Field measurements were made under both summer and winter conditions, and no moisture contribution to the thermal conductivity was noted.

KEY WORDS: insulation systems, thermal performance, spray-applied insulations, moisture, thermal insulation

Spray-applied insulation systems have many desirable features. They are well-suited to retrofit applications; they provide complete coverage for building components alleviating thermal bridging; and they can be applied conveniently and economically. Interior spray-applied systems are protected on their outer surfaces by the nearly impermeable building envelope, but their inner surfaces are generally exposed to the inside air. Water vapor in the inside air of a building diffuses into the insulation and condenses in regions of the insulation where the temperature is below the dew point. Such condensation is particularly severe under conditions of high inside humidity and low outside temperature. Free water in insulation systems lowers their thermal resistance and may possibly have an adverse effect on their mechanical properties.

Fiberglass and cellulose are the materials most commonly spray applied as interior insulation systems. In the present study, the effects of moisture on these systems have been studied in both laboratory and field environments. The laboratory tests were designed to determine the thermal performance of spray-applied specimens while the specimens are maintained under controlled, but realistic, conditions; their free surfaces were exposed to air at 30°C with 10 to 50% relative

[1]Assistant professor of chemical engineering and professor of physics and atmospheric science, respectively, Center for Insulation Technology, Drexel University, Philadelphia, PA 19104.
[2]Member of technical staff, Singer Co., Cherry Hill, NJ 08003.

humidity while their substrate surfaces were maintained at 10°C. The field measurements were made under similar but uncontrolled conditions. The free surfaces of the insulation systems were exposed to air at normal indoor temperature and humidity, and the substrate surfaces were exposed to the typical outdoor conditions of the Philadelphia area. During the winter, the surface conditions of the field specimens were therefore similar to the surface conditions of the laboratory specimens. The emphasis of this paper is on the correlation of field and laboratory results. Additional experimental results obtained in the laboratory under more severe conditions of temperature and humidity will be published separately.

Literature Survey

Moisture intrusion and its negative effect on the thermal resistance of materials has long been known as a major problem in the design and operation of insulation systems. Most of the experimental and theoretical groundwork on moisture effects in insulation was done by Luikov [*1,2*], starting in the 1930s, and although the experimental aspects are now somewhat dated, Luikov's theory still forms the basis for a number of current mathematical models. The first major experimental investigations of the effects of moisture on thermal conductivity were those of Jespersen [*3*] and Joy [*4*]. Their results seemed to indicate that a small amount of moisture in insulation would greatly increase its thermal conductivity. For example, for a moisture content of 2% by volume, the thermal conductivity of glass-fiber insulation was found to increase by 100 and 200% above its dry value, according to Jespersen's and Joy's results, respectively.

Subsequent experiments in which thermal conductivity has been measured versus moisture content show a much more modest rise in thermal conductivity [*5–7*]. The high values for the thermal conductivity are probably related to the test methods employed by Jespersen and Joy. If a thermal gradient is placed across moist insulation, the moisture begins to migrate toward the cold side by vapor diffusion. This vapor movement has a profound effect on the heat transfer characteristics, and the measured thermal conductivity is not indicative of the steady-state value. The effect of moisture redistribution on heat transfer in insulations has been noted by a number of researchers [*5–11*]. Recent work in this area was done by Langlais et al [*5*] and Thomas et al [*8*]. In the former study, specimens of mineral fiberboard were sprayed on one side with water, sealed in a polyethylene bag, and placed in a guarded hot plate with the wetted surface on the hot side. The thermal conductivity was then determined as a function of time. The results showed a sharp increase in conductivity in the first 2 h and a "quasi-steady-state" value at about 5 h. The value was not a true steady-state value because the conductivity continued to decrease slowly with time. The magnitude of the quasi-steady-state conductivity increased rapidly as the moisture content was increased from 0 to 3% by volume. When the wetted surface was placed on the cold side, however, almost no change in conductivity occurred with time and moisture content. This seems to indicate that the rapid variations in thermal conductivity values were due to moisture diffusion and phase change.

Using a similar experimental technique, Thomas et al [*8*] measured the moisture and temperature distributions, as well as the thermal conductivity, versus time. Starting with fiberboard uniformly wetted to about 10% by volume, their results showed a redistribution of moisture toward the cold side and a corresponding rapid increase in the conductivity. This increase was followed by a gradual decrease in conductivity over the 200-h test period. Bomberg and Shirtliffe [*9*] found similar moisture migration and heat transfer characteristics in their study on aerated concrete and fiberboard.

Knab et al [*7*] exposed a variety of insulation specimens to both humidity and temperature gradients and determined their moisture gains by weighing them and their thermal conductivity with a heat flow metering apparatus. Their results for glass-fiber insulations showed a gradual increase in conductivity with moisture content (2.5% increase in conductivity for a 1% by volume moisture content).

In their study, Tobiasson and Ricard [*6*] utilized an experimental technique very similar to that of Knab et al [*7*] and obtained comparable results. Therefore, the data obtained by Jespersen and Joy, which show a rapid increase in conductivity with small increases in moisture content, are probably the result of an initial nonequilibrium moisture distribution.

Three studies that investigated moisture in insulation under dynamic boundary conditions were done by Powell and Robinson [*10*], Hedlin [*11*], and Tye and Spinney [*12*]. Hedlin sealed specimens of wetted roofing insulation in plastic and then placed them in a test house where the temperatures and heat flows for each specimen were monitored over a seven and one-half to eight-month period. Even under dynamic conditions, the results for conductivity versus moisture content were comparable to those obtained by investigators using steady-state methods. The main difference was the observation that there were two distinct values for conductivity at each moisture content—one at high temperature difference and one at low. Hedlin's results also showed a redistribution of moisture depending on the season but did not indicate how the season affected the conductivity.

In the Powell and Robinson study, the specimens were placed in a laboratory test chamber where one surface was exposed to normal indoor conditions and the other to simulated diurnal cycles for both winter and summer. The heat flows were measured by heat flux transducers placed on each specimen. The results showed only slight increases in thermal conductivity during the winter conditions but large increases during the summer. This was due to the contribution of vapor migration to the heat transfer as the thermal gradient reversed direction from day to night during the summer. During the winter conditions, the gradient did not change direction, and the moisture accumulated on the cold side.

Tye and Spinney studied the effect of moisture on loose-fill cellulose under both steady-state and dynamic conditions. The cellulose in ceiling and wall constructions was subjected to temperature and humidity gradients over a 57-day period. The specimens showed little reduction in thermal resistance, especially when cyclic conditions were used. These studies indicate the importance of the moisture distribution in insulation, as well as the amount of moisture.

Experimental Procedures

Laboratory Experiments

The insulation specimens used in the laboratory experiments were spray-applied fiberglass (46 kg/m^3) and cellulose (89 kg/m^3). Using manufacturer's specifications, each insulation was sprayed into a specimen holder consisting of an aluminum plate and four Plexiglas sides. The one surface left exposed was leveled by sawing to make a uniform surface. The size of the test specimens was 25 by 25 cm and 5 cm thick. The specimen holders were placed horizontally into the test assemblies (Fig. 1) with constant-temperature water chambers resting on the top of the metal plates and the open surfaces exposed to air chambers. The air entering the air chamber was maintained at a constant humidity and temperature. A Beckman Instruments humidity probe was used to monitor the humidity to within 1% relative humidity, and a proportional controller was used to maintain the temperature. Each water chamber had two heat flux transducers (25 mm in diameter) embedded in its lower plate, and a sheet of ethylene propylene diene monomer rubber (EPDM) was placed between the bottom of the water bath and the specimen chamber. The heat flux transducers were calibrated by replacing the specimen with a standard expanded polystyrene (EPS) specimen and applying temperature differences ranging from 15 to 30°C. To monitor the temperature distribution in the specimens, five thermocouples were placed at various depths inside each test specimen while they were being sprayed. The sensors, thermocouples, and probe were connected to a microcomputer that processed and recorded the data. The specimen holders would be removed and weighed on a triple-beam balance to measure the moisture weight gain.

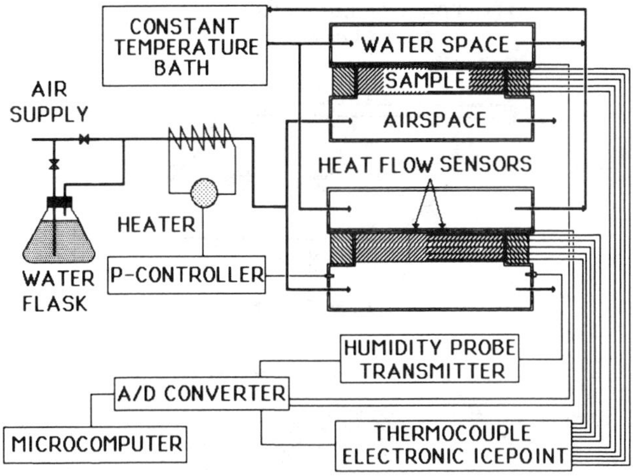

FIG. 1—*Laboratory experimental design.*

Before the test run was started, the specimens were dried by passing air of 10% relative humidity into the air chambers. The difference between the moisture content at 10% and 0% relative humidity is small, so a constant weight was assumed to be the dry weight of the specimen. The water chambers were then set at 10°C, and the conditions in the air chambers were set at 30% relative humidity in the first run and at 50% relative humidity in the second and at a constant temperature of 30°C for both runs. Once every 24 h, the specimens were quickly removed from the test assemblies, weighed to determine their moisture contents, and returned to the test assemblies. The temperature, humidity, and heat flow were monitored every half hour throughout the run. The run was continued until there was no further moisture gain by the specimens. To study the drying rates, the humidity in the air chambers was then lowered to 10% until the specimens were dried to their initial weight.

Field Experiments

The insulation specimens used in the field measurements were spray applied to the undersides of corrugated, galvanized steel decks mounted on EPS frames (frame/deck units). The specimens were 1.22 m square and, nominally, 5 cm thick, with densities of 75 and 90 kg/m^3 for the fiberglass and cellulose materials, respectively. The frame/deck units were tested initially inside the laboratory and were then assembled into outdoor modules and tested outside on an adjacent roof. A schematic diagram of the test configuration is shown in Fig. 2. The metal decks are continuously exposed to the outdoor weather of the Philadelphia area while laboratory air is continuously circulated underneath the insulated roof systems. The relative humidity and temperature of the laboratory air and the outside air were continuously recorded with hygrothermographs.

The thermal resistance values for the systems were measured periodically with the temperature-controlled test plate shown in Fig. 2. Three heat flux transducers are attached to the plate underneath a single sheet of EPDM. The test plate is placed onto the corrugated surface, an EPS lid is placed over the test plate, and the test plate is either heated or cooled to an equilibrium temperature. High and low-temperature testing are typically performed during the summer and winter, respectively, so as not to alter unnecessarily the physical condition of the insulation specimens. The upper corrugated deck temperatures, the lower insulation surface temperatures, and the heat flows from the three heat flux transducers are recorded at half-

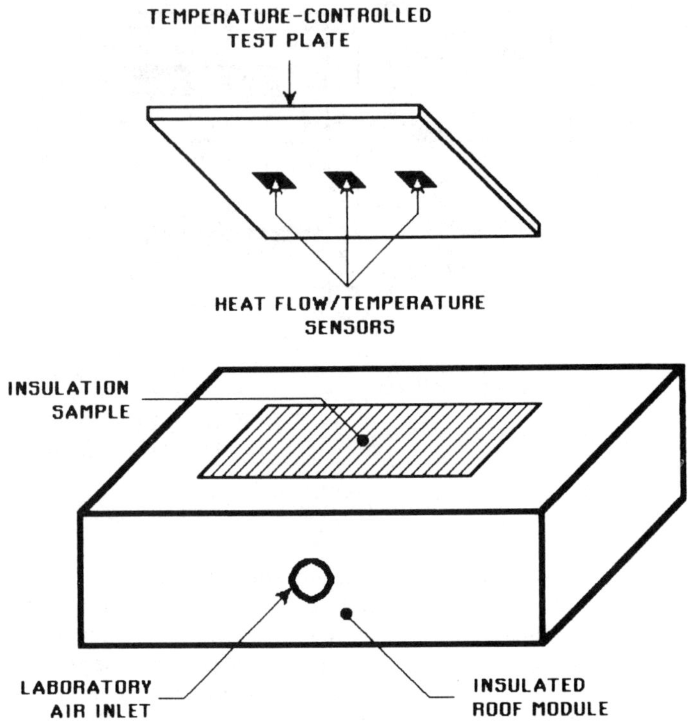

FIG. 2—*Schematic of field configuration.*

hourly intervals for a 3-h period after equilibrium is achieved. The calibration of the sensors and the application of the test plate to nonuniform insulation specimens has been described [*13,14*].

Results

The moisture gains in volume percent for fiberglass and cellulose insulation materials are shown in Figs. 3 and 4, respectively. For air conditions of 30% relative humidity, the moisture gain rate and steady-state values for both materials are almost identical with the final gain for fiberglass and cellulose, reaching 1.57 and 1.70% by volume, respectively. For air conditions of 50% relative humidity, however, the steady-state moisture gain for the fiberglass increases to 5% and that for the cellulose to about 9% by volume. (The cellulose had not quite reached steady state when the test run ended.) When the humidity is reduced to 10%, both insulations dry very rapidly, which indicates good recoverability.

The apparent thermal conductivities as a function of time (and therefore moisture content) are shown in Figs. 5 and 6 for fiberglass and cellulose, respectively. The conductivities for both specimens are essentially constant with time and moisture gain for the 30% relative humidity run but increased by about 50% for the 50% relative humidity run. The conductivity values tend to oscillate slightly, and the lines through the data points represent average values. Because of the accuracy of heat flux transducers, the absolute values for the conductivities are accurate to about 10%. But, since the calibration would be the same for each reading, the relative values can be considered very precise. At the onset of drying, the apparent thermal conductivity is observed to decrease abruptly, as shown in Figs. 5 and 6. These changes are transient effects associated with the drying process.

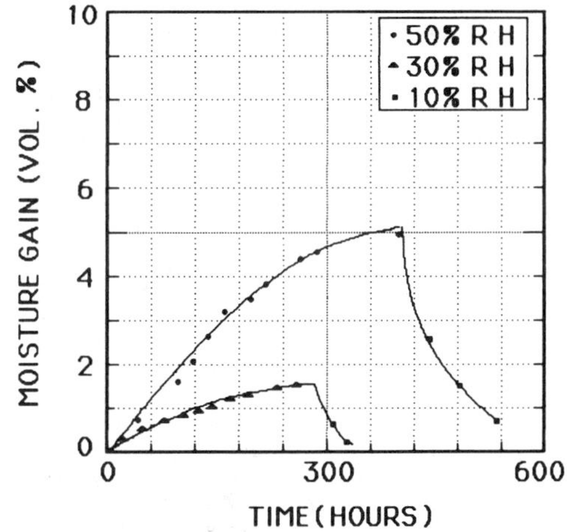

FIG. 3—*Moisture gain of spray-applied fiberglass.*

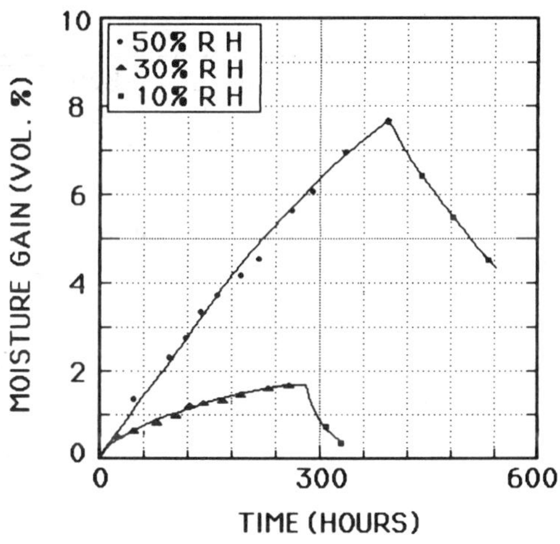

FIG. 4—*Moisture gain of spray-applied cellulose.*

The results of the field experiments are summarized on Table 1. Measurements were made on each insulation system five times over a period of one year. The thicknesses of the insulation in the two systems were measured at both the crests and the troughs of the substrate corrugations; both averages are shown on Table 1. The thermal resistance values were calculated using the expression $R = f\Delta T/Q$, where $f = 1.12$ is a correction factor, ΔT is the average temperature difference across the insulation system, and Q is the average heat flow as calculated from the three heat flux transducers. The factor, f, is calculated from repeated calibrations, as discussed previously [*14*]. Using an average insulation thickness, the average apparent thermal conductivities of the fiberglass and cellulose specimens may be calculated to be 0.0392 and 0.0503 W/m · K, respectively, at average insulation temperatures of 39.1 and 40.4°C, respectively. The average conductivities were 0.0342 and 0.0421 W/m · K at average insulation temperatures of 6.8 and 6.6°C. No systematic variation of thermal resistance with time was noted.

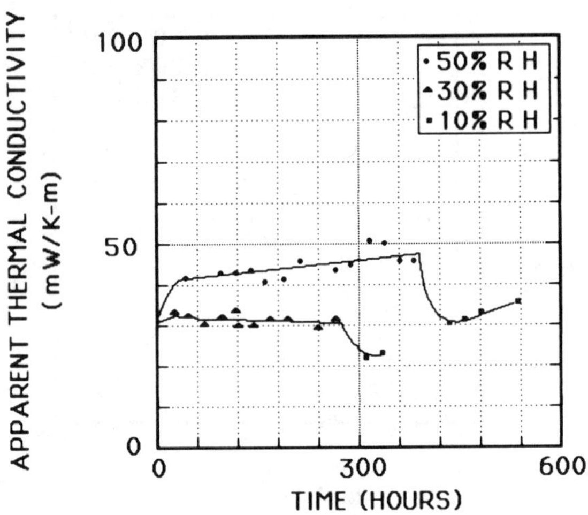

FIG. 5—*Apparent thermal conductivity of spray-applied fiberglass.*

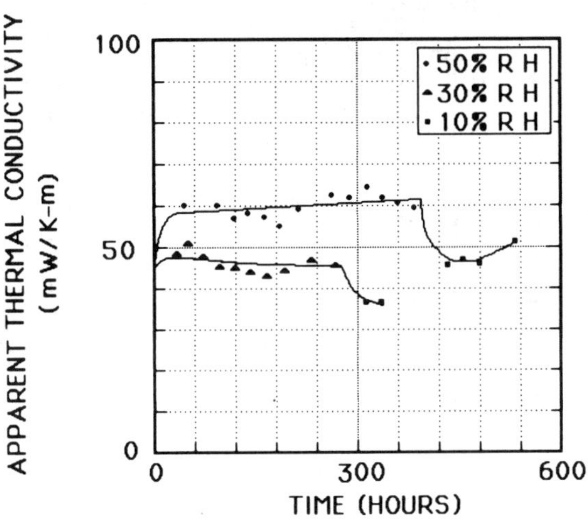

FIG. 6—*Apparent thermal conductivity of spray-applied cellulose.*

The field tests were performed under relatively uncontrolled conditions in which the relative humidity and temperature of the air passing under the insulation systems vary diurnally as well as seasonally. This air comes from the laboratory, which is heated during the winter and cooled during the summer. The relative humidity of the air rises during rainy periods but, in general, shows only gradual day-to-day variation. On a seasonal basis, the monthly average relative humidity of the inside air varied from a low of 22% in December 1983 to a high of 53% in June 1984; the humidity levels were not recorded when the first two test data of Table 1 for both the fiberglass and cellulose systems were acquired. The weekly average relative humidities during the weeks preceding the cold tests of December 1983 and March 1984 were 23 and 34%, respectively, and the weekly average relative humidity during the week preceding the hot tests of June 1984 was 47%.

TABLE 1—*Test data for insulation spray applied to the undersides of corrugated metal decks.*

Descriptions[a]	Date	Thickness,[b] cm	Temperature Difference, °C, ΔT	Heat Flow, W/m²				Thermal Resistance, R[c]		Mean Temperature, °C, T_m
				Q_1	Q_2	Q_3	$\bar{Q}$	$m^2 \cdot K/W$	$ft^2 \cdot h \cdot °F/Btu$	
FG-I	6/83	5.1 to 6.2	20.3	13.5	18.3	16.0	15.9	1.43	8.12	38.6
FG-O	11/83	5.1 to 6.2	27.1	18.4	20.9	23.4	20.9	1.45	8.22	36.8
FG-O	12/83	5.1 to 6.2	17.0	10.4	13.4	13.0	12.3	1.56	8.83	8.0
FG-O	3/84	5.1 to 6.2	14.6	8.3	9.5	9.9	9.2	1.77	10.03	5.5
FG-O	6/84	5.1 to 6.2	26.2	17.9	21.7	20.2	19.9	1.47	8.36	41.9
CE-I	7/83	5.6 to 7.2	18.9	20.3	13.5	18.9	17.6	1.20	6.84	39.4
CE-O	11/83	5.6 to 7.2	25.0	24.0	19.5	24.3	22.6	1.24	7.04	40.2
CE-O	12/83	5.6 to 7.2	16.8	12.8	11.2	13.7	12.6	1.50	8.51	7.0
CE-O	3/84	5.6 to 7.2	22.2	16.5	14.3	18.1	16.3	1.53	8.67	6.2
CE-O	6/84	5.6 to 7.2	26.2	23.2	18.2	23.7	21.7	1.35	7.68	41.8

[a]FG = fiberglass; CE = cellulose; I = inside laboratory; O = outside laboratory.
[b]Average minimum and average maximum insulation thickness under a corrugated deck.
[c]R value = $1.12\ \Delta T/\bar{Q}$.

Discussion

Dew point calculations indicate that for air conditions of 30% relative humidity and 30°C and a plate temperature of 10°C, a condensation plane (a plane along which the relative humidity reaches 100%) will occur in the insulation specimens very near the plate. This is not the case when the relative humidity of the air is raised to 50%. A condensation plane then occurs near the middle of the insulation, with a large portion of the specimen volume on the cold side experiencing condensation. In the case of 30% relative humidity, since little of the specimen is experiencing normal condensation, a large part of the moisture gain is due to capillary condensation and adsorption. In porous and fibrous media, as a result of reduced vapor pressures in the macrospaces and microspaces, condensation can occur before the bulk air stream has reached 100% relative humidity. This is commonly referred to as capillary condensation. The larger moisture content for the cellulose is probably due to its higher density (89 kg/m^3 compared with 46 kg/m^3 for fiberglass) and to the existence of micropores in the cellulosic fibers. Both conditions provide greater surface area per unit volume for the physical adsorption of moisture and promote the occurrence of capillary condensation. The moisture gain results from this study cannot be compared directly with the previously cited studies for two main reasons: (1) the insulation materials studied previously were not spray applied, and (2) the boundary conditions were different.

The only study that may be comparable is the work by Knab et al [*7*], in which the specimens were exposed to air conditions of 38°C and 100% relative humidity on one side and 10°C and 25% relative humidity on the other. For 5-cm glass-fiber insulation, the moisture gain after 20 days was about 10% by volume, which is comparable to the value of 5% by volume obtained in this study under less stringent conditions. No data are available on spray-applied cellulose, but some results have been obtained with loose-fill cellulose by Tye and Spinney [*12*]. In their study, in which the insulation was exposed to a temperature gradient of 38°C and a relative humidity of 40%, the loose-fill cellulose showed a moisture gain of 10% by volume, which is comparable to the gain of 9% for the cellulose specimen in this study.

As mentioned in the Results section, the apparent thermal conductivity increased by as much as 50% from the initial value. This was in conjunction with moisture gains of 5 and 9% by volume for fiberglass and cellulose. These results differ from those of some previous works [*3,4*] that show a greater effect of moisture gain on the conductivity and others that show a lesser effect [*6,7,12*], but the type of materials, boundary conditions, and method of conductivity measurement differ from those of earlier studies. In general, the higher thermal conductivity values were obtained under transient conditions while the lower values were obtained under near-equilibrium conditions. For glass-fiber insulation materials, Knab et al [*7*] and Tobiasson and Ricard [*6*] showed increases in thermal conductivity of 15 and 25%, respectively, for moisture gains of 5% by volume. Tye and Spinney [*12*] found that the thermal conductivity of cellulose increased by only 15% for a moisture gain of 10%.

In most of the earlier studies, the specimen was wetted, then placed in an apparatus that subjected it to a thermal gradient to determine its thermal resistance. During the measurement, the moisture would redistribute through the specimen in order to reach a new equilibrium state. This movement of moisture by vapor diffusion has a profound effect on the apparent heat transfer characteristics of the material (see Literature Survey).

In the 30% relative humidity run of the present study, the rates and steady-state values of moisture gain were small, and thus the contributions of the water vapor migration and condensed water to the apparent conductivities were small. But, in the 50% relative humidity run, the moisture gain rate was almost three times that for the 30% relative humidity conditions. There was, therefore, an appreciable increase in the apparent conductivities of the specimens because of the vapor diffusion contribution. The gradual rise in the conductivities was due to condensation of water in a portion of the insulation volume near the cold plate. As steady state was approached (just prior to drying), both specimens showed decreases in their conductivities as a result of the reduction in the vapor diffusion and condensation inside them. During the drying phase, the opposite process occurs—the accumulated moisture vaporizes in the specimens and diffuses toward the air chamber. As a result of this convective heat transfer, a sharp decrease occurred in the apparent thermal conductivity. As the moisture content returns to the initial value, the conductivity begins to return to its initial value.

The significance of this effect is in how it relates to methods of measurements in which moisture redistribution may be going on during the test. Thus, because of the possibility of moisture migration and condensation inside a specimen, the manner in which the moisture is gained and the conductivity is measured can play a major role in determining the values obtained for the thermal conductivities.

Only the low-temperature field test data acquired in December 1983 and March 1984 can be compared to the laboratory data. During these tests, the substrate sides of the insulation systems were maintained at low temperatures. The mean insulation temperatures were 7°C, compared with mean insulation temperatures of 20°C for the laboratory experiments; the temperature conditions are therefore somewhat more severe than those in the laboratory environment. Humidity levels during the cold tests were not as high as those employed in the laboratory tests at 50% relative humidity. Within experimental error, the thermal resistance values of March 1984 at 34% relative humidity were the same as those of December 1983 at 23% relative humidity. Higher humidity values, which could produce a lowering of the thermal resistance values, were not present during these low-temperature tests. Higher indoor air humidity values were observed during the spring and summer, but the outdoor temperatures are higher during these periods. As a result, the field test data acquired to date show no influence of indoor air relative humidity.

Conclusions

The field test data are consistent with the laboratory test data. Indoor air moisture does not influence the thermal resistance of spray-applied fiberglass or cellulose insulation systems at the relative humidity levels and temperatures encountered in the field experiments. Higher humidity levels and lower substrate temperatures, of course, do reduce the thermal performance of these systems, as shown in the laboratory 50% relative humidity experiments. The values of apparent thermal conductivity obtained in both the 30% relative humidity laboratory experiments and field experiments are also in good agreement. The low-temperature field tests gave an average value of 0.034 and 0.042 W/m · K at 6.8 and 6.6°C for fiberglass and cellulose, respectively, whereas the laboratory experiments gave values of 0.033 and 0.046 W/m · K at 20°C. The laboratory thermal conductivities at 50% relative humidity were somewhat higher as a result of the moisture transport contribution.

Acknowledgments

The authors would like to thank Thomas Edmunds and Jon Larson for their help with the experimental measurements, and CertainTeed Corporation for its continued financial support.

References

[1] Luikov, A. V., *Heat and Mass Transfer in Capillary Porous Bodies*, Pergamon Press, Elmsford, NY, 1966.

[2] Luikov, A. V., "Systems of Differential Equations of Heat and Mass Transfer in Capillary-Porous Bodies," *International Journal of Heat and Mass Transfer*, Vol. 18, 1975, pp. 1–14.

[3] Jespersen, H. B., *Journal of the Institute of Heating and Ventilation Engineering*, Vol. 21, No. 216, Aug. 1953, pp. 157-174.

[4] Joy, F. A., *Symposium on Thermal Conductivity Measurements and Applications of Thermal Insulations, ASTM STP 217*, American Society for Testing and Materials, Philadelphia, 1957, pp. 65-80.

[5] Langlais, C., Hyrien, M., and Klarsfeld, S., "Influence of Moisture on Heat Transfer Through Fibrous-Insulating Materials," in *Thermal Insulation, Materials, and Systems for Energy Conservation in the '80s, ASTM STP 789*, American Society for Testing and Materials, Philadelphia, 1983, pp. 563-581.

[6] Tobiasson, W. and Ricard, J., "Moisture Gain and Its Thermal Consequences for Common Roofing Insulations," *Proceedings*, 5th Conference on Roofing Technology, National Bureau of Standards and National Roofing Contractors Association, Gaithersburg, MD, April 1979, pp. 4-16.

[7] Knab, L. I., Jenkins, D. R., and Mathey, R. G., *The Effect of Moisture on the Thermal Conductance of Roofing Systems*, Building Science Series No. 123, National Bureau of Standards, Gaithersburg, MD, April 1980.

[8] Thomas, W. C., Bal, G. P., and Onega, R. J., "Heat and Moisture Transfer in a Glass Fiber Roof-Insulating Materials," in *Thermal Insulation, Materials, and Systems for Energy Conservation in the '80s, ASTM STP 789*, American Society for Testing and Materials, Philadelphia, 1983, pp. 582-601.

[9] Bomberg, M. and Shirtliffe, C. J., "Influence of Moisture and Moisture Gradients on Heat Transfer Through Porous Building Materials," in *Thermal Transmission of Insulations, ASTM STP 660*, American Society for Testing and Materials, Philadelphia, 1978, pp. 211-233.

[10] Powell, F. J. and Robinson, H. E., *The Effect of Moisture on the Heat Transfer Performance of Insulated Flat-Roof Constructions*, Building Science Series 37, U.S. National Bureau of Standards, Gaithersburg, MD, Oct. 1971.

[11] Hedlin, C. P., "Effect of Moisture on Thermal Resistance of Some Insulations in a Flat Roof under Field-Type Conditions," in *Thermal Insulation, Materials, and Systems for Energy Conservation in the '80s, ASTM STP 789*, American Society for Testing and Materials, Philadelphia, 1983, pp. 602-625.

[12] Tye, R. P. and Spinney, S. C., "A Study of the Effects of Moisture Vapour on the Thermal Transmittance Characteristics of Cellulose Fibre Thermal Insulation," *Journal of Thermal Insulation*, Vol. 2, 1979, pp. 175-196.

[13] Larson, D. C. and Corneliussen, R. D., "Thermal Testing of Roof Systems," in *Thermal Insulation, Materials and Systems Energy Conservation in the '80s, ASTM STP 789*, American Society for Testing and Materials, Philadelphia, 1983, pp. 400-412.

[14] Larson, D. C., "Thermal Resistance Measurements with Heat Flow Sensors—Application to Spray-Applied Insulation Systems," in *Building Applications of Heat Flux Transducers, ASTM STP 885*, American Society for Testing and Materials, Philadelphia, 1985, pp. 129-139.

Heinz R. Trechsel,[1] Paul R. Achenbach,[2] and Spencer Conklin[3]

Field Study on Moisture Problems in Exterior Walls of a Masonry Housing Development on the Coast of the Gulf of Mexico

REFERENCE: Trechsel, H. R., Achenbach, P. R., and Conklin, S., **"Field Study on Moisture Problems in Exterior Walls of a Masonry Housing Development on the Coast of the Gulf of Mexico,"** *Thermal Insulation: Materials and Systems, ASTM STP 922,* F. J. Powell and S. L. Matthews, Eds., American Society for Testing and Materials, Philadelphia, 1987, pp. 371-393.

ABSTRACT: Like many other buildings in the warm and humid climate of the Gulf Coast, family quarters at the Pensacola Naval Air Station, Florida, have experienced serious moisture and mildew problems. Earlier attempts to correct the situation have not been totally successful.

Following an exploratory study in the spring of 1982, measurements of temperatures, relative humidities, air infiltration, and water leakage were taken, and other tests were conducted on several dwelling units over the period from September 1982 to May 1983. During the same period, two surveys of individual houses were conducted to determine the extent of the moisture problem and to document occupant practices that might influence or aggravate the problems.

This report summarizes the individual test and survey results. The results indicate that rainwater penetration through cracks in the masonry walls and at windows is a major source of the moisture observed in the gypsum board and the thermal insulation. Inadequate ventilation, particularly within the bedrooms, and possibly capillary rise of moisture from the foundations may also have contributed to the moisture problem. The report discusses these conclusions and lists possible remedial measures.

KEY WORDS: buildings, family housing, hot humid climates, masonry, moisture, moisture problems, thermal insulation

A brief field visit to the Naval Air Station (NAS) in Pensacola, Florida, in the spring of 1982 [*1*] revealed serious moisture problems in masonry family housing units. Although problems were also found in other developments, the ensuing investigation centered mainly on the Corry Field Development, and this report discusses only the work related to Corry. One problem was moist and waterlogged wallboard; the other problem was mildew on walls and in furnishings, such as on mattresses of beds placed against walls and on rugs. The initial visit did not uncover a clear-cut cause of the moisture problems. However, condensation as a direct result of air conditioning in summer did not appear to be the only or even the most significant cause of the moisture problems, as moist and waterlogged gypsum wallboard was in evidence during the visit to an unoccupied house (Corry Unit 2364 B) in early April, in which the air conditioning had been turned off since the fall of 1981. Accordingly, in this study all likely causes were investigated.

[1]Consultant, H. R. Trechsel Associates, Germantown, MD 20874.
[2]Consultant, McLean, VA 22101.
[3]Architect (retired), Naval Civil Engineering Laboratory, Port Hueneme, CA 93043.

Description of Housing Units

All the Corry Field units are three-bedroom, one-and-one-half bath, semidetached one-story houses with rather large living room and kitchen/dinette combinations. Each unit also has a carport. There is provision for a dryer in the kitchen, with an outside exhaust opening. Both kitchen and baths have exhaust fans. The bathroom fans are controlled by manual switches that can be operated only when the light switch is on. Construction is exterior exposed concrete block walls with 50-mm (2-in.) (nominal) furring, 25-mm (1-in.) cellular board (expanded polystyrene) insulation, and gypsum board finish, as shown in Fig. 1. Except over the entrance doors, there are no gutters, and rainwater drains from the roof onto the sandy ground about 0.6 m (2 ft) in front of the walls. The gable roofs have soffit ventilation and gable vents. Ceiling insulation is 150 mm (6 in.), with a vapor retarder. Each dwelling unit is served by a gas central hot-air heating system, using outside air for combustion, and an electric central air conditioning system. The houses contain about 111 m^2 (1200 ft^2) of floor area and an outside storage room. The houses were built in about 1972. Figure 2 is a photograph of a Corry Field house, and Fig. 3 provides a floor plan. Most of the tests described in this report were conducted on Corry House No. 2363 A.

An effort was made previously to eliminate the moisture problem. The original gypsum board and fiberglass thermal insulation were removed, a coating was applied to the concrete block wall between the furring strips, and new expanded polystyrene board thermal insulation and gypsum board were installed. The purpose of the coating on the concrete block wall was to act as a vapor retarder to reduce the moisture flow from the concrete block into the insulation and the gypsum board. Two types of coatings were used: a clear sealant and a black asphaltic coating. The quality of the clear sealant application could not be determined, but an inspection of several segments of asphaltic-coated wall indicated that the application of the coating left many pinholes open. Since the coatings were applied only between the furring strips, moisture could still enter at the strip locations. In any case, the previous effort did not eliminate the moisture problems.

Experimental Work

Discounting active roof leaks, for which no evidence was found, there are five basic possible causes for the observed moisture problems:

(*a*) rainwater and sprinkler-water leaks through masonry or at windows,
(*b*) rising damp from water leakage at the footing,
(*c*) condensation of moist indoor air in winter or moist outdoor air in summer,
(*d*) inadequate ventilation and heat in winter, and
(*e*) inadequate summer dehumidification.

It is possible that the problems result from more than one of these causes. To identify the causes, or to estimate the contribution of individual causes, a series of tests, observations, and surveys was devised. These were conducted in September, November, January, February/March, and May; thus essentially covering all the seasons.

The tests and observations consisted of measuring air temperatures and relative humidities indoors, outdoors, and within wall cavities and concrete block surface temperatures (exposed and internal); determining moisture content and water vapor permeance of wall materials; measuring air infiltration; observing drainage patterns on the site and water seepage into walls below grade; water-spray testing at windows and masonry wall segments; and conducting occupant surveys to determine occupant practices and the extent and location of moisture problems.

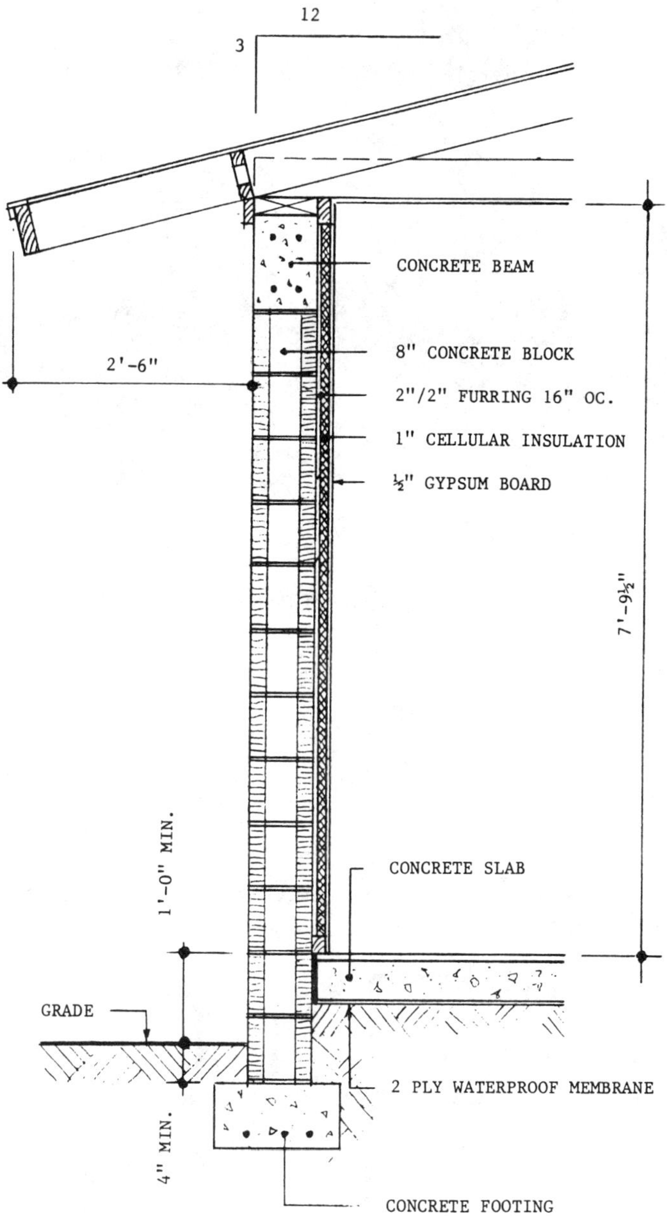

FIG. 1—*Wall section.*

FIG. 2—*Photograph of housing unit.*

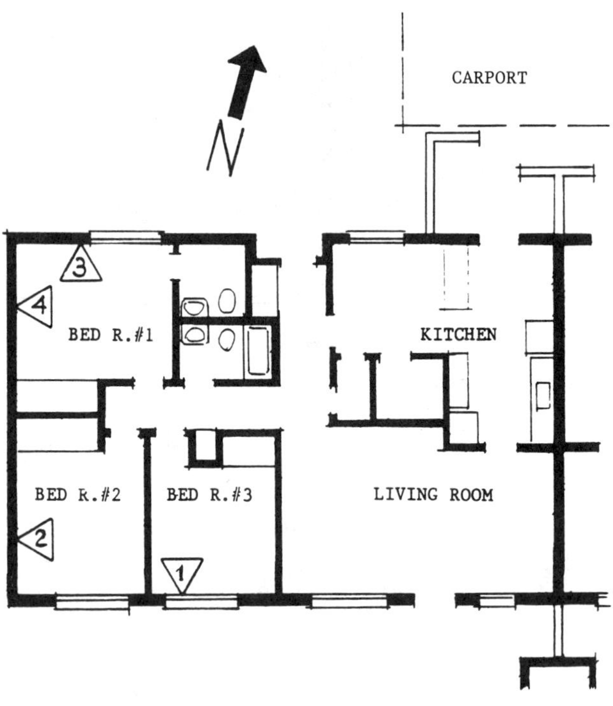

FIG. 3—*Floor plan of Housing Unit 2363 A.*

Temperatures and Relative Humidities

The temperatures and relative humidities were measured inside the house, out of doors, and within several of the wall cavities. During the first series of tests (in September 1982), temperature and relative humidity readings were typically taken at approximately 6-h intervals during the 24-h day; this was later discontinued. It appeared that taking readings twice daily, approximately during the hours of highest and lowest temperature, was sufficient.

The instruments used were the Humicap HMP 15 piercing probe and Humicap HMI 31 digital indicator. The instruments were periodically calibrated with the Humicap HMK 11 probe calibrator. Figure 4 shows the probe and indicator. The probe gave reliable service, except that in saturated air the readings exceeded 100% relative humidity. This is attributed to condensation forming on the sensor itself. Inaccuracies above 99% relative humidity appear irrelevant to the conclusions.

For measuring the temperatures and relative humidities within concrete block cavities, holes were drilled from the building interior into the block cavity. A larger hole was also drilled through the gypsum board and the insulation board. When the probe was not inserted, the hole in the block was sealed with a cork stopper, and the hole in the gypsum board was covered with tape.

The temperature and moisture measurements were taken in walls with and without a rainshield. The rainshield was constructed of 2 by 4 framing with polyethylene sheeting. The rainshield was approximately 0.6 m (2 ft) in front of the wall and had openings of approximately 100 mm at the top and bottom to allow air currents. The rainshields were subject to deterioration and vandalism, and they were not effective during the later part of the test period (February to May 1983).

The probe locations (stations) are shown in Fig. 5. Probes 1, 2, and 3 were the original locations [Probes 2 and 3 having both a high top (A) and a low bottom (B) location within the same vertical block cavity]. Probes 4A and B and 3C and D were added later. Because both temperature and relative humidity were virtually identical for the top and bottom stations, the bottom Stations 2B, 3B, and 4B were discontinued after an initial number of readings. Stations 1, 1A, 2, and 2B were behind the rainshield. The wall outside Probes 3 (A, B, C, and D) and 4 (A and B) was exposed to the weather.

Moisture Content of Wall Materials

In September, November, and February, wall material specimens were collected and analyzed for water content. The tests were conducted by oven drying the specimens at 82°C (180°F) until a constant weight was obtained.

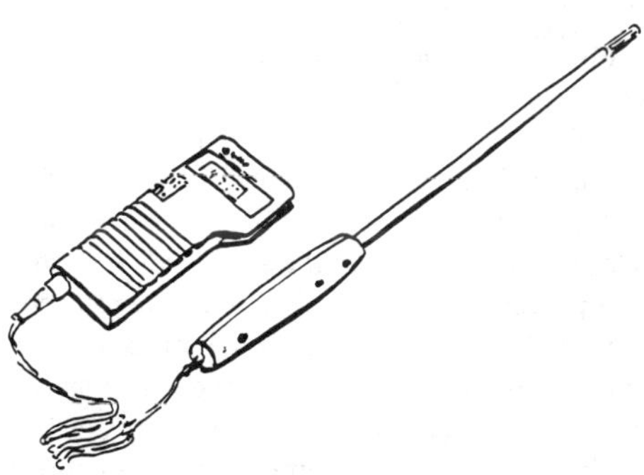

FIG. 4—*Temperature and relative humidity measurement instrument.*

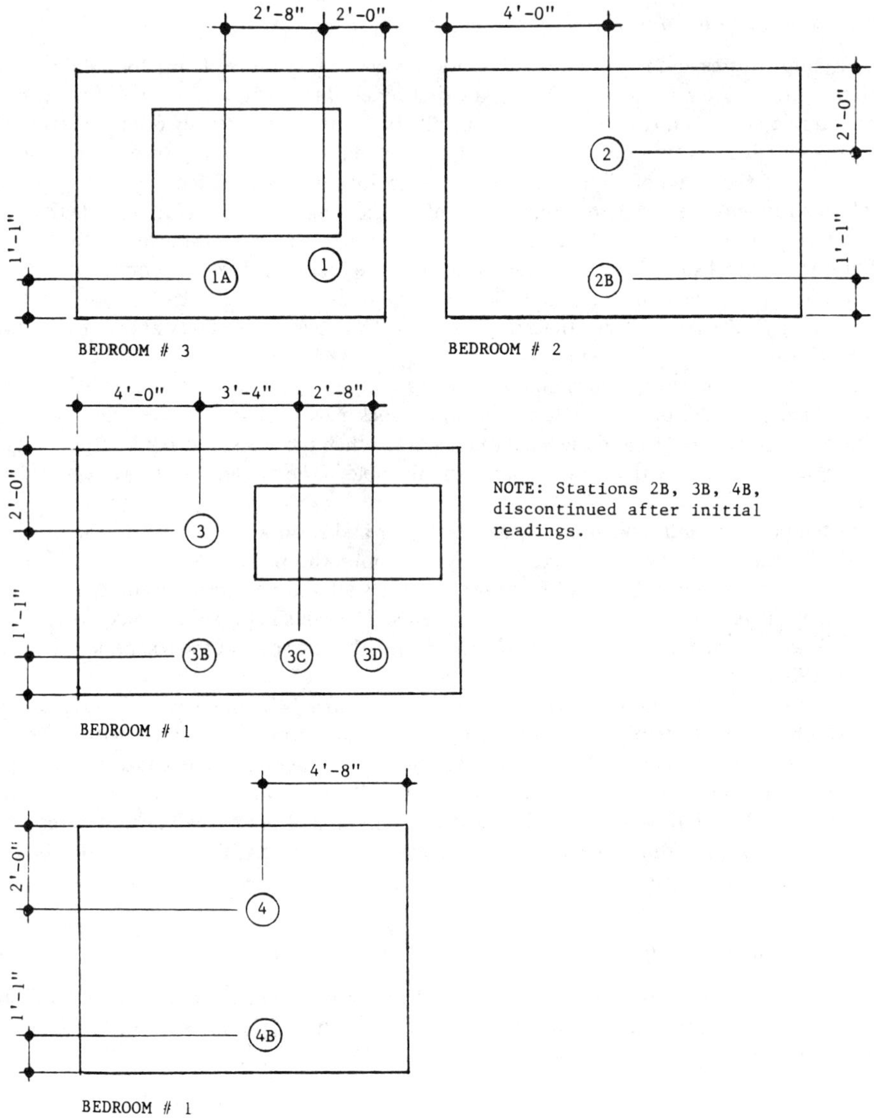

FIG. 5—*Station locations and wall elevations.*

Concrete block inside face shell and gypsum board specimens from Stations 1 and 4 were analyzed. In November, additional specimens were tested. Near Station 1, specimens were collected from the gypsum board and both the outside and inside face shell on the concrete block to allow the establishment of a profile through a wall behind the rainscreen. Gypsum board specimens were collected from near Station 2, also behind the rainscreen. Near Station 3 (exposed to the weather) a series of specimens was taken high in the wall: one specimen of concrete block each from the outside face shell, from the web, and from the inside face shell, and one specimen from the gypsum board, allowing the establishment of a moisture profile through the wall. Specimens were also taken of gypsum board and of the interior face shell of the concrete block directly below the aforementioned specimens near Station 3 to provide information regarding stratification of moisture within the wall. Another specimen of gypsum board was taken near Station 4.

Air Infiltration Tests

All tracer gas tests were conducted in accordance with ASTM Method for Determining Air Leakage Rate by Tracer Dilution Test (E 741-83). The tracer gas used was sulfur hexafluoride (SF_6) [*2,3*].

The tests included both whole-house infiltration tests and tests in individual bedrooms, with the doors both open and closed. In the bedroom tests, the tracer was released into the furnace return air intake when the equipment fan was operating. The fan was then shut off prior to and during the sampling in the bedrooms. In addition to these tests in Corry No. 2363 A, container sample tests were conducted in 25 selected occupied houses. The latter tests were to be used to determine the correlation, if any, between houses that exhibited certain air infiltration characteristics and those with a high incidence of moisture problems.

Ground Drainage Tests and Observations of Standing Water During Rain

The normal water table at Corry had been investigated previously by Public Works personnel and found to be several feet below grade. However, some question remained as to how fast water would drain during and after a heavy rain. Therefore, a simple test was devised. This called for flooding the ground with a garden hose, discontinuing the water supply, and measuring the rate of drainage by means of a dipstick in a perforated pipe sunk into the ground. This test was conducted twice in November: in the first test, water was supplied for $^1/_2$ h to simulate a short rain; in the second test, water was applied for approximately 12 h, simulating a long-lasting rainfall. No effort was made to measure the water supplied in either test.

On 28 Feb. 1983, a heavy rain provided an opportunity to observe the drainage patterns of the ground around some of the houses. The rainfall on that date was recorded at the nearby air station as 60 mm (2.41 in.). [During all of 1982 and the first two months of 1983, this level of daily rainfall was exceeded only twice—on 27 June 1982 with 104 mm (4.14 in.) and on 30 July 1982 with 66 mm (2.65 in.).] The pattern of standing water was drawn on prepared schematic drawings of the houses. The houses were selected based on the results of the occupant survey, so that about half the houses observed had moisture problems.

Water Leakage Observation at Footing

Because of a concern that water might leak into the concrete block cavities below grade, two approximately 3 m (10-ft) trenches were dug in January along the wall and footing outside Stations 1 and 3. One of the trenches was behind a rainscreen and the other was not. The trenches were approximately 300 mm (1 ft) deep and 350 mm (14 in.) wide; they exposed the footings and the block wall below grade (Fig. 6). The wall and footing were then well cleaned of dirt and loose debris.

At two locations (Stations 3C and 3D), water was inserted into the wall cavities through ports drilled into the block from inside the house to observe whether it would drain out at the joint between footing and wall. As it rained, the level of water in the trenches was observed and the rate of drainage noted.

The footing and wall below grade were then cleaned again, loose mortar was chiseled out, the joint between the footing and the wall was patched as needed to provide a smooth and clean surface, a heavy coat of asphaltic waterproofing material was applied to the top of the footing and the wall up to about 50 mm (2 in.) above grade, and the trenches were filled with earth.

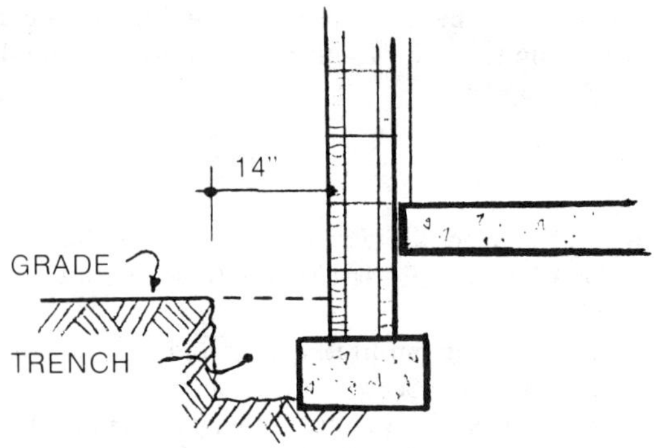

FIG. 6—*Trench at the foot of the wall.*

Relative humidity readings were taken in the wall cavities at the trenches (Stations 1 and 1A, and 3C and 3D) before digging, during the period when the trenches were open, and after parging and waterproofing.

Water Spray Tests

Because moisture problems appeared more frequently in the vicinity of, and particularly below, windows than in windowless walls and because visual inspection indicated deficiencies in glazing gaskets and weatherstripping, an initial water-hose test was conducted in November and repeated in February. That test consisted of randomly directing the full stream of a garden hose against the window and its frame and observing water leakage on the inside. The test was conducted on the windows of the smaller bedrooms.

To investigate the potential of water wetting the concrete block through hairline cracks in the deteriorated paint finish, the absorption of water at the hairline cracks was tested in February by depositing drops of water with an eye dropper at the cracks and observing capillary action, if any.

These pilot tests were then followed in May with tests conducted according to ASTM Test for Water Penetration of Exterior Windows, Curtain Walls, and Doors by Uniform Static Air Pressure Difference (E 331-83) modified. During the test, the time to leakage and the location and type of leakage were recorded. The test was conducted on two windows of Corry House 2363 A and on two windows of Corry House 2399 B. The tests were started at zero indoor/outdoor pressure differential. The pressure was then set at 25 Pa (0.10 in. of H_2O) and raised to 75 Pa (0.3 in. of H_2O). The pressure was maintained across the entire building envelope by means of a "fan door." All gypsum board and insulation were removed below and at the jambs of the window so that any water penetration through the wall could be observed.

The same tests were also conducted on two windowless wall areas in House 2363 A and on one wall area in House 2399 B. All gypsum board and insulation were stripped from wall areas [approximately 1.8 m (6 ft) wide and room height] to observe any leakage or moisture seepage. In addition, viewing ports were cut into the concrete block from the house interior to give visual access to the block cavities.

Occupant Surveys

A total of three occupant surveys were conducted. Two in-depth surveys covered a total of 80 dwelling units. The third survey was short, to confirm specific data, and covered 6 houses.

The purpose of the two larger surveys was to accomplish the following:

(*a*) establish the extent of the moisture and mildew problem in NAS housing units;
(*b*) identify the location and types of problems;
(*c*) identify houses with and without problems so that (1) air infiltration tests could be conducted to establish a correlation between infiltration rates and problems and (2) a visual observation of drainage patterns around buildings could establish additional correlations; and
(*d*) determine occupant behavioral and heating and air conditioning equipment management patterns that might point to correlations with the incidence of moisture problems.

The first survey, conducted in September 1982, covered 26 dwelling units [*4*]. The second survey (February 1983) covered 54 units, using a questionnaire that was revised based on the experience with the first survey [*5*]. The third survey (April 1983) was conducted to obtain data on the incidence of moisture problems in 5 specific houses that had been found to have a serious drainage problem but had not been included in the two earlier surveys. A much simplified questionnaire was used in this survey.

Results and Discussion of Experiments

Temperatures and Relative Humidities in Cavity of Concrete Blocks

The temperatures and relative humidity measured in the cavities of the concrete block used in the wall construction provided information on the warming effect of the sun on the walls of the house and the resulting influence on moisture movement in the wall when these measurements were compared with the indoor and outdoor temperatures and relative humidities. Since the dew point is the best measure of the vapor pressure and the direction of moisture movement by diffusion within a building construction, dew points were determined from dry-bulb temperatures and relative humidity.

Table 1 shows the dry bulb and dew point temperatures at Stations 1, 2, and 3 and indoors and outdoors at several times of day during the period 17–19 Sept. 1982, when summer conditions prevailed. Most of the time the dry-bulb temperatures were higher than either the indoor or outdoor temperature. This is attributed to the warming of the concrete block by the sun during the day and the heat storage effect of concrete block during the night. Exceptions to this relationship between cavity and outdoor temperature occurred in about one third of the observations, mostly before the sun struck the outside wall.

More often than not, the dew point temperatures in the concrete block cavities were significantly higher than either the indoor or outdoor dew points and sometimes higher than the indoor and outdoor dry-bulb temperatures. In some cases the air in the cavity was saturated at a temperature higher than either the indoor or outdoor temperature. This strongly indicates that the concrete block contained moisture that was evaporated by solar heat to produce dew point temperatures in the cavity above the indoor and outdoor temperatures. When these conditions existed, water vapor would be moved by diffusion in both the indoor and outdoor directions from the concrete block cavity, and since the temperature of the insulation and gypsum board inside the concrete block was typically below the cavity dew point temperature, condensation would occur on these materials. Based on the data in Table 1, the conditions for condensation on the insulation and gypsum board existed from 12:30 P.M. until 10:30 P.M. on 17 September at Stations 1 to 3 and at some stations from 9:30 A.M. to 10:30 P.M. on 18 September. Before sunup on 17 September the dew point temperatures of the insulation and gypsum board may have been slightly higher than the cavity dew point temperature, in which case some condensation could have reevaporated and been reabsorbed by the concrete block. These moisture conditions existed in the concrete block even though the rainfall during the first half of September

TABLE 1—*Dry-bulb and dew point temperatures in the concrete block cavities, indoors and outdoors for Corry Unit 2363 A under summer conditions.*

Time	Indoors	Station 1, Bottom	Station 2 Top	Station 2 Bottom	Station 3 Top	Station 3 Bottom	Outdoors
		DRY-BULB TEMPERATURE, °F,[a] 17 SEPT. 1982					
4:30 A.M.	78	79	79	80	80	79	76
12:30 P.M.	81	96	91	89	86	88	91
4:45 P.M.	77	86	95	93	91	88	89
10:30 P.M.	79	83	85	85	84	83	78
		DEW POINT TEMPERATURE, °F, 17 SEPT. 1982					
4:30 A.M.	65	79	74	76	80	76.5	74
12:30 P.M.	62	92.5	81.5	81	86	86	77.5
4:45 P.M.	58.5	86	95	93	89	88	77
10:30 P.M.	56.5	83	81.5	81	84	81	76.5
		DRY-BULB TEMPERATURE, °F, 18 SEPT. 1982					
6:00 A.M.	78	77	78	78	78	78	75
9:30 A.M.	76	81	79	79	79	79	80
6:30 P.M.	79	88	84	84	82	81	80
10:30 P.M.	77	80	81	81	81	80	79
		DEW POINT TEMPERATURE, °F, 18 SEPT. 1982					
6:00 A.M.	59.5	77	72	72	75.5	75.5	71.5
9:30 A.M.	56.5	81	74	73.5	78.5	79	76.5
6:30 P.M.	57	88	81	80.5	82	81	77
10:30 P.M.	63.5	80	76	76	80.8	77	77
		DRY-BULB TEMPERATURE, °F, 19 SEPT. 1982					
9:30 A.M.	78	81	79	79	79	78	79
1:30 P.M.	79	91	78	77	88	80	84
		DEW POINT TEMPERATURE, °F, 19 SEPT. 1982					
9:30 A.M.	70.5	81	73	73	78	76.5	76.5
1:30 P.M.	65	81	75.5	74	86	80	80.5

[a]To convert from Fahrenheit to Celsius divide (degrees Fahrenheit −32) by 1.8.

was approximately half the normal rainfall for Pensacola, and both Stations 1 and 2 had been protected by the rainscreen since 27 August.

Table 2 shows similar dry-bulb temperature, relative humidity, and dew point temperature data at five stations in the concrete block cavities, indoor and outdoors, for the period 17–20 Jan. 1983, when winter conditions prevailed. During the winter period, the outdoor temperatures were observed to be from 5.5 to 14°C (42 to 57°F). The cavity temperatures, except for those at all the stations on 17 January, were between the indoor and outdoor temperatures in both clear and rainy weather. Except for Station 2, on 18 and 20 January, the dew point temperatures in the concrete block cavities were always higher than the indoor and outdoor dew point temperatures, which indicates that the concrete block still contained moisture. However, condensation probably did not occur on the insulation and gypsum board during the period 18–20 January, because their temperatures were probably higher than the cavity dew point temperatures.

Temperatures and relative humidities were also observed in the concrete block cavities and indoors and outdoors during the period 18 Feb. through 2 March 1983. Outdoor temperatures

TABLE 2—*Dry-bulb (DB), relative humidity (RH), and dew point (DP) temperatures in the concrete block cavities, indoors and outdoors, for Corry Unit 2363 A under winter conditions.*

	Date and Time of Day														
	17 Jan. 1983, 11:30 A.M.-Noon			18 Jan. 1983, 9:15-9:30 A.M.			19 Jan. 1983, 1:00-2:00 P.M.			20 Jan. 1983, 10:00-10:30 A.M.			20 Jan 1983, 2:00-3:00 A.M.		
Station	DB	RH	DP	DB	RH	DP	DB	RH	DP	DB	RH	DP	DB	RH	DP
Indoors	70	31	29	68	38	41.5	69	33	39	69	40	44	65	52	47
Outdoors	57	35	22	42	46	24	44	66	34	48	89	46	48	86	44
1	72	100	72	56	88	53	52	92	50	51	90	48	54	92	52
1A	...	...	...	...	...	...	55	93	53	53	91	50	54	91	51.5
2	56	65	45	50	56	35	53	63	41	54	62	41.5	52	69	42
3	54	92	52	46	91	44	50	89	47	51	89	48	51	91	49
4	53	95	51.5	46	92	44	51	88	47	51	89	47	51	88	48
Weather	sunny			sunny			light rain			heavy rain			light rain		

were in the range of 5 to 11°C (41 to 52°F) on 26 February but more typically in the early spring range of 13 to 19°C (55 to 67°F) during the remainder of this period. Table 3 presents the data for 25 February, when the outdoor temperature ranged from 11°C (52°F) at 9 P.M. to 17°C (63°F) at about 2 P.M. and the weather was sunny. During this period, water was evaporating in the house at a rate to simulate an occupancy of four persons, producing an indoor relative humidity in the range of 62 to 76%.

On 25 February, when the outdoor temperature ranged from 11 to 17°C (52 to 63°F), the dew point temperatures in the cavities were always higher than the outdoor dew point temperatures, indicating that moisture would be moving toward the exterior of the house. For most of the day, the cavity dew point was higher than the outdoor dry-bulb temperature, indicating that condensation could occur in the exterior shell of the concrete block. Furthermore, the dew point temperatures at Stations 1 and 1A from 10:30 A.M. to 4:30 P.M. were higher than the indoor dry-bulb temperature, indicating that condensation could occur on or in the insulation and gypsum board. The indoor dew point temperature was equal to, or higher than, the cavity dew point temperatures during the forenoon at Stations 2, 3, 3C, 3D, and 4, but the reverse was typical in the afternoon when the sun shone on the walls at these stations. The indoor dew point temperatures were only marginally above the cavity dry-bulb temperatures for a short time around 11 A.M. on 25 February at Stations 3, 3C, and 3D, indicating that condensation in the concrete block caused by diffusion of indoor moisture was minimal. These observations indicate that the moisture was moving back and forth within the wall at different times of the day, causing condensation under certain conditions and leaving the wall entirely at other times. At Stations 1, 1A, 3, 3C, 3D, and 4 the relative humidities in the concrete block cavities were near saturation throughout the day.

On 16 May 1983, one set of readings was taken to indicate the conditions that might exist in late spring. During that test, the dew point temperatures in the cavities were higher than the outdoor and indoor dew point temperature, indicating that moisture was moving from the concrete block cavity toward both the exterior and the interior.

Table 4 gives an overview of the seasonal swings of dry-bulb, relative humidity, and dew point temperatures indoors, within the concrete block cavity, and outdoors. It should be noted that except for the period of 16 to 19 November [when the outdoor dew point was 0.5 degree Celsius (one degree Fahrenheit) higher], the dew point temperature within the block cavity was higher by 2.2 to 14 degrees Celsius (4 to 25 degrees Fahrenheit) than either the indoor or outdoor dew points. It can also be noted that the outdoor dew point temperature was higher than the indoor dew point temperature in September and November but that a reversal occurred then and the indoor dew point temperature became higher than the outdoor temperature but below the cavity dew point temperature in January, February, and May.

The totality of the temperature and relative humidity observations in the cavities of the concrete block and indoors and outdoors strongly indicates that there was always enough moisture in the concrete block to allow the sun's heat to raise the dew point temperatures above the indoor and outdoor dew points and frequently above the indoor and outdoor dry-bulb temperatures. Thus, the condensation observed in the insulation and the wetting and softening of the gypsum board is attributed principally to the presence of water in the concrete blocks and the effect of the sun's heat on the concrete blocks that seldom, if ever, dry out.

Moisture Content of Wall Materials

The moisture content of wall materials was measured to provide information on the following:

(*a*) actual moisture content,
(*b*) change of moisture content resulting from installation of the rainscreen,

TABLE 3—*Dry-bulb (DB), relative humidity (RH), and dew point (DP) temperatures in the concrete block cavity, indoors and outdoors, for Corry Unit 2363 A on 25 February 1983.*

	7:00–8:00 A.M.			10:30–11:10 A.M.			1:50–2:25 P.M.			4:05–4:30 P.M.			8:15–9:00 P.M.		
Station	DB	RH	DP	DB	RH	DP	DB	RH	DP	DB	RH	DP	DB	RH	DP
Indoors	70	65	57.5	71	76	63	74	62	60	73	67	61.5	69	64	56.5
Outdoors	55	85	51	57	60	43	63	41	39	60	39	35	52	50	34
1	61	93	59	77	102	77	85	104	85	81	99	81	67	103	67
1A	61	94	59.5	77	102	77	87	105	87	81	102	81	67	101	67
2	60	73	51.5	65	71	55	73	76	65	80	83	74.5	69	82	63.5
3	59	94	57	60	97	59	66	91	63	64	95	62.5	60	94	58.5
3C	59	97	58	61	95	59.5	63	96	62	65	95	63.5	59	100	59
3D	58	98	57.5	60	97	59	62	97	61.5	66	93	64	58	101	58
4	59	91	56	63	90	60	70	104	70	79	103	79	59	96	58
Weather		partly cloudy, strong wind			sunny			sunny			sunny			clear	

TABLE 4—*Average dry-bulb (DB), relative humidity (RH), and dew point (DP) temperatures indoors, within concrete block cavities, and outdoors during the six observation periods from September to May.*

	16–20 September			21–24 September			16–19 November			17–21 January			18–23 February			16 May		
	DB	RH	DP	DB	RH	DP	DB	RH	DP	DB	RH	DP	DB	RH	DP	DB	RH	DP
Indoor	79	56	62	67	49	47	72	60	58	69	39	43	70	63	57	76	84	71
Cavity	84	90	82	74	92	72	71	85	66	55	86	51	65	94	64	77	92	75
Outdoor	81	84	76	69	62	56	72	82	67	48	61	37	59	66	48	74	86	70

(*c*) moisture content as a function of time,
(*d*) distribution of moisture through the wall from the outside to the inside, and
(*e*) distribution of moisture vertically in the wall.

The results of the tests are given in Table 5. An examination of the values shows that the concrete block contained from 2.5 to 10.3% water by weight for an average of 6.2% and that the gypsum board contained from 0.5 to 35.2% of water by weight for an average of 14.2%. The large variations, although partially spread over time, indicate a relative local distribution of very high and very low moisture content of wall materials. This indicates that the source of moisture or the mechanism of its transport is also localized and suggests that either leakage of liquid water or of moist air are primary sources of the moisture in the materials.

Air Infiltration Tests

The data collected by S Cubed (La Jolla, CA) [*2,3*] in November 1982 are summarized in Table 6. Additional container test results on occupied houses are shown on Table 7. Reference is made to Table 9 and to the Conclusions section in the discussion of the findings.

In general, the air leakage rates were found to be very low. With the furnace fan off, they ranged from close to zero in individual bedrooms with doors closed to 0.1 air changes per hour (ACH) in bedrooms with doors open. With the fan running, an infiltration rate of 0.2 ACH for the whole house was observed in the unoccupied Corry No. 2363 A. In occupied Corry units, the measurements showed levels of 0.16 to 0.53 ACH. Assuming a family of four, this would correspond to a ventilation rate of 10.8 to 36 m^3/h/person (6.4 to 21.2 ft^3/min/person). [The American Society of Heating, Refrigerating, and Air Conditioning Engineers (ASHRAE) Ventilation Standard 62-73 establishes a minimum of 8.3 m^3/h/person (5 ft^3/min/person).] These rates are for the house as a whole. If only the bedrooms are considered, the 0.1 ACH measured with open doors would result in only 2.1 m^3/h (1.3 ft^3/min) total or, with two people in the room, in only 1.1 m^3/h (0.67 ft^3/min) per person. (With doors closed this rating would be even lower.)

TABLE 5—*Moisture content of wall materials, percent by weight.*[a]

Station	Location	Material	Month of Test			
			August	September	November	January
1	Inside	concrete block	3.1	NA	2.5	NA
	Outside	concrete block		6.4[b]	4.7	NA
	Interior	gypsum board	15.3	23.3[c]	17.5	NA
2	Interior	gypsum board	. . .	NA	13.3	NA
3	Outside high	concrete block	. . .	NA	6.8	NA
	Web high	concrete block	. . .	NA	10.3	8.5
	Inside high	concrete block	. . .	NA	9.1	2.3
	Interior high	gypsum board	. . .	NA	2.0	12.8
	Interior low	gypsum board	. . .	NA	0.5	NA
	Inside low	concrete block	. . .	NA	8.6	NA
4	Interior	gypsum board	13.6	13.8	12.2	. . .

[a]The tests were conducted by Pensacola Testing Laboratory, Inc. except for January tests, which were conducted by H. R. Trechsel Associates. Specimen locations are near but not directly at the measurement stations. Specimens collected during successive months are from nearby but not exactly the same location.
[b]Average of two readings at 3.1 and 9.7.
[c]Average of three readings of 21.4, 35.2 and 13.25%.

TABLE 6—*Summary of measured air change rates in Corry Field No. 2363 A and 2381 B in November.*[a]

House	Occupied/Unoccupied	Remarks	Air Change Rate
2363 A	unoccupied	[b]	0.2
2363 A	unoccupied	bedrooms only[c] doors open	0.1 average
2363 A	unoccupied	bedrooms only[c] doors closed	0
2381 B	occupied	[d]	0.3

[a]Data from S-Cubed. The wind speed for all the tests was below 5 mph.
[b]HVAC fan operating.
[c]Fan shut off.
[d]Container method test.

TABLE 7—*Air change rates measured by container method in selected occupied houses at Corry Field.*[a]

House	Rate
2303 B	0.53
2304 A	0.26
2305 A	0.19
2305 B	0.49
2308 B	0.26
2310 A	0.34
2311 A	0.49
2313 A	0.21
2316 B	0.23
2317 B	0.16
2319 B	0.26
2322 B	0.17
2331 A	0.22
2338 A	0.25
2339 B	0.36
2351 B	0.25
2352 A	0.40
2368 A	0.25
2372 A	0.27
2373 B	0.18
2381 B	0.21
2381 B	0.21
2386 A	0.22
2397 A	0.29
2398 A	0.28

[a]Data by S Cubed analysis.

Ground Drainage Tests and Observations of Standing Water During Rain

The results of the drainage tests conducted in November, given in Table 8, show that the sandy ground drained well after the simulated short rains and drained to a level below the top of the footing in about 1$^1/_2$ h after the simulated 12-h rain. However, in both tests the surface of the ground showed some standing water during the $^1/_2$-h and 12-h period while the water was supplied.

TABLE 8—*Results of ground drainage tests (November).*

Short Test ($^1/_2$-h wetting)		Long Test (12-h wetting)	
Time (after water shutoff), min (except where noted)	Drop in Water Level, in.[a]	Time (after water shutoff), min (except where noted)	Drop in Water Level, in.
0	saturation, standing water	0	saturation, standing water
5	1	5	0
10	$1^1/_2$	10	0
15	2	15	0
25	6		
40	$8^1/_2$ (below top of footing)	30	$1^1/_4$
6 h and 40 min	15	1 h and 45 min	10 (below top of footing)

[a]1 in. = 25.4 mm.

Observations made in February during and after the heavy rain indicated that out of the 20 houses observed, 12 showed substantial standing water for several hours at or near the house walls, 6 showed moderate amounts of standing water, and only 2 showed no standing water at all. The correlation, or lack thereof, between these observations and reported moisture and mildew problems is discussed in the Conclusions section. Figure 7 shows a house classified as having "heavy" standing water.

Water Leakage Observation at Footing

Shortly after the trenches were dug, rainfall on two days [19 January—16 mm (0.62 in.); 20 January—32 mm (1.28 in.)] filled the trenches to above the original grade level. They drained partially only after more than an hour, thus confirming the findings of the drainage tests conducted in November and the observations during the heavy rain on 28 February. The trench at Station 1, behind the rainscreen, filled as fast as the trench at Station 3 without a rainscreen.

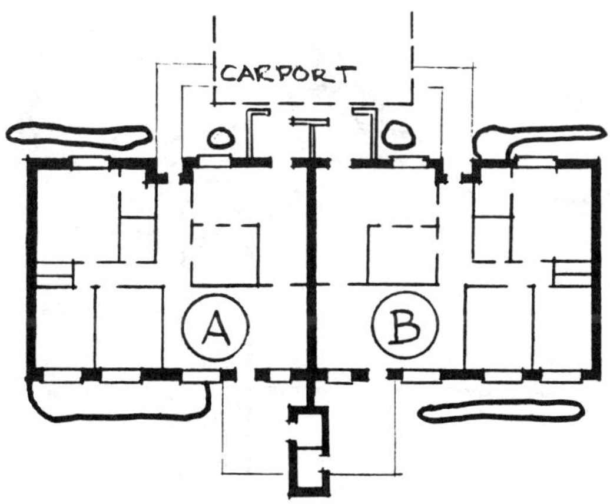

FIG. 7—*Diagram of standing water near Houses 2310 A and 2310 B.*

After the trenches were dug, the footing and wall below grade were exposed to air between rainfalls, but this did not result in a lowering of the relative humidity of the air inside the affected wall cavities over the period from November to February.

Moderate amounts of water introduced into the wall cavities at Stations 3C and 3D drained to the outdoors at the bottom of the concrete block wall. This indicates that water could also enter the wall cavities from outside, whenever rain causes the presence of standing water near the building wall.

After a waterproof coating was applied to the footing and wall below grade, the relative humidity inside the wall cavity did not change over the period from early March to mid-May. Whether or not it might have changed over a longer period is not known.

Water Spray Tests

Water Pilot Tests—In the initial water-hose test in November it appeared that water leaked not only through the window itself but also into the block wall. Similar results were obtained in February. During both the November and February tests, water was found to be leaking into the wall below windows in three modes: (1) water leaked both through the weatherstripping and around the glazing gaskets, overflowed the interior sill of the window channel onto the ceramic tile stool, and leaked by capillary action into the joint between the aluminum window frame and the stool (Fig. 8); (2) it followed the first path through the window, but then leaked into the wall from the sill at the window corner and at the jamb where the tile grout had deteriorated; (3) leaking was observed on the interior at a lower corner of the precast window sill. The origin of this third leak could not be determined.

The careful caulking of the joint between the aluminum window frame and the interior tile stool, and the patching and caulking of the joint between the jamb and the stool, appeared to block the leakage at these points and to stop the leakage into the wall, although window leakage continued and water continued to flow over the stool and down the interior face of the wall below the window.

The aluminum screen frame blocked the water flow from the weep holes in the window channels. This contributed to the overflow of the window channel, and the amount of overflow decreased substantially after the screen frames were notched.

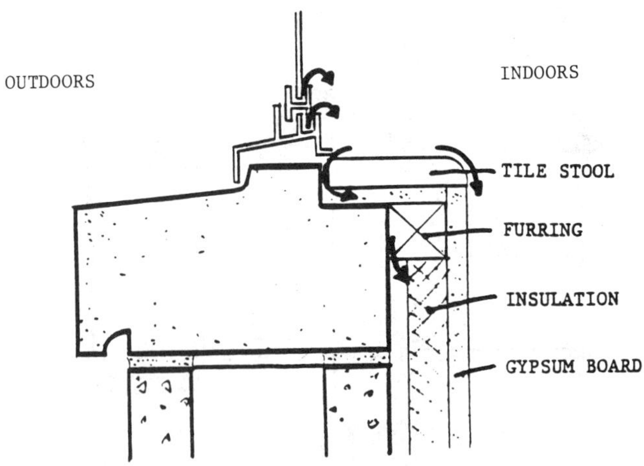

FIG. 8—*Leakage paths of water at the window sill.*

The eye-dropper water tests of capillary action at hairline cracks in the deteriorated paint showed that most cracks "took in" water. Thus, water can enter the concrete blocks even if neither air nor water pressure is present.

Modified ASTM Method E 331-83 Tests—The tests on the windows essentially corroborated the results of the water-hose tests. Substantial window leakage was observed at the same places as in the earlier tests. It was also found that the water, after leaking through the windows, found its way into the wall below, specifically into the space between the insulation board and the concrete block. The gross water leakage is attributed by the testing agency to loose glazing at fixed lights, weather-seal shrinkage, deteriorated sill corner sealant, and lack of proper sill drainage.

During the tests on the solid masonry walls, water leakage into the concrete block cavity was observed through the viewing ports cut into the concrete block. On the southeast corner of Corry No. 2363 A, the concrete block cavity filled up to a level 300 mm (12 in.) above the exterior grade and some 450 mm (18 in.) above the footing, and a small amount of water flowed from underneath the baseboard onto the floor. On the northwest corner of No. 2363 A, moisture became apparent on the inside at the mortar joints, and on the southeast corner of Corry No. 2399 B, water was seen dripping inside the concrete block cavity. In that same location, the insulation board, when removed prior to the test, was wet on its exterior (the side facing the block wall). This condition was identical to that observed on Corry No. 2363 A during the initial visit on 3 April 1982.

The tests indicate that the source of the moisture within the concrete wall cavity could be outside rainwater infiltrating the window and cracks in the exterior concrete block, deteriorating mortar joints, and hairline cracks in the paint.

Occupant Surveys

Of the 200 dwelling units at Corry Field, 86 houses were surveyed. Of this sample, 26 (30%) had current, past, or potential moisture problems and 41 (48%) had mildew problems. A total of 56 houses (66%) had mildew or moisture problems, or both, and only 29 (34%) had neither moisture nor mildew problems.

Of 28 rooms with identified moisture problems, 26 (93%) were bedrooms, and 2 (7%) were kitchen/dinette spaces. No problems were found in living rooms. Moisture problems were also found to exist in bathrooms and mechanical rooms. These are not included in these numbers because they are outside the scope of this study.

Of the 28 moisture problems, 17 (61%) were under or next to windows, and 11 (39%) were either on walls without windows or at some distance from the window.

Of the eleven problems not near a window, two (18%) were leaks below baseboards, four (36%) were along wall/ceiling joint, two (18%) were low in corners, and three (27%) were on blank walls that were not near ceilings, corners, or baseboards.

Several occupants mentioned that the moisture problems appeared or worsened after substantial rainfall. Several also complained about cold bedrooms in winter.

According to the two larger surveys, there were essentially no differences between homes with moisture problems and those without. In both groups, about two thirds of the homes had mildew problems. Thermostat settings were similar, as was the use of air conditioning and natural ventilation. Exhaust-fan use was about the same. No homes with moisture problems had aquariums; eight with no moisture problems had aquariums. The presence of house plants was similar for both groups. The incidence of water from sprinklers touching outside walls was almost equal. Almost all the occupants in both groups noted that in winter condensation formed on the inside of windows. Discoloration on the exterior of outside walls occurred both where moisture problems existed and where they did not. Occupant complaints of cold bedrooms in winter appeared to be independent of whether or not the houses had moisture or mildew problems, or both.

Conclusions

Section 4 discusses individual test results and observations. In this section, the tests will be related to one another.

The most significant findings were the results of the dew point, relative humidity, and temperature measurements; the water-spray tests; the air infiltration tests; and the surveys. Calculations show that, except for some days in January and February, the dew point within the concrete block cavity was higher than that of either the outside or inside ambient air. This indicates that moisture is present within the concrete block wall itself. It also indicates that this moisture, as a general rule, did not move into the wall by water vapor pressure from either outdoors or indoors, because the pressure gradients slope toward both the indoor and outdoor face of the wall, away from the block cavity. This moisture must have entered in liquid form. Four modes of such liquid water infiltration were observed.

The water-spray test indicated that under appropriate conditions, such as high winds and concurrent rain, substantial liquid water can enter the wall. It was also shown that water leakage at windows can seep into the wall between the insulation board and the concrete block wall and that water can penetrate the deteriorated paint film through capillary action at the many hairline cracks. Observations of water leakage into the wall at the footing indicates another mode in which the water can infiltrate the block wall. Although there are a few days during which moisture could also be transferred into the wall from the interior through vapor diffusion, it appears that the massive moisture accumulation within the wall is due primarily to the four modes of liquid water entry outlined previously. Furthermore, vapor diffusion from the interior would not have caused the water to form on the exterior side of the insulation, as was found in April 1982 and May 1983.

During the occupant survey, it was found that almost two thirds of the moisture problems occurred below or adjacent to windows. This suggests that window and sill leakage are major factors causing the problems, and the window tests confirmed this. Since moisture problems were also found on walls without windows, a second mode—either leakage through the wall or at the footing—may also be active in all or most of the problem areas under windows. Furthermore, the rainscreen installed to protect the window and wall did not allow the walls to dry out, although a reduction in moisture content was observed. Based on available data, it is not possible to state with certainty whether exterior wall leakage or capillary rise of moisture from the footing is the predominant cause of the moisture problem observed in walls not near a window. In some locations, such as the one where standing water was observed in the concrete block cavity, rainwater penetration through the wall must be the major cause. In other areas water could also enter the wall cavity below grade. However, observations of standing water at or near building walls do not correlate well with reported moisture problems.

A major finding of the occupant survey was that, from a total of 28 moisture problems, all but 2 occurred in bedrooms (excluding problems related to bathtubs and equipment rooms). The windows in all the rooms are of the same type and make, and the wall construction is uniform throughout the dwelling units. As can be seen from the floor plans on Figs. 3 and 8, the living and kitchen/dinette rooms are moderately sheltered from wind-driven rain. Also, the concrete rear patio in front of the living room and the concrete court at the kitchen entrance would reduce the possibility of water entering the wall from below grade. It is difficult, however, to believe that the modest sheltering of the living/kitchen/dining rooms could have almost entirely prevented moisture problems in these rooms.

The air infiltration tests showed that all the houses in which measurements were made had very low infiltration rates. This was true based on both the tracer gas and pressurization tests.

The tests also showed that the bedrooms have even lower air change rates than the house as a whole during periods when the heating or cooling fan was not operating, particularly with the bedroom doors closed. The occupant survey indicates that the bedrooms also appeared to be difficult to heat. The combination of very limited ventilation and low heat is a potential reason for the moisture and mildew problems appearing almost exclusively in the bedrooms. Both tend to inhibit the natural drying out of the gypsum board moistened by the water vapor driven toward the interior when the concrete block cavity is wet.

Table 9 compares the results of various tests and investigations for a few representative houses and indicates the lack of a clear correlation between standing water, air infiltration rate, and incidence of moisture and mildew problems. Because of this lack of correlation, the predominant causes for the moisture problems in any given location cannot be identified at this time.

In summary, and based on the field tests performed, it appears that the primary cause of moisture problems in the Corry Field houses is accumulation of moisture in the concrete block, which tends to remain year-round in certain locations of the walls. Sunshine on the walls raises the temperature and dew point in the cavities sufficiently to cause moisture condensation on the outer side of the insulation and on the gypsum board and to wet the wooden furring strips.

TABLE 9—*Summary of occupant survey, infiltration rate, and observation of standing water on the ground.*

House	Moisture Problem	Mildew Problem	Infiltration Rate	Standing Water
2303 B	no	no	0.53	NA
2304 A	no	yes	0.26	NA
2305 A	yes	no	0.19	yes
2305 B	yes	yes	0.49	some
2308 B	no	yes	0.26	NA
2310 A	yes	no	0.34	yes
2311 A	yes	no	0.49	yes
2311 B	no	no	NA[a]	some
2313 A	no	no	0.21	NA
2316 B	no	yes	0.23	NA
2317 B	yes	no	0.16	yes
2319 A	yes	no	NA	yes
2319 B	yes	no	0.26	NA
2322 B	no	no	0.17	NA
2331 A	no	yes	0.22	NA
2338 A	yes	yes	0.25	NA
2339 B	no	yes	0.36	NA
2351 B	no	no	0.25	yes
2352 A	yes	yes	0.40	some
2352 B	no	no	NA	yes
2368 A	no	yes	0.25	NA
2372 A	yes	no	0.27	NA
2373 B	yes	no	0.18	NA
2380 B	no	no	NA	yes
2381 A	no	yes	0.38	some
2381 B	yes	no	0.21	some
2386 A	no	yes	0.22	NA
2397 A	no	yes	0.29	NA
2398 A	yes	no	0.28	yes
2398 B	yes	no	NA	yes

[a]NA = not available.

Evidence was found during the field studies and tests that the concrete blocks were being wetted, the drying of the gypsum board was being prevented, or mold and mildew were formed through the actions and interactions of one or more of the following processes:

(*a*) rain or sprinkler-water leakage through the glazing gasket and window weatherstripping onto the sill and down into the furring space and over the wall surface;
(*b*) rain or sprinkler water entering through hairline cracks in the paint, deteriorated mortar joints, and cracks in the concrete blocks;
(*c*) groundwater seepage into concrete wall cavities below grade above the foundation;
(*d*) inadequate wintertime ventilation;
(*e*) inadequate heat distribution to the bedrooms; and
(*f*) lack of summertime moisture control in the bedrooms.

For each of these causes and contributing factors, the following remedial measures are possible:

(*a*) repairing or replacing the windows;
(*b*) sealing all the major cracks and coating the exterior of the house with a class of paint or waterproof coating that has good elasticity and stable curing properties;
(*c*) cleaning and sealing the joint between the footing and the concrete block wall, and parging and sealing the wall below grade and to 50 mm (2 in.) above grade level, and installing gutters, downspouts, and splash blocks, regrading the property for better surface drainage, and installing drain tiles and gravel around the foundation;
(*d*) increasing the wintertime ventilation rate, particularly in bedrooms;
(*e*) increasing the heat input to the bedrooms; and
(*f*) reducing the indoor relative humidity in summer.

Acknowledgments

While the authors accept all responsibility for any shortcomings in the research conducted, the work described in this paper could not have been completed without the dedicated support and assistance of others.

Throughout the field study, the technical support from W. W. Reinschmidt, director of the Engineering Division, and O. H. Taffe, Jr., manager of the Architectural Branch at the NAS Public Works Center in Pensacola, Florida, was invaluable. J. J. Pastucha, N. C. Hansen, and W. S. Kotich of the NAS provided effective liaison with the occupants and assisted with procuring necessary adjunct services. J. Williams of the NAS Weather Station provided detailed weather data required for the analysis of seasonal variations of observations, and J. King and R. Tinsley of the Naval Civil Engineering Laboratory, Port Hueneme, provided very valuable support in connection with measurement technology and instrumentation.

P. Lagus of S-Cubed, of LaJolla, California, collected all the air infiltration data; R. F. Miller of Architectural Testing Inc., York, Pennsylvania, conducted the water leakage tests; and J. Elder, research psychologist of Gaithersburg, Maryland, prepared the survey documents and conducted the surveys.

To all of these, the authors express their most sincere thanks and appreciation.

References

[*1*] "Investigation of Moisture Problems in Navy Public Works Center Housing Units, Pensacola Florida," final report to the Navy Public Works Center, Pensacola, FL, under Contract No. N62467-82 C-2045, by H. R. Trechsel Associates, Germantown, MD, May 1982.

[2] Lagus, P., "Building Air Leakage Tests and Measurements," final report to the Naval Construction Battalion Center, Port Hueneme, CA, under Contract No. N62583/83 M T140, August 1983.

[3] Lagus, P. L. and King, J. C., "Air Leakage and Fan Pressurization Measurements in Selected Naval Housing," *Measured Air Leakage of Buildings, ASTM STP 904,* American Society for Testing and Materials, Philadelphia, 1986, pp. 5–16.
[4] Elder, J., "Survey of Occupants of Lexington Terrace Housing and Corry Housing," Contractor's Report to Naval Air Station Civil Works Center, Gaithersburg, MD, 4 Nov. 1982.
[5] Elder, J., "Survey of Occupants of Corry Housing," Contractor's Report to the Naval Civil Engineering Laboratory, Gaithersburg, MD, 1 May 1983.

Per Ingvar Sandberg[1]

Thermal Resistance of a Wet Mineral Fiber Insulation

REFERENCE: Sandberg, P. I., "**Thermal Resistance of a Wet Mineral Fiber Insulation,**" *Thermal Insulation: Materials and Systems, ASTM STP 922,* F. J. Powell and S. L. Matthews, Eds., American Society for Testing and Materials, Philadelphia, 1987, pp. 394–404.

ABSTRACT: A strategy for assessing the thermal resistance of a wet insulation material has been developed. A computer program has been developed to calculate the effects of moisture flow and phase changes. Laboratory experiments in a guarded hot plate apparatus with moist mineral fiber insulation have been conducted. There was satisfactory agreement between the computer calculations and the experimental results.

The practical thermal resistance of mineral fiber insulation in two applications has been studied using the aforementioned strategy:

1. Insulation with accidental high moisture content caused by leakage, rain penetration, and so on. A typical example is insulation on the exterior of basement walls.
2. Insulation with trapped moisture inside impermeable vapor barriers. This situation is sometimes found in roof insulation.

A conclusion was that phase changes in some cases were important energy transfer mechanisms and therefore must be considered.

KEY WORDS: condensation, fibrous material, guarded hot plate, heat transfer, moisture migration, thermal conductivity, relative humidity, phase change, thermal insulation

For a long time we have been aware that moisture in thermal insulation affects the insulating ability of the material. However, our knowledge in this area is, in many respects, sparse and contradictory. The reason for this is that many different test methods have been used and the results have sometimes been wrongly interpreted. The thermodynamic processes during a test of a wet material are complicated, and the meaning of the concept of "thermal conductivity of a wet material" is not quite clear.

To get a basis for classifying insulating materials in Sweden, a research project is being carried out at the National Testing Institute. The aim of the project is to review existing information and to propose test procedures.

This paper reports some general aspects of thermal resistance of wet materials and also a more detailed analysis of heat and moisture transfer in wet mineral wool.

Theory and Strategies to Assess Thermal Resistance in Practice

For the following discussion, it is convenient to divide heat flow into three components:

1. Heat flow caused by a temperature gradient in a condition of moisture equilibrium, that is, no moisture transfer. This flow is affected by various mechanisms, among them being the moisture content of the material.

[1]Head, Division of Building Physics, National Testing Institute, S-501 15 Borås, Sweden.

2. Convective heat transfer by moisture flow. Water vapor and water each carries its own enthalpy. It should be observed that the enthalpy of water vapor differs from the enthalpy of water with the latent heat of vaporization.
3. Heat transfer resulting from phase changes.

Simulation of all these kinds of moisture effects in a test procedure is unrealistic and unnecessary. The effects of Types 2 and 3 depend entirely on the occurrence and size of moisture transfer in the material, and if these effects are allowed during a test, it is difficult to assess a material property or a building component property. There will also be a great risk of overestimating these types of effects.

For these reasons the author has suggested the following strategy:

(*a*) to determine the effects of Type 1 (thermal conductivity of the wet material) by *testing,* and

(*b*) to determine the effects of Types 2 and 3 (effects of moisture flow and phase changes) by *calculations or estimations based on experience.*

The result of Strategy *a* is "thermal conductivity of wet material." This is a material property that has to be combined with Strategy *b.* This combination yields something we may call "thermal resistance of a wet building component." This resistance value, which will represent an average over, say, a year or a heating season, varies not only with the material and its thickness but also with the boundary conditions.

The purpose of the test is primarily to measure the thermal conductivity of the wet material (Type 1 effect). If possible, heat flow due to moisture flow and phase changes (Types 2 and 3) should be negligible during the test. If this is not possible, it is necessary to estimate Type 2 and 3 effects by supplementary measurements and calculations and to correct the test results accordingly.

Calculation of Type 2 and 3 effects is needed in the case of change in moisture content or moisture content distribution. Only net changes during the period studied are of interest. Consequently, daily oscillations, for example, have no effect on the annual mean value as long as the moisture content is the same at the beginning and end of the period. There is, however, one important exception: If moisture migrates in one direction in the vapor phase and then back in the liquid phase, Type 3 effects may be considerable even if no net change in the moisture content occurs. Compare *d* and *e* in Fig. 1.

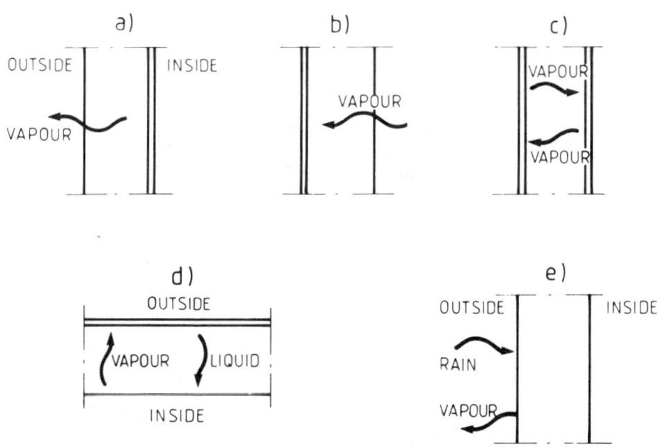

FIG. 1—*Some examples in which moisture effects may have to be taken into consideration.*

Figure 1 illustrates some examples in which Type 2 and 3 effects may have to be taken into consideration:

(*a*) Drying of the initial moisture content. The moisture content decreases. The heat of vaporization is taken mainly from the inside, which causes an increased heat flow at the wall's inner surface, and the thermal resistance of the wall seems to deteriorate.

(*b*) Condensation against a cold outer surface. Heat is released in the condensation zone but is then mainly lost to the outside air. A slight rise in temperature at the outer surface will occur, and the thermal resistance seems to be improved.

(*c*) Periodic moisture flowing between the surfaces in a closed building element. Heat will be liberated and absorbed alternately at the inner and outer surface, and the net effect over several periods is negligible. Daily oscillations, for example, are not of interest when a heating season is studied. During a laboratory test, however, which may last for only a couple of hours and constitute only part of a period, the effects may be significant, and they must be taken into consideration when the test results are evaluated. This is why effects of Type 2 and 3 should be avoided or carefully considered during the test. The effects during the test may be considerable while the effects averaged over a longer period are negligible.

(*d*) Vapor flow in one direction and liquid flow in the other. During the winter, moisture in a roof construction may be transported upwards in vapor phase by diffusion and back downwards in the liquid phase by capillary suction or the action of gravity. The heat of vaporization is taken mainly from the inside while the heat of condensation is lost to the outside. Although the moisture content is the same, the thermal resistance of the roof is reduced.

(*e*) The same principle as in Example *d*. Liquid (rain) hits the wall and is absorbed. When the wall dries out again, the heat of vaporization is taken partly from the inside. The thermal resistance of the wall seems to deteriorate.

Thermal Resistance of Wet Mineral Wool

Until recently much of our knowledge in Scandinavia about moisture influence on the thermal conductivity of mineral wool has been based on measurements by Jespersen [*1*]. He used a stationary method with a hot plate, a cold plate, and heat flow meters on each side of the test specimen. The results for mineral wool are shown in Fig. 2. At about 1% by volume the thermal conductivity rises steeply from ~0.035 to ~0.065 W/m · K. It is difficult to find a physical explanation for this sudden increase in thermal conductivity.

The sudden increase can be explained only by phase changes in the material. In that case, the thermal conductivity reported by Jespersen is not a material property but an "apparent thermal

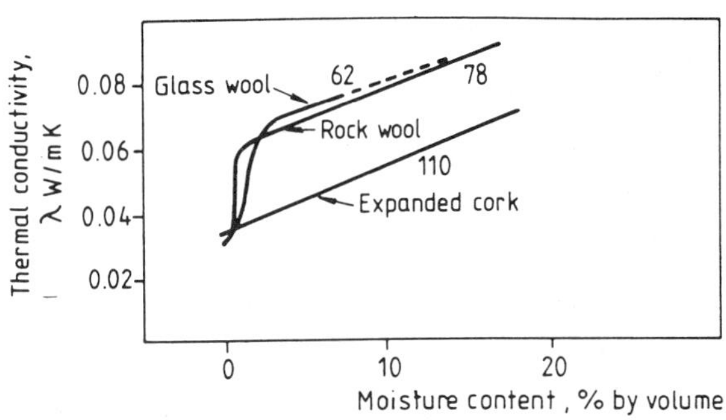

FIG. 2—*Thermal conductivity as a function of moisture content according to Jespersen* [1]: *rock wool, 78 kg/m*^{3}*; glass wool, 62 kg/m*^{3}*; expanded cork, 110 kg/m*^{3}.

conductivity," which is dependent not only on the moisture content but also on the moisture flow and the boundary conditions.

In Jespersen's tests, the moisture flow causes a heat sink (evaporation) at the warm side and a heat source (condensation) at the cold side. Consequently, the heat flow recorded by the heat flow meters at the test specimen's surfaces is much higher than the Type 1 heat flow in the material.

Apparently many of those who have used Jespersen's results have made the mistake of basing heat flow calculations in mineral wool on measurements carried out with a very special moisture situation that is seldom encountered in practice. This situation requires a permanent supply of moisture at the warm side. In reality, the amount of moisture is limited, and in such permeable materials as mineral wool, the moisture under the influence of a temperature gradient is rapidly concentrated to a thin layer on the cold side. A similar opinion has been expressed by Langlais et al [*2*] and Thomas et al [*3*].

The moisture effects of Types 1, 2, and 3 must be separated and the assessment of the thermal resistance handled differently depending on the particular moisture situation. For that reason, a computer program has been developed for studying the thermal resistance under different moisture conditions. The computer model has been verified by comparison with laboratory tests (see Ref *4*).

Guarded Hot Plate Measurements on Wet Material

Measurements were made in a guarded hot plate apparatus with a measurement area of 1.0 by 1.0 m. The test arrangement was done according to that shown in Fig. 3. The density of the mineral fiberboard was ~250 kg/m^3.

Between the hot plate and the top piece of mineral fiberboard, a cloth with a known amount of water was placed. A dry cloth was placed between the cold plate and the bottom piece. During testing, the moisture migrated from the top cloth to the bottom cloth.

The heat flow from the hot plate was continuously measured and registered as mean values over 0.5 h. Thermocouples, three on each level, were placed between the mineral fiberboard pieces and on the board surfaces in contact with the cloths. The averages on each level over 3 h were registered.

The heat flow from the hot plate and the boundary temperatures are shown in Fig. 4.

The computer program was used to simulate the same test, and the agreement was found to be good. The difference between the laboratory test and the computer model was less than 0.5 W/m^2 in heat flow and less than 0.5°C in temperature.

At the end of the test (when a state of moisture equilibrium had been established) the vapor

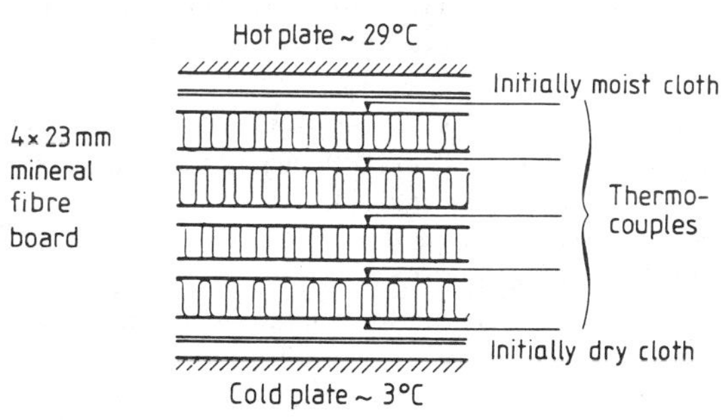

FIG. 3—*Test configuration.*

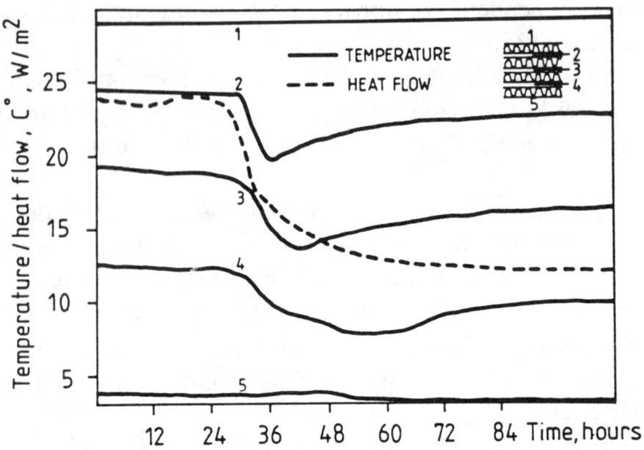

FIG. 4—*Heat flow and temperatures during the test.*

concentration was the same in all the board pieces, and the relative humidity (RH) could be calculated:

Piece No.	RH	λ_{dry}[a]	$\lambda_{end\text{-}of\text{-}test}$
1	~20 to 30	≃ 0.040	0.041
2	~30 to 43	≃ 0.040	0.042
3	~43 to 66	≃ 0.040	0.040
4	~66 to 100	≃ 0.040	0.040

[a] λ = thermal conductivity, in watts per metre degrees kelvin.

There is evidently no measurable tendency toward higher thermal conductivity for those pieces with high relative humidity. Moisture effects on Type 1 heat flow are negligible for mineral wool in the hygroscopic range. This experience has also been confirmed in several similar tests.

Accidental High Moisture Content

A mineral fiber insulation that is fully saturated with water will dry out in two stages. During the first stage, which will last for a couple of hours, gravity action will cause a drainage of the material. The rate of draining and the amount of water retained after drainage depends to a great extent on the orientation of the insulation. In a vertical position, the water drains quickly to a water content of 3 to 5% by volume except in the lower 0.1 to 0.2 m, where the material remains saturated [*5,6*].

During the second stage, which will last from a couple of days to a couple of months, the drying mechanism is vaporization of water and diffusion of water vapor out of the material. If there is an impervious surface layer on the cold side (for example, a vapor retarder) preventing the drying, however, the vapor will condense here. The major part of the moisture will soon be concentrated to a thin layer at the surface.

During the drying (or redistribution) of moisture in the second stage, heat flows and temperatures in the material are affected by phase changes. These must be considered when thermal conductivity design values are assessed.

A typical example of a mineral fiber insulation exposed to accidental high moisture contents is insulation on the outside of basement walls. The concept of mineral fiber as an external insulation and a drainage layer is now well known in many countries [*5, 7–9*]. Let us use this building component as an illustration of how the aforementioned strategy for determining the thermal resistance of a wet material may be applied.

In 1983, approval of a type of basement wall with an external mineral fiber insulation was due for revision and extension. Up to then, the design value for the thermal conductivity of the mineral fiber had been 0.060 W/m · K. The manufacturer was of the opinion that this value was too high and wanted it reconsidered. Together with the manufacturer, the National Testing Institute analyzed the hygrothermal performance of the insulation with the following results:

1. The insulation is almost continuously subjected to a temperature gradient, and the material is very permeable to vapor diffusion. Consequently, there is a very strong potential for drying.
2. The material is very slightly hygroscopic; at 90% relative humidity, the equilibrium moisture content is less than 0.3% by volume.
3. The normal condition for the material is therefore "almost dry," and moisture effects on Type 1 heat flow are negligible.
4. Type 2 effects (convective heat transfer) can be estimated (see Ref *10*) to be less than 0.5% of the total heat flow and may also be neglected.
5. Type 3 effects (phase changes) must be considered. The problem is that there is very little knowledge about accidental wetting by rain, melting snow, or high groundwater level. Lindkvist and Mattsson [*11*] measured moisture contents in eight houses on one or two occasions. The moisture content was below 0.2% by volume in 35 out of 38 test specimens.

As a design basis, the annual extent of accidental wetting was fixed at ten "light" wettings (ordinary rain) to 0.5% by volume and one "major" wetting (melting snow or heavy rain causing total saturation) to 3.8% by volume. The first figure (0.5%) was based on field measurements [*11*] and the second (3.8%) on laboratory tests [*5,6*].

Assuming a temperature difference of 10°C across the insulation and a thickness of 0.07 m, the heat flow through the warm surface of the insulation was calculated for the two cases of accidental wetting (see Fig. 5). In Fig. 5 the heat flow is expressed by an "apparent thermal conductivity," λ_a

$$\lambda_a = \frac{q \cdot d}{\Delta\Theta}$$

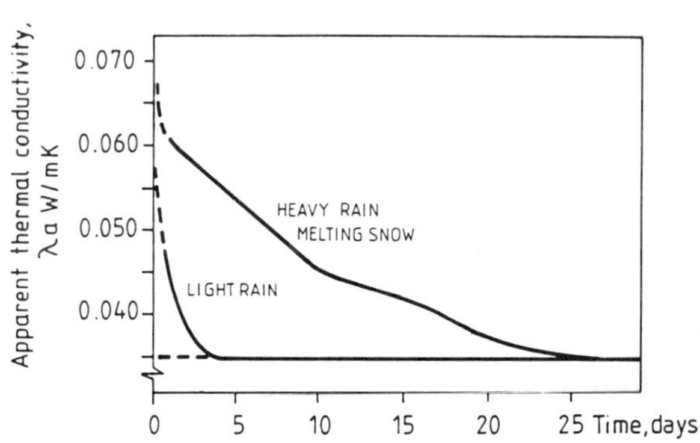

FIG. 5—*Calculated values of the apparent thermal conductivity as a function of time.*

where

λ_a = apparent thermal conductivity, W/m · K,
q = heat flow through the warm surface of the mineral wool, W/m^2,
d = thickness = 0.07 m, and
$\Delta\Theta$ = temperature difference = 10°C.

The thermal conductivity for the dry material is 0.035 W/m · K.

After the "light" rain, the thermal conductivity is affected for 4 days and the average value of λ_a during this period is ~0.046 which is 31% above the "dry" value. After the "heavy" rain, λ is affected for 25 days, and the average is ~0.047, 35% above the dry value. Considering these effects, the annual mean λ_a can be estimated as

$$\bar{\lambda}_a = \frac{40 \times 1.31 + 25 \times 1.35 + 300 \times 1.00}{365 \times 0.035 = 0.037 \text{ W/m} \cdot \text{K}}$$

Consequently, the moisture effects may justify an increase in thermal conductivity ($\Delta\lambda_{\text{moisture}}$) of about 0.002 W/m · K. The heat of vaporization, which was taken from the inside, will be recovered in the ground just outside the insulation when the water vapor condenses. This means that the estimate of $\Delta\lambda_{\text{moisture}} = 0.002$ W/m · K is on the safe side. On the other hand, it may be argued that the mean value should be calculated only for the heating season, which means a slightly higher $\bar{\lambda}_a$ ($\Delta\lambda_{\text{moisture}} \simeq 0.003$).

Approval of the basement wall type was extended (see Fig. 6). After consideration of scatter in production, moisture, compression from earth pressure, aging, and other factors, the insulation was assigned a design value of 0.042 W/m · K.

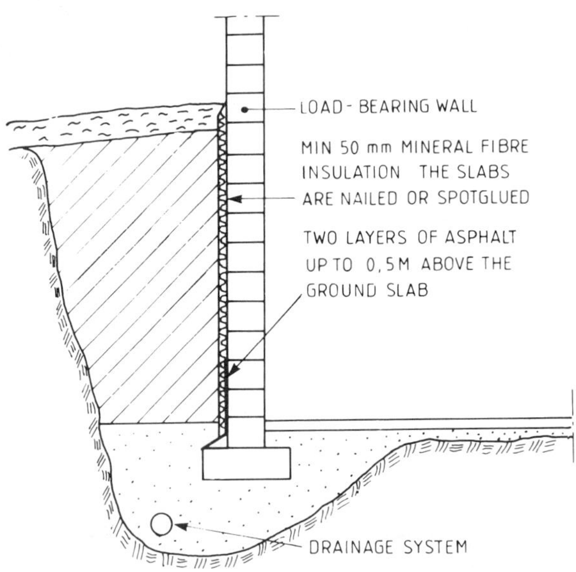

FIG. 6—*Approved type of basement wall.*

Insulation with Trapped Moisture Inside Impermeable Vapor Barriers

In a closed building element, moisture may flow periodically between the surfaces. Heat will be released and absorbed alternately at the inner and outer surface, and the net effect is negligible over several periods. This situation may occur during autumn or spring in roof insulation with a vapor barrier on the inside and a roofing membrane on the outside. Figure 7 shows the calculated heat flow at the inner surface and the apparent thermal conductivity in insulation 0.10 m thick, containing 6% by volume of moisture, with an inside temperature of 20°C and an outside surface temperature cycling sinusoidally between −10 and +30°C. The average temperature difference is then 10°C, and the dry thermal conductivity is assumed to be 0.035 W/m · K, which means that the average heat flow should be

$$\bar{q} = \frac{0.035 \times 10}{0.1 = 3.5 \text{ W/m}^2}$$

This is also, as expected, the average of the heat flow calculated in Fig. 7, although there are variations between 10 and −7 W/m^2.

Two factors influence the heat flow response at the inner surface—effects of phase changes and the thermal diffusivity. To separate these effects, another calculation was carried out with the latent heat of vaporization, $R = 0$. The results are shown in Fig. 8. As can be seen from this figure, the moisture acts as a "negative thermal capacity"—the amplitude increases, and the time lag decreases.

The weather conditions with constant cycling will last for only limited periods. In reality a "cycling equilibrium" is never reached. Therefore, further calculations modeling the annual performance of the insulation have been carried out. The following conditions were assumed: thickness, 0.1 m; moisture content, w, 6% by volume; indoor temperature, 20°C; and thermal conductivity, $\lambda = 0.035 + (0.035/10)w$. This is a moisture effect on Type 1 heat flow, meaning that λ is doubled at 10% by volume moisture content. This correction, which is in accordance with results from Hedlin [*12*], Bomberg and Shirtliffe [*13*], and Knab et al [*14*], seems considerable, but we have to bear in mind that the moisture in mineral wool tends to gather in thin layers and therefore affects only very limited parts of the material. A moisture content of 6% in a 0.1-m-thick insulation may be concentrated in a 6-mm-thick layer.

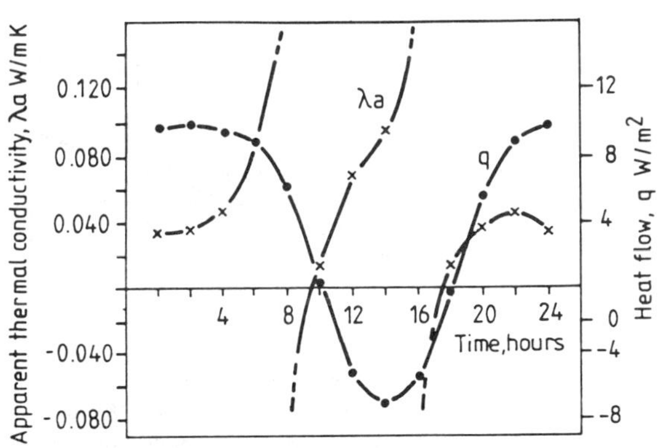

FIG. 7—*Calculated values of the apparent thermal conductivity and heat flow over 24 h.*

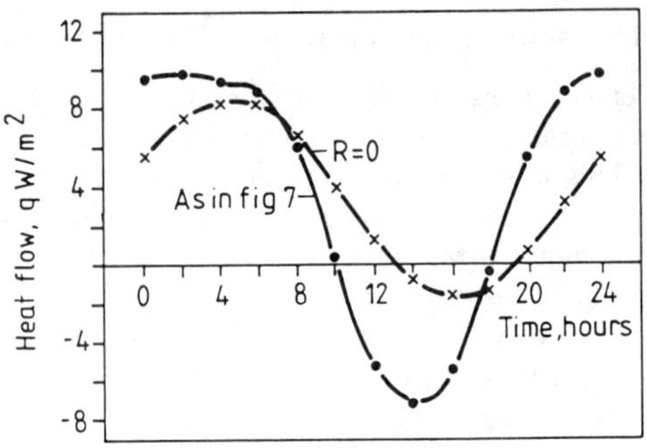

FIG. 8—*Calculated values comparing heat flow at the inner surface with and without phase changes considered.*

The external surface temperature has been expressed as a sine function over the year. To this has been added a sine function over the day simulating the daily oscillation

$$\Theta_e = 12 + 10 \sin(\omega_2 t) + [6 + 4 \sin(\omega_2 t)] \times \sin(\omega_1 t)$$

where

Θ_e = external surface temperature, °C,
t = time, s,
$\omega_1 = 2\pi/24$ h, and
$\omega_2 = 2\pi$/year.

The results of the calculation (Fig. 9) are shown as ten-day averages of heat flow as a function of the average temperature difference. The straight line represents the corresponding values for dry insulation. Field tests reported by Hedlin [*12*] gave similar results (Fig. 10).

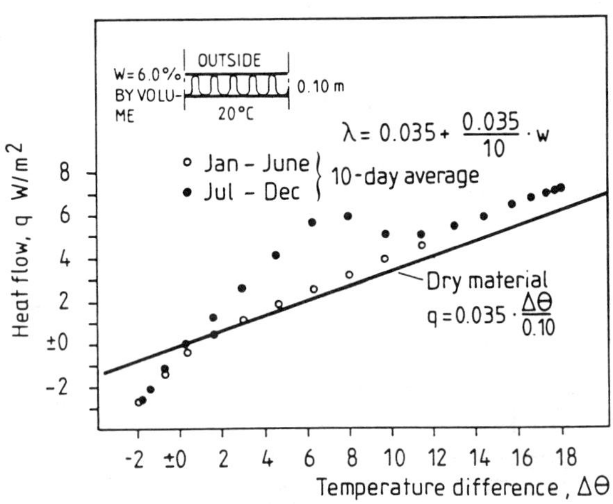

FIG. 9—*Calculated values of the heat flow at the inner surface of roof insulation as a function of surface-to-surface temperature difference.*

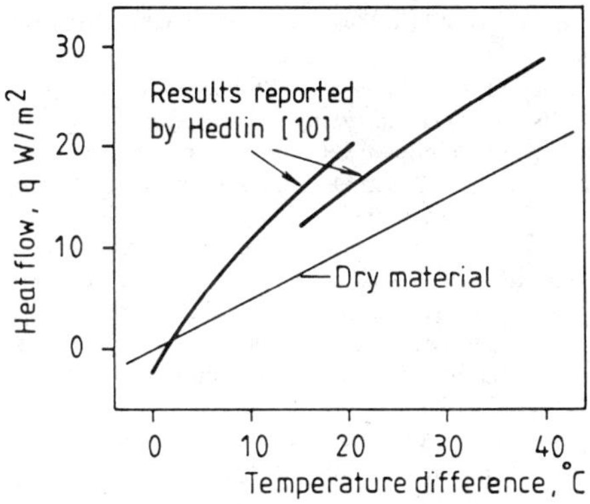

FIG. 10—*Heat flow versus temperature difference for glass fiber at 6% moisture content by volume (results from field measurements by Hedlin* [12]*).*

The annual mean value of the heat flow in the dry insulation would be 2.8 W/m². The calculated mean value for the moist insulation is around 3.3 W/m², which is ~18% higher than the dry value.

During the summer season vapor diffuses from the outer parts of the insulation and condenses at the inner surface, causing an extra inflow of heat. It is seldom possible to utilize this extra heat, and in many cases the heat may be a disadvantage. It would, therefore, be more correct to restrict the discussion to the period with a net heat flow from inside to outside, that is, the heating season. During this period, $\bar{\lambda}_a$ is ~0.045 W/m · K, which is ~28% above the dry value. This calculation is made with an arbitrary set of conditions to illustrate the moisture effects, and it is not a proposal for the size of $\Delta\lambda_{moisture}$.

Conclusions

Two examples have illustrated a suggested strategy for assessing the thermal resistance of wet insulation. These examples have indicated the importance of considering the large-scale and long-term moisture movements in the material. This requires a computer model and knowledge of the moisture boundary conditions and the moisture properties of the material.

Moisture in a mineral fiber insulation is very mobile, and phase changes are therefore often an important energy transfer mechanism.

References

[1] Jespersen, H. B., "Fugtige materialers varmeledningstal" (in Danish), *VARME*, No. 3, Copenhagen, 1952.

[2] Langlais, C., Hyrien, M., and Klarsfeld, S., "Moisture Migration in Fibrous Insulating Materials under the Influence of a Thermal Gradient and Its Effect on Thermal Resistance," in *Moisture Migration in Buildings, ASTM STP 779*, American Society for Testing and Materials, Philadelphia, 1982, pp. 191–206.

[3] Thomas, W. C., Bal, G. P., and Onega, R. J., "Heat and Moisture Transfer in a Glass Fiber Roof-Insulating Material," in *Thermal Insulation, Materials and Systems for Energy Conservation in the '80s, ASTM STP 789*, American Society for Testing and Materials, Philadelphia, 1983, pp. 582–601.

[4] Sandberg, P. I., "Thermal Resistance of a Wet Mineral Fibre Insulation," paper presented at the Conseil International du Batiment W40 meeting, Louvain, Belgium, 1983.

[5] Bomberg, M., "Some Performance Aspects of Glass Fiber Insulation on the Outside of Basement Walls," in *Thermal Insulation Performance, ASTM STP 718*, American Society for Testing and Materials, Philadelphia, 1980, pp. 77–91.

[*6*] "Drainage Ability of Mineral Wool" (in Swedish), unpublished Laboratory Report No. 22, Gullfiber AB, Billesholm, Sweden.

[*7*] Tao, S. S., Bomberg, M., and Hamilton, J. J., "Glass Fiber as Insulation and Drainage Layer on Exterior of Basement Walls," in *Thermal Insulation Performance, ASTM STP 718*, American Society for Testing and Materials, Philadelphia, 1980, pp. 57–76.

[*8*] Elmroth, A. and Höglund, I., "New Basement Wall Designs for Below-Grade Living Space" (in Swedish), Byggförlaget, Stockholm, Sweden, 1971.

[*9*] Deacon, P. C., "Glass Fiber as a Draining Insulation System for the Exterior of Basement Walls," in *Thermal Insulation, Materials and Systems for Energy Conservation in the '80s, ASTM STP 789*, American Society for Testing and Materials, Philadelphia, 1983, pp. 413–434.

[*10*] Sandberg, P. I., "Thermal Conductivity of Wet Materials. Some Personal Reflections," *Studies in Building Physics*, Report TVBH-3007 Division of Building Technology, Lund Institute of Technology, Lund, Sweden, 1981.

[*11*] Lindkvist, H. and Mattsson, B., "Moisture Protection and Thermal Insulation of Basement Walls" (in Swedish), working paper from the Division of Building Technology, Royal Institute of Technology, Stockholm, Sweden, 1972.

[*12*] Hedlin, C. P., "Effect of Moisture on Thermal Resistance of Some Insulations in a Flat Roof Under Field-Type Conditions," in *Thermal Insulation Materials and Systems for Energy Conservation in the '80s, ASTM STP 789*, American Society for Testing and Materials, Philadelphia, 1983, pp. 602–625.

[*13*] Bomberg, M. M. and Shirtliffe, C. J., "Influence of Moisture and Moisture Gradients on Heat Transfer Through Porous Building Materials," in *Thermal Transmission Measurements of Insulation, ASTM STP 660*, American Society for Testing and Materials, Philadelphia, 1978, pp. 211–233.

[*14*] Knab, L. I., Jenkins, D. R., and Mathey, R. G., "The Effect of Moisture on the Thermal Conductance of Roofing Systems," NBS Building Science Series 123, National Bureau of Standards, Washington, DC, 1980.

Gerald E. Sherwood[1]

Condensation Potential in Wood-Frame Walls

REFERENCE: Sherwood, G. E., "**Condensation Potential in Wood-Frame Walls,**" *Thermal Insulation: Materials and Systems, ASTM STP 922,* F. J. Powell and S. L. Matthews, Eds., American Society for Testing and Materials, Philadelphia, 1987, pp. 405–417.

ABSTRACT: As a result of steadily rising energy costs, construction practice for light-frame wood structures has changed over the past few years. The use of 152-mm (6-in.)-thick walls and application of high-*R*-value (thermal resistance value), low-permeance sheathings to 102-mm (4-in.) walls has caused concern for the moisture patterns that may occur in walls. To observe actual moisture patterns and the potential for condensation, test structures were constructed near Madison, Wisconsin, and near Gulfport, Mississippi, for exposure of eight types of insulated wall panels at controlled indoor conditions and typical outdoor weather conditions. Panels were instrumented with moisture sensors and tested without (Phase 1) and with (Phase 2) penetrations (electrical outlets) in the indoor surface.

Continuous inside vapor retarders effectively prevented cold weather condensation in all the panels. Installation of an electrical outlet changed moisture patterns in both the cold winter climate and the hot, humid summer climate. Although condensation occurred for limited time periods in some panels at both test sites, the moisture content of framing did not rise to critical levels.

This paper should be useful to building designers, builders, and building code officials.

KEY WORDS: condensation, moisture control, vapor retarder, air leakage, wood-frame walls, foam sheathing, thermal insulation

High-efficiency thermal insulation systems for wood-frame residential construction have become essentially standard for many parts of the country in recent years. These systems include rigid foam wall sheathing, foil-backed foam wall sheathing, or nominal 152-mm (6-in.) wall studs with 152-mm (6-in.) insulation batts. All of these walls have higher *R* values (measurement of resistance to heat loss), and foam sheathings have lower perm values (measurement of the water vapor movement rate through a material) than previously used wall constructions. Theoretically, all of these systems should result in within-wall moisture patterns different from—and perhaps in excess of—those of conventional walls with nominal 102-mm (4-in.) studs and wood or wood-base sheathing materials.

Excessive moisture in wall cavities can have several detrimental effects. It may decrease the effectiveness of the cavity insulation [*1*]. If the cavity remains wet for extended periods coincident with warm temperatures in the wall, wood structural components may decay. Under winter conditions, outdoor temperature and indoor humidity are the critical variables, since indoor moisture is moving toward the drier outdoors and may condense on the sheathing or siding [*2*]. Under summer conditions, indoor temperature and outdoor humidity are the critical variables, since outdoor moisture is moving toward the drier air-conditioned space and condenses on the gypsum board or vapor retarder when placed on the back of gypsum board.

The potential for these detrimental effects can be assessed based on measurements of moisture levels at various locations in walls exposed on one side to a complete annual cycle of out-

[1]Engineer, U.S. Forest Products Laboratory, Madison, WI 53705.

door weather conditions and on the opposite side to indoor conditions with controlled temperature and humidity.

The objective of the research described in this report was to evaluate the potential detrimental effects of moisture accumulation in wall cavities in both a cold climate [*3*] and a hot, humid climate with a long air-conditioning season. The cold climate location was Madison, Wisconsin, with average monthly temperatures from December to February ranging from −7 to −4°C (19 to 24°F) and frequent lows of about −21°C (−5°F). The hot, humid climate location was Gulfport, Mississippi, where average monthly temperatures from June to August range from 27 to 28°C (80 to 83°F), with highs frequently approaching 38°C (100°F). Average relative humidities at Gulfport during summer months are 85% at 4 A.M. (the coolest time of the day) and 64% at 1 P.M. (the warmest time of the day).

The described work is part of an ongoing program of thermal/moisture research at the Forest Products Laboratory (FPL)[2] to determine the potential for condensation in walls. Because all variables could not be considered in a single study, additional studies are planned in both controlled laboratory tests and field observations of complete houses.

Background

The results of previous research at FPL on moisture condensation in walls have been summarized [*2*]. General recommended practice applies mostly to cold climates, but there is concern for how warm the winter must be to eliminate the need for a vapor retarder on the inside face of the wall. There is also concern that an outside vapor retarder may be needed during hot, humid summers to reduce moisture movement to the interior face of the wall. Closed-cell foam sheathings or foil-backed foam sheathings act as outside vapor retarders and could reduce moisture movement toward the inside in the summer.

A recent study by Burch and Treado [*4*] showed that, for certain conditions, condensation occurred as a thin film on cold surfaces and had minimal effect on the rate of heat transfer because it did not wet the insulation. However, wet insulation has been found in walls after prolonged periods of condensation. In some cases the condensation runs to the bottom of the wall cavity, saturating the sole plate as well as the lower few inches of insulation.

Moisture also reduces the thermal resistance of wood and wood products. A method for estimating that reduction is presented in the *Wood Handbook* [*5*]. More serious effects of moisture on wood are dimensional changes and the potential for decay, although this author is not aware of documented reports of extensive decay in wood-frame walls resulting from condensation. Such decay is a greater threat in warm climates than in cold climates because decay fungi require temperatures above 4°C (40°F) for growth [*5*]. The only problems generally found—and those most visible—are mildew and peeling or blistering paint.

Previous cold climate studies [*6, 7*] have shown the increased potential for condensation with high indoor humidities when outdoor winter temperatures are low. As more airtight houses result in higher indoor humidities, an even greater potential for condensation may be expected. Air-conditioning studies have been conducted in the relatively mild climate of Athens, Georgia [*8*], but no documented studies from hot, humid climates are available. Although laboratory tests have included condensation studies, the actual moisture patterns through the cross section of a variety of walls exposed to outdoor conditions are needed to evaluate the effect of construction changes. This can best be accomplished by testing exposure structures in more than one climate to include the effect of climate on moisture patterns.

Also, two conditions are necessary to evaluate the effects of diffusion versus air movement. This is best accomplished in two phases: one with no penetrations in the vapor retarder, and a second with an electrical outlet penetrating the vapor retarder.

[2]The laboratory is maintained in cooperation with the University of Wisconsin, Madison, WI.

Materials and Methods

Exposure Structures

Two structures were built (one near Madison, Wisconsin, the other near Gulfport, Mississippi) for the purpose of exposing test walls to outdoor weather conditions on one side while exposing the opposite side to typical indoor conditions. The buildings were long and narrow, 2.4 m (8 ft) wide by 14.6 m (48 ft) long, with the long axis running east-west for maximum exposure of north and south walls (Fig. 1). The center 2.4-m (8-ft)-long section was an instrument room. The remaining length of the building was partitioned every 1.2 m (4 ft), resulting in ten 1.2 by 2.4-m (4 by 8-ft) rooms (Fig. 2) [*3,9*] connected by doors in partitions. The only exterior door was in the instrument room. Support for the roof and ceiling was provided by partitions, so that exterior wall panels could be removed and replaced while the building remained intact. Wall panels were completely instrumented during fabrication and then installed by lag bolting them to partitions. Identical panels were installed on north and south walls for extremes of exposure. Both the ceiling and floor were insulated with 305-mm (12-in.) (R-38) glass-fiber batts.

Rooms were individually heated by a resistance-type electric heater and individually cooled by a window-type air conditioner mounted in the floor. Humidification was achieved by a vaporizing-type humidifier in each room during the heating season to maintain a minimum relative humidity (RH) of 40%. The humidity was not controlled during the air-conditioning season. Heaters were controlled by wall thermostats to maintain a temperature between 19 and 21°C (67 and 70°F). Air conditioners were set to cycle on at 26°C (79°F) and off at 24°C (76°F). Ceiling fans operated when either the heater or the air conditioner was running.

End rooms were considered buffers rather than test rooms. This left eight identical rooms in each building for test and comparison purposes.

Test Panels

For this study, all the test panels had 13-mm (1/2-in.) gypsum board on the inside and 11 by 305-mm (7/16 by 12-in.) primed hardboard lap siding on the outside. The siding was painted after the panels were fabricated. Full-thickness glass-fiber insulation was placed in each wall cavity. The pair of panels for one room was framed with 2 by 6 studs at 610-mm (24-in.) spacing; all the other panels were framed with 2 by 4 studs at 406-mm (16-in.) spacing. The primary variables were the sheathing material and the vapor retarder (Fig. 2). In addition, one panel with foil-backed polyisocyanurate foam was vented at the top by a corrugated plastic strip between the top plate and sheathing. Polystyrene sheathing was installed in 0.6 by 1.2-m (2 by 4-ft) sections; all other sheathings were in 1.2 by 2.4-m (4 by 8-ft) sections. Sheathing materials included 13-mm (1/2-in.) fiberboard, 13-mm (1/2-in.) plywood, 25-mm (1-in.) extruded polystyrene foam, and 25-mm (1-in.) foil-backed glass-fiber reinforced polyisocyanurate foam. Only two types of vapor retarders were used: 6-mil polyethylene film, continuous over the face of the framing, or asphalted kraft paper backing on blanket insulation stapled between studs. Although the asphalted kraft paper could be installed by the recommended method of lapping all joints over studs, in practice it is often stapled between studs with no laps. That method was followed to simulate typical field conditions.

Calculated *R* values for each wall panel are shown in Fig. 2. In addition, the sheathing surface was the plane with the greatest potential for condensation in winter. When condensation forms as ice, it is more likely to accumulate to critical amounts than when it forms as water; therefore, the outdoor temperature for which freezing would occur on the warm side of the sheathing is shown.

Each test panel was instrumented with wooden moisture sensors at eleven locations in the wall (Fig. 3). A thermocouple was also placed at each moisture sensor location. At heights of 0.3

FIG. 1—*Experimental condensation study structure near Madison, Wisconsin. The building near Gulfport, Mississippi, is identical.*

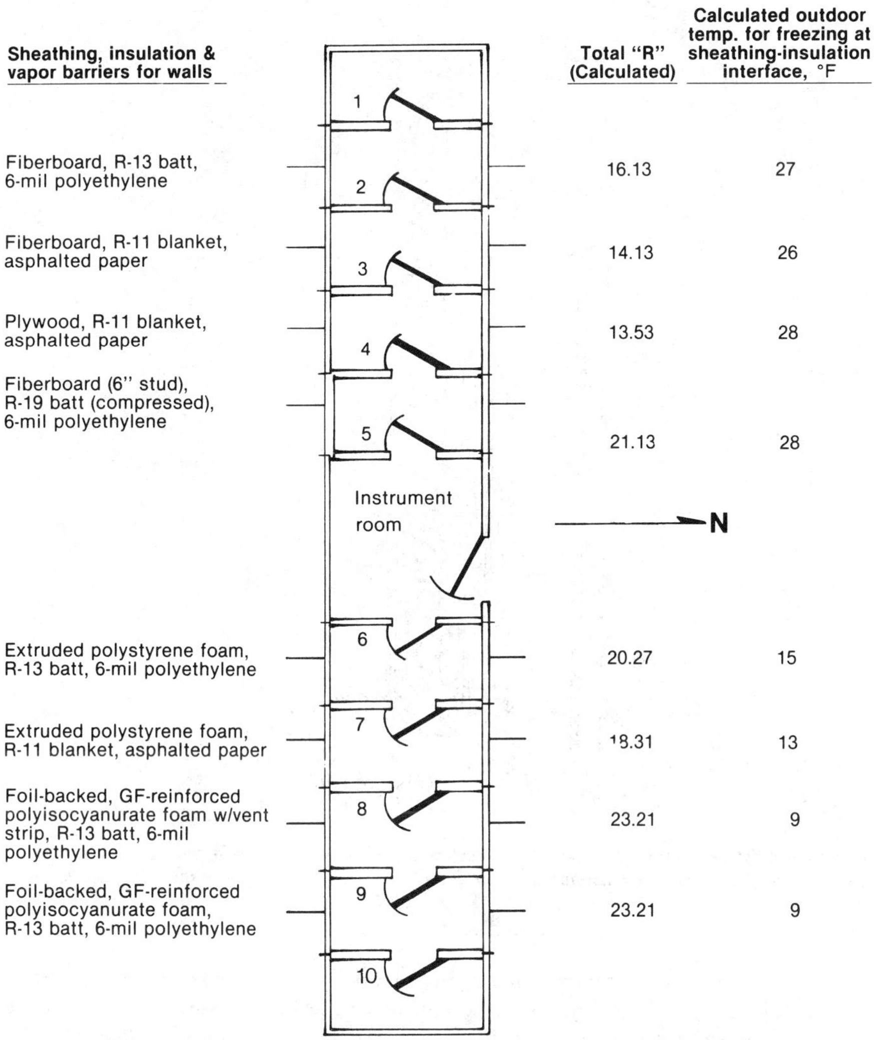

FIG. 2—*Plan of the experimental structure showing the variables of construction of each wall panel [3]. Note that both* R *values and interface temperatures are based on calculation methods shown in the* ASHRAE Handbook of Fundamentals *[9].*

and 2.1 m (1 and 7 ft) above the floor, moisture measurements were made at the siding-sheathing interface, at the sheathing-insulation interface, at the center of the cavity insulation, and in the adjacent stud. Sensors were also located in the center of the top plate, the center of the sole plate, and between the siding and the sheathing at the mid-height of the wall. Since the purpose of the study was to monitor the moisture content (MC) of wood components, no moisture sensor was placed at the vapor retarder interface. Brief periods of condensation could have occurred there and been undetected unless the condensation affected the MC of insulation or ran down to the sole plate. Lead wires from all these data points were brought into the room through the vapor retarder and gypsum board at two points (0.3 and 2.1 m above the floor). The punctures in the vapor retarders were caulked around each wire individually.

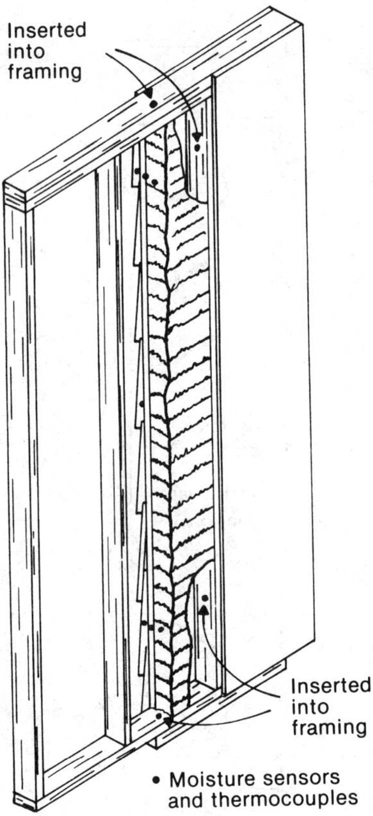

FIG. 3—*Moisture sensor locations in each test panel. Four sensors are in the framing: two are in the center of insulation, two at the insulation/sheathing interface, and three at the sheathing/siding interface.*

All the test panels were without open punctures in the gypsum board or vapor retarder for the first year (Phase 1) of the study. In the second year of testing (Phase 2), a standard duplex electrical outlet was installed in each wall panel to observe the effect of air leakage into the wall cavity. In conventional construction, joints around windows or at baseboards and other discontinuities in the vapor retarder may result in additional leakage. For this study the electrical outlet was selected to provide air leakage for comparison purposes.

After installation of test panels, all the joints with floor, ceiling, and partitions were caulked. On the outside, vertical joints between panels were caulked, and the joint between the floor framing and the bottom edge of the wall panel was caulked. Six-mil polyethylene taped to each face of the partitions extends out between adjoining panels to prevent transfer of moisture between panels.

Data Acquisition

Moisture Content and Temperature

The MC of wooden sensors was measured at various locations in the walls (see Fig. 3) as well as in the rooms and outdoors. The sensors were selected wood elements in which electrical resistance changed with the MC of the wood. Wooden sensors thus selected typically have an error no greater than ±2% MC for readings in the range of 9 to 20% MC. Construction and details of the operation of this sensor are given by Duff [*10*]. Determination of the MC beyond these limits was less accurate because of difficulties in measuring extreme ranges of resistance, and beads of condensed water were often present on surfaces at sensor readings of 20% or higher.

To measure the very high resistance inherent in the sensor effectively and to transmit data to the logger accurately, amplifiers were located as close to each sensor as was practical; their output was connected to the data logger and calibrated.

Temperature measurements were made at each wooden sensor with a type-T (copper-constantan) thermocouple and used for the temperature corrections. The resistance readings were adjusted for temperature and species to provide the MC of wood at 21°C (70°F) [*10*].

Data Recording

All of the moisture and temperature data were digitized and recorded on cassette tape using a multichannel, programmable data logger. Readings were made three times a day—at 1 A.M., 9 A.M., and 5 P.M.—in the Gulfport building. Because of data logger equipment problems in the Madison building, hand readings were made only three times a week through much of the test period.

Results

In both the cold climate and the hot, humid climate major changes occurred in the moisture patterns between Phase 1 (no penetrations) and Phase 2 (with penetrations). The installation of electrical outlets that penetrated the vapor retarder permitted air movement through the wall cavity in both directions, resulting in generally higher moisture levels. In both geographic locations moisture conditions during winter were more severe in the north walls than in the south walls because of the more severe exposure, so these were selected for presentation of data. During the air-conditioning season, moisture conditions were more severe in some south panels, so those were selected for presentation of summer conditions.

There was only one case of moisture levels in framing that would create a decay potential; this was in the hot, humid summer location and existed for only a four-day period. All of the walls in the hot, humid summer location were disassembled and examined for signs of condensation or deterioration. There was no deterioration of materials in any of the wall panels. Signs of moisture such as streaking and water stains generally verified the test data. Examination revealed that Panel 7S had been damaged during installation. The broken sheathing allowed outdoor air to enter the wall cavity near the bottom, so data from this panel was excluded from the results. The results for the two study locations are discussed in more detail under separate headings.

Cold Winter Climate

The moisture levels for critical locations in each test wall panel are shown in Table 1. During both Phase 1 and Phase 2, condensation occurred only during the coldest weather—in January, February, or March—when there was little danger of decay. All the walls were completely dry by early April when outdoor temperatures began to rise above freezing. Condensation occurred at some location in every wall panel over the two-year period, but none of the panels had an accumulation that would create a serious problem. Moisture levels are reported at only two locations in the walls because these were the points of greatest potential for condensation, and they proved to be the only points where major changes occurred. The specific findings were as follows:

1. No condensation was observed during Phase 1 in walls with a continuous polyethylene vapor retarder, regardless of the type of sheathing used.

2. North walls with fiberboard or polystyrene sheathing and only asphalted paper backing on glass-fiber insulation (no punctures) stapled between studs had condensation on the sheathing for a limited time (no more than six weeks).

3. Where condensation occurred in walls with fiberboard sheathing, it initially formed on the back of siding and later on the sheathing. Some moisture also passed through horizontal joints in polystyrene sheathing and condensed on siding.

4. A cold-side vapor retarder, such as the glue joint in plywood sheathing, reduced the hazard of condensation at the sheathing-siding interface without unduly increasing the cavity MC.

5. In Phase 2, condensation formed on the sheathing behind electrical outlets in all north-facing walls with batt insulation of R-13 and R-19. No localized condensation formed behind outlets in walls with R-11 blanket insulation. This was apparently because of more air movement through the less-dense insulation, which distributed moisture to other parts of the wall.

6. Condensation formed on sheathing near the top of walls with electrical outlets only where sheathing had a low insulating value.

7. Based on moisture loss from the rooms, vent strips at the top of walls with high-R, low-permeance sheathing resulted in greater air leakage with no apparent benefit in moisture control.

8. For both years and all constructions, all the data points showed the MC to be below 11% by early April, which was near the end of the heating season.

9. The MC of framing did not increase significantly at any time during the two-year study.

Hot, Humid Summer Climate

The moisture levels for locations that exhibited the most change in each test wall panel are shown in Tables 2 and 3. Only the probe with the highest MC on a panel is shown. The MCs at all the data points were consistently higher than those in the cold winter climate. All framing MCs were about 11% during Phase 1 but rose to about 14% after the vapor retarder was penetrated with electrical outlets. Some framing MCs in the south panel with 152-mm (6-in.) studs, fiberboard sheathing, and polyethylene vapor retarder rose to about 16% during the summer when condensation occurred in that panel. The framing MC at one location in that panel rose above the 20% level for about four days during the summer of Phase 2 but quickly returned to 16%. Examination showed no signs of deterioration. Moisture levels are reported for the two locations in walls that had the greatest potential for condensation. These locations are different for summer and winter conditions.

1. No condensation was detected in any of the walls during the first winter (Phase 1, without penetrations).

2. The only wall with sustained condensation during the first summer (Phase 1, without penetrations) was the wall with 152-mm (6-in.) studs and fiberboard sheathing.

3. The MCs at all points in all walls increased from about 11% to about 14% when the walls were penetrated by an electrical outlet (Phase 2).

4. Although some walls had periods of high MC during the second *winter* (Phase 2, with penetrations), no extended periods of condensation were recorded. The only room that had extended periods of condensation during the second *summer* (Phase 2, with penetrations) was that with the wall with 152-mm (6-in.) studs and fiberboard sheathing.

5. The only wall showing an increase in framing MC was the wall with 152-mm (6-in.) studs, which had an MC of about 16% at the end of the summer in both Phase 1 and Phase 2.

TABLE 1—*Maximum moisture content of wooden probes at key points in north-facing panels at the Madison, Wisconsin, site.*[a]

Panel No.	Insulation-Sheathing Interface						Sheathing-Siding Interface					
	January		February		March		January		February		March	
	Phase 1	Phase 2	Phase 1	Phase 2	Phase 1	Phase 2	Phase 1	Phase 2	Phase 1	Phase 2	Phase 1	Phase 2
2N	low	cond	low	cond	low	cond	low	cond	low	cond	low	cond
3N	low	high	mod	cond	low	high	high	cond	cond	cond	cond	cond
4N	high	high	low	cond	high	high	low	high	mod	cond	mod	high
5N	low	cond	low	cond	low	cond	mod	cond	mod	cond	low	cond
6N	mod	cond	high	cond	low	cond	high	high	mod	high	mod	mod
7N	high	cond	cond	mod	cond	mod	low	cond	mod	high	mod	mod
8N	mod	cond	low	cond	mod	cond	low	low	low	mod	low	low
9N	low	cond	high	high	low	mod	low	low	low	mod	low	low

[a]Low = <12% MC; mod (moderate) = 12 to 16%; high = 16 to 20%; cond (condensation) = >20%. Phase 1 is without penetrations in the vapor retarder; Phase 2 is with penetrations in the vapor retarder.

TABLE 2—*Maximum moisture content of wooden probes at key points in north-facing panels during winter at the Gulfport, Mississippi, site.*[a]

Panel	Insulation-Sheathing Interface						Sheathing-Siding Interface					
	December		January		February		December		January		February	
No.	Phase 1	Phase 2	Phase 1	Phase 2	Phase 1	Phase 2	Phase 1	Phase 2	Phase 1	Phase 2	Phase 1	Phase 2
2N	low	mod	low	mod	low	mod	low	mod	low	mod	low	mod
3N	low	mod	low	mod	low	mod	low	mod	low	mod	low	mod
4N	low	mod	low	mod	low	mod	low	mod	low	mod	low	mod
5N	low	mod	low	mod	low	mod	low	mod	low	mod	low	mod
6N	low	mod[b]	low	mod[b]	low	mod[b]	low	mod	low	mod	low	mod
7N	low	mod	low	mod	low	mod	low	mod	low	mod	low	mod
8N	low	mod	low	mod	low	mod	low	mod	low	mod	low	mod
9N	low	high	low	high	low	high	low	mod	low	mod	low	mod

[a]Low = <12% MC; mod (moderate) = 12 to 16%; high = 16 to 20%; cond (condensation) = >20%. Phase 1 is without penetrations in the vapor retarder; Phase 2 is with penetrations in the vapor retarder.

[b]These conclusions were not replicated on south-facing wall panels. The reasons for these differences are speculative, which indicates the need for further testing.

TABLE 3—*Maximum moisture content of wooden probes at key points in south-facing panels during summer at the Gulfport, Mississippi, site.*[a]

Panel No.[b]	Insulation					Framing				
	June		July		August[c]	June		July		August[c]
	Phase 1	Phase 2	Phase 1	Phase 2	Phase 1	Phase 1	Phase 2	Phase 1	Phase 2	Phase 1
2S	low	mod	low	mod	low	low	mod	low	high	low
3S	low	mod	low	mod	low	low	mod	low	mod	low
4S	low	mod	low	mod	low	low	mod	low	mod	low
5S	cond[d]	mod	cond[d]	cond[d]	cond[d]	low	mod	low	cond[d]	mod
6S	low	mod[d]	low	mod[d]	low	low	mod	low	mod	low
8S	low	mod	low	mod	low	low	mod	low	mod	low
9S	low	high[d]	mod	mod[d]	mod	low	mod	low	mod	low

[a]Low = <12%; mod (moderate) = 12 to 16%; high = 16 to 20%; cond (condensation) = >20%. Phase 1 is without penetrations in the vapor retarder; Phase 2 is with penetrations in the vapor retarder.

[b]Panel 7S was damaged during installation, which was not apparent until the structure was dismantled after exposure.

[c]Lightning damage prevented obtaining data for August Phase 2.

[d]Range of humidity was not replicated on opposite-facing walls. The reasons for differences were not resolved, which indicates the need for further study.

Conclusions

These conclusions apply only to conventional construction and indoor conditions of 19 to 21°C (67 to 70°F), 40% relative humidity during winter, and 24 to 26°C (76 to 79°F) during summer. Higher indoor humidities, which may occur as a result of construction moisture, extremely tight construction, or major indoor moisture sources, will increase the condensation potential. Lower air-conditioning temperatures will increase the summer condensation potential. While specific results are limited to the geographic location of the test building, the conclusions from the cold winter climate study are applicable to much of the upper Midwest and Northeast of the United States, and conclusions from the hot, humid summer climate are applicable to much of the southeastern United States. Condensation potential increases with severity of climate in both cases.

Asphalted paper backing on insulation stapled between studs does not provide adequate vapor retarder protection in cold climates to prevent condensation in the wall cavity or streaking of the siding where a permeable sheathing is used. A continuous 6-mil polyethylene vapor retarder can control winter condensation in insulated walls even where low-permeance sheathing is used. Puncturing the vapor retarder, as with an electrical outlet, completely changes moisture patterns in the wall during both winter and summer and results in condensation on the sheathing behind the electrical outlet in cold weather.

In all of the types of construction observed, both with and without penetrations (outlets), condensation in the wall cavity during winter forms on the back of the siding or on the back surface of the sheathing and does not wet the bulk of the cavity insulation. Low-permeance foam sheathings present no greater cold-weather condensation hazard in winter than do the other types of sheathing studied, and they appear to be beneficial in reducing moisture movement into wall cavities during summer. Vent strips at the top of walls with high-*R*, low-permeance sheathing produce no apparent benefit in moisture control.

While conditions that would promote decay in wood framing do not appear to be a danger in winter, moisture levels can be high enough in most panels to produce significant dimensional changes in thin panel products or long strips of siding. Winter condensation is not a decay hazard at either of the geographic locations in any of the wall constructions tested. Although summer condensation may wet the insulation in high-*R*-value walls with compressed cavity insulation and with low resistance to moisture movement near the outside face, the potential for deterioration of materials is minor. Current moisture movement theory does not explain why condensation occurred in these walls, which points up a need for further study of the mechanisms of moisture movement in walls. There is no high potential for decay in any of the materials of any of the walls tested.

Acknowledgments

This research was conducted in cooperation with the American Hardboard Association, Dow Chemical U.S.A., Jim Walter Research Corporation, and the U.S. Department of Housing and Urban Development. The author acknowledges the valuable contributions of Doug Wheeler [Forest Products Laboratory (FPL) staff engineering technician] in instrumenting the test structures and providing electronic expertise throughout the two-year study. Thanks also go to Terry Niedziela (formerly FPL staff carpenter) for his creative solutions to problems during the construction, instrumentation, and installation of the test panels.

References

[*1*] Joy, F. A., "Thermal Conductivity of Insulation Containing Moisture," in *Thermal Conductivity Measurements and Applications of Thermal Insulations, ASTM STP 217,* American Society of Testing and Materials, Philadelphia, 1957, pp. 65–80.

[*2*] Anderson, L. O. and Sherwood, G. E., "Condensation Problems in Your House: Prevention and Solution," Agriculture Information Bulletin 373, Forest Products Laboratory, Madison, WI, 1974.

[*3*] Sherwood, G. E., "Condensation Potential in High Thermal Performance Walls—Cold Winter Climate," Research Paper FPL 433, U.S. Department of Agriculture, Forest Service, Forest Products Laboratory, Madison, WI, 1983.

[*4*] Burch, D. M. and Treado, S. J., "A Technique for Protecting Retrofitted Wood-Frame Walls from Condensation Damage," *A.S.H.R.A.E. Transactions,* Vol. 84, Part I, American Society of Heating, Refrigerating, and Air-Conditioning Engineers, New York, 1978.

[*5*] U.S. Department of Agriculture, Forest Service, Forest Products Laboratory, *Wood Handbook—Wood as an Engineering Material,* Handbook No. 72. U.S. Department of Agriculture, Washington, DC, revised 1974.

[*6*] Sherwood, G. E., "Paint as a Vapor Barrier," Research Paper FPL 319, U.S. Department of Agriculture, Forest Service, Forest Products Laboratory, Madison, WI, 1978.

[*7*] Sherwood, G. E. and Peters, C. C., "Moisture Conditions in Walls and Ceilings of a Simulated Older Home During Winter," Research Paper FPL 290, U.S. Department of Agriculture, Forest Service, Forest Products Laboratory, Madison, WI, 1977.

[*8*] Duff, J. E., "The Effect of Air-Conditioning on the Moisture Conditions in Wood Walls," Research Paper SE 78, U.S. Department of Agriculture, Forest Service, Southeastern Forest Experimental Station, Asheville, NC, 1971.

[*9*] *ASHRAE Handbook of Fundamentals,* American Society of Heating, Refrigerating, and Air-Conditioning Engineers, New York, 1977.

[*10*] Duff, J. E., "A Probe for Accurate Determination of Moisture Content of Wood Products in Use," Research Note FPL-0142, U.S. Forest Service, Forest Products Laboratory, Madison, WI, 1966.

Roofs

Wayne Tobiasson,[1] *Alan Greatorex,*[1] *and Doris Van Pelt*[1]

Wetting of Polystyrene and Urethane Roof Insulations in the Laboratory and on a Protected Membrane Roof

REFERENCE: Tobiasson, W., Greatorex, A., and Van Pelt, D., **"Wetting of Polystyrene and Urethane Roof Insulations in the Laboratory and on a Protected Membrane Roof,"** *Thermal Insulation: Materials and Systems, ASTM STP 922,* F. J. Powell and S. L. Matthews, Eds., American Society for Testing and Materials, Philadelphia, 1987, pp. 421–430.

ABSTRACT: When subjected to a sustained temperature gradient in the presence of moisture in laboratory wetting tests, urethane and expanded polystyrene roof insulations accumulate enough moisture to reduce their insulating ability significantly. Extruded polystyrene is quite resistant to moisture in such tests. But the vapor drive is not as great in actual roofs, and it may reverse direction, thereby seasonally drying the insulation. To determine how well the laboratory tests could predict the wetting rate of insulation in actual protected membrane roofs, extruded and expanded polystyrene and urethane insulations were installed in a protected membrane roof in Hanover, New Hampshire. After three years of exposure, little moisture had accumulated in the extruded polystyrene, and it still retained essentially all of its initial insulating ability. Moisture progressively accumulated in 16-kg/m^3 (1-lb/ft^3) and 30-kg/m^3 (1.9 -lb/ft^3) expanded polystyrene insulations, and at the end of the test they retained only about 30 and 40% of their initial thermal resistance, respectively. The urethane accumulated enough moisture to reduce its insulating ability to about 30% of its dry value. The laboratory tests provided a valuable indication of the potential long-term moisture gain of these insulations when installed in protected membrane roofs in cold regions.

KEY WORDS: roofs, protected membranes, thermal insulation, cellular plastics, urethane, styrene, moisture, wetting, thermal conductivity, thermal resistance

In a *protected membrane* roof, the membrane is placed below the insulation for protection. There it is not affected by most of the temperature variations, solar effects, and mechanical abuse that exposed membranes are subjected to. The insulation, which is usually loose laid, is protected from the sun and from being blown off by the wind by a ballast of stones or concrete pavers, as shown in Fig. 1. The insulation boards above the membrane are in a relatively harsh environment. All the surfaces of the boards are bathed in water during a rain, and moisture may remain between the ballast and the insulation and between the insulation and the membrane for some time thereafter. If the deck is "dead level" or if dirt that enters the system blocks the drainage channels between boards, the insulation may sit in ponded water for long periods.

An old protected membrane (PM) roof is shown in Fig. 2. The bark membrane is protected by a layer of insulating sod. In today's protected membrane roofs, the bark has been replaced by bituminous built-up membranes or "single-ply" membranes made of rubber, plastic, or "rubberized" asphalt, and the sod has been replaced with extruded polystyrene insulation and ballast. Extruded polystyrene insulation is remarkably resistant to moisture.

[1]Research civil engineer, and civil engineering technicians, respectively, U.S. Army Corps of Engineers, Cold Regions Research and Engineering Laboratory, Hanover, NH 03755-1290.

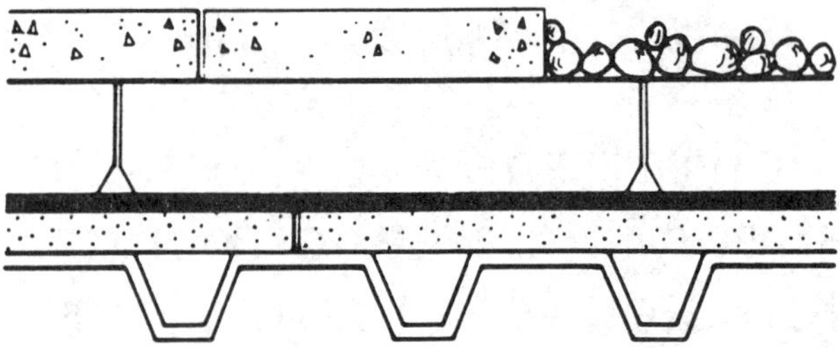

FIG. 1—*Cross section of a protected membrane (PM) roof.*

FIG. 2—*An old protected membrane roof with a bark membrane below sod insulation.*

Field Experience

The U.S. Army Cold Regions Research and Engineering Laboratory (CRREL) has been interested in PM roofs for many years [*1–6*]. In 1972, a large portion of the roof over CRREL's Hanover, New Hampshire, laboratory was reroofed with an elastomeric membrane (EPDM), 89-mm-(3½-in.) thick extruded polystyrene insulation, and 51-mm (2-in.)-thick concrete pavers. We believe it is the oldest "rubber"-protected membrane in the United States.

The protected membrane concept is patented [*7*]. One reason for CRREL's recent work on these roofs is that the patent expires in 1985 and several new protected membrane systems are being marketed. Less expensive, *expanded* polystyrene (EPS) insulation (also known as beadboard) and urethane insulation have been considered for use in protected membrane roofs. A urethane protected membrane roof was marketed briefly a few years ago, but it was withdrawn because of moisture problems.

EPS insulation has been substituted for extruded polystyrene insulation as a cost-saving measure in some protected membrane roofs. In 1982, Tobiasson examined one such roof in Alaska three years after it was constructed. Two boards were weighed. The 100-mm (4-in.)-thick, 0.76 by 1.22-m (2½ by 4-ft), 30-kg/m^3 (1.9-lb/ft^3) EPS boards should have weighed about 3 kg (6 lb) dry. One board weighed 19 kg (42 lb) and the other 34 kg (75 lb). The laboratory studies discussed in this paper indicated that the EPS in that roof had lost about half of its insulating ability.

Laboratory Tests

Equipment and Procedure

Insulations have been tested in the laboratory for moisture gain by immersing them in water [*8*] and by subjecting them to a thermally induced vapor pressure gradient [*9–12*]. Cellular plastic insulations such as polystyrenes and urethanes become wet much faster in the latter type of test than by immersion. Since insulations in use are subjected to temperature gradients, laboratory tests with temperature gradients appear to be appropriate. We see little value in isothermal immersion tests for determining the wetting characteristics of insulations in roofs.

In order to evaluate urethane and styrene insulations for use in protected membrane roofs, 25-mm (1-in.)-thick, 300-mm (11¾-in.)-square specimens were sealed with two coats of vapor barrier coating on their sides and top, then subjected to hot, wet conditions [29°C (85°F), 100% relative humidity] below and cool conditions [4°C (40°F), 75% relative humidity] above, as shown in Fig. 3. This represents a severe boundary condition conducive to wetting, since no drying can occur from the sealed top surface, but vapor is driven upward toward the cool top from the dripping-wet bottom. The increase in water content with time for 32-kg/m^3 (2.0-lb/ft^3) urethane (URE), 16-kg/m^3 (1.0-lb/ft^3) expanded polystyrene (EPS-1), 30-kg/m^3 (1.9-lb/ft^3) expanded polystyrene (EPS-2), and 38-kg/m^3 (2.4-lb/ft^3) extruded polystyrene (EXT) specimens is shown in Fig. 4. Data points are shown for the EPS-2 specimen only. The data fit for the other three specimens was similar.

In 400 days, the URE and EPS-1 and EPS-2 specimens had a moisture content exceeding 30% by volume, but even after 1800 days (that is, about five years), the EXT specimen had a volumetric moisture content of less than 10%.

Periodically during this long test, each specimen was removed from the wetting apparatus for a short time, wrapped in a sheet of 0.0013-mm-(0.0005-in.) thick plasticized polyvinyl chloride, and placed in a thermal conductivity instrument, which was maintained with its top plate at about 4°C (40°F) and its bottom plate at about 29°C (85°F). This reproduced the same top and bottom temperatures as in the wetting apparatus. Measurements were made in accordance with the ASTM Test for Steady-State Thermal Transmission Properties by Means of the Heat Flow Meter (C 518-76). The requirements of this standard were met, except that the specimens were not dry because the purpose of the study was to test wet specimens, and six successive readings for all wet specimens did not always yield thermal resistance values that agreed within 1%.

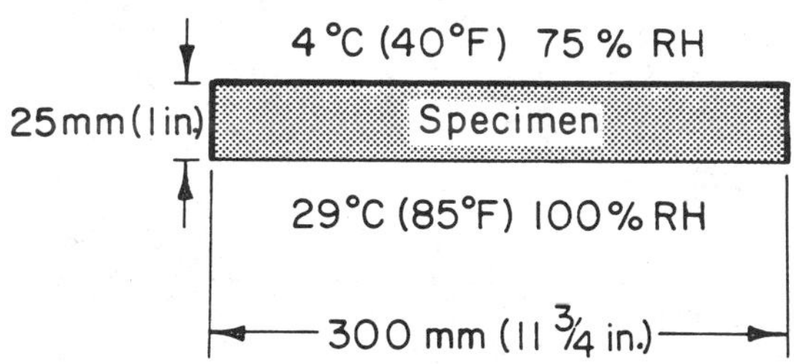

FIG. 3—*Laboratory wetting conditions.*

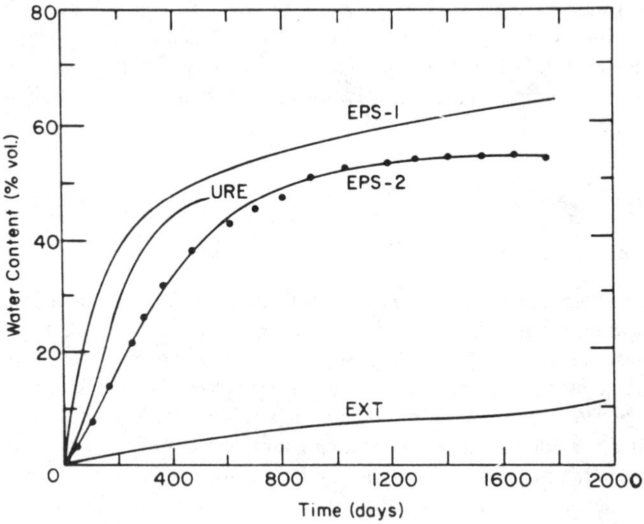

FIG. 4—*Increase in water content with time for laboratory specimens.*

As a specimen takes on water, its thermal resistance decreases. The ratio of its wet thermal resistance to its dry thermal resistance is a measure of its thermal efficiency. We call this efficiency ratio, expressed as a percentage, the thermal resistance ratio (TRR). The decay in TRR as the specimens took on moisture is shown in Fig. 5. Each of the three specimens (URE, EPS-1, and EPS-2) is represented by the single solid curve. The curve below is for the EXT specimen. That curve has a solid upper portion; then, at a TRR of about 68%, the curve becomes dashed. Because the EXT specimen took on very little water even after five years in the laboratory wetting apparatus, the laboratory results only provide data for that EXT specimen up to a volumetric water content of about 11%. The EXT curve was extended by testing pieces

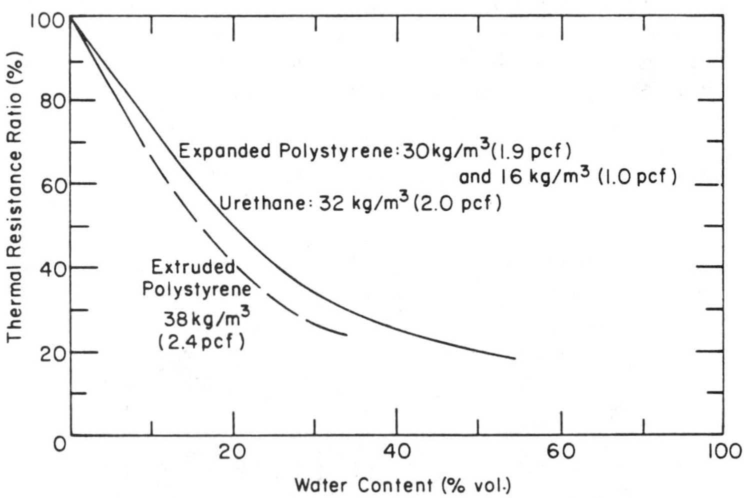

FIG. 5—*The decay in thermal resistance for laboratory specimens.*

of similar insulation removed from the aforementioned CRREL protected membrane roof eleven years after it had been installed.

When we disassembled that roof in 1983, we found that most of the insulation was dry and had retained more than 80% of its initial *R* value. However, a few boards were noticeably heavier than the rest. These specimens were not uniformly wet (Fig. 6). We measured their thermal resistance and their moisture content.

The laboratory tests demonstrated that in an environment with a severe wetting potential, extruded polystyrene is much more likely to retain its thermal resistance than expanded polystyrene or urethane.

Insulation above a membrane in a protected membrane roof is in a relatively harsh wetting environment, but that environment is not as harsh as the one produced in our laboratory tests. We conducted an exposure test to determine how well these same insulations will perform on an actual protected membrane roof that is not sealed on top or subjected to a sustained one-way vapor drive.

Exposure Tests

In 1980, 305 by 305-mm (12 by 12-in.) specimens of urethane, extruded polystyrene, and expanded polystyrene were placed in the CRREL protected membrane roof (Fig. 7). The membrane was often wet in this area because the lip of the nearby drain was a little higher than the membrane here, and dirt had accumulated within the surrounding protected membrane system, retarding drainage.

The type and thickness of each exposure specimen are shown in Fig. 8. Note that the higher-density, expanded polystyrene specimen (EPS-2) is in one layer while all other specimens are in two layers, with the top layer designated *T* and the bottom layer *B*. The EPS-2 specimen was

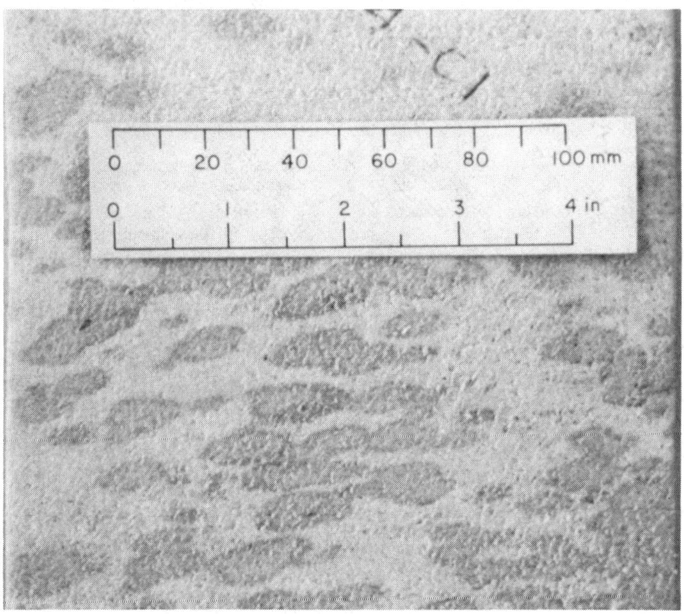

FIG. 6—*Section through a wet board of extruded polystyrene after eleven years of exposure in New Hampshire.*

FIG. 7—*Exposure test specimens installed in the CRREL protected membrane roof.*

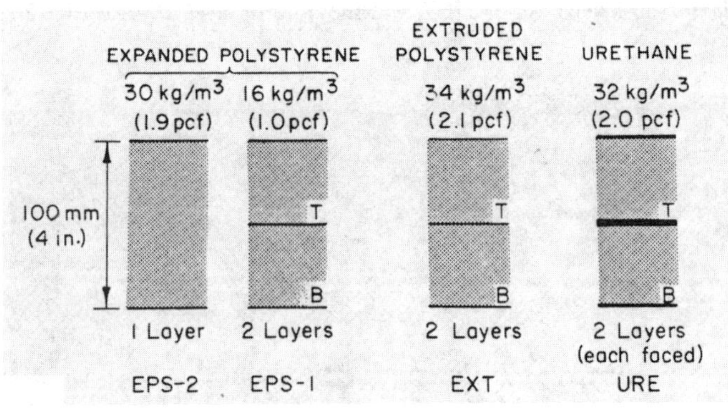

FIG. 8—*The type and thickness of the exposure test specimens.*

the same material used in the aforementioned Alaskan protected membrane roof. The top and bottom urethane specimens were each faced with an asphaltic felt.

The sides of insulation boards are not sealed when they are used in roofs. However, since such boards are much larger than these specimens, we sealed two adjacent sides of each exposure specimen. The remaining two sides were unsealed to represent the exposed edges and corner of a typical roof insulation board.

A 610 by 610-mm (2 by 2-ft) concrete paver was placed directly on top of the specimens. Periodically, over a three-year period, the paver was moved aside, and the specimens were weighed.

The rates of wetting for the EPS-2 specimen and the *top* layer of the other three specimens are shown in Fig. 9. There is no significant seasonal variation in the rate of wetting. As in the

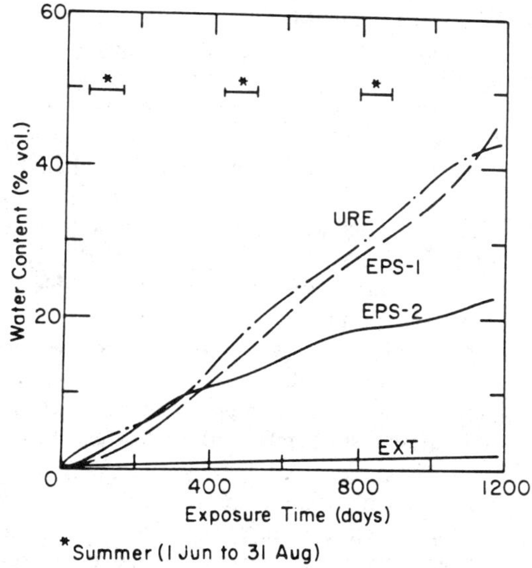

FIG. 9—*Increase in water content with time for the* top *layer of the exposure test specimens.*

laboratory test, the urethane and expanded polystyrene specimens became quite wet, and the extruded polystyrene specimen remained relatively dry. The rate of wetting for each exposure specimen was less than that of the corresponding laboratory specimen.

The rates of wetting for the EPS-2 specimen and the *bottom* layer of the other three exposure specimens are shown in Fig. 10. There is a noticeable reduction in wetting rate during the summer for the EPS-1 and URE specimens. Again, the EXT specimen remained relatively dry while the others took on a significant amount of water.

Some of the difference in performance between the EPS-1 and EPS-2 specimens is due to the

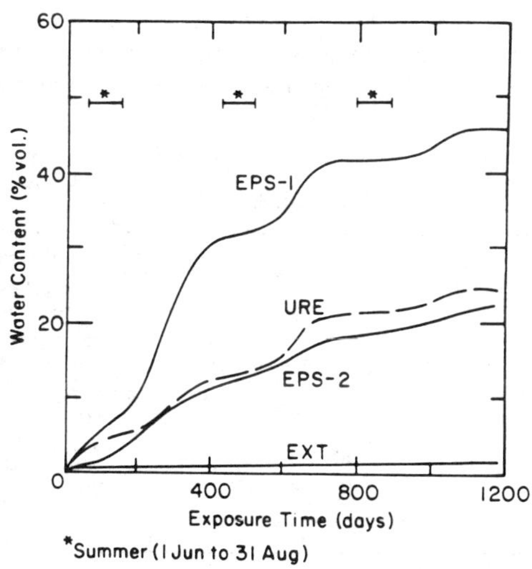

FIG. 10—*Increase in water content with time for the* bottom *layer of the exposure test specimens.*

difference in density (EPS-2, which is about twice as dense as EPS-1, is somewhat more resistant to water). However, a portion of the difference is also related to the difference in thickness (EPS-2 is twice as thick as EPS-1).

We calculate that, over the course of a year, the laboratory test vapor drive would average about five times the average vapor drive to which the exposure test specimens were subjected. This helps explain why the laboratory specimens wet more quickly than the exposure specimens did.

After 1178 days of exposure, the specimens were removed from the roof, surface-dried with paper towels, wrapped in a film of plasticized polyvinyl chloride, then tested in a heat flow meter thermal conductivity instrument. Results are presented in Table 1.

Comparing laboratory test and exposure test findings, we conclude that the laboratory test gives a reasonable prediction of field performance in cold regions. Extrapolating these findings to the 10 to 20 or more years of performance desired of a roof insulation, it seems clear that only extruded polystyrene insulation should be used in protected membrane roofs where water can pond on the membrane and pavers rest directly on the insulation, thereby retarding upward drying.

The positive effects of a sloped membrane and ventilation of the top surface of the insulation in a protected membrane roof have been demonstrated in Canada [*13*]. In order to assess these factors, numerous full-board exposure specimens of expanded and extruded polystyrene were installed in two layers in the CRREL protected membrane roof during the summer of 1983 (Fig. 11). About a year later, the specimens were reweighed. All the extruded polystyrene boards had taken on very little moisture. Where the membrane was dead level and no ventilation was provided between the pavers and the insulation, the EPS had a TRR as low as 48%. Where the membrane was on a slope of about 1:50 (1/4 in./ft) and crushed rock was used as ballast, the TRR of the EPS was above 84%. Obviously, membrane slope and surface ventilation are important. However, since the EPS lost up to 16% of its insulating ability in one year, we are not very optimistic about its use even in well-sloped, well-ventilated protected membrane roofs. The specimens installed in 1983 will be reweighed periodically over the next several years to define their long-term wetting characteristics better.

Conclusions

A laboratory wetting test has been developed that subjects insulation specimens sealed on their edges and cold side to thermally induced vapor pressure gradients. The test provides a valuable indication of the wetting behavior of insulations used in protected membrane roofs in

TABLE 1—*Thermal resistance ratio (TRR) of exposure test specimens at the end of the test.*

Insulation	Density, kg/m³ (lb/ft³)	Thermal resistance ratio after 1178 Days of Exposure, %	
		Top Layer	Bottom Layer
Extruded polystyrene	34 (2.1)	100	100
Expanded polystyrene	30 (1.9)	41[a]	
Expanded polystyrene	16 (1.0)	26	32
Urethane	32 (2.0)[b]	13	44

[a]Only one layer.
[b]Facings are not included.

FIG. 11—*Styrene test boards installed in the CRREL protected membrane roof in 1983.*

cold regions. These laboratory tests, plus exposure tests and field experience, indicate that *extruded* polystyrene insulation can be used above the membrane in protected membrane roofs, but *expanded* polystyrene (EPS) and urethane insulation should not be used since they are likely to become wet and lose much of their insulating ability.

The rate of wetting for insulation in a protected membrane roof can be reduced by providing the membrane with slope to drain and configuring the ballast so that the upper portions of the insulation can air dry.

References

[*1*] Schaefer, D., "Protected Membrane Roofing System Installation at Hanover, New Hampshire," Special Report 215, U.S. Army Cold Regions Research and Engineering Laboratory, Hanover, NH, 1974.

[*2*] Aamot, H. W. C. and Schaefer, D., "Protected Membrane Roofs in Cold Regions," CRREL Report 76-2, U.S. Army Cold Regions Research and Engineering Laboratory, Hanover, NH, 1976.

[*3*] Schaefer, D., "Water Absorption of Insulation in Protected Membrane Roofing Systems," CRREL Report 76-38, U.S. Army Cold Regions Research and Engineering Laboratory, Hanover, NH, 1976.

[*4*] Schaefer, D., Larsen, E. T., and Aamot, H. W. C., "Observation and Analysis of Protected Membrane Roofing Systems," CRREL Report 77-11, U.S. Army Cold Regions Research and Engineering Laboratory, Hanover, NH, 1977.

[*5*] Aamot, H. W. C., "Mid-Winter Installation of Protected Membrane Roofs in Alaska," CRREL Report 77-21, U.S. Army Cold Regions Research and Engineering Laboratory, Hanover, NH, 1977.

[*6*] Schaefer, D., "Installation of Loose-Laid Inverted Roof System at Ft. Wainwright, Alaska," Special Report 77-18, U.S. Army Cold Regions Research and Engineering Laboratory, Hanover, NH, 1977.

[*7*] Best, J. S., Roof Construction and Method Thereof, U.S. Patent No. 3,411,256, 1968.

[*8*] Kaplar, C. W., "Moisture and Freeze-Thaw Effects on Rigid Thermal Insulations," Technical Report 249, U.S. Army Cold Regions Research and Engineering Laboratory, Hanover, NH, 1974.

[*9*] Dechow, F. J. and Epstein, K. A., "Laboratory and Field Investigations of Moisture Absorption and Its Effects on Thermal Performance of Various Insulations," *Thermal Transmission Measurements of Insulation, ASTM STP 660,* American Society for Testing and Materials, Philadelphia, 1978, pp. 234–260.

[*10*] Hedlin, C. P., "Moisture Gains by Foam Plastic Roof Insulations Under Controlled Temperature Gradients," *Journal of Cellular Plastics,* Vol. 13, No. 5, September/October 1977, pp. 313-319.

[*11*] Tobiasson, W. N. and Ricard, J., "Moisture Gain and Its Thermal Consequence for Common Roof Insulations," *Proceedings, 5th Conference on Roofing Technology,* April 1979, National Roofing Contractors Association, Oak Park, IL, 1979, pp. 4-16.

[*12*] Knab, L. I., Jenkins, D. R., and Mathey, R. G., "The Effect of Moisture on the Thermal Conductivity of Roofing Systems," NBS Building Science Series 123, National Bureau of Standards, Gaithersburg, MD, 1980.

[*13*] Hedlin, C. P., "Moisture Content in Protected Membrane Roof Insulations—Effect of Design Features," *Roofing Systems, ASTM STP 603,* American Society for Testing and Materials, Philadelphia, 1976, pp. 36-50.

Walter J. Rossiter, Jr.,[1] *and Robert G. Mathey*[1]

Risk of Blistering of Built-Up Roofing Membranes Applied to Polyurethane Foam Insulation

REFERENCE: Rossiter, W. J., Jr., and Mathey, R. G., **"Risk of Blistering of Built-Up Roofing Membranes Applied to Polyurethane Foam Insulation,"** *Thermal Insulation: Materials and Systems, ASTM STP 922,* F. J. Powell and S. L. Matthews, Eds., American Society for Testing and Materials, Philadelphia, 1987, pp. 431–448.

ABSTRACT: The use of polyurethane foam insulation boards in low-sloped roof constructions has increased in recent years to upgrade their thermal efficiency. Built-up membranes applied directly over polyurethane foam insulation may experience greater incidence of blistering than membranes applied on other roofing insulation boards. The bubbling of hot asphalt during its application to the polyurethane insulation board can cause voids in the asphalt layer, which eventually can grow into blisters. This paper presents the results of a study to investigate the effect of application of hot asphalt directly onto polyurethane insulation boards. In addition, it discusses causes of blistering and methods for using polyurethane foam insulation boards in built-up roofing systems so as to reduce the risk of blistering.

Polyurethane foam insulation specimens, typical of those used in the late 1970s in the United States for built-up roofing construction, were included in the study. Most of the insulations contained a facing material. Hot asphalt was applied to the insulations with facings to determine the extent of off-gassing. Hot asphalt was also applied to insulations without facings, to facings removed from insulation, and to oven-dried insulations and facings to investigate the effect of the moisture content of the insulations and facings on the off-gassing.

In other tests, oven-dried and moist ("room dry") felts were used in the preparation of built-up membranes over polyurethane boards. Delamination of the roofing specimens indicated many large voids in the asphalt layer between the polyurethane insulation and the first ply of felt. These voids were considered to increase the risk of blistering in built-up membranes applied directly on the polyurethane foam insulation. The data support previous suggestions to apply a thin layer of rigid insulation to the polyurethane foam before the membrane is applied.

KEY WORDS: blistering, built-up membranes, cellular plastic, foam, insulation, moisture, off-gassing, polyurethane, roofing, thermal insulation

The total area of the nation's low-sloped roofing is about 2.3 billion square metres (25×10^9 ft^2) [*1*]. In 1976, it was estimated that an average decrease of 0.1 in the U factor of these low-sloped roofs would save nearly 20 million barrels of crude oil annually [*2*]. More recently in 1981, it was reported that an annual reduction in energy consumption of about 34 million barrels of crude oil might be achieved by increasing the R value from 7 to 20 [*3*]. Clearly, on the national level, a compelling incentive exists to increase the thermal efficiency of low-sloped roofs. On the individual level, increased thermal efficiency of low-sloped roofs can mean substantial reductions in heating and cooling costs for building owners. Griffin [*4*], Cash [*5*], and Courville and Kolb [*6*] have discussed the economic impact of increasing the thermal efficiency of low-sloped roofing systems.

[1]Research chemist and materials research engineer, respectively, Building Materials Division, Center for Building Technology, National Bureau of Standards, Gaithersburg, MD 20899.

A result of the upgrading of the thermal efficiencies of low-sloped roofs in the 1970s was an increase in the use of rigid cellular plastic (foam) insulation boards [7]. Types of foam plastic insulations used at that time included polyurethane, polyisocyanurate, polystyrene, and the composite boards whereby an inorganic insulation such as perlite or fibrous glass was laminated to one or both faces of a polyurethane or polystyrene board. According to data from Project Pinpoint surveys by the National Roofing Contractors Association (NRCA), the foam plastic insulations were used in about 7% of low-sloped roofing installations in 1974, while in 1979 they were installed in 20% of the jobs surveyed [7].

The increased use of the foam plastic insulations did not occur without concern that these materials might have an adverse effect on bituminous built-up membrane performance. The 1979 NRCA Project Pinpoint data indicated that 38% of all roofing problems reported in the survey involved roof assemblies that included foam plastic insulation boards [7]. This paper presents the results of a study concerning one aspect of the performance of roofing systems using rigid polyurethane foam insulation: the blistering of built-up roofing membranes applied directly on polyurethane foam boards.

According to the ASTM Definitions of Terms Relating to Roofing, Waterproofing, and Bituminous Materials (D 1079-83a) a blister is a raised portion of a roofing membrane resulting from local internal pressure. Although blisters may not leak initially, they may shorten the service lives of built-up membranes. The membrane may be weakened because of ply delamination, and also the blister may be readily punctured by foot traffic or other sources of load application and by impact. Aggregate on the membrane surface may slide from the raised blister, exposing membrane felts to the deleterious effects of the weathering environment.

Background

Concerns that built-up roofing membranes applied directly over polyurethane foam insulation boards might experience a high frequency of blistering date back to the mid-1970s. The early concerns centered on the possibility that the fluorocarbon blowing agent in the cells of the foam may have resulted in membrane blistering. In 1976, Mirra [8] reported that experimental test roofs installed in 1972 and 1973 with urethane-glass fiber composite insulation boards experienced membrane blistering more frequently than control test roofs, which contained only glass fiber insulation board. He indicated that these blisters were predominantly flat and randomly located on the insulation boards and over joints between boards. He further stated that they were initially located between the top surface of the insulation board and the bottom ply of the built-up membrane. In addition, he reported that the blisters developed within 30 to 60 days after roof installation. In Mirra's study, the cause of the blistering was attributed to the large differential thermal movement between the polyurethane foam and the built-up membrane. The role of the fluorocarbon blowing agent released from the foam on the growth of the blisters was investigated during the study. Fluorocarbon gas pressurization of initial voids made during hot asphalt application was ruled out as a mechanism for blister growth.

Mirra's conclusion [8] attributing the blistering problem to differential thermal movement was not accepted by all segments of the roofing industry. In August 1978, NRCA informed its membership of the possibility of an increased incidence of blistering of built-up roofing membranes over some polyurethane and polyurethane-composite roof insulation boards [9]. This NRCA bulletin indicated that the blistering problem was associated with material characteristics and field factors concerning off-gassing (vapor release) from the insulation boards. NRCA recommended that, to minimize the risk of blistering, a thin layer of an insulation such as fiberboard, perlite, or fibrous glass be installed over the polyurethane foam on which the built-up membrane would be fabricated. The 1978 NRCA bulletin was superseded in 1979 by a second bulletin [10] that contained a similar recommendation for the use of thin insulation board to separate the polyurethane foam and the membrane. In addition, the second bulletin also recommended that, as an option to the separation layer, a base ply could be installed on the foam insulation to allow for venting.

What prompted the NRCA bulletins were observations that release of a gas (off-gassing) from the insulation boards occurred when hot bitumen was applied to their surfaces. This phenomenon has been referred to as a bubbling, frothing, or foaming of the hot applied bitumen. In spite of the concerns over the phenomenon, few reports describing the process appeared prior to 1980. When hot asphalt is applied to the facing material on the surface of the insulation board, many small gas bubbles evolve, which either rise through the asphalt or are trapped in it [*11,12*]. Adjacent small bubbles may link together and produce large voids. These voids may ultimately grow into blisters. The evolved gas has been reported to be water vapor [*11,12*]. One report indicated that the source of the water vapor was moisture present on or near the surface of the insulation [*12*]. Another report attributed the foaming or bubbling of the applied asphalt principally to moisture in the facing sheet on the insulation [*11*]. In contrast to the bubbling observed when hot asphalt is applied to polyurethane foam insulation, no bubbling has been observed in asphalt applied to some conventional insulation materials [*12*]. In 1981, Porcher [*13*] reviewed the NRCA studies on the causes of bubbling and frothing of hot asphalt applied to polyurethane foam. He hypothesized that moisture on the surface of the insulation is vaporized during application of the hot asphalt. Since the insulation has relatively low permeability, the released moisture vapor rises through the asphalt, and bubbling and frothing occur. Porcher recommended that the NRCA recommendations [*9,10*] be followed, to avoid the possibility of blistering of membranes applied directly on polyurethane foam insulation.

Mechanism of Blister Formation

A mechanism for blister formation in built-up roofing membranes was proposed by Warden [*14*] approximately 25 years ago. A review of the subject has been done by Griffin [*15*]. Laaly et al [*16*] recently reported the results of laboratory tests investigating some parameters affecting blister formation. The key element in blister formation is a void within the plies of the membrane (or between the membrane and the insulation board substrate), and the expansion of air, moisture, or other gases trapped within the void. The diurnal temperature cycle to which the roof is exposed also plays a role. Early in the day, as the sun strikes the roof, the membrane warms, causing the bitumen to soften and be more susceptible to flow. In addition, the gas entrapped in the void increases in pressure as the membrane warms. Since the membrane has low permeability, the gas cannot readily escape from the void. If the gas pressure is greater than the bond strength between the membrane plies, the void will increase in size, and a blister will begin to grow. Later in the day, as the rays of the sun no longer strike the roof, the membrane cools. At this time, the bitumen stiffens, and with the decrease in temperature, the pressure in the enlarged void is reduced. However, since the bitumen has stiffened, the enlarged void does not return to its original size but remains expanded. The cooled gas develops a negative pressure, which may draw air or moisture into the void throughout the night until equilibrium is reached with the ambient pressure. Subsequently, the cycle repeats itself, and the blister continues to grow.

Two key factors should be noted concerning this mechanism of blister formation. First, moisture need not be entrapped in the system for a blister to grow. It is only necessary to have a void with air in the system. Second, the system with the void need not be completely closed for air leakage into the void to occur during nighttime cooling.

Objective and Scope

This paper describes the results of a two-phase study to investigate the causes of bubbling and frothing of hot asphalt when applied directly to the surfaces of polyurethane foam insulation boards. In the laboratory phase, observations were recorded concerning the effect on the bubbling of the insulation type, the moisture content of the insulation, and the asphalt temperature. The effect of the bubbling on producing voids in the asphalt layer between the insulation board and the built-up membrane was also investigated. In the field phase of the study, gases in built-up membrane blisters were sampled and analyzed for composition.

At the beginning of the study, possible causes of the bubbling and frothing of the hot asphalt were listed, as follows:

(*a*) release of moisture from the surface of the foam insulation,
(*b*) release of moisture from the facing on the insulation board,
(*c*) release of air, fluorocarbon, or other gases from within the cells of the foam, and
(*d*) release of gaseous decomposition products produced when hot asphalt strikes the surface of the foam or facing.

Experimental Procedure

Foam Insulation Specimens

Fourteen polyurethane-type, rigid insulation board specimens (Nos.1 through 14) were included in the study (Table 1). These proprietary insulations were generally typical of those used in built-up roofing construction in the late 1970s and were supplied by manufacturers' representatives. Table 1 includes a listing of the types of boards and facings, board thicknesses, and densities of the foam. The types of boards were polyurethane foam, polyisocyanurate foam, and composites in which the polyurethane foam was laminated to an inorganic board insulation, such as perlite or fibrous glass. The descriptions of the foam specimens as either polyurethane or polyisocyanurate were those given by the manufacturer. The facing types were bituminous felts, aluminum foil laminated to kraft paper, and a glass mat. One polyurethane foam was unfaced.

Asphalts and Felts

The asphalts in the study were specified as Type I and Type III, as designated in the ASTM Specification for Asphalt Used in Roofing (D 312-78). Organic and asbestos felts were specified according to the ASTM Specification for Asphalt-Saturated Organic Felt Used in Roofing and Waterproofing (D 226-82) and the ASTM Specification for Asphalt-Saturated Asbestos Felts Used in Roofing and Waterproofing, (D 250-84), respectively. Asbestos felts were included in the study, since they were used to fabricate built-up membranes in the late 1970s.

Application of Hot Asphalt to the Insulation Boards and Components

To observe the effect of the application of hot asphalt on polyurethane foam insulation boards, laboratory test specimens having dimensions of 150 by 150 mm (6 by 6 in.) were prepared. The test specimens to which hot asphalt was applied included foam insulations with facings intact in the condition "as-received" from the manufacturers' representatives, foam insulations with facings removed and in the as-received condition, foam insulations with facings removed and dried at 85°C (185°F) in a laboratory vacuum oven, facings separated from the insulation boards and in the as-received condition, and delaminated facings that had been dried at 97°C (207°F) in a laboratory oven. Heated asphalts were applied to the test specimens using a large ladle. The application temperatures of the asphalt were 177, 200, and 232°C (350, 312, and 450°F), although most tests were conducted at the middle temperature. Heating of the

TABLE 1—*Polyurethane foam insulation specimens.*

Sample No.	Insulation Board Type	Facing Type[a]	Thickness,[b] mm	in.	Foam Density,[c] kg/m³	lb/ft³	Foam Moisture, % by mass[d]	Facing Moisture, % by mass[e]
1	polyurethane foam	felt	35	1.4	29.3	1.8	3.6	3.8
2	polyurethane foam	felt	38	1.5	32.8	2.0	11.8	5.6
3	polyurethane foam	felt	38	1.5	27.7	1.7	3.9	2.2
4	polyurethane foam	felt	33	1.3	30.1	1.9	3.1	3.0
5	polyurethane foam[f]	felt	37	1.4	31.1	1.9	2.9	2.9
6	polyurethane foam	aluminum[h]	22	0.9	40.0	2.5	6.3	...[i]
7	polyurethane foam	none	51	2.0	36.3	2.3	1.4	...[j]
8	Isocyanurate foam[f]	felt	32	1.3	38.1	2.4	2.4	0.7
9	PUF/perlite composite	felt	45	1.8	28.9	1.8	2.8	2.0
10	PUF/perlite composite	felt	45	1.8	33.1	2.1	5.2	2.0
11	PUF/perlite composite	felt	48	1.9	34.4	2.1	1.8	1.7
12	PUF/perlite composite	felt	40	1.6	32.7	2.0	3.2	2.6
13	PUF/perlite composite	felt	45	1.8	30.2	1.9	3.5	1.7
14	PUF/FG composite[g]	glass mat	45	1.8	33.4	2.1	5.0	3.3

[a]Composite boards have a facing on one side only.
[b]Total thickness of the board.
[c]Density of the polyurethane component only.
[d]Percentage by mass of the dry foam in the as-received condition.
[e]Percentage by mass of the dry felt in the as-received condition.
[f]Glass fibers were incorporated in the foam.
[g]FG indicates fibrous glass.
[h]The aluminum foil was laminated to kraft paper, which was adhered to the foam.
[i]Not determined, since the facing was not readily removable.
[j]No facing on the foam specimen.

asphalt was accomplished using a laboratory hot plate, which maintained the asphalt temperature to ±5°C (9°F).

In the case of the composite insulation boards, the inorganic layer was intact on the polyurethane foam when hot asphalt was applied to the facing on the foam surface. For most of the composite insulation boards, hot asphalt was not applied to the surface of the inorganic insulation layer. In a few cases, as controls, hot asphalt was applied to the inorganic insulation layer in its as-received condition. This provided the basis for comparison of the extent of bubbling of hot asphalt applied to the inorganic insulations and the polyurethane foam.

Moisture Content of the Foam Insulations

With one exception, three cylindrical foam specimens (without facings) having diameters of about 21 mm (0.8 in.) were cut with a cork borer from the foam insulation specimens. Four replicate determinations of moisture content were made for insulation board Specimen 1. The lengths of the cylindrical specimens corresponded to the thicknesses of the foam in the insulation board specimens. Each cylindrical foam specimen was placed in a tared glass weighing jar, and the specimen mass was determined using an analytical balance with a sensitivity of 0.1 mg. The weighing jars containing the foam specimens were placed in an oven with forced air convection from about 8:00 A.M. to 5:00 P.M. for three consecutive days. The temperature of the oven was 97 ± 3°C (207 ± 5°F). This temperature was selected so that the maximum would not be above 100°C (212°F). During the night, the foam specimens were allowed to cool and remain in a dessicator containing calcium chloride dessicant. The mass of each foam specimen was determined each morning before it was replaced in the oven. Any loss in mass due to heating was recorded as moisture content. The percentage of volatile loss was calculated as the percentage by mass of the oven-dried foam specimens as follows

$$\text{Percent moisture content (by mass)} = \frac{100(W_1 - W_2)}{W_2}$$

where

W_1 = mass of original cylindrical specimen, and
W_2 = mass of specimen after oven drying.

Moisture Content of the Insulation Board Facings

Three specimens of insulation board facings, with dimensions of 150 by 150 mm (6 by 6 in.), were cut and separated by peeling with a spatula from the insulation board specimens. Any foam that remained on the underside of the separated facings was removed by rubbing with steel wool. The separated facings were weighed using an analytical balance having a sensitivity of 0.1 mg. Each facing specimen was suspended by wire for 22 h in an oven with forced air convection and then placed in a dessicator containing calcium chloride dessicant. The oven temperature was 97 ± 3°C (207 + 5°F). After cooling, the mass of each facing specimen was again determined. The heating and weighing process was then repeated twice. The moisture content was considered to be equivalent to the percentage by mass of the volatiles lost from the facing specimens during heating in the oven. The moisture content was calculated as for the foam specimens.

In three cases (Specimens 2, 8, and 9), the moisture contents of the felt facings were determined according to the procedure described in the ASTM Test for Water in Petroleum Products and Bituminous Materials by Distillation (D 95-83). It was found that the measured moisture content using this ASTM method differed by between 15 and 30% from that estimated using the oven-drying technique.

Preparation and Delamination of Built-Up Roofing/Insulation Specimens

Roofing specimens with dimensions of 460 by 460 mm (18 by 18 in.), and consisting of either four plies of organic perforated felt or four plies of asbestos perforated felt and a polyurethane-perlite composite insulation board (Specimen 9), were prepared using hot ASTM Type III asphalt. For all specimens, the composite insulation board was in the as-received condition. Two specimens (one with each type of felt) were fabricated from felts that contained some moisture ("room dry") as conditioned at ambient laboratory temperature and humidity. These room dry felts contained about 1.4 and 3.0% moisture by mass for the asbestos and organic felts, respectively. Also, two specimens (one with each type of felt) were prepared with four plies of felt, which were oven dried for about 20 h at 93 to 99°C (200 to 210°F). The asphalt was heated to about 205 to 220°C (401 to 428°F) and applied to the felt facing of the composite insulation board specimen. The asphalt application took about 20 to 25 s to cover the surface of the insulation. Immediately after application of the asphalt, a ply of felt was placed onto the hot asphalt. The felt was pressed in place ("broomed") by hand using an industrial-type floor brush (with the handle removed). In general, brooming of each ply of felt was carried out for about 15 to 30 s. Next, additional layers of hot asphalt and plies of felt were applied in turn (sandwich fashion) until a four-ply roofing specimen had been prepared. In addition to the four-ply specimens, one two-ply specimen was prepared in a similar manner. This specimen consisted of organic perforated felts and a sandwich board having polyurethane foam with perlite boards laminated on both faces. The felts and insulation were both in the room dry condition.

All of the completed roofing specimens were placed in a freezer overnight at −40°C (−40°F). Immediately after they were removed from the freezer, the plies of felt were delaminated from the insulation board and each other using a metal spatula.

Thermogravimetric Analysis of Polyurethane Foam Specimens

Polyurethane foam specimens cut from the insulation board specimens were subjected to thermogravimetric analysis. The foam specimens (about 2.4 to 3.3 mg) were heated from ambient room temperature to 250°C (482°F) in nitrogen gas at a rate of 5°C/min (9°F/min). The nitrogen flow rate was 75 mL/min. The analyzer instrument was evacuated and then filled with purified nitrogen prior to each run.

Analysis of Gases in Roofing Blisters

Sealed glass ampoules of 40-cm^3 capacity and fitted with rubber septa were used to contain and transport gas samples removed from blisters in existing built-up membranes fabricated on polyurethane insulation boards. The buildings were located in the southwestern United States, and the ages of the roofing ranged from about one to five years. The gas samples were taken by National Bureau of Standards (NBS) research personnel. On a roof, just prior to removal of a gas sample from a blister, 20 cm^3 of air was removed from a glass ampoule using a 20-cm^3 gas-tight syringe with a 13-mm (0.5-in.)-long, Size 26 removable needle. The roof blister was penetrated using an 89-mm (3.5-in.)-long, Size 13 needle. The asphalt in the built-up membrane provided a tight seal around the needle as it penetrated the membrane. This needle was attached to the gas-tight syringe, and 20 cm^3 of gas from the blister was drawn into the syringe. The Size 13 needle was replaced with a Size 26 needle to prevent damage to the rubber septa of the ampoules. The 20 cm^3 of gas from the blister was injected into the glass ampoule. Since 20 cm^3 of air had previously been removed from the ampoule, it contained one part blister gas sample and one part air at the time of analysis. This dilution was taken into consideration in calculating the composition in mole percent of the gases in the blisters. The ampoules containing blister gas samples were returned to the NBS laboratories for gas analysis by mass spectroscopy and by gas chromatography for the fluorocarbon blowing agent.

Approach and Discussion of Results

Application of Asphalt to Foam Test Specimens

In an initial experiment, the heated ASTM Type I and Type III asphalts were applied to the as-received insulation specimens with the facings intact on the foam. Since Specimen 7 did not have a facing (see Table 1), this foam was not included in the initial experiment. In all cases, when the hot bitumen was applied to the facing on the insulation specimen, vigorous bubbling and frothing of the asphalt were observed. The bubbling generally began 2 to 3 s after asphalt application and ended about 1 to 1½ min after application. The pattern of the bubbling of the asphalt was generally the same for all the foam specimens. Numerous small bubbles were seen on the asphalt surface, many of which would coalesce to form larger bubbles in the asphalt. As the heated asphalt cooled and its viscosity increased, small voids created by the bubbling were fixed in the asphalt surface, leaving it with a "crater" or "pock-marked" appearance. Figure 1 is an example of the bubbling and frothing resulting from the application of hot asphalt to a foam insulation specimen. Although the overall pattern of the bubbling phenomenon was about the same for most of the foam specimens, some specimens were observed to produce more active bubbling than others. The pattern of bubbling was quite different in the case of Foam Specimen 6. This foam had a facing consisting of perforated aluminum foil with kraft paper laminated on its underside. When hot asphalt was applied to the aluminum foil facing on this foam insulation, the bubbling was seen only at the perforations in the facing, indicating that it was probably arising from the foam and not the facing. This observation was inconsistent with a previous report [*13*] that moisture in the facing was the cause of bubbling.

The temperature of the asphalt had a noticeable effect on the bubbling. In general, as the asphalt temperature increased, the bubbling appeared to be more active. Attempts were not made to quantify the amount of bubbling as a function of temperature. No differences in the extent of bubbling were observed for the two different types of asphalt. However, upon cooling, for a given asphalt temperature, the voids resulting from the bubbling that occurred in the Type III asphalt were more apparent than those that appeared in the Type I asphalt. The asphalt was still warm when bubbling and frothing ceased, and subsequent flow would result in a partial smoothing of the asphalt surface. Since the Type III asphalt provided more resistance to flow, because of its higher viscosity, the voids in this type of asphalt surface were evidently held in place more than for the Type I asphalt.

In some experiments, the temperature at the interface of the foam and felt was recorded as a function of time. A copper-constantan thermocouple, embedded in the center of the specimen at the interface of the felt facing and foam, and a digital thermocouple indicator were used to measure the temperature. Figure 2 shows an example of plots of interface temperatures versus time for replicate measurements made during the application of asphalt to a foam specimen. The temperature of the heated asphalt in the container was about 200°C (392°F). The temperature of the interface was observed to rise almost immediately upon application of the hot asphalt to the felt facing. In about 30 s, it had reached about 95°C (203°F), and a maximum temperature of 118°C (245°F) was recorded from about 1½ to 2 min after asphalt application. The temperature of the interface was still above room temperature 10 min after asphalt application, at which time the temperature measurements were no longer recorded. In general, the bubbling in the hot applied asphalt had ceased about the time the interface temperature had reached a maximum.

In another series of experiments, the felt facings were removed from the as-received foam insulation specimens. Then asphalt at about 200°C (392°F) was applied directly to the surface of the foam. With two exceptions, application of the hot asphalt to the foam surface of the specimens without facings also resulted in bubbling and frothing of the asphalt. In most of these cases, the bubbling and frothing was so active that a crackling sound accompanied the off-

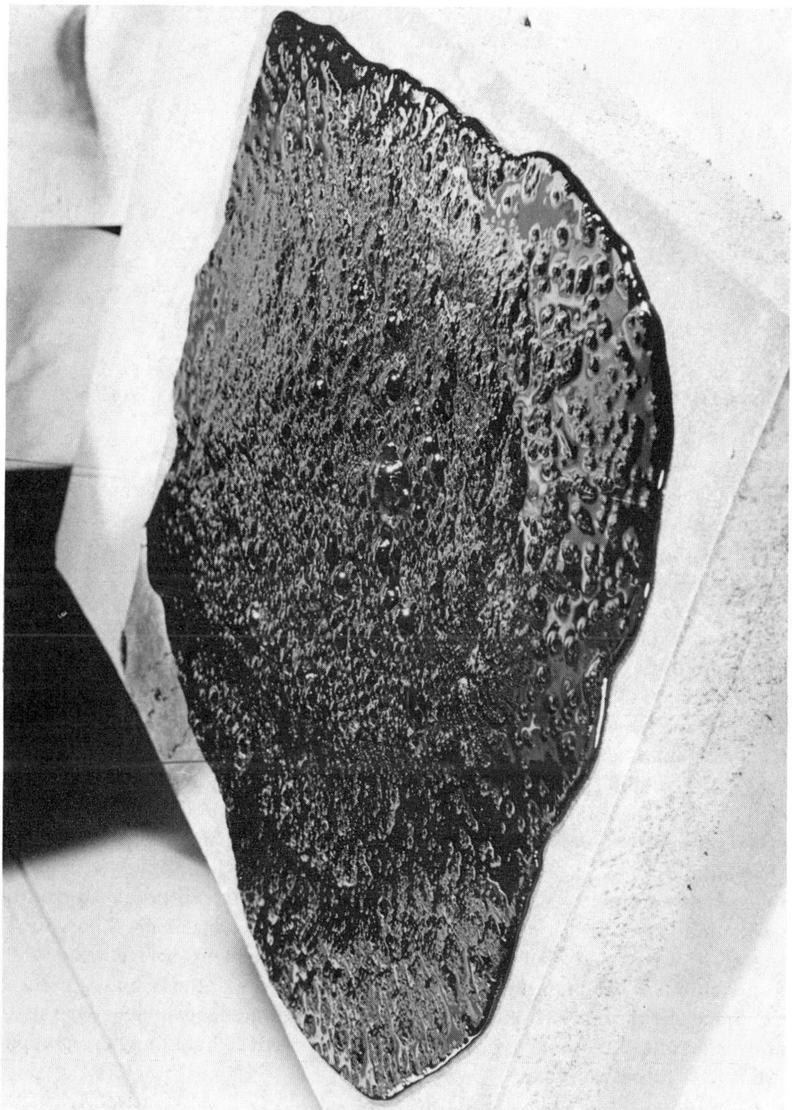

FIG. 1—*Bubbling and frothing of hot asphalt, which occurs when it is applied to a polyurethane foam insulation board.*

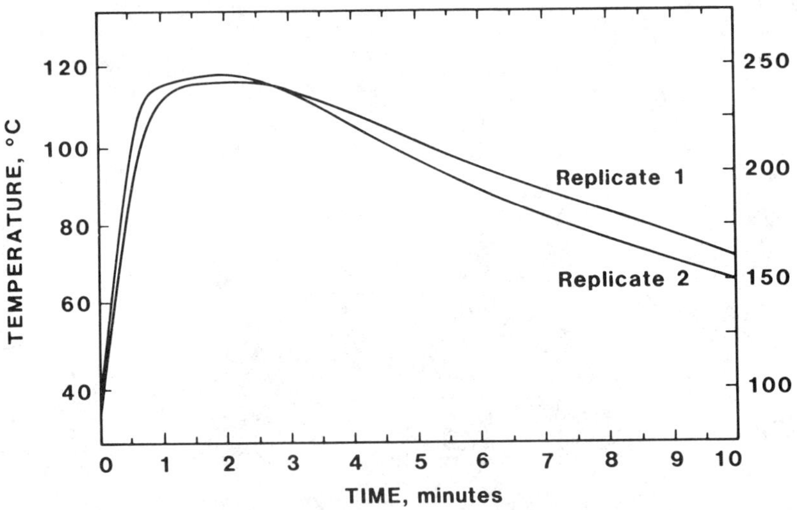

FIG. 2—*Plot of the temperature at the interface of the felt facing and polyurethane foam insulation versus time during which hot asphalt, at about 200°C (392°F), was applied to the felt facing.*

gassing phenomenon. The pattern of the bubbling was comparable to that observed when asphalt was applied to the felt facings intact on the foam surface. However, when asphalt was applied directly to the foam surface, the resulting bubbling and frothing were more active than when it was applied to the felt facings on the foam. This was consistent with the foam being a contributing cause of the off-gassing. However, when asphalt at 200°C (392°F) was applied directly to the surface of the foam, the surface temperature may have risen above the decomposition temperature of the foam. Polyurethane foams may start to decompose at temperatures between 150 and 200°C (302 and 392°F). The contribution to bubbling resulting from thermal decomposition of the polyurethane because of direct application of the hot asphalt onto the unfaced foam was not ascertained.

Specimens 7 and 8 exhibited only minor amounts of bubbling and frothing when hot asphalt was applied directly to their unfaced surfaces. In these cases, the bubbling was not at all comparable to that resulting from application of hot asphalt to the other foam specimens. In general, relatively few bubbles were observed in the asphalt, and no crackling sound accompanied the little bubbling that took place. As previously mentioned, Specimen 8 produced considerable bubbling when hot asphalt was applied directly to the felt-faced specimen. Thus, the lack of bubbling observed when hot asphalt was applied directly to the foam surface without felt indicated that the felt facing, and not the foam, was the major cause of the bubbling in the case of Specimen 8. Since this specimen (No. 8) was the only isocyanurate foam included in the study, it should not be generalized that such types of foams do not produce bubbling during application of asphalt directly on their surfaces.

A series of tests was also conducted to apply hot asphalt to the as-received felts that were removed from the insulation specimens. The aluminum/kraft paper on Specimen 6 was not included in this series of tests, since this facing could not be removed from the foam in one piece.

Application of asphalt at 200°C (392°F) to the as-received felts without foam produced variable results. The felt from Specimen 8 produced considerable bubbling, which was characterized as being typical of that observed when hot asphalt was applied to moist roofing felt. This observation was consistent with the observation described earlier that, in the case of Specimen

8, the felt facing was the major source of the bubbling. For Specimens 2, 3, 5, and 11, a minor amount of bubbling was observed when the hot asphalt was poured on felts. In these cases, the bubbling was considerably less vigorous than that observed when asphalt was applied to the same felts intact on the respective foam specimens. This indicated that for these specimens the presence of the felt had some contribution to, but was not the sole source of, the off-gassing. Felts removed from Specimens 1, 4, 9, 10, 13, and 14 produced essentially no bubbling of the hot asphalt. This indicated that for these specimens the felts were not contributors to the off-gassing observed when asphalt was applied to the felt-faced specimens.

Felt facings were removed from foam specimens and subjected to laboratory drying. The percentage of moisture by mass in the felts, as represented by the loss in mass after heating in the oven, is given in Table 1. It is apparent that the moisture content (percentage by mass) ranged from 0.7 to 5.6%. No relationship between the moisture in the felt and bubbling during hot asphalt application was developed in this study. Application of asphalt at 200°C (392°F) to the surfaces of the dried felts showed no significant bubbling of the asphalt for any of the specimens. Thus, for those felts that produced bubbling of the hot applied asphalt in the as-received condition, the bubbling was attributed to moisture (or other volatiles) present in the felts. Consistent with this observation, in practice, if the moisture content of the felt facings on foam boards was increased in relation to that determined for the as-received specimens, then the contribution of the moisture in the felts to the off-gassing would be expected to increase in comparison with that observed for many specimens tested in this study.

Asphalt at 200°C (392°F) was also applied directly to the unfaced polyurethane foam surfaces of specimens that were vacuum-oven dried. Specimens 6 and 8 were not included in this test. When hot asphalt was applied to Specimen 7, essentially no bubbling was observed. (This specimen also exhibited no bubbling when hot asphalt was applied to the surface of the as-received specimen.) For all other specimens, bubbling, lasting about 40 to 60 s and having various degrees of intensity, was observed when the hot asphalt was applied to their surfaces. This observation was consistent with one report [*17*] that, when hot asphalt was applied to an oven-dried foam specimen, gas was evolved during asphalt application. In the present study, for many of the specimens, the bubbling was observed to be slight. In others, considerable bubbling was observed. Nevertheless, for all the foam specimens, it was less active than when hot asphalt was applied to the surfaces of the corresponding as-received foams without facings. For the most part, the crackling sound produced upon application of hot asphalt to the as-received foams was absent when asphalt was applied to the vacuum-oven-dried specimens. These observations indicate that off-gassing still occurred after removal of moisture (or other volatiles) from the specimens. However, since the off-gassing was seen to be less active for dried specimens, it appeared that moisture (or other volatiles) contributed to some degree to the off-gassing of the as-received foams.

The moisture contents of the polyurethane foam portion of the insulation specimens for the as-received condition are given in Table 1. With one exception, the moisture contents ranged from 1.4 to 6.3% by mass. No relationship between the percentage of moisture in the foam insulation and the amount of bubbling was determined.

Thermogravimetric Analysis

Figure 3 presents the results of thermogravimetric analysis on selected foam specimens (Nos. 2, 8, and 9). The results indicate that the three specimens tested lost mass after being heated in nitrogen and showed varying degrees of thermal stability. The mass loss generally commenced at between 90 and 140°C (194 and 284°F), depending on the specimen, and in this range was about 1 to 4% for the three specimens. Mass loss increased as the temperature was raised. At 250°C (482°F), the mass losses for the three specimens—Nos. 2, 8, 9—were about 16, 7, and 23%, respectively.

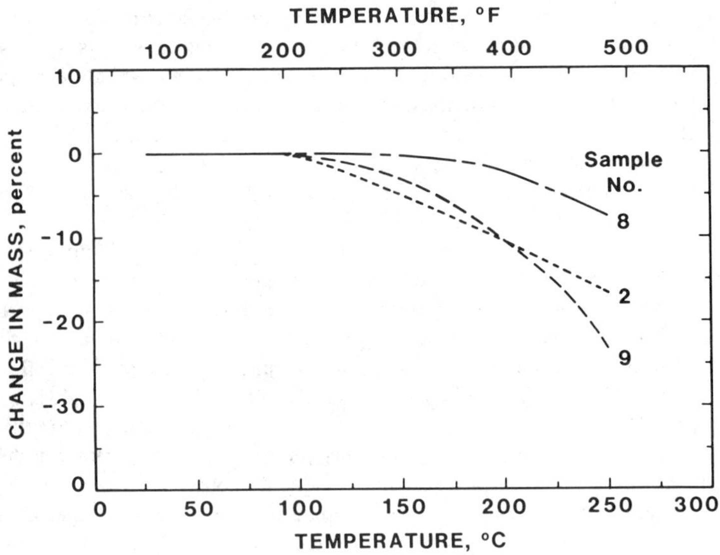

FIG. 3—*Results of thermogravimetric analyses of selected foam specimens (Nos. 2, 8, and 9).*

As was shown previously in Fig. 2, the maximum measured temperature of the foam/felt interface reached slightly less than 120°C (248°F) when hot asphalt at 200°C (392°F) was applied to the felt surface. The 120°C (248°F) temperature was then about midrange of the temperatures for which thermogravimetric analysis indicated the beginning of mass loss of the selected foam specimens. At temperatures about 100°C (212°F), moisture or other volatile materials may be driven from the foam. It has been reported that polyurethane foams are thermally stable at temperatures up to about 110°C (230°F), which is within the temperature range normally encountered in building use [*18*]. However, beyond 110°C (230°F), rupture of the cell walls of the foam may occur and be accompanied by loss of volatile components in the cells [*18*]. As previously indicated, many of these foams may start to decompose at temperatures between 150 and 200°C (302 and 392°F) [*19*].

In this study, as is evident from Fig. 3, rapid loss of mass did not in general begin to occur until the temperature reached about 140 to 150°C (284 to 302°F). This finding was consistent with the statements in the literature concerning the onset of thermal decomposition [*19*]. The results of the thermogravimetric analysis, coupled with the temperature measurements given in Fig. 2, provided evidence that at the asphalt temperatures employed in this study, thermal decomposition of the polyurethane foam polymer probably did not occur, and thus it did not contribute to the bubbling observed on the addition of hot asphalt to the faced surface of the foam boards. Although thermal decomposition of foam might occur and result in bubbling if the asphalt application temperature were sufficiently hot, the results of the study have indicated that thermal decomposition is not a prerequisite for off-gassing during asphalt application.

It is not improbable that release of air, fluorocarbon, or other gases from within the cells at the surface of the polyurethane foam might also be a contributing cause. Polyurethane foams are closed cell foams. When the hot asphalt strikes the foam surface, the gas in the cells heats and exerts pressure within the cells. If the temperature of the asphalt is sufficiently high, the pressure in the cells can increase to the point where the weak cell walls are ruptured [*18*]. At this point, the gaseous content of the cells would be released, causing bubbling within the hot asphalt. If cell wall rupture were a contributing mechanism, then foams that have relatively strong cell walls (and also are dry) should not undergo bubbling when hot asphalt is applied to their surfaces. A study of the off-gassing of the polyurethane foam insulations, as related to the strength and structure of the cell walls was beyond the scope of this study, but should be the subject of further study. Two foams (Nos. 7 and 8) in this present study did not produce active bubbling when hot asphalt was applied to their surfaces. The reasons for this observation were not investigated. Nevertheless, the densities of foam Specimens 7 and 8 (Table 1) were, with one exception, greater than the densities of the other specimens. Greater density could contribute to improved cell wall strength.

Delamination of Built-Up Roofing/Insulation Specimens

During the preparation of the four-ply membrane/insulation specimens, when hot asphalt was applied to the felt facing on the polyurethane foam insulations, extensive bubbling resulted. The first felt (or bottom ply) of the membrane specimen was placed onto the bubbling layer of hot asphalt (Fig. 4) and pressed in place using a hand broom. The brooming technique pushed the first ply down into the hot asphalt. However, after the brooming was stopped and hot asphalt was being applied to the top surface of the first ply of felt (in preparation of adding the second ply to the membrane specimen), the first ply of felt was pushed upwards because of the pressure created by the continued off-gassing of the polyurethane foam insulation. The second ply of felt was set into the hot asphalt and again pushed into place with the broom. The broom-

FIG. 4—*A ply of felt being placed onto the bubbling layer of hot asphalt.*

ing of the second ply also pushed the membrane flat onto the surface of the polyurethane insulation board. The third and fourth plies of felt were put in place in a similar manner, although the bubbling appeared to have stopped by this time.

During preparation of the membrane specimens, vigorous bubbling was observed only when hot asphalt was poured on the insulation boards. When the asphalt was applied to some of the felts used in membrane preparation, minor bubbling occurred, but it was noticeably less than that arising from the insulation boards. This off-gassing from the felts occurred primarily with both room-dry asbestos and organic felts, although a few bubbles were seen for oven-dried organic felts.

All the specimens were delaminated in the laboratory for observation. In all cases, extensive and relatively large voids were found in the interfacial asphalt layer (between the membrane and insulation board facing). These voids generally ranged in size from about that of a dime (17 mm, or 2/3 in., in diameter) to that of a nickel (21 mm, or 4/5 in., in diameter). In a few cases, the voids were estimated to be about 1300 to 1900 mm^2 (2 to 3 in.^{2}) in area. Figure 5 shows an example of the voids found at the interface of the foam insulation and built-up membrane in one specimen.

Delamination of the plies of felt for the four specimens indicated that some interply asphalt layers had a few small voids, particularly in the case of the room-dry felts. Other interply asphalt layers contained no voids at all. Where interply voids were observed, their size was usually less than 3 mm (1/8 in.) in diameter, although a few were as large as 6 mm (1/4 in.) in diameter.

This experiment of fabricating membrane specimens on polyurethane foam boards demonstrated that the off-gassing resulting when hot asphalt is applied directly to the polyurethane foam insulation board can increase the risk of blistering of the membrane. The large number of relatively large voids in the interfacial layer of asphalt was attributed to the off-gassing. These voids were found in the interfacial layer of asphalt in the laboratory specimens although each applied felt ply was broomed into the layer of asphalt onto which the felt was placed. In a similar experiment, fabrication of a two-ply membrane on a perlite board produced no bubbling of the applied asphalt, and no voids were observed in the interfacial asphalt layer.

In this laboratory study, the membrane specimens were fabricated in sandwich fashion (Fig. 6*a*). Thus, all the voids attributed to off-gassing of polyurethane foam were found in the interfacial layer of asphalt. This might indicate that, in practice, blisters resulting from the voids would be at the interface of the insulation and membrane. However, in practice, most built-up roofing membranes are currently fabricated in shingle fashion (Fig. 6*b*). With a specimen fabricated in shingle fashion, some of the voids created by off-gassing may occur at the overlap of the plies (Fig. 6*b*). The possibility exists that the voids at the ply laps could grow between plies to form interply blisters.

Composition of the Gas Sampled from Roof Blisters

The results of the gap sampling of twelve blisters in existing built-up membranes fabricated on polyurethane insulation boards are given in Table 2. Although water vapor is difficult to quantify by the methods employed, it was estimated that each sample contained between 1 to 1.5 mole percent of water vapor.

As is evident from Table 2, the major gaseous components of the blisters were nitrogen and oxygen, with some argon, carbon dioxide, and fluorocarbon 11. In general, the gas composition was comparable to that of ordinary air, although in some blisters the oxygen content was less than that of ordinary air. The presence of fluorocarbon 11 (which was generally 1 mole percent or less) was attributed to its release from the foam insulation. Since the gases in the blisters were analyzed a year or more after membrane fabrication, it could not be determined whether the fluorocarbon entered voids during membrane construction, because of off-gassing from the foam, or at a later time, because of diffusion from the foam. In investigating the causes of these

FIG. 5—*Voids observed at the interface of the foam insulation and built-up membrane.*

blisters, in many cases, it was not possible to assign a cause. In a few cases, skips (areas of felt without asphalt) in the applied asphalt were seen, which could have been initial voids that subsequently grew into blisters.

Summary and Conclusions

This study was conducted to investigate the bubbling that results when hot asphalt is applied to the surface of polyurethane foam insulation boards. Commercially available insulation

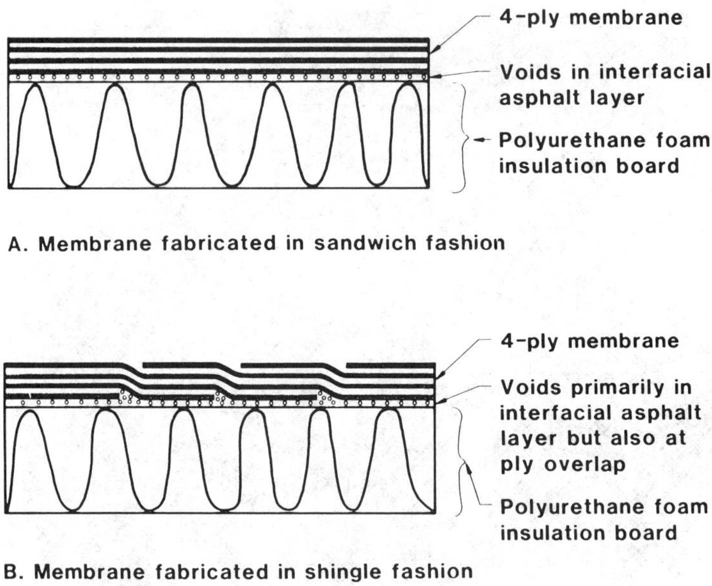

FIG. 6—*Voids occurring in the interfacial asphalt layer, as shown for membranes fabricated in sandwich fashion* (a) *and in shingle fashion* (b).

TABLE 2—*Composition of the gas within roof blisters.*

	Composition, mole %, by Specimen No.[b]											
Gas Component[a]	1	2	3	4	5	6	7	8	9	10	11	12
N_2	81.1	80.2	79.0	84.6	79.9	84.2	80.3	81.7	77.9	86.0	85.7	81.3
O_2	17.5	18.4	19.8	13.6	19.0	14.6	17.7	17.0	19.8	11.8	11.4	17.3
Argon	0.9	0.9	0.9	0.9	0.9	0.9	0.9	0.9	0.9	0.9	0.9	0.9
CO_2	0.30	0.08	0.16	0.60	0.11	0.16	0.14	0.10	0.24	1.2	1.5	0.18
Fluorocarbon 11	0.25	0.42	0.20	0.36	0.05	0.10	0.96	0.3	1.2	0.06	0.50	0.34

[a]Water was estimated to be in the range of 1 to 1.5 mole percent for all the specimens.
[b]Error was estimated to be ±20%.

boards, the majority of which had felt facings, were obtained, and hot asphalt was applied directly to their surfaces. This generally resulted in extensive bubbling. Little bubbling was observed for one specimen. Based on the results of the study, the following conclusions were drawn.

Cause of Bubbling

It was postulated that the observed bubbling might be due to (1) the release of moisture from the surface of the foam insulation, (2) the release of moisture from the facing on the foam, (3) the release of air, fluorocarbon, or other gases within the cells of the foam, or (4) the release of gaseous decomposition products from the foam. Based on the results of the study, the cause of the bubbling appears to be a combination of Reasons 1, 2, and 3, depending on the circumstances. In some cases, moisture on the surface of the foam or within facings was the cause or at least a contributor. Release of air, fluorocarbon, and other gases from within the cells of the foam was also considered to be a contributor to the off-gassing. This conclusion was based on the observation that vacuum-oven-dried foam without facings produced some bubbling during application of hot asphalt to their surfaces. Definitive evidence on the release of air, fluorocarbon, and other gases from within cells was not obtained in this study. At temperatures above 110°C (230°F), rupture of the cell walls may occur and result in the loss of volatile components in the cells. For the asphalt temperatures in the study, thermal decomposition of the foam was eliminated as a major contributor to off-gassing, since the off-gassing was shown to occur at relatively low temperatures (120°C, or 248°F, or less). These temperatures are less than those that result in thermal decomposition, which usually begins at temperatures above 150°C (302°F).

With regard to moisture in the facings as a cause of off-gassing, the foam specimens with the facings removed were shown to produce bubbling when asphalt was applied. Moreover, in many cases, the facings removed from the foam produced no bubbling when asphalt was applied. Thus, the facings in such cases could not be considered a cause of the off-gassing from the faced insulation boards. In contrast, for one specimen, the foam with its felt facing removed produced little bubbling, but the felt facing itself produced vigorous bubbling. In some cases, minor bubbling from the removed felt facings was observed and attributed to moisture (or other volatiles) in the facings. In these cases, if the felt facings were on foam boards, the moisture in them would contribute to the off-gassing.

With regards to moisture on the surface of the foam, this was considered to be a contributing factor but not the sole cause of the off-gassing. Drying of foam specimens in a vacuum oven did not eliminate the bubbling, but did reduce its extent.

Risk of Blistering

During the study, four-ply built-up membrane specimens were fabricated on polyurethane foam insulation boards using hot asphalt and felts. Delamination of these membrane/insulation specimens indicated that extensive and relatively large voids were formed at the membrane-insulation board interface because of off-gassing. Since blistering has been attributed to the growth of voids incorporated in the layer of applied asphalt, the results of the study have indicated that off-gassing of the polyurethane insulation boards increases the risk of blistering of built-up roofing membranes applied directly onto the boards. Experience has shown that not all voids in built-up roofing membranes result in blisters [*20*], and also that many built-up roofing membranes over polyurethane foam have performed satisfactorily without excessive blistering. In this regard, in a field test conducted by the Roof Insulation Committee of the Thermal Insulation Manufacturers Association (RIC/TIMA), a roof containing 14 different insulation boards, including polyurethane foam, fibrous glass, and perlite, and having three different built-up membranes directly on the insulation showed essentially no blistering after one year of service [*21*]. However, because application of built-up roofing directly on a polyurethane foam board can result in an increased risk of blistering, steps need to be taken to minimize the risk.

The recommendation of NRCA [*10*], given in conjunction with TIMA, is that a thin layer of rigid board insulation be applied with hot asphalt to the polyurethane foam before membrane application, or that a base ply be installed to the polyurethane to allow for venting. The data from this study support the part of the recommendation dealing with the application of the rigid board. The use of a base ply to minimize blistering was not addressed in this study. When a rigid board insulation is used, the membrane will not be applied directly to the bubbling layer of asphalt resulting from off-gassing of the polyurethane foam insulation. If voids are present in this asphalt between the thin layer of rigid board insulation and the polyurethane, blistering will be inhibited at this juncture because of the rigidity of this layer of insulation.

Acknowledgments

This investigation was conducted under the Tri-Service Building Materials Investigational Program and was jointly sponsored by the U.S. Army Corps of Engineers Headquarters; U.S. Navy, Naval Facilities Engineering Command; and U.S. Air Force, Air Force Engineering and Services Center. The authors appreciate the support of these services. The authors wish to thank their colleagues at the National Bureau of Standards who assisted in the study: Jessie Hairston provided assistance in the preparation and testing of laboratory specimens; Mike Manka conducted thermogravimetric analyses of foam insulation specimens; William Dorko provided analyses of the gaseous composition of blisters in built-up roofing membranes.

References

[1] Busching, H. W., Mathey, R. G., Rossiter, W. J., Jr., and Cullen, W. C., "Effect of Moisture in Built-Up Roofing—A State-of-the Art Literature Survey," Technical Note 965, National Bureau of Standards, Gaithersburg, MD, July 1978.

[2] "Good Roofs Save Energy," *The Roofing Spec,* July 1976, pp. 13–15.

[3] Robinson, J. N., "The Assessment of Roofing Research, An Interim Report," ORNL/TM-7640, Oak Ridge National Laboratory, Oak Ridge, TN, July 1981.

[4] Griffin, C. W., "Impact of Roof Insulation Thickness on Life-Cycle Costing of Built-Up Roofing Systems," Paper No. 18, *Proceedings,* Symposium on Roofing Technology, National Bureau of Standards and National Roofing Contractors Association, Rosemont, IL, September 1977, pp. 134–143.

[5] Cash, C. G., "Optimization of the Thermal Resistance of Roof Insulation in Reroofing," Paper No. 24, *Proceedings,* Symposium on Roofing Technology, National Bureau of Standards and National Roofing Contractors Association, September 1977, pp. 208–214.

[6] Courville, G. E., and Kolb, J. O., "Economic Analysis of Insulation Materials Used in Low-Slope, Built-Up Roof Systems," Report ORNL/TM-9004, Oak Ridge National Laboratory, Oak Ridge, TN, June 1984.

[7] "Pinpoint Finds Urethane Problems," *The Roofing Spec,* May 1980, pp. 28–31.

[8] Mirra, E., "Blistering Phenomenon and Urethane Insulation," *Roofing/Siding/Insulation,* December 1976, pp. 58–61, 71.

[9] Interim Alert Bulletin No. 4, The National Roofing Contractors Association, Oak Park, IL, 10 Aug. 1978.

[10] Bulletin No. 7, The National Roofing Contractors Association, Oak Park, IL, 1 March, 1979.

[11] "Tests Show Urethane Blistering Due to Moisture Release," *The Roofing Spec,* April 1980, pp. 10 and 11.

[12] "Blisters Over Urethane," *The Roofing Spec,* May 1980, pp. 18–24, 26.

[13] Porcher, H. "Blistering Over Urethane," *Proceedings,* Sixth Conference on Roofing Technology, Gaithersburg, MD, National Bureau of Standards, National Roofing Contractors Association, 30 April–1 May, 1981, pp. 46–49.

[14] Warden, W. B., "A Theory of the Mechanism of Blistering" *A Study to Improve Bituminous Built-Up Roofs,* Monograph No. 1, Building Research Institute/National Academy of Sciences/National Research Council, Washington, DC, 1960, pp. 14–26.

[15] Griffin, C. W., *Manual of Built-Up Roof Systems,* 2nd ed., McGraw-Hill, New York, 1982, pp. 284–303.

[16] Laaly, H. O., Dutt, O., and Feldman, R. F., "Interply Adhesion and Blistering of Bituminous Built-Up Roofing Membranes," *Durability of Building Materials,* No. 2, 1985, pp. 209–228.

[17] "Discussion of Session No. 15," *Proceedings of the Second International Symposium on Roofs and Roofing,* J. O. May, Ed., Vol. II, Society of the Chemical Industry, London, 1981, p. 149.

[18] Roux, G., Balme, A., Ghaleb, M., and Rivero, J., "Les Isolants Thermiques en Polyurethanne Cellulaire: Methodes d'Identification, Durabilité," *Cahiers du Centre Scientifique et Technique du Batiment,* No. 1927, May 1984, p. 8.

[19] Fabris, H. J., "Thermal and Oxidative Stability of Urethanes" *Advances in Urethane Science and Technology,* K. C. Frisch and S. L. Reegan, Eds., Vol. 4, Technomia, Westport, CT, pp. 89–111, 1976.

[20] Jennings, F. D. and Jennings, P. E., "Preventing Blistered Built-Up Roofs," *Proceedings of the Second International Symposium on Roofs and Roofing,* J. O. May, Ed., Vol. I, Society of the Chemical Industry, London, 1981, pp. 368–376.

[21] "TIMA Test: BUR Is Blister Free," *Roofing/Siding/Insulation,* February 1982, p. 112.

George E. Courville, [1] *Kenneth W. Childs,* [1] *Donald J. Walukas,* [1] *Philip W. Childs,* [1] *and Edwin I. Griggs* [2]

An Apparatus for Thermal Performance Measurements of Insulated Roof Systems

REFERENCE: Courville, G. E., Childs, K. W., Walukas, D. J., Childs, P. W., and Griggs, E. I., **"An Apparatus for Thermal Performance Measurements of Insulated Roof Systems,"** *Thermal Insulation: Materials and Systems, ASTM STP 922,* F. J. Powell and S. L. Matthews, Eds., American Society for Testing and Materials, Philadelphia, 1987, pp. 449–459.

ABSTRACT: The U.S. Department of Energy is conducting thermal performance measurements on low-slope roofs with a recently developed field test apparatus at Oak Ridge National Laboratory (ORNL). The apparatus accommodates four 1.2 by 2.4-m (4 by 8-ft) test specimens and includes the measurement capabilities for specimen temperatures, temperature gradients, heat flows, and moisture content. A weather station characterizes outdoor weather conditions. Tests underway include (1) validation of a roof surface temperature model developed to study the effects of wet insulation, (2) measurement of temperature distributions and heat transfer in high *R* value roofs, and (3) validation of an analysis of the effectiveness of high reflectance surfaces. Preliminary experimental results are presented and correlations between experiment and modeling are discussed.

KEY WORDS: roofs, thermal performance, *R* value, roof reflectance, dynamic performance, thermal insulation

In situ measurements of thermal performance of roof systems are complicated by difficulties with placement of temperature and heat flow sensors in already-built structures, with calibrating these devices, and with putting in place an adequate data acquisition system. To overcome these difficulties and provide accurate measurements of the thermal performance of roof systems, the U.S. Department of Energy has established a roof thermal research apparatus (RTRA) at Oak Ridge National Laboratory (ORNL) [*1*]. The goals of this work are to provide reliable thermal and hygric measurements of well-characterized roof systems under field conditions, to validate mathematical models that accurately describe roof thermal performance so that they can be extrapolated to all environments and will be appropriate for a wide range of potential roof materials, and to substantiate energy-saving attributes and identify potential issues related to energy-efficient roof systems.

The RTRA has been designed to address many of the energy-related research needs identified in a recent report entitled "Assessment of Roofing Research" [*2*]. These needs include moisture detection techniques, in-place *R* value measurements, effects on thermal performance of color, reflective coatings, roof mass, and moisture migration. Accurate performance measurements of roof systems with different types and amounts of insulation will demonstrate the effectiveness of insulated roof decks in reducing building energy useage. This will provide baseline data for

[1]Staff scientists and senior technician, respectively, Oak Ridge National Laboratory, Oak Ridge, TN 37831.

[2]Department of Mechanical Engineering, Tennessee Technological University, Cookeville, TN 38505.

determining the economic optimum thickness of insulation and for evaluating the relative performance of different insulation systems under field conditions. Measurements from the RTRA will also be available for correlating the results of field measurements with laboratory measurements and with analytic simulations. Although preliminary efforts will be focused on one-dimensional problems, more complex issues such as the effect on heat transfer of various joint configurations will be considered.

Techniques for measuring the amount and location of moisture within roof systems are also being developed for use at the RTRA. This will allow detailed studies of the effect of moisture on thermal performance and will provide data on moisture movement within roofs.

The Building

The apparatus is shown in Fig. 1. It is housed in a 2.4-m-wide by 7.9-m-long by 2.7-m-high (8-ft-wide by 26-ft-long by 9-ft-high) building with a concrete slab-on-grade floor and concrete block walls. The roof has a conventional built-up roof (BUR) center section with two 1.2 by 2.4-m (4 by 8-ft) test panels on either side. A walkway near the roof level is mounted along the back and both ends of the building so that inspection of the panels is facilitated.

The interior temperature of the RTRA is controlled by a through-the-wall heat pump, and a humidifier is available to control humidity during the winter. Plans include the addition of exterior wall test panels to study exterior wall insulations.

Test Panel Construction

Test panels are assembled by roofing mechanics and instrumented by project staff in a building near the test facility. Each experimental panel is designed to approximate a "real" roof segment. Panels are mounted in special steel angle frames to facilitate transportation and installation. Conventional wood nailers are mounted in the frames for membrane and perimeter flashing attachment. Steel decks are being used in current panels, although other types of decks can be accommodated and will be tested at a later date. The initial experiments are concerned primarily with the properties of board insulations, which are covered with conventional three-ply BUR membranes. Other insulations and other membranes will be incorporated into later tests.

A computer analysis of installed test panels has shown that edge temperature effects are negligible within the central 0.6 by 1.8-m (2 by 6-ft) region of each test panel. This has been confirmed by thermocouple measurements and surface infrared scanning. Thus, thermal experiments carried out in this region will be typical of field conditions.

Material Characterization

Insulation layouts are designed around separate 0.9 by 1.5-m (3 by 5-ft) sections placed in the center of the panel. Precise laboratory density and thermal conductance measurements are carried out on these sections. Full-thickness steady-state conductance measurements are made using a unique hot screen apparatus in use at ORNL [*3*]. This apparatus has an accuracy of 1.5% traceable to standard reference materials issued by the National Bureau of Standards. When heat flow measurements are required, heat flux transducers are embedded in the insulation stack. Heat flux transducers are calibrated simultaneously with these laboratory conductance measurements.

Instrumentation

Most temperature measurements are made using copper-constantan thermocouples. These measurements typically are made at the panel surface, at various points within the interior of a

FIG. 1—*The U.S. Department of Energy roof thermal research apparatus at Oak Ridge National Laboratory.*

panel, and at several elevations within the building. Comparative temperature measurements of the panel surface can also be made with a two-dimensional infrared scanner. This equipment includes a camera, a cathode ray tube (CRT) monitor, a video and audio tape recorder, and a playback unit. Both monochromatic and color capabilities are available and are used.

Moisture measurements are currently made with a set of load cells (sensitivities of 0.25%) for continuous weighing of the test panels (four load cells are used to determine the total panel weight and the center of gravity of a panel). Qualitative moisture measurements are obtained by using an infrared imager or a capacitance-based device. Plans for instrumentation include a quantitative, electrostatic capacitance probe (under development).

A conventional recording weather station is available for characterization of the outdoor climatic conditions. Specific measurements include outdoor ambient temperature, relative humidity, barometric pressure, wind speed, and wind direction. Other quantities measured include sky radiation, solar incidence, and outdoor panel surface reflectance.

Data Acquisition System

A major feature of the data acquisition system is the use of multiplexing units into which many sensors are inputted and out of which a single line moves data to the data acquisition unit in the control building. The current system, which is expandable, can handle 100 inputs.

All measurement transducers input to basic transducing/multiplexing units, each of which can handle 20 channels. The information from five units is again multiplexed and sent over one wire to a data acquisition system (DAS). This system performs some of the data reduction, for example, making hourly averages and applying engineering units. The DAS is hard-wired to a DEC PDP 11/34 computer for data storage. This unit, in turn, is accessible from several IBM personal computers, with which further data reduction and analyses can be performed. The personal computers also function as terminals for even larger mainframes available at ORNL, which can perform data reduction and analysis on a larger scale.

Experiments

The RTRA has been operational since July 1984. Three experiments are currently underway. These are (1) validation of an ORNL mathematical model for predicting roof surface temperature with wet insulation under the membrane, (2) comparison of the thermal performance of roof systems with different insulation densities (thermal storage effects), and (3) investigation of the energy-savings potential of high reflectance surfaces.

Additional work will include validation of a comprehensive thermal/hygric/structural mathematical model under development at the University of Illinois [*4*], continuance of mass effect studies (extending to different deck systems and to insulating concrete), and analysis of the effect of wet insulation and of misting systems on thermal performance.

Results

Roof Surface Temperature

ORNL is carrying out an analysis of various factors that have influenced the use of infrared imagers in detecting moisture in roof systems [*5*]. In such work, the relevant parameter is the difference in membrane surface temperature over wet and dry (less wet) insulation sections. This difference depends on the amount of moisture, the inside/outside temperature differences, the solar isolation, wind, and several other variables. The ORNL analysis is an attempt to quantify these effects so that their influence on surface temperature can be predicted. The RTRA is used to validate the results of the analysis.

Two nearly identical test panels are placed side by side in the RTRA. Both are BUR membranes (no ballast) over nominal R-2.6 (15) fibrous glass board insulation on a metal deck. Provisions have been made for varying the water content (only total water, not distribution, can be monitored at this time). Figures 2, 3, and 4 show some preliminary results. Figure 2 is a comparison of the measured surface temperature on a dry panel with the calculated temperature, using the ORNL model. Two points should be made. First, the overall agreement is good, particularly during evening hours, which are most relevant for this study because infrared viewing is carried out during these times. Second, the disagreement during the day has been shown to be caused mostly by local wind effects during times when the surface-to-air temperature difference is large. Two possible causes are under investigation. One is a difference in wind speed at the surface compared with measured wind speed 1.2 m (4 ft) above the surface, and the other is an error in the algorithm for the surface convective heat transfer coefficient.

Figure 3 compares the temperatures on the two adjacent panels. Note that they follow one another closely, especially during the evening hours. The next step in this work is to insert known amounts of water into one of the panels and compare measured and predicted values of surface temperature, particularly in the evening hours. Agreement will constitute validation of the ORNL model, which provides confidence in its subsequent use to generate guidelines of infrared scanning.

In a first attempt, a known amount of water was inserted into the insulation of Panel 1, and Panel 2 was kept dry. Water is introduced into Panel 1 through a pipe that was flashed into one corner of the panel. The tube extends to the bottom layer of insulation. Lateral spread of the water is checked by monitoring changes in the center of gravity of the panel. Figure 4 shows the results. Here, thermocouples are placed between insulation boards on both panels, and the temperature differences between boards are recorded. Before one panel is wetted, the ratio of temperature differences for similarly placed thermocouples in the two panels should equal unity because, as was previously shown, the panels are similar. This is confirmed by experiments, as can be seen in the small graphs in the upper right of Fig. 4. When water is added to Panel 1 (5% by volume), the effect is to drive down the temperature difference across the lowest board in Panel 1. Note that the other temperature differences are not affected. This result suggests that the bulk of the added water moved to the bottom of the panel, and therefore has little effect on the temperature at the surface. Subsequent tests will involve different amounts and distribution of water in the panel.

Mass Effects of Roof Insulation

Analysis of heat storage effects has not progressed far enough to provide quantitative results. Preliminary results, shown in Fig. 5, suggest the complexity of the problem and also illustrate the hazard in attempting to use simple, steady-state equations for calculating heat flow. In this case, a vertical array of thermocouples is installed in a four-board stack of fibrous glass under BUR (each board is 0.02 m thick). The points connected by the solid lines represent the measured temperature as a function of position in this stack at several times during a day (1 Sept. 1984). The dashed line connects temperatures measured at the top and bottom of the stack. In a steady-state calculation ($Q = \Delta T/R$), the dashed temperature distribution is assumed. The figure clearly indicates that substantial errors exist between this assumed temperature distribution and the actual distribution. Also note, for example, that at 9:00 P.M. and at midnight the steady-state assumption predicts heat flow from the building, whereas the actual temperature gradient near the inner surface suggests that heat is flowing from the insulation into the build-

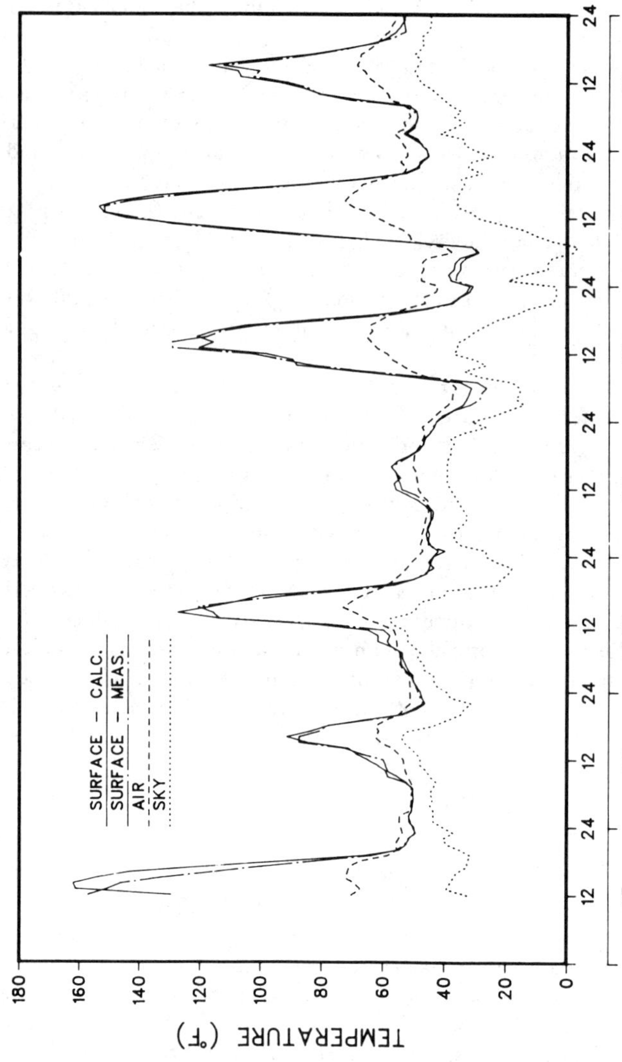

FIG. 2—*Measured versus calculated roof surface temperatures. The test panel was nominal R-2.6 (15) fibrous glass with a three-ply BUR membrane. The surface coating was an asphalt flood coat.*

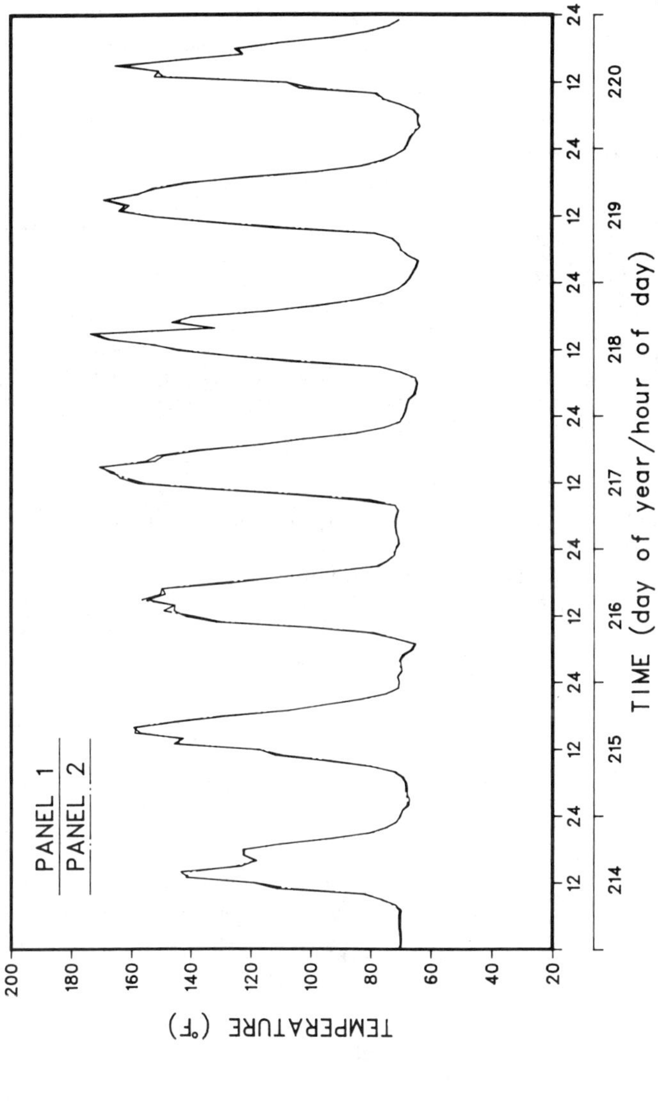

FIG. 3—*Measured roof surface temperatures on two adjacent test panels over a seven-day period.*

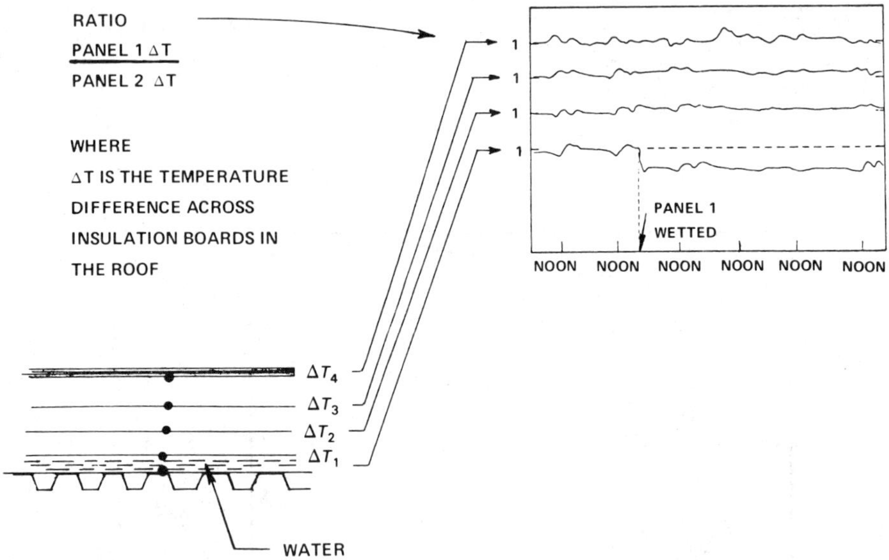

FIG. 4—*Effect of wetting the insulation of one of a pair of similar test panels. Before wetting, the ratio of similarly placed thermocouples was unity; after wetting, changes in the thermocouple readings suggest that most of the water moved to the bottom of the system.*

ing (as well as to the air). The reverse of this is true at noon. This illustrates the heat storage property of the insulation.

One expression commonly used to characterize the thermal performance of *in situ* building components involves the quantity

$$R_n \equiv \sum_{i=1}^{n} \Delta T_i \Big/ \sum_{i=1}^{n} q_i$$

in which R_n is the ratio of the average temperature

$$\frac{1}{n} \sum_{i=1}^{n1} \Delta T_i$$

to the net heat transfer

$$\frac{1}{n} \sum_{i=1}^{n1} \Delta q_i$$

after n hours [*6*]. It has been shown that R_n approaches the steady-state R value of the component when n is large compared with the component time constant, measured in hours, and where the temperature variation is periodic. This calculation can be readily carried out with data from RTRA. Figure 6 shows the result from a one-week period in October 1984 for a roof

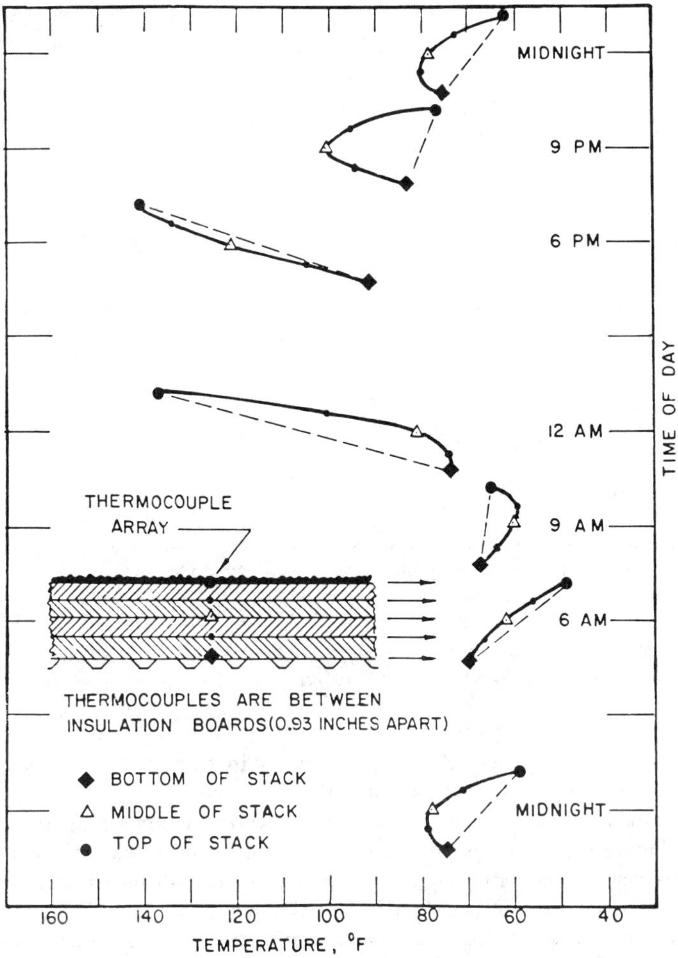

FIG. 5—*A series of temperature versus vertical position measurements taken at different times during the day. The measurements were taken between 0.2-m fibrous glass insulation boards. The dashed lines connect temperature readings at the top and bottom of the insulation stack. They represent the temperature distribution if the roof temperature is assumed to be in a steady state.*

system with a measured steady-state R value of 14.7 ft^2 · h · °F/Btu. Here, the time variation of R_n shows strong variations initially, with damped oscillations converging to a finite value. This limiting value, 14.3 ft^2 · h · °F/Btu, is within 3% of the laboratory value. Work on this effect is continuing in order to assess the significance of different roof parameters and different environmental conditions.

Thermal Effect of Surface Absorptance

ORNL has also initiated a study of the effects of variations in solar reflectance on the thermal performance of roof systems. This study has three objectives: (1) to validate and apply the ORNL surface temperature model to predict the impact of different reflectances under arbitrary climatic conditions, (2) to use the DOE2.1 whole building mathematical model to assess

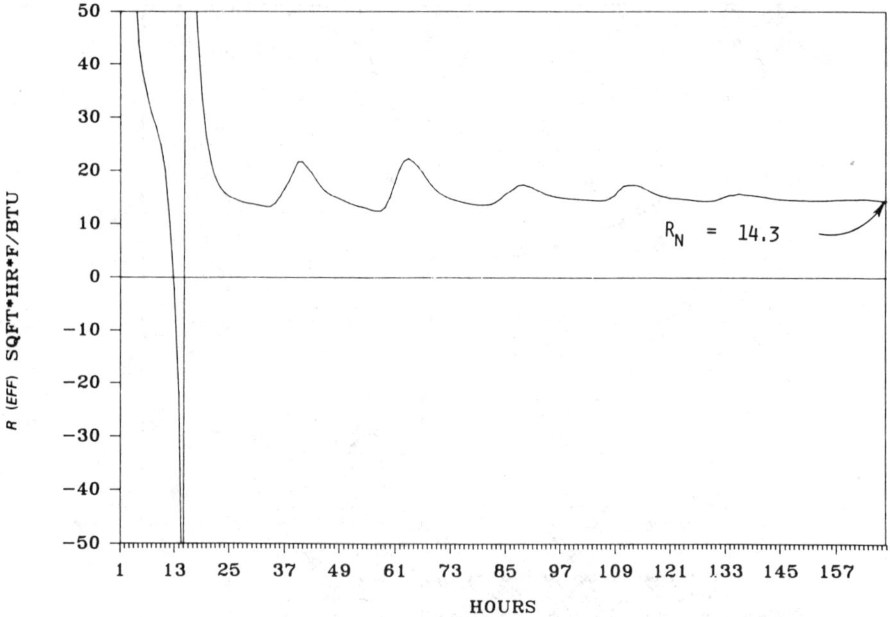

FIG. 6—R_{eff}, *defined as the ratio of the average temperatures to the net heat flow after* n *hours, for 16–22 Oct. 1984. The curve is a cumulative history of this quantity.*

the impact of varying reflectance on the cost of operating roofs, and (3) to supplement existing data on the solar reflectance and thermal emittance of a range of roof systems.

The experimental apparatus is being used to validate certain aspects of the surface temperature model. In Figure 7 we have plotted temperature versus time for two roof sections, one with an asphalt flood-coat surface (reflectance $\simeq 5\%$) and the other with aluminum flakes sprinkled in the flood coat (reflectance $\simeq 50\%$). Note that the aluminum flakes clearly result in a decrease in daily temperatures. Also note that the long wave emittance also is decreased for these flakes because the evening temperatures on the treated surface are above those on the untreated surface.

Summary

Oak Ridge National Laboratory has established a roof thermal research apparatus for carrying out thermal and hygric experiments on sections of low-sloped roofs. The test panels are exposed to a controlled-temperature interior space and to the prevailing East Tennessee exterior environment. They are well instrumented and all data is stored and is aided in the analysis by computer systems.

Current experiments include studies of the effect of wet insulation on membrane temperature, thermal storage phenomena in built-up roof insulation, and the effects of varying surface reflectance on roof thermal performance.

Acknowledgment

The research for this paper was sponsored by the Office of Building Energy Research and Development, Building Systems Division, U.S. Department of Energy, Under Contract DE-AC05-840R21400 with Martin Marietta Energy Systems, Inc.

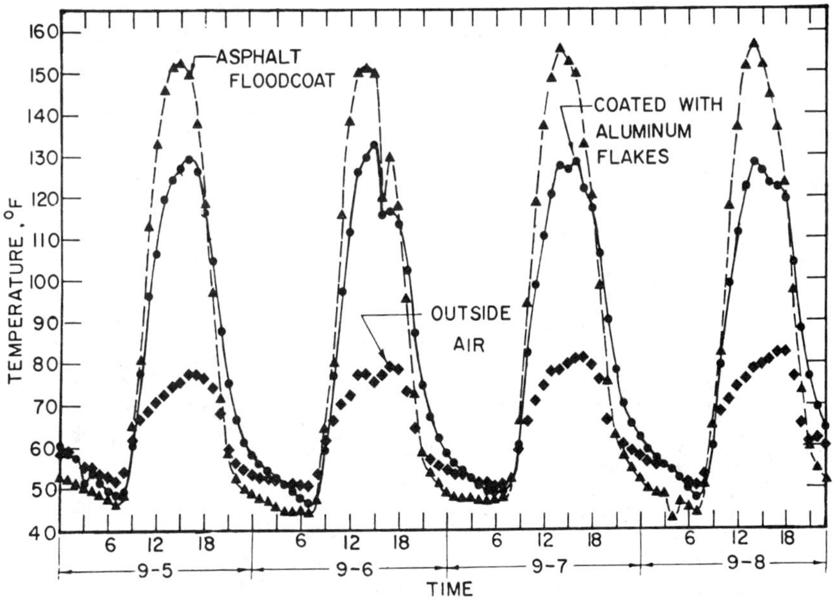

FIG. 7—*A comparison of surface temperature measurements for a high-absorptance surface (asphalt flood coat) and a low-absorptance surface (aluminum flakes) blown into a flood coat.*

References

[1] Courville, G. E., Sanders, J. P., Sr., and Childs, P. W., "Dynamic Thermal Performance of Insulated Metal Deck Roof Systems," *Proceedings*, Third ASHRAE/DOE/BTECC Conference on Thermal Performance of the Exterior Envelopes of Buildings, Clearwater Beach, FL, 2-5 Dec. 1985.

[2] Robinson, J. N., "The Assessment of Roofing Research," ORNL/TM-7640, Oak Ridge National Laboratory, Oak Ridge, TN, July 1981.

[3] Graves, R. S., Yarbrough, D. W., and McElroy, D. L., "Apparent Thermal Conductivity Measurements by an Unguarded Technique," *Proceedings*, 18th International Conference on Thermal Conductivities, Rapid City, SD, 2-6 Oct. 1983, Plenum Press, New York, 1983.

[4] Dempsey, B. J. and Haber, R. B., "Mathematical Modeling of Whole Roof System Performance," University of Illinois, Urbana, IL, June 1984.

[5] Childs, K. W. and Courville, G. E., "An Investigation of Factors Influencing Infrared Moisture Surveys Using a Mathematical Model," *Proceedings of the Society of Photo-Optical Instrument Engineers*, Vol. 446, 1984, p. 82.

[6] Flanders, S. N., "Time Constraints on Measuring Building R-Values," CRREL Report 80-15, U.S. Army Cold Regions Research and Engineering Laboratory, Hanover, NH, June 1980.

Industrial Applications

Vernon Batdorf[1]

Anomalous Behavior of Water Vapor Retarders Applied to Spray-Applied Polyurethane Foam Insulation on Low-Temperature Outdoor Storage Tanks

REFERENCE: Batdorf, V., "**Anomalous Behavior of Water Vapor Retarders Applied to Spray-Applied Polyurethane Foam Insulation on Low-Temperature Outdoor Storage Tanks,**" *Thermal Insulation: Materials and Systems, ASTM STP 922,* F. J. Powell and S. L. Matthews, Eds., American Society for Testing and Materials, Philadelphia, 1987, pp. 463–474.

ABSTRACT: The diffusion of water vapor through permeable retarders applied over thermal insulation in low-temperature service is a function of vapor pressure differentials. For permeable insulation materials of a fibrous nature, with an installed vapor retarder on the warm side, the annual intake of water by vapor diffusion is readily calculated.

It has been observed that for sprayed-on polyurethane foam insulation, the actual intake is substantially less than calculated. Field data from eight and eleven-year-old installations of spray-applied polyurethane foam, covered with vapor-retardant mastic on large storage tanks containing agricultural liquid ammonia held at −33°C (−28°F), show that the calculated level of water intake was not attained and that the insulation material continues to perform without significant change in thermal transmission.

The author hypothesizes that the cellular nature of the insulation restricts and changes the diffusion mechanism and that the absence of free air spaces within the insulation attenuates moisture diffusion paths.

KEY WORDS: vapor retarder coatings, polyurethane foam, water vapor diffusion, thermal insulation, foam service life

The amount of water vapor diffused through retarder mastics or coatings applied over thermal insulation in low-temperature service is of paramount importance to the insulation design engineer. Moisture absorbed by the insulation will result in a significant loss in thermal resistance of the insulation, which in turn will greatly increase the operative cost of cooling. If moisture permeates into the insulation over a long period of time, sufficient moisture may build up to cause complete failure of the insulation system and corrosion of the substrate. To prevent this from happening, moisture vapor retarder coatings are currently being used to protect insulation in low-temperature service. Typically, these moisture vapor retarders will have a permeance of less than 0.1 perm when tested at 37.8°C (100°F), 90% relative humidity differential, in a Mocon Permatran-W1.[2] This method gives close agreement with permeance values obtained by either the ASTM Test for Water Vapor Transmission Rate of Sheet Materials using a Rapid Technique for Dynamic Measurement (E 398-83) or the ASTM Test for Water Vapor Trans-

[1]Technical director, Foster Products Division, H. B. Fuller Co., Vadnais Heights, MN 55110.

[2]The Permatran-W1 employs an infrared sensor to measure the amount of water vapor diffusion through a test film. The instrument is manufactured by Modem Controls, Inc., Minneapolis, MN.

mission of Materials Method E at (E 96-80) (both at 37.8°C [100°F], 90% relative humidity differential). However, from calculations based on the low permeance of vapor retarders for tanks operating below 0°C (32°F), one can demonstrate that there could be a significant amount of moisture pickup, enough to cause a drastic reduction in thermal resistance of the insulation within a few years.

The experience of H. B. Fuller Co. has shown that when a chlorosulfonated polyethylene (CSPE) rubber (Hypalon)[3] based moisture vapor retarder mastic or coating is used over polyurethane foam insulation in low-temperature service, the insulation system continues to operate with relatively unimpaired thermal efficiency for at least eleven years. In several installations, the author has observed that the actual diffusion of moisture is substantially less than that which could be calculated for vapor retarders with the given service conditions. To document this experience, service data were obtained from the examination of two large storage tanks containing agricultural liquid ammonia held at −33°C (−28°F) and stored at atmospheric pressure [*1*].

The tanks are 52 m (170 ft) in diameter by 34 m (110 ft) high and hold up to 27 200 metric tons of liquid anhydrous ammonia for use as fertilizer. Each of these single-wall steel tanks was specified to be insulated with a 7.6-cm (3-in.)-thick seamless layer of spray-applied urethane foam, designed to maintain a heat transfer rate of 15.8 W/m^2 (5 $Btu/h \cdot ft^2$). The CSPE mastic was spray applied onto the foam in two coats. Both tanks have been in continuous service, with the liquid ammonia level varying with usage, normally at an 5.5 to 6.0-m (18 to 20-ft) level during the summer months, and near full during the winter months. The tank at Garner, Iowa, put into service in 1973, had been emptied to 1350 metric tons two or three times during its service life, always in April or early May for about two weeks before being refilled. No known time of emptying had occurred for the Westchester, Iowa, tank since it was put into service in 1976. The lowest liquid level ever noted was at 1.5 m (5 ft), in the fall of 1980. Both tanks are refilled by pipeline throughout the year as needed.

These tanks are now eleven and eight years old, respectively, and are presently operating with good thermal efficiency. It appears that the cellular nature of the urethane foam insulation restricts and changes the diffusion mechanism experienced with fibrous insulation, so that in the absence of any free air spaces within the insulation, the moisture diffusion is attenuated. This would not be the case in a fibrous-type insulation system, such as mineral wool or fibrous glass, because of the continuous network of open interstices that allow for the free movement of moisture.

Procedure

Data were obtained on these two operating agricultural ammonia storage tanks in Iowa, one near Garner (90 miles north of Des Moines in North Central Iowa), in service for eleven years, and the other in Westchester (75 miles south of Waterloo in Southeast Iowa), in service for eight years. No repair work had been needed on the tank at Westchester. The roof insulation on the tank at Garner had been recoated in 1978 because of hail damage. There are some frost blisters in the foam on the walls of this tank, caused by moisture entry after the hail damage on the roof, and some tearing of the mastic on the walls caused by falling ice, which had blown off the top of the tank. Both tanks were visited in June 1984.

Up to 20.3-cm (8-in.)-square cutouts of foam and retarder were removed down to within 1.25 cm (0.5 in.) from the metal tank wall and immediately sealed in 0.07-mm (2.7-mil) polyethylene freezer bags to be tested for moisture content. The specimens were weighed within 4 h of removal from the Garner tank, and within 24 h of removal from the Westchester tank. No condensate was observed within the bag during removal of the specimens for testing, and moisture

[3]Hypalon is a trademark of E.I. du Pont de Nemours and Co.

loss through evaporation was assumed to be negligible. The condition of the foam left on the tank wall behind the cutout was also closely observed. To determine the thermal efficiency of the insulation system, readings were made around the tank using a Linear Laboratories Model C-700 ThermoFlow thermometer. The meter measures both surface temperature and net heat flow without contact. Accurate readings for the surface temperature of the mastic and heat absorption of the tank could be made only on the shaded side of the tank. The tanks were then observed for water seepage around the perimeter near the bottom, and examined for overall condition. The insulated tank walls were checked by hand pressure for the presence of soft, moisture-laden areas.

Test Data

When the foam specimens were removed from the tank, ice was found in the foam at a distance as much as 1.3 cm (0.5 in.) away from the tank wall. Even though there was ice in the residual foam next to the tank wall, the foam remained well-bonded to the tank surface and could not be pulled loose. Typically, the urethane foam broke at the frost line, and the removed piece was observed to be damp or wet for a distance of as much as another 1.3 cm (0.5 in.) into the foam. Outward beyond this wet area, the foam appeared to be dry. Examination of the cross sections of the foam specimens revealed a uniform closed-cell structure, with no delamination observed between the built-up layers. Tables 1 and 2 list the actual sprayed thickness of the foam specimens as cut out, their dry density, and their moisture content. The dry densities of the urethane foam are well within the range of 32 to 48 kg/m^3 (2.0 to 3.0 lb/ft^3) specified for this type of work.

Using the ThermoFlow thermometer on the shaded tank wall surface, heat transfer rates of approximately -9.5 to -15.8 W/m^2 (-3 to -5 Btu/h · ft^2) were obtained. These rates are within the design criterion. The temperature of the mastic surface at Garner was approximately 15 to 17°C (60 to 62°F), with an ambient air temperature of 18°C (66°F). At Westchester, the mastic surface temperature was measured to be 27°C (81°F), with an ambient air temperature of 28°C (83°F). Using the observed heat transfer rates, foam thickness, and temperature differential across the foam, a calculated thermal conductivity of 0.020 to 0.030 W/m · K (0.14 to 0.21 Btu · in./h · ft^2 · °F) is obtained. This range of thermal conductivity is representative of a good-quality sprayed polyurethane insulation.

In the laboratory, the CSPE mastic was stripped from the foam specimens, which were removed from both tanks and tested for permeance. The CSPE mastic was well bonded to the

TABLE 1—*Condition of urethane foam removed from the ammonia tank at Garner, Iowa.*

Wall Location	Thickness, cm (in.)	Dry Density, kg/m^3 (lb/ft^3)	Moisture Content	
			Location in Foam	Weight %[a]
North	12.0 (4.75)	38.4 (2.4)	inner 3.8 cm	2.5
			outer 7.6 cm	1.5
Southwest	7.0 (2.75)	38.4 (2.4)	inner 4.4 cm	18.3[b]
			outer 2.5 cm	2.5
East	8.5 (3.30)	41.6 (2.6)	inner 4.1 cm	1.7
			outer 3.2 cm	1.2

[a]The percentage of moisture in the foam specimens removed, not including from 0.6 to 1.3 cm of foam next to the tank wall. The percentage is based on the weight of the dry foam.

[b]The foam broke loose within the frost area on these specimens, so the inner section contained some frost, which resulted in a higher moisture level.

TABLE 2—*Condition of urethane foam removed from the ammonia tank at Westchester, Iowa.*

Location on Tank	Thickness, cm (in.)	Dry Density, kg/m^3 (lb/ft^3)	Moisture Content	
			Location in Foam	Weight %[a]
North wall	11.4 (4.50)	33.6 (2.1)	inner 2.2 cm	5.9[b]
			outer 8.3 cm	2.2
West wall	10.2 (4.00)	36.8 (2.3)	inner 3.2 cm	2.9
			outer 6.4 cm	1.2
East wall	13.3 (5.25)	36.8 (2.3)	inner 3.8 cm	43.0[b]
			outer 8.6 cm	1.9
Roof	6.4 (2.50)	56.0 (3.5)	inner 2.5 cm	.6
			outer 3.9 cm	2.6

[a]The percentage of moisture in the foam specimens removed, not including from 0.6 to 1.3 cm of foam next to the tank wall. The percentage is based on the weight of the dry foam.

[b]The foam broke loose within the frost area on these specimens, so the inner section contained some frost, which resulted in a higher moisture level.

urethane foam, and a layer of urethane foam was left on the back side of the mastic after forcing delamination. The results of water vapor transmission rate (WVTR) testing on these films are listed in Tables 3 and 4. Except for one case, the permeance values were all much higher than would be expected for the CSPE mastics. This normally would be at 0.046 metric perm (0.07 perm), tested in a Permatran W1 at 37.8°C (100°F), 90% relative humidity differential, at the typical coverage of 0.76-mm (30-mil) dry film thickness. However, on these specimens, the foam surface was irregular, varying from a coarse orange peel to almost a popcorn texture. These irregularities in the foam surface were replicated on the back side of the film surface, causing difficulty in obtaining a positive seal with the rubber sealing ring used in WVTR test cups; therefore, some local environmental moisture was likely to have influenced the results. Also, some pinholes observed in the tested mastic specimens would have increased the permeance substantially. Pinholes can be caused by blow holes or crevices in the foam, air entrapment on application, too much mastic applied in one coat, or some thin spots in the mastic applied over this rough surface. Obviously, the presence of these pinholes in the film should have made the moisture content in the installed foam even higher than that calculated at the normal permeance of sound CSPE vapor retarder mastic.

In order to calculate the amount of water that should have diffused into the foam, data were obtained on the climatic conditions of temperature and relative humidity at the locations of these tanks. These data were provided from recorded average monthly humidity conditions in

TABLE 3—*Condition of the CSPE mastic[a] on the ammonia tank at Garner, Iowa.*

Location on Tank	Film Thickness, mm (mil)	Pinholes in Film[b]	Permeance, metric perm (perm)[c]
North	1.0 to 1.8 (40 to 70)	none	0.127 (0.193)
Southwest	0.7 to 1.9 (30 to 75)	many	0.990 (1.500)
East	0.7 to 1.0 (30 to 40)	few	0.380 (0.580)

[a]Foster Monolar mastic.

[b]Detected by the use of a spark tester; none of the pinholes were visible to the eye.

[c]Tested at 37.8°C (100°F), 90% relative humidity differential, using a Mocon Permatran W1.

TABLE 4—*Condition of the CSPE mastic*[a] *on the ammonia tank at Westchester, Iowa.*

Location on Tank	Film Thickness, mm (mil)	Pinholes in Film[b]	Permeance, metric perm (perm)[c]
North wall	1.0 to 1.3 (40 to 50)	none	0.037 (0.056)
West wall	0.8 to 1.1 (30 to 45)	several	0.178 (0.270)
East wall	0.6 to 0.8 (25 to 30)	several	0.077 (0.117)
Roof	1.0 to 1.3 (40 to 50)	several	0.230 (0.350)

[a]Foster Monolar mastic.
[b]Detected by the use of a spark tester; none of the pinholes were visible to the eye.
[c]Tested at 37.8°C (100°F), 90% relative humidity differential, using a Mocon Permatran W1.

the close vicinity of these tanks, as well as the normal temperature for each month of the year. The data obtained are presented in Appendixes I and II, along with a calculation of the total water vapor transmission over the current lifetime of each tank as expected for the given conditions.

Discussion of Results

A comparison of the calculated water absorption of the foam with the measured water content in the cut-out specimens is presented for both tanks in Tables 5 and 6. In all cases, the actual water content of the foam is much below that which a simple calculation would theoretically suggest to be present. This is true for the typical coating permeance of 0.046 metric perm (0.07 perm) for a pinhole-free CSPE mastic film, as well as for the measured permeance. Except for the three cases noted on Tables 1 and 2, where the removed foam specimens actually included some foam in the frost area, the moisture content was very low, and typical of what might be expected for uncoated foam itself exposed to a humid environment. Foam specimens removed from the job site and predried at 60°C (140°F), showed a 2.6% weight increase after six days of exposure in a 100% relative humidity environment. This indicates that there was no liquid moisture in the foam cells on the tanks beyond the frost line. Where water was observed in the foam, it was always within 2.51 cm (1 in.) away from the tank wall and was suspected to have been ice that melted after the specimens were removed. In Fig. 1, an approximate temperature profile across the foam thickness is shown, indicating the moisture content at various

TABLE 5—*Comparison of the calculated and actual water pickup by urethane foam on the ammonia tank at Garner, Iowa.*

Wall Location	Theoretical Water Pickup, % of Dry Foam Weight[a]	Theoretical Water Pickup at 0.07 perm, % of Dry Foam Weight[b]	Actual Water Pickup, % of Dry Foam Weight
North	43.8	15.7	1.76
Southwest	571.6	26.7	2.85
East	161.6	20.1	1.48

[a]Based on the total calculated water vapor transmission for the eleven years of service, using the actual permeance of the CSPE mastic, the average vapor pressure differential for the eight-month period of positive vapor drive, and the calculated weight of the urethane foam per square foot of surface area, determined from the measured density and thickness. See Appendix I for the calculations.
[b]Using the typical permeance for a CSPE mastic on a flat, pinhole-free unweathered free film.

TABLE 6—*Comparison of the theoretical and actual water pickup by urethane foam on the ammonia tank at Westchester, Iowa.*

Wall Location	Theoretical Water Pickup, % of Dry Foam Weight[a]	Theoretical Water Pickup at 0.07 perm, % of Dry Foam Weight[b]	Actual Water Pickup, % of Dry Foam Weight
North wall	11.3	14.2	2.62
West wall	54.6	14.2	1.40
East wall	17.8	10.7	4.00
Roof	...	...	1.90

[a]Based on the total calculated water vapor transmission for the eight years of service, using the actual permeance of the CSPE mastic, the average vapor pressure differential for the seven-month period of positive vapor drive, and the calculated weight of the urethane foam per square foot of surface area, determined from the measured density and thickness. See Appendix II for the calculations.

[b]Using the typical permeance for a CSPE mastic on a flat, pinhole-free unweathered free film.

levels across the foam thickness. The graph was plotted from temperature readings made in the foam insulation on a return trip to Garner on 4 Oct. 1984. A 0.16-cm (1/16-in.) diameter thermocouple probe was used with an Omega 450 AET thermometer, Type E. The moisture content was determined by removing a 1.9-cm (0.75-in.)-diameter core, and cutting off 0.64-cm (0.25-in.) slices for weight loss measurements. A laboratory test is currently being conducted which shows a similar temperature profile on an insulation system constructed to simulate the tank service conditions. The figure demonstrates that most of the foam volume is still dry. Even though the inner 1.30 cm (0.50 in.) contains frost, it is functioning at a lower thermal conductivity than the dry foam, which is partly due to the lower temperatures. This conductivity change may also be due to the ice buildup forming an inner wall in the foam cells, which further restricts heat flow, or to a partial vacuum in these cold foam cells resulting from condensation of moisture and Freon vapor. Foam specimens that were dried at 60°C (140°F) before being immersed in water for seven days at room temperature were found to have an average water absorption of 50.7% by weight. This indicates that the foam does have a much higher capacity for water absorption than that which actually took place. Water absorption by sprayed polyurethane foam can typically vary from a 10% to a greater than 100% weight gain by this test method, depending on the foam quality and density.

Conclusions

The test data demonstrate an anomalous behavior between fibrous and cellular insulations, disclosed by comparing actual and calculated moisture diffusion through moisture vapor retarder mastics in low-temperature service. This anomaly has not previously been documented, and the reasons are not easily understood, involving complexities inherent in this type of insulation system. Since there is both a vapor pressure and a temperature gradient across the insulation system, one must deal with the combined effect of moisture and heat flow; neither is independent of the other [*2*]. The moisture movement is complicated by its changing state (vapor, liquid, solid) and the necessity of accounting for the associated latent heat emitted in changing phases on cooling, the latent heat transfer being restricted by the insulating foam. The thermal conductivity of the urethane foam does not remain constant across the temperature range but will decrease with the temperature even where ice is formed in the foam cells, as indicated by the steeper slope of the temperature gradient across the foam thickness given in Fig. 1. This lower thermal conductivity for the foam near the tank wall, apparently from a layer of ice buildup on the cell walls, but not filling the cells, helps to prevent further ice buildup outward in the foam,

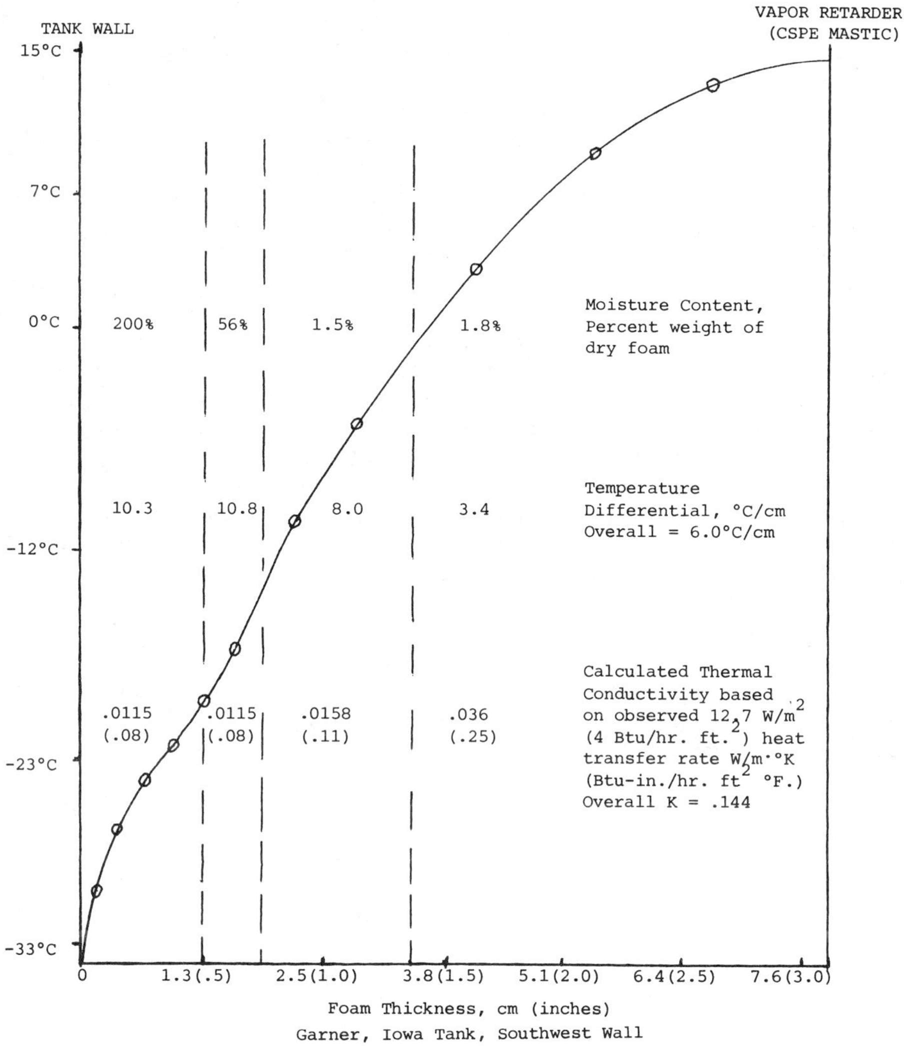

FIG. 1—*Typical temperature-moisture profile across the urethane foam thickness.*

as the temperature differential would not decrease across the foam with ice buildup. Also, the thermal conductivity of the urethane foam will increase with age in the outer layers, because of a slow diffusion of the Freon out of the foam cells and a diffusion of air and moisture vapor back into the same cells.

The moisture permeance of the urethane foam probably is not constant over the temperature range, but will decrease with temperature as the foam becomes harder and more resistant to moisture. The moisture permeance has been observed also to decrease with increasing mean relative humidity [*3*]. This decrease may be explained by a blocking of vapor diffusion paths within the urethane foam structure, due to a capillary condensation of moisture vapor at high humidity conditions.

Moisture permeance through vapor retarder mastics or coatings may be much slower when used on a foam insulation system than that obtained on free films. The closed cell foam restricts vapor to a cell-by-cell movement, resulting in a moisture vapor pressure differential across the foam. Therefore, the vapor pressure at the back side of the retarder may be less than at the exterior only by that difference corresponding to the temperature drop induced by the thermal resistance of the retarder. Since the thermal resistance of a thin retarder is very low, there will be very little temperature or vapor pressure differential across the retarder. The result is a much lower amount of moisture diffusion into the outer foam than would be predicted by calculations based on the total vapor pressure differential across the whole insulation system. In contrast, in "open" fibrous-type insulation systems, the insulation will transmit pressure differentials quickly. Thus, calculations of water vapor transfer based on the differential vapor pressure dictated by the vessel temperature with respect to the ambient air conditions will be in agreement with test data on free films. It also has been observed by others [*3*] that unprotected urethane foam will continue to pick up moisture until saturated when a differential vapor pressure is maintained across the urethane foam.

There remain unanswered questions as to the diffusion rate of moisture vapor into polyurethane foam in low-temperature service when protected with a vapor retarder mastic or coating. It is suggested that computer modeling of these phenomena could establish better criteria for system design. This study indicates that this type of retarder mastic does provide adequate protection for polyurethane foam thermal insulation and remains effective for many more years than simple calculations would predict.

Acknowledgments

This study was performed with the consent and help of Farmland Industries, Inc., of Kansas City, Missouri, the owner of the tanks. The author wishes to acknowledge the help of Harvey Spangler, terminals manager for Farmland Industries, Inc.

APPENDIX I

Calculation of Water Vapor Pressure for Each Month at Garner, Iowa

Month of Year	Relative Humidity at 6 A.M., %[a]	Relative Humidity at 6 P.M., %[a]	Temperature, °C, (°F)[b]	Vapor Pressure at 6 A.M., in. Hg[c]	Vapor Pressure at 6 P.M., in. Hg[c]
January	79	72	−6.7 (19.9)	0.09	0.080
February	85	73	−4.8 (23.4)	0.11	0.095
March	86	67	3.8 (38.8)	0.21	0.160
April	83	57	9.3 (48.7)	0.29	0.260
May	78	53	16.4 (61.5)	0.43	0.290
June	80	53	21.8 (71.2)	0.62	0.410
July	81	55	24.5 (76.1)	0.74	0.500
August	84	58	23.2 (73.7)	0.71	0.490
September	84	61	18.6 (65.4)	0.53	0.380
October	78	55	12.3 (54.2)	0.32	0.230
November	82	66	2.8 (37.1)	0.18	0.150
December	83	72	−3.7 (25.3)	0.12	0.100

[a]Monthly average for a seven-year period at Des Moines, Iowa, 90 miles due south of Garner, Iowa.
[b]Monthly normal temperature for 1931 through 1960 at Des Moines, Iowa.
[c]1 in. Hg = 25.4 mm Hg.

Calculations

The vapor pressure of water at 0°C (32°F) was 4.57 mm Hg (0.18 in. Hg). It was taken as the inner pressure, as most of the specimens broke off at the frost line when removed.

The average vapor pressure for the eight-month period of March through October (245 days), the months of positive vapor drive having a vapor pressure greater than 4.57 mm Hg (0.18 in. Hg), was 10.34 mm Hg (0.407 in. Hg).

The differential vapor pressure was 5.77 mm Hg (0.407 − 0.18 = 0.227 in. Hg).

The total water vapor transmission (WVT) was calculated as follows:

$$\text{perms} \times 24\ \text{h} \times 245\ \text{days} \times 11\ \text{years} \times 0.227\ \text{in. Hg}$$

or

$$\text{WVT} = (\text{perms} \times 14682)\ \text{grains water/ft}^2$$

or

$$\text{WVT} = (\text{perms} \times 2.096)\ \text{lb}$$

or

$$\text{WVT} = (\text{perms} \times 0.9507)\ \text{kg},\ (\text{metric perms} \times 0.6265)\ \text{kg}$$

The ratio of the calculated water pickup to the percentage of dry foam weight was

$$\frac{(\text{perms} \times 2.096)\ \text{lb water/ft}^2 \times 100}{\text{foam dry density (lb/ft}^3) \times \text{thickness (ft)}}$$

APPENDIX II

Calculation of Water Vapor Pressure for Each Month at Westchester, Iowa

Month of Year	Relative Humidity at 6 A.M., %[a]	Relative Humidity at 6 P.M., %[a]	Temperature, °C, (°F)[b]	Vapor Pressure at 6 A.M., in. Hg[c]	Vapor Pressure at 6 P.M., in. Hg[c]
January	80	75	−7.8 (17.9)	0.08	0.075
February	84	75	−5.9 (21.4)	0.10	0.090
March	87	74	0 (32.0)	0.16	0.130
April	86	61	8.5 (47.3)	0.28	0.200
May	85	58	15.5 (59.9)	0.44	0.300
June	86	61	21.2 (70.1)	0.64	0.450
July	88	63	23.2 (73.8)	0.74	0.530
August	92	63	22.0 (71.7)	0.72	0.490
September	90	64	17.2 (62.9)	0.52	0.370
October	84	63	10.9 (51.7)	0.32	0.240
November	84	71	1.5 (34.7)	0.17	0.140
December	83	77	−4.9 (23.2)	0.10	0.100

[a]Monthly average for a seven-year period at Waterloo, Iowa, 75 miles north of Westchester, Iowa.
[b]Monthly normal temperature for 1931 through 1960 at Waterloo, Iowa.
[c]1 in. Hg = 25.4 mm Hg.

Calculations

The average vapor pressure for the seven-month period of April through October (214 days), the months of positive vapor drive, having a vapor pressure greater than 4.57 mm Hg (0.18 in. Hg) was 11.32 mm Hg (0.4457 in. Hg).

The differential vapor pressure was 6.75 mm Hg (0.4457 − 0.18 = 0.2657 in. Hg).

The total water vapor transmission (WVT) was calculated as follows:

$$\text{perms} \times 24 \text{ h} \times 214 \text{ days} \times 8 \text{ years} \times 0.2657 \text{ in. Hg}$$

or

$$\text{WVT} = (\text{perms} \times 10917) \text{ grains water/ft}^2$$

or

$$\text{WVT} = (\text{perms} \times 1.558) \text{ lb}$$

or

$$\text{WVT} = (\text{perms} \times 0.7068) \text{ kg, (metric perms} \times 0.4658) \text{ kg}$$

The ratio of the calculated water pickup to the percentage of dry foam weight was

$$\frac{(\text{perms} \times 1.558) \text{ lb water/ft}^2 \times 100}{\text{foam dry density (lb/ft}^3) \times \text{thickness (ft)}}$$

References

[1] H. B. Fuller Company, "Largest Spray-On Urethane Job Insulates Four Anhydrous Ammonia Fertilizer Tanks," *NICA Outlook,* April 1974, pp. 32–34.

[2] "Moisture in Building Construction," in *1981 ASHRAE Handbook of Fundamentals,* American Society of Heating, Refrigerating, and Air Conditioning Engineers, Atlanta, GA, 1981, Chapter 21, pp. 21.1–21.22.

[3] Paljak, I., "Condensation in Slabs or Cellular Plastics," *Materiaux et Constructions,* Vol. 6-N, No. 31, 1973, pp. 53–56.

Materials

Norman M. P. Low[1]

Material Degradation of Thermal Insulating Mineral Fibers

REFERENCE: Low, N. M. P., "**Material Degradation of Thermal Insulating Mineral Fibers,**" *Thermal Insulation: Materials and Systems, ASTM STP 922,* F. J. Powell and S. L. Matthews, Eds., American Society for Testing and Materials, Philadelphia, 1987, pp. 477–492.

ABSTRACT: Inorganic mineral fibers of the slag wool and rock wool type and of the glass fiber type were found to exhibit physical disintegration and morphological modifications on the fiber surface when these fibrous products were exposed to environments in which an excessive water medium and heating at temperatures up to 100°C were maintained for a period of several months. The physical disintegration transformed the fine, threadlike fibers into small segments and granules. Scanning electron microscopic examinations showed that test specimens of several slag-rock fibers displayed a rough, eroded fiber surface after the fibers had been immersed in water and heated to temperatures below 60°C for about three months, while similar fibers displayed the development of a reticular network structure, or "honeycomb morphology," on the fiber surface after they had been immersed in water and heated to temperatures between 85° and 100°C for 14 days. Analyses of the water solution and of the disintegrated fibers suggest that these modifications of the fibrous products are caused by chemical leaching and hydration processes. Thermal conductivity measurements showed that the disintegrated fibers have thermal insulating values about 15% lower than those of normal, threadlike fibers.

KEY WORDS: mineral wool, glass fibers, slag wool, rock wool, thermal insulating materials, durability, material degradation, moisture problems, humidity effects, thermal insulation

Energy conservation has prompted extensive use of thermal insulating materials in building envelopes to impede the loss of heat energy. Both organic and inorganic thermal insulators are used extensively in the construction industry. Inorganic mineral fibers, such as glass fibers, slag wool, and rock wool fibers, are the most common thermal insulators. These inorganic fibers are generally considered to be durable products for insulating applications because they are able to withstand high operating temperatures, they are chemically stable, and they are also believed to be nonabsorbent of water. In spite of these special qualities, information in the literature indicates that the action of heat and humidity could have detrimental effects on these inorganic fibers. It has been reported that mineral wool fibers can become fragile and the fibrous form can be destroyed if fibers are exposed to heat and humidity [*1*]. The effect of heat and humidity on the material behavior of inorganic mineral fibers has been studied in a recent research project. This paper presents and discusses some of the experimental results.

Experimental Procedure

Two types of mineral fibers were used in the present study. First, slag-rock wool fibers, and second, glass fibers. Both of these insulation materials are commercially available in the form of loose pouring fibers and they are manufactured and marketed by four Canadian producers. The

[1]Research assistant professor, Centre for Building Studies, Concordia University, Montreal, Quebec, Canada H3G 1M8.

test specimens prepared from the slag-rock wool fibers were identified as Specimens 1, 2, and 3, and the test specimen prepared from the glass fibers was identified as Specimen 4.

Two groups of experiments have been conducted. The first was carried out by immersing about 50 g of loose fibers in distilled water in a 1-L airtight glass container so that the fibers were constantly in contact with the liquid water. Each test specimen was placed in an individual container. The second group of experiments was carried out with slightly modified procedures. About 100 mL of distilled water was first placed in the bottom of a similar-size container, and a layer of clean glass cullet about 1.0 cm thick was then placed in the water. About 50 g of the mineral fibers were then placed on top of the glass cullet so that the fibers were not in direct contact with the water inside the container. Each test specimen was also placed in an individual container. Each glass container was closed with a lid that had a metal wire strap to hold it to the container so that water evaporation from the container was minimized. The containers were then placed in an electric oven and heated at six different temperature levels: 25, 45, 60, 85, 92, and 100°C. Containers kept at temperatures above 85° were heated for a period of two weeks, containers kept at temperatures above 25°C were heated for a period of about three months; and containers kept at 25°C were heated for a period of about eight months.

During the heating process, each container was briefly removed from the oven about once a week to examine the integrity of the fibers inside the container. For those containers in the first group, the liquid water and fibers were agitated by hand for about 60 s before the containers were returned to the oven for the continuing heating process. No such manual agitation was carried out in the containers for the second group. At the end of the heating process, a small quantity of the moist fiber mass was removed from each container and dried at 85°C for 24 h. The surface features of the fibers were then examined by scanning electron microscopy (SEM).

The pH value of the water solution before and after the fiber immersion process was measured at ambient temperature. Thermogravimetric analysis[2] was performed on two test specimens. The first specimen consisted of the normal slag-rock wool fibers from Specimen 1, which had not been exposed to any water treatment. The second specimen consisted of the disintegrated fibers from Specimen 1, which had been exposed to the heat-water treatment process at 100°C for six days.

The thermal conductivity of two sets of specimens was measured employing a heat flow meter[3] which conforms to the ASTM Test for Steady-State Thermal Transmission Properties by Means of the Heat Flow Meter (C 518-76). The first set of specimens consisted of the normal slag-rock wool fibers from Specimens 1 and 2, which had not been exposed to the water-heat treatment process. The second set of specimens consisted of the disintegrated fibers, also from Specimens 1 and 2, which had been exposed to the heat-water treatment process and subsequently oven dried. The test specimens were prepared by a hand pouring method. Test specimens with five different fiber densities within the range of 45 to 215 kg/m^3 were tested.

Results

Mineral fibers of either the glass fiber type or the slag-rock wool type generally appear in the form of fine, threadlike filaments. The surface of each individual fiber is smooth and free from defects, as shown in the scanning electron micrograph in Fig. 1. The fibrous form and defect-free surface characteristics remained unchanged when fibers of either slag-rock wool or glass fibers were heated from ambient temperature up to 500°C for a period 5 to 6 h and subsequently cooled back down to the ambient temperature in a water-free environment. This dry heating

[2]The thermogravimetric analysis (TGA) measurements were made with the Du Pont 1090 differential thermal analyzer, made by E. I. du Pont de Nemours and Co., Wilmington, DE.

[3]The heat flow meter apparatus used was the R-Matic system, made by the Dynatech, R/D Co., Cambridge, MA. It has a test accuracy of about ±2%.

FIG. 1—*Scanning electron photomicrograph of the smooth, defect-free fiber surface of normal mineral fibers.*

treatment was observed to cause no modifications of the fibers. However, when water medium, in the form of either vapor or liquid, was introduced to the fibers and the fiber-water system was heated at the low temperature levels selected for the present study, both slag-rock wool fibers and glass fibers exhibited physical changes. The extent of this physical change varied depending on the type of the fibers and also on the water medium to which the fibers were exposed. The overall experimental results are summarized in Table 1. The material change and the property change of the mineral fibers are discussed separately in the following sections.

Physical Change

The first group of experiments was carried out by immersing the test fibers in distilled water and heating them to the various temperature levels. This simple water-heat exposure treatment caused the fiber to disintegrate, transforming the fine, threadlike fibers into small segments and granules when the water-fiber mixtures were manually agitated. It was noted that the time period in which the fiber disintegration process developed varied depending on the heating temperatures. For example, Specimens 1 and 2 of the slag-rock wool fibers partially disintegrated into small segments after the fibers had been immersed in the water at 25°C for a period of eight months, while Specimen 3 of the slag-rock wool fibers and Specimen 4 of the glass fiber exhibited no physical change at all. These observations were made by visual inspection of the fibers in the containers. When the temperature of the water was increased to 45°C, Specimens 1 and 2 of the slag-rock wool fibers also disintegrated into small segments, but this occurred after four months of heating. When the temperature of the water solution was further increased to 60°C, the fiber disintegration process occurred more readily. Specimens 1 and 2 of the slag-rock wool

TABLE 1—*Observation of material disintegration in mineral fibers after exposure to water and heat treatments at various temperatures.*

Heating Temperature, °C	Exposure Time, days	Detection of Material Disintegration			
		Vapor Phase		Liquid Water	
		Specimen No.	Form of Change	Specimen No.	Form of Change
25	240		no change	1,2,	disintegration with manual agitation
45	120	1,2	surface change	1,2,	disintegration with manual agitation
60	90	1,2,3	disintegration with finger rubbing	1,2,	disintegration with manual agitation
75	40	1,2,3	disintegration with finger rubbing	1,2,	disintegration with manual agitation
85	27	1,2,3	disintegration with finger rubbing	1,2,	disintegration with no agitation
94	14	1,2,3	disintegration with finger rubbing	1,2,3	disintegration with no agitation
100	14	1,2,3	disintegration with finger rubbing	1,2,3	disintegration with no agitation
100	120	4	glass fiber partial disintegration with finger rubbing	4	glass fibers, no disintegration with manual agitation

fibers disintegrated into granules after three months of heating. At 85°C, similar fiber disintegration was also observed, but it occurred in an even shorter period of only 27 days.

The combination heat-water treatment caused no visible change in Specimen 3 of the slag-rock wool fibers or Specimen 4 of the glass fibers for heating temperatures up to 85°C. However, when the temperature was raised to the 92 to 100°C range, all three specimens of the slag-rock fibers partially disintegrated into granular form after only 6 days of heating, and a complete disintegration occurred after 14 days of heating. Glass fibers from Specimen 4 exhibited no physical change even after the fibers had been immersed in the water and heated to 100°C for a period of as long as four months. The physical disintegration of Specimens 1 and 2 of the slag-rock wool fibers that occurred at the various temperatures is shown in Fig. 2. There is a simple correlation between the heating temperatures and the durations in which the fibers had been immersed in the water solution for the fiber disintegration process (Fig. 3).

Immersion of slag-rock wool fibers in water solution in conjunction with the low-temperature heating was found not only to cause a physical disintegration of the elongated fibers but also to produce morphological modifications on the fiber surface. Close examination by scanning electron microscopic technique of the small segments of the disintegrated fibers from Specimens 1 and 2 that had been immersed in water and heated at 45°C for four months showed a rough, eroded fiber surfaces (Fig. 4). On the other hand, close examination of the small segments of the disintegrated fibers from all specimens (Specimens 1, 2, and 3) that had been immersed in water and heated at 100°C for 14 days showed a reticular network structure surrounding the fiber stem, or a "honeycomb morphology," as shown in Fig. 5. The physical changes shown in

FIG. 2—*Physical change of the slag-rock wool fibers after immersion in water and heating at various temperature levels.*

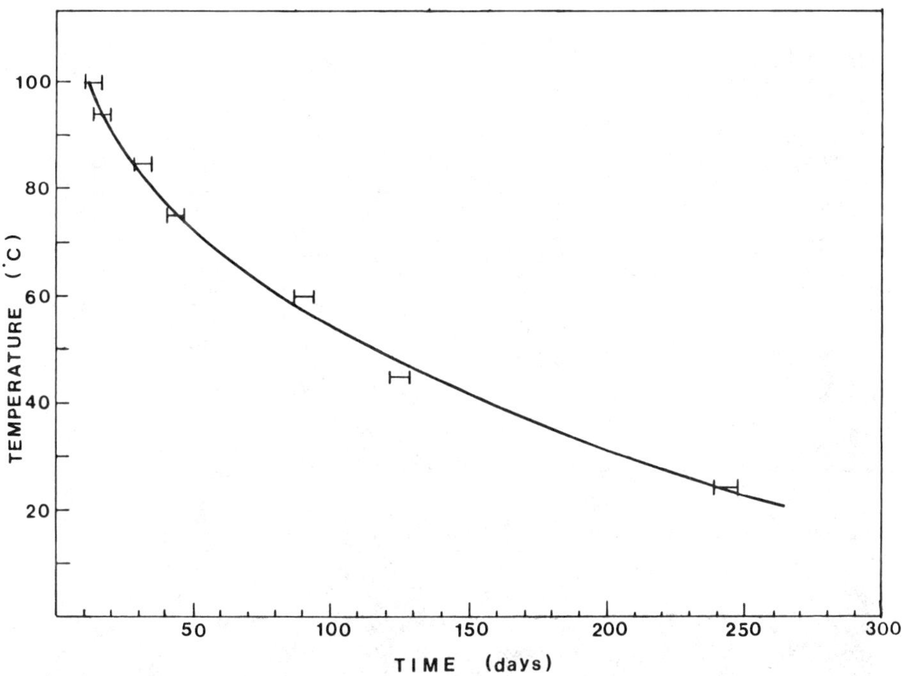

FIG. 3—*A simple correlation between the heating temperature of the water solution and the durations in which the fibers have been immersed in the water solution for the fiber disintegration process in slag-rock wool fibers.*

Figs. 4 and 5 evidently indicate that a major material degradation process occurred in the slag-rock wool fibers after they had been immersed in water solution with low-temperature heating.

The second group of experiments exposed the mineral fibers to water vapor, which was generated by evaporation of the liquid water at the bottom of the glass container while it was being heated to the various temperature levels. This simple exposure subjecting slag-rock wool fibers and glass fibers to a heat-water vapor environment also caused physical change and microstructural modifications in the fibers, but the changes were observed in a different manner. After exposure to water vapor at 25°C for eight months, slag-rock wool fibers from all three specimens and glass fibers from Specimen 4 displayed no material change or surface modification, as determined by visual inspection and SEM examinations. However, slag-rock wool fibers from all three specimens disintegrated into a granular form when the moist fibers were gently rubbed with the fingers after the fibers had been exposed to water vapor and heated to temperatures between 45 and 100°C for periods similar to those used in the first group of experiments. The fiber disintegration produced by hand rubbing was more discernible for those slag-rock wool fibers that had been heated at the higher temperature range (85 to 100°C). When normal fine slag-rock wool fibers that had not been exposed to the heat-water vapor treatment were rubbed by fingers, the fibers generally remained intact. After exposure to water vapor and heat treatment at 100°C for a period of four months, a portion of the glass fibers from Specimen 4 also disintegrated partially into the granular form during the hand rubbing process.

Microstructural modifications of the two types of mineral fibers after exposure to heat-water vapor treatments displayed different features. For example, after four months of exposure to water vapor at 45°C, slag-rock wool fibers from Specimens 1 and 2 displayed lumpy spots or

FIG. 4—*Scanning electron photomicrograph of the rough, eroded fiber surface of the slag-rock wool fibers immersed in water and heated to 45°C for four months.*

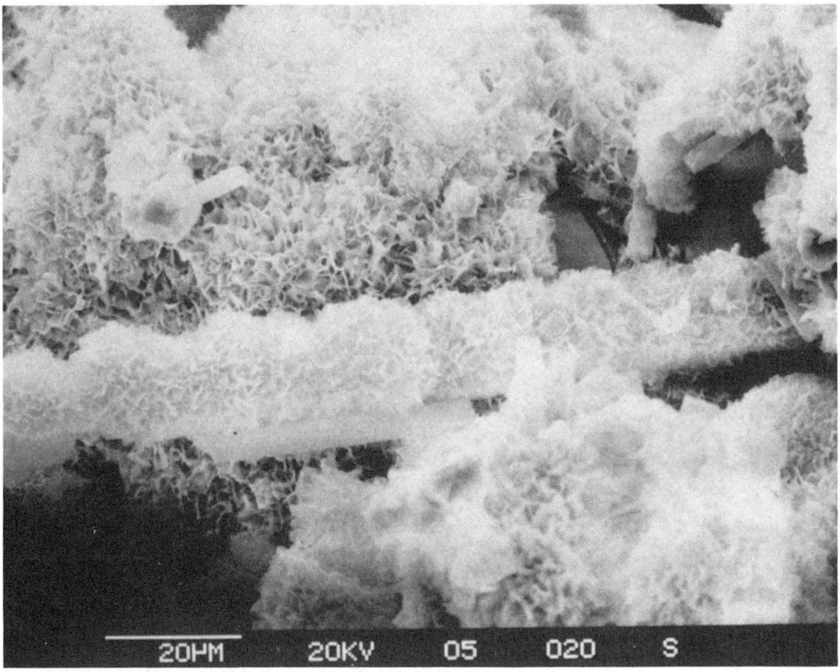

FIG. 5—*Scanning electron photomicrograph of the reticular network structure, or honeycomb morphology, on the surface of the slag-rock wool fibers immersed in water and heated to 100°C for 14 days.*

deposition on the surface of the fiber stem, as shown in the SEM micrographs in Fig. 6. After three months of exposure to water vapor at 60°C, slag-rock wool fibers from Specimens 1 and 2 also showed the presence of foreign substance on the fiber surfaces and the formation of fractures on the fiber stems, as shown in Fig. 7. A recticular network structure or "honeycomb" morphology, was also observed in those fibers that had been exposed to water vapor and heated to temperatures above 85°C after a period of 40 to 50 days. After exposure to water vapor at 100°C for four months, glass fibers from Specimen 4 displayed very different features. Circular nodules were observed clinging to the fiber stem, as shown in the photomicrograph in Fig. 8. It is evident that exposure of the two types of mineral fibers to water vapor with low-temperature heating resulted in material degradation and fiber surface modifications similar to those developed during the water immersion process.

Chemical Change

The nature of the morphological change on the fiber surface observed in the slag-rock wool fibers suggests that this modification could have been caused by some chemical reaction. To establish a basis for such a possibility, the pH value of the water solution before and after the fiber immersion process was carefully measured at ambient temperature. The pH value of the distilled water prior in contact with the mineral fibers was near neutral (about 6.8) and increased to the 10 to 12 range after the fibers had been immersed in the distilled water. This

FIG. 6—*Scanning electron photomicrograph of the lumpy spots or deposition on the surface of the slag-rock wool fibers exposed to water vapor and heated to 45°C for four months.*

change in pH value of the water solution occurred in the slag-rock wool fibers and glass fibers from all four specimens. Figure 9 shows a plot of the variation in pH as a function of time for the four test specimens.

It is evident that the distilled water became alkaline in the fiber immersion process. It was noted that the change in the pH value of the distilled water was unaffected by either the length of the period in which the fibers were immersed or the temperature within the 25 to 100°C range to which the fibers were heated. The change in pH value of the water solution was also unaffected by a preheating treatment of the mineral fibers up to 500°C for several hours prior to immersion in the distilled water. The preheating treatment before immersion was intended to remove any organic resin material that might have been sprayed on the fibers by the manufacturer as a binding agent to hold the fibers together during the fiber production stage. The removal of this organic matter did not appear to alter the pH value of the solution, as it again rose to the 10 to 12 range. It also remained in the 10 to 12 range when the fiber-water system was frequently exchanged with fresh distilled water after each pH measurement. It was noted that slag-rock wool fibers from Specimens 1 and 2, when subjected to this frequent water exchange, disintegrated readily into small segments with manual agitation, although the fibers had not been heated above room temperature.

Property Change

Thermogravimetric Analysis—TGA measurement was carried out on two test specimens. The first specimen consisted of the normal, threadlike slag-rock wool fibers from Specimen 1 that had not been exposed to any water medium. The second specimen consisted of small segments of the disintegrated slag-rock wool fibers from Specimen 1 that had been heated in water

FIG. 7—*Scanning electron photomicrograph of the foreign substance and fracture on the fiber stem of slag-rock wool fibers exposed to water vapor and heated to 60°C for three months.*

at 100°C for 14 days. All the fibers used for these TGA measurements were oven dried at 85°C for 24 h before the specimen preparation. The two test specimens were analyzed separately in the differential thermal analyzer within the temperature range of 25 to 1000°C with a heating rate of 3 degrees Celsius per minute. Figure 10 shows the thermogram of the weight change as a function of the heating temperatures in the TGA measurements. Curve A shows the weight change for the normal fibers, and Curve B shows the weight change for the disintegrated fibers. It is evident that the normal slag-rock fibers exhibited no weight change within the temperature range, while the disintegrated fibers showed a gradual weight loss as the temperature increased. Maximum weight loss of about 18% was attained at about 1000°C. Within the temperature range of 680 to 710°C, the thermograms showed an abrupt and large weight loss rate for the disintegrated slag-rock fibers from Specimen 1 (Fig. 11). This abrupt and large rate of weight loss is found to be quite similar to that which occurs during dehydration of a number of ceramic minerals [*2*].

Thermal Conductivity Measurement—Thermal conductivity measurements were carried out on two sets of test specimens at a mean temperature of about 25°C and a temperature differential across the test specimen of about 24°C. The first set of test specimens was prepared with the normal slag-rock wool fibers from Specimens 1 and 2; the second set was prepared with the small segments and granules of the disintegrated fibers from the same specimens. Each specimen was prepared by hand pouring the fibers into a specimen container 25 mm thick and 310 by 310 mm. Several fiber density values in the range of 45 to 215 kg/m^3 were measured. The measurements showed that the effective thermal resistance of a 25-mm-thick barrier formed by normal slag-rock wool fibers with a fiber density of about 130 kg/m^3 is 0.65 (m^2 · °C/W). For a similar thickness and fiber density with the use of the small segments and granules of the disin-

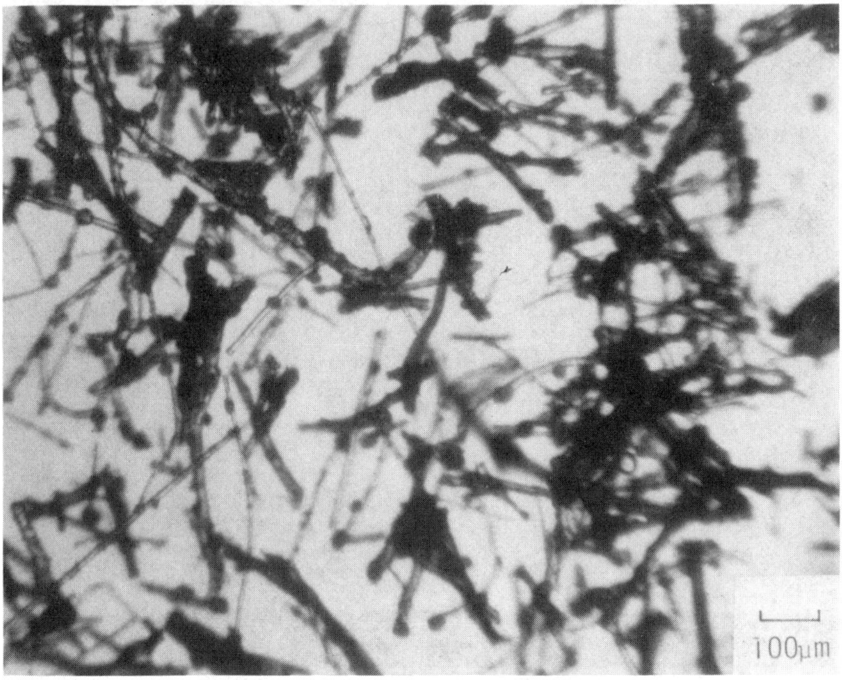

FIG. 8—*Optical photomicrograph of the circular nodules on the glass fiber stems exposed to water vapor and heated to 100°C for four months.*

tegrated slag-rock wool fibers, the thermal resistance of the barrier decreased to 0.50 ($m^2 \cdot °C/W$). Figure 12 shows the variation of the thermal resistance as a function of fiber density. Curve A is the thermal resistance of the normal slag-rock wool fibers, and Curve B is that of the disintegrated fibers. It is evident that the insulation barrier rendered by the disintegrated fibers is about 15 to 20% lower than that rendered by the normal, threadlike fibers. It must be pointed out that the larger difference in thermal resistance value at the low fiber density values within the 45 to 100-kg/m^3 range is attributed to the air space above the fibers in the container. A significant portion of the container space was unfilled by the small segments and granules of the disintegrated fibers because of the settlement of the fiber granules in the container.

Discussion

The present study has observed several experimental results. The pH value of the water solution was changed by the fiber immersion process. The surface characteristics of the slag-rock wool fibers were modified, and a physical desintegration occurred in both slag-rock wool fibers and glass fiber as a result of exposure of the fibers to a water medium under a low-temperature heat treatment. These material and physical changes could have been caused by some chemical reactions between the water medium and the surface of the mineral fibers. Two possible mechanisms may be responsible for these materials degradation processes; they can be interpreted according to the following considerations.

Surface erosion and the formation of fracture on the fiber stem in those fibers that had been heated at temperatures up to 60°C may be a consequence of a chemical leaching process. When

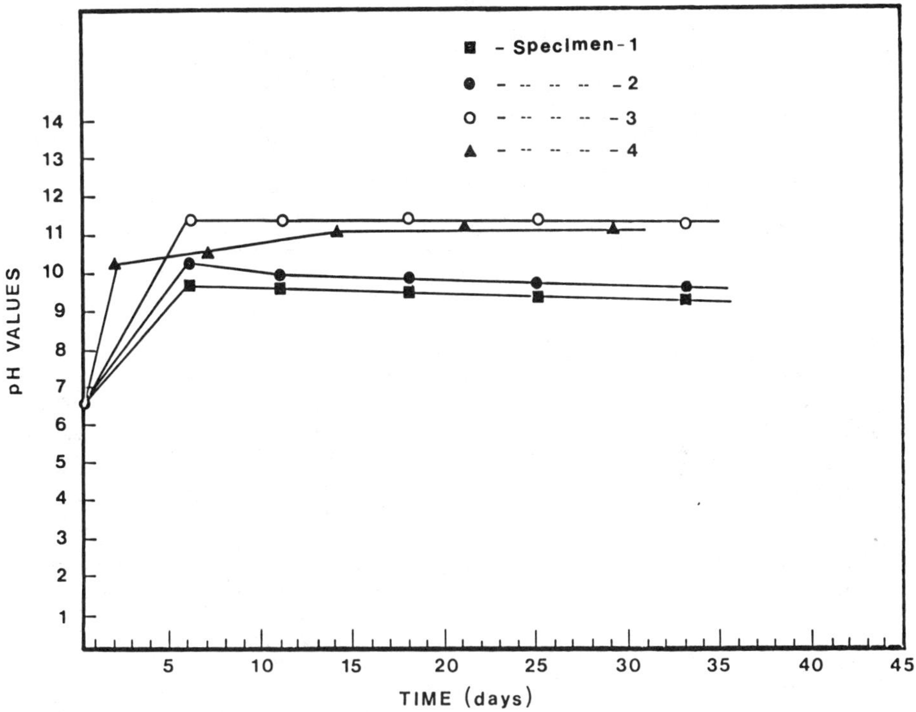

FIG. 9—*Modification of the pH values of distilled water by immersion of glass fibers and slag-rock wool fibers:* ■—*slag-rock wool fibers from Manufacturer A;* ●—*slag-rock wool fibers from Manufacturer B;* ○—*slag-rock wool fibers from Manufacturer C;* ▲—*glass fibers from Manufacturer D.*

the water medium is in contact with the mineral fiber surface, certain chemical species are removed from the fibers and dissolved in the water solution, thus changing the neutral water solution to an alkaline solution. Under the continual removal and accumulation of these leached-out components, some of the components begin to settle on some parts of the fibers that are not affected by the water chemical reaction, thus leading to the formation of the foreign substance and lumpy spots on the fibers. On the other hand, the removal of chemical species from the fibers would result in the formation of weak spots or stress concentrators in the fiber stem. Movement of the fibers by external vibration or by manual agitation would lead to fracture and subsequent disintegration of the fibrous form in the weakened spots, thus transforming the long, fine fibrous solids into small segments and granules.

The formation of the reticular network structure, or honeycomb morphology, on the surface of those fibers that have been immersed in water and heated to temperatures above 85°C evidently develops from a different chemical process. It is known that when portland cement grains come into contact with the water medium, a hydration process occurs, and the calcium silicate compound in the cement become hydrated. This hydration often causes the formation of a reticular network structure on the surface of the calcium silicate hydrates [*3*]. A close comparison of the physical appearance of the reticular network structure in the slag-rock wool fibers and that in the calcium silicate hydrates in portland cement finds them to be very similar. Second, silicate compounds, which are the sources contributing to the formation of the honeycomb structure, are present in both products. Based on these similarities, it seems not unreasonable

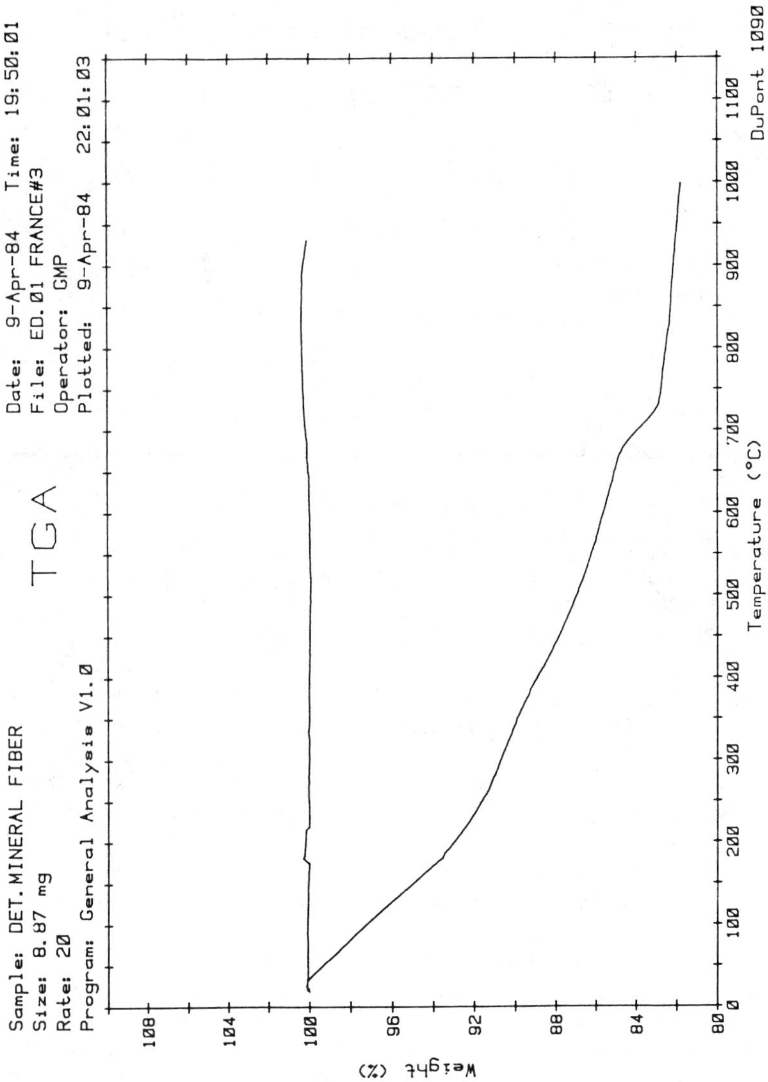

FIG. 10—*Thermogram of weight change as a function of heating temperatures in the thermogravimetric analysis measurement of normal slag-rock wool fibers (Curve A) and of the disintegrated slag-rock wool fibers immersed in water and heated to 100°C for 14 days (Curve B).*

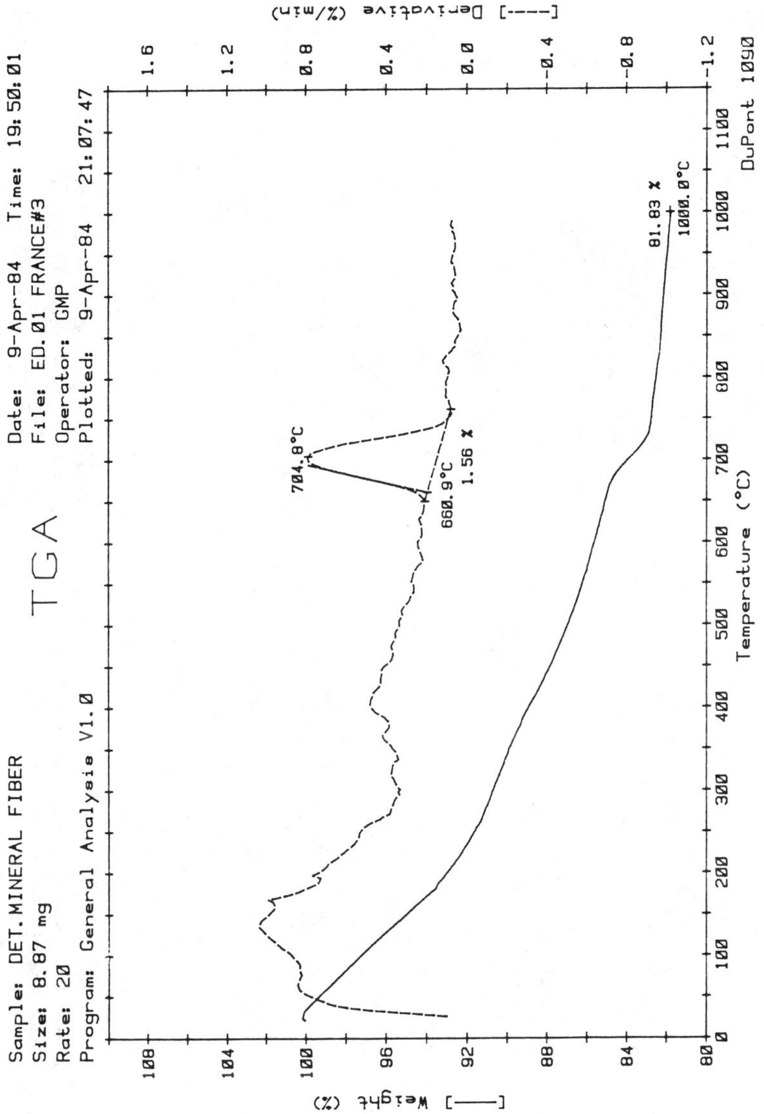

FIG. 11—*Thermogram of the abrupt and large rate of weight change in the thermogravimetric analysis measurement of disintegrated slag-rock wool fibers, a possible indication of the thermal dehydration process.*

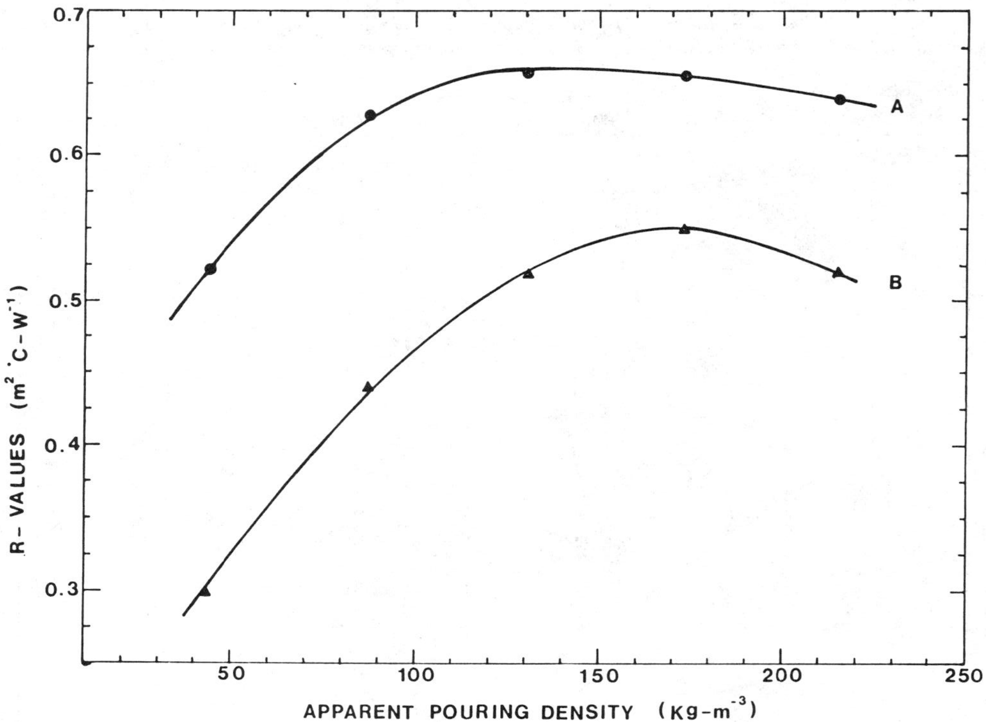

FIG. 12—*Variation of the thermal resistance of a 25-mm-thick test specimen that contained fibers of various fiber densities for the normal, threadlike slag-rock wool fibers (Curve A) and the small segments and granules of the disintegrated slag-rock wool fibers (Curve B).*

to suggest that a chemical hydration process occurs on the slag-rock wool fibers when they come into contact with the water medium and is activated by heat treatment at temperatures within the 85 to 100°C range.

The mechanism of these chemical hydration and leaching processes taking place between the inorganic slag-rock wool fibers and the water medium is not a totally unrealistic explanation in view of the chemical sources used to produce this type of inorganic fibrous product. According to information in the literature [*4*] mineral wool fibers of the slag-rock wool type are fabricated primarily from mechanical mixtures of metallurgical slags and natural rocks, which contain the main ingredients of silica (SiO_2), alumina (Al_2O_3), calcia (CaO), and magnesia (MgO) with a small proportion of other oxides such as iron oxide (Fe_2O_3) and the alkalies sodium oxide (Na_2O) or potassium oxide (K_2O). In order to control the melting temperature or the viscosity of the molten mixture, foreign oxide components are often added to the starting components. Furthermore, sulfur is frequently present in the metallurgical slags and natural rocks, and it is also invariably present in the coke, which is the major fuel source in the mineral wool fiber production process employing the conventional cupola furnace technique. Intermixing of all these chemical ingredients with homogenity in the melting process and in the fiberization of the molten mass would become difficult. Many of the fibrous solids produced would contain various chemicals with very little chemical homogeneity. It is quiet possible that some silicate-type and other related chemical compounds would be formed during the high-temperature melting and subsequent cooling processes. When these silicates and related compounds are in contact with the water medium and are activated by the heating process, the initiation of the leaching and hydration processes is not impossible.

The observation of a higher thermal conductivity or lower thermal insulation value in the small segments and granules of the disintegrated fibers is consistent with the nature of these particulate materials. If the disintegrated fibers have indeed become hydrated, the short fibers would contain a number of the hydrated water molecules in the fiber structure. The presence of water in the fiber structure would therefore enhance the heat conduction mode through the fiber mass. Second, small segments and granules of the disintegrated fibers generally have a greater tendency to settle in the container during the hand pouring process. More settling of the short fibers would reduce the overall pore space in the fiber mass. It is generally known that a decrease in the porosity of a particulate medium in the presence of air increases the thermal conductivity of the granulate material [*5*]. Based on these two considerations, the presence of the hydrated forms in the fiber structure and the reduction of porosity in the fiber mass would contribute to a higher thermal conductivity value in the disintegrated fibers of the slag-rock wool fibers, as are observed in the heat flow meter measurement.

Conclusion

Results of the present study show that when inorganic mineral fibers are in contact with water medium, chemical processes develop. If the water medium is in the form of liquid, certain chemical species are removed from the surface of the fibers and dissolved in the water solution, thus altering the pH value of neutral water into that of an alkaline solution. This leaching action was observed to cause slag-rock wool fibers to erode on their surfaces. The erosion and leaching processes also lead to fracture formation on the fiber stems. As a result, fiber disintegration develops, which transforms the fine, threadlike fibers into small segments and granules with manual agitation. When heating is elevated to temperatures above 85°C, the chemical reaction between the fibers and the water medium is activated to become a hydration process, which leads to the formation of a reticular network structure, or honeycomb morphology, on the fiber surfaces. Thermal conductivity measurements show that the small segments and granules of the disintegrated fibers have lower thermal insulation values than the normal, threadlike fibers.

Whether this material degradation process in the mineral fibers observed in the laboratory study could occur in similar materials and products in the field is uncertain; this possibility is yet to be investigated. However, the results of the present study could raise an awareness about the durability of inorganic mineral fibers under environmental influences. Furthermore, this experimental study could augment the many studies of the durability of conventional building materials and components documented in the literature [*6*].

Acknowledgment

This research project was supported in part by the Natural Sciences and Engineering Research Council of Canada under Grant No. A1192.

References

[*1*] Lejeune, G., "L'etude du Vieillissement des Laines Minerals," *Chimie et Industrie*, Vol. 73, No. 6, p. 1167.
[*2*] Kingery, W. D., *Introduction to Ceramics*, Wiley, New York, 1960, p. 336.
[*3*] Mindess, S. and Young, J. F., *Concrete*, Prentice-Hall, Englewood Cliffs, NJ, 1981, p. 96.
[*4*] Low, N. M. P., "A Critical Evaluation of Mineral Wool Fiber Manufacturing Processes," Technical report prepared under Contract No. OSQ77-00044, for the Canada Centre for Minerals and Energy Technology, Department of Energy, Mines and Resources, Ottawa, Ontario, Canada, 1977.
[*5*] Pratt, A. W., "Heat Transmission in Low Conductivity Materials," *Thermal Conductivity*, Vol. 1, R. P. Tye, Ed., Academic Press, New York, 1969, p. 301.
[*6*] Sereda, P. J. and Litvan, G. G., Eds., *Durability of Building Materials and Components, ASTM STP 691*, American Society for Testing and Materials, Philadelphia, 1980.

David L. McElroy,[1] *David W. Yarbrough,*[1] *and Ronald S. Graves*[1]

Thickness and Density of Loose-Fill Insulations After Installation in Residential Attics

REFERENCE: McElroy, D. L., Yarbrough, D. W., and Graves, R. S., **"The Thickness and Density of Loose-Fill Insulations After Installation in Residential Attics,"** *Thermal Insulation: Materials and Systems, ASTM STP 922,* F. J. Powell and S. L. Matthews, Eds., American Society for Testing and Materials, Philadelphia, 1987, pp. 493-505.

ABSTRACT: Limited information exists on the extent of settling of loose-fill attic insulations as a function of time after installation. This study reports thickness and density measurements in attics immediately after installation and periodically for two years following. Installations of fiberglass A, rock wool, fiberglass B, perlite, and cellulose in 44 attics in Ohio, Washington, and Tennessee were examined periodically.

The thickness, t, at fixed locations was measured using probes, rulers attached to ceiling joists, and polystyrene floats guided by rulers. The density, ρ, was determined using a cylindrical cutter to define a specimen of known volume that was weighed on site. The observed values of t and ρ were correlated with time after installation, and manufacturers' descriptions of products including thermal resistance expectations.

As part of the analysis of cellulosic insulation behavior, a series of laboratory tests of the blower-cyclone-shaker technique were conducted, and the findings from these tests are reported.

KEY WORDS: loose-fill attic insulations, settling, density, *R* value, *in situ* insulation density, *in situ* insulation thickness, cellulosic insulation settling test, cellulosic insulation density test, thermal insulation

Nomenclature

a, b, c	Parameters in Eq 1
k_a	Apparent thermal conductivity, $W/m \cdot K$
R	Thermal resistance, $m^2 \cdot K/W$
t	Thickness at a given time, t_0, initial value, m
ρ	Density at a given time, kg/m^3
ρ_0	Density, initial value
$\rho_{(BCS)}$	Density test results from blower-cyclone-shaker
ρ_{label}	Density derived from label information
mean	Mean of variable X,

$$\frac{1}{n} \sum_{i=1}^{n} x_i$$

[1]Metallurgist, chemical engineer, and engineer, respectively, Physical Properties Group, Metals and Ceramics Div., Oak Ridge National Laboratory, Oak Ridge, TN 37831.

σ_x Standard deviation,

$$\sqrt{\frac{1}{n-1}\sum_{i=1}^{n}(x_i-\bar{x})^2}$$

$\sigma_{\bar{x}}$ Standard deviation of the mean, $\sigma_x/\sqrt{n}$

The thermal resistance, R value, is calculated by dividing thickness by apparent thermal conductivity, k_a, and for loose-fill thermal insulations k_a depends on a number of factors including type of fiber or particle, thickness, temperature, and density. Thermal test results [ASTM Test for Steady-State Thermal Transmission Properties by Means of the Guarded Hot Plate (C 177-85); ASTM Test for Steady-State Thermal Transmission Properties by Means of the Heat Flow Meter (C 518-85)] provide k_a, from which R values are calculated. Loose-fill thermal insulation is marketed on the basis of R value, and a typical product label indicates simultaneous values for the minimum thickness and the minimum weight of insulation per unit area to achieve the R value. In addition, the area that can be insulated to a specific R value with the contents of the package is given. The manufacturer's specifications for attic usage is intended to account for any settling that occurs after the insulation is installed so that the R value does not fall below the specific value. If settling occurs, then the R value decreases from the initial value.

The decrease in R value with settling can be calculated from k_a, which is conventionally expressed as a function of density, as shown in Eq 1[2]

$$k_a = a + b\rho + \frac{c}{\rho} \tag{1}$$

The parameters a, b, and c in Eq 1 are given in Table 1 [*1–4*] for four loose-fill materials. These parameters were obtained by applying the method of least squares to thermal test results. This information can be combined with Eq 2 and the definition of R value to derive expressions for R value as a function of thickness and the rate of change of R value with thickness, given initial

TABLE 1—*Constants for the calculation of apparent thermal conductivity of loose-fill insulations as a function of density at a mean temperature of 297 K (75°F).*

Material	a	b	c	Ref
Fiberglass A	2.5956E-2[a] (0.1800)[b]	4.5010E-5 (0.0050)	3.8990E-1 (0.1688)	[*1,2*]
Rock wool	7.5561E-3 (0.0524)	2.2145E-4 (0.0246)	9.4473E-1 (0.4090)	[*3*]
Cellulose	3.8300E-2 (0.2656)	4.4470E-5 (0.00494)		[*4*]
Fiberglass B	1.5862E-2 (0.1100)	1.7914E-4 (0.0199)	4.3934E-1 (0.1902)	[c]

[a]Density, in kilograms per cubic metre; k_a, W/m · K; E-2 indicates power of ten, 10^{-2}.
[b]Numbers in parentheses are density, in pounds per cubic foot; k_a, Btu · in./ft^2 · h · °F.
[c]D. J. McCaa, CertainTeed Corp., private communication to R. S. Graves and D. L. McElroy, 13 April 1984.

[2]See the Nomenclature for identification of variables.

values for density and thickness (that is, a given weight of insulation per unit area). Equation 2 does not account for mass loss or gain due to water absorption by the insulation.

$$t_0\rho_0 = t\rho \tag{2}$$

$$R = \frac{\rho_0 t_0 t^2}{a\rho_0 t_0 t + b(\rho_0 t_0)^2 + ct^2} \tag{3}$$

$$\frac{dR}{dt} = \frac{a(\rho_0 t_0)^2 t^2 + 2b(\rho_0 t_0)^3 t}{[a\rho_0 t_0 t + b(\rho_0 t_0)^2 + ct^2]^2} \tag{4}$$

Equation 4 shows that dR/dt will be positive for the material parameters in Table 1. This means that a decrease in thickness will result in a decrease in R value. Table 2 contains values for dR/dt calculated with Eq 4 for products at thicknesses and densities for $R = 5.28$ $m^2 \cdot K/W$ ($R = 30$ $ft^2 \cdot h \cdot °F/Btu$). The tabulated dR/dt values represent the initial rate of R value decrease with thickness decrease. The R value as a function of thickness can also be calculated for a product installed at ρ_0 and t_0 using Eq 3. Figure 1 shows typical curves of R versus thickness for four loose-fill products installed at a weight per unit area required for $R =$ 5.28 $m^2 \cdot K/W$.

Figure 1 shows the variation of R value with thickness at constant mass per unit area. These curves demonstrate the need to know how the thickness of an installed loose-fill insulation changes with time, given that the initial installation meets the manufacturer's specifications for minimum thickness and minimum weight of material per unit area. The need to account for thickness loss is recognized by the Federal Trade Commission [5], and loose-fill cellulosic insulations are required to account for settling, with a laboratory test as a predictor [6]. The current work was undertaken to observe over long periods the thickness and density of installed loose-fill insulations in residential attics in order to determine if settling occurs.

Two sets of *in situ* density and thickness data have been collected. The first set of attic insulation densities collected in Knoxville, Tennessee; Tampa, Florida; Denver, Colorado; and Portland, Oregon, included installed materials up to ten years old [7,8]. This data set contains densities above typical label values for the products and shows the need for observations starting

TABLE 2—*Calculated values for the initial rate at which* R *value decreases with decrease in thickness.*

Material	Typical ρ_0, kg/m³ (lb/ft³)	t_0 for $R = 5.28$, m (in.)[a]	dR/dt, per cm (per in.)
Fiberglass A	10.7 (0.67)	0.332 (13.07)	0.068 (0.98)
Rock wool	28.8 (1.80)	0.247 (9.72)	0.093 (1.34)
Cellulose	36.0 (2.25)	0.211[b] (8.31)	0.261 (3.76)
Fiberglass B	13.8 (0.86)	0.265 (10.43)	0.083 (1.19)

[a] $R = 5.28$ $m^2 \cdot K/W$ corresponds to $R = 30$ $ft^2 \cdot h \cdot °F/Btu$.

[b] Loose-fill cellulose is normally installed at a greater thickness to compensate for settling.

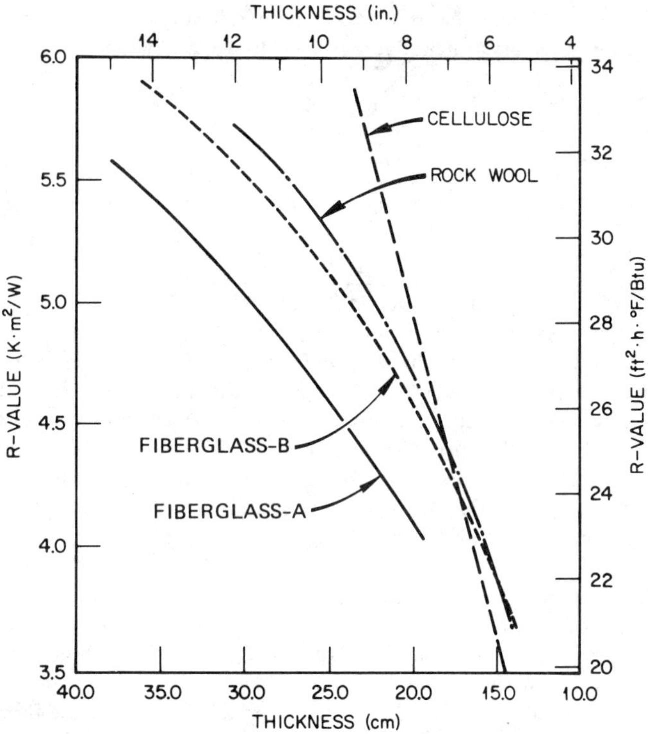

FIG. 1—R *value as a function of thickness for installations with mass of insulation required for* R = *5.28* $m^2 \cdot K/W$ *at the thickness and densities shown in Table 2.*

with installation since it was not possible to determine whether the insulations had settled or had been installed at high density. The second set of data, therefore, contains measurements of thickness and density for periods up to 2.3 years starting with installation. This second set of observations includes five types of loose-fill insulation that were routinely installed by insulation contractors. Residences in Bucyrus, Ohio; Columbia, Tennessee; and Seattle, Washington, were included in this part of the project.

Measurement of *In Situ* Thickness and Density

The densities of installed loose-fill insulation were determined by weighing known volumes of insulation. The volume of material for a density determination was calculated from an average of five depth measurements obtained with a 3.2-mm-diameter brass probe and the 0.3-m-diameter metal cylindrical cutter used to isolate insulation specimens. This material was removed from the cylinder and weighed to ±1.0 g using an electronic balance (Fisher Scientific Co., 10 lb/4536 g, Model EWO-4010). This method is the same as was used in an earlier study by A. D. Little, Inc. (unpublished report) except for the substitution of the electronic balance for a spring scale.

Insulation thicknesses were measured by five methods, two of which were eventually discarded. Thicknesses were periodically determined at marked attic locations using the average of five probe depths. Floats made from manila file folders and 2.54-cm-thick expanded polystyrene board were positioned on joist-mounted rulers and allowed to rest on the installed insulation to provide thickness as a function of time. The 21.6 by 27.9-cm folders and 30 by 122-cm polystyrene boards caused compressions of less than 5.0 mm [*9*]. The two measurement schemes that were found to be less satisfactory involved fixing a long wire horizontally in the attic at a known distance above the attic floor or mounting a wire grid perpendicular to the attic floor. These latter two methods interfered with installation and were discontinued because of inconvenience. These measurement techniques have been described in more detail by McElroy et al [*10*].

Thickness and Density Variations with Time

This section summarizes the data obtained in Bucyrus, Columbia, and Seattle using the techniques previously described. In each case, initial values for thickness and density were obtained within three days of installation. The initial and final values for density and thickness of the products tested are contained in Tables 3 through 6. Measurements were made at intermediate times and continued until the thickness changes between measurements were judged to be negligible, thus indicating that a final density had been achieved.[3]

Two houses in the Columbia segment of the project were partially insulated with loose-fill perlite. The perlite insulation was installed at a density of 48 kg/m^3 and a thickness of 0.2524 m

TABLE 3—*Summary of thickness and density observations for fiberglass A.*

Location	Type of Measurement	Number of Sites	Initial	Final	Change in Density, %
Seattle, WA, 5 attics, 616 days	cylinder, kg/m^3	10	15.55	16.00	2.89
	probe, m	10	0.3183	0.3111	2.31
	float, m	10	0.3297	0.3251	1.41
	ruler, m	25	0.3381	0.3332	1.47
	wire, m	15	0.3251	0.3215	1.12
	R value				
	K · m^2/W		6.38	6.34	−0.63 change in *R*
	°F · ft^2 · h/Btu		36.2	36.0	
	average increase in density = 1.71%				
	average value for final density (1.0171) (15.55) = 15.82 kg/m^3				
	label density = 11.53 kg/m^3				
	ratio of final density to label density = 1.37				
Bucyrus, OH, 8 attics, 831 days	cylinder, kg/m^3	16	11.19	12.94	15.64
	probe, m	16	0.4204	0.3947	6.51
	float, m	4	0.4202	0.4060	3.55
	ruler, m	7	0.4250	0.4025	5.59
	R value				
	K · m^2/W		6.88	6.83	−0.73 change in *R*
	°F · ft^2 · h/Btu		39.1	38.8	
	average increase in density = 9.48%				
	average value for final density (1.095) (11.19) = 12.25 kg/m^3				
	label density = 10.73 kg/m^3				
	ratio of final density to label density = 1.14				

[3]A listing of thickness and density data at individual attic sites can be obtained from the authors.

TABLE 4—*Summary of thickness and density observations for rock wool.*

Location	Type of Measurement	Number of Sites	Initial	Final	Change in Density, %
Seattle, WA, 5 attics, 616 days	cylinder, kg/m^3	10	50.72	50.24	−0.95
	probe, m	10	0.1877	0.1833	2.40
	float, m	10	0.2103	0.2103	0.00
	ruler, m	21	0.2135	0.2127	0.38
	wire, m	15	0.1959	0.1981	−1.11
	R value				
	K · m^2/W		5.44	5.43	−0.18 change in *R*
	°F · ft^2 · h/Btu		30.9	30.8	
	average increase in density = 0.09%				
	average value for final density (1.001) (50.72) = 50.77 kg/m^3				
	label density = 30.44 kg/m^3				
	ratio of final density to label density = 1.67				
Columbia, TN, 6 attics, 406 days	cylinder, kg/m^3	12	46.88	43.72	−6.74
	probe, m	12	0.2703	0.2646	2.15
	float, m	12	0.2733	0.2674	2.21
	ruler, m	30	0.2810	0.2726	3.08
	R value				
	K · m^2/W		7.27	7.10	−2.34 change in *R*
	°F · ft^2 · h/Btu		41.3	40.3	
	average increase in density = 0.97%				
	average value for final density (1.0097) (46.88) = 47.33 kg/m^3				
	label density = 27.14 kg/m^3				
	ratio of final density to label density = 1.74				

to provide an *R* value of 5.28 m^2 · K/W (30 ft^2 · h · °F/Btu). The perlite insulation did not decrease in thickness (settle) over the duration of the project (406 days).

The data in Tables 3 through 6 are arranged to show changes in density or thickness determined by each of the techniques used. Average changes in the density of the loose-fill products were computed by weighting the percentage change obtained for each measurement technique by the number of sites represented. Thickness and density ratios are interchangeable using Eq 2. It is recognized that a detailed data analysis that takes into account stratification of the data and the differences in measurement uncertainty inherent in each of the techniques could be undertaken. Detailed analysis of these results could result in minor changes in the reported averages but will have no impact on the conclusions. An Appendix contains a listing of mean values, standard deviations, and standard deviation of means for the reported measurements. The standard deviation of the mean is an indication of the uncertainty to be associated with the reported mean value for a variable.

The attics in Bucyrus insulated with fiberglass A and cellulose started near label values for density and show settling. The fiberglass A material was installed at an average density of 11.19 kg/m^3 (0.698 lb/ft^3), which is near the label value of 10.73 kg/m^3 (0.670 lb/ft^3). The average final density value for the fiberglass A insulation was 12.25 kg/m^3 (0.765 lb/ft^3), and this represents a 9.5% increase in density and a corresponding decrease in thickness. The final density was 14% above the value calculated from label information. The data in Table 2 can be used to show that a 9% density increase for a fiberglass A insulation installed at a thickness of 0.332 m and a density of 10.7 kg/m^3 to provide R = 5.28 m^2 · K/W would result in a predicted 3.8%

TABLE 5—*Summary of thickness and density observations for cellulose.*

Location	Type of Measurement	Number of Sites	Initial	Final	Change in Density, %
Bucyrus, OH, 8 attics, 831 days	cylinder, kg/m^3	16	35.99	46.08	28.04
	probe, m	16	0.2144	0.1744	22.94
	ruler, m	43	0.2290	0.1880	21.81
	R value				
	K · m^2/W		5.64	4.58	−18.79 change in *R*
	°F · ft^2 · h/Btu		32.0	26.0	
	average increase in density = 23.4% average value for final density (1.234) (35.99) = 44.41 kg/m^3 label density = 39.57 kg/m^3 ratio of final density to label density = 1.12				
Seattle, WA, 5 attics, 616 days	cylinder, kg/m^3	10	47.52	53.13	11.80
	probe, m	10	0.2156	0.1868	15.42
	float, m	9	0.2159	0.1963	9.98
	ruler, m	25	0.2236	0.2014	11.02
	wire, m	15	0.2262	0.2072	9.17
	R value				
	K · m^2/W		5.49	4.91	−10.56 change in *R*
	°F · ft^2 · h/Btu		31.2	27.9	
	average increase in density = 11.2% average value for final density (1.112) (47.52) = 52.84 kg/m^3 label density = 42.77 kg/m^3 ratio of final density to label density = 1.24				

decrease in *R* value. The *in situ* thickness and density data exhibit real variations, which influence the calculated initial and final *R* values. These data yielded a calculated initial *R* value of 6.88 m^2 · K/W (39.1°F · ft^2 · h/Btu) and a final *R* value of 6.83 m^2 · K/W (38.8°F · ft^2 · h/Btu). Both exceed the target *R* value, and the decrease in *R* value is less than the predicted change because of the real variations in the attics.

The cellulose insulation in Bucyrus was installed at 36.0 kg/m^3 (2.25 lb/ft^3), and this was the label value for as-blown material. The average final density for the material was 44.4 kg/m^3 (2.77 lb/ft^3) while the final density indicated by the label was 39.1 kg/m^3 (2.44 lb/ft^3). The *R* values obtained for these conditions, using the data from Tables 1 and 2, are 5.64 m^2 · K/W (32.0 ft^2 · h · °F/Btu) for the measured initial thickness and density, 4.58 m^2 · K/W (26.0 ft^2 · h · °F/Btu) for the measured final thickness and density, and 5.15 m^2 · K/W (29.2 ft^2 · h · °F/Btu) using the label values for final thickness and density.

The fiberglass A, rock wool, cellulose, and fiberglass B insulations in Columbia and Seattle were installed at densities above the label value. Settling was negligible for the rock wool installations and the Seattle fiberglass B. The Columbia fiberglass B had an average final density 4% above the initial value of 16.39 kg/m^3 (1.023 lb/ft^3), and the Seattle cellulose average final density was 52.8 kg/m^3 (3.296 lb/ft^3).

The *in situ* density and thickness measurements indicate settling in every case, but the changes were quite small in the case of the high-density rock wool and Seattle fiberglass B. Except for the products tested in Bucyrus, the insulations were installed at densities significantly above the label value; this limits the conclusions that can be made about settling of these products when installed at the label value. The data show that mechanical equilibrium occurred on the average at densities above the label value for the products tested.

TABLE 6—*Summary of thickness and density observations for fiberglass B.*

Location	Type of Measurement	Number of Sites	Initial	Final	Change in Density, %
Columbia, TN, 2 attics, 429 days	cylinder, kg/m^3	4	16.39	17.33	5.74
	probe, m	4	0.3186	0.3107	2.54
	float, m	4	0.3191	0.3076	3.74
	ruler, m	10	0.3191	0.3051	4.59
	R value				
	$K \cdot m^2/W$		7.00	6.88	−1.71 change in *R*
	$°F \cdot ft^2 \cdot h/Btu$		39.7	39.1	
	average increase in density = 4.27%				
	average value for final density (1.043) (16.39) = 17.09 kg/m^3				
	label density = 13.73 kg/m^3				
	ratio of final density to label density = 1.24				
Seattle, WA, 5 attics, 616 days	cylinder, kg/m^3	10	16.07	14.70	−8.53
	probe, m	10	0.2987	0.2790	7.06
	float, m	10	0.3069	0.3058	0.36
	ruler, m	25	0.3085	0.3058	0.88
	wire, m	15	0.3133	0.3051	2.69
	R value				
	$K \cdot m^2/W$		6.68	6.56	−1.80 change in *R*
	$°F \cdot ft^2 \cdot h/Btu$		37.9	37.2	
	average increase in density = 0.73%				
	average value for final density (1.0073) (16.07) = 16.19 kg/m^3				
	label density = 13.73 kg/m^3				
	ratio of final density to label density = 1.18				

Average *R* values for each set of test results have been calculated using Eq 3 and the constants given in Table 1. The calculated *R* values are based on the initial density measurements obtained with the cylinder technique and a weighted average of the thickness measurements. The final *R* values were calculated using the average value for the final density and a weighted average of the final thickness measurements. The calculated *R* value changes are not entirely consistent with the reported changes in density or thickness. The inconsistency stems from the fact that the initial *R* value calculation is directly related to the cylinder density measurement while the density change calculations include all the available data.

Laboratory Determination of Settled Density

Test methods for the settled density of mineral fiber loose-fill insulations have been under discussion within ASTM Committee C-16 on Thermal Insulation, but consensus has not been achieved. Laboratory tests that use mechanical vibration to settle loose-fill insulations give results that depend on the frequency, amplitude, and duration of the test [*11*]. Therefore, test parameters that yield densities like those reported in Tables 3, 4, 5, and 6 could be identified.

Loose-fill cellulosic products are tested for settling by a method known as the blower-cyclone-shaker (BCS) test [*6*]. The BCS test was studied in detail to determine factors affecting test results and to establish a link between BCS results and field data. The results of work done with the BCS have been summarized in interim reports by Graves et al [*12*] and McElroy et al [*13*].

The BCS study shows that factors not clearly identified in the test procedure significantly affect the results for settled density. A correlation between the blown or initial density of a material prepared for BCS testing and the BCS result for settled density has also been demonstrated, and this suggests a major change in the way the BCS test is used.

The evaluation of the BCS test procedure done in conjunction with an intralaboratory study showed that factors such as the mechanical condition of the shaker, the dimensions and location of the receiving beaker, and handling of material before testing measurably affect the test results. The entries in Table 7 show the magnitude of the changes in BCS settled density due to changes in the experimental apparatus and procedure.

Figure 2 shows a correlation between BCS settled density and the density of the blown product before the shaker test is performed. The figure also shows data for the settled density of the

TABLE 7—*Summary of results of BCS test procedure study.*

Variable	Change in Settled Density, kg/m^3
Shaker condition, old to new	4.0 to 4.8
Receiver height, 20.32 to 30.48 cm	3.2
Cyclone to receiver gap, 3.0 to 8.6 cm	4.8
Preblow material	1.3
Screen extended below cyclone	0.8
Voltage to blowers	1.6 to 3.2
Change of plate from 75 to 25 g	1.9
Change in relative humidity from 50 to 70%	0.0
Change in receiving beaker inside diameter from 14.0 to 15.2 cm	1.8

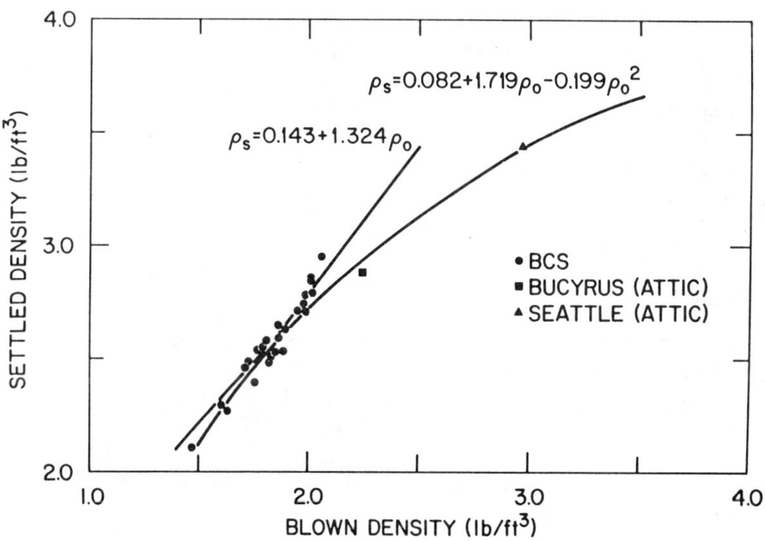

FIG. 2—*Correlations between blown density and settled density obtained from BCS and attic measurements.*

loose-fill cellulosic products that were installed in attics. The attic results from Bucyrus agree to better than 10% with the overall correlation between blown and settled density. The correlation constructed on the assumption that the field test data and the laboratory BCS data can be grouped together shows some curvature, and this is reasonable because high-density loose fills will clearly resist settling. Equation 5 relates BCS settled density to blown densities in the range of 24.0 to 33.6 kg/m³ (1.5 to 2.1 lb/ft³). Equation 6 covers a broader range of blown densities and includes the field test results.

$$\rho_{(BCS)} = 2.2906 + 1.324\,\rho_{0(BCS)} \tag{5}$$

$$\rho_{(BCS)} = 1.314 + 1.719\,\rho_{0(BCS)} - 0.0124\,\rho^2_{0(BCS)} \tag{6}$$

These data suggest that the BCS procedure is meaningful only if the product installed density matches the product density as blown for the BCS test. Differences between the two blown densities can occur because of differences in the equipment used to blow the insulation. A correction should be applied to the BCS settled density to obtain the predicted final density. The correction can be obtained from Eq 5 or 6. If Eq 5 is used, then the predicted final density is given by Eq 7. If Eq 6 is used, then the predicted final density is given by Eq 8.

$$\rho = \rho_{(BCS)} + 1.324[\rho_{0(label)} - \rho_{0(BCS)}] \tag{7}$$

$$\rho = \rho_{(BCS)} + [1.719 - 0.0124(\rho_{0(label)} + \rho_{0(BCS)})] \times [\rho_{0(label)} - \rho_{0(BCS)}] \tag{8}$$

Conclusions

The attic field test results for loose-fill mineral fiber insulations show final densities from 12 to 74% above densities derived from label information. With one exception, however, these materials were installed at densities above the label value, and the increases in density with time were either zero or small.

The attic field test results for loose-fill cellulosic insulations suggest that the BCS test for final density can be used to predict performance if the test variables are carefully controlled. Differences between densities of material blown for testing and blown in attics must be taken into account, and appropriate corrections have been formulated.

The field test results show that rock wool insulations installed at 46.9 kg/m³ (2.9 lb/ft³) and above did not settle. Fiberglass B installed at 16.1 kg/m³ (1.0 lb/ft³) settled about 4% in the worst case. Loose-fill fiberglass A installed near the label density settled about 9%, and this represents a decrease in R value of about 4%.

The loose-fill perlite that was tested did not settle.

Acknowledgments

This research was sponsored by the Office of Building and Community Systems, U.S. Department of Energy, under Contract DE-AC05-840R21400 with the Martin Marietta Energy Systems, Inc. The authors appreciate the participation of Richard Leuthold, president of Fiberchem Company, in the Bucyrus segment of the project; Lael Peterson, Thermoguard Insulation Company, in the Seattle segment; and Carl Haas, Tennessee Valley Authority, in the Columbia segment. The encouragement of W. Gerken and E. C. Freeman, program managers with the U.S. Department of Energy, is also gratefully acknowledged.

APPENDIX

Calculated Standard Deviations and Means for the Field Study Data Obtained in Bucyrus, Ohio; Columbia, Tennessee; and Seattle, Washington

Material and Measurement	Initial				Final			
	N	Mean	σ_x	$\sigma_{\bar{x}}$	N	Mean	σ_x	$\sigma_{\bar{x}}$
				SEATTLE				
Fiberglass A								
Cylinder, kg/m^3	10	15.55	1.51	0.48	10	16.00	1.15	0.36
Probe, m	30	0.3183	0.0132	0.0024	50	0.3111	0.0163	0.0023
Float, m	10	0.3297	0.0130	0.0041	10	0.3251	0.0133	0.0042
Ruler, m	25	0.3381	0.0182	0.0036	25	0.3332	0.0184	0.0037
Wire, m	15	0.3251	0.0227	0.0059	15	0.3215	0.0220	0.0057
Rock wool								
Cylinder, kg/m^3	10	50.72	4.35	1.38	10	50.24	4.61	1.46
Probe, m	30	0.1877	0.0158	0.0029	50	0.1833	0.0146	0.0021
Float, m	10	0.2103	0.0122	0.0039	10	0.2103	0.0126	0.0040
Ruler, m	21	0.2135	0.0154	0.0034	21	0.2127	0.0163	0.0036
Wire, m	15	0.1959	0.0165	0.0043	15	0.1981	0.0176	0.0045
Cellulose								
Cylinder, kg/m^3	10	47.52	2.39	0.76	10	53.13	3.34	1.06
Probe, m	30	0.2156	0.0169	0.0031	50	0.1868	0.0170	0.0024
Float, m	9	0.2159	0.0131	0.0044	9	0.1963	0.0130	0.0043
Ruler, m	25	0.2236	0.0139	0.0028	25	0.2014	0.0134	0.0027
Wire, m	15	0.2262	0.0233	0.0060	15	0.2072	0.0207	0.0053
Fiberglass B								
Cylinder, kg/m^3	10	16.07	1.47	0.46	10	14.70	1.15	0.36
Probe, m	30	0.2987	0.0174	0.0032	50	0.2790	0.0218	0.0031
Float, m	10	0.3069	0.0246	0.0078	10	0.3058	0.0239	0.0076
Ruler, m	25	0.3085	0.0211	0.0042	25	0.3058	0.0211	0.0042
Wire, m	15	0.3133	0.0289	0.0075	15	0.3051	0.0317	0.0082

Calculated Standard Deviations and Means for the Field Study Data Obtained in Bucyrus, Ohio; Columbia, Tennessee; and Seattle, Washington—*Continued.*

Material and Measurement	Initial				Final			
	N	Mean	σ_x	$\sigma_{\bar{x}}$	N	Mean	σ_x	$\sigma_{\bar{x}}$
				BUCYRUS				
Fiberglass A								
Cylinder, kg/m^3	16	11.19	1.17	0.29	16	12.94	1.92	0.48
Probe, m	48	0.4204	0.0385	0.0056	80	0.3947	0.0365	0.0041
Float, m	4	0.4202	0.0221	0.0110	4	0.4060	0.0193	0.0097
Ruler, m	7	0.4250	0.0200	0.0076	7	0.4025	0.0169	0.0064
Cellulose								
Cylinder, kg/m^3	16	35.99	2.30	0.58	16	46.08	1.88	0.47
Probe, m	48	0.2144	0.0164	0.0024	80	0.1744	0.0118	0.0013
Ruler, m	43	0.2290	0.0153	0.0023	43	0.1880	0.0128	0.0020
				COLUMBIA				
Rock wool								
Cylinder, kg/m^3	11	46.88	3.36	1.01	12	43.72	5.32	1.54
Probe, m	60	0.2703	0.0221	0.0029	60	0.2646	0.0211	0.0027
Float, m	12	0.2733	0.0208	0.0060	12	0.2674	0.0195	0.0056
Ruler, m	30	0.2810	0.0179	0.0033	30	0.2726	0.0187	0.0034
Fiberglass B								
Cylinder, kg/m^3	4	16.39	2.27	1.14	5	17.33	1.58	0.71
Probe, m	20	0.3186	0.0207	0.0046	20	0.3107	0.0167	0.0037
Float, m	4	0.3191	0.0101	0.0051	4	0.3076	0.0151	0.0076
Ruler, m	10	0.3191	0.0153	0.0048	10	0.3051	0.0160	0.0051

References

[1] Bomberg, M. and Solvason, K. R., *How to Ensure Good Thermal Performance of Blown Mineral Fibre Insulation in Horizontal and Vertical Installations*, NRC-C Report No. 167, National Research Council-Canada, Ottawa, Ontario, Canada, Aug. 1980, Table 11.

[2] Wilkes, K. E. and Rucker, J. L., "Thermal Performance of Residential Attic Insulations," *Energy and Buildings*, Vol. 5, No. 4, 1983, pp. 263–277.

[3] Bomberg, M. and Solvason, K. R., *How to Ensure Good Thermal Performance of Blown Mineral Fibre Insulation in Horizontal and Vertical Installations*, NRC-C Report No. 167, Aug. 1980, Table 11. The Equation for Table 10 values was adjusted to describe label values for ten rock wool products.

[4] Shirtliffe, C. J. and Bomberg, M. in *Blown Cellulose Fiber Thermal Insulations: Part 2—Thermal Resistance, ASTM STP 660*, American Society for Testing and Materials, Philadelphia, 1978, pp. 104–129.

[5] Federal Trade Commission, "Labeling and Advertising of Home Insulation," Rule 16 CFR Part 460, Federal Trade Commission, Washington, DC, 1978.

[6] "Federal Specification, Insulation Thermal (Loose-Fill for Pneumatic or Poured Application):Cellulosic or Wood Fiber," Specification GSA HH-I-515D, General Services Administration, Washington, DC, 15 June 1978.

[7] Yarbrough, D. W., Wright, J. H., McElroy, D. L., and Scanlan, T. F., "Settling of Loose-Fill Insulations Due to Vibration," in *Thermal Insulation, Materials, and Systems for Energy Conservation in the 80's, ASTM STP 789*, American Society for Testing and Materials, Philadelphia, 1983, pp. 703–714.

[8] Yarbrough, D. W., McElroy, D. L., and Harris, W. W., *Properties and Testing of Loose-Fill Cellulosic Insulation*, ORNL/TM-6433, Oak Ridge National Laboratory, Oak Ridge, TN, February 1982, pp. 19, 21.

[9] Yarbrough, D. W. and Wright, J. H., *Reduction in the Thermal Resistance (R-value) of Loose-fill Insulation and Fiberglass Batts Due to Compression*, ORNL/Sub-7715/2, Oak Ridge National Laboratory, Oak Ridge, TN, April 1981.

[10] McElroy, D. L., Yarbrough, D. W., and Graves, R. S., "An In-Situ Study of Attic Loose-Fill Thermal Insulation in Residential Applications," *Thermal Performance of the Exterior Envelopes of Buildings II*, ASHRAE SP 38, American Society of Heating, Refrigerating, and Air Conditioning Engineers, Atlanta, GA, 1983, pp. 343–357.

[11] Yarbrough, D. W. and Wright, J. H., *The Effect of Vibrations on the Density of Loose-Fill Insulations*, ORNL/Sub-7715/3, Oak Ridge National Laboratory, Oak Ridge, TN, May 1981.

[12] Graves, R. S., McElroy, D. L., and Yarbrough, D. W., "Interim Report on an Intercomparison of the Blower-Cyclone-Shaker Technique," paper presented at an ASTM Committee C16 on Thermal Insulation Meeting, San Antonio, TX, 16–20 April 1984.

[13] McElroy, D. L., Yarbrough, D. W., and Graves, R. S., "Interim Report on Attic Field Test of Loose-Fill Cellulosic Insulations," paper presented at an ASTM Committee C16 on Thermal Insulation Meeting, Philadelphia, PA, 19–21 Sept. 1983.

Marion Hollingsworth, Jr. [1]

Thermal Testing of Reflective Insulations

REFERENCE: Hollingsworth, M., Jr., **"Thermal Testing of Reflective Insulations,"** *Thermal Insulation: Materials and Systems, ASTM STP 922,* F. J. Powell and S. L. Matthews, Eds., American Society for Testing and Materials, Philadelphia, 1987, pp. 506–517.

ABSTRACT: ASTM hot box tests have traditionally been specified for the determination of thermal performance of reflective insulations designed for building applications. These methods require that specimens be representative of the construction to be investigated; however, ASTM committees have yet to standardize the details of specimen mounting. In building applications, multilayer reflective insulations are almost always mounted between framing members (joists or studs), and it is logical to duplicate this construction in hot box evaluations. A variant procedure followed under California regulations allows the nailing flanges of adjacent insulation pieces to be fastened together to form a blanket large enough to stretch across the hot box with no framing members in the test area.

This paper presents a comparison of hot box data showing that such tests, where no framing members are present, can result in the overstatement of *R* values by as much as nearly 70%, in comparison with data from specimens mounted in the normal manner between framing members, even though the thermal resistance of the framing members has been calculated out. Thus, it appears that significant lateral heat transfer occurs between the insulation and the framing members, and, therefore, the total system must be tested if the results are to be realistic.

KEY WORDS: reflective insulation, multilayer insulation, thermal insulation, thermal resistance, *R* value, thermal conductance, thermal performance, guarded hot box

Multilayer reflective insulations have, for many years, occupied a small but significant portion of the total market of thermal insulation used in buildings. In spite of this long history, much is still not known concerning their use and evaluation. A recent report by Yarbrough [*1*] gives a comprehensive assessment of the status of these products and points to those areas where controversy exists and additional knowledge is needed. It must be noted that this paper addresses only the narrow subject of the procedures used in hot box evaluations of thermal performance. For a broader discussion of these products and the use of other possible evaluation techniques, the reader is referred to Yarbrough [*1*].

Product Description and Usage

These products generally consist of two or more layers of aluminum foil, or aluminum foil laminated to one or both sides of kraft paper, which are joined at their long edges to form nailing flanges. They are shipped flattened in rolls. Proprietary and often ingenious designs provide for expansion to form the required air spaces between layers when the products are stretched laterally between building framing members. It is the series of reflective surfaces limiting the transfer of radiant energy across adjoining air spaces that reduces the overall heat transfer and thus allows these products to serve as thermal insulations. Since thermal perfor-

[1]Retired from Research and Development Division, Owens-Corning Fiberglas Corp., Technical Center, Granville, OH 43023.

mance is highly dependent on the nature of the surroundings, which exchange radiation with the product surfaces, the total system of product plus surroundings must be considered when describing thermal performance.

While these products can significantly reduce radiant transfer, conduction and convection within the air spaces may still occur, and it is because of convection effects that the thermal performance is sensitive to the orientation and direction of heat flow. Thus, it is more than accidental that the largest market for multilayer reflective insulations is in horizontal ceiling applications in warmer climates, where the buoyancy of heated air prevents downward convection and higher effective thermal resistances are achieved. These horizontal applications of multilayer reflective insulations, both for heat flow up and for heat flow down, are the subject of this paper. Vertical wall applications, where convection effects can be more severe, are not addressed.

Test Procedures

Since the dependence of thermal performance on the surroundings is generally recognized, proposals for thermal testing of reflective insulations invariably require large-scale specimens that duplicate, as nearly as is practical, the size, orientation, and other conditions of the installation being investigated. Traditionally, these proposals have used hot box apparatus, such as is described in either the older guarded box method, the ASTM Test for Steady-State Thermal Performance of Building Assemblies by Means of a Guarded Hot Box (C 236-80), or the newer calibrated box method, the ASTM Test for Thermal Performance of Building Assemblies by Means of a Calibrated Hot Box (C 976-82). Recently, the use of field test data, either from instrumented houses or from special outdoor laboratory buildings, where comparisons can be made more reliably, has been advocated (see Ref *1*), but it is laboratory hot box testing that is the subject of this paper.

Even when there is agreement to use hot box tests, the proposals often disagree on the details of specimen mounting and data interpretation, both of which can greatly affect the results. In a previous paper [*2*], this author briefly discussed some of the proposals made for thermal testing of reflective insulations.

Probably the most common procedure is to construct a test panel with standard framing members at standard spacings and to mount the insulation product between them in the normal manner. The faces of the panel are then closed with plywood or gypsum board sheets. Test results for such panels incorporating framing members in the test area directly represent performance of an actual installation, including the heat transfer through the framing members, and thus meet the requirements of the ASTM hot box methods that specimens be representative of the construction to be investigated. From such data, it may be desirable to calculate out the direct transfer through the framing members to arrive at an estimation of the thermal resistance of the insulated cavity alone. One common procedure for this calculation is the parallel path methodology [*3*] of the American Society of Heating, Refrigerating, and Air Conditioning Engineers (ASHRAE), using published R values for wood [*4*]. Although this procedure strictly applies to situations where there is no lateral heat transfer between the two paths, it may be used in this case to obtain a reasonable approximation. This value can be considered the effective R value of the insulation, but one must remember that it strictly applies only to the product when mounted in the same manner, in the same surroundings, and under the same conditions used during the test. Thus, it is important for such products, whose performance is highly dependent on their surroundings and the details of mounting, that derived R values be based on tests which duplicate normal end-use conditions as closely as possible.

A second procedure that has been advocated by some reflective insulation manufacturers is to use a test panel consisting of only the peripheral frame and the facing cover sheets, with no framing members inside. In this case, adjacent pieces of the insulation are joined at their edges by fastening their nailing flanges together (by stapling, glueing, or taping). The resulting large blanket is stretched across the frame and fastened at its edges. Proponents of this procedure argue that the results obtained represent the thermal performance of the reflective insulation alone, and thus can be compared directly with results quoted for competing mass insulations, such as fiberglass or rock wool, without the uncertainty of additional calculations.

Data from such tests, however, may not be representative of performance for two reasons. The first is that the exact configuration and position in the cavity is different for specimen pieces mounted to each other than it is for those mounted between framing. The other reason is that without the adjacent framing members, any increase in heat flow due to two or three-dimensional interaction will be reduced, and therefore quoted thermal resistances will be artificially high. The need to consider multidimensional heat flow effects in these products has been pointed out by Yarbrough [*1*]. The prime objective of the work reported here was to obtain data so that a direct comparison of the results of these two test procedures—that is, tests in panels with and without framing members—could be made.

Standards and Regulations

The requirements of thermal insulations are usually given in standard specifications, which are incorporated, in whole or in part, into codes and adopted by regulatory agencies. Such standards, to be meaningful, must not only specify the product properties but must give, in sufficient detail to avoid ambiguity, the test procedures to be followed in determining the specified properties. In the case of reflective insulations, unfortunately, there has been no agreement on the details of testing for thermal performance, the most important property of a thermal insulation. As a consequence, ASTM Subcommittee C16.21 on Reflective Insulations, a subcommittee of Committee C-16 on Thermal Insulation, has been unable to produce an acceptable specification standard.

Despite the lack of agreement on exact test procedures, some regulatory bodies have adopted regulations for these products. Notable among these is the Federal Trade Commission (FTC) [*5*], which specifies the use of ASTM Test C 236 (or ASTM Test C 976) at a mean temperature of 23.9°C (75°F) and with a temperature differential of 16.7°C (30°F). Since no details of the test panel and specimen mounting or of the data interpretation are specified, a wide range of *R*-values can result, depending on the configuration tested.

Another notable regulation is that of the State of California [*6*], which required tests in accordance with ASTM Test C 236 in a panel with a nominal 2-by-6 wooden frame, actually 38 by 140 mm (1.5 by 5.5 in.), with 19-mm (3/4-in.) plywood sheets on both sides, and states that "the resultant thermal performance shall be based upon the insulation only." Testing conditions were set at a 23.9°C (75°F) ± 1.1°C (2°F) mean temperature, with at least a 22.2°C (40°F) temperature difference. These California regulations were written to permit testing in panels without framing members in the test area. Note that the temperature difference requirements of the FTC and the State of California are mutually exclusive, thus further compounding the differences expected under the two sets of regulations.

Laboratory Tests

Test Plans

Pursuant to the narrow objective of providing only a comparison of the two test configurations—one with normally spaced framing members (joists) and the other with no framing members—it was felt that tests on a limited number of products would be sufficient. Fortunately, test reports for the configuration with no framing members in the test area were already publicly available [*7*], and it was thus necessary only to obtain data for the same products but mounted between framing members. To avoid criticism of the validity of the comparisons, the new tests would be run in the same guarded hot box and under the same nominal test conditions as the older tests.

Apparatus and Test Panels

The apparatus used for all data reported here is an ASTM Test C 236 guarded hot box, accredited by the National Voluntary Laboratory Accreditation Program (NVLAP) of the U.S. Department of Commerce (administered by the National Bureau of Standards), taking a 2.4 by 2.4-m (8 by 8-ft) specimen and with a 1.5 by 1.5-m (5 by 5-ft) metering area. The test panel was built up of nominal 2-by-6, actually 38 by 140-mm (1.5 by 5.5-in.), joists spaced 610 mm (24 in.) apart on center and surrounded by a 2-by-6 perimeter frame, as shown in Fig. 1. (The earlier tests without the interior members used a perimeter frame only.) Figure 1 also shows the location of the twelve exterior surface thermocouples and the two cover screws on each side in the test area. Corresponding air temperature thermocouples were also located on each side. No interior thermocouples were installed, nor were convection barriers used inside the test panels for either set of tests (with or without interior framing). Note that two choices were made that tend to produce a lower total heat flow, and therefore a higher apparent thermal resistance,

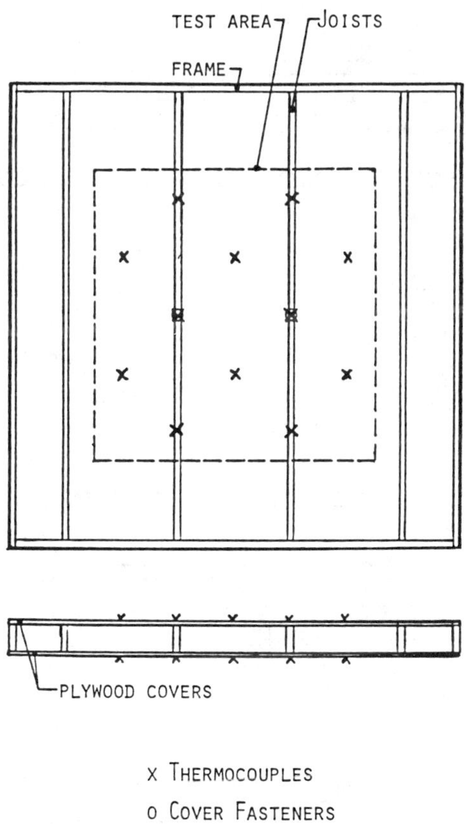

FIG. 1—*Test panel.*

than would occur in normal applications. The first is the use of only two cover fasteners on each side in the test area. Normal applications of facing materials (roof decking, ceiling drywall, and so on) would use fasteners spaced 180 to 300 mm (7 to 12 in.) apart, thus introducing additional metallic paths for heat transfer. The second feature reducing heat flow is the positioning of only two joists in the 1.5-m (5-ft) test width, which is less than the two for every 1.2 m (4 ft) in normal applications. An alternative layout would have placed one joist at the centerline, thus introducing three joists in the test area, which would have been more than normal.

Test Specimens

Test specimens for the new tests (specimens mounted between joists) were from new samples obtained through normal distribution outlets, since the samples used for the previous tests (without joists) were presumed to have been discarded. One product from each of three different manufacturers was chosen to match those in the available data from tests without joists, thus allowing direct comparisons. The nominal configurations of these products during the hot box tests are shown in Fig. 2. It should be noted that Product A includes one more layer, and therefore one more air space, than the other two, and and because of this would be expected to exhibit superior performances. Examination of the test data discussed later (see Table 1), however, shows no consistent advantage for Product A. Two observations may provide the explanation. It was noted at the time of installation that even with special attention Product A failed to

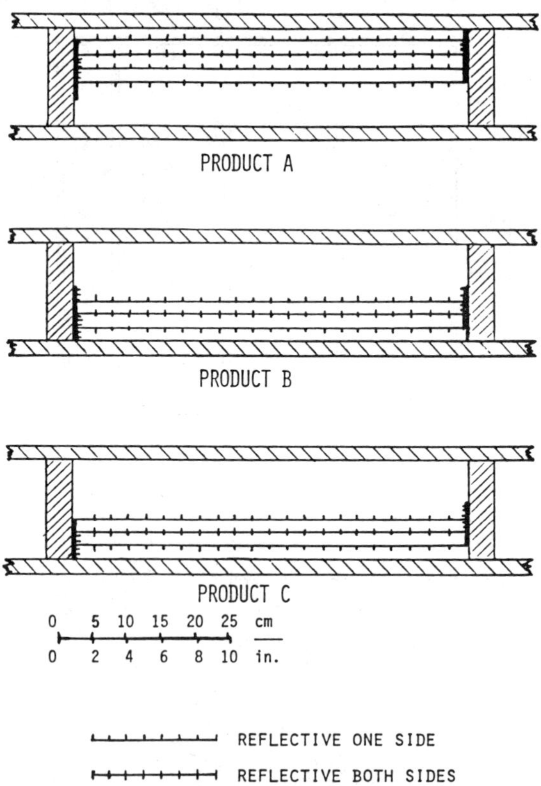

FIG. 2—*Test configurations.*

give full air space deployment, apparently because of inadequate stiffness at the edges. And it was discovered some time after the tests and calculations were complete, that the single-sided inner layer in Product A was reversed in manufacture (as shown in Fig. 2), so that one air space was completely nonreflective. It is not known whether this defect occurred in the specimen used for the earlier tests, and no attempt was made to allow for it in the calculations reported in this paper.

Test Results and Discussion

All the tests were run following the standard procedures of ASTM Test C 236 and using nominal air temperatures of 37.8°C (100°F) on the warm side and 10°C (50°F) on the cold side. Tests were run with the specimen horizontal, with temperatures adjusted for both heat flow up and heat flow down. Results of tests on both configurations (with and without joists) are given in Table 1, where all thermal resistance values have been adjusted to exclude the resistance of the two plywood cover sheets, and thus represent the effective resistance between the inner faces of the cover sheets. The resistance values given in Table 1 give a direct measure of the thermal performance of the total system, including the framing members (joists) when used, and thus are valuable in determining the typical structure heat loss. It is important to note that the thermal resistance of the insulations that do not contain the interior framing joists is significantly greater than that for those in which the reflective insulation is mounted in the normal manner between framing members.

Also included in Table 1 are calculated resistance values through the insulation path alone for the tests in which the joists were present. This calculation uses the ASHRAE parallel path concept [*3*] that heat transfer through the two paths—one through the wood joists and the other through the cavities containing the insulation—is parallel and independent and driven by the same temperature differences. These calculated values give an effective R value for the insulation as installed, which, since the direct heat transfer through the joists has been eliminated, can more logically be compared with that of other insulations. It is important to note in Table 1 that the data from the tests without joists overstate the R values by amounts ranging from 23 to 80%, in comparison with those obtained from the tests with the insulation mounted between joists.

Calculation Techniques—Other Configurations and Temperatures

Reflective insulation data from hot box tests such as those in Table 1 strictly apply only to the exact configuration tested and to the exact conditions of the test. Often data are needed for a configuration or for conditions that differ from those of the available test data. In such cases, an assessment must be made of the practicability of conducting new tests versus the risk of errors in calculations for the new configuration (or new conditions) based on the existing data. Sometimes it is impossible to achieve the new conditions with available test apparatus, and a calculation procedure must be used. Thus, a suitable calculation procedure for extending test data to other configurations and to other temperature conditions is desirable.

Calculation Procedure

The calculation procedure used in this paper, which has been used in deriving some of the data submitted for the certification of reflective insulation products, is based on the ASHRAE tables for the thermal conductance and resistance of a plane air space [*8*]. Since these published values are for ideal specimens (plane and parallel surfaces, no air leakage, and so on), a factor is introduced to provide for increased heat transfer due to the nonideal characteristics inherent in commercial multilayer reflective insulations. The assumption is made that the same factor

TABLE 1—*Hot box test results in SI and inch-pound units.*

Result	Product A, Heat Flow Direction		Product B, Heat Flow Direction		Product C, Heat Flow Direction	
	Down	Up	Down	Up	Down	Up
SI Units						
Specimens mounted between 38 by 140-mm joists, 610 mm on center						
Warm air temperature, °C	38.4	38.3	38.4	38.3	38.4	38.3
Warm surface temperature, °C[a]	37.2	36.5	37.3	36.3	37.4	36.5
Cold surface temperature, °C[a]	10.7	10.8	10.6	10.7	10.6	10.7
Cold air temperature, °C	9.9	9.9	9.9	9.9	9.9	9.8
Heat flux, average, W/m^2	11.17	17.13	10.57	17.92	10.38	16.53
Measured average *R* value, K · m^2/W[b]	2.04	1.18	2.20	1.11	2.25	1.23
Calculated insulation *R* value, K · m^2/W[b,c]	2.11	1.18	2.29	1.09	2.34	1.23
Specimens mounted together—no joists						
Warm air temperature, °C	37.6	37.5	38.0	37.9	37.5	37.4
Warm surface temperature, °C[a]	36.8	36.6	37.2	36.6	36.6	36.0
Cold surface temperature, °C[a]	10.1	10.5	10.9	11.2	10.5	10.8
Cold air temperature, °C	9.7	9.9	10.6	10.4	10.0	9.9
Heat flux, average, W/m^2	8.08	10.73	7.92	12.71	8.14	12.43
Measured average *R* value, K · m^2/W[b]	2.98	2.11	2.98	1.67	2.89	1.71
Percent overstatement, %[d]	42	80	31	53	23	38

Inch-Pound Units

Specimens mounted between 2 by 6 joists, 24 in. on center						
Warm air temperature, °F	101.1	101.0	101.1	101.0	101.1	101.0
Warm surface temperature, °F[a]	99.0	97.7	99.2	97.4	99.3	97.7
Cold surface temperature, °F[a]	51.3	51.4	51.1	51.3	51.1	51.2
Cold air temperature, °F	49.9	49.9	49.9	49.9	49.9	49.6
Heat flux, average, Btu/h · ft²	3.54	5.43	3.35	5.68	3.29	5.24
Measured average *R* value, °F · h · ft²/Btu[b]	11.6	6.7	12.5	6.3	12.8	7.0
Calculated insulation *R* value, °F · h · ft²/Btu[b,c]	12.0	6.7	13.0	6.2	13.3	7.0
Specimens mounted together—no joists						
Warm air temperature, °F	99.6	99.5	100.4	100.3	99.5	99.4
Warm surface temperature, °F[a]	98.3	97.9	98.9	97.9	97.8	96.8
Cold surface temperature, °F[a]	50.2	50.9	51.7	52.2	50.9	51.4
Cold air temperature, °F	49.4	49.8	51.0	50.8	50.0	49.9
Heat flux, average, Btu/h · ft²	2.56	3.40	2.51	4.03	2.58	3.94
Measured average *R* value, °F · h · ft²/Btu[b]	16.9	12.0	16.9	9.5	16.4	9.7
Percent overstatement, %[d]	42	80	31	53	23	38

[a]Surface temperatures measured on the outer surfaces of plywood covers. These are simple averages of temperatures from the twelve thermocouples, six over the joists and six over the cavities (see Fig. 1).

[b]Thermal resistances (*R* values) are between the inner surfaces of plywood covers obtained by subtracting ASHRAE values for the two plywood covers [0.164 K · m²/W (0.93°F · h · ft²/Btu) each] from *R* values measured between the outer surfaces.

[c]Insulation *R* values are for the path through the insulated cavity alone. The effect of the direct heat transfer through the joists was calculated out using an ASHRAE parallel path calculation.

[d]The percent overstatement is the percentage by which the *R* values obtained from tests with no joists overstate performance compared with those obtained from tests with the reflective insulation mounted between joists, but with the direct heat transfer through the joists calculated out.

(which is a multiplier of the air space conductance) applies equally to all air spaces within the insulation system under all conditions of use. The factor is unique for a particular product, product width, direction of heat flow, and mounting situation (and thus is different when mounting is between joists than when there are no joists). The factor can be obtained from calculations of the existing test configuration by adjusting its value until the calculation produces the measured total thermal resistance. It is then used in a similar calculation to predict performance under the new conditions. In such calculations, the thermal resistances for any exposed surfaces are added without modification by the multiplying factor.

Air Space Correlation

The air space conductance and resistance values are dependent on the direction of heat flow, the air space thickness, the temperature difference, the mean temperature, and the effective emittance (which, for reflective insulations, is taken to be 0.05). Since the values of the factor, the temperatures of the layers, and the heat flux are all interrelated in the calculation, it is necessary to use an iterative process to determine their magnitude. This requires interpolation of the ASHRAE air space table entries, which is conveniently done with a computer correlation. The correlation used for the calculation data reported in this paper was produced in-house, from the National Bureau of Standards data [*9*] underlying the ASHRAE tables, and is unpublished. Another such correlation has been published by Yarbrough [*1*]. These and other correlations of these data may differ slightly, and thus small differences in calculated factors and predicted resistances may occur.

Calculation Examples

The calculation procedure was used first to determine the multiplying factors and then to predict the thermal performance of the products when installed approximately 76 mm (3 in.) below a roof deck with the bottom face exposed. Such installations are typical in many commercial building applications. The multiplying factors determine from the test data reported here are given in Table 2. For the specimens mounted between joists, the resistance used was that through the insulation path, that is, with the effect of the direct heat transfer through the joists removed by the parallel path calculation. It is interesting to note that, for the specimens tested without joists, the multiplying factors are small, averaging only 1.14 (that is, 14% greater heat flow than the ASHRAE prediction) for heat flow down and near unity for heat flow up. For the specimens mounted between joists, the multiplying factors are much larger. For heat flow down, the average increase in heat flow over the ASHRAE prediction is 49%; for heat flow up, it is 57%. Note that the multiplying factors for Product A, mounted between joints, are significantly higher than those for the other two products. The probable explanation for this is the lack of full air space deployment and the manufacturing defect of the inner layer reversal noted

TABLE 2—*Factors necessary to multiply ASHRAE air space conductances to achieve hot box measured thermal performance.*

	Product A, Heat Flow Direction		Product B, Heat Flow Direction		Product C, Heat Flow Direction	
Specimen	Down	Up	Down	Up	Down	Up
Specimens between joists	1.716	1.841	1.396	1.530	1.356	1.344
Specimens together—no joists	1.213	0.995	1.107	0.994	1.108	0.969

earlier. These imperfections were not considered in the calculations. It is not clear whether they were also present in the earlier tests (with no joists), since the multiplying factor is higher, in relation to the other products, for heat flow down but not for heat flow up.

The calculated thermal resistances for this exposed application are given in Table 3. In these calculations, the fixed temperatures of 38°C (95°F) on the warm side and 12.8°C (55°F) on the cold side were taken to be at the underneath surface of the plywood roof deck and at the bottom (exposed) surface of the insulation. ASHRAE values [*10*] for the downward-facing bottom surface thermal resistance of 0.232 K · m²/W (1.32°F · h · ft²/Btu) for heat flow up and 0.801 K · m²/W (4.55°F · h · ft²/Btu) for heat flow down were used (not modified by the factor). The comparisons show that, on the average for heat flow down, the calculated thermal resistances are 21% greater when based on tests with the insulation mounted together, with no joists used, than when based on tests with the insulation mounted between joists. For heat flow up, the same comparison yields an average of 46%, with values ranging as high as 67%.

Discussion and Conclusions

The most important conclusion evident from this study is that the heat flow through multilayer reflective insulations is significantly less when the insulations are tested alone, that is,

TABLE 3—*Insulation R-values calculated for installations with bottom surface exposed.*[a]

Result	Product A, Heat Flow Direction		Product B, Heat Flow Direction		Product C, Heat Flow Direction	
	Down	Up	Down	Up	Down	Up
SI UNITS (temperatures of 35°C and 12.8°C)						
Based on tests of specimens mounted between joists						
Insulation spaces, K · m²/W	1.80	0.88	1.89	0.76	1.89	0.86
Bottom air film, K · m²/W	0.80	0.23	0.80	0.23	0.80	0.23
Total *R* value, K · m²/W	2.60	1.11	2.69	0.99	2.69	1.09
Based on tests of specimens mounted together—no joists						
Insulation spaces, K · m²/W	2.54	1.63	2.42	1.18	2.32	1.19
Bottom air film, K · m²/W	0.80	0.23	0.80	0.23	0.80	0.23
Total *R* value, K · m²/W	3.34	1.86	3.22	1.41	3.12	1.42
Percent overstatement, %	29	67	20	42	16	30
INCH-POUND UNITS (temperatures of 95°F and 55°F)						
Based on tests of specimens mounted between joists						
Insulation spaces, °F · h · ft²/Btu	10.20	4.99	10.74	4.32	10.75	4.88
Bottom air film, °F · h · ft²/Btu	4.55	1.32	4.55	1.32	4.55	1.32
Total *R* value, °F · h · ft²/Btu	14.8	6.3	15.3	5.6	15.3	6.2
Based on tests of specimens mounted together—no joists						
Insulation spaces, °F · h · ft²/Btu	14.44	9.24	13.77	6.68	13.15	6.77
Bottom air film, °F · h · ft²/Btu	4.55	1.32	4.55	1.32	4.55	1.32
Total *R* value, °F · h · ft²/Btu	19.0	10.6	18.3	8.0	17.7	8.1
Percent overstatement, %	29	67	20	42	16	30

[a]Temperatures were taken at the bottom surface of the deck and at the bottom surface of the insulation.

fastened together without framing members between adjacent strips, than when the same products are mounted in the normal manner between framing joists. This statement is true even though the direct heat transfer through the joists has been calculated out using parallel path calculation methodology. It is thus apparent that thermal resistance *R* values obtained from tests on such insulations mounted without the framing members will overstate the *R* value performance and will thus be misleading in estimating building heat loss or in making comparisons with other thermal insulations.

Exactly why these products mounted between joists exhibit a significantly greater heat flow than can be explained by direct transfer through the joists is not fully understood, but the phenomenon is probably due to two-dimensional (or three-dimensional) heat transfer. In effect, the probable transfer of heat between the two parallel paths (through the joists and through the insulated cavities) increases the total transfer beyond that which would occur if the two paths were independent. Thus, the joists would contribute to the total heat flow an amount in excess of the direct transfer that would occur through them if only one-dimensional heat conduction existed. Although such multidimensional effects could occur with all thermal insulations, it seems logical that they would be especially large for the reflective insulations because of the highly conductive lateral layers. In contrast to this, the competing mass insulations are relatively much more uniform in all directions, and any multidimensional heat transfer effects are relatively insignificant.

A secondary conclusion evident from the multiplying factors necessary in calculating new conditions is that the heat flow through these commercial reflective insulations is usually significantly greater than that predicted by the ASHRAE tables [*8*]. This corroborates, to some degree, the conclusions of an earlier paper [*2*], but the effects were much larger for vertically mounted specimens.

Recommendations

The major recommendation coming from this study is that the thermal resistance values for reflective insulations (and for other highly anisotropic insulations) be derived from tests of the product mounted in the manner in which it will be used, but with any direct conduction transfer through mounting members calculated out. In this manner, the resulting effective *R* values will more nearly represent the installed performance of the product and will be more suitable for comparison with the *R* values of other insulations.

A further recommendation is that similar and more detailed studies be conducted to ascertain more thoroughly the existence and magnitude of these multidimensional heat transfer efects in insulation systems.

Acknowledgments

The author wishes to express his appreciation to the Mineral Insulation Manufacturers Association (MIMA), which sponsored the tests on the specimens mounted between joists. Thanks also go to W. J. Sanders and others at the Butler Manufacturing Company Research Center, who conducted the hot box tests; to T. E. Weast of the Midwest Research Institute, who prepared the test reports and provided guidance on the calculation procedures; to M. B. Gettings of the Owens-Corning Fiberglas (OCF) Technical Center, who performed the computer calculations for the factors and new configuration *R* values; and to others at MIMA and OCF for their guidance and criticism of this work.

References

[*1*] Yarbrough, D. W., "Assessment of Reflective Insulations for Residential and Commercial Applications," Report ORNL/TM-8891, Oak Ridge National Laboratory, Oak Ridge, TN, 1983.

[2] Hollingsworth, M., Jr., "Experimental Determination of the Thermal Resistance of Reflective Insulations," *Transactions of the American Society of Heating, Refrigerating, and Air Conditioning Engineers,* Vol. 89, Part 1B, 1983, pp. 568–578.

[3] "Design Heat Transmission Coefficients," 1981 *ASHRAE Handbook of Fundamentals,* American Society of Heating, Refrigerating, and Air Conditioning Engineers, Atlanta, GA, 1981, Chapter 23, pp. 23.2–23.3.

[4] 1981 *ASHRAE Handbook of Fundamentals Volume,* American Society of Heating, Refrigerating, and Air Conditioning Engineers, Atlanta, GA, 1981, Chapter 23, Table 3A, p. 23.17.

[5] "Labeling and Advertising of Home Insulation," Trade Regulation Rule, 16CFR 460, Bureau of Consumer Protection, Federal Trade Commission, Washington, DC, 1978, Section 460.05 (b).

[6] "California Insulation Quality Standards," State Energy Resources Conservation and Development Commission, State of California, Sacramento, CA, 1981, Section 1553 (b)(2).

[7] "Test Reports in Support of Applications for Product Listing in the CEC Directory of Certified Insulation Material," California Energy Commission (CEC), Sacramento, CA, March 1982.

[8] 1981 *ASHRAE Handbook of Fundamentals,* American Society of Heating, Refrigerating, and Air Conditioning Engineers, Atlanta, GA, 1981, Chapter 23, Table 2, pp. 23.12–23.13.

[9] Robinson, H. W. and Powlitch, F. J., "The Thermal Insulating Value of Air Spaces," Housing Research Paper 32, U.S. Housing and Home Finance Agency, Washington, DC, 1957.

[10] 1981 *ASHRAE Handbook of Fundamentals,* American Society of Heating, Refrigerating, and Air Conditioning Engineers, Atlanta, GA, 1981, Chapter 23, Table 1, p. 23.12.

R. P. Tye[1] *and C. F. Baker*[2]

Development of Experimental Data on Cellular Plastic Insulations Under Simulated Winter Exposure Conditions

REFERENCE: Tye, R. P. and Baker, C. F., "**Development of Experimental Data on Cellular Plastic Insulations Under Simulated Winter Exposure Conditions,**" *Thermal Insulation: Materials and Systems, ASTM STP 922,* F. J. Powell and S. L. Matthews, Eds., American Society for Testing and Materials, Philadelphia, 1987, pp. 518–537.

ABSTRACT: An experimental program on expanded and extruded polystyrene insulating materials has been undertaken to investigate the effects of moisture gain on thermal performance. The investigation had two major objectives:

(*a*) to assist in the development of test methods and techniques that address more realistically the subject of moisture effects on the properties of thermal insulations; and

(*b*) to develop a credible data base for typical polystyrene materials when subjected to a set of combined temperature, moisture gradient, and continuous time exposure parameters comparable to the most severe that could be foreseen for the continental United States.

In addition, more limited measurements were made on a phenolic foam product, and limited measurements of thermal performance of the materials were made at several mean temperatures encompassing those normally encountered in applications involving the building envelope. Finally, the result of an interlaboratory study on certain specimens from one of the investigations is presented. This was undertaken both to confirm the absolute thermal property values and to illustrate that the ASTM Test for Steady-State Thermal Transmission Properties by Means of the Heat Flow Meter (C 518-76) is capable of discriminating between very small incremental changes in thermal performance of the exposed specimens.

The results, which indicate relatively small effects under these conditions, are discussed in relation to those obtained in other studies involving moisture effects and to the probable behavior in the normal environment. The viability of the approach as a suitable standard testing technique is also discussed.

KEY WORDS: apparent thermal conductivity, expanded polystyrene, extruded polystyrene, moisture thermal performance, thickness effect, thermal insulation

The use of cellular plastic insulating board for roofing applications has been steadily increasing during the past decade, and considerable growth potential clearly exists in the plastics industry [*1*]. The major materials used include both expanded molded and expanded extruded polystyrenes, polyurethanes and isocyanurates, and, more recently, closed-cell phenolic foams. Detailed information on the physical, mechanical, and thermal properties of these materials is available in appropriate material specification sheets and descriptive literature. However, the information can be classified as being obtained under ideal laboratory conditions and established primarily to facilitate comparisons of insulations in the marketplace. In most cases relevant data for performance under actual conditions are not available to the architect and de-

[1]Engineering director, Dynatech Scientific Inc., Cambridge, MA 02139.
[2]Program manager, Fiber Materials, Inc., Biddeford, ME 04005.

signer, and this is the information that is considered most necessary [*2*] if realistic energy conservation goals are to be realized.

Following these assessments and the subsequent development of the National Program Plan [*3*], a number of conferences and symposia devoted solely to insulation performance have been held. These include the ASTM [*4*], joint ASTM/U.S. Department of Energy (DOE) [*5,6*], American Society of Heating, Refrigerating and Air-Conditioning Engineers (ASHRAE)/DOE [*7,8*], and National Bureau of Standards/National Roofing Contractors Association (NBS/NRCA) [*9,10*] meetings where information on field performance has been highlighted in many instances. However, there is still an urgent need to obtain more basic information on both materials and systems in order to interpret individual field performance information better for the more general case or for particular applications. Among the most significant factors related to this subject of performance of insulations, especially in roofs, are the effects of moisture and moisture transfer.

When installed in a roofing system, a thermal insulation rarely encounters standard test conditions but is subject to changes of temperature and humidity that, in some cases, can be severe. Should the insulation, under these conditions, absorb and retain moisture, its thermal performance will be lowered, and other properties could be affected adversely. Such changes will have significant effects on the overall heat losses in any building containing the insulation and could influence the choice of materials for energy-efficient designs.

These factors illustrate the need to determine the properties of materials that have been subjected to conditions simulating those expected in real life. This type of technical information is necessary to confirm that a material can be chosen and that it will perform in accordance with manufacturers' recommendations. Additionally, the need to produce this information necessitates the development of new and revised techniques and procedures. All of these will assist the overall goals of providing adequate data on the performance of materials, the improvement of testing criteria, and the development of standards, codes, and similar documents. These, in turn, will result in the design, construction, and use of safe, energy-efficient structures.

Two studies on moisture effects on the properties of thermal insulations used in roofing applications have been carried out by the U.S. Army Cold Regions Research and Engineering Laboratory (CRREL) [*11*] and the National Bureau of Standards (NBS) [*12*], respectively. In both cases, specimens of different thermal insulation materials were conditioned by developing a temperature and moisture gradient across the surfaces and studying the changes in the thermal resistance of the specimens. The temperature and humidity gradients—29.5°C, 100% relative humidity to 4.4°C, 75% relative humidity for the CRREL study, and 38°C, 100% relative humidity to 10°C, 25% relative humidity for NBS—were chosen to ensure that the specimens experienced an environment that contained excessive water, but they were not really representative of conditions that would be experienced in applications. As a result, the data were of considerable scientific value in illustrating the effects of water on the thermal properties and providing empirical relationships for deriving thermal properties from moisture contents, however, because of the high moisture conditions, the results were less applicable to the practical environmental conditions an insulation would encounter.

As a result of the foregoing, the Society of the Plastics Industry decided in 1982 to undertake an investigation of expanded and extruded polystyrene, having three major objectives:

1. To develop a credible data base on typical commercially available polystyrene insulation boards when subjected to a set of continuous humidity and temperature-gradient parameters simulating severe winter conditions in the continental United States. The physical, mechanical, and thermal properties would be evaluated before and after exposure to selected conditions representative of a common winter condition with an internal face of the insulation subjected to 21 ± 1°C/42% relative humidity and the external face to −7°C/saturation. These conditions established a vapor pressure differential of 4.9 mm Hg across the insulation from the warm side

to the cold side. They created the potential of having both dew and freeze points within the specimens. An exposure time of 120 days at this vapor pressure differential was chosen as being one that would produce "maximum" effects, which in all probability, would not be exceeded *in situ.* The particular set of exposure conditions was the same as that which had been used previously for a study on the influence of moisture on the performance of cellulose insulation [*13*].

2. To assist in the development of test methods and techniques that more realistically address the subject of moisture effects being addressed within ASTM Committee C-16 on Thermal Insulation, and in particular by a task force of ASTM Subcommittee C 16.31 on Chemical and Physical Properties, which is responsible for developing test methods for physical properties.

3. To establish thermal performance data for the materials at two mean temperatures encompassing those normally encountered in building applications. Such information is required both for inclusion in future materials specifications such as ASTM Specification for Preformed, Cellular Polystyrene Thermal Insulation (C 578-83), code, and regulatory documents, and to enable architects and designers to improve modeling and design criteria for producing more energy-efficient buildings.

Following this study on the two forms of polystyrene, the Koppers Co. commissioned a similar study on a commercially available faced, closed-cell cellular phenolic product.

This paper describes the development of the test criteria and testing methodology and presents and discusses the results obtained on the three types of cellular plastic insulation board materials.

Details of Test Materials

1. *Block molded polystyrene*—This consisted of two different lots of a nominal 16-kg/m^3 molded expanded polystyrene (EPS) board designated Mat I and Mat II. They were supplied from different sources and were both described as typical of commercially available materials.

2. *Extruded polystyrene*—This consisted of two different lots of commercially available boards supplied by two manufacturers and designated Mat B and Mat P.

In both of these cases, each lot consisted of up to ten boards of three nominal thicknesses. Sufficient boards of each thickness were chosen randomly from each lot to provide material from which all necessary specimens could be prepared both for initial characterization and for the subsequent conditioning, coupled with preconditioning and postconditioning property measurements. Sufficient numbers of specimens were prepared by cutting, using the hot wire technique, so that actual test specimens and replicates were again chosen randomly.

Mat I, Mat B, and Mat P were characterized in all three thicknesses and Mat II in one thickness only, except for the apparent thermal conductivity (thermal resistance). In addition, the effects of edge and face sealing of the EPS Mat I and Mat II materials with an impervious tape simulating protective coverings or vapor retarders were also evaluated. This was not studied for the extruded Mat B or Mat P materials.

3. *Phenolic foam*—This consisted by two lots of commercially available products nominally 30 mm thick, with corrugated paper and a regularly pierced reflective coating on one face, and a heavier paper coating on the other face.

An original lot, 102X-488, was characterized in the preconditioned state and placed in the exposure chamber. However, because of a mechanical failure in the chamber pumping system, testing had to be suspended midway through the planned 120-day cycle. This interruption necessitated starting with a new lot identified as 63X-271. This second material had a slightly higher density than the original because the corrugated facing material had been changed in the manufacturing process. Hence, the results of preexposure measurements in this work apply to Lot

102X-488 parts, while results pertinent to postexposure conditions apply to Lot 63X-271 parts. Scanning electron and optical microscopic examinations of the actual foam indicated they were very similar, and the effects of any potential differences between the two lots would be insignificant insofar as the results of this study are concerned.

Characterization

Specimens of the as-received materials that had been stored at standard laboratory conditions for at least 48 h were used to characterize the physical and mechanical properties of the materials according to current ASTM tests to ensure that they were representative of the respective product.

Since thermal performance was the major parameter being evaluated, detailed information is provided for this parameter. Details of the test conditions are given in Table 1.

All measurements of apparent thermal conductivity, λ_{app}, were undertaken on 30-cm-square specimens using a Dynatech Model TCHM-R4, Rapid-k, heat flow meter apparatus in accordance with ASTM Test for Steady-State Thermal Transmission Properties by Means of the Heat Flow Meter (C 518-76). Prior to, during, and after the measurement programs, the instrument was calibrated at each thickness of specimen and at each temperature using both the NBS SRM 1450b and other national standards laboratory traceable reference materials and transfer standards. By calibrating at the same thickness as those of the test materials, the overall accuracy of the method was maintained by use of the appropriate calibration constant determined for that thickness. In addition, prior to the first study on the molded EPS materials, a series of tests was conducted by four different industry laboratories on the same six 25-mm-thick specimens of one of the materials; this was done to establish partial verification of the accuracy and reliability of the tests being undertaken. Details on this interlaboratory study are given in the appendix.

The mean temperature condition at 7°C was chosen specifically so that it could be used subsequently for direct comparison of the effect of exposure on the property. For this condition, the hot face was maintained at approximately 21°C and the cold face at approximately −7°C, that

TABLE 1—*Test matrix for thermal properties measurements.*

	Mean Temperature, °C					
	Preexposure				Postexposure	
Material and Thickness	−10	7	24	43	7	24[a]
Mat I, 25 mm (1 in.)	2	2	4	2	4	3
51 mm (2 in.)	2	2	4	2	4	...
100 mm (4 in.)	2	2	4	2		
Mat II, 25 mm (1 in.)	...	2	4	...	4	3
51 mm (2 in.)	...	2	8	2	8	...
100 mm (4 in.)	...	2	4	2	4	...
Mat B and P, 25 mm (1 in.)		4	4	1	4	...
51 mm (2 in.)		4	4	1	4	...
75 mm (3 in.)		4	4	1	4	...
Faced phenolic, 30 mm (1.2 in.)		8	8	2	4	4

[a]Specimens were further conditioned for 72 h at 38°C followed by 72 h at 60°C or constant weight after the original exposure. This was done to show that there may have been some moisture in the original "preconditioned" material.

is, the temperature levels for the exposure of the specimens. The matching of test conditions to those of the exposure was considered an essential element in measuring the thermal conductivity of specimens containing even small quantities of moisture.

Thus, after 120 days, specimens would be taken from the chamber, wrapped in plastic, and inserted immediately into the instrument with the plates already established at the same two temperatures. Under these circumstances, steady-state conditions are reached almost immediately. No higher temperature conditions would be imposed, and thus any water in the specimen would remain fixed in position. The amount and effect in any section could be examined in detail once the apparent thermal conductivity had been evaluated.

The term "apparent thermal conductivity" is used to draw attention to the fact that, in low-density materials at any particular mean temperature, this property can vary with changes in parameter such as temperature difference and specimen thickness. This is because radiation is a significant contribution to the overall heat transmission characteristics [*14*]. Thus, each specimen will have unique thermal conductance and resistance values. In practice, the thickness effects for the current materials are relatively small, particularly for the higher densities.

Results and Discussion

The collected results for the apparent thermal conductivity are given in Table 2. The results for the various other physical and mechanical properties indicate that all the materials were homogeneous, reproducible, and within the range expected for the particular product type. Any small differences observed could be related directly to density or differences in thickness. It was also observed that the dimensional stability as measured by ASTM Test for Response of Rigid

TABLE 2—*The apparent thermal conductivity of preexposed materials.*

Specimen	Test Thickness, mm	Test Density, kg/m^3	λ_{app}, W/m · K			
			−10°C	7°C	24°C	43°C
Mat I						
1B1	29.0	16.0	0.0364	0.0382	0.0403	0.0426
1B2	28.4	16.0	0.0365	0.0382	0.0408	0.0429
1B3	28.2	16.3			0.0400	
1B4	27.9	16.0			0.0400	
2A1	53.6	15.7	0.0365	0.0392	0.0410	0.0442
2A2	53.6	15.4			0.0413	
2A3	53.6	16.2			0.0411	
2A4	53.6	15.6			0.0410	
2D1	53.3	15.3	0.0368	0.0392	0.0413	0.0442
2D2	53.8	15.4			0.0417	
2D3	53.1	15.7			0.0413	
2D4	53.6	15.4			0.0413	
4B1	104.1	15.5	0.0371	0.0403	0.0418	0.0447
4B2	104.6	15.1	0.0377	0.0400	0.0411	0.0446
4B3	104.1	16.6			0.0423	
4B4	105.2	15.7			0.0418	
Mat II						
1-4	25.9	16.3		0.0365	0.0393	0.0423
1-11	25.4	16.6			0.0390	
1-18	25.1	16.5		0.0371	0.0393	
1-23	25.9	16.6			0.0392	

TABLE 2—*Continued.*

Specimen	Test Thickness, mm	Test Density, kg/m^3	λ_{app}, W/m · K			
			−10°C	7°C	24°C	43°C
2-9	54.1	16.6		0.0368	0.0397	0.0429
2-10	53.6	16.5		0.0372	0.0403	
2-11	54.9	17.8			0.0392	
2-14	54.6	16.0		0.0372	0.0403	0.0433
2-17	54.9	16.6			0.0400	
2-19	53.6	16.2		0.0371	0.0400	
2-20	54.4	17.4			0.0397	
2-22	53.8	17.4			0.0398	
4-8	105.9	16.3			0.0405	
4-11	105.9	16.8			0.0408	
4-13	105.4	17.0		0.0380	0.0405	
4-23	105.7	15.8		0.0382	0.0405	0.0439
Mat B						
B1-1	25.6	31.0		0.0267	0.0284	
B1-2	25.6	31.0		0.0264	0.0277	
B1-3	25.6	30.7		0.0266	0.0280	0.0294
B1-4	25.6	31.0		0.0267	0.0281	
B2-1	51.0	27.8		0.0267	0.0290	0.0316
B2-2	51.0	27.8		0.0266	0.0287	
B2-3	50.8	27.7		0.0264	0.0286	
B2-4	51.0	28.0		0.0266	0.0293	
B3-1	76.5	29.9		0.0268	0.0291	
B3-2	76.5	29.6		0.0268	0.0286	
B3-3	76.5	29.6		0.0266	0.0284	
B3-4	76.7	29.8		0.0267	0.0286	
Mat P						
P1-1	25.6	30.1		0.0245	0.0276	
P1-2	25.4	30.2		0.0245	0.0274	
P1-3	25.4	30.1		0.0245	0.0274	
P1-4	25.4	30.1		0.0248	0.0277	0.0299
P2-1	50.8	27.2		0.0262	0.0287	
P2-2	50.3	27.5		0.0260	0.0283	
P2-3	49.8	27.4		0.0262	0.0294	0.0310
P2-4	49.8	27.4		0.0261	0.0286	
P3-1	75.2	26.2		0.0258	0.0283	
P3-2	75.2	26.1		0.0260	0.0287	
P3-3	75.9	26.1		0.0258	0.0284	
P3-4	75.9	26.1		0.0255	0.0283	
Faced phenolic 102x-455						
K-1	31.8	56.0		0.0157	0.0165	
K-2	32.3	56.2		0.0156	0.0163	
K-3	31.8	55.5		0.0164	0.0170	0.0176
K-4	32.5	55.4		0.0157	0.0166	
63x-271						
A	30.0	67.8		0.0163	0.0169	
B	30.0	68.3		0.0160	0.0173	
C	30.0	65.4		0.0162	0.0169	0.0176
D	30.0	67.8		0.0162	0.0172	

Cellular Plastics to Thermal and Humid Aging (D 2126-75), namely 70°C and 90% relative humidity and −29°C and ambient relative humidity, each for periods of seven days and longer—were much lower than the limits allowed in appropriate materials specifications.

Some further comments can be made concerning the different thermal conductivity results, especially for the block-molded polystyrene material. The results given in Table 2 show that the values for Mat II are of the order of 3% below those for Mat I for comparable thicknesses of each material. This difference may be due to the higher density of the Mat II product. In addition, the results indicate clearly the existence of the so-called thickness effect in low-density materials. In the present case, at 24°C, the overall effect is of the order of 3.7% for both expanded materials over the thickness range of 25 to 100 mm (1 to 4 in.). For the extruded polystyrene materials, the thickness effect is still apparent; however, the magnitude is lower. This can be ascribed to both the higher densities of these products and possible differences in cell size and cell gas content.

The results also indicate a decrease in *R* value of approximately 14% between 7 and 43°C for the molded materials. In addition, there is a small but distinct difference in behavior between the two materials, with the change for Mat I close to 12% and that for Mat II 16%. The reasons for this difference are not known, but it could be due to a combination of factors such as bead size precursors and density. Similar small differences have been noted previously for cellular plastics [*15*]. For the extruded materials, the results indicate a decrease in *R* value per unit thickness between 7 and 43°C of some 12 and 20% for Mat B and P, respectively. This difference in behavior can also be due to the aforementioned cell size and gas factors. For both types of cellular polystyrene, the overall change in unit *R* value is consistent with that documented in an earlier study on insulation for building applications [*2*].

Details of Exposure Study

Experimental Details

To simulate conditions typical of a winter exposure in the northern continental United States, a special test chamber was designed and constructed. Details of the construction are shown in Figs. 1 and 2. The chamber consists essentially of a large box structure containing a lower "warm" and upper "cold" section, in which the temperature and humidity are controlled continuously at reasonably constant levels for long periods of time. The chambers were separated horizontally by a specimen holder section containing 36 equal-sized holes into which 300 by 300-mm specimens of different thicknesses could be tightly fitted for the exposure. The specimen holder was designed so that one surface of each specimen was flush with the surface of the holder facing the cold side conditions with the remainder of the specimen exposed to the warm side conditions, as shown in Fig. 2. The whole conditioning box was situated in a temperature-controlled room. The temperature and humidity conditions on the lower side were adjusted by means of controlled heaters and a suitable choice of salt solution. For the present investigation, the temperature was maintained at 21 ± 1°C and a relative humidity of 42% obtained with a saturated salt solution of zinc nitrate heptahydrate. Under these conditions, no condensation of vapor occurred on the warm side of the test specimens. The upper, cold side was maintained at −7 ± 1°C using a circulated controlled-temperature chilled antifreeze solution pumped continuously through the cooling coil system.

For the exposure cycle, the specimens were weighed to 0.1 g and then introduced into the exposure chamber on a staggered time basis. This was done so that the postexposure thermal and moisture content tests could be undertaken immediately after the 120-day period and the moisture would not have time to migrate within or out of the specimen. Each specimen was exposed for a total period of 120 days. At weekly intervals during this period, each specimen was

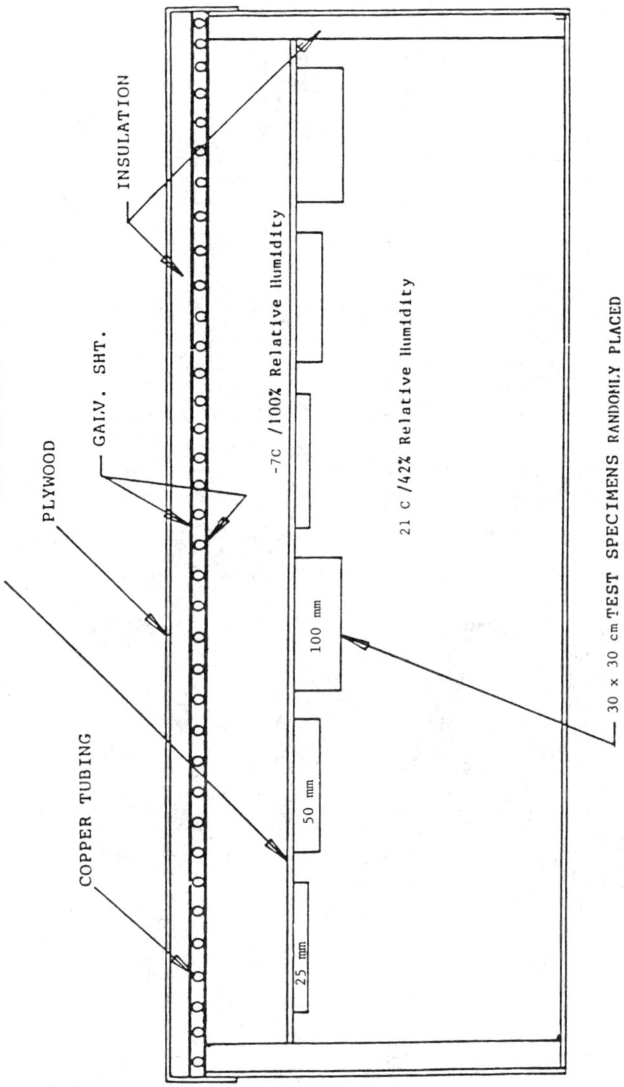

FIG. 1—*Schematic of the conditioning chamber.*

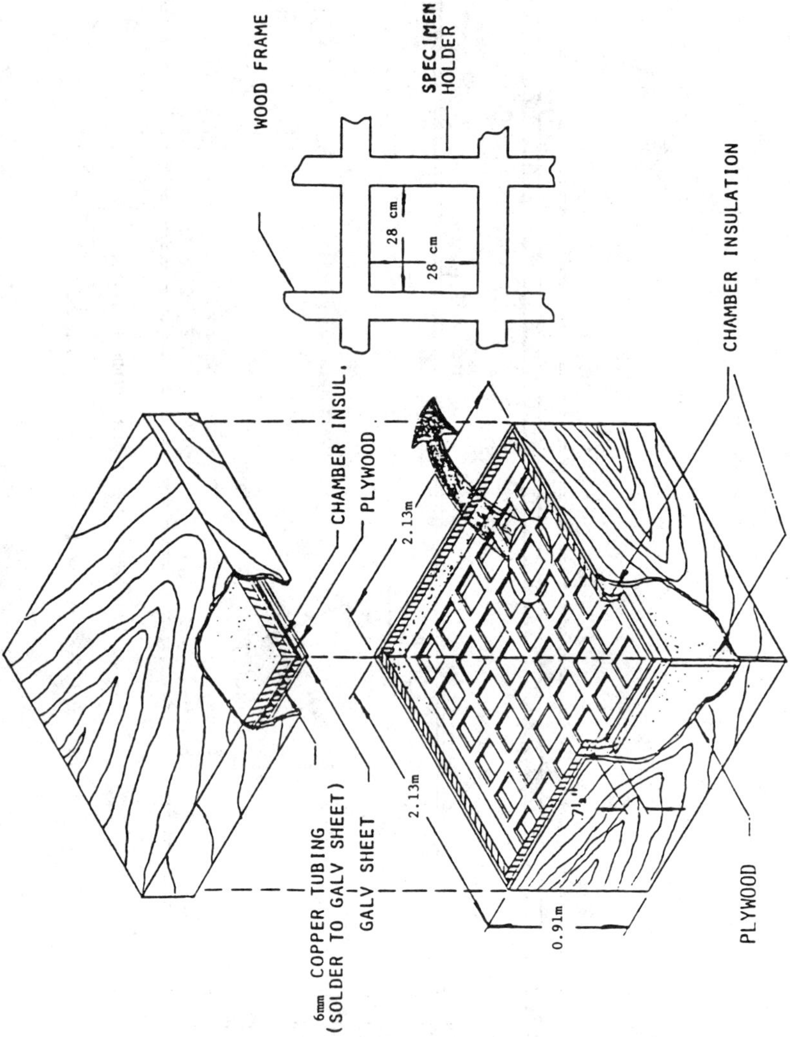

FIG. 2—*Details of the specimen holder in the chamber.*

removed from the holder and weighed, an operation that took a few minutes, during which time a dummy insulation specimen was placed in the vacated slot.

At the end of the 120-day period, each specimen was removed from the chamber, weighed, and wrapped in a thin film of plastic while moved from the chamber to the thermal conductivity test apparatus. The plastic was removed, and the specimen was placed into the apparatus with the hot and cold plates established at the temperatures of the exposure chamber, and a measurement was undertaken in accordance with ASTM Method C 518-76, except that now the specimens were not in the laboratory-conditioned state. Randomly selected specimens were then conditioned to constant weight and measured once again at 24°C.

Immediately after the thermal measurement at 7°C, selected specimens of the polystyrene materials were examined only for moisture content and distribution through the specimen. A 51-mm block was cut from the still cold specimen and sliced horizontally into thirds with a scalpel, and each piece was weighed and then dried to constant weight in a vacuum desiccator.

An attempt was also made to measure the temperature distribution within a specimen of each thickness of polystyrene immediately after the exposure. For this, a fine-gage sheathed temperature probe was inserted into the foam to different depths. A second technique, drilling a small hole into the foam and using an unsheathed thermocouple, was also tried, since the first technique proved difficult because of the foam's resistance to penetration. Neither technique proved successful since the measured temperatures in each case rose rapidly to essentially 24°C, and no real gradient could be found. This was probably because of the effects of direct heat conduction along the high-conductivity probe and thermocouple wires heating the specimen locally and disrupting the real temperature distribution.

The remaining 25-mm-thick specimens were then conditioned at standard laboratory conditions for a period of one week and, following this, the compressive and flexural strengths were measured in a manner similar to the measurement of the as-received specimens. Finally, a brief optical and scanning electron microscopic study was undertaken on random specimens of the preexposed and postexposed materials. Specimens of each thickness of material were observed before exposure by optical stereomicroscopy in the ×5 to ×50 magnification range and by scanning electron microscopy in the ×20 to ×200 range to establish a baseline against which specimens after exposure could be judged. Fracture surfaces from mechanical test specimens were used for these observations as they provide a more representative view of the material. After conditioning, specimens of each size material were observed using the technique noted earlier. In addition a comparison of specimens before and after conditioning was made visually.

Material Parameters

Block-Molded Polystyrene—The results for the preconditioned materials indicated that either could have performance characteristics representative of a commercially available 16-kg/m^3 EPS material used for roofing insulation applications. The authors decided to choose equal lots of each material for exposure and further testing.

Sixteen specimens in four groups of 4 covering the thickness range were chosen randomly from each of the two materials to make a total population of 32 specimens.

The study was designed to include both unsealed specimens and those having sealed faces or edges, or both. Four replicate specimens of each form were to be measured, and the number of spaces available in the exposure chamber was limited; a decision was made to concentrate more effort on the 50-mm-thick material since this is currently the most widely used thickness. Thus, a total of 16 specimens 50-mm thick were exposed, together with 8 each of the 25-mm and 100-mm-thick materials.

To simulate the influence of the vapor-retarding effects of a roof membrane, selected specimens of each thickness were sealed carefully on the edges only and on the cold face and edges with a low-permeance aluminum foil adhesive tape. In addition, for the 50-mm-thick material only, the tape was attached to the cold face and one set of specimens was left unsealed. Table 3 contains the test matrix and the identification of the randomly chosen specimen sets of each form.

Extruded Polystyrene—A total of 24 specimens were chosen for exposure in the chamber, 12 of each material divided evenly among the three thicknesses. No sealing of edges or faces was carried out.

Phenolic Foam—Four specimens only were exposed for the 120-day period. Each specimen retained its original perforated facings since these were representative of the product used in a roofing application. No edge sealing was attempted.

TABLE 3—*Test matrix for exposure of EPS specimens.*

Form	Test Specimen Identification		
	25-mm Material	50-mm Material	100-mm Material
Sealed edges	1-4, 11, 18, 23	2A1 to A4	4-8, 11, 12, 23
Sealed cold face and edges	1B1 to B4	2-9, 10, 17, 19	4B1 to B4
Sealed cold face	...	2-11, 14, 20, 22	...
Unsealed	...	2D1 to D4	...
Total number of specimens	8	16	8

Results After Exposure

Moisture Gain and Physical and Thermal Properties

Table 4 contains details of the average moisture gain of each group of specimens of each material, and Table 5 shows the distribution of moisture through the different polystyrene materials. Tables 6 and 7 contain comparisons of the preexposed and postexposed mechanical and thermal properties, respectively. Measurements of apparent thermal conductivity at 24°C of dried specimens after exposure are shown in Table 8.

Visual Examination

Visual and microscopic examination of the specimens indicated that the exposure conditions had little or no effect on the materials under study. Flexure failures before and after exposure were also compared visually to determine any changes in the cross section of the material. No differences were observed in any of the specimens. Optical stereomicroscopy in cross section confirmed the lack of difference between preexposed and postexposed specimens. Photomicrographs taken by scanning electron microscopy further confirmed these findings, showing well-bonded structures, uniformity, and no major voids or imperfections or any sign of damage or alteration in form in the conditioned specimens.

Discussion of Results

Moisture Pickup

A comparison of the experimental data for the preexposed and postexposed specimens of the two polystyrene insulations and the phenolic foam materials indicates that exposure to sim-

TABLE 4—*Average moisture pickup of specimens exposed to simulated winter conditions.*

Specimen and Thickness, in. (mm)	Form	Weight Gain, weight %
Mat II, 1″ (25)	sealed edges	0.2
Mat I, 2″ (51)	sealed edges	0.135
Mat II, 4″ (100)	sealed edges	0.15
Mat I, 1″ (25)	sealed cold face and edges	7.5[a]
Mat II, 2″ (51)	sealed cold face and edges	0.175
Mat I, 4″ (100)	sealed cold face and edges	0.2
Mat II, 2″ (51)	sealed cold face	0.15
Mat I, 2″ (51)	unsealed	0.125
Mat B, 1″ (25)	unsealed	0.0
2″ (51)	unsealed	0.45
3″ (15)	unsealed	0.8
Mat P, 1″ (25)	unsealed	0.0
2″ (51)	unsealed	0.4
3″ (75)	unsealed	0.425
Faced phenolic	faced	−3.5

[a]One specimen only trapped a considerable amount of liquid water during a power failure, and the results are not considered typical. The other three specimens of this group had <0.2% gain.

ulated winter conditions had a very limited effect on the thermal and mechanical properties of either material. Under the conditions of this study, the total moisture pickup of the block-molded and extruded polystyrene specimens was less than 0.2 and 1% by weight, respectively, in every case, with the largest amount of moisture tending generally to be toward the center of the material. An examination of Table 8 indicates clearly that the moisture levels involved in the study are, in all cases, less than 0.55 and 1.4% by weight for the block-molded and extruded polystyrenes, respectively. In addition, it was less than 1% in the majority of sections of all the specimens. The moisture appeared to be uniformly distributed across a specimen in the direction parallel to the cold faces. However, the moisture contents did vary through the thickness, always being lowest in the one-third section adjacent to the warm side and generally higher in the cold one-third section. This would be expected, since the differential vapor pressure driving force due to the combined humidity and temperature differentials would allow moisture to migrate from the warm surface through the material until it condenses in the center third, thus slowing further movement. The fact that the amounts of water seen in Table 5 are higher than those in Table 4 indicates that some moisture was probably present in the original materials.

Further examination of the results of Table 5 for the 50-mm-thick block-molded polystyrene specimens yields the unexpected results that the use of a vapor retardant membrane for sealing the cold surfaces or edges, or both, appears to have little effect on the overall pickup and retention of moisture. The unsealed specimens had a total moisture gain of 0.24%, whereas the two specimens that had one face sealed gained 0.14 and 0.13%. However, the lowest moisture gain is seen to occur for the specimen that had only the edges sealed. The amounts of moisture involved are too small to explain the specific reasons for these differences and why there was not a larger retention of moisture for the specimens with both edges and one face sealed. Assuming

TABLE 5—*Distribution of moisture in polystyrene specimens after exposure.*

Specimen and Form	Moisture Content, weight %		
	Cold Third	Middle Third	Warm Third
Block-molded material			
1-18 sealed edges	0.18	0.01	0.01
2-A3 sealed edges	0.09	0.01	0.03
4-13 sealed edges	0.34	0.18	0.04
1-B3 sealed face and edges	0.12	0.98	0.0
2-17 sealed face and edges	0.25	0.13	0.01
4-B3 sealed face and edges	0.40	0.26	0.13
2-22 sealed face	0.27	0.14	0.01
2-D3 unsealed	0.52	0.20	0.01
Extruded material			
B1-1	0.33	0.40	0.32
B1-2	0.48	1.00	0.45
B1-3	0.41	0.41	0.36
B1-4	0.37	0.42	0.35
B2-1	0.42	0.55	0.33
B2-2	0.48	0.42	0.37
B2-3	0.51	0.60	0.49
B2-4	0.46	0.48	0.42
B3-1	0.49	0.77	0.41
B3-2	0.60	1.01	0.55
B3-3	0.52	0.91	0.45
B3-4	0.66	0.83	0.59
P1-1	0.41	0.22	0.44
P1-2	0.40	0.62	0.34
P1-3	0.54	0.88	0.52
P1-4	1.00	1.41	1.02
P2-1	0.43	0.80	0.32
P2-2	0.45	0.24	0.26
P2-3	0.46	0.80	0.44
P2-4	0.54	0.86	0.44
P3-1	0.51	0.40	0.34
P3-2	0.84	0.29	0.26
P3-3	0.55	1.32	0.24
P3-4	0.52	1.09	0.42

that the vapor retarder used was efficient, the weight gains for the specimens having a sealed face or edges, or both, can be calculated using the following equation

$$W = uAt(\Delta p/\ell)$$

where W is the total mass of absorbed moisture, A and ℓ are area of cross section and thickness, respectively, t is the elapsed time, u is the average permeability of the specimen, and Δp is the water vapor pressure difference across the specimen.

For the sealed specimens, the actual moisture gains were considerably less than those that could be calculated. The reasons for this discrepancy, other than the possibility that the particular vapor retarder and its form of application were ineffective, are still unknown.

For the phenolic material there was an actual loss in weight, as illustrated for one specimen in

TABLE 6—*Comparison of mechanical properties before and after exposure.*

Material and Thickness, mm	Compressive Strength, mPa		Flexural Strength, mPa	
	Exposed	Preexposed	Exposed	Preexposed
Block-molded polystyrene				
Mat I, 25	0.793 ± 0.014	0.758 ± 0.021	2.090 ± 0.006	1.840 ± 0.011
Mat II, 25	0.903 ± 0.021	0.958 ± 0.021	2.090 ± 0.006	2.075 ± 0.028
Extruded polystyrene				
Mat B, 25	2.482 ± 0.069	2.323 ± 0.172	8.350 ± 0.345	7.743 ± 0.421
50	2.689 ± 0.214	2.503 ± 0.193	5.736 ± 0.303	5.268 ± 0.145
75	4.254 ± 0.048	3.971 ± 0.207	3.328 ± 0.310	5.288 ± 0.214
Mat P, 25	1.710 ± 0.165	1.696 ± 0.096	9.039 ± 0.621	9.046 ± 0.359
50	1.958 ± 0.124	1.793 ± 0.069	6.426 ± 0.296	6.419 ± 0.165
75	2.144 ± 0.186	2.027 ± 0.138	4.992 ± 0.089	5.171 ± 0.110
Phenolic foam, 30	2.972 ± 0.200	1.641 ± 0.352	8.356 ± 1.689	7.136 ± 2.172

TABLE 7—*Comparisons of apparent thermal conductivity of preexposed and postexposed specimens at 7°C.*

Specimen and Form	λ_{app}, W/m · K		
	Exposed	Preexposed	Difference, %
Block-molded Polystyrene			
1-4 sealed edges	0.0374	0.0365	2.3
1-11	0.0372		
1-18	0.0367	0.0371	−1.1
1-23	0.0374		
2A1	0.0395	0.0392	0.7
2A2	0.0398		
2A3	0.0395		
2A4	0.0400		
4-8	0.0395		
4-11	0.0388		
4-13	0.0388	0.0380	2.3
4-23	0.0394	0.0382	3.1
1B1 sealed face and edges	0.0381	0.0382	<0.5
1B2	0.0381	0.0382	<0.5
1B3	0.0380		
1B4	0.0378		
2-9	0.0388	0.0368	5.5
2-10	0.0385	0.0372	3.5
2-17	0.0387		
2-19	0.0387	0.0371	4.3
4B1	0.0410	0.0403	1.8
4B2	0.0411	0.0400	2.9
4B3	0.0413		
4B4	0.0405		

TABLE 7—*Continued.*

Specimen and Form	λ_{app}, W/m · K		
	Exposed	Preexposed	Difference, %
Block-molded Polystyrene			
2-11 sealed edges	0.0381		
2-14	0.0388	0.0372	4.3
2-20	0.0375		
2-22	0.0384		
2D1 unsealed	0.0395	0.0392	0.7
2D2	0.0398		
2D3	0.0395		
2D4	0.0397		
Extruded polystyrene			
B1-1 unsealed	0.0273	0.0267	2.2
B1-2	0.0268	0.0264	1.6
B1-3	0.0273	0.0266	2.7
B1-4	0.0270	0.0267	1.1
B2-1	0.0274	0.0267	2.7
B2-2	0.0273	0.0266	2.7
B2-3	0.0270	0.0264	2.2
B2-4	0.0276	0.0266	3.8
B3-1	0.0274	0.0268	2.2
B3-2	0.0273	0.0268	1.6
B3-3	0.0277	0.0264	4.9
B3-4	0.0278	0.0267	4.3
P1-1	0.0253	0.0243	2.9
P1-2	0.0255	0.0243	4.1
P1-3	0.0254	0.0243	3.5
P1-4	0.0254	0.0248	2.3
P2-1	0.0264	0.0262	0.5
P2-2	0.0264	0.0260	1.7
P2-3	0.0265	0.0262	1.1
P2-4	0.0265	0.0261	1.6
P3-1	0.0265	0.0258	2.8
P3-2	0.0265	0.0260	2.2
P3-3	0.0265	0.0258	2.8
P3-4	0.0263	0.0255	2.8
Faced phenolic			
A	0.0154	0.0168	−5.6
B	0.0154	0.0160	−3.7
C	0.0156	0.0162	−3.7
D	0.0156	0.0162	−3.7

Fig. 3. This, again, could be due to the fact that moisture was present during the original manufacturing and that the average humidity level in the chamber was lower than the ambient laboratory condition under which baseline values were determined.

Mechanical Properties

The exposure to simulated winter conditions produced no detrimental effects on the mechanical properties of any of the materials. The compressive strengths and the flexural strengths

TABLE 8—*Apparent thermal conductivity of dried material at 24°C.*

	λ_{app}, W/m · K		
Specimen	Postexposure	Preexposure	Difference, %
Mat 1			
1B1	0.0391	0.0402	3.0
1B2	0.0398	0.0408	2.5
1B4	0.0388	0.0400	3.0
Mat II			
1-4	0.0388	0.0394	1.5
1-11	0.0384	0.0390	1.5
1-23	0.0385	0.0392	1.9
Faced phenolic			
A	0.0172	0.0169	−1.7
B	0.0169	0.0173	2.6
C	0.0169	0.0169	0
D	0.0170	0.0172	0.8

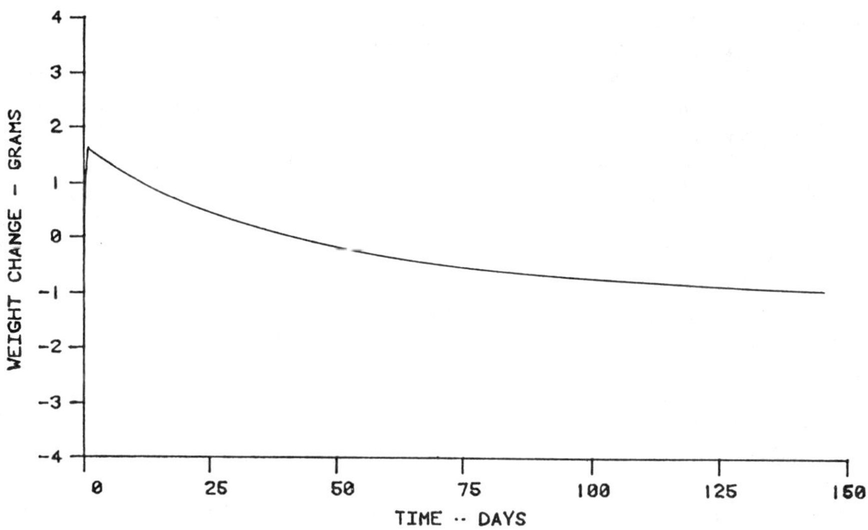

FIG. 3—*Change in the moisture content of phenolic foam specimen with conditioning time.*

showed minor increases after exposure, compared with before exposure, as can be seen in Table 6. However, these increases tend to fall within the variability of the tests and the materials themselves, and thus no definite conclusions as to property improvements should be drawn.

Thermal Performance

Thermal performance was the major parameter studied, and the results are discussed separately for each material.

Block-Molded Polystyrene—An examination of the λ_{app} (thermal conductance, since the same specimens were used) of the exposed specimens shown in Table 7 indicates that, in the majority of cases, a small increase in value is obtained because of the influence of the absorbed moisture. However, Specimens 1B1 to 1B4 indicated a very minimal change. The values for the preexposed specimens, where measured, are also shown in the table for comparison. Not all of the specimens in this state were evaluated at the respective 7°C mean temperature. However, the complete results for the preexposed materials indicate considerable uniformity of property, so that the values obtained at one mean temperature could be accepted as being representative of a particular material for that thickness.

The overall average increase in thermal conductance for all specimens, irrespective of thickness and form, where the property was measured on both preexposed and postexposed specimens, was 2.2%. An examination of the collected results for the 51-mm specimens indicate a maximum increase in thermal conductance of 5.5% and a minimum of less than 1%. The only inconsistency with the results is that increases of the order of 1% for the unsealed specimens are less than would be expected for the higher moisture content on these specimens. Since the preexposure and postexposure measurements were carried out on the same specimens, using identical techniques, the relative difference for each specimen is real.

It should be borne in mind that the differences in results obtained in this study are of the same order as the overall absolute accuracy of the heat flow meter method. It is possible that the absolute values of λ_{app} could vary by up to 1% since this was the order of the maximum variation seen in the calibration constant of the instrument, as measured for the respective thicknesses at various times during the two sets of measurements. Since these occurred at least 120 days apart and, for Material I, at least 150 days apart, it is believed that daily changes during either of the sets of measurements would be less than 1%. The effects of these changes are seen to be minimal on the overall comparative results, and thus they will not affect the general conclusions that can be drawn from the comparative values.

The results shown in the appendix confirm that the particular method and apparatus used are capable of providing highly reproducible and accurate results and that the calibration was, in fact, reliable. Furthermore, the measurement undertaken on the six thinnest exposed specimens after subsequent drying indicated small improvements in λ_{app} of 1.5 to 2.9%. It will be noted that the total amount of moisture obtained from drying the specimens was higher than that gained during the exposure. This is a further indication that moisture was present in the original material prior to exposure. Consequently, it would be expected that the thermal conductivity for a totally dried specimen would be lower.

Extruded Polystyrene—A review of Table 8 shows that λ_{app} of the specimens evaluated increased only to a limited extent after exposure to simulated winter conditions for 120 days. The average increase for all materials was 2.5%, which is consistent with the results for the other materials.

With the exception of Material P (the 25-mm specimen), λ_{app} tended to increase with specimen thickness for both materials. This correlates with the moisture pick up of specimens, as shown in Table 5: the thicker the specimen, the greater the moisture pickup.

Phenolic Foam—The major observations for this material are that it does not appear to pick up or retain moisture, and thus the thermal performance appears to be unaffected by the 120-day exposure to simulated winter conditions. In fact, there is a small but significant (~4%) improvement in performance, which can be attributed to the small decrease in moisture content from that in the as-received material. Furthermore, there appears to be no "aging effect" for this fluorocarbon blown material. Although the product had facers, they were highly permeable, and thus gas exchange similar to that seen for similarly faced fluorocarbon blown polyurethanes and isocyanurates would be possible [*16,17*]. This did not occur, which would indicate that there is a difference in the aging characteristics of the different materials. Further study of this phenomenon over longer periods of time but without the imposition of moisture gradients is required for phenolic foams.

An overall comparison of these results with those of the two earlier studies is interesting. The objectives of these studies were somewhat different. Furthermore, the amounts of moisture involved in them were very much greater because of the higher vapor pressure differentials being used. These were some five and ten times greater for the CRREL and NBS studies, respectively, than for the present investigation. The results obtained in the present study show that the reduction in thermal resistance of two of the three materials due to moisture pickup during the simulated winter conditions was, in general, less than 5%. This result is consistent with the results of the NBS study, which showed that one can expect a 4 to 5% reduction in thermal resistance for small moisture gains of less than 2% by volume.

This behavior, combined with the facts that the mechanical properties remained comparable and moisture absorption and retention was minimal, would lead to the conclusion that if these materials are subjected in use to conditions similar to those used in this study—namely, 21°C/42% relative humidity, and −7°C/saturation—they will perform as designed. These conditions were selected to approximate a winter day in the northern continental United States, and thus there is no reason to believe the materials would not be just as effective elsewhere in the United States.

Summary

A set of differential temperature and humidity environmental criteria and a testing methodology have been proposed and established experimentally to simulate winter conditions in the northern United States. They have been used to investigate the effects of moisture on the thermal and other properties of three cellular plastic insulation materials. The results of this investigation confirm that the method of imposing a combined moisture and temperature gradient across insulation specimens for a given time period is a viable one for studying the effects of moisture in a more realistic manner than present test methods involving isothermal or static parameters and that it can be used for other sets of conditions.

The results for the three materials indicate that none of the materials was damaged in any way. Small but significant changes in thermal properties were seen with all three materials, and these can be attributed to appropriate increases and decreases in moisture contents.

APPENDIX

Round-Robin Study

A round-robin study of specimens from the Society of the Plastic Industry (SPI) project was undertaken by the EPS manufacturers on random specimens selected by Energy Materials Testing Laboratory (EMTL). The following laboratories performed the work:

American Hoechst Corporation
R&D Laboratory
Leominster, MA

BASF Corporation
R&D Laboratory
Parsippany, NJ

ARCO Polymers, Inc.
R&D Laboratories
Newtown Square, PA

Energy Materials Testing Laboratory
Division of Fiber Materials, Inc.
Biddeford, ME

The purpose of the round robin was to confirm the level of λ values (calibration) and the capability of the thermal testing apparatus to discriminate between very small incremental changes in thermal performances of exposed specimens. The results are shown in the following table.

SPI Project No. 6443: Round-robin testing of five 25-mm-thick specimens according to ASTM Method C 518-76 by four laboratories using 300-mm-square apparatus

	λ_{app} at 24°C, W/m · K					
Specimen No.	Laboratory A	Laboratory B	Laboratory C	Laboratory D	Average	Range
1-32	0.0398	0.0397	0.0392	0.0392	0.0395	0.0006
1-16	0.0401	0.0397	0.0395	0.0392	0.0397	0.0009
1-12	0.0403	0.0404	0.0400	0.0398	0.0401	0.0006
1-22	0.0404	0.0404	0.0400	0.0398	0.0401	0.0006
1-1	0.0405	0.0405	0.0401	0.0400	0.0403	0.0005
Average	0.0403	0.0401	0.0398	0.0397		
Range	0.0007	0.0008	0.0009	0.0008		

All the apparatus was calibrated with 25.4-mm thickness of high-density fiberglass board.

The following conclusions can be drawn from the results:

1 Interlaboratory differences were in the range of 1 to 2%.
2 Specimen differences were 2% minimum.
3 All four laboratories ranked the five specimens in the same order from low to high.

Conclusion

The thermal test apparatus showed reliable calibration. Thermal test results are highly reproducible and accurate. Test data in the Final Report showing preexposure/postexposure thermal performance differences in the range of 1 to 5% appear to be reliable.

References

[1] *Chemical Marketing Reporter,* 10 April 1976, p. 7.

[2] Tye, R. P., Ashare, E., Guyer, E. C., and Sharon, A. C., "An Assessment of Thermal Insulation Materials and Systems for Building Applications," Report BNL-50862, U.S. Department of Energy, Washington, DC, June 1978.

[3] *The National Program Plan for Building Thermal Envelope Systems and Insulating Materials,* P. R. Achenbach, Ed., ORNL/Sub-7973/1, U.S. Department of Energy, Washington, DC, March 1982.

[4] *Thermal Transmission Measurement of Insulation, ASTM STP 660,* American Society for Testing and Materials, Philadelphia, 1978.

[5] *Thermal Insulation Performance, ASTM STP 718,* American Society of Testing and Materials, Philadelphia, 1980.

[6] *Thermal Insulation, Materials and Systems for Energy Conservation in the 80's, ASTM STP 789,* American Society of Testing and Materials, Philadelphia, 1983.

[7] "Thermal Performance of the Exterior Envelope of Buildings," ASHRAE SP 28, American Society of Heating, Refrigerating and Air-Conditioning Engineers, Atlanta, GA, 1981.

[8] "Thermal Performance of the Exterior Envelope of Buildings II," ASHRAE SP 38, American Society of Heating, Refrigerating and Air-Conditioning Engineers, Atlanta, GA, 1983.

[9] *Proceedings,* Sixth Conference on Roofing Technology, 30 April–1 May 1981, sponsored by the National Bureau of Standards and the National Roofing Contractors Association, Chicago.

[10] *Proceedings,* Seventh Conference on Roofing Technology, 14–15 April 1983, sponsored by the National Bureau of Standards and the National Roofing Contractors Association, Chicago, 1983.

[11] Tobiasson, W. and Ricard, J., "Moisture Gain and Its Thermal Consequence for Common Roof Insulations," *Proceedings,* Fifth Conference on Roofing Technology, the National Bureau of Standards and the National Roofing Contractors Association, Chicago, 1979.

[12] Knab, L. I., Jenkins, D. R., and Mathey, R. G., "The Effect of Moisture on the Thermal Conduc-

tance of Roofing Systems," NBS Building Science Series 123, U.S. Department of Commerce, Washington, DC, April 1980.

[*13*] Tye, R. P. and Spinney, S. C., "Study on the Effects of Moisture Vapour on the Thermal Transmittance Characteristics of Cellulose Fibre Insulation," *Journal of Thermal Insulation,* Vol. 2, 1979, pp. 175–196.

[*14*] Larkin, B. K. and Churchill, S. W., *Journal of the American Institute of Chemical Engineers,* Vol. 5, No. 4, 1959, p. 467.

[*15*] Tye, R. P., *Proceedings,* 13th International Congress of Refrigeration, Washington, DC, Institut International du Froid, Paris, Vol. I, 1971, pp. 371–378.

[*16*] Norton, F. J., *Journal of Cellular Plastics,* Vol. 3, 1965, p. 23.

[*17*] Norton, F. J., *Journal of Cellular Plastics,* Vol. 18, 1982, p. 300.

Walls and Windows

Douglas M. Burch,[1] *George N. Walton,*[1] *Kevin Cavanaugh,*[1] *and Betty A. Licitra*[1]

Effect of Wall Mass on the Annual Heating and Cooling Loads of Single-Family Residences for Five Selected Climates

REFERENCE: Burch, D. M., Walton, G. N., Cavanaugh, K., and Licitra, B. A., **"Effect of Wall Mass on the Annual Heating and Cooling Loads of Single-Family Residences for Five Selected Climates,"** *Thermal Insulation: Materials and Systems, ASTM STP 922,* F. J. Powell and S. L. Matthews, Eds., American Society for Testing and Materials, Philadelphia, 1987, pp. 541–566.

ABSTRACT: The space heating and cooling loads for a typical house and a high-solar-gain house containing partition walls and interior furnishings are simulated using a computer program called the Thermal Analysis Research Program (TARP). Separate computer runs are carried out for the following wall constructions: insulated wood frame, insulated masonry with mass on the exterior, and insulated masonry with insulation sandwiched between interior and exterior mass. The reductions in annual space heating and cooling loads achieved in typical houses with masonry wall construction compared with identical houses with lightweight wood-frame wall construction are derived as a function of five climates. Insulation credits for energy conservation standards to account for reductions in space conditioning loads for masonry houses are investigated. The presence of interior mass features was observed to cause the typical house to approach a "static high-mass limit" that coincided with steady-state theory. Under such a condition, wall mass was found to have a small effect on annual space heating and cooling loads.

KEY WORDS: thermal mass, energy conservation standards, whole-building performance, steady-state theory, internal heat gain utilization, thermal insulation

The National Bureau of Standards (NBS) recently carried out field studies [*1,2*] using six one-room test cells in Gaithersburg, Maryland, to investigate the effect of wall mass on space heating and cooling loads. The test cells were extensively instrumented, and their space heating and cooling loads were monitored over a one-year period.

The study [*1*] pertaining to space heating found that wall mass did not have a measurable effect on space heating loads during the cold part of the winter in Gaithersburg, Maryland. However, during mild spring heating days, when the internal heat gains caused the indoor temperature to rise above the thermostat set temperature, a significant thermal mass effect was observed. The heavyweight masonry and log buildings consumed less space heating energy than identical[2] lightweight buildings that had equivalent thermal resistance in their building envelopes. Wall mass was found to be more effective when placed inside, rather than outside, the wall insulation. The study [2] pertaining to space cooling found that wall mass had a significant effect on the space cooling loads of these test buildings during the entire summer season in Gaithersburg, Maryland.

[1]Mechanical engineers, engineering student, and computer programmer, respectively, Center for Building Technology, National Bureau of Standards, Gaithersburg, MD 20899.

[2]The buildings were identical except for their wall constructions.

While the field study results conclusively demonstrated the existence of a thermal mass effect, they were found to have limited applicability to real houses because the test cells were small, the solar gains through windows were small, and the top surfaces of the floors were insulated.

A limited series of tests was conducted with a partition wall installed in two of the test cells. However, both the field measurements and the computer predictions indicated that the effects of a partition wall in the test cells was very small. This was because the direct solar gains through windows did not enter the test buildings during periods when a thermal mass effect would normally be expected. The present study examines the effect of partition walls and interior furnishings in a typical house and a high-solar-gain house under conditions when direct solar gains contribute in a normal way to the internal heat gains.

Comparison of TARP to Measured Data

In an effort to extend the results of the NBS field measurements [*1,2*] to typical houses, a computer program, called the Thermal Analysis Research Program (TARP), was tested by comparing predicted and measured space heating and cooling loads for the six test cells used for the field studies. This computer program was subsequently used to investigate the effect of wall mass on space heating and cooling loads in a typical house and a high-solar-gain house as a function of climate.

Description of Computer Program

TARP is a computer program for predicting the space heating and cooling loads of a building under a dynamic set of boundary conditions. The program uses a detailed heat-balance method for determining space heating and cooling loads from the predicted heat losses and heat gains. Unlike most contemporary computer programs, TARP permits the evaluation of the multiroom situation by simultaneously solving the heat-balance equations for several independent zones. The computer algorithms are based partly on subroutines from the Building Loads Analysis and System Thermodynamics (BLAST) computer program. For the TARP predictions of the test cells, a detailed description of the building, including the heat-transfer properties for the materials comprising the building envelope and its interior mass features, an operation schedule for the building, and hourly outdoor climate data were specified for the program. In addition, an empirical infiltration correlation for each of the test cells was specified for TARP. Further information on TARP may be found in Ref *3*.

Description of Test Cells

The test buildings were small, one-room test cells identical except for the wall constructions, which were as follows:

No. 1—insulated lightweight wood frame,
No. 2—uninsulated lightweight wood frame,
No. 3—insulated masonry with interior insulation (mass on exterior),
No. 4—uninsulated masonry,
No. 5—log, and
No. 6—insulated masonry with insulation sandwiched between the interior and exterior masses.

A photograph of one of the test cells is presented in Fig. 1. These buildings were operated with a fixed thermostat setting and exposed to winter heating, spring heating, and summer cooling climates.

FIG. 1—*A photograph of one of the test cells.*

Results of Comparisons

Measured hourly outdoor climate data for the six test cells for 13-day winter and summer periods were entered into TARP. Comparisons between the measured and the computer-predicted sensible space heating loads for the last 62 h of the winter period are given in Fig. 2. The outdoor climate data are given in Fig. 2*a;* sensible heating load comparisons for Buildings 1, 3, and 6 are given in Fig. 2*b*, *c*, and *d*, respectively. Similar agreement was obtained for the other test buildings. A similar set of sensible space cooling load comparisons for the last 72 h of the summer period is given in Fig. 3.

Percentage differences between predicted and measured loads for the six test cells during the winter and summer periods are summarized in Table 1. For the winter period, TARP predicted cumulative space heating loads within a root-mean-square (RMS) average of 11%. For the summer period, it predicted cumulative space cooling loads within an RMS average of 25%. These percentage differences, in an absolute sense, appear to be large. Perhaps the strongest case for the validity of TARP in relation to the current study is that, during climate periods when a thermal mass effect was experimentally observed, the program predicted the correct relative cumulative space conditioning loads—that is, the ranking of the test cells and the relative magnitudes of the predicted thermal mass effects were the same as those for the actual test cells. This is illustrated by a bar chart of the cumulative space heating loads for the spring comparison period, given in Fig. 4. A discussion of experimental errors and uncertainties that contribute to the percentage differences follows.

The differences between predicted and measured space heating and cooling loads were believed to be due to the following factors: uncertainties in the data entered into the program, simplifying approximations used in the algorithms of TARP, and experimental error in the measured space heating and cooling loads.

A complete description of the building envelopes including the heat-transfer and physical properties for the materials comprising their envelopes was entered into TARP. In many cases, heat-transfer properties were obtained from handbooks and may have differed from actual values by as much as $\pm 10\%$. Variability in the construction and the presence of thermal anomalies such as compressed blanket insulation may have introduced additional uncertainty. Another factor introducing uncertainty was that the amount of framing for Test Cells 1 through 4 was not precisely known. It was estimated from photographs taken during the construction of the test cells.

A series of outdoor temperatures was specified as input to TARP and served as a boundary condition. A study of the microclimate surrounding the test cells indicated that the outdoor temperature at different locations varied at any instant within the range of 1 degree Celsius (2 degrees Fahrenheit). The outdoor temperature was considered to be uncertain by this amount.

During periods requiring space heating, the indoor temperature was specified as a constant that corresponded to the average of the hourly indoor temperature measured at the midheight level. The effects of time fluctuations of the indoor temperature as a result of thermostat control and vertical temperature gradients were not taken into account in TARP.

Even though TARP is one of the most sophisticated computer programs, it nonetheless contains many simplifying assumptions. A few of the more important ones are a uniform indoor temperature, one-dimensional heat transfer, and constant heat-transfer properties. The indoor temperature specified in TARP was the midheight value. The indoor temperature within the test cells was observed to be stratified by as much as 4 degrees Celsius (8 degrees Fahrenheit) during certain periods. This caused the floor and ceiling heat-transfer surfaces to be modeled with slightly inexact boundary condition temperatures. The one-dimensional heat-transfer assumption resulting in imprecise predictions of heat transfer at corners and within the earth beneath the slab-on-grade floors. Regarding the constant property assumption, mean temperature has been shown to have a small effect on the thermal conductances of building materials.

Other simplifying assumptions that may have affected the predicted results include isotropic diffuse solar radiation and constant ground reflectance.

Regarding experimental error, the heating loads involved the measurement of electricity that were believed to be accurate to within ±0.25%. On the other hand, the sensible cooling loads were computed as the product of a mass flow rate of air through an air conditioner and an integrated temperature rise across the unit. The air flow measurement was believed to be uncertain by about ±5 to 8%. Including the uncertainty in the temperature difference measurement, the overall uncertainty in the cooling load measurement could have been as large as ±12%.

In view of these uncertainties and simplifying assumptions, the level of agreement between predicted and measured space heating and cooling loads was considered to be good. The level of agreement is comparable and in most cases better than that for other similar computer programs cited in the literature [*4–7*].

Simulation of Residences

The TARP computer simulations were carried out for a typical house and a high-solar-gain house. Both of these houses were selected to have high thermal resistance in their building envelopes consistent with current energy conservation practice. These houses are described next.

Typical House

The typical house was a wood-frame rambler with a floor area of 110 m^2 (1180 ft^2). This house was similar to the house specified in the Hastings' report [*8*]. It had a pitched roof and ventilated attic with a thermal resistance, R, of 3.3 $m^2 \cdot K/W$ (19 $h \cdot ft^2 \cdot °F/Btu$) in the ceiling insulation. The wall constructions analyzed included insulated wood frame, insulated masonry with mass on the exterior, and insulated masonry with mass on both sides of the insulation. A description of these wall constructions is given in Table 2. The wall constructions were the same as those for the field test cells, except that the overall thermal resistances were made to be identical by slightly adjusting the thermal conductivities of the wall insulations.

For the typical house, the windows were double pane and had a surface area of 13.1 m^2 (141 ft^2), or 12% of the floor area. For each orientation, the ratio of window area to gross wall area was constant. The roof overhung the windows by 0.61 m (2.0 ft) along the north and south walls.

The floor consisted of 0.025-m (1-in.) wood and insulation [$R = 1.9$ $m^2 \cdot K/W$ (11 $h \cdot ft^2 \cdot °F/Btu$)] placed over a ventilated crawl space. A floor plan and elevation are shown in Figs. 5*a* and 5*b*.

The heat-transfer properties for the components of the building envelope are summarized in Table 3. For the analysis, the rate of infiltration was taken to be constant at one volume change per hour.[3]

High-Solar-Gain House

The high-solar-gain house was identical to the typical house, with the exception that the window area was increased to 16.4 m^2 (176 ft^2), or 15% of the floor area, and the entire window opening was placed on the south wall.

[3]During the review of this paper, it was pointed out that this value is somewhat on the high side for new construction.

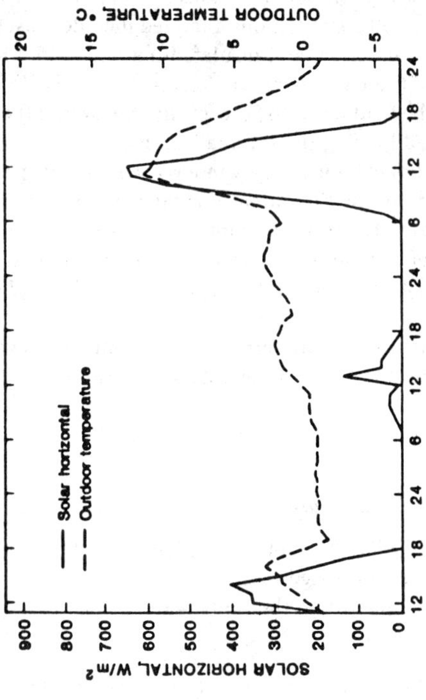

A. Outdoor climate

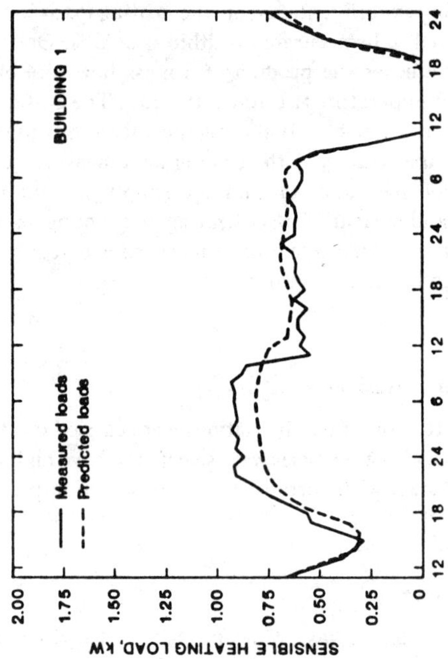

B. Wood-frame test cell

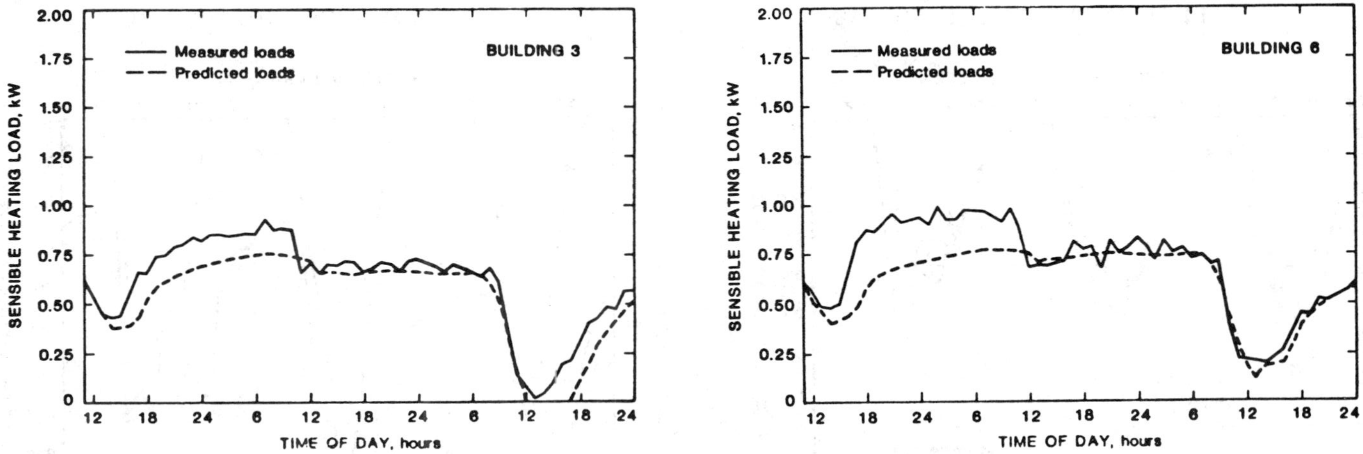

C. Masonry test cell with mass on the exterior

D. Masonry test cell with mass on the interior

FIG. 2—*Comparison of predicted and measured sensible heating loads for a 62-h winter period.*

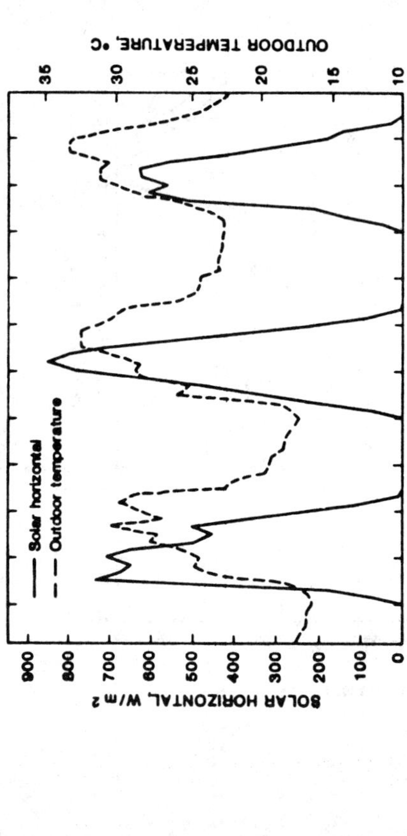

A. Outdoor climate

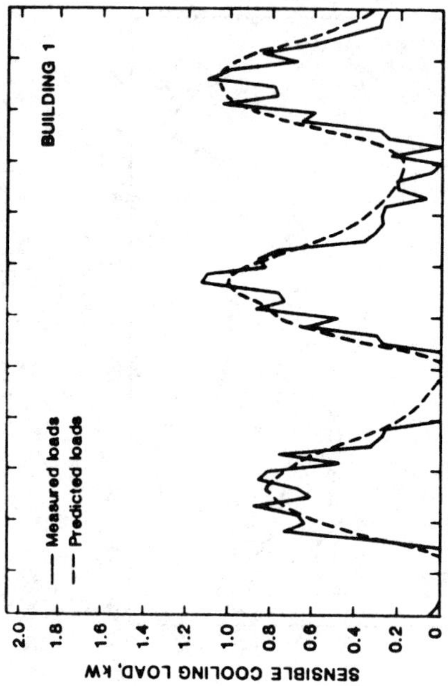

B. Wood-frame test cell

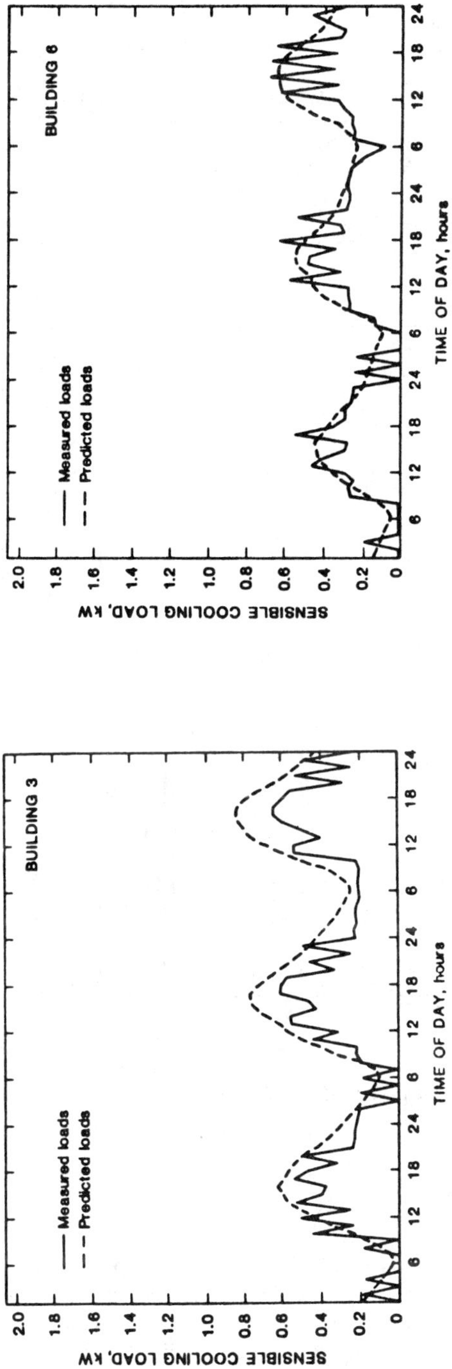

C. Masonry test cell with mass on the exterior

D. Masonry test cell with mass on the interior

FIG. 3—*Comparison of measured and predicted sensible cooling loads for a 72-h summer period.*

TABLE 1—*Summary of differences between predicted and measured space conditioning loads, in percentages.*

Test Cell Description	Winter Cumulative Load	Winter Peak Load	Summer Cumulative Load	Summer Peak Load
Insulated wood frame	−2.7	−13	+16	−11
Uninsulated wood frame	+6.0	−5.0	+7.8	—4.9
Insulated masonry (mass on outside)	−13	−19	+37	+29
Uninsulated masonry	+20	−8.4	+36	+23
Log	+5.3	−9.9	+22	+21
Insulated masonry (mass on inside)[a]	−12	27	+18	−6.3
Root-mean-square average	11	15	25	18

[a]Insulation was sandwiched between the interior and exterior masses.

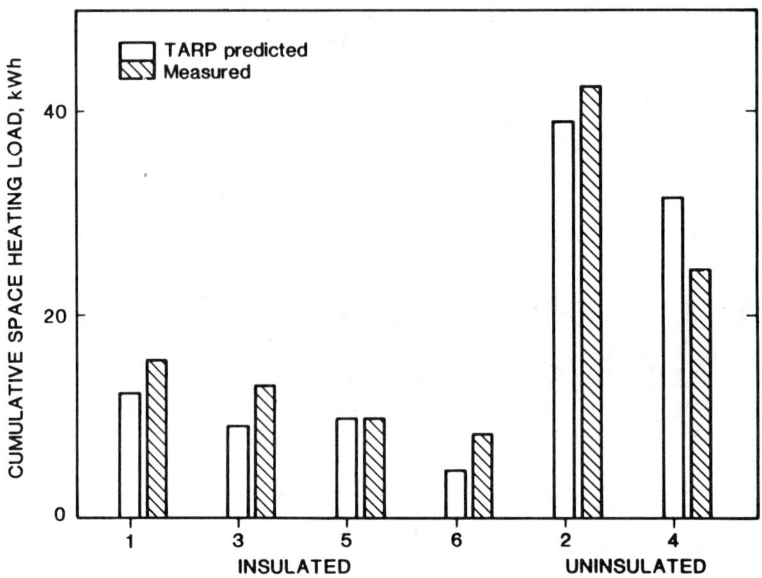

FIG. 4—*Bar chart of the cumulative space heating loads for the spring period.*

Description of Computer Model

Each house was simulated as three zones: a living space, an attic, and a crawl space. Space heating and cooling were provided only for the living space. The air temperature within each zone was treated as being well mixed. Partition walls and interior furnishings were included as surfaces within the living space. The radiant exchange among the surfaces within each zone was computed by the mean-radiant-temperature network method [*9*]. This method is equivalent to putting all surfaces on a hypothetical sphere, permitting each surface to have some view of every

TABLE 2—*Description of wall constructions.*[a]

Construction Type	Description
Insulated lightweight wood frame	0.5-in. gypsum board 0.002-in. polyethylene film 2 by 4 studs placed 16 in. on center with R-11 blanket insulation installed between the studs 5/8-in. exterior plywood
Insulated masonry with interior insulation (mass on outside)	0.5-in. gypsum board 0.002-in. polyethylene film 2-in.-thick extruded polystyrene insulation placed between 1 1/2-in.-wide wood furring strips placed 24 in. on center 1/4-in. air space 4-in. two-core hollow concrete block at 105 lb/ft^3 4-in. face brick
Insulated masonry with insulation sandwiched between inside and outside mass	0.5-in. plaster 8-in. two-core hollow concrete block at 105 lb/ft^3 3 1/2-in. perlite insulation in cavity 4-in. face brick

[a]1 in. = 0.0254 m.

other surface. Of contemporary computer programs, TARP is one of the few that handles the radiation exchange among interior surfaces and envelope surfaces.

A special series of computer runs was carried out in which the living space of the typical house was modeled using six zones instead of one. This model included convective exchange between the zones that occurred through connecting doors and passageways. The relative effects of wall mass for the six-zone model coincided closely with those for the one-zone model. Since modeling the living space using one zone instead of six zones predicted very nearly the same relative thermal mass effects and required approximately one fourth as much computer time, it was used in the analysis.

Each of the partition wall surfaces consisted of 0.05 by 0.10-m (2 by 4-in.) framing with 0.013-m (1/2-in.) gypsum board attached at opposite sides. The surface area of the partition walls was identical to the actual partition walls of the Hastings' house.

Interior furnishings were modeled as a 0.05-m (2-in.) thick slab of wood having a surface area of 86.6 m^2 (932 ft^2). The total weight of interior furnishings was 3200 kg (7000 lb), and its specific heat was taken to be 1200 J/kg · K (0.29 Btu/lb · °F). The weight of interior furnishings included not only the weight of movable furniture but also fixed furnishings such as appliances and fixtures.

For the computer simulations, the thermostat was set at 20°C (68°F) for space heating and 24°C (76°F) for space cooling. Within a 4.4 degrees Celsius (8.0 degrees Fahrenheit) range between the setpoints, space conditioning was not provided. A constant internal load of 8.1 W/m^2 (0.75 W/ft^2) of floor was used to simulate heat release associated with occupancy.

Results

Using climate data from Weather Year for Energy Calculations (WYEC) [*10*] magnetic computer tapes, annual space heating and cooling loads were predicted for the following cities: Madison, Wisconsin; Lake Charles, Louisiana; Washington, D.C.; Los Angeles, California; and Charleston, South Carolina. These cities were selected to represent the climates of the

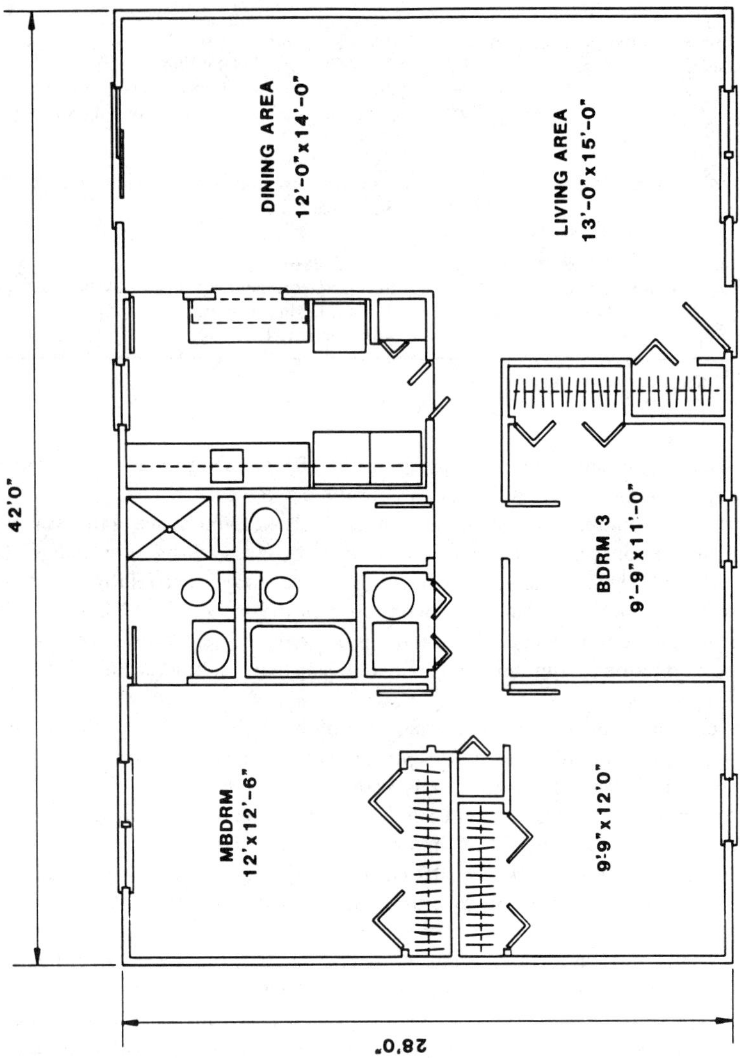

FIG. 5a—*Floor plan for the typical ranch house.*

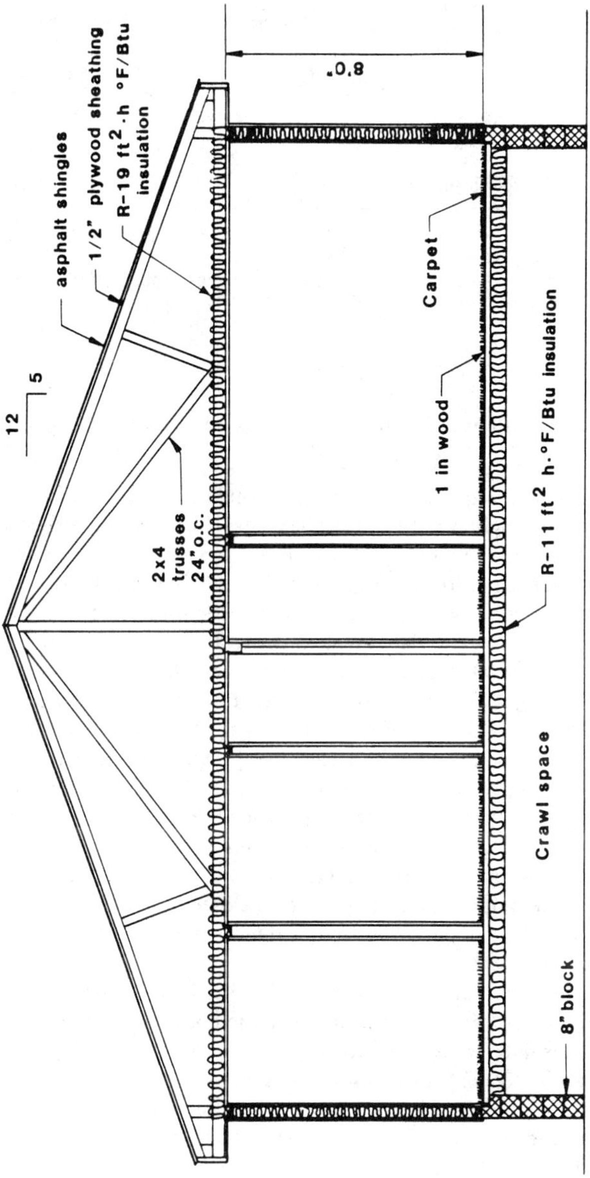

FIG. 5b—*Elevation for the typical ranch house.*

TABLE 3—*Heat-transfer properties for the components of the typical house.*

Component	Surface Area, A, m^2 (ft^2)	Thermal Transmittance, U, $W/m^2 \cdot K$ ($Btu/h \cdot ft^2 \cdot °F$)	$U \cdot A$ Product, W/K ($Btu/h \cdot °F$)
Glazing	13.1 (141)	2.75 (0.485)	36.1 (68.4)
Walls	89.1 (959)	0.471 (0.083)	42.0 (79.6)
Floor/crawl space[a]	...	...	23.4 (44.4)
Ceiling/attic[a]	...	...	32.2 (61.1)
Door	1.87 (20.1)	1.62 (0.285)	3.01 (5.7)
Infiltration[b]	...	...	88.2 (167.3)
Overall envelope heat-transfer coefficient, K			224.9 (426.5)

[a]Calculated from a three-node model that included an air node, a bottom surface node, and top surface node. The bottom and top surfaces exchanged heat by radiation.

[b]Calculated from the relation $H_I = \rho \cdot V \cdot I \cdot C_p$, where ρ is the density of air, V is the volume of the living space, I is the rate of infiltration, and C_p is the specific heat of air.

northern, Gulf Coast, mid-Atlantic, southern West Coast, and southern regions of the United States, respectively.

Typical House—Weekly average winter heating loads expressed in kilowatt hours per day for the typical house with lightweight wood-frame wall construction exposed to Washington, D.C., climate are plotted as a function of average outdoor temperature in Fig. 6. The special case of interior surfaces without thermal storage is given in Fig. 6*a*. Here the interior surfaces participated in the radiant exchange within the living space, but they did not store heat. Note that the majority of the heating loads are correlated by a linear relationship obtained by the method of least squares. As the outdoor temperature becomes warmer, the heating loads rise above this linear relationship. The larger heating loads during mild heating periods are caused by an inability of the typical house to utilize all of its internal heat gains. The indoor temperature rose above the thermostat set temperature, and thermal energy was rejected to the environment. In extreme cases of overheating, the indoor temperature actually rose above the setpoint for space cooling, and cooling loads occurred. The effect of adding thermal heat capacity to the interior surfaces is illustrated in Fig. 6*b*. Here it is seen that adding thermal heat capacity to the interior surfaces caused the typical house to utilize a larger portion of its internal heat gains. As a result, the heating loads near the balance-point temperature closely approached the linear relationship. The term balance-point temperature denotes the average outdoor temperature at which the space heating load vanishes.

The linear relationship functions as a "static high-mass limit"; that is, the space heating loads approach this limit as thermal mass is added to the house. Heating load correlations for the typical house with insulated masonry wall constructions are given in Fig. 7*a* for the case of mass on the outside and in Fig. 7*b* for the case of mass on the inside. These results indicate that wall mass has a small effect on space heating loads in Washington, D.C., because the presence of interior surfaces caused the space heating loads to approach closely the static high-mass limit. Under this condition, additional mass provided in the walls had a small effect. The linear relationship is shown to coincide closely with steady-state theory.

The space heating load, Q_h, is given by the relation

$$Q_h = K \cdot (T_b - T_0) \tag{1}$$

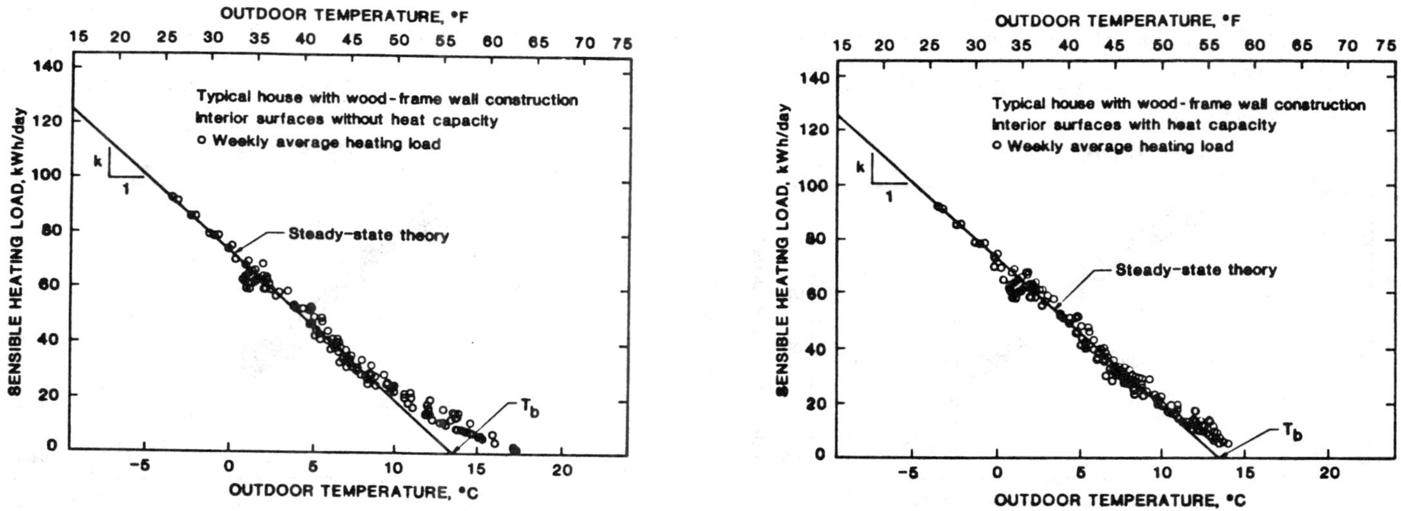

a. Interior surfaces without heat capacity

b. Interior surfaces with heat capacity

FIG. 6—*Heating load correlations for a typical house with wood-frame wall construction located in Washington, DC.*

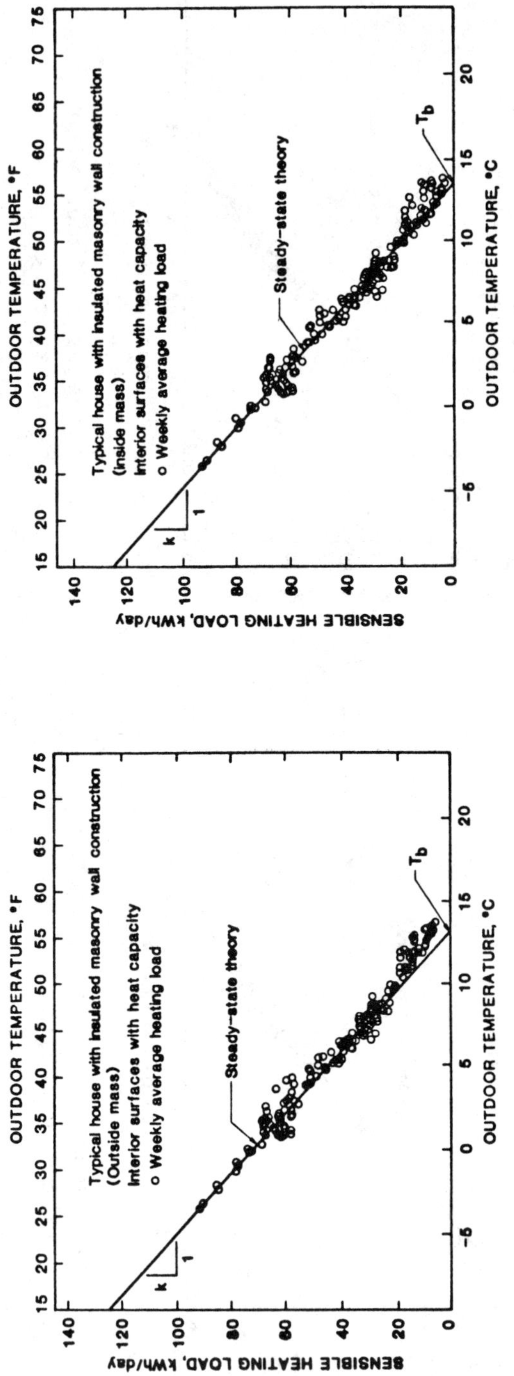

a. Mass on outside

b. Mass on inside

FIG. 7—*Heating load correlations for the typical house with masonry wall constructions located in Washington, DC.*

where T_0 = average outdoor temperature, in degrees Celsius (degrees Fahrenheit). The overall envelope heat-transfer coefficient, K, is given by

$$K = \sum_{i=1}^{N} U_i \cdot A_i + G_i \tag{2}$$

where

$U_i \cdot A_i$ = product of thermal transmittance and surface area for the ith component, W/K (Btu/h · °F), and
G_i = heat-transfer coefficient due to infiltration, W/K (Btu/h · °F).

The balance-point temperature, T_b, may be predicted by the relation

$$T_b = T_i - \frac{Q_i + Q_s - Q_e}{K} \tag{3}$$

where

T_i = average indoor temperature, °C (°F),
Q_i = internal heat generation associated with occupancy, W (Btu/h),
Q_s = solar gain, W (Btu/h), and
Q_e = heat transfer to the earth, W (Btu/h).

Convergence to steady-state theory in the limit as thermal mass is added is another check on the validity of TARP. Good agreement with steady-state theory is observed. The slope of the linear portion of the heating load correlation agrees within 0.8% of the predicted value using the steady-state theory given in Table 3. The balance-point temperature, T_b, agrees within 0.4 degree Celsius (0.7 degree Fahrenheit) when the solar gain, Q_s, is computed by the relation

$$Q_s = \tau \cdot \mathrm{SC} \sum_{i=1}^{4} A_i \cdot H_i \tag{4}$$

Here A_i is the window surface area for the ith orientation, H_i is the incident solar radiation for the ith orientation, τ is the coefficient of solar transmission for standard double-strength glass, and SC is a seasonal average shading coefficient for double-pane windows. Relation 4 neglects solar loading of opaque portions of the building envelope.

Reductions in annual heating load achieved by using masonry instead of wood-frame wall construction for the five climates analyzed are summarized in Table 4. Here it should be emphasized that the houses compared have equivalent steady-state envelope heat-transfer coefficients, K.

From Table 4, it is evident that inside mass is considerably more effective than outside mass. That is, reductions in annual space heating loads are about four to five times larger. Both the percentage and the absolute reductions for cases of inside and outside mass are larger in the mild climate regions because the majority of the heating days are distributed near the balance-point temperature. At an electric rate of $0.06 per kilowatt hour, the largest observed absolute reductions of 294 kWh for Charleston, South Carolina, represents an annual monetary savings in heating cost of about $18. This reduction is 9.9% of the annual heating load.

Reductions in annual space cooling loads for the typical house are summarized in Table 5. Here the same trends are apparent; that is, absolute and percentage reductions are largest in mild cooling climates, and mass is more effective when it is positioned inside rather than outside

TABLE 4—*Comparison of reductions in annual space heating loads achieved by masonry wall construction and wood-frame wall construction in a typical house.*

		Annual Heating Loads, 10^3 kWh			Reductions			
					Mass Outside		Mass Inside	
Region	City	Wood Frame	Masonry (Mass Outside)	Masonry (Mass Inside)	kWh	%	kWh	%
Northern	Madison, WI	15.7	15.6	15.5	41	0.26	158	1.0
Mid-Atlantic	Washington, DC	8.01	7.98	7.88	26	0.33	135	1.7
Gulf Coast	Lake Charles, LA	1.94	1.88	1.71	55	2.9	227	11.7
Southern West Coast	Los Angeles, CA	0.651	0.591	0.416	60	9.2	235	36
Southern	Charleston, SC	2.98	2.88	2.68	93	3.1	294	9.9

TABLE 5—*Comparison of reductions in annual space cooling loads achieved by masonry wall construction and wood-frame wall construction for a typical house.*

		Annual Cooling Load, 10^3 kWh			Reductions			
					Mass Outside		Mass Inside	
Region	City	Wood Frame	Masonry (Mass Outside)	Masonry (Mass Inside)	kWh	%	kWh	%
Northern	Madison, WI	3.07	2.97	2.79	108	3.5	286	9.3
Mid-Atlantic	Washington, DC	6.14	6.06	5.96	88	1.4	188	3.1
Gulf Coast	Lake Charles, LA	10.4	10.3	10.2	105	1.0	255	2.4
Southern West Coast	Los Angeles, CA	3.80	3.61	3.33	193	5.1	469	12.3
Southern	Charleston, SC	8.27	8.16	8.01	105	1.3	258	3.1

the wall insulation. As in the case of reductions in space heating load, the absolute reductions convert into rather small monetary savings. For instance, at an electric rate of \$0.06 per kilowatt hour and for air-conditioning equipment with a seasonal average coefficient of performance (COP) = 2.4, the largest reduction of 469 kWh represents an annual monetary savings in cooling costs of about \$12. This reduction is 12.3% of the annual cooling load.

High-Solar-Gain House—Reductions in space heating loads for the high-solar-gain house achieved by using masonry instead of wood-frame wall construction for the five geographic locations are summarized in Table 6. The reductions are consistently higher than those for the typical house given in Table 4. A similar summary of reductions in space cooling loads for the high-solar-gain house is given in Table 7. Here it can be seen that the percentage reductions are generally only slightly above those for the typical house given in Table 5. However, the absolute reductions are considerably larger than those for the typical house because the absolute cooling loads in the high-solar-gain house are larger. These results indicate that wall mass has a larger effect on space heating and cooling loads in houses with larger solar gains. It should be pointed out that special actions would normally be taken to minimize summer solar gains in a high-solar-gain house, and whole-house ventilation instead of vapor-compression air conditioning would probably be used to provide space cooling during mild summer periods.

Reductions in Wall Thermal Resistance

Background

It has been suggested that prescriptive levels of wall thermal resistance specified in energy conservation standards should be reduced in masonry houses to account for reductions in space heating and cooling energy achieved by wall mass [*11*]. The reduction in wall thermal resistance has been called ΔR.

This section summarizes the results of a computer analysis to determine ΔR values for the typical masonry houses with partition walls and interior furnishings located in the five different climate regions. Two cases are considered. The first case is based on reductions only in annual space heating load. Here the ΔR values cause the typical masonry residences to have annual space heating loads equivalent to those of the typical wood-frame house. The second case is based on the combined reductions in both annual space heating and cooling loads. Here the ΔR values cause the typical masonry residences to have annual space heating and air-conditioning loads (including both space heating and cooling) equivalent to those of the typical wood-frame house.

The reductions in wall thermal resistance, ΔR, were determined by an iterative procedure. A sequence of TARP computer simulations was made with decreasing thickness of wall insulation until the annual load for the typical house with masonry wall construction was equal to that for the typical house with wood-frame wall construction. The ΔR value was computed by taking the difference in thermal resistances between the original masonry wall and the modified masonry wall.

Based on Equivalent Annual Space Heating Loads

Reductions in wall thermal resistance are summarized in Table 8 and are plotted with respect to a heating degree-days base of 18°C (65°F) in Fig. 8. The ΔR values are less than or equal to 0.23 $m^2 \cdot K/W$ (1.3 $h \cdot ft^2 \cdot °F/Btu$) for all climatic regions except for the case of the typical masonry house with wall mass inside located in the mild climates of the Gulf Coast, southern West Coast, and southern regions. Such ΔR values are considered to be small. A thermal resistance of 0.23 $m^2 \cdot K/W$ (1.3 $h \cdot ft^2 \cdot °F/Btu$) is equivalent to that of about 0.01-m (0.4-in.) glass-fiber blanket insulation.

TABLE 6—*Comparison of reductions in annual space heating loads achieved by masonry wall construction and wood-frame wall construction for a high-solar-gain house.*[a]

		Annual Heating Loads, 10^3 kWh			Reductions			
					Outside Mass		Inside Mass	
Region	City	Wood Frame	Masonry (Mass Outside)	Masonry (Mass Inside)	kWh	%	kWh	%
Northern	Madison, WI	14.6	14.5	14.2	87.9	0.60	363	2.5
Mid-Atlantic	Washington, DC	7.23	7.16	6.88	61.5	0.85	343	4.7
Gulf Coast	Lake Charles, LA	1.77	1.69	1.41	73.8	4.2	359	20.3
Southern West Coast	Los Angeles, CA	0.545	0.476	0.227	68.9	12.6	318	58.3
Southern	Charleston, SC	2.64	2.51	2.11	125	4.8	526	19.9

[a]Fenestration is 15% of the floor area, and it is all placed on the south wall.

TABLE 7—*Comparison of reductions in annual space cooling loads achieved by masonry wall construction and wood-frame wall construction for a high-solar-gain house.*[a]

		Annual Cooling Loads, 10^3 kWh			Reductions			
					Outside Mass		Inside Mass	
Region	City	Wood Frame	Masonry (Mass Outside)	Masonry (Mass Inside)	kWh	%	kWh	%
Northern	Madison, WI	3.79	3.65	3.35	144	3.8	442	11.7
Mid-Atlantic	Washington, DC	7.03	6.91	6.66	120	1.7	366	5.2
Gulf Coast	Lake Charles, LA	11.8	11.7	11.3	146	1.2	478	4.0
Southern West Coast	Los Angeles, CA	6.33	6.11	5.60	223	3.5	735	11.6
Southern	Charleston, SC	9.79	9.60	9.18	185	1.9	607	6.2

[a]Fenestration is 15% of the floor area, and it is all placed on the south wall.

TABLE 8—*Reductions in wall thermal resistance, ΔR, for a typical house with masonry wall construction (based on equivalent annual space heating loads).*

Region	City	ΔR, $m^2 \cdot K/W$ ($h \cdot ft^2 \cdot °F/Btu$)	
		Mass Outside	Mass Inside
Northern	Madison, WI	0.021 (0.12)	0.079 (0.45)
Mid-Atlantic	Washington, DC	0.021 (0.12)	0.11 (0.65)
Gulf Coast	Lake Charles LA	0.12 (0.69)	0.51 (2.9)
Southern West Coast	Los Angeles, CA	0.23 (1.3)	0.83 (4.7)
Southern	Charleston, SC	0.17 (0.98)	0.48 (2.7)

Based on Equivalent Annual Space Conditioning Loads Including Both Heating and Cooling

Reductions in wall thermal resistance, ΔR, are summarized in Table 9. With the exception of those for Los Angeles, these ΔR values are larger than when reductions only in space heating loads are considered (see Table 8). For the typical masonry house with wall mass outside, the ΔR values are larger than 0.23 $m^2 \cdot K/W$ (1.3 $h \cdot ft^2 \cdot °F/Btu$) for the mild climate regions (that is, the Gulf Coast and southern region). For the typical masonry house with wall mass inside, the ΔR values are larger than 0.23 $m^2 \cdot K/W$ (1.3 $h \cdot ft^2 \cdot °F/Btu$) for all the climate regions except Los Angeles. Note that the ΔR values for Los Angeles are large negative values. A negative ΔR value means that the wall resistance of the masonry house had to be increased instead of decreased to achieve an equivalent annual space conditioning load. An explanation for the peculiar behavior for Los Angeles follows.

The majority of the space conditioning loads for Los Angeles are space cooling loads that occur very near the balance-point temperature. For this condition, the typical house gained heat from the internal heat gains and lost heat by means of conduction through its building envelope. In cooler climates, a reduction in wall resistance increases the conductive losses through the building envelope. In warmer climates, a reduction in wall resistance has the opposite effect of increasing conductive gains. Therefore, when the typical house is operated in a mild cooling climate such as Los Angeles, ΔR adjustments have a reverse effect to that in a hot cooling climate.

Caveats and Cautions

The effect of thermal mass in houses is dependent on the amount of internal heat gains including the solar gains, the thermal resistance of the building envelope, and the amount of thermal storage provided by the partition walls, interior floors/ceilings, and interior furnishings. Since these factors vary considerably from one house to the next, the effect of thermal mass will likewise vary considerably among the housing stock. Therefore, the ΔR values given in this paper, in a strict sense, are applicable only to the particular typical house used for the analysis. Moreover, the insulation levels used for the houses analyzed were not changed with respect to climate. It is common practice for houses located in cold climates to be fitted with more insulation than houses located in warm climates.

This study did not include the effect of wall mass on the part-load efficiency of heating, ventilating, and air-conditioning (HVAC) equipment. Another study in preparation indicates that

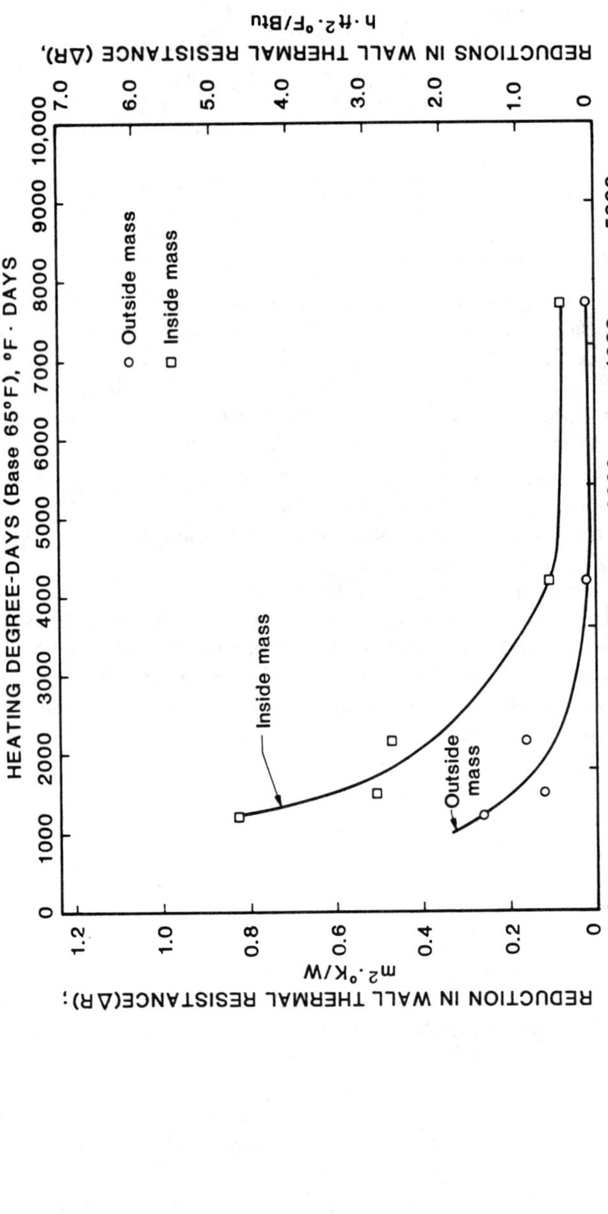

FIG. 8—*Reductions in thermal resistance, ΔR, plotted as a function of heating degree-days.*

TABLE 9—*Reductions in wall thermal resistance, ΔR, for a typical house with masonry wall construction (based on equivalent annual space heating and air-conditioning loads).*

Region	City	ΔR, $m^2 \cdot K/W$ ($h \cdot ft^2 \cdot °F/Btu$)	
		Mass Outside	Mass Inside
Northern	Madison, WI	0.088 (0.5)	0.23 (1.3)
Mid-Atlantic	Washington, DC	0.088 (0.5)	0.26 (1.5)
Gulf Coast	Lake Charles LA	0.28 (1.6)	0.63 (3.6)
Southern West Coast	Los Angeles, CA	−3.3[a](−19.0)	$-\infty$ ($-\infty$)
Southern	Charleston, SC	0.32 (1.8)	0.70 (4.0)

wall mass has a small effect on peak space heating and cooling loads in single-family residences. These results suggest that wall mass will have a small effect on the part-load efficiency of HVAC equipment.

Summary and Conclusions

A computer program called TARP was used to predict the space heating and cooling loads of a typical house and a high-solar-gain house. Each of these houses contained representative interior surfaces including partition walls and interior furnishings.

Weekly space heating loads were correlated with respect to average outdoor temperature. The presence of partition walls and interior furnishings caused the space heating load correlations of the typical house to approach a linear relationship that coincided with steady-state theory. Under this condition, the presence of additional mass, such as wall mass, was found to have a small but beneficial effect on space heating and cooling loads, except for climates where the house operated predominantly near its balance-point temperature.

Similarly, wall mass was found to have a small effect on space cooling loads in the typical house with partition walls and interior furnishings.

Regarding insulation credits in energy conservation standards, the reductions in wall thermal resistance, ΔR, in typical masonry houses that provide equivalent annual space conditioning loads for both heating and cooling were found to range from 0.09 (0.5) to 0.70 $m^2 \cdot K/W$ (4.0 $h \cdot ft^2 \cdot °F/Btu$). When only reductions in annual space heating load were considered, the ΔR values for the typical masonry house with wall mass on the outside were found to be less than or equal to 0.23 $m^2 \cdot K/W$ (1.3 $h \cdot ft^2 \cdot °F/Btu$) for the five climate regions considered. This ΔR value corresponds to about 0.01-m (0.4-in.) glass-fiber blanket insulation. Considerably larger ΔR values were observed for cases with wall mass on the inside.

Thermal mass effects were shown to be relatively more important in instances in which the internal heat gains were large, the mass was located inside the insulation, and the house operated predominantly near its balance-point temperature.

References

[1] Burch, D. M., Krintz, D. F., and Spain, R. S., "The Effect of Wall Mass on Winter Heating Loads and Indoor Comfort—An Experimental Study," *A.S.H.R.A.E. Transactions*, Vol. 90, Pt. 1, 1984, pp. 94–121.

[2] Burch, D. M., Davis, K. L., and Malcolm, S. A., "The Effect of Wall Mass on the Summer Space Cooling of Six Test Buildings," *A.S.H.R.A.E. Transactions*, Vol. 90, Pt. 2, 1984, pp. 5–21.

[3] Walton, G. N., *Thermal Analysis Research Program—Reference Manual*, NBSIR 83-2655, National Bureau of Standards, Washington, DC, 1983.

[*4*] Arumi-Noe, F. and Burch, D. M., "DEROB Simulation of the NBS Thermal Mass Test Buildings," *A.S.H.R.A.E. Transactions*, Vol. 90, Pt. 2, 1984, pp. 254–274.
[*5*] Burch, D. M., Peavy, B. A., and Powell, F. J., "Comparison Between Measured and Computer-Predicted Hourly Heating and Cooling Energy Requirements for an Instrumented Wood Frame Townhouse Subjected to Laboratory Tests," *A.S.H.R.A.E. Transactions*, Vol. 81, Pt. 2, 1975, pp. 70–88.
[*6*] Judkoff, R., Wortman, D., and Burch, J., "Measured Versus Predicted Performance of the SERI Test House: A Validation Study," paper presented at the National Heat Transfer Conference, Seattle, WA, July 1983.
[*7*] Anderson, B., Bauman, F., and Kammerud, R., "Verification of BLAST by Comparison with Direct Gain Test Cell Measurements," Report No. LBL-10619, Lawrence Berkeley Laboratory, Berkeley, CA, November 1980.
[*8*] Hastings, S. R., "Three Proposed Typical House Designs for Energy Conservation Research," NBSIR 77-1309, National Bureau of Standards, Washington, DC, October 1977.
[*9*] Carrol, J. A., "An 'MRT Method' of Computing Radiant Energy Exchange in Rooms," *Proceedings*, 2nd Systems Simulation and Economics Analysis Conference, San Diego, CA, 1980, pp. 903–917.
[*10*] Crow, L. W., "Development of Hourly Data for Weather Year for Energy Calculations (WYEC)," *A.S.H.R.A.E. Journal*, Vol. 23, No. 10, October 1981, pp. 37–41.
[*11*] Goodwin, S. E. and Catani, M. J., "The Effects of Mass on Heating and Cooling Loads and on Insulation Requirements of Buildings in Different Climates," *A.S.H.R.A.E. Transactions*, Vol. 85, Pt. 1, 1979, pp. 869–884.

Robert P. Bowen[1] *and K. Richard Solvason*[1]

A Calorimeter for Determining Heat Transmission Characteristics of Windows

REFERENCE: Bowen, R. P. and Solvason, K. R., "**A Calorimeter for Determining Heat Transmission Characteristics of Windows,**" *Thermal Insulation: Materials and Systems, ASTM STP 922,* F. J. Powell and S. L. Matthews, Eds., American Society for Testing and Materials, Philadelphia, 1987, pp. 567-581.

ABSTRACT: Heat loss through windows is receiving an increasing amount of attention from consumers, manufacturers, and building designers. The technology is changing and should lead to substantial improvements in thermal performance. To assist in the evaluation of existing as well as new products, the Division of Building Research, National Research Council of Canada (DBR/NRCC), has designed and built a new calorimeter for use in their Environmental Test Facility.

This paper presents an approach for determining the heat transmission characteristics of windows using the radiative and convective components of heat transfer from the calorimeter to the specimen surface. Through calibration tests using materials of known conductance, expressions for the radiation exchange between the calorimeter surfaces and the specimen and the calorimeter air to the specimen can be determined. These expressions are then used when testing a window to calculate an equivalent surface temperature for the window. This approach is necessary because of the difficulty in determining an average surface temperature on glass.

Although the approach can be used with conventional calorimeter designs, uncertainties are introduced because of the changing calorimeter characteristics for different sizes of specimens. To reduce the uncertainties, the new calorimeter was designed to have a constant convection coefficient and uniform temperature radiation field. The calibration program is discussed, and some preliminary results are given.

KEY WORD: windows, thermal performance, laboratory measurement, conductance, guarded hot boxes, calorimeters, thermal insulation

Consumers and manufacturers are placing more emphasis on the thermal performance of windows since they can constitute a substantial heat loss in winter. Consumers are demanding improved thermal performance, and manufacturers are responding with new techniques and products. A test to determine the heat transmission characteristic is required to assess the relative thermal merits of window systems.

Presently, two ASTM test methods are used for determining the heat transmission of windows: the ASTM Test for Steady-State Thermal Performance of Building Assemblies by Means of a Guarded Hot Box (C 236-80) and the ASTM Test for Thermal Performance of Building Assemblies by Means of a Calibrated Hot Box (C 976-82). These standards, however, were not developed with windows in mind. Establishing the overall coefficient of heat transfer (*U* value) or thermal conductance (*C* value) of a window presents a number of unique problems that must be accounted for in the test method and test apparatus calibration. These problems include measurement of glass temperature, establishing the heat flow through the wall surrounding the

[1]Research officer and senior research officer, respectively, National Research Council of Canada, Division of Building Research, Ottawa, Ontario, Canada K1A 0R6.

window specimen, and determining the effect of specimen size on apparatus calibration and on possible air leakage through the specimen.

Ad hoc tests using the Division of Building Research (DBR), National Research Council of Canada (NRCC), Environmental Test Facility indicated a 1.3-degree-kelvin difference in thermocouple measurements of the same point on a glass surface resulting simply from varying the attachment to the glass. The attachment techniques compared were different lengths of 12-mm-wide aluminum tape and 12-mm-wide cloth tape. The method providing the highest reading was with 75-mm cloth tape only, and the lowest reading was obtained with 75-mm aluminum tape only. Figure 1 presents the isotherms on the room side surface of a sealed unit subjected to a 56-degree-kelvin temperature difference. This is an indication of how difficult it would be to establish a single value for the room-side glass surface temperature by surface thermocouple measurements.

The objective of this paper is to present an approach for determining the heat transmission characteristics of windows and to discuss a new calorimeter design that will overcome some of the uncertainties associated with conventional calorimeter designs used with guarded hot box test facilities.

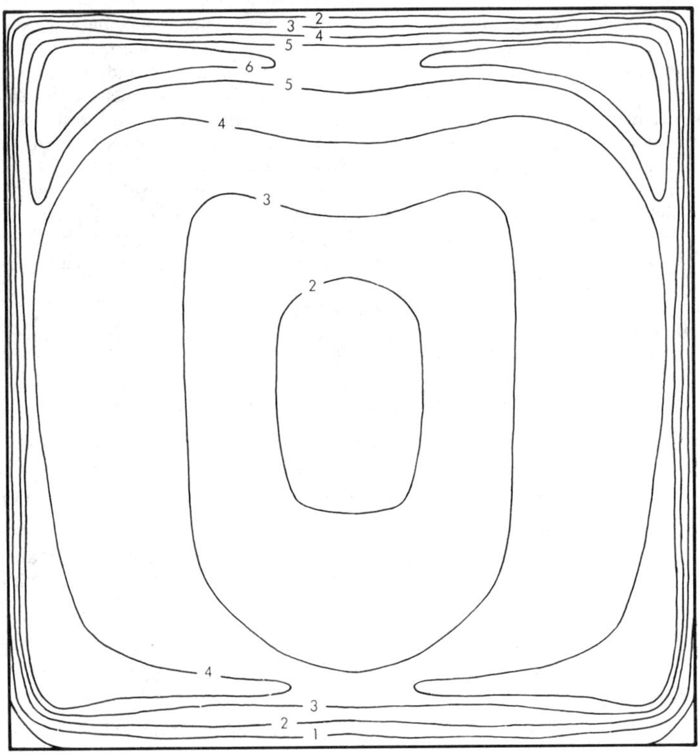

HERMETICALLY SEALED GLAZING UNIT
TRIPLE-GLAZED 2 - 6 mm AIR SPACES
WEATHER SIDE: -35°C; ROOM SIDE: 21°C
ISOTHERMS AT 1°C INTERVALS

FIG. 1—*Room side surface temperature profile of hermetically sealed glazing unit.*

Theory

The heat transfer through a window, Q_s, can be represented as

$$Q_s = UA\Delta T \tag{1}$$

where

U = the overall coefficient of heat transfer, in watts per square metre kelvin,
A = the area of the specimen, in square metres, and
ΔT = the temperature difference across the specimen, in degrees kelvin.

The U value cannot be measured directly but can be derived from measured values of Q_s, A, and ΔT.

The Q_s can be determined by using a calorimeter or metering box placed over the room side of the window and support wall (Fig. 2). The total energy input to the calorimeter is measured after steady-state conditions are reached and represents the total loss from the calorimeter, Q_T. This total heat transfer can be expressed as

$$Q_T = Q_b + Q_m + Q_s \tag{2}$$

where

Q_b = heat flow through the calorimeter walls, and
Q_m = heat flow through the support wall for the specimen (referred to as the mask loss).
The net heat loss or gain through the calorimeter walls is minimized or made equal to zero by

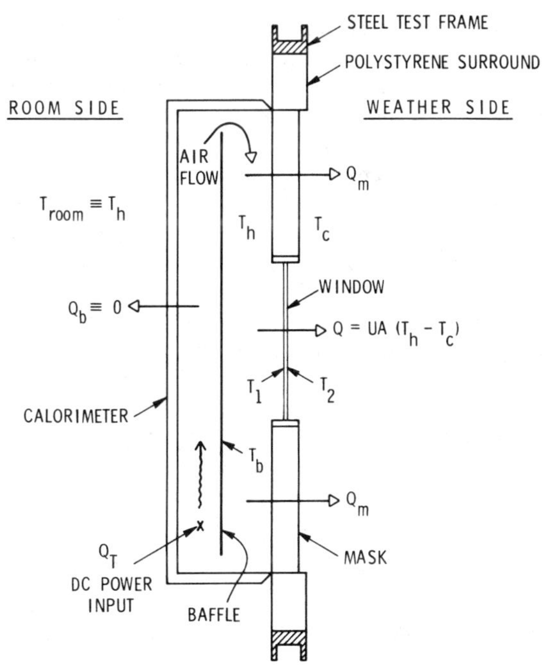

FIG. 2—*Conventional configuration for calorimeter mounted on the room side of the specimen and support wall.*

controlling the guard box air temperature, so that the mean temperature difference across the calorimeter walls is practically zero. The heat flow through the specimen is then

$$Q_s = Q_T - Q_m \tag{3}$$

Q_m is determined from calibration runs on the mask.

The determination of ΔT is difficult because the heat flow through the specimen is transferred from the calorimeter to the specimen surface by a combination of radiation and convection. Therefore, the calorimeter surface temperatures as well as the mean air temperature must be determined to define the inside temperature to which the specimen is exposed. Care must also be taken in determining the weather side temperature. The following is a detailed discussion of this problem including a technique for determining the window surface temperature for use in calculating the heat transmission for a window.

The U values commonly used by designers and specifiers for windows are those recommended by the American Society of Heating, Refrigeration and Air-Conditioning Engineers (ASHRAE) [*1*]. Table 1 summarizes values for the overall coefficient of heat transfer, U value, and overall thermal resistance, R value, using data from Chapter 23, Table 8, of the *ASHRAE Handbook of Fundamentals* ($R = 1/U$). The values were calculated using the surface film coefficients assumed for natural convection on the room side [$f_i = 8.29$ W/(m$^2 \cdot$ K)] and forced convection on the weather side of the window [$f_0 = 34.1$ W/(m$^2 \cdot$ K)]. The combined resistance provided by the inside and outside air film, $1/f_i + 1/f_0$, is 0.15 (K $\cdot$ m^2/W). Compared with the total R values, the inside and outside air film represents a significant component of the thermal resistance assumed for current window designs. Since the air films generated by a test setup will be facility dependent, they may not match the values used by ASHRAE or be the same from one apparatus to another. The air films will also vary for a given apparatus depending on the size of the specimen.

Two different approaches to the air films are possible. One is to establish a test procedure that would involve adjusting the equipment to obtain an air film resistance similar to that of ASHRAE, and thereby yield an overall thermal transmission value, the U value. The other approach is to determine the air film resistance for a particular facility or test setup and determine a C value. The C value is the heat transfer between defined surfaces (that is, no boundary air films). In the latter case, standard air film values could be added to obtain a U value.

Either approach would be satisfactory if the mean surface temperature of the glass could be measured accurately. The following factors indicate the difficulty in establishing a mean value for the specimen surface temperature from thermocouple measurements: two-dimensional heat flow at the perimeter of the glazing, the glass deflection associated with barometric changes, thermocouple contact error, and disturbance of the air film caused by the many thermocouple wires required to adequately establish the temperature profile.

TABLE 1—*Thermal resistance,* R *value, of hermetically sealed glazing.*

Glazing	U Value, W/(m$^2 \cdot$ K)	R Value, K $\cdot$ m^2/W
Double		
6-mm air space	3.3	0.30
13-mm air space	2.8	0.36
Triple		
6-mm air space	2.2	0.45
13-mm air space	1.8	0.56

In the early stages of development of guarded hot box test facilities at the Division of Building Research [*2*], it was shown that the relationships describing radiant and convective components of the inside surface coefficient could be used to determine an equivalent inside surface temperature for nonisotropic walls. The heat exchange, Q_s, between the test specimen and the guarded hot box can be expressed as

$$Q_s = Q_r + Q_c \tag{4}$$

where

Q_r = the radiant heat exchange between the calorimeter surfaces and the specimen surface, and
Q_c = the convective heat exchange between the calorimeter air and the specimen surface.

In a gray enclosure, the radiation exchange between two surfaces can be expressed in the form [*3*]

$$Q_r = A_1 \mathfrak{F}_{1b} \sigma (T_b^4 - T_1^4) \tag{5}$$

where

A_1 = area of Surface 1,
$\mathfrak{F}_{1b}$ = the overall interchange factor for radiation from Surface 1 to Surface b,
σ = the Stefan-Boltzmann constant, and
T_b, T_1 = the absolute temperature of Surfaces b and 1.

For one source (guarded hot box), one sink (test wall), and no interconnecting surfaces, the $\mathfrak{F}_{1b}$ is obtained from the following:

From McAdams (Eq 4-33)

$$\frac{1}{A_1 \mathfrak{F}_{1b}} = \frac{1}{A_b}\left(\frac{1}{e_b} - 1\right) + \frac{1}{A_1}\left(\frac{1}{e_1} - 1\right) + \frac{1}{A_1 F_{1b}}$$

Solving for $\mathfrak{F}_{1b}$

$$\mathfrak{F}_{1b} = \frac{1}{\dfrac{A_1}{A_b}\left(\dfrac{1}{e_b} - 1\right) + \left(\dfrac{1}{e_1} - 1\right) + \dfrac{1}{F_{1b}}}$$

Then, if the geometric view factor, F_{1b}, equals 1

$$\mathfrak{F}_{1b} = \frac{1}{\dfrac{A_1}{A_b}\left(\dfrac{1}{e_b} - 1\right) + \dfrac{1}{e_1}} \tag{6}$$

This expression for radiation heat exchange can be used to approximate the radiation exchange to a window specimen if the box surface temperatures remain uniform and the interchange between the box and the mask is definable. If not, the geometric view factor would not be one, and a more detailed analysis would be required. This will be addressed in more detail in a subsequent paper discussing the calibration of the window calorimeter. The convection component is approximated by

$$\frac{Q_c}{A_1} = q_c = C(T_h - T_1)^B \tag{7}$$

where

q_c = convective heat transfer per square metre,
T_h = air temperature in the box, and
B, C = constants determined for the box.

The constants can be determined over a range of heat flows by calibration with specimens of known conductance similar to that of double or triple glazing.

Equations 5 and 7 can also be written as

$$\frac{Q_r}{A_1} = q_r = h_r(T_b - T_1) \tag{8}$$

$$q_c = h_c(T_h - T_1) \tag{9}$$

where

q_r = radiant heat exchange per square metre,
$h_r = \mathfrak{F}_{1b}\sigma(T_b^2 + T_1^2)(T_b + T_1)$, and
$h_c = C(T_h - T_1)^{B-1}$.

When $T_h \neq T_b$, the sum of the two coefficients, h_r and h_c, provides the inside film coefficient, f_i. If $T_h = T_b$, the total heat transfer to the specimen can be expressed as

$$Q_s = f_i(T_b - T_1) \tag{10}$$

If the constants for the box are known, specimens of nonuniform conductance, such as windows, can be evaluated. Measuring Q_t, T_h, and T_b, a mean specimen surface temperature, T_1, can be found by using Eqs 4, 5, and 7. Once T_1, is calculated, the inside surface coefficient can be determined from Eqs 8 and 9.

The outside air film can be provided by forced convection. Previous unpublished development work on a test method to measure the condensation potential of metal window frames led to an understanding that local coefficients adjacent to the framing would affect the frame heat loss and thus the inside surface temperature. This work led to the design of the DBR wind machine (Fig. 3), which produces a very uniform and repeatable weather side air film. The air film can be established during calibration and used to calculate the outside surface temperature, T_2, by using measured outside air temperatures, T_c. Unlike the room side temperatures, the weather side box temperatures in the view of the window (wind machine tubes and back wall) remain close to that of T_c (within approximately 0.1 degree Celsius).

Problems with a Conventional Calorimeter

Initial work on the development of a test method for the *U* value of windows (see the Appendix) showed that the conventional type of calorimeter design was not suitable. The main concern is the inability to maintain a uniform and definable calorimeter surface and air temperature. Placing a window in a wall onto which the calorimeter is mounted produces a large difference between the heat transfer from the calorimeter to the window and to the support wall. This difference causes the temperature fields of T_b and T_h, used in the preceding discussion, to be nonuniform. For example, in the test on the sealed double-glazing unit with a 55-degree-

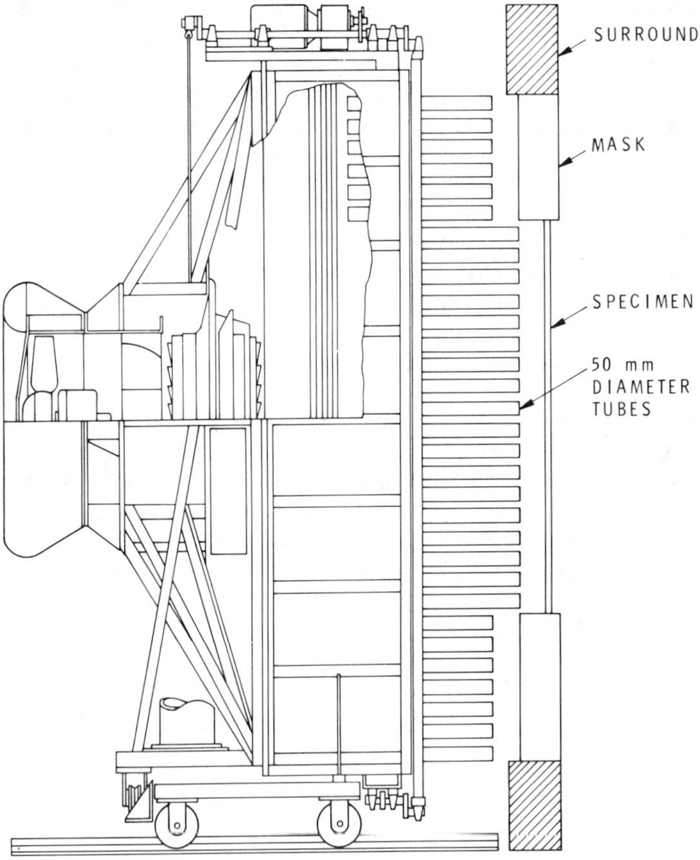

FIG. 3—*Wind machine. Air is blown at the face of the specimen through tubes spaced 115-mm on centers. The bank of tubes rotate so that each tube is moved in a 76-mm-diameter circle.*

kelvin temperature difference from calorimeter air to weather side air, T_b, as measured on the baffle immediately in front of the window, was 6.9 degrees kelvin lower than that measured directly in front of the support wall. A systematic error, $\Delta T_1 = 1$ K, in establishing the weighted mean temperature of the baffle, T_b, results in an error of approximately 7% for the constant, C, in Eq 6 and a 6% error in the value determined for the inside film, f_i.

The establishment of the calorimeter air temperature, T_h, can also introduce systematic errors. The air stream that flows down over the window and up behind the baffle is reheated by electric heaters running the full width of the calorimeter. Since electric heaters provide constant flux over their length, a horizontal gradient in the air caused by the convective losses from the air stream as it falls over the window and surrounding mask will not be eliminated as the air flows over the heater.

DBR Window Calorimeter—Design and Construction

From experience using the calorimeter designed for testing walls, and consideration of the preceding approach for windows, it was clear that if an accurate, repeatable, and reproducible test procedure was to be developed, the uncertainty in the values used for the surface temperatures on the baffle in view of the window, T_b, and the air temperature, T_h, had to be reduced. Also, consideration had to be given to the different sizes of windows that might be tested and the resulting impact on the temperature gradients associated with T_b and T_h.

To meet these requirements, a new calorimeter was designed for measuring the heat transmission characteristics of windows with the following features:

(*a*) a constant and uniform temperature baffle, and
(*b*) a constant and uniform temperature convection heater.

To achieve a constant and uniform temperature on the baffle, an evaporating/condensing panel using Refrigerant 12 was designed (Fig. 4). Three condensing panels (744 by 2632 mm) were connected together to form a baffle 2232 mm wide by 2632 mm high for the new calorimeter. An electric tubular heater is contained in a pipe reservoir connected to the panels for boiling the refrigerant.

The constant and uniform temperature convection heater was also obtained by evaporating and condensing the refrigerant (Fig. 5). The heater was constructed of two fin tubes connected to a copper pipe reservoir, which contained a second tubular heater of the same power rating as the one in the pipe reservoir of the baffle. The baffle and convection heaters were selected to have as short a cold end as possible and be capable of maintaining a 65-degree-kelvin temperature difference across a specimen 2.44 by 2.44 m with a U value of 0.35 W/(m^2 · K). The constant temperature baffle and convection heater were installed in a new calorimeter box constructed as shown in Fig. 6.

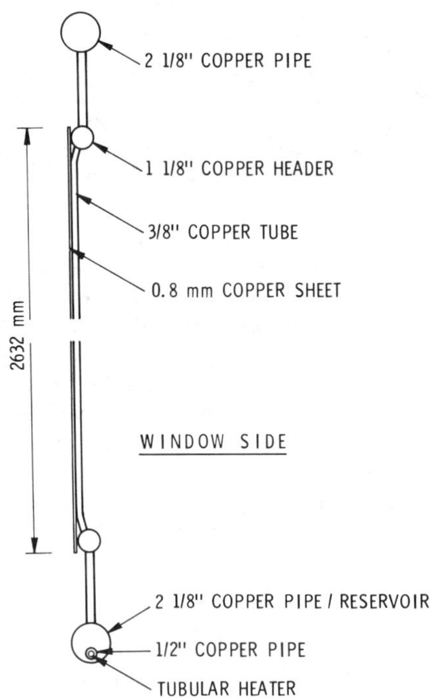

FIG. 4—*Uniform temperature baffle.*

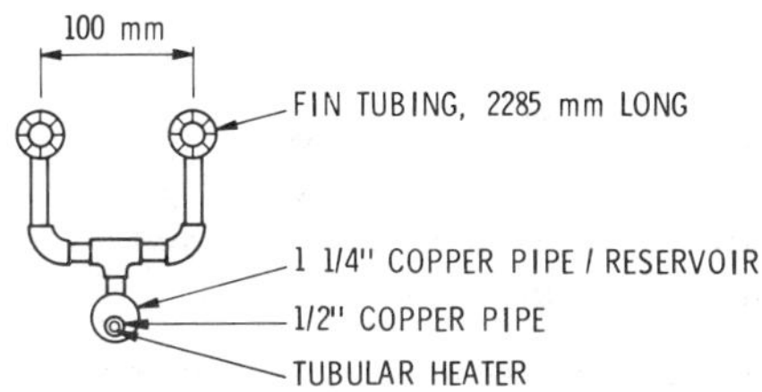

FIG. 5—*Uniform temperature convection heater.*

FIG. 6—*Window calorimeter.*

Calibration

The objectives of the calibration program currently in progress are as follows:

(*a*) to check the total power input measurements of the window calorimeter,
(*b*) to determine the uniformity of the baffle and air temperatures, T_b and T_h, and
(*c*) to determine *B* and *C* constants (see Eq 7) for a range of power requirements and for different sizes and orientations of calibration specimens.

Using a homogeneous wall constructed of material with known conductivity, tests were conducted using the window calorimeter. The power and ΔT across the specimen at a number of different room side and weather side temperature conditions were measured. The thermal conductance of the specimen determined using the window calorimeter and the values obtained with the wall calorimeter are shown in Table 2.

Having established the ability to determine total power input to the calorimeter, Q_T, the support wall for the window specimens was built without openings and tested to determine the conductance and mean temperature dependence. Calibration specimens simulating very simple windows 1.0 by 1.6 m and 1.0 by 0.8 m will be used to establish the values of *B* and *C* over the expected range of power requirements for the calorimeter. During these tests the performance of the baffle and convection heater will be established.

Expanded polystyrene was used for the core of the wall since it represented a material that had a stable conductivity with time and could be sealed to maintain a low moisture content. To protect the polystyrene from mechanical damage and to facilitate a modular construction with a spline connecting system, 17 mm of plywood was laminated to both sides, using a water-based contact cement. The wall was made in sections to allow initial calibration and subsequent interchanging or substitution of wall sections for calibration specimens. The nominal conductance of the wall is 0.326 W/(m^2 · K) at 0°C mean temperature.

The first calibration specimens were constructed of 11-mm extruded polystyrene laminated to 3-mm glass on each side. Thermocouples could be used to measure the temperature at the glass/polystyrene interface since problems discussed earlier regarding measurement of glass temperature at an air/glass interface would not be a factor. Thermocouples were placed between the glass and the polystyrene (Fig. 7). The tips were soldered to 25 by 38-mm copper shims and then glued to the glass surface before the glass was glued to the polystyrene. After testing with glass surfaces on both sides, the room side surface will be modified to provide a low-emissivity surface. The low-emissivity surface reduces q_r relative to q_c, so that errors associated with it should have little effect on q_c.

The calibration procedure involves the measurement of Q_T, T_b, T_1, T_2, T_h, and T_c. The mask loss is calculated from surface temperature measurements on the mask and the conductivity

TABLE 2—*Comparison of the DBR wall calorimeter and window calorimeter for a 100-mm expanded polystyrene specimen 2.44 by 2.44 m, with latex coatings.*

Mean Temperature of Specimen, °C	Thermal Conductance, W/(m^2 · K)	
	Wall Calorimeter	Window Calorimeter
6.39	0.340	0.339
−1.02	0.331	0.330
−8.09	0.323	0.322

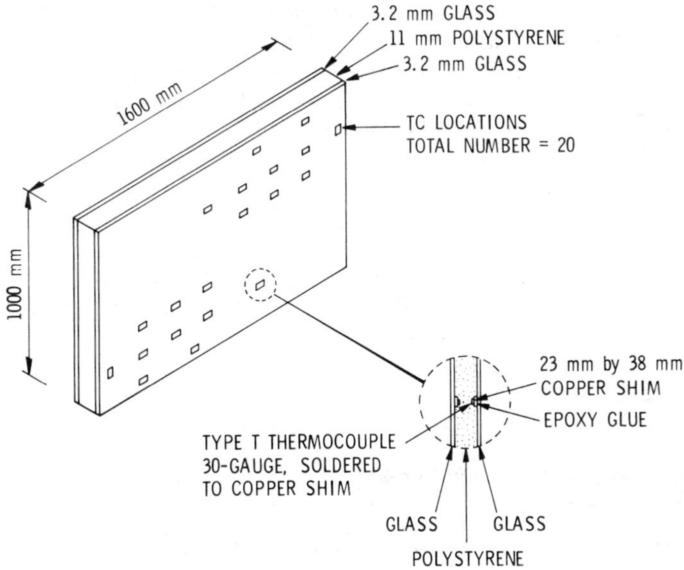

FIG. 7—*Calibration specimen.*

relationship established from the tests referred to earlier. With the calibration specimen in the mask, a new term must be added to Eq 3

$$Q_s = Q_T - Q_m - Q_e \tag{11}$$

where Q_e is the additional heat loss around the edge of the calibration specimen, which is not included in Q_m. This loss will be a function of the perimeter length of the calibration specimen, position in the mask, the method for holding the calibration specimen in the mask, and local film coefficients. Two techniques will be considered for determining the value of Q_e: measurement using the calibration specimen as a heat flow meter and measurement by finite element analysis. The former is a preferred option because an analytical approach requires a number of assumptions regarding local film coefficients at the calibration specimen edge.

Once Q_s is determined for each test condition and set up, Q_r is calculated (Eq 5) using T_b, T_1, and the measured surface emissivities of the baffle and window and q_c established from Eqs 4 and 7. By using calibration specimens with both a high and low surface emissivity on the room side surface, B and C can be established over a broad range of q_c and $(T_h - T_1)$.

The calorimeter is currently being calibrated. Initial testing has shown that with the 1.6 by 1.0-m specimen in the 2.4 by 2.4-m wall and 22°C room side/−35°C weather side temperatures, the difference between the maximum and minimum thermocouple reading on the baffle is only 0.2 degree kelvin. The calorimeter air stream temperatures at similar elevation are constant to within 0.1 degree kelvin.

Future Plans

The work to date has demonstrated that the constant and uniform temperature baffle and convection heater does provide more uniform and definable characteristics for the calorimeter than the existing wall calorimeter. Once the calorimeter calibration is completed, a program of evaluating different sealed window units and windows will be undertaken. The objective of the program is to develop a standard procedure for determining a nominal thermal conductance for windows using the window calorimeter. Factors to be addressed will include the effect of specimen size on results, reporting of results based on rough opening area or glazed area, and the effect of glass deflection on the *U* values arrived at for sealed units.

Acknowledgment

This paper is a contribution from the Division of Building Research, National Research Council of Canada, and is published with the approval of the director of the division.

APPENDIX

Window Tests Using DBR Wall Calorimeter: Apparatus and Test Setup

The DBR environmental chamber and calorimeter for wall heat flow measurements are shown in Figs. 8 and 9. The calorimeter box is constructed of isocyanurate foam covered inside and outside with fiberglass and polyester resin. A row of electric heaters is located between a baffle and the back wall just above the bottom of the baffle, so that the baffle shields the test specimen from direct radiation from the heaters. The heated air flows by natural convection up between the baffle and the back wall, through the gap at the top, down between the baffle and the test assembly, and back through the gap at the bottom. A thermopile with 64 uniformly spaced junctions is used to measure the mean temperature difference across the walls of the calorimeter. The output from the thermopile is used with a differential controller to control the temperature of the room side guard box surrounding the calorimeter box to maintain a near-zero mean temperature difference across the calorimeter box walls.

A steel test frame was used to construct a wall in which to mount the test windows. A cross section of the wall and test window is shown in Fig. 10. All the joints in the plywood were sealed with caulking to prevent air leakage through the test assembly.

The conductivity of the support wall and a 25-mm painted plywood calibration specimen was established in a guarded hot plate apparatus. The calorimeter constants were determined using the calibration specimen prior to testing the windows. For the tests with the plywood calibration specimen, the same conditions were maintained as for the tests with the windows; that is, the weather side was adjusted and maintained at three nominal temperatures, −7, −21, and −35°C, and the calorimeter box was maintained close to 21°C with the heat input held constant after equilibrium had been reached. The heat input to the room side chamber was controlled so that the average temperature difference across the calorimeter walls was practically zero. This zero difference ensured that the heat exchange between the calorimeter and the room side chamber was negligibly small.

The heat transfer from the calorimeter heaters to the test wall was by natural convective air flow and radiation transfer from the calorimeter surfaces. The multinozzle wind machine located in the weather side chamber was used to provide a uniform air film over the weather side surface.

Upon completion of the calibration tests, the plywood blank was removed and the test window installed. A compressible sponge weatherstrip was placed between the stop and glass. Caulking was then applied to seal the stops to the frame and the glass to prevent air leakage. Thermocouples were installed using aluminum tape on the room side and weather side glass surfaces. The calorimeter was installed and operated as in the calibration run with the measurements taken after equilibrium was reached at each weather side temperature.

The *U* values established from the tests for double and triple-sealed units with nominal 6-mm air spaces and test temperatures of approximately 21°C room side and −35°C weather side

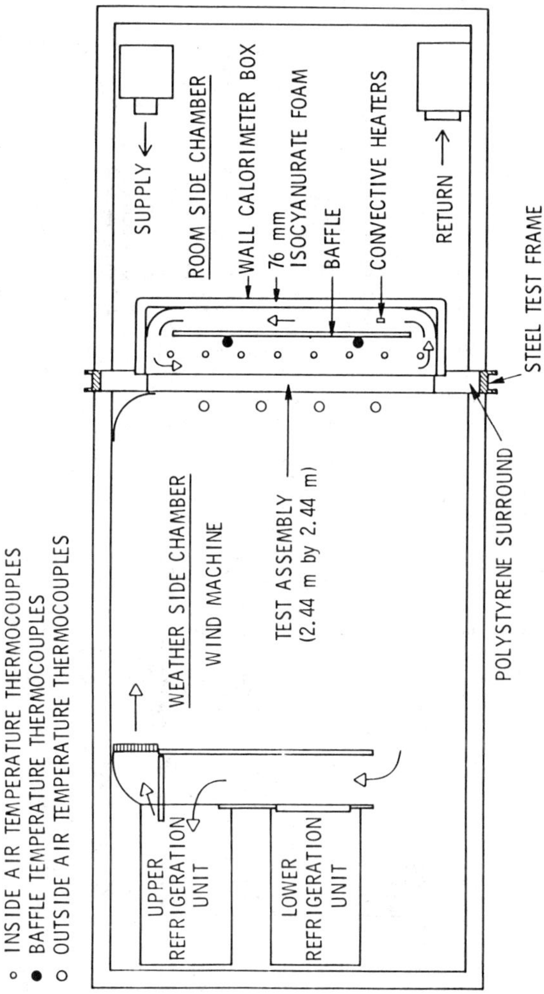

FIG. 8—*DBR/NRCC environmental test facility, vertical section.*

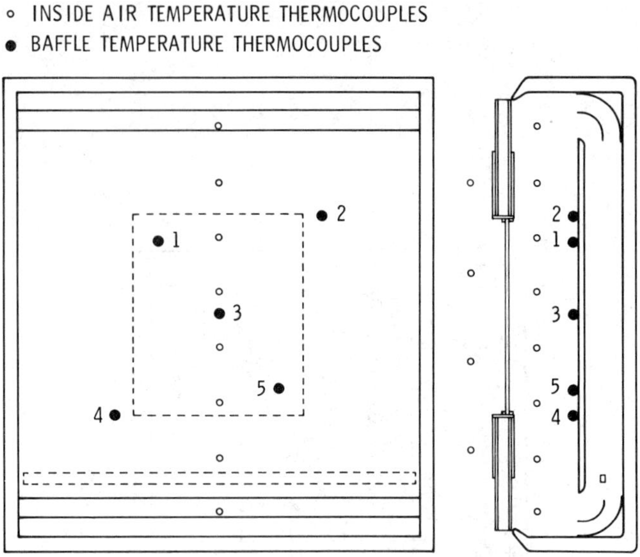

FIG. 9—*DBR/NRCC wall calorimeter box, front view and side section.*

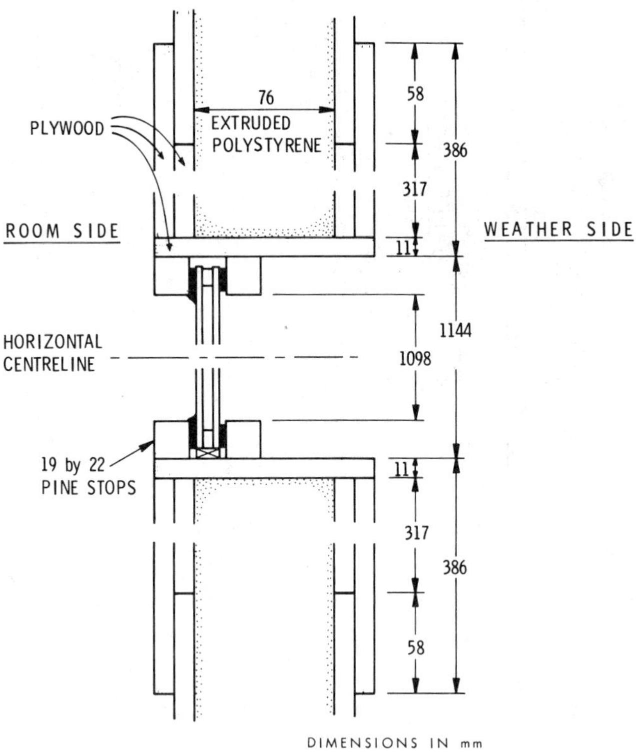

FIG. 10—*Test wall with sealed glazing unit, vertical section.*

TABLE 3—*High and low baffle temperatures for the plywood calibration specimen and the double- and triple-glazed sealed glazing units.*

Specimen	Room Side Temperature, °C	Weather Side Temperature, °C	Baffle Temperatures, °C High Thermocouple No. 2	Low TC No. 3	ΔT, K, No. 2 − No. 3
Plywood	21.6	−7.0	21.3	18.2	3.1
	21.7	−21.0	21.3	17.0	4.3
	21.6	−35.1	21.1	16.0	5.1
Double-glazed unit	21.6	−6.6	21.4	17.4	4.0
	21.6	−20.9	21.4	15.9	5.5
	21.7	−34.5	21.3	14.4	6.9
Triple-glazed unit	22.3	−7.0	22.1	18.9	3.2
	21.6	−21.1	21.3	17.1	4.2
	21.5	−35.1	21.1	15.9	5.2

were 3.45 and 2.41 W/($m^2 \cdot$ K), respectively. It should be noted that the absolute numbers were not as important as the technique and problems associated with testing of windows.

The first area of uncertainty was in establishing the baffle temperature, T_b. For wall testing, the baffle temperature remains uniform even when walls with high conductance studs are used. With windows, however, there is a large area of high conductance, which causes temperature gradients on the baffle in front of the window. Figure 9 shows the thermocouple locations on the baffle and the view of the window on the baffle. Table 3 summarizes the high and low baffle temperatures for the test conditions specified. The worst case resulted in a temperature difference of 6.9 degrees kelvin between thermocouple Locations 2 and 3.

Although glass temperatures were measured, they were not used in the calculations. These thermocouples were used only to provide an indication of the temperature gradients on the glass. The effect of thermocouple attachment was investigated by comparing two thermocouples having the same reading and similar location. The attachment was varied by using different lengths of aluminum and black cloth tape and cloth tape on top of the aluminum tape. The temperature measured by the thermocouples for the different attachments varied by 1.3 degrees kelvin.

References

[*1*] *ASHRAE Handbook of Fundamentals,* American Society of Heating, Refrigerating, and Air-Conditioning Engineers, New York, 1981, Table 8, Chapter 23.

[*2*] Brown, W. C., Solvason, K. R., and Wilson, A. G., *A.S.H.R.A.E Transactions,* Vol. 66, 1961, pp. 561–577.

[*3*] McAdams, W. H., *Heat Transmission,* 3rd ed., McGraw-Hill, New York, 1954, pp. 72–76.

William R. Strzepek[1]

Thermal Resistances of Various Concrete Masonry Wall Constructions Incorporating Rigid Plastic Foam Insulation

REFERENCE: Strzepek, W. R., **"Thermal Resistances of Various Concrete Masonry Wall Constructions Incorporating Rigid Plastic Foam Insulation,"** *Thermal Insulation: Materials and Systems, ASTM STP 922,* F. J. Powell and S. L. Matthews, Eds., American Society for Testing and Materials, Philadelphia, 1987, pp. 582–598.

ABSTRACT: Series of thermal transmission tests were conducted on large wall specimens, which measured 300 by 400 cm (118 by 157 in.), using a calibrated hot box test apparatus. The wall specimens were constructed with concrete masonry units and insulated using various interior insulation systems incorporating extruded and block molded expanded polystyrene (EPS) insulation. An exterior insulated finishing system incorporating extruded EPS insulation was also evaluated. The interior and exterior insulation systems included various methods of attachment of the insulation and finish. The impact of these attachment methods on thermal resistance was studied. Thermal resistance values for the various wall constructions were determined for two different mean temperatures. For some constructions R values were calculated using a standard heat transfer methodology commonly followed throughout the construction industry. The calculated system R values were generally only slightly greater than the measured R values, but for some insulation systems the calculation method significantly overestimated the thermal resistance of the wall construction.

KEY WORDS: calibrated hot box testing, full-scale thermal testing, concrete masonry walls, thermal insulation, cellular plastic, heat transmission, thermal resistance

The need to conserve energy, and subsequently the need for insulation, is readily acknowledged by most people associated with the construction industry. However, construction technology has progressed to the point where it is no longer acceptable to add insulation indiscriminately to structures. In addition to the obvious economic considerations and other pertinent factors, the effectiveness of the insulation when installed in a particular system is also an important consideration.

There are many insulating materials and systems to choose from when approaching a given application whether it is, for example, an attic, a concrete masonry wall, or a crawl space that is being insulated. Since the insulating materials and the systems in which they are used—that is, the means of attaching the insulation and finishing material, if required—can vary widely, it is reasonable to expect that the thermal resistances of the systems may vary significantly as well.

In the study discussed in this paper, various commercially available insulation systems were applied to an uninsulated wall constructed with concrete masonry units. The insulating material used was generally extruded expanded polystyrene (EPS) foam sheathing. Only the manner in which the extruded EPS foam insulation and gypsum wallboard were attached to the concrete block wall differed from one system to the next. Block molded EPS sheathing was installed in

[1]Senior research engineer, Dow Chemical U.S.A., Granville, OH 43023.

one test, so that the effect of using a different insulating material in a given system could be examined.

As a result, it was possible to investigate the impact of the design of a particular system on its thermal resistance. Some designs incorporated highly conductive elements, such as mechanical fasteners and metal furring channels, which acted as thermal short circuits and thermal bridges when these materials penetrated the insulation. Such factors as the number of mechanical fasteners used and the furring channel configuration were examined. Also, some tests were conducted to address the issue of repeatability. Some measured results were then compared with calculated values obtained using a heat transfer methodology commonly followed throughout the construction industry.

Description of Test Apparatus and Experimental Procedure

Test Facility

The test facility and the calibrated hot box test apparatus have been described in previously published work by Palfey and Contreras [*1,2*]. As is shown in Figs. 1 and 2, the calibrated hot box was located in a controlled environment room. This room was maintained at the same temperature as the metering chamber. Test procedures and guidelines presented in the ASTM Test for Thermal Performance of Building Assemblies by Means of a Calibrated Hot Box (C 976-82) were followed.

Test specimens measuring 300 by 400 cm (118 by 157 in.) were constructed in portable frames and instrumented with surface thermocouples in a general assembly area. Frames were transported into the test room and then positioned and sealed between the metering (hot) and climatic (cold) chambers. Temperature-controlled chilled air was circulated through insulated ducts to the climatic chamber from an external refrigeration evaporator coil.

Instrumentation and Control

The primary control system, which was located in the control room adjacent to the calibrated hot box, was used to control or monitor the following test parameters:

(*a*) hot and cold side air temperatures,
(*b*) hot and cold side wall surface temperatures,
(*c*) air infiltration mass flow,
(*d*) hot side relative humidity, and
(*e*) energy consumption.

Transducer signals were monitored with a digital voltmeter and interfaced with the computer peripheral where data were collected for processing. The temperatures of both chambers were controlled with three-stage internal electric heaters. A small circulating fan was used to distribute air through the vaned plenum chamber on each side of the test wall and to maintain a uniform air flow along the surfaces of the test wall. As indicated in Fig. 2, the curtain of air flowed in the direction of natural convection, that is, down on the hot side and up on the cold. The high-emittance plenum panels on each side opposite the test wall were adjustable to control air velocity.

Temperatures were sensed with Type T copper-constantan thermocouples. Four sets of 48 thermocouples were uniformly distributed on the surfaces of the test wall and in the adjacent air curtains, so that those on the wall were directly opposite those in the air. Thermocouples were not placed in direct contact with fasteners. Heat transfer through the metering chamber wall was sensed with nine pairs of differential thermocouples in series. The sensing thermocouples were located on the inside and outside wall surfaces of the seven-sided chamber. Energy input to

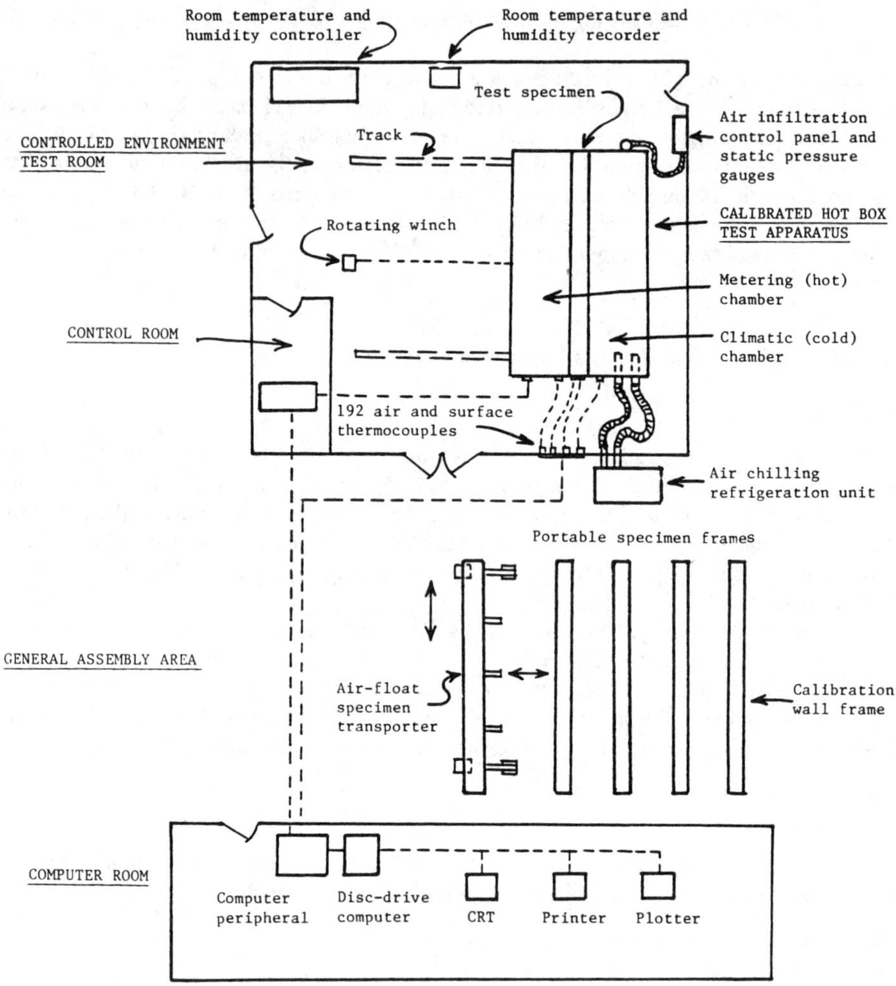

FIG. 1—*Diagram of the calibrated hot box test facility.*

the metering chamber heaters and fan was measured with transducers that produce signals proportional to the watt-hours of energy used. One signal was equivalent to 0.125 W · h of energy consumed.

All the test data were collected and stored on magnetic computer disks. All test parameters, including 192 temperature points, were scanned at 15-min intervals and stored as data blocks. Each disk stored 940 data blocks, which was equivalent to 235 h of testing. Block data could be reviewed on a cathode ray tube monitor, or a hard copy could be printed. Computer programs were used to determine the test wall thermal transmission properties.

Metering Chamber Calibration

The metering chamber was calibrated so that the net heat flow through the test wall constructions and the insulated test frame could be determined. The calibration wall consisted of two

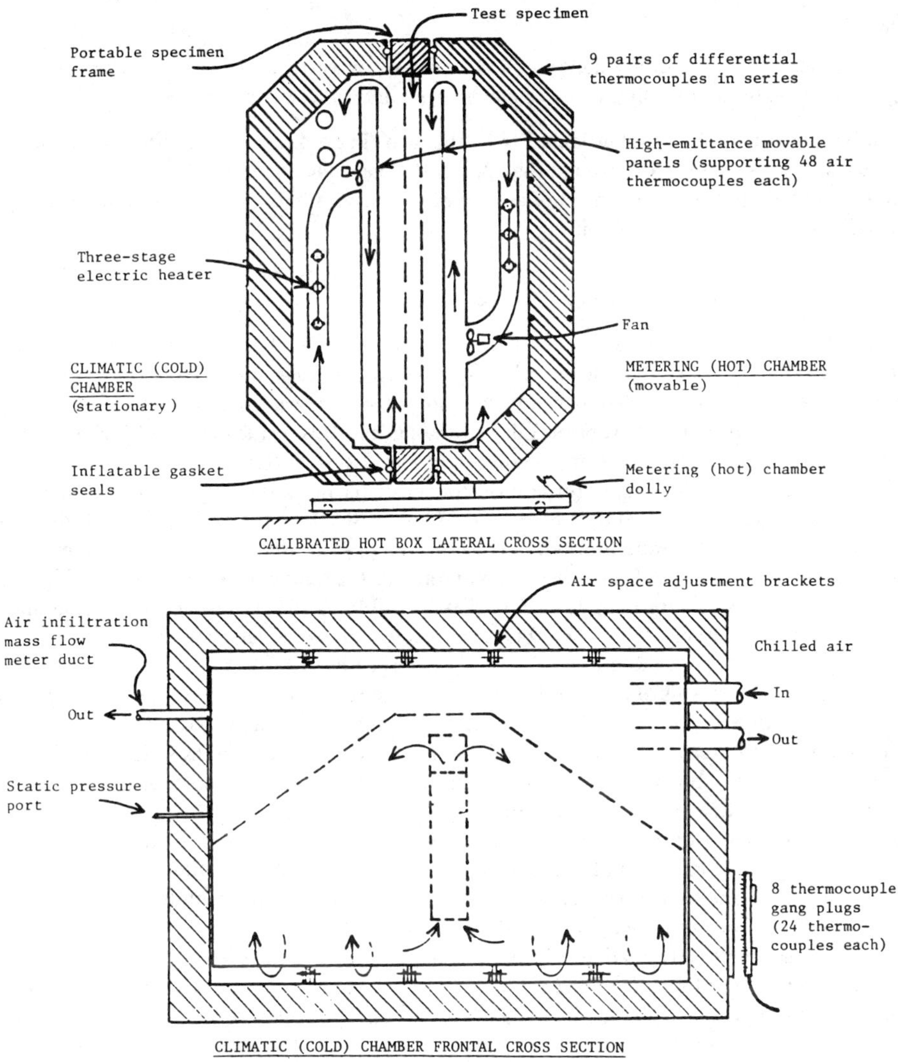

FIG. 2—*Cross sections of the calibrated hot box test apparatus.*

plies of aged, high-density 2.54-cm (1.00-in.) extruded expanded rigid polystyrene thermal insulation covered with air-impervious aluminum sheet facings. The facings were painted with a high-emittance black paint.

After establishing a temperature difference across the calibration wall and reaching thermal equilibrium, all internal air and surface temperatures were maintained constant while the temperature outside the box was varied. Total power input to the metering chamber was plotted as a function of thermopile output, E, which reflected the temperature difference across the walls of the metering chamber. The slope of this line was used as the correction factor, M, in the expression

$$q = Q - ME$$

where

q = net heat flux (including flanking heat transfer), in watts, and
Q = total heater and fan input to the metering chamber, in watts.

Flanking heat transfer was considered to be the heat that did not pass through the concrete masonry wall constructions or through the walls of the metering chamber but through the insulated test frame. Specific comments on flanking heat transfer relative to this study are included in the Discussion of Results section.

Test Conditions and Data Reduction

The wall constructions were tested vertically in a heating mode; that is, the interior insulation systems were placed adjacent to the metering chamber, which was maintained at a high temperature relative to the climatic chamber, and the exterior system was positioned opposite the climatic chamber. This regimen was not followed in one test; this test is discussed further in the sections on Description of Wall Constructions and Discussion of Results.

The designed operating temperature ranges for the metering chamber and climatic chamber were 0 to 60°C (32 to 140°F) and 25 to −40°C (77 to −40°F), respectively. In this study data were collected for mean temperatures of 24°C (75°F) and 4°C (40°F). For the tests in which a wall mean temperature of 24°C (75°F) was maintained, the temperatures of the metering and climatic chambers were approximately 38°C (100°F) and 10°C (50°F), respectively. To obtain a wall mean temperature of 4°C (40°F), the hot and cold sides were maintained at temperatures of approximately 21°C (70°F) and −12°C (10°F), respectively.

Velocities of the curtains of air that flowed along the surfaces of the test walls were controlled and normally ranged from 0.1 to 0.2 m/s (0.2 to 0.4 mph). The variation in air temperature within both the hot and cold side air curtains was aproximately 2 degrees Celsius (4 degrees Fahrenheit). The variation in temperature of a given hot or cold side wall surface ranged up to 9 degrees Celsius (16 degrees Fahrenheit) for the uninsulated concrete block wall, and the variation for the insulated walls was approximately 5 degrees Celsius (9 degrees Fahrenheit). The air pressure within the metering and climatic chambers was controlled to minimize the pressure differential across the wall constructions. As a result, the measured air infiltration mass flow rates through the wall specimens were negligible in all cases.

Some test parameters scanned at 0.25-h intervals for a 12-h period after a steady-state condition was attained are plotted in Fig. 3 for one of the wall constructions evaluated in this study. The maximum percentage change in energy transfer during successive 4-h periods was within the 1% tolerance recommended in ASTM Method C 976-82. At steady-state conditions, the thermal resistance of a particular wall construction was calculated from

$$q = Q - ME$$

$$R = \frac{(t_1 - t_2)A}{q}$$

where

q = net heat flux (including flanking heat transfer), W,
Q = total heater and fan input to the metering champter, W,
M = thermopile constant for the metering chamber walls (correction factor), W/V,
E = thermopile output for metering chamber walls, V,
t_1 = mean temperature of wall surface (hot), K,
t_2 = mean temperature of wall surface (cold), K,

A = area of wall construction normal to heat flow, m^2, and
R = thermal resistance, $K \cdot m^2/W$.

Description of Wall Constructions

Four different interior insulation systems were evaluated in addition to the uninsulated concrete masonry wall construction; namely, the drywall laminate system, the Thermo-Stud[2] system, the metal Z-furring channel system, and the Temp Guard Insulation Furring (TGIF)[3]

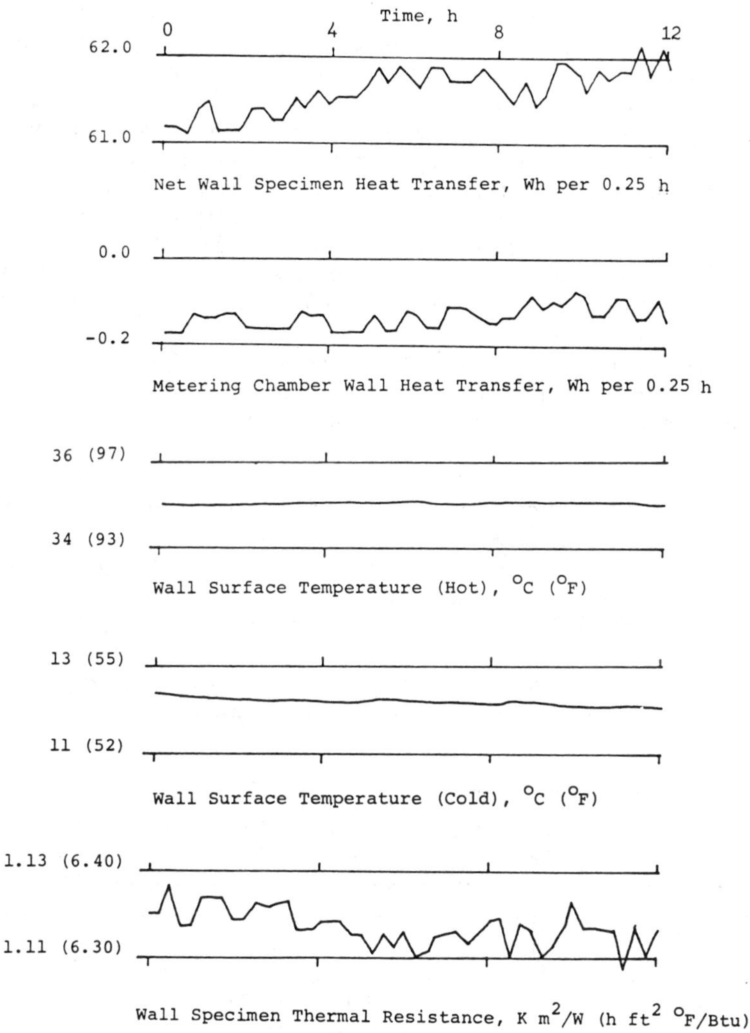

FIG. 3—*Calibrated hot box performance parameters for the drywall laminate system (first variation, 27 January–3 February 1983) at a wall system mean temperature of 24°C (75°F).*

[2]Trademark of W. R. Grace & Co.
[3]Manufactured by Judkins Associates Inc.

system. In addition, one exterior insulated finishing system was examined. In some cases several different variations of a given system were tested. All the systems and most of the variations were commercially available in the marketplace. System variations not commercially available were included to investigate specific areas of interest, such as the effect of the number of mechanical fasteners on thermal performance.

Construction Materials

The two-core blocks used in the study had a nominal thickness of 15 cm (6 in.), and the width and length were approximately 20 cm (8 in.) and 40 cm (16 in.), respectively. The density of the concrete in the block was 2070 kg/m^3 (129 lb/ft^3), and the thermal conductivity of the concrete was 1.82 W/m · K (12.6 Btu · in./h · ft^2 · °F) when measured at a mean temperature of 24°C (75°F) in accordance with the ASTM Test for Steady-State Thermal Transmission Properties by Means of the Guarded Hot Plate (C 177-76$^\epsilon$). Both the density and thermal conductivity of the concrete were measured by Dynatech R/D Company.

The nominal 2.54-cm (1.00-in.) extruded EPS sheathing used in the concrete masonry test walls was all from the same lot, that is, it was assumed that all the material had virtually the same physical properties. Insulating boards measured approximately 122 by 244 cm (48 by 96 in.), but they were cut to a 61-cm (24-in.) width when required for the installation of a particular system. The nominal density of the material was 32 kg/m^3 (2 lb/ft^3), and the thermal resistance at a mean temperature of 24°C (75°F) was 0.86 K · m^2/W (4.90 h · ft^2 · °F/Btu). This value was representative of the material at the time the concrete masonry wall constructions were tested.

The nominal 2.54-cm (1.00-in.) block molded EPS sheathing, which was used in only one variation of the Thermo-Stud system, had nominal dimensions of 61 by 244 cm (24 by 96 in.) and a nominal density of 16 kg/m^3 (1 lb/ft^3). The thermal resistance of the material was 0.61 K · m^2/W (3.68 h · ft^2 · °F/Btu) when measured at a mean temperature of 24°C (75°F). The paper-faced gypsum wallboard had a nominal thickness of 1.27 cm (0.50 in.) and a width of 122 cm (48 in.). In some tests, boards 244 cm (96 in.) in length were used, in other tests, boards 305 cm (120 in.) in length were installed.

Screw-type masonry fasteners with diameters of 0.48 cm (0.19 in.) were used in various lengths to attach furring channels to the block wall when required. Self-tapping drywall screws with a diameter of 0.38 cm (0.15 in.) and lengths of 2.54 cm (1.00 in.), and for some tests 3.18 cm (1.25 in.), were used to attach the wallboard when required. Additional information concerning all these materials is provided in manufacturers' literature and appropriate references, such as the 1981 volume of the *ASHRAE Handbook of Fundamentals*.

Wall System Installation

With each variation of each system the drywall, insulation, and furring were removed from the block wall following each test, except where noted. The same concrete block wall served as the base for each system. Holes for the masonry fasteners had to be predrilled through the furring channels (and the insulation in most cases) into the concrete masonry. When the walls were stripped after completion of a given test series, these holes were filled with mortar or a caulking material before the next variation was installed. Comments concerning the effect of this practice on the test results follow in the Discussion of Results section.

The insulation and gypsum wallboard were installed vertically or horizontally, depending on the system; and both the drywall and foam sheathing often had to be trimmed to fit the dimensions of the test frame. Good construction practices were followed. All the joints were tightly butted. The joints in the drywall were either covered with a cotton-cloth, vinyl-coated adhesive tape or taped and spackled, and the heads of fasteners in the field of the drywall were covered with small pieces of the same adhesive tape. All the perimeter edges where the test wall butted against the frame were caulked on both sides of the test wall. Penetrations into the test wall at the collection point for the surface thermocouples (which was adjacent to the thermocouple gang plugs shown in Fig. 2) were also caulked.

One surface of the concrete block wall was sealed with a commercially available surface bonding cement. The coating was approximately 0.48 cm (0.19 in.) thick and was applied to minimize air flow through the block wall during the tests. As mentioned previously, this wall served as the base wall for all the tests. The interior insulation systems and the exterior insulated finishing system were applied to the unsealed surface of the wall.

Drywall Laminate System

The first variation of this system consisted of extruded EPS sheathing attached to the unsealed surface of the block wall with a commercially available mastic adhesive and gypsum wallboard attached to the surface of the foam insulation with the same adhesive. This insulation and drywall were installed vertically. The drywall was sufficiently long to extend the full height of the test wall, which was 300 cm (118 in.). This variation represented the ideal case with no mechanical fasteners or furring, and thus no thermal short circuits.

The second variation differed from the first in that 5.72-cm (2.25-in.) masonry fasteners were screwed through the drywall and insulation into the block after holes were predrilled. Fasteners were installed along the vertical edges of the drywall on 40-cm (16-in.) centers and along the vertical centerline of the boards on 60-cm (24-in.) centers. A total of 82 masonry screw anchors was used. For the third variation, additional masonry screw anchors were installed so that fasteners were located on 20-cm (8-in.) and 30-cm (12-in.) centers at the perimeter and centerline of the boards, respectively. This represented a total of 152 fasteners.

The fourth variation was similar to the first variation except that 2.54 by 5.08-cm (1.00 by 2.00-in.) wood furring strips were attached horizontally at the top and bottom of the test wall with 4.78-cm (1.88-in.) masonry fasteners on approximately 60-cm (24-in.) centers and mastic adhesive. A total of 16 fasteners was used. The extruded EPS sheathing was placed between the furring and attached to the wall with a mastic adhesive. No air space existed within the system. The drywall was adhesively attached to the foam surface and mechanically anchored to the wood furring with self-tapping drywall screws on 20-cm (8-in.) centers.

For the fifth variation, four horizontal wood furring strips were located on 122-cm (48-in.) centers vertically beginning at the base of the wall. Thirty-two masonry fasteners were used to attach the strips. The extruded EPS insulation was installed horizontally between the wood furring strips, and the drywall was also installed horizontally. To attach the foam insulation and drywall, a mastic adhesive was used. Each board of drywall was also mechanically fastened to the wood furring strips with drywall screws on 20-cm (8-in.) centers at the horizontal joints.

Thermo-Stud System

This system consisted of block molded EPS foam sheathing, which measured 61 cm (24 in.) in width, attached to the concrete block wall with 25-gage, U-shaped metal furring channels and covered with gypsum wallboard. The eight Thermo-Stud furring channels, which measured 4.14 cm (1.63 in.) in width, were installed vertically on 60-cm (24-in.) centers. The channels had serrated edges, which were pressed into the exposed surface of the insulation, and 5.72-cm (2.25-in.) masonry fasteners were installed on 60-cm (24-in.) centers to attach the furring to the wall. A total of 56 screw anchors was used. The gypsum wallboard was installed vertically and mechanically fastened into the channels of 20-cm (8-in.) and 30-cm (12-in.) centers at the perimeter (that is, vertical joints) and the centerline, respectively. A second variation of the system was also examined. Extruded expanded polystyrene foam sheathing, which measured 122 cm

(48 in.) in width, was substituted for the block molded EPS material. The number of fasteners used to secure the furring channel and the drywall remained unchanged.

To minimize flanking effects, a gap measuring approximately 5 cm (2 in.) was left between the extreme left and right channels and the test frame. The channels were installed so that they straddled vertical joints in the insulating boards where they occurred. The serrated edges of the Thermo-Stud furring channel projected approximately 1.60 cm (0.62 in.) from the base of the channel at the peaks and approximately 0.50 cm (0.20 in.) in the troughs. Therefore, the edges of the channel, which projected into the foam sheathing, did not come into direct contact with the concrete block wall. The approximate frequency of the serrations was one per 2.54 cm (1.00 in.). An isometric drawing of the Thermo-Stud furring channel is included in Fig. 4.

Metal Z-Furring Channel System

The metal Z-furring channel system consisted of extruded EPS sheathing, which measured 61 cm (24 in.) in width, installed between 24-gage Z-shaped metal furring channels and covered with drywall. The seven channels were installed vertically on 61-cm (24-in.) centers. The 1.91-cm (0.75-in.) flange was fastened directly to the concrete block with 31.8-cm (1.25-in.) masonry fasteners on 61-cm (24-in.) centers. A total of 42 masonry fasteners was installed. The drywall was secured to the 3.18-cm (1.25-in.) flange with drywall screws on 40-cm (16-in.) centers at the vertical joints and 60-cm (24-in.) centers at the centerline. Eighty-two drywall screws were installed.

A second variation of this system was also tested. In this case, the wall construction remained unchanged except for the installation of additional drywall screws. Screws were added so that the resulting attachment pattern for the drywall was 20 cm (8 in.) and 30 cm (12 in.) on center at the perimeter and centerline, respectively. As a result, the total number of drywall fasteners was increased to 153.

In this test series the extreme left and right channels were located at least 15 cm (6 in.) from the sides of the test frame to minimize flanking heat loss. Also, that portion of the channel passing between the insulating boards contained slots intended to reduce thermal transmission. See Fig. 4 for an isometric drawing of the metal Z-furring channel.

Temp Guard Insulation Furring (TGIF) System

The TGIF system consisted of extruded EPS sheathing attached to the block wall with mechanically fastened TGIF strips and covered with drywall. The eight strips were installed vertically on 60-cm (24-in.) centers and secured to the wall with 5.72-cm (2.25-in.) masonry fasteners installed on 60-cm (24-in.) centers. A total of 56 screw anchors was used. The drywall was installed vertically and fastened to the furring strips on 20-cm (8-in.) and 30-cm (12-in.) centers at the vertical joints and the centerline, respectively.

To minimize flanking heat loss, the same practice was followed with regard to placement of the strips as was used for the Thermo-Stud furring channels. Also, the strips were installed so that they straddled vertical joints in the insulating boards where they occurred. The TGIF furring strip was 4.14 cm (1.63 in.) in width and fabricated from 25-gage, galvanized steel. Small tabs were punched into it approximately every 13 cm (5 in.) to support the strip on the surface of the foam sheathing until it was mechanically fastened to the substrate. The TGIF strip was essentially flat except for two ribs, 0.30 cm (0.12 in.) in depth, which ran the length of the strip and the edges, which were turned and projected approximately 0.30 cm (0.12 in.) from the base of the strip, as shown in Fig. 4. Hence, a small air space existed between the insulation and the drywall in this system.

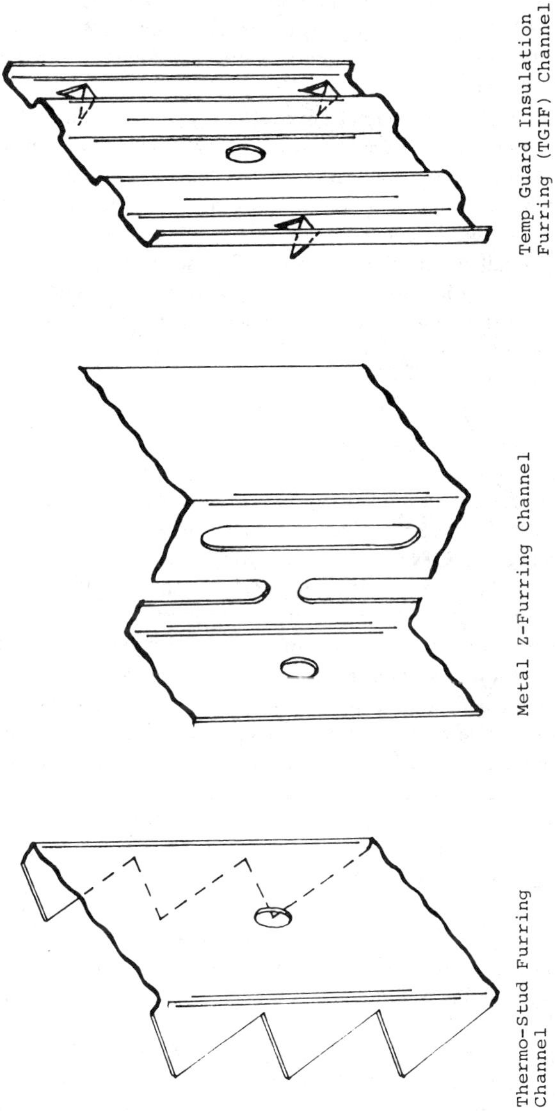

FIG. 4—*Isometric drawings of the metal furring channels evaluated in the study (drawings are not to scale).*

Exterior Insulated Finishing System

Three variations of this system were tested. The first variation consisted of extruded EPS foam sheathing attached to the unsealed surface of the concrete masonry wall with a mastic adhesive and covered with a cementitious plaster finish. Before the cementitious finish was applied, an alkali-resistant fiberglass mesh was stapled to the foam sheathing. The attachment pattern for the staples, which were 0.97 cm (0.38 in.) in length, was approximately one per 0.14 m^2 (1.5 ft^2). A coat of acrylic latex was applied over the mesh and foam sheathing, and then the portland-cement-based plaster was applied in two coats. The total thickness of the finish was approximately 0.64 cm (0.25 in.).

For the second variation, 49 nail-type masonry fasteners, with a diameter of approximately 0.51 cm (0.20 in.) and a length of 3.81 cm (1.50 in.), were installed in a relatively regular array across the field of the test wall. Holes approximately 3.81 cm (1.50 in.) in diameter were drilled through the cementitious finish to the mesh, and pilot holes for the nail anchors were drilled through the mesh and foam into the concrete block wall. The fasteners with 2.54-cm (1.00-in.)-diameter washers were installed, and the holes were filled with additional plaster.

The third variation was identical to the second variation except that the test wall was reversed, that is, the insulated finishing system was adjacent to the metering chamber. This point is addressed in more detail in the Discussion of Results section.

Discussion of Results

Thermal resistances for the various wall constructions presented in the tables and figures represent surface-to-surface R values; in other words, the values are for the wall only and do not include air film resistances. In this study, flanking heat transfer was not measured. The apparent maximum flanking heat transfer was calculated to be less than 3% of the net heat flux, q. The flanking heat transfer should have been very nearly the same for the various insulated wall constructions, provided the location of the insulation in the test specimens relative to the metering chamber was the same. Consequently, the flanking heat transfer for interiorly and exteriorly insulated systems may have differed, and this disparity is addressed in a subsequent paragraph.

It should therefore be recognized that the thermal resistances for the wall constructions presented in this paper are relative R values in that comparison of the values indicates the difference in thermal performance between the systems. The R values shown are not intended to be absolute R values: flanking heat transfer and measurement error associated with the test apparatus have not been taken into consideration. The author feels that the relative thermal resistances presented are, in most cases, within 5% of the absolute R values, but no data are included to substantiate this statement.

Table 1 gives measured thermal resistance values for the concrete block wall insulated with several interior insulation systems at mean temperatures of 4°C (40°F) and 24°C (75°F). As expected, the data showed that, when thermal short circuits and thermal bridges in the form of mechanical fasteners and furring were incorporated in the insulation systems, the R values of the systems decreased.

Also, in every case the measured thermal resistances of the insulated wall systems were greater at a mean temperature of 4°C (40°F) than at 24°C (75°F). This was largely due to the well-established fact that thermal insulation characteristics of foam plastic materials improve as mean temperature decreases. This phenomenon was indirectly reflected in the thermal resistance data for the uninsulated block wall, which remained constant despite the change in mean temperature.

As mentioned previously, the first variation of the drywall laminate system served as the ideal case in that all components of the system were adhesively bonded together, thereby eliminating thermal short circuits. The R value of this variation was 1.11 $K \cdot m^2/W$ (6.32 $h \cdot ft^2 \cdot °F/Btu$)

TABLE 1—*Thermal resistance data for the concrete block wall with the interior insulation systems.*

Wall Description	Thermal Resistance, $K \cdot m^2/W$ ($h \cdot ft^2 \cdot °F/Btu$), at Two Wall System Mean Temperatures	
	4°C (40°F)	24°C (75°F)
Uninsulated	0.17 (0.94)	0.17 (0.94)
Drywall laminate system with no mechanical fasteners or wood furring (first variation)	1.16 (6.59)	1.11 (6.32)
Drywall laminate system with mechanical fasteners and wood furring at top and bottom of wall (fourth variation)	1.14 (6.45)	1.08 (6.14)
Drywall laminate system with mechanical fasteners and wood furring on 122-cm centers (fifth variation)	1.05 (5.98)	1.01 (5.76)
Thermo-Stud system with block molded EPS sheathing (first variation)	0.86 (4.89)	0.82 (4.64)
Thermo-Stud system with extruded EPS sheathing (second variation)	1.10 (6.24)	1.05 (5.97)
Metal Z-furring channel system with gypsum wallboard fastened on 40-cm and 60-cm centers at the perimeter and field, respectively (first variation)	0.99 (5.63)	0.94 (5.37)
Metal Z-furring channel system with gypsum wallboard fastened on 20-cm and 30-cm centers at the perimeter and field, respectively (second variation)	0.99 (5.64)	0.93 (5.30)
Temp Guard Insulation Furring (TGIF) system	1.16 (6.61)	1.10 (6.27)

when measured at a mean temperature of 24°C (75°F). In the fourth variation of the drywall laminate system, in which wood furring and mechanical fasteners were added, the *R* value decreased to 1.08 $K \cdot m^2/W$ (6.14 $h \cdot ft^2 \cdot °F/Btu$). When additional furring and fasteners were installed, the thermal resistance was further reduced to 1.01 $K \cdot m^2/W$ (5.76 $h \cdot ft^2 \cdot °F/Btu$).

The metal Z-furring channel system exhibited the worst thermal performance with a measured *R* value of 0.93 $K \cdot m^2/W$ (5.30 $h \cdot ft^2 \cdot °F/Btu$) at a 24°C (75°F) mean temperature. This was expected, given that the metal Z-furring channel represented the most severe thermal bridge evaluated in the study. Note that seven furring channels were incorporated in this wall construction in comparison with eight for both the Thermo-Stud and TGIF systems. When the number of drywall screws used to attach the gypsum wallboard to the metal Z-furring channel was increased from 82 to 153, the *R* value of the system appeared to decrease from 0.94 (5.37) to 0.93 $K \cdot m^2/W$ (5.30 $h \cdot ft^2 \cdot °F/Btu$) at a mean temperature of 24°C (75°F). However, because that decrease was not observed at a mean temperature of 4°C (40°F), and because the decrease was small, it was questionable whether the apparatus was able to measure the effect of the additional drywall fasteners.

The interior insulation system with the best thermal performance relative to the ideal case was the TGIF system, which had a measured system thermal resistance of 1.10 $K \cdot m^2/W$

(6.27 h · ft^2 · °F/Btu) at a 24°C (75°F) mean temperature. The second variation of the Thermo-Stud system, which was similar in design to the TGIF system, had an *R* value of 1.05 K · m^2/W (5.97 h · ft^2 · °F/Btu) at a 24°C (75°F) mean temperature. System design features that probably contributed to the difference between these similar systems were the small, 0.30-cm (0.12-in.) air space between the insulation surface and the drywall in the case of the TGIF system and the difference in the means of securement of the two furring channels to the surface of the insulation—that is, tabs versus serrated edges.

By comparing the thermal resistances of the first and second variations of the Thermo-Stud system, the impact of using a different material could be examined. The *R* value of the first variation, which incorporated molded EPS sheathing, was 0.82 K · m^2/W (4.64 h · ft^2 · °F/Btu) at a mean temperature of 24°C (75°F). The difference between this value and that of the second variation with extruded EPS sheathing was greater than the difference in *R* values of any two systems. This suggested that material selection could have a greater impact on thermal performance than system design.

Figure 5 illustrates the effect that masonry fasteners had on the thermal performance of the drywall laminate system. When 82 masonry fasteners were installed uniformly across the test wall surface, the *R* value was reduced from 1.11 (6.32) to 1.08 K · m^2/W (6.12 h · ft^2 · F/Btu) at a mean temperature of 24°C (75°F). When the number of fasteners was approximately doubled, the *R* value was further reduced to 1.04 K · m^2/W (5.92 h · ft^2 · °F/Btu). Similar decreases were measured in tests conducted at a mean temperature of 4°C (40°F).

Table 2 shows measured thermal resistance values for the concrete block wall with the exterior insulated finishing system at mean temperatures of 4°C (40°F) and 24°C (75°F). The first variation served as the ideal case in which no thermal short circuits were present. The *R* value of this variation at a mean temperature of 24°C (75°F) was 0.99 K · m^2/W (5.64 h · ft^2 · °F/Btu). When the 49 mechanical fasteners were added, the *R* value decreased as expected to 0.92 K · m^2/W (5.21 h · ft^2 · °F/Btu). A similar decrease was measured at a 4°C (40°F) mean temperature, but the system *R* values were greater, as anticipated.

For the third variation, no changes were made in the wall construction, but the test wall was reversed. The insulation system, which had been adjacent to the climatic (cold) chamber, was in this test exposed to the metering (hot) chamber. The measured system *R* value increased noticeably although the insulation was most likely maintained at a slightly higher mean temperature, which would have resulted in a somewhat lower *R* value for the material. This effect could also be observed to some degree in the difference between the thermal resistances of the ideal cases for the interior and exterior systems.

The location of the insulation in the test specimen relative to the metering chamber could have resulted in increased flanking losses in the second variation, and therefore could have helped account for the disparity between the second and third variations. Also, the location of the relatively massive concrete block wall relative to the metering chamber could have had some effect on thermal transmission through the test wall. It was unclear why the relative performance of the variations at a 4°C (40°F) mean temperature was different from that at 24°C (75°F). More testing needs to be performed to fully investigate these phenomena.

Tables 3*a* and *b* summarize thermal resistance data collected, respectively, on the uninsulated concrete block wall and on the wall insulated with the drywall laminate system with no thermal bridges. Repeatability of results for the uninsulated concrete block wall over a period of approximately two years was very good. The tests performed in November 1981 were the first to be run in this study, and those conducted in December 1983 were the last. The data indicated that the *R* value of the block wall did not change over the course of the study, although holes were periodically drilled into the wall for installation of masonry fasteners and then filled before installation of the next system.

The data presented in Table 3*b* did not provide as good an indication of repeatability as those

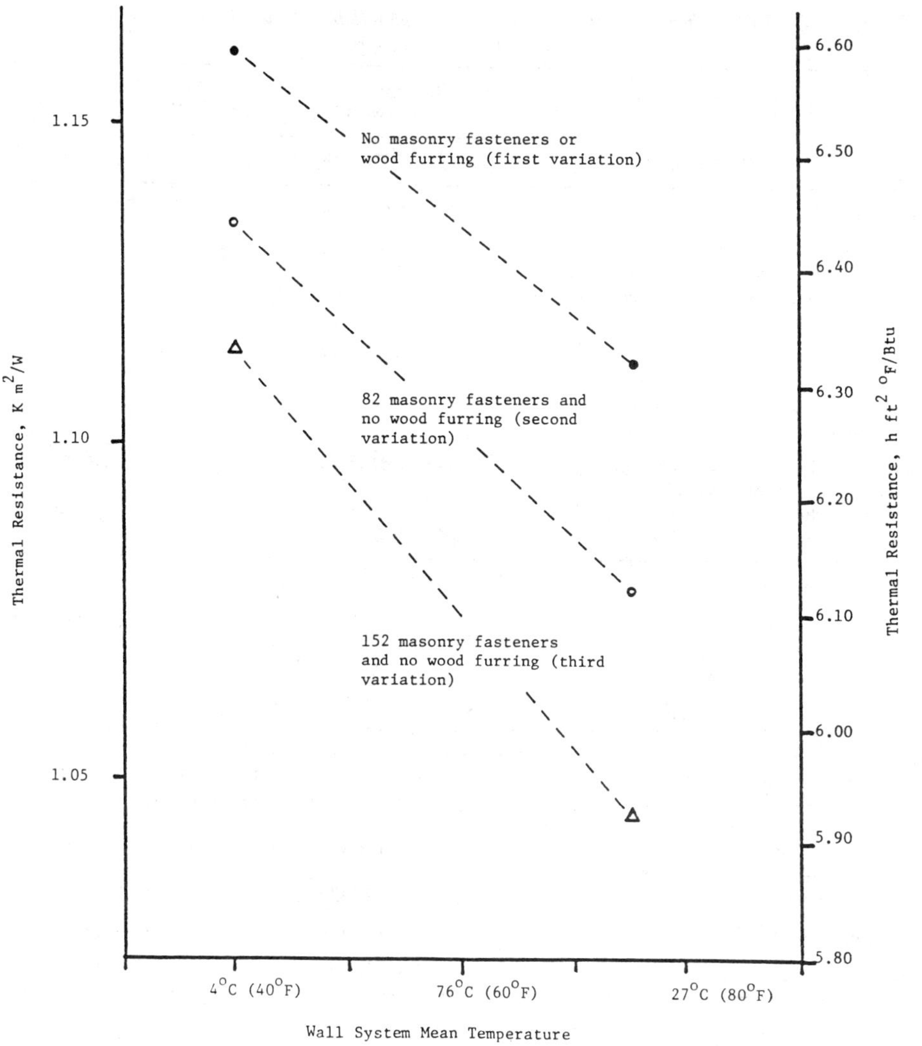

FIG. 5—*Thermal resistance versus mean temperature for the concrete block wall insulated with the drywall laminate system using different attachment patterns.*

in Table 3*a* because workmanship was also a factor. In the 1982 tests the adhesive application proved unsatisfactory. The insulation system separated from the concrete block wall, and an air space up to 0.97 cm (0.38 in.) wide developed. The space would tend to increase the *R* value of the system, as the data indicated. The difference in measured values was nevertheless, less than 5%.

In the 1982 tests the evaluation at a mean temperature of 24°C (75°F) was continued for a second 12-h period without removing the test wall from the apparatus. The measured *R* value increased from 1.15 to 1.16 K · m²/W (6.53 to 6.60 h · ft² · °F/Btu). This could have reflected

TABLE 2—*Thermal resistance data for the concrete block wall with the exterior insulated finishing system.*

Wall Description	Thermal Resistance, K · m²/W (h · ft² · °F/Btu), at Two Wall System Mean Temperatures	
	4°C (40°F)	24°C (75°F)
Uninsulated	0.17 (0.94)	0.17 (0.94)
Extruded EPS sheathing with no masonry fasteners; insulation system adjacent to climatic chamber (first variation)	1.05 (5.94)	0.99 (5.64)
Extruded EPS sheathing with masonry fasteners; insulation system adjacent to climatic chamber (second variation)	1.02 (5.77)	0.92 (5.21)
Extruded EPS sheathing with masonry fasteners; insulation system adjacent to metering chamber (third variation)	1.04 (5.93)	0.96 (5.48)

TABLE 3*a*—*Summary of thermal resistance data for the uninsulated concrete block wall.*

Test Date	Thermal Resistance, K · m²/W (h · ft² · °F/Btu), at Two Wall System Mean Temperatures	
	4°C (40°F)	24°C (75°F)
5-10 Nov. 1981	0.17 (0.94)	0.17 (0.94)
3-6 March 1983	0.16 (0.92)	0.16 (0.92)
25 Nov.-13 Dec. 1983	0.17 (0.94)	0.17 (0.94)

TABLE 3*b*—*Summary of thermal resistance data for the concrete block wall insulated with the drywall laminate system with no mechanical fasteners or wood furring (first variation).*

Test Date	Thermal Resistance, K · m²/W (h · ft² · °F/Btu), at Two Wall System Mean Temperatures	
	4°C (40°F)	24°C (75°F)
1-9 May 1982	1.21 (6.88)	1.15 (6.53)
3 May 1982		1.16 (6.60)
27 Jan.-3 Feb. 1983	1.16 (6.59)	1.11 (6.32)

the increasing impact of the growing air space, but most likely the apparatus was not sensitive enough to measure such an insignificant effect. In any event, the difference between the two values, which was less than 1%, was negligible.

Following a heat transfer calculation methodology described in the 1981 volume of the *ASHRAE Handbook of Fundamentals* and using thermal transmission data given in the same volume for a mean temperature of 24°C (75°F), an *R* value of 1.12 $K \cdot m^2/W$ (6.35 $h \cdot ft^2 \cdot {}^\circ F/Btu$) was calculated for the first variation of the drywall laminate system, which incorporated no mechanical fasteners or wood furring. When specific data for the extruded EPS foam insulation and the concrete masonry wall evaluated in this study were substituted for the general data presented in the ASHRAE handbook, a value of 1.11 $K \cdot m^2/W$ (6.29 $h \cdot ft^2 \cdot {}^\circ F/Btu$) was obtained. The difference between the calculated values and the measured value of 1.11 $K \cdot m^2/W$ (6.32 $h \cdot ft^2 \cdot {}^\circ F/Btu$) was less than 1%.

However, unlike this variation of the drywall laminate system, other commercially available systems evaluated included furring and mechanical fasteners of some type (which are not commonly considered in *R* value calculations). In an effort to simulate common practice in the marketplace, no attempt was made to incorporate the effect of thermal short circuits in the calculations performed for this study; and other factors, such as the slots on the metal Z-furring channel, were also not included. The inaccuracy introduced by this practice was examined by comparing measured data to calculated *R* values. Depending on the specific thermal design of the system, the inaccuracies were found to be as great as 20% of the calculated *R* value.

In cases where the furring was placed on the surface of the insulation—for example, the TGIF and Thermo-Stud systems—the measured *R* values were approximately 1 to 6% less than the calculated value of 1.12 $K \cdot m^2/W$ (6.35 $h \cdot ft^2 \cdot {}^\circ F/Btu$). When wood furring and mechanical fasteners where incorporated in the drywall laminate system in Variations 4 and 5, the measured *R* values were up to 10% below the calculated value.

For the systems incorporating metal Z-furring channels, the measured *R* values were nearly 20% less than calculated value. A second calculation based on the more sophisticated zone method described in the 1981 ASHRAE handbook was performed to better account for the presence of the metal Z-furring channel. The calculated *R* value was 0.63 $K \cdot m^2/W$ (3.55 $h \cdot ft^2 \cdot {}^\circ F/Btu$), which indicated that the effect of the metal Z-furring channel in this type of wall construction was greatly overestimated using this methodology.

Conclusions

The manner in which an insulation system is attached to a concrete masonry wall can have a significant impact on the thermal resistance of the system. Some commercially available systems are better designed than others in terms of thermal performance, providing greater *R* value for a given thickness and type of insulation. The effects of the use of various mechanical fastener attachment patterns and different furring strips and channels (that is, the effects of system design) can be measured, and they can be significant. And the *R* value of the insulating material can have an even greater impact on thermal performance than system design.

A well-designed system from a thermal viewpoint, such as the TGIF system, can achieve an *R* value within 1% of that attained under ideal conditions in which no thermal short circuits or thermal bridges are present. On the other hand, the *R* value of a system with relatively poor thermal design, such as the metal Z-furring system, can be nearly 20% less than that of the ideal case.

In addition, it can be difficult to calculate accurately the thermal resistance of a system with a relatively poor thermal design, whereas inaccuracies in *R* value calculations for well-designed systems are not as great.

Acknowledgments

Deserving special recognition are George S. Deverman and Darrell W. Hollandsworth, both of Dow Chemical U.S.A., for their efforts in constructing the wall test specimens, conducting the tests, and collecting the data.

References

[*1*] Palfey, A. J., "Thermal Performance of Low Emittance Building Sheathing," *Journal of Thermal Insulation,* Vol. 3, January 1980, pp. 129–141.

[*2*] Contreras, A. G. and Palfey, A. J., "Thermal Resistances of Insulated Brick Veneer Walls with Reflective and Nonreflective Air Spaces," *Thermal Insulation, Materials, and Systems for Energy Conservation in the '80s, ASTM STP 789,* American Society for Testing and Materials, Philadelphia, 1983, pp. 373–383.

John V. Alderson,[1] Alan Guy,[1] Barry Justin,[1] and Gerald Shaw[1]

Performance of a Group of Well-Insulated Solar Houses in the United Kingdom

REFERENCE: Alderson, J. V., Guy, A., Justin, B., and Shaw, G., **"Performance of a Group of Well-Insulated Solar Houses in the United Kingdom,"** *Thermal Insulation: Materials and Systems, ASTM STP 922,* F. J. Powell and S. L. Matthews, Eds., American Society for Testing and Materials, Philadelphia, 1987, pp. 599–612.

ABSTRACT: The monitoring of 14 single-story, one-bedroom senior citizens' houses in the northwest region of the United Kingdom is described. Nine of the 14 dwellings were built with very high levels of insulation (in comparison with the then existing 1976 building regulations) and with a form of Trombe wall to provide passive solar gains. The remaining five buildings, of the same internal layout as the others, were insulated only to the minimum specified standards and contained no special solar features. This group was used as a control for comparative purposes. The houses were wholly electrically heated using an off-peak low-tariff-storage heating system. Some 250 sensors, mainly resistance thermometers, were distributed within and on various parts of the structure of the dwellings for monitoring structural thermal behavior. Also, a local weather station was installed on site for climatic measurements. These were interfaced by means of an on-site data logger and land transmission line to a remote control computer at Pilkington Brothers, Research & Development Laboratories some 30 km (20 miles) away. Data were collected every hour, excluding the inevitable breakdowns, for a period of some three years. The resulting energy savings associated with the various insulation and solar measures were determined and used to provide estimates of cost-effectiveness. The most effective measures were extra ceiling insulation, cavity wall insulation, and double glazing.

KEY WORDS: domestic dwellings, thermal performance monitoring, insulation measures, solar walls, cost-effectiveness, thermal insulation

In 1978, a small estate of 14 single-bedroom, one-story houses for senior citizens was completed at Bebington, Merseyside, in the northwest of the United Kingdom. This estate was the result of a collaborative project involving links between academia (Loughborough University of Technology), a major U.K. housing association (Merseyside Improved Houses), and industry (Pilkington Brothers). The scheme was developed as an experiment to meet the following objectives:

(*a*) to design low-energy dwellings incorporating a balanced mix of insulation, double glazing, thermal mass, and means for passive solar utilization;
(*b*) to achieve public sector planning, design, and financial approval for the scheme;
(*c*) to monitor the technical performance of the houses and the efficacies of the various energy-saving measures, and
(*d*) to monitor the social acceptability of the scheme, that is, occupant satisfaction.

Detailed aspects of the design and overall philosophy of the scheme are given elsewhere [*1,2*].

[1]Senior technologists, Pilkington Brothers, Research & Development Laboratories, Lathom, Ormskirk, Lancashire L40, 5UF, England.

The monitoring exercise carried out in relation to objective *c* was jointly funded by the Energy Technology Support Unit (ETSU) of the U.K. Department of Energy, and Pilkington Brothers, the latter being contracted to carry out a two-year program (later extended) of monitoring, analysis, and reports. This paper summarizes the main results obtained and the conclusions drawn. (For conciseness, much of the detail and subsequent analysis of the measured data has been omitted. Some of this is described in earlier work concerning preliminary aspects of the measurements [*3,4*], although full details are available from the final report [*5*] prepared for the preceding contract.)

Summary Details of the Houses

Nine of the 14 houses were built with high levels of insulation, are double glazed throughout, and have a double-glazed Trombe-type wall incorporated as a passive solar feature. The insulation (glass fiber) is applied above ceiling level (100 mm), below the roof deck (80 mm), in the wall cavities (100 mm), and beneath the upper concrete screed deck of the solid floors (50 mm). The houses are also draught sealed. A plan layout of a "solar" house is given in Fig. 1.

The remaining five houses are of similar layout but have no special solar features and are single glazed. At construction they were insulated to meet the minimum requirements specified by the (1976) building regulations. (Subsequent to the initial phase of the monitoring investigation, the house insulation levels were improved.) One of the houses was draught sealed to the same level as the solar houses. This group of traditional houses was used as a control for comparative purposes.

The arrangement and orientation of the houses is shown schematically in Fig. 2. The site is set out with two main terraces of five houses (one solar, one traditional) facing approximately 17° east of south. The other four solar houses comprised two semidetached pairs facing southeast and southwest (see Fig. 2). A view of the two terraces is given in Fig. 3.

Both sets of houses have all-electric heating systems consisting of fan-assisted thermostatically controlled storage heaters, sized and distributed to meet the expected requirements of the occupants. These operated using a low-tariff off-peak electricity supply, taking in energy overnight and releasing this during the following day according to demand.

In the solar houses, the solar wall consists of 225 mm of dense (2400 kg/m^3) engineering "black" (for solar absorption purposes) brick. This is faced by conventional double glazing some 500 mm away. Any warm air in the formed solar wall cavity may be allowed to circulate by natural convection along four ceiling ducts to living room, kitchen, and bedroom areas of the houses (see Fig. 4). On dull days and at night during the heating season, air circulation and heat loss can be reduced by closing off vents at the bottom of the solar wall. In summer, unwanted air circulation can be prevented in a similar manner, and the wall cavity also vented to the outside (see Fig. 4). Additionally, an external "Dutch" blind is provided for summer shading if required (see, for example, Fig. 4, summer conditions, for schematic of blind usage).

Design Heat Loss Characteristics

Various physical properties were measured on structural components (for example, bricks, blocks) of the houses. These included densities, thermal conductivities, specific heats, and solar absorptances. On the basis of these and additional data, design U values and specific heat loss coefficients for the various structural elements of the houses were calculated. Table 1 gives a summary of the calculated design U values.

The design specific heat loss coefficients for the pair of terraces (assuming a relative air exchange rate of 0.5 volumes per hour for each) were calculated to be 445 W/K for the solar terrace and 725 W/K for the traditional terrace. The floor area of each terrace is 251 m^2.

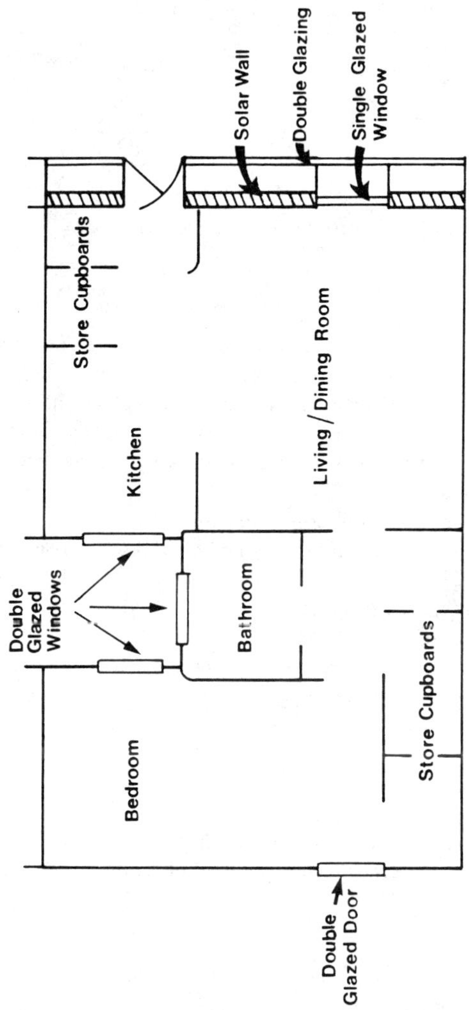

FIG. 1—*Layout of a solar-assisted house.*

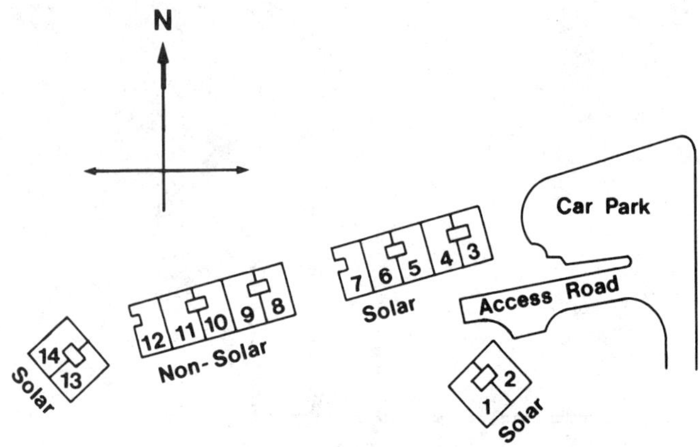

FIG. 2—*Acorn Close, Bebington, site plan.*

Instrumentation

The estate was instrumented with about 270 permanent detectors, mainly platinum resistance thermometers for measurement of structural and internal air temperatures, but also including surface-mounted heat flux transducers, removable globe thermometers, and thermistor-bead-type anemometers (positioned in the solar wall ducts). For the measurement of external conditions, a complete weather station was installed on the roof of the southeast facing pair of semidetached houses.

All the sensors were connected by underground cabling to a 300-channel data-logger situated in a brick hut on site. The logger was, in turn, linked by post office telephone line (dedicated) to a remote computer at the Pilkington Brothers Research & Development Laboratories some 25 miles distant. The computer was programmed to initiate data collection, transmission, and storage on hard disk for subsequent analysis, usually on an hour-by-hour basis.

Three separate electricity meters were provided for each house to measure individual space heating, domestic hot water, and total energy consumption. Readings of these were taken manually at weekly intervals.

Short-Term Experiments

Apart from the long-term regular collection of data, additional measurements were taken over shorter periods using a variety of techniques and instrumentation. These included the following.

Thermographic Survey of the Solar Houses

To check for any thermal defects in the structure of the solar houses (for example, possibly arising from inadequately installed insulation), a thermal-imaging camera was used to "view" the site. This was carried out at night during winter (to eliminate solar radiation effects), with external air temperatures of around 0°C.

No obvious faults in the structure were detected, although the thermal bridging effect of the concrete floor raft was clearly apparent, and there appeared to be some evidence of warm air leakage around the external flap closures in the solar wall. The effect of different living room and bedroom area temperatures was also noticeable, as was the presence of electric storage heaters that were internally positioned adjacent to external walls viewed by the equipment.

FIG. 3—*A view of five of the low-energy bungalows with the five traditional dwellings in the background.*

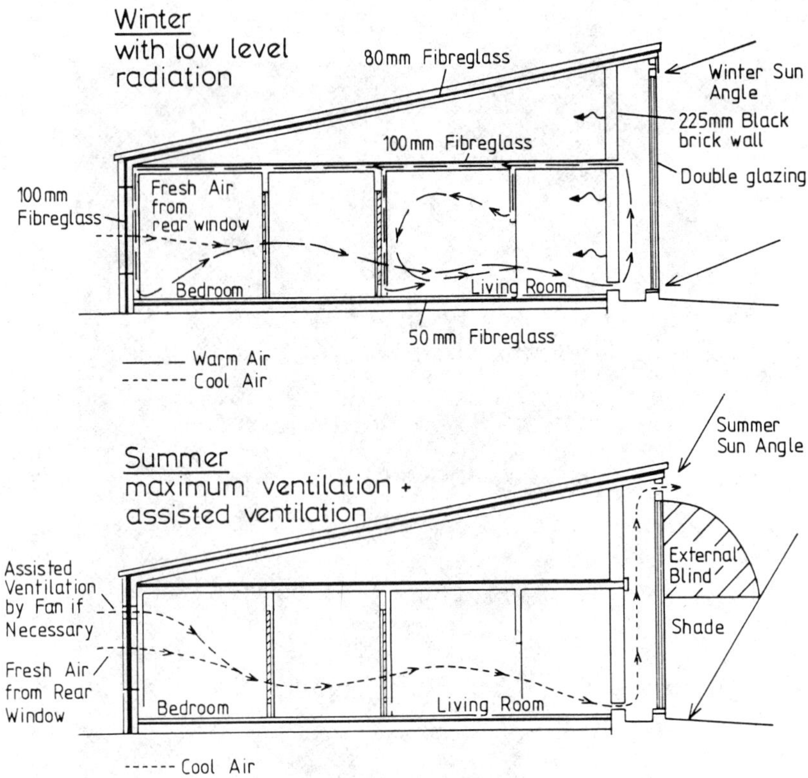

FIG. 4—*Acorn Close—solar heating and ventilation systems.*

TABLE 1—*Design U values, in watts per metre square kelvin.*

Element	Solar	Traditional
Floor	0.33	0.6
Ceiling plus roof	0.17	0.4
Rear and end walls	0.3	1.0
Patio walls	0.27	1.0
Solar wall	1.5	...
Front wall (traditional)	...	1.0
Double glazing	2.8	...
Single glazing	...	5.7

Experimental Measurement of House Heat Loss Coefficients

Some short-term (over three days or so), quasi-steady measurements of house heat losses were made by heating a house interior (with electric fan heaters) to a controlled high temperature (here, 27°C). These were carried out over periods of reasonably settled overcast external conditions, with the house vacated, all windows closed, and with the house interior atmosphere well-stirred (using fans) to achieve as uniform an interior temperature distribution as possible.

The energy consumed by the fans and heaters was logged hourly, and a house specific heat loss coefficient, L, in watts per kelvin, was calculated from the relation

$$L = \frac{P}{\Delta T} \tag{1}$$

where

P = the average power input (in watts) over the eventual measurement period (some two to three days, following a similar time for equilibrating purposes), and

ΔT = the average internal to external temperature difference (in degrees kelvin) over the same period.

Because of occupancy factors, it proved possible to measure only three houses. The resulting calculated house heat loss coefficients are shown in Table 2. The measurement of House 2 was carried out prior to occupation, when the structures of the houses were unlikely to have fully dried out. This probably accounts for the high value found for that test. The remainder of the measurements were carried out well over a year after initial occupation. For these, resulting levels of loss coefficient agreed quite well with the calculated design value (see Table 2).

Ventilation Exchanges

Ventilation exchanges were measured under a variety of conditions on some of the houses using a conventional tracer-gas decay technique [here based on continuous infrared (IR) absorption measurement of internal atmospheric carbon dioxide (CO_2) after initial injection of the pure gas]. A summary of some measured average exchange rates is given in Table 3. These relate to conditions in which all the external doors and windows were kept closed, with tests being carried before and some time after occupation.

The preoccupation tracer-gas decay measurements were complemented by carrying out pressurization tests (for example, Ref *6*) to provide a comparative assessment of the relative "leakiness" of each house. The results of these tests (see Ref *5* for full details) were qualitatively in line with ventilation exchange behavior (see Table 2). As a group, the solar houses were found to be more leaky than the traditional ones, the draught-proofed traditional house showing the least leakage.

On the basis of the results of Table 2, it was concluded that if no deliberate ventilation was applied, a reasonable level of ventilation exchange to assume for any calculation purposes was

TABLE 2—*Measured house specific heat loss coefficients.*

House	Measured L, W/K	Design Value, W/K
1	91	93[a]
2	136	93
3	90	93
	110[b]	

[a]The design value is that calculated for an end-terrace or semidetached house (all the houses tested being of this type), with an assumed relative air exchange rate of 0.5 volumes per hour.

[b]Repeat measurement at a later date.

TABLE 3—*Measured air exchange rates.*

Test Details	No. of Measurements	Average Exchange Rate, volumes/h
	PREOCCUPATION	
Solar	6	0.47
Traditional	3	0.43
Draught-proofed traditional	5	0.40
	POSTOCCUPATION	
Solar, 1 year	3	0.47
Solar, 3 years	1	0.42

0.5 volume changes per hour. Subsequent observations over the three years of monitoring indicated that window opening by the occupants was infrequent and usually occurred for only short periods. It seems reasonable to assume, therefore, that the preceding level was representative of typical practical exchange rates over the monitoring period.

Shading of the Solar Wall

A major factor influencing solar wall performance is the amount of solar radiation absorbed at the front surface of the black brick wall. This depends not just on the brick solar absorptance but also on the solar transmission characteristics of the glass facade of the wall and internal/external shading effects. While glass transmission losses were readily calculable, the losses resulting from shading were not as easy to estimate, since, first, the glazing bars, overhang, and reveal structures were fairly complex and, second, trees, nearby buildings, and other structures shaded particular houses to different extents.

A decision was made, therefore, to measure the solar wall shading rather than compute it. This was done by the following two means:

(*a*) photographing individual house fronts throughout the day at various times of the year; and

(*b*) mounting a solarimeter vertically for a limited period (five weeks) on the brick face of a house solar wall, and measuring integrated energy collections for comparison with step *a*. [Radiation in the plane of the wall was calculated from corresponding measured global and diffuse horizontal values, and glass transmission losses (30%) were then allowed for.]

The results of both these procedures were in reasonable agreement. On the average, the amount of solar energy actually received at the front face of the brick wall was estimated to be between 35 and 40% of that available on a corresponding unobstructed surface (depending on the location and orientation of the particular wall involved).

Convective Contribution from Solar Wall Ducts

The convective contribution from solar wall ducts was investigated using a CO_2 tracer gas technique, which measured the time taken, under various conditions of solar radiation, for a slug of gas injected at one end of a solar wall duct to appear at the other end.

The earlier measurements of this nature produced some unexpected (at that time) observations. On occasions when, according to the design predictions, the prevailing weather conditions should have produced a convective flow into each solar wall duct, this was not always detected. Indeed, sometimes the opposite occurred. Further investigation showed that this was due to adverse air circulations taking place around the solar wall duct systems, driven by temperature differences between livingroom/kitchen and bedroom areas. In the absence of solar radiation, these circulations cooled the bedroom areas and, by doing so, were effectively self-sustaining.

Because the adverse flows resulted in needless cooling of the interiors of the houses, steps were taken to prevent this by fitting nonreturn flaps in the solar wall ducts where outflow into the solar wall cavity occurred. It was confirmed that this measure eliminated the adverse flows, and so the flaps were fitted to each solar house.

The slug injection technique, coupled with duct air inlet and outlet temperature measurements, was used to estimate the energy contribution by convection through the solar wall ducts. This was found to be small (see the next section).

Long-Term Monitoring—Results of Analysis

Analysis of the long-term collected data was carried out mainly on the following four time scales:

(*a*) approximately monthly, to examine quasi-steady performances of building components,
(*b*) weekly, for statistical correction of energy requirements with internal and external environmental conditions,
(*c*) daily, to relate day-to-day effects of variations in solar insolation and ambient temperatures, and
(*d*) hourly, mainly to investigate more rapid dynamic response behavior.

As a result of these analyses, it proved possible to relate space heating requirements for the two types of dwelling after accounting for occupancy differences. In addition, the separate contributions to energy savings of the various measures incorporated in the dwellings were determined. These are now discussed.

Savings in Space Heating Energy Consumption

One of the main objectives of the monitoring was to relate typical overall energy savings of a solar house over its traditional counterpart. Figure 5 compares the average measured weekly space heating demands for the solar (lower histogram) and traditional houses in the two terraces. (Note—the marked reduction in demand for the traditional terrace for the 1980–1981 heating season was a direct result of cavity-filling the walls of these with blown glass fiber in August 1980).

In simple terms, over the twelve months from September 1979 to September 1980, the average traditional house used 6500 kWh (23.4 GJ) for space heating compared with 3250 kWh (11.7 GJ) for the average solar house. However, the number of occupants in the dwellings and their living patterns significantly affected energy demands. At Bebington, it was found that energy consumption variations of up to nearly 100% occurred because of this.

To attempt to compensate for such individual differences, the results for each house were normalized to a mean house temperature of 18.3°C and 1.5 occupants per house. The resulting average space heating demands for the houses were then found to be somewhat lower than those just cited, and an estimated normalized mean energy balance for each type of dwelling is shown in Fig. 6.

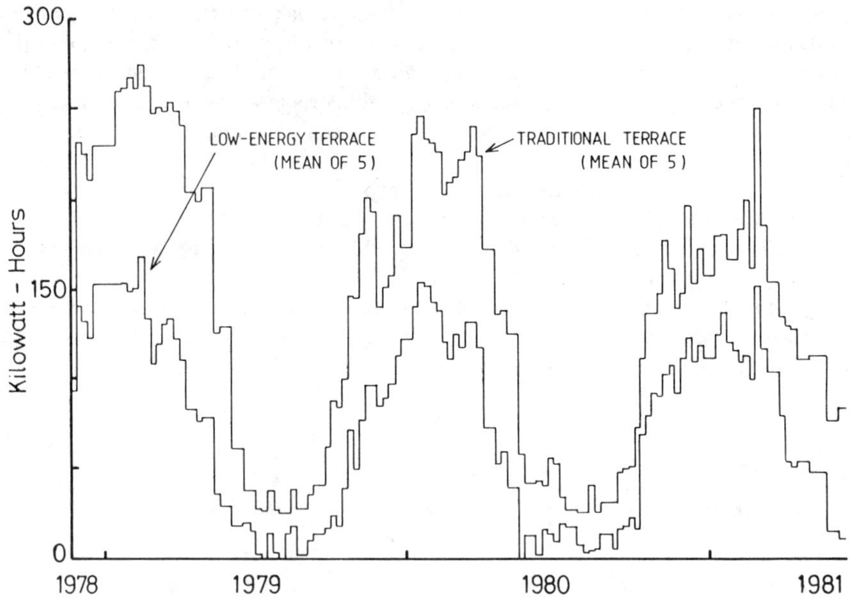

FIG. 5—*Acorn Close, Bebington, weekly space heating.*

From Fig. 6 it may be inferred that (for the normalized occupation) the combined effect of the extra insulation measures and the solar wall in the solar houses led to average space heating demands of approximately half those of the traditional houses and a reduction in the total annual energy requirement of around 4000 kWh (14.5 GJ).

Energy Savings from the Solar Wall

Because of the somewhat novel use of the Trombe-type solar wall (at least in the United Kingdom), estimation of the solar contribution to space heating and its comparison with more conventional energy saving methods was of considerable interest. The measured U value of the solar wall (estimated from multiple regression analysis of wall heat flows) was 1.55 $W/m^2 \cdot K$, which agreed quite well with the design value (see Table 1). The analysis also indicated that, on the average, approximately 22% of the daily available solar insulation in the plane of the wall was useful in offsetting solar wall conduction losses.

The combination of these factors with the results of the measurement of the wall convective contributions indicated that about 18% of the total energy savings was due to replacing the traditional front wall with the Trombe wall. Only about one tenth of this saving came from the convected warm air, the remainder resulting from conducted gains. Although facilities also existed for drawing air from the insulated loft space into the bedrooms, these were initially found by the occupants to have no noticeable effect and were subsequently never used.

In practical terms, the results for the solar wall meant that an opaque, insulated wall with a U value of around 0.65 $W/m^2 \cdot K$ or better would have been more beneficial for energy saving than the particular solar wall used in the dwellings. Nevertheless, the solar wall did make a contribution to energy savings, and there appeared to be obvious design possibilities for improvement on this (for example, by reducing shading or increasing absorber area).

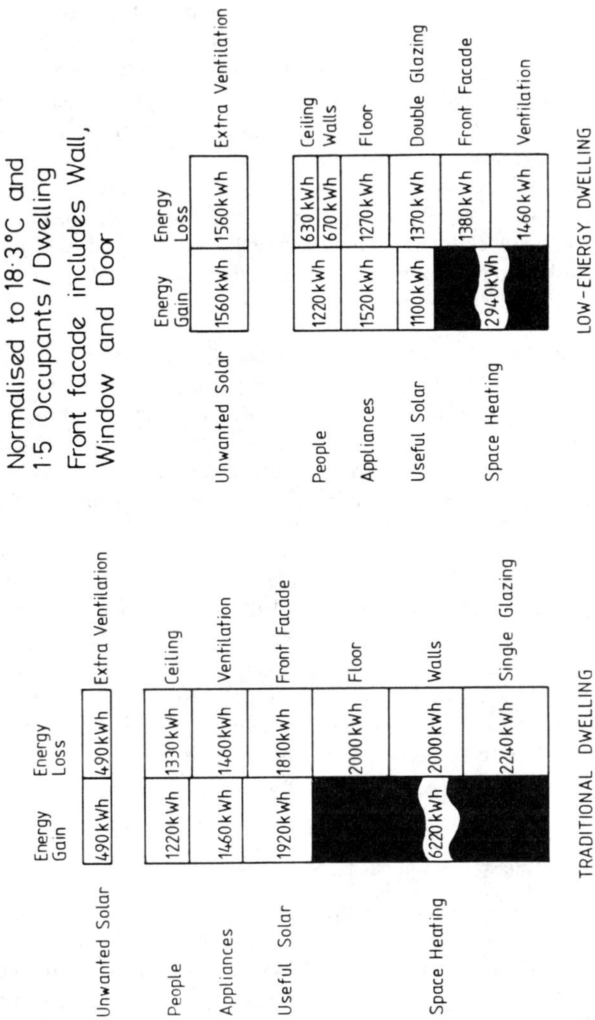

FIG. 6—*Annual normalized energy balance.*

Energy Savings Resulting from Extra Insulation

Ceilings—The traditional houses were insulated at ceiling level with 50 mm of glass fiber mat. Analysis showed that about 16% of the total energy savings resulted from the higher insulation levels in the ceilings (100 mm) and roofs (80 mm) of the solar houses. However, further calculations indicated that the same savings could have been achieved at lower cost by positioning all the additional insulation at ceiling level.

Walls—The external walls of the traditional houses were of brick/cavity/block construction with a nominal design U value of 1 $W/m^2 \cdot K$ (see Table 1). The corresponding measured estimated value (obtained in a similar manner to that for the solar wall) of 0.97 $W/m^2 \cdot K$ was in good agreement with this. The analysis also showed, however, that conducted heat losses were influenced by solar radiation. For example, the effective U value over a heating season for a front wall of the traditional terrace was estimated to be 0.89 $W/m^2 \cdot K$. (Note—after the cavity-filling operation of August 1980, the average measured U value decreased to 0.43 $W/m^2 \cdot K$. Assuming a mean cavity width of 50 mm, this was roughly equivalent to a blown insulation thermal conductivity of 0.034 $W/m^2 \cdot K$, which was well within the claims of the installers.)

Overall, the analysis showed that some 26% of the total energy savings of the solar houses was due to the cavity insulation (100 mm) of the rear and end walls.

Floors—Both types of houses had solid floors, although in the solar houses these were insulated over the full floor area (50-mm glass fiber board beneath 75-mm cement screed). Although measurements were made using heat flux transducers (movable) and also by taking temperature differences across the floor insulation, these proved insufficient for precise evaluation of energy savings, as the floor heat flows were quite complex. The best estimate that could be made was that the energy savings resulting from the floor insulation measured did not exceed 18% of the total savings.

Double Glazing—The traditional houses were single glazed throughout, whereas all the external glazing of the solar houses consisted of sealed, double-glazed units. All the window frames were aluminum and contained no thermal break.

No on-site measurements were made of glazing transmittances, and U values of 5.7 and 2.8 $W/m^2 \cdot K$ were assumed for single and double glazing, respectively. Assessments of energy savings based solely on these values may well be overestimated, particularly as any effects of drawing curtains at night have been ignored. The savings resulting from the double glazing were therefore evaluated as not more than 22% of the total.

Cost-Effectiveness of the Energy-Saving Measures

The effectiveness of the various energy-saving measures was determined from the annual energy savings and an assumed annual increase in real fuel price of 3% over the cost in 1978.

The low-energy terrace of five dwellings cost $40 000 (£ 21 000) more to build than the traditional terrace and saved electricity worth $315 (£ 164) per year at that time, giving a discounted pay-back period of 53 years for the total energy-saving package. The marginal capital costs of each individual measure were more difficult to assess, as the prices for some of the items in the agreed final account seemed unrepresentative [for example, $8.80 (£ 4.6) per square metre for sealed, double-glazed units]. Where there were clearly good grounds for revising the cost of an item, as in the preceding example, this was done using commercial price guides, and a discounted pay-back period was then calculated for each individual measure.

These calculations led to pay-back periods of between 5 and 10 years for extra ceiling insulation (130 mm extra), external cavity wall insulation, and the double glazing. The pay-back period for the installed floor insulation was at least 26 years, whereas that for the solar facade was 80 years, decreasing to 62 years if the ducting and vent controls required to produce the very small convective contribution were eliminated.

Social Aspects

Although the work described in this report was primarily concerned with the energy aspects of the houses, the experimenters found it impossible to ignore the influence of the occupants who had almost as significant an effect on the performance of the houses as did the conservation measures. The manner in which the residents operated and understood the heating system and its controls, the way some used the solar cavity to grow plants, and the frequency of use of the solar wall blinds all had significant effects on the energy performance.

It was, of course, important to know whether the solar houses, in addition to having reduced energy consumptions, provided accommodation that was at least as pleasant and manageable as the traditional counterparts. Although this aspect of the work formed no part of the authors' brief, interacting closely with the occupants over a long period of time nevertheless provided a clear impression that the occupants found the houses pleasant and comfortable and no particular trouble to manage. Inevitably, there were minor complaints, particularly about winter drafts from the internal louvred doors, night-time rustlings from the solar ducts, and so on. However, these are problems that could be overcome with simple design changes.

Conclusions

The objective of producing low-energy public sector senior citizens' dwellings, satisfactory both to the occupants and the housing management, has been largely met by the scheme described in this paper.

The most cost-effective elements in the package of installed energy-saving measures consisted of extra insulation to the ceilings, cavity wall insulation, and double glazing.

Energy savings were also achieved by the double-glazed Trombe wall, although in this case the level of savings and the cost-effectiveness are much more dependent on specific design details. Room for improving the energy savings and decreasing the capital cost exists; however, even if the savings were doubled, the additional capital cost would need to be reduced 14-fold to give a discounted pay-back period comparable with the most cost-effective insulation measures.

The complex and sometimes unexpected interactions between the building, the heating systems, and the occupants indicate that development of more closely integrated designs could save even more energy at lower cost without sacrificing human comfort or unduly limiting the occupants' freedom to choose their own living conditions. However, that remains a challenge for the future.

Acknowledgments

The writers are indebted to a considerable number of colleagues, participants, and collaborators who provided much valuable input and assistance during the course of this work. In particular, grateful acknowledgment is given to A. Leff for administering the variety of on-site instrumentation; to Merseyside Improved Houses, their architects, and especially H. W. Evans, for their constant interest and support; to the designers of the houses, P. Greenwood and Dr. H. Ward; to the U.K. Department of Energy for the substantial financial support for the monitoring program and the staff of the Energy Technology Support Unit for their guidance and encouragement; to the occupants of the houses for their cooperation and understanding; and finally to the directors of Pilkington Brothers and A. S. Robinson, director of Group Research and Development, with whose permission this paper is published. The work was carried out under ETSU Contract No. E/5A/CON/1003.

References

[1] Greenwood, P. and Ward, H., "Solar Houses for the Elderly," *Proceedings,* Conference C19, International Solar Energy Society, United Kingdom Section, London, April 1979.

[2] Evans, H. W., "Rented Housing—Why Solar Energy?" *Proceedings,* Conference C29, International Solar Energy Society, United Kingdom Section, April 1982.

[3] Justin, B., Guy, A. G. and Shaw, G., "Monitoring Solar Walls in Occupied Houses," *Proceedings,* Conference C24, International Solar Energy Society, United Kingdom Section, October 1980.

[4] Justin, B., Alderson, J. V., Guy, A. G., and Shaw, G., "Some Experience Gained from Monitoring Occupied Houses," *Proceedings,* International Solar Energy Society Congress, Brighton, U.K., Vol. 3, August 1981, p. 1911.

[5] Alderson, J. V., Guy, A. G., Justin, B., Leff, A. and Shaw G., "The Monitored Behavior of Occupied Low-Energy Bungalows," Final Report, ETSU-S-1003, Pilkington Brothers PLC, Lathom, U.K., 1983.

[6] Dickson, D. J., "Methods of Measuring Ventilation and Leakage Rates of Houses," ECRC Report M1419, Electricity Council Research Centre, Capenhurst, U.K., April 1981.

Air and Moisture

Robert Kane[1] and Guy Titley[1]

Case Histories of Moisture Monitoring in Residential Walls

REFERENCE: Kane, R. and Titley, G., "**Case Histories of Moisture Monitoring in Residential Walls,**" *Thermal Insulation: Materials and Systems, ASTM STP 922,* F. J. Powell and S. L. Matthews, Eds., American Society for Testing and Materials, Philadelphia, 1987, pp. 615–629.

ABSTRACT: There is an unjustified concern within the building industry that the use of an exterior low water vapor permeance foam sheathing design allows an accumulation of moisture between the vapor barrier and the exterior sheathing.

In this study the authors measured the moisture content in the wood studs of seven occupied homes, in four Canadian cities, with a low water vapor permeance sheathing (extruded polystyrene), with or without a noninsulating sheathing, such as plywood, fiberboard, or chipboard, used on the exterior of the house. The degree days ranged from 4068 to 5920°C, and the relative humidity ranged from 71 to 83.5%. In all cases the actual moisture content of the studs was measured with Delmhorst moisture elements and then read with Delmhorst moisture meters. The results show a similarity in moisture concentration patterns regardless of the exterior sheathing installed. The effects of construction practices and local relative humidities are discussed.

The practical usefulness of actual moisture studies of real homes, done in a variety of climates, will be most interesting to today's architects and builders.

KEY WORDS: double vapor barriers, field studies, low permeance sheathings, moisture content, moisture research, thermal insulation

There is an unjustified concern in the building industry that the use of an exterior low water vapor permeance foam sheathing design causes accumulation of moisture between the vapor barrier and the exterior sheathing. Research done by the U.S. Department of Agriculture [*1*], Owens-Corning Fiberglass [*2*], and others [*3–5*] has shown that low permeance foam sheathings present no greater cold weather condensation hazard than the other types of sheathing studied, and therefore, the concern is unjustified.

Incorrect assumptions have led to confusion within the construction industry about the use of insulated, low water vapor permeance foam as an external sheathing on wood-frame constructed homes. In this study the authors measured the moisture content in the wood studs of seven occupied homes in four Canadian cities. Each home had a low water vapor permeance sheathing (extruded polystyrene), with or without a noninsulating sheathing, such as plywood, fiberboard, or chipboard, installed on the exterior of the house. (See Table 1 for the permeance values of common exterior sheathing materials.) Recognizing that there are many uncontrollable variables in this type of study, extreme conditions were chosen for each site. In Winnipeg, Manitoba, a high heating degree-day region using conventional construction was examined; in London, Ontario, a high interior relative humidity was studied in a moderate heating degree-day region; in Regina, Saskatchewan, an energy-efficient design was studied in a high heating degree-day region; and last, in St. John's, Newfoundland, a moderate heating degree-day region was examined in a maritime climate with a high relative humidity.

[1]Senior research engineer and group leader, respectively, Dow Chemical Canada Inc., Sarnia, Ontario, Canada N7T 7K7.

TABLE 1—*Permeance of exterior sheathing materials.*

Product or Material	Thickness, mm	Metric, ng/Pa · s · m^2	Imperial, perm
Polyurethane (foil-faced)	25.4	1.5	0.025
Polyethylene film	(4 mil)	5.0	0.08
Exterior plywood	9.5	20.0	0.35
Extruded polystyrene	38.1	23.0	0.40
Exterior plywood	12.7	27.0	0.47
Extruded polystyrene	25.4	35.0	0.60
Building paper		58 to 290	1.0 to 5.0
Molded polystyrene	38.1	76 to 193	1.3 to 3.3
Phenolic (kraft-faced)	28.6	320	5.5
Gypsum sheathing	12.7	2330	40
Fiberboard	12.7	2330 to 5830	40 to 100
Rigid glass fiber	25.4	1750	30

The true measure of moisture performance for any building material is how well it performs against Mother Nature. It is important to understand the minimum conditions required for fungi growth when evaluating moisture content in wood stud data. Fungi growth causes wood studs to rot. This growth requires free water in cell cavities [*6*], which is present because the cell walls are already saturated. The saturation point is reached when the moisture content is within the range of 25 to 32%. Correspondingly, fungi growth requires a minimum temperature of 4°C, with the optimum temperature being about 27°C. Adequate oxygen is always present in conventional buildings; therefore, fungi growth could occur in the summer months when the weather is warm, but only if the moisture content is excessive. Moisture damage occurs when the design allows for an accumulation of moisture that results in excessive levels during the warm months when fungi are most active.

The ultimate measure of performance, then, is to ensure that the moisture content is less than 25% when the temperature is greater than 4°C.

The results showed that the moisture concentration in the wood studs was generally less than 15% in the homes studied and always less than the 25% required for fungi growth, regardless of the sheathing installed. As suggested by others [*7,8*], external relative humidity and airtightness were shown to be important variables in controlling the moisture content in wood studs. The water vapor permeance of the walls studied was shown to be an insignificant factor in controlling the moisture content in the wood stud.

Technical Setup

Instrumentation Details

Delmhorst moisture elements were installed in the wood studs during the construction of several occupied homes in Canada. The homes were located in Winnipeg, London, Regina, and St. John's. Each test site represented a specific actual condition: Winnipeg—cold winters; London—high interior relative humidity; Regina—energy-efficient design; St. John's—high exterior relative humidity. For more information on the homes chosen, see the section entitled Construction Details. The length of the studies, the amount of data collected, and the type of data collected were specific to each test site. (See Table 2 for the weather conditions.) A Delmhorst moisture meter (Model J-1) was used to collect readings, and interior wall cavity temperature corrections were according to Cech and Pfaff [*9*]. The readings were recorded in weight percentages.

TABLE 2—*Weather conditions of four locations studied.*

Location	Degree-Days, Centigrade	Average Relative Humidity, %			
		1 A.M.	7 A.M.	1 P.M.	7 P.M.
Winnipeg	5889	80	81	64	69
Regina	5920	78	80	61	65
London	4068	85	86	68	75
St. John's	4804	88	86	76	84

The electrical resistance of wood is so highly sensitive to moisture content that a change in moisture content from 7 to 20% results in a 10 000-fold decrease in resistance [*2*]. The Delmhorst moisture meter data are accurate to within 0.5 to 2.0% moisture concentration, as the actual moisture concentration varies from minimum values of 5% through the fiber saturation (25 to 32%) [*10*].

The elements were installed by drilling 6-mm-diameter holes to a depth of 12 mm at a 45° angle to the stud face. Delmhorst probes were then installed with a snug fit by forcing the plastic collar into the 6-mm hole. A more detailed investigation was conducted in St. John's, Newfoundland, where Delmhorst moisture elements were used to measure the moisture content in the stud and also the humidity at the batt/sheathing interface. Thermocouples were also installed in St. John's to measure the temperature at the batt/sheathing interface. The moisture elements were generally installed on the leeward side of the buildings near a high-relative-humidity location such as a bathroom or kitchen.

Construction Details

Winnipeg—A six-year study was conducted to examine the long-term effects of installing a low permeance foam sheathing on the exterior of a residential building. Winnipeg was chosen because the housing authorities in Canada advised that locations with cold winters would test the worst scenario. Today, all authorities in Canada agree that buildings in high-relative-humidity regions are more susceptible to moisture damage than those in the cold, dry areas in the Midwest. Each of the Winnipeg homes was built using typical 1976 construction methods (see Table 3). A vapor barrier was installed without any attempt to seal punctures or joints, it can therefore be referred to as a vapor barrier, but it cannot be considered an effective air barrier. Two Delmhorst elements were used to measure the moisture concentration in the studs in each of the four houses. The heating degree-days for Winnipeg are 5889 Celsius [*11*].

Three wall designs were included for comparison:

(*a*) traditional construction using fiberboard sheathing as a control site,
(*b*) extruded polystyrene sheathing because of the low permeability (two homes), and
(*c*) extruded polystyrene over plywood because of the low permeance combination (plywood has a lower water vapor permeability than extruded polystyrene).

The home was electrically heated.

London—A five-year study was conducted to test a high-relative-humidity building with a low permeance foam sheathing installed externally (see Table 4). High interior relative humidities placed a high moisture stress on the building structure [*12*]. An indoor pool was chosen to ensure that this high moisture stress was indeed continuously placed on the wall design. Four

TABLE 3—*Winnipeg construction details.*

Construction Detail	Home No. 1	Home No. 2	Home No. 3	Home No. 4
Style	single story	split level	single story	split level
Heating	hot-air gas	electric forced air	oil furnace	oil furnace
Wall construction[a]	12-mm gypsum 2-mil polyethylene RSI 2.1 batts (R12) 12-mm fiberboard 1-layer paper stucco finish	12-mm gypsum 2-mil polyethylene RSI 2.1 batts (R12) 9-mm plywood RSI 1.3 (R7.5) extruded polystyrene 1-layer paper siding	12-mm gypsum 2-mil polyethylene RSI 2.1 batts (R12) RSI 1.3 (R7.5) extruded polystyrene 1-layer paper brick exterior	12-mm gypsum 2-mil polyethylene RSI 2.1 batts (R12) RSI 1.3 (R7.5) extruded polystyrene 1-layer paper brick exterior
Probe No.	1, 2	1, 2	1, 2	1, 2
Probe orientation	W, N	N, N	NE, SW	N, S
Room[b]	DR, H	H, bath	BR, K	BR, K

[a]Extruded polystyrene was installed in 600 by 2400-mm dimensions. Plywood and fiberboard were installed in 1200 by 2400-mm dimensions.
[b]DR = dining room; H = hall; bath = bathroom, BR = bedroom; K = kitchen.

TABLE 4—*London construction details.*

Construction Detail	Characteristics			
Style	home with indoor pool			
Heating	hot-air gas			
Wall construction[a]	cedar polyethylene RSI 3.5 batts (R20) RSI 1.3 (R7.5) extruded polystyrene 12-mm fiberboard building paper horizontal furring cedar siding			
Probe No.	1	2	3	4
Probe orientation	NE	NE	NW	NW
Room[b]	IP	IP	IP	IP

[a]Extruded polystyrene was installed in 600 by 2400-mm dimensions.
[b]IP = indoor pool.

Delmhorst moisture elements were used to measure the moisture concentration in the studs of the indoor pool building enclosure. The heating degree-days for London are 4068 Celsius [*11*].

Regina—A one-year study was conducted to test the effects of a properly installed interior polyethylene vapor barrier with a low permeance foam sheathing installed externally (see Table 5). An energy-efficient home was chosen with RSI 3.5 (R20) glass fiber batts and RSI 1.7 (R10) extruded polystyrene. A continuous interior polyethylene air/vapor barrier was installed (see the Discussion section for a more detailed explanation of continuous air/vapor barrier). Three Delmhorst moisture elements were used to measure the moisture concentration in the studs of the energy-efficient home chosen. The heating degree-days for Regina are 5920 Celsius [*11*].

TABLE 5—*Regina construction details.*

Construction Detail	Characteristics		
Style	split level		
Heating	gas forced air		
Wall construction[a]	12-mm gypsum polyethylene (air/vapor barrier) RSI 3.5 batts (R20) RSI 1.7 (R10) extruded polystyrene building paper siding		
Probe No.	1	2	3
Probe orientation	S	E	W
Room[b]	bath,	K	BR

[a]Extruded polystyrene was installed in 600 by 2400-mm dimensions.
[b]Bath = bathroom; K = kitchen; BR = bedroom.

Newfoundland—A one-year study was conducted to see the effects of the high relative humidities in maritime regions such as Newfoundland on walls with a low permeance foam sheathing installed externally. Since the beginning of the moisture monitoring studies in 1976, many researchers have examined the cause and effect relationship that results in moisture damage. It has been shown that the highest percentage of moisture damage occurs in the high exterior relative humidity regions such as Newfoundland or the interior of British Columbia [*7*]. This study was initiated in March 1982 in St. John's, Newfoundland, to determine the effects of low permeance sheathings in a region of this type. A single dwelling was chosen to examine the effects of three different sheathing combinations (see Table 6). The single dwelling approach was used to eliminate uncontrolled variables such as interior relative humidity and weather exposure.

The house chosen showed typical signs of interior moisture problems caused by high internal relative humidity. Condensation was found on the windows and mildew on the interior drywall. The house had an original vapor barrier installed which could not be classified as an air/vapor barrier. The wind-driven rain side of a house in St. John's faces northeast. It is considered to be the side exposed to the highest moisture stress. The moisture content in the studs was monitored, as well as the wall relative humidity and the wall temperature of the north face of the house, which was exposed to this high moisture stress. The north face was divided into three sections: chipboard sheathing over studs, extruded polystyrene sheathing over studs, and an extruded polystyrene retrofit over chipboard and studs. Delmhorst moisture elements were used to measure the moisture concentration in the studs. Other Delmhorst moisture elements were used to measure the relative humidity within the wall, and thermocouples were used to measure the temperature within the wall of each of the three sections. The measurements were taken from the middle stud section of each of the three divisions defined here (see Fig. 1 for details). The heating degree-days for St. John's are 4804 Celsius [*11*].

Results

Winnipeg

The Winnipeg study showed that the moisture concentrations in the studs with chipboard, extruded polystyrene, and extruded polystyrene over plywood cycle with the seasons (see Fig. 2). The highest moisture concentrations were found as winter progressed, and the lowest moisture

TABLE 6—*St. John's construction details.*[a]

Construction Detail	Wall Section No. 1	2	3
Style	two story	two story	two story
Heating	electric baseboard	electric baseboard	electric baseboard
Wall construction[b]	12-mm gypsum polyethylene RSI 2.1 batts (R12) 12-mm chipboard exterior siding	12-mm gypsum polyethylene RSI 2.1 batts (R12) RSI 1.3 (R7.5) extruded polystyrene exterior siding	12-mm gypsum polyethylene RSI 2.1 batts (R12) RSI 1.3 (R7.5) extruded polystyrene over 12-mm chipboard exterior siding

[a]Note that this study includes only one house.
[b]Extruded polystyrene was installed in 600 by 2400-mm dimensions.

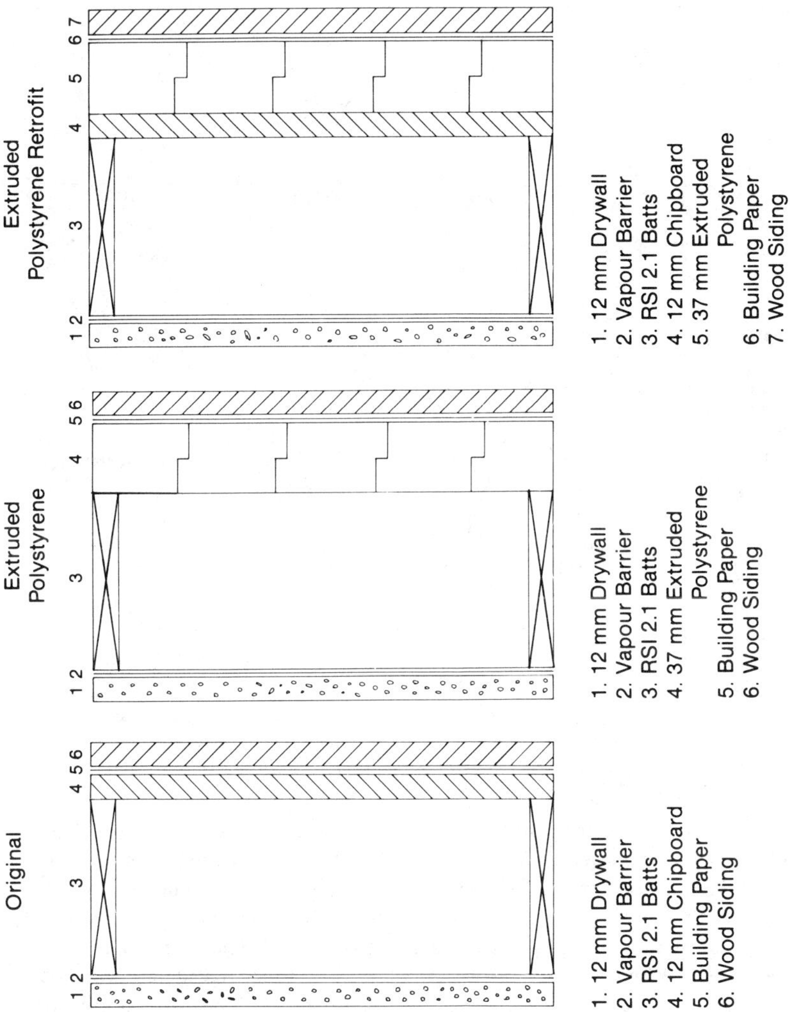

FIG. 1—*Wall construction (St. John's, Newfoundland).*

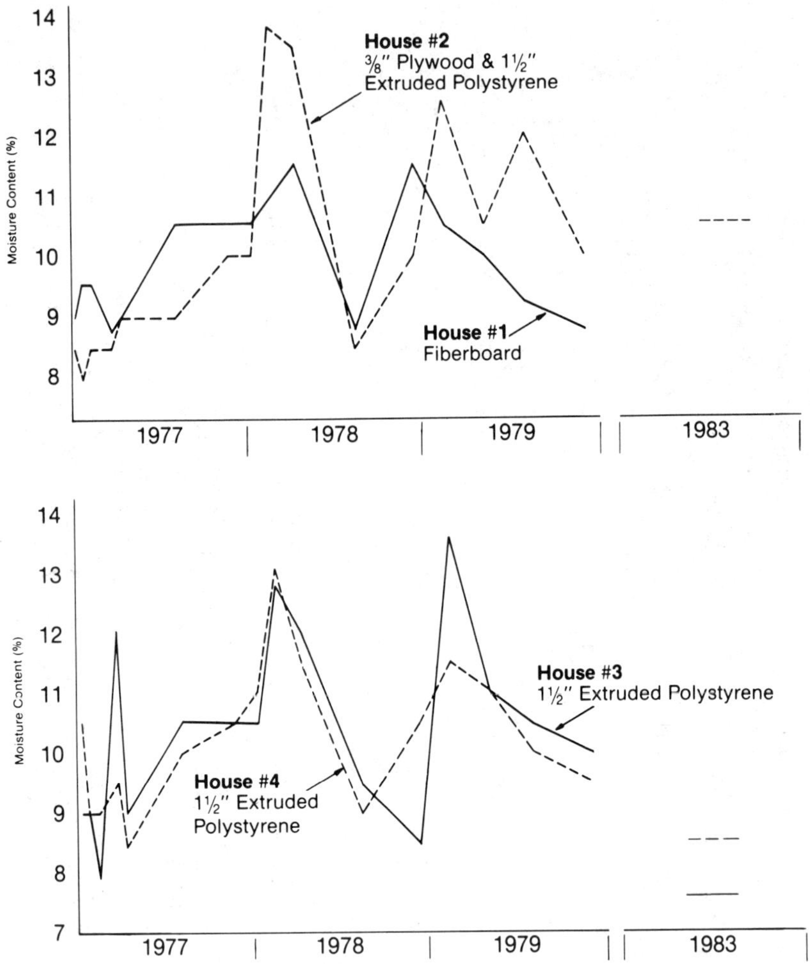

FIG. 2—*Stud moisture content versus time (Winnipeg, Manitoba).*

concentrations were found during the summer. The moisture concentrations in the studs ranged from 8 to 14%, which was well below the 25 to 32% required for fungi growth. The moisture data from the fiberboard, extruded polystyrene, or extruded polystyrene over plywood homes were generally within ±2%. Even the unvented house with electric heat and the very low permeance combination of extruded polystyrene and plywood had results comparable to those of the other test sites in Winnipeg. The results from all four houses showed no signs of an accumulation of moisture in the stud after six years. The weight percent moisture concentrations at the four test sites started at summer levels of 10, 9, 11, and 10 and ended at summer levels of 9, 11, 8, and 9, respectively. The summer moisture levels for House No. 2 increased from 1977 to 1979; however, a check in 1983 showed no increase in summer moisture from 1979 to 1983.

London

The London study was of a home with an indoor pool. Indoor pools ensure high interior relative humidities. The actual interior relative humidities were not measured. High interior relative humidities produced a higher than normal moisture stress on the wall design. The moisture concentrations in the studs ranged from 10 to 22%, which is still below the levels required for fungi growth. The results also show that the wall was able to recover from the high initial concentrations (see Table 7).

Regina

The Regina study showed that an energy-efficient house design with an air/vapor barrier installed on the warm side had moisture concentrations in the stud well below the levels required for fungi growth. The moisture concentrations in the stud ranged from 7 to 9%. These moisture concentrations were very low and did not show any signs of cycling with the seasons (see Fig. 3).

Newfoundland

The results of the Newfoundland study showed that the weather variations and the high relative humidity of this coastal region definitely affected the performance of the wall design, as can be seen in Fig. 4. The moisture concentrations of the three wall sections examined were higher than those from drier regions. The data also showed that the moisture concentrations were independent of the wall design selected.

TABLE 7—*Stud moisture content versus time, (London, Ontario).*

	Northeast		Northwest	
Date	Top Plate, %	Bottom Plate, %	Top Plate, %	Bottom Plate, %
10/78	16	18	17	19
11/78	17	19	16	22
12/78	18	15	17	20
6/79	17	11	15	13
12/79	12	10	17	16
6/82	10	11	7	10
7/83	10	10	17	17

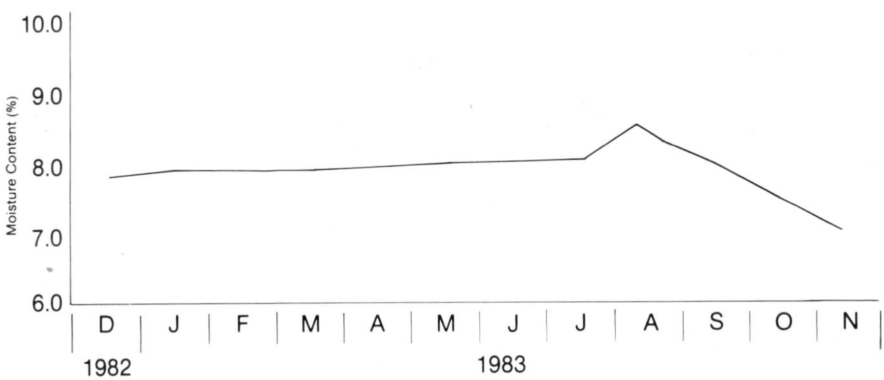

FIG. 3—*Stud moisture content versus time (Regina, Saskatchewan).*

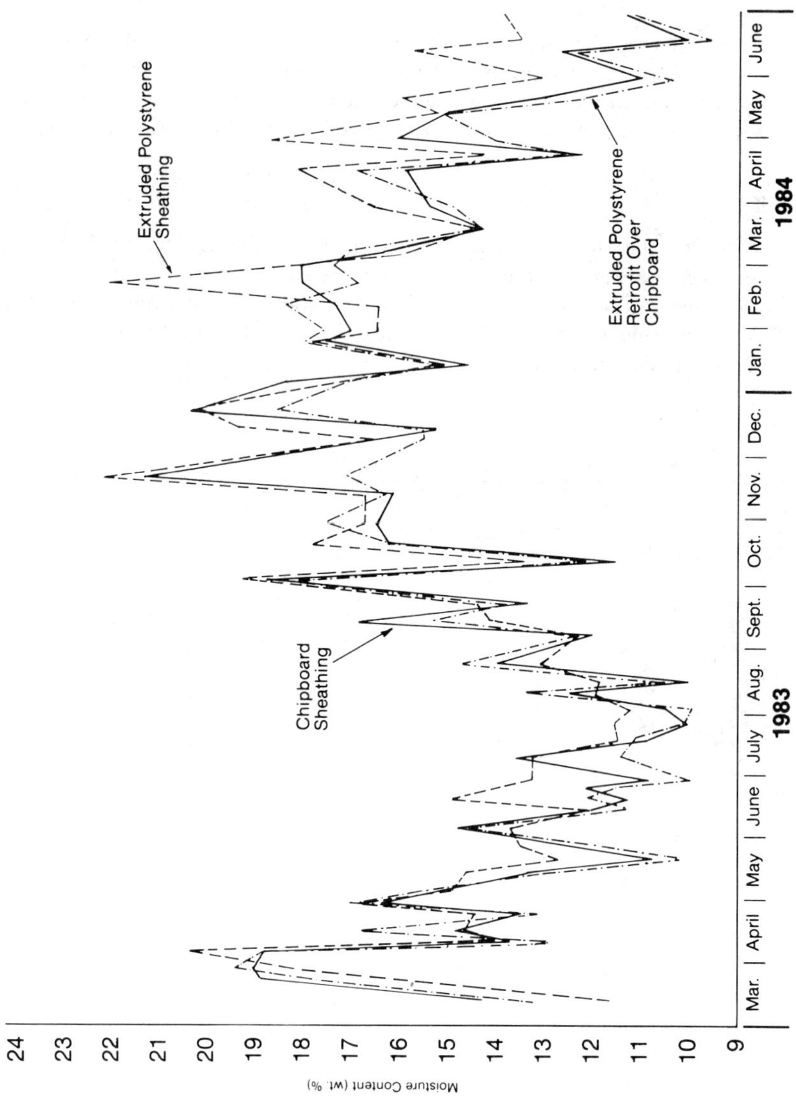

FIG. 4—Stud moisture content versus time (St. John's, Newfoundland).

A comparison of the data in Figs. 4 and 5 shows that the relative humidity within the wall varied daily and this, in turn, controlled the moisture concentration of the studs. The moisture concentration of the section using low permeance foam sheathing was equivalent to that of the section using higher permeance chipboard sheathing.

In Fig. 6, the temperature at the batt/sheathing interface exhibited a 3 to 6-degree-Celsius increase when low permeance foam sheathing was used in place of chipboard sheathing.

Discussion

The practice of installing low permeance foam sheathings on the exterior of a building has caused some technical people to hypothesize that moisture could accumulate between the polyethylene vapor barrier and the exterior sheathing. These problems have not materialized, and the hypothesis has remained unsupported by research or practical data. It is estimated that approximately 750 000 homes in North America have been sheathed with low permeance foam sheathings, and there has not been one documented case of moisture damage caused by low permeance foam sheathing. This practical evidence has also been supported by several research reports [*1–5*].

A point of interest found by Sherwood and substantiated in this study is that homes with an air/vapor barrier, such as the Regina home studied, control moisture cycling better than homes with only a vapor barrier. Air/vapor barriers such as the one installed in the Regina house are continuous and require special treatment at the sill plate, around the windows and doors, around electric outlets, and at the polyethylene joints. Some other designs allow for a distinction between the air barrier and the vapor barrier.

Regardless of the design and materials chosen, the importance of installing a proper air barrier to control moisture cannot be overemphasized. However, the results of this study and others [*1–5*] clearly show that normal building practices will perform satisfactorily when low water vapor permeance foam sheathings are installed externally.

In homes without an air barrier, a moisture cycling pattern can be found. It is best shown in the Winnipeg homes. The moisture content in the studs is higher in the winter and lower in the summer. The winter peaks are caused by the condensation of the moisture contained in the warm interior air, as it exfiltrates outward through the wall on the leeward side of the house.

As spring turns to summer, the warm, dry air from the exterior or the interior travels through the wall by infiltration or exfiltration, thereby drying the wood studs as the moisture content in the air reaches equilibrium with the moisture content in the wood studs.

An explanation for the excellent moisture control performance of extruded polystyrene is demonstrated in Timusk and Lischkoff [*13*]. They found that exterior insulating sheathings performed significantly better than noninsulating sheathings, when the amount of condensed water was controlled. As shown in Fig. 6, the batt/sheathing interface temperature for the walls sheathed with extruded polystyrene is 3 to 6 degrees Celsius warmer than the wall sheathed with chipboard sheathing. These warmer batt/sheathing interface temperatures can increase the capacity of air to contain moisture by over 100%, depending on the interface temperature, as shown in Fig. 7.

Conclusions

1. Extruded polystyrene used as external sheathing does not cause moisture accumulation in the wood studs.

2. Extruded polystyrene used as external sheathing will warm the batt/sheathing interface, thereby controlling the amount of condensation.

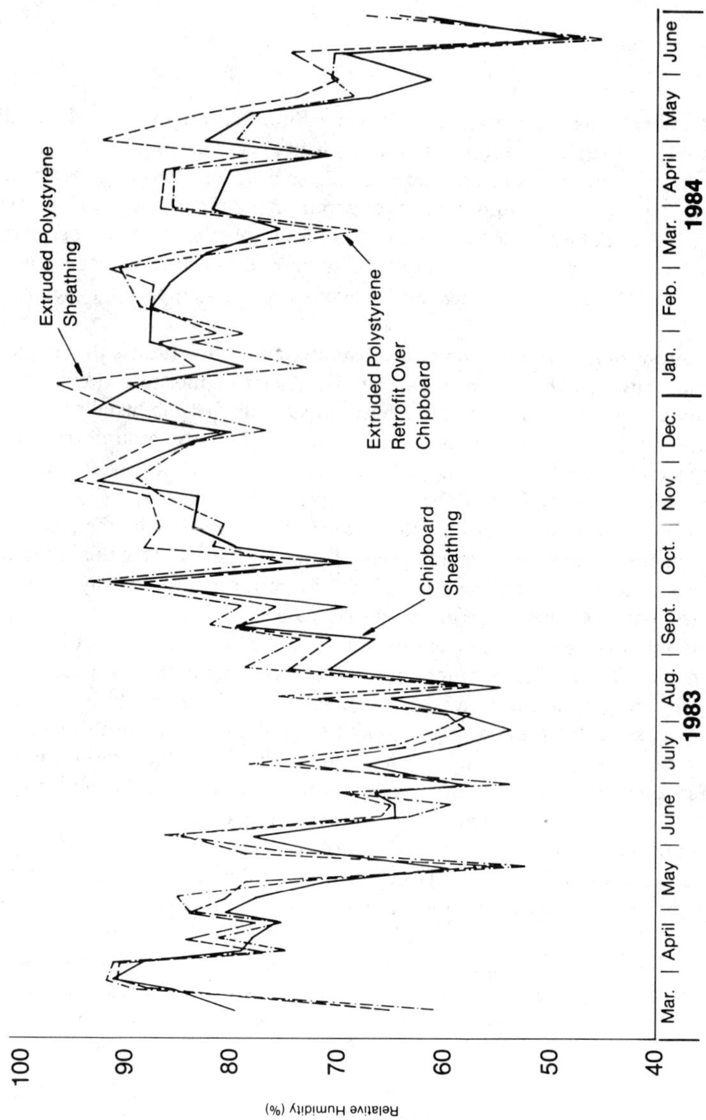

FIG. 5—*Wall relative humidity versus time (St. John's, Newfoundland).*

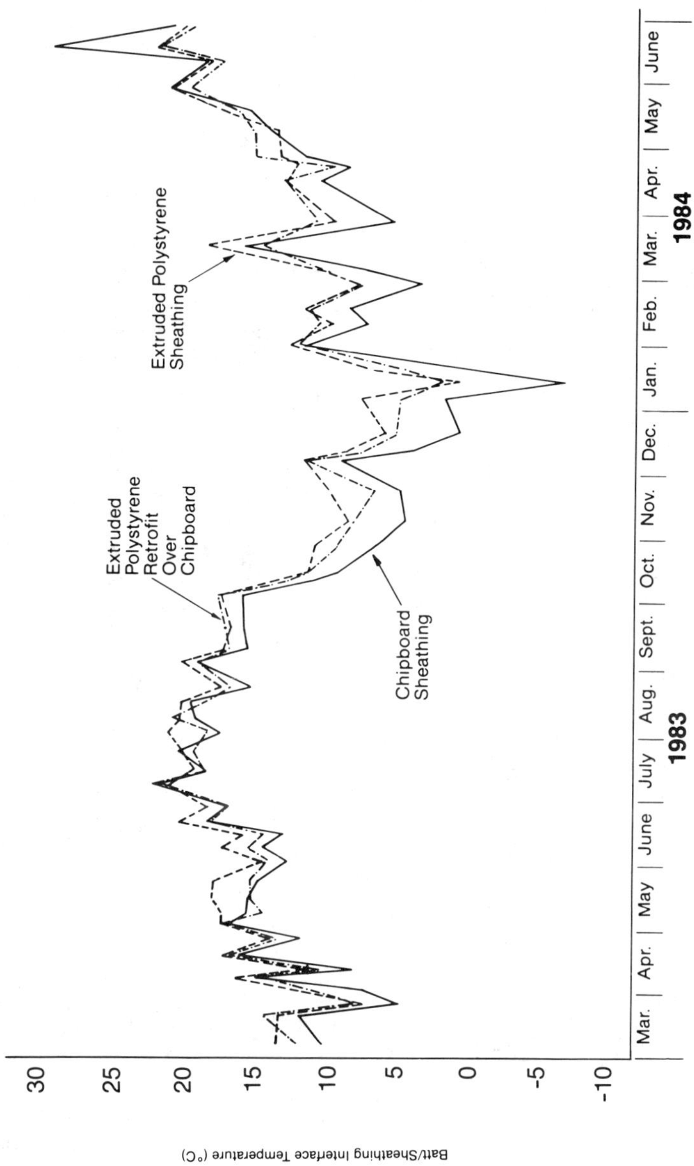

FIG. 6—*Batt/sheathing interface temperature versus time (St. John's, Newfoundland).*

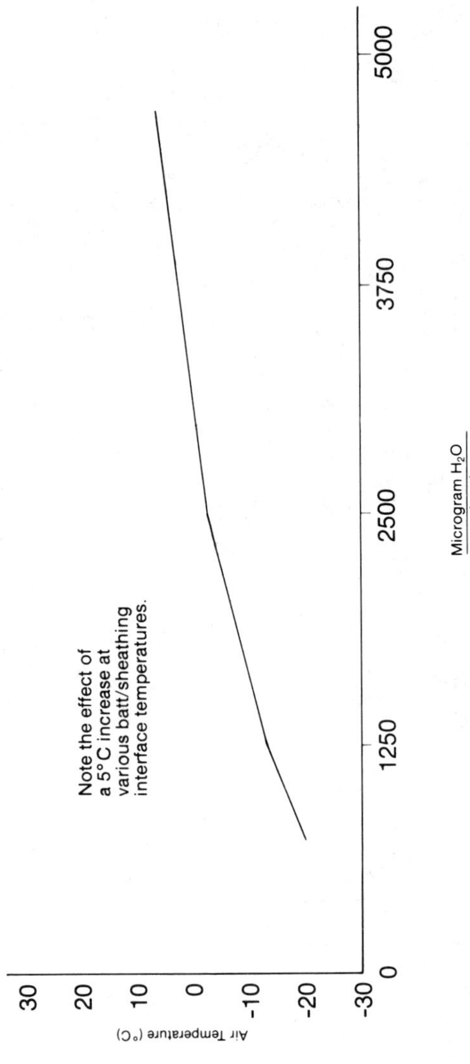

FIG. 7—*Temperature versus moisture capacity at 100% relative humidity.*

References

[*1*] Sherwood, G. E, "Condensation Potential in High Thermal Performance Walls—Cold Winter Climates," Forest Products Laboratory of the U.S. Department of Agriculture, Madison, WI, May 1983.

[*2*] Luft, F. J., "A Study of Moisture Migration and Accumulation in Residential Stud Walls," paper presented at the American Society of Heating, Refrigerating, and Air Conditioning Engineers/Department of Energy Conference, Owens-Corning Fiberglas, Toledo, OH, December 1982.

[*3*] "A Field Investigation to Measure Moisture Content of Building Materials Used in Conjunction with Extruded Polystyrene Exterior Sheathing," National Association of Home Builders, Research Foundation Inc., Rockville, MD, March 1978.

[*4*] Burch, D. M., Treado, S. J., and Contreras, A. G., "The Use of Low-Permeability Insulation as an Exterior Retrofit System—A Condensation Study," *A.S.H.R.A.E. Transactions,* Vol. 85, Part 2, 1979.

[*5*] Wang, F. S., "Comparative Studies of Vapor Condensation Potentials in Wood Framed Walls," paper presented at the American Society of Heating, Refrigerating, and Air Conditioning Engineers/Department of Energy Conference in Orlando, FL, December 1979.

[*6*] "Canadian Wood Construction—Structure and Properties," CWC SP-1, Canadian Wood Council, Ottawa, Ontario, Canada, 1978.

[*7*] "Moisture Induced Problems in NHA Housing," Canada Mortgage and Housing Corp., Toronto, Ontario, Canada, June 1983.

[*8*] "Humidity, Condensation and Ventilation in Houses," National Research Council of Canada, Building Science Insight, Ottawa, Ontario, Canada, 1983.

[*9*] Cech, M. Y. and Pfaff, F., "Moisture Content Correction Tables for Resistance Type Moisture Meters," Forestry Technical Report No. 7, Canadian Forestry Service of the Department of the Environment, Ottawa, Ontario, Canada, 1975.

[*10*] "Owner's Manual—Moisture Detectors for Wood," Delmhorst Instrument Co., Towaco, NJ, 1978.

[*11*] "Climatic Information for Building Design in Canada," Supplement No. 1 to the National Building Code of Canada, 30-year averages, 1977.

[*12*] Eyre, D., "A Dissertation of Residential Energy Conservation and Moisture Damage," Saskatchewan Research Council, Saskatoon, Saskatchewan, Canada, June 1984.

[*13*] Timusk, J. and Lischkoff, J. K., "Moisture and Thermal Aspects of Insulated Sheathings," *Proceedings,* Second CSCE Conference on Building Science and Technology, Canadian Society for Civil Engineering, Waterloo, Ontario, Canada, November 1983.

Peter Cleary[1] *and Robert Sonderegger*[2]

A Method to Predict the Hour-by-Hour Humidity Ratio of Attic Air

REFERENCE: Cleary, P. and Sonderegger, R., **"A Method to Predict the Hour-by-Hour Humidity Ratio of Attic Air,"** *Thermal Insulation: Materials and Systems, ASTM STP 922,* F. J. Powell and S. L. Matthews, Eds., American Society for Testing and Materials, Philadelphia, 1987, pp. 630–638.

ABSTRACT: Long-term measurements of attic wood moisture content have shown that at least 50 kg of water may be stored in roof sheathing members over the course of the winter, to be released in the spring. Overlaid on this seasonal cycle is a daily variation driven by temperature changes in the attic. A dynamic model is required to predict the resulting attic air humidity ratio. Hourly moisture flows in a typical attic—from ventilation and from the wood—are examined, and a simple first-principles dynamic mathematical model to relate them is developed. The model predicts the hour-by-hour attic air humidity ratio. The results are compared with measured data. The implications for attic ventilation rates at different levels of attic insulation are discussed.

KEY WORDS: attic air, humidity, moisture, modeling, thermal insulation

An attic is usually naturally ventilated, its temperature and relative humidity being determined both by the weather and by air, heat, and moisture transport from the living space. The aim of ventilation is to prevent condensation and limit moisture accumulation by surface adsorption in winter and to reduce cooling loads in summer. Moisture enters the attic from the house (by both convection and diffusion), and mixes with the usually drier ventilation air. During the day, the attic is heated by solar radiation. At night, the roof sheathing cools down by radiating heat to the sky; its temperature can drop below the outside air temperature.

The sheathing was thought to be an inert surface. Thus, if its temperature were to drop below the attic air dew point temperature, condensation would form. It was believed that good ventilation could prevent condensation, since more ventilation would result in a generally drier air mixture in the attic.

Recent studies have found that the wood in an attic is not inert but that it adsorbs and desorbs moisture continually. Ford [*1*], Harrje et al. [*2*], and Cleary [*3*] investigated houses in New Jersey and California and showed year-long cycles in which wood dried in the summer and adsorbed moisture in the winter. This cycle is illustrated in Fig. 1 for a single-family unoccupied house in Oroville, California [*3*]. Approximately 60 kg of water was adsorbed during the winter. To investigate the driving forces behind this cycle, the attic was monitored over the six-month period from December 1983 to May 1984.

Experimental Procedure

Oroville is located in the northeast Sacramento Valley, approximately 120 km northwest of Sacramento itself. The winter is mild. Chico, about 30 miles away in the same climate zone, has a 30-year average minimum temperature in January of 2.2°C [*4*].

[1]Staff scientist, Lawrence Berkeley Laboratory, University of California, Berkeley, CA 94720.
[2]President, Morgan Systems Inc., Berkeley, CA 94707.

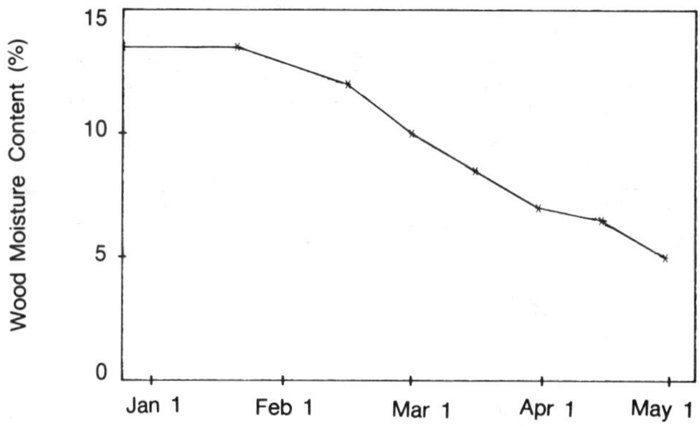

FIG. 1—*Measured moisture content of the west roof sheathing at the Oroville test house, January to May 1984.*

The house was a single story, 7.9 by 7.9 m, with a gable roof sloped at a 33.7° angle to the horizontal. Ventilation was by approximately 0.3 m^2 of soffit vents. There was no ceiling vapor retarder, and the attic floor was insulated with 0.15-m-thick fiberglass batts. Parameters measured continuously at the site included outside dew point, wind speed and direction, attic sheathing temperature at four points, wood electrical resistance at three points, and attic air dew point. Periodic measurements of the attic ventilation rate were made by sulfur hexafluoride (SF_6) injection and decay.

The concentration of moisture in the attic and outside air was measured by means of aspirated chilled-mirror hygrometers. Roof sheathing temperatures were measured inside the attic with calibrated solid-state sensors, two terminal-integrated circuits that produced a current of 1 μA per degree kelvin. The sensors were epoxied to copper disks, and the copper disks were nailed on the undersurface of the sheathing. Wood moisture content (kilograms of water per kilogram of dry wood) was derived from measurements of the electrical resistance between pairs of electrodes inserted 10 mm into the wood. Details of the method may be found in Ref *3*. It was not possible to remove wood samples from the attic to check the results, but research by Burch and coworkers at the National Bureau of Standards [*5*] has given an accuracy of ±1.4% for wood resistance measurements.

There were no sources of moisture in the house except that caused by the periodic visits of researchers (for example, showers, washing), approximately once every two weeks. Air flow between the house and the attic was small; on the one occasion when it was measured by SF_6 injection into the attic, a flow of 4 m^3/h from the attic to the house was found.

Results

The ventilation rate was not measured continuously. A number of measurements were made at different wind speeds and an average value of an empirical area, V, was found, where V is defined by $R = V \times S$. R is the ventilation rate, in cubic metres per second, S is the wind speed, in metres per second, and V is an empirical constant, in square metres. V was found to be 0.023 m^2; the physical area is approximately 0.3 m^2. Temperature differences between the house, the attic, and outside were not found to have a significant effect on the ventilation rate.

Figure 2 shows the roof sheathing temperature for three days in January 1984. The outdoor and attic air humidity ratios for the same period are shown in Fig. 3. There is a distinct 24-h periodicity to the data. The same periodicity can be seen in Ford's research [*6*], for a New Jersey

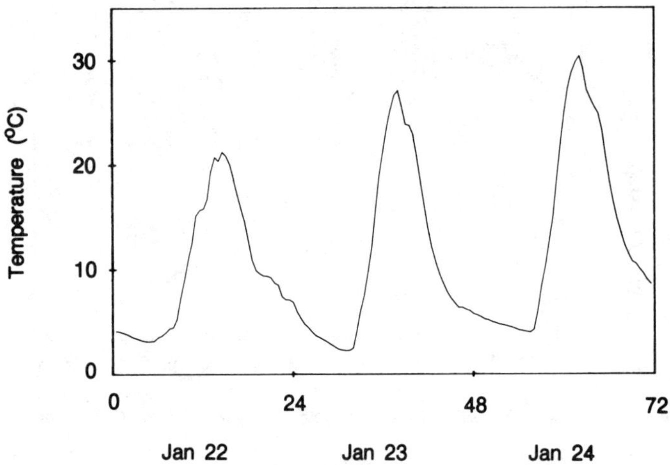

FIG. 2—*Measured temperatures at the underside of the roof sheathing of the Oroville test house, 22–24 Jan. 1984. This is an average value from four sensors, two on each slope. The horizontal axis indicates the number of hours.*

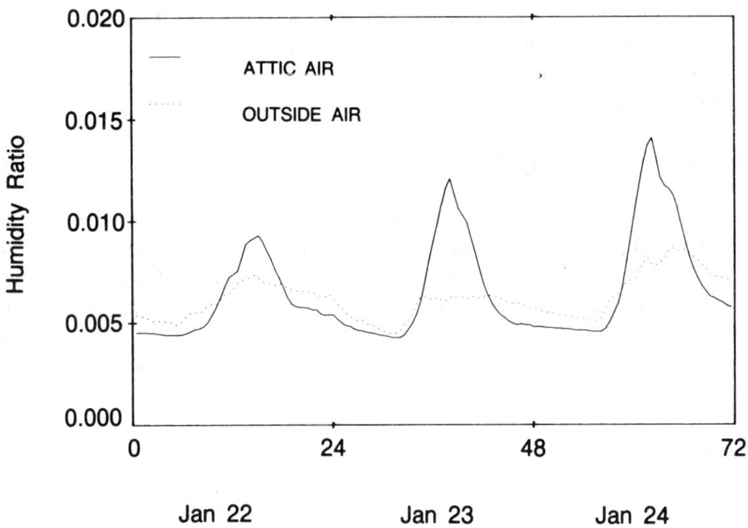

FIG. 3—*Measured humidity ratios for attic air and outside air at the Oroville test house, 22–24 Jan. 1984. The horizontal axis indicates the number of hours.*

attic in September 1980, and a similar pattern was observed by Burch and coworkers in an environmental chamber study [5].

It was postulated that the source of the peaks in the humidity ratio was moisture desorbed from the roof sheathing, caused by warming from solar radiation. The wood in the attic weighs approximately 1100 kg and in mid-January contained an estimated 124 kg of water. Thus, a fractional variation in the percentage of moisture content in the wood can change the air humidity ratio by more than 100%.

This dynamic flow of moisture contrasts with the classic picture, in which the attic humidity ratio is equal to that of the outside air unless condensation occurs on some surface inside the attic. The amount of water desorbed from the wood can be calculated by a mass balance for water entering and leaving the attic. (In an occupied house, a third term would have to be added for transport into and out of the living space.) Assuming that the wood is the sole source of moisture and that the attic air is perfectly mixed, the mass balance gives

$$m = M(W_{\text{attic}} - W_{\text{outside}}) \tag{1}$$

where

m = rate of water flow from the wood, in kilograms per second,
M = (dry) mass flow rate of ventilation air, in kilograms per second,
W_{attic} = attic air humidity ratio, dimensionless, and
W_{outside} = the outside air humidity ratio, dimensionless.

The calculated water flow rate for this period is shown in Fig. 4. It can be seen that the flow peaks just after noon each day and that, during the night, the attic actually absorbs water from the ventilation air. The magnitude of the flows is rather large, up to 2 kg/h. This may be compared with a typical moisture generation rate for an occupied house of 0.4 kg/h [7].

Theoretical Model

A simple model has been developed to predict the flow of water into the wood. For a more complete analysis of moisture and heat flow, see Ref *8*. Following standard models [*9*], the flow of water from the wood is given by

$$m = kA(W_s - W_{\text{attic}}) \tag{2}$$

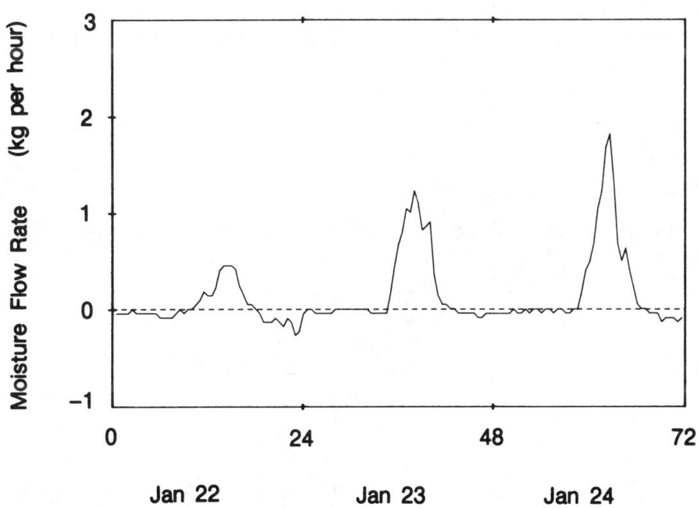

FIG. 4—*Moisture flow for the attic of the Oroville test house, 22–24 Jan. 1984. The values are calculated from a mass balance for water in the ventilation air. The horizontal axis indicates the number of hours.*

where

m = flow of water, in kilograms per second,
k = transfer coefficient, in kilograms per square metre per second,
A = transfer surface area, in square metres,
W_s = humidity ratio of the air surface film, dimensionless, and
W_{attic} = humidity ratio of the attic air, dimensionless.

We assume that the air film above the wood is in equilibrium with the surface of the moist wood and that the wood moisture content is uniform. This should hold for slow drying or wetting. Assuming a Lewis relationship between heat and mass transfer of 1.0 [*10*], the relation between transfer coefficient is

$$h_c = kC_p \tag{3}$$

where

h_c = convective heat transfer coefficient, in watts per square metre degree Celsius, and
C_p = specific heat of moist air, in joules per kilogram degree Celsius.

A constant value of 0.0014 $kg/m^2 \cdot s$ for k was used in this study. The surface humidity ratio may be found from data on wood properties. (For example, Table 3-4 of Ref *11* gives the moisture content of wood at various temperatures and relative humidities. It is said to apply to any species of wood.) If it is assumed that the wood is homogeneous, and that the surface film humidity ratio is a function solely of temperature and wood moisture content, these data can be used to find W_s. The data from Table 3-4 of Ref *11* were transformed into humidity ratios for various combinations of temperature and wood moisture content, and an empirical correlation was made to the data. A good fit was found of the form

$$W_{\text{surface}} = (b + cu + du^2 + eu^3)e^{T/a} \tag{4}$$

where

T = wood temperature, in degrees Celsius,
u = weight of water in the wood divided by the dry weight of the wood, dimensionless,
a = 15.8°C,
b = −0.0015,
c = 0.053,
d = −0.184, and
e = 0.233.

The term for water flow may be eliminated from Eqs 1 and 2, giving an equation for the attic air humidity ratio of

$$W_{\text{attic}} = \frac{\dfrac{AkW_s}{M} + W_{\text{outside}}}{1 + \dfrac{Ak}{M}} \tag{5}$$

This equation predicts attic humidity ratio as a function of wood area (sheathing area and truss area), ventilation rate, outside humidity ratio, and wood surface film humidity ratio (itself a function of wood moisture content and temperature). The unitless parameter, Ak/M, gives a measure of whether the attic is dominated by the wood surface humidity ratio (Ak/M far greater than 1) or by the outside air humidity ratio (Ak/M far less than 1). For the three-day period examined here, wind speeds were low, and Ak/M was approximately 10; thus, the attic was surface dominated.

The wood in the roof consists of two main components, the roof sheathing and the roofing trusses. It was assumed that all the roof sheathing was at one temperature and moisture content and all the roof trusses were at another temperature and moisture content. The predicted and measured attic air humidity ratio are compared in Fig. 5. The measured wood moisture content was almost constant for the three-day period. Daily values were 12.9, 12.8, and 12.8% for the west sheathing, 9.6, 9.4, and 9.4% for the truss. Fixed values of 12.8 and 9.4% were used for the prediction. Good agreement is seen between measured and predicted attic air humidity ratio, although the prediction is always lower than the measurement at night.

Sensitivity Studies

To see how errors in attic ventilation rate and wood moisture content might affect the results, two sets of sensitivity runs were made.

The effect of errors in the wind speed/ventilation rate correlation was investigated by varying the wind speed by a factor of two above and below the measured value—that is, varying the ventilation rate by a factor of four. The results are shown in Fig. 6. It can be seen that the effect on the prediction is minor. As discussed earlier, the attic air humidity ratio was dominated by the wood surface humidity ratio not by the ventilation rate.

The wood moisture content was varied by 2% above and below the measured values of 12.8% for the sheathing and 9.4% for the trusses. The results are shown in Fig. 7. It can be seen that this variation in wood moisture content causes approximately a 25% change in the predicted attic air humidity ratio.

Discussion

Guidelines for attic ventilation are in a state of flux. The 1981 *ASHRAE Handbook of Fundamentals*, published by the American Society of Heating, Refrigerating and Air Condi-

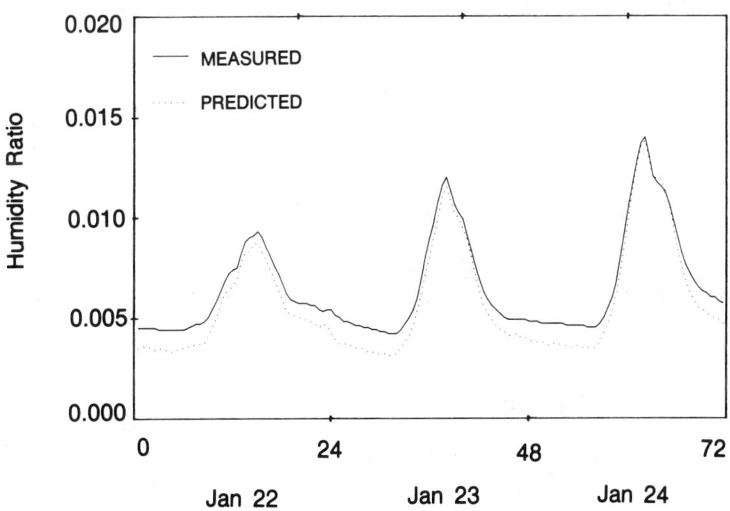

FIG. 5—*Measured and predicted attic air humidity ratios for the Oroville test house, 22–24 Jan. 1984. The horizontal axis indicates the number of hours.*

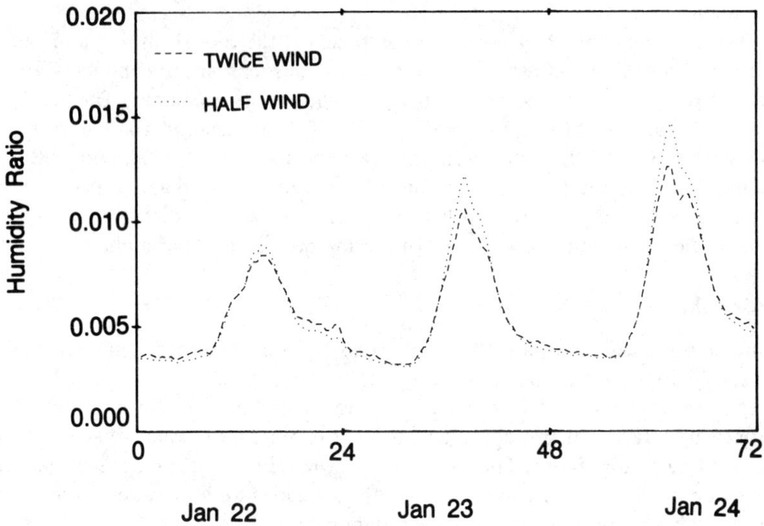

FIG. 6—*Predicted attic air humidity ratios using twice and half the measured wind speed for the Oroville house, 22–24 Jan. 1984. The horizontal axis indicates the number of hours.*

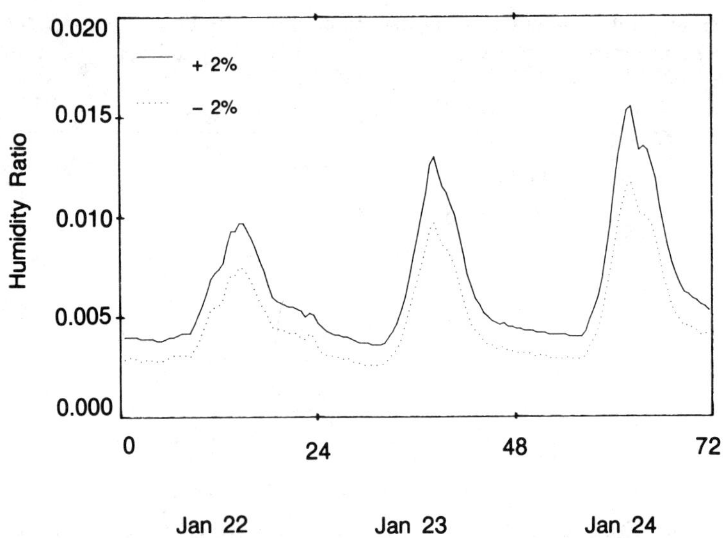

FIG. 7—*Predicted attic air humidity ratios using wood moisture content values 2% greater and 2% less than the measured values for the Oroville house, 22–24 Jan. 1984. The horizontal axis indicates the number of hours.*

tioning Engineers (ASHRAE) [*12*], gives only what it calls "past practice" and notes that it may not apply to new construction.

In the classic picture of an attic, the roof sheathing is inert and does not store moisture. The measurements described previously indicate that there is, in fact, a constant flow of moisture into and out of the wood in an attic. In the attic of the house that was monitored, flow rates of as high as 2 kg/h were measured. Current attic ventilation guidelines do not allow for the possibility of moisture storage and release and may either underestimate or overestimate the amount of ventilation needed for moisture control.

By chance, the old guidelines appear to work well for attics with low levels of insulation. Before this model can be used to develop guidelines for well-insulated attics, much work must be done; little is known about moisture flow from house to attic, and attic ventilation rates are not well characterized. However, the model could be used for sensitivity studies, given a range of moisture flows and attic ventilation rates.

A more general problem is that the purpose of winter attic ventilation is now unclear. It was thought that ventilation removed the moisture that leaked into the attic from the house. In the Oroville case, the house was not occupied and all the moisture that was stored in the wood of the attic came from the ventilation air. Thus, more ventilation may not help.

The old rationale of preventing the sheathing temperature reaching the dew point is inadequate; a new rationale for ventilation is required. Two obvious candidates are (1) prevention of wood moisture content levels above some threshold and (2) prevention of conditions known to be conducive to fungal or bacterial growth.

Conclusion

Measurements show that there is considerable flow of water into and out of the roof sheathing and trusses of a residential attic. This is in contrast to the classic picture of attics and agrees well with studies carried out recently by other researchers.

A simple model has been presented which treats the moisture flow in an attic as a function of wood temperature, wood moisture content, outside humidity ratio, and attic ventilation rate. The model is shown to predict the attic humidity ratio well.

Acknowledgments

The authors are grateful for the assistance of their colleagues at the Energy Performance of Buildings Group at the Lawrence Berkeley Laboratory.

This work was supported by the assistant secretary for Conservation and Renewable Energy, Office of Building Energy Research and Development, Building Systems Division of the U.S. Department of Energy under Contract No. DE-AC03-76SF00098.

References

[*1*] Ford, J. K., "Heat Flow and Moisture Dynamics in a Residential Attic," Master of Science in Engineering thesis, Princeton University, Princeton, NJ, 1981.

[*2*] Harrje, D. T., Gibson, R. G., Jacobson, D. I., Dutt, G. S., and Hans, G., "Field Measurements of Seasonal Wood Moisture Variations in Residential Attics," Report No. 173, Center for Energy and Environmental Studies, Princeton University, Princeton, NJ, January 1984.

[*3*] Cleary, P., "Moisture Control by Attic Ventilation—An In-Situ Study," Report No. LBL-18062, Lawrence Berkeley Laboratory, University of California, Berkeley, CA, July 1984.

[*4*] Climatology of the United States No. 81 (By State), "Monthly Normals of Temperature, Precipitation, and Heating and Cooling Degree Days 1951–1980, California," National Oceanic and Atmospheric Administration, Asheville, NC, September 1982.

[*5*] Burch, D. M., Lemay, M. R., Rian, B. J., and Parker, E. J., "Experimental Validation of an Attic Condensation Model," *A.S.H.R.A.E. Transactions,* Vol. 90, Part 2A, 1984, p. 59.

[6] Ford, J. K., "Humidity Measurement and Modeling in a Residential Attic," *Moisture Migration in Buildings, ASTM STP 779,* American Society for Testing and Materials, Philadelphia, 1982, pp. 160–190.

[7] *Environmental Control Principles,* American Society of Heating, Refrigerating and Air Conditioning Engineers, New York, 1977, p. 82.

[8] Kohonen, R. and Maatta, J., "Transient Analysis of the Thermal and Moisture Physical Behaviour of Building Constructions," Report No. 168, Technical Research Center of Finland. Reproduced with permission by U.S. National Technical Information Service as Report No. PB83-227454.

[9] Kays, W. M. and Crawford, M. E., *Convective Heat and Mass Transfer,* McGraw-Hill, New York, 1980.

[10] *ASHRAE Handbook of Fundamentals,* American Society of Heating, Refrigerating and Air Conditioning Engineers, Atlanta, GA, 1981, p. 3.10.

[11] *Wood Handbook,* Forest Products Laboratory, Forest Service, U.S. Department of Agriculture, Government Printing Office, Washington, DC, 1974.

[12] *ASHRAE Handbook of Fundamentals,* American Society of Heating, Refrigerating and Air Conditioning Engineers, Atlanta, GA, 1981, p. 21.15.

David Jacobson,[1] *David T. Harrje,*[1] *and Gautam S. Dutt*[1]

Measurement of Air Leakage Properties of Common Residential Insulating Materials

REFERENCE: Jacobson, D., Harrje, D. T., and Dutt, G. S., "**Measurement of Air Leakage Properties of Common Residential Insulating Materials,**" *Thermal Insulation: Materials and Systems, ASTM STP 922,* F. J. Powell and S. L. Matthews, Eds., American Society for Testing and Materials, Philadelphia, 1987, pp. 639-648.

ABSTRACT: A methodology for the measurement of air leakage properties of common insulating materials is presented. The method involves measuring air flows through a test ceiling section with cracks of various widths and covered by different insulating materials. Corresponding pressure drops are measured, and pressure-leakage curves are fit through the points. Leakage flows through different insulating materials at specific pressure differences are compared. The results indicate that loose-filled cellulose retards flow somewhat better than fiberglass batts and much better than blown-in fiberglass and vermiculite. The study also investigated several combinations of insulation materials in which one type of insulation has been added on top of another, as in a typical ceiling reinsulation retrofit. Improvements in the form of reduced air leakage are also observed in these tests. Our measurements indicate that the choice of insulation material can greatly influence the air leakage of houses and thus affect energy use, ventilation, and indoor air quality.

KEY WORDS: insulation, air leakage, energy conservation, retrofit, residential insulation, thermal insulation

Adding insulation is one of the most common conservation strategies for many homeowners. The insulation is primarily intended to reduce conductive/radiative heat transfer through the building envelope, and its cost-effectiveness as a retrofit measure is computed accordingly. Typically, comparisons of the properties of the candidate insulating materials concentrate on the thermal resistance, or *R* value. This paper examines another possible impact of adding insulation—reduction in air leakage. In this study, the air leakage properties of common insulating materials used in walls and ceilings were measured in a laboratory setting. We examined how air "leaks" or flows through insulation by measuring the flow in a specially constructed apparatus. This apparatus simulated a typical leakage site covered with insulation such as might be experienced on the attic floor/living space ceiling. In addition to the laboratory measurements, a limited number of field studies were conducted, and these demonstrate the measurable reduction in overall house air leakage following the addition of cellulose wall insulation.

Fluid Mechanics of Air Leakage

Air infiltration is caused by pressure differences induced across the envelope of the building. These pressures are created by wind and inside-outside temperature difference (the latter being referred to as "stack effect") and are approximated by the following relationships [*1*]

[1]Research assistant, senior research engineer, and research engineer, respectively, Center for Energy and Environmental Studies, Princeton University, Princeton, NJ 08544.

$$P_w = 1/2 \rho_o v^2 \tag{1}$$

$$P_s = (\rho_o - \rho_i)h = \rho_i \frac{h(T_i - T_o)}{T_o} \tag{2}$$

where

P_w = wind-induced pressure difference, Pa,
P_s = temperature-induced pressure difference, Pa,
ρ_o = outside air density, kg/m^3,
ρ_i = inside air density, kg/m^3,
v = wind speed, m/s,
g = gravitational acceleration, 9.8 m/s^2,
T_i = indoor temperature, K,
T_o = outdoor temperature, K, and
h = distance to neutral level, m—positive if above the neutral level and negative if below.

The flow, Q, through a building envelope is directly related to these induced pressure differences and may be described by an expression of the form [*1*]

$$Q = C(\Delta P)^n \tag{3}$$

where

Q = flow rate, m^3/s,
P = pressure difference Pa, and
n, C = constants.

An expression of this kind has a basis in fluid mechanics. The parameter, n, generally takes on a value of 0.5 for orifice flow, fully developed turbulent flow, or when entrance effects prevail. In this case, the flow is proportional to the square root of the pressure difference.

For laminar flow, n is 1.0, with flow being directly proportional to the pressure difference. Associated with the n values found in these idealized situations are the corresponding values of C, which in some cases have been developed analytically. An example of such an expression is the Hagen-Poiseuille equation for developed laminar flow in pipes [*2*].

Unfortunately the flow paths in a house do not tend to be as geometrically quantifiable as that of simple duct or orifice flow. Flow paths often have many bends and area contractions and expansions. Even if the geometries could be specified easily, it would be impossible to find all the leakage sites. Air flows in buildings are typically characterized by the general forms of Eq 3. The parameter, n, is empirically determined and usually falls between 0.5 and 1.0, often about 0.65.

To solve for the flow exponent, n, and C, we first take the natural log of both sides of Eq 3

$$\ln Q = n \ln \Delta P + \ln C \tag{4}$$

The natural logs of flow and pressure difference are then linearly regressed to determine the slope, n, and the intercept, $\ln C$.

Testing Apparatus to Simulate Air Leakage

A small test box [0.33 m^3 (11.8 ft^3) in volume] was constructed using framing lumber covered with a 0.1-mm (4-mil) polyethylene sheet (Fig. 1). The top of the frame was fitted with a card-

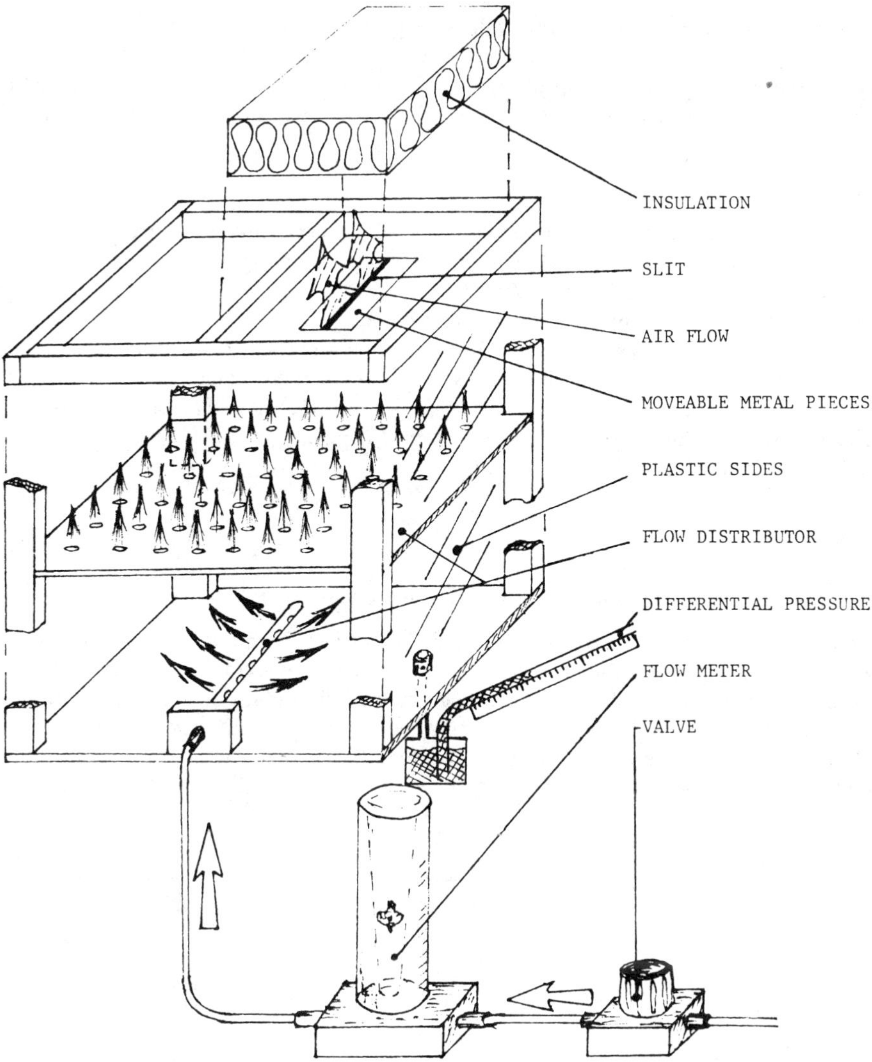

FIG. 1—*Test box for measuring the effect of insulation on crack air leakage, exploded view.*

board sheet, coated with a plastic film to make it airtight. The cardboard has a rectangular hole cut in it and is covered with two movable thin sheet metal pieces taped in place to create a 0.32-m (12.75-in.) long crack of various widths. The top compartments were constructed to simulate two wall or attic floor sections [0.24 m (16 in.) on-center], into which the insulation was placed over the leakage site.

The bottom of the box has two plates, above and below the framing. The top air distribution plate has 1.3-cm (1/2-in.) holes drilled through it (see Fig. 1). Air from a laboratory compressed-air supply flows in through a plastic tube inserted between the two plywood plates, and the holes help distribute the air flow evenly as it makes its way to the slit. The air flow was determined using a rotometer with a full-scale flow rate of approximately 3.28 m^3/h (1.93 ft^3/min). The flow was recorded as percentages of the full-scale value. The measurement of the absolute flow rate is unnecessary because the dimensions of the slit and the test box are arbitrary. The pressure difference between the interior of the box and the outside is measured by an inclined-tube manometer connected to a perforated copper tube inside the frame, as shown in Fig. 1.

Experimental Procedure

The metal pieces were taped in place to provide the uniform width crack used to simulate an actual building shell crack. The widths used in this experiment were chosen to be 3.2 mm (1/8 in.), 1.6 mm (1/16 in.), and 0.8 mm (1/32 in.).

A specified thickness of insulating material was then placed in the compartment above the crack. Flow measurements were taken at 10% increments of the maximum flow. At each flow rate, the pressure drop across the insulation system was recorded. Values for C and n were then calculated as described previously from regressions of pressure-flow characteristics. In our experiment, extremely high correlations were obtained for the regressions of $\ln Q$ against $\ln \Delta p$. The R^2 values exceeded 0.98 in 90% of the cases.

Results

Single Insulation Choices, Representing New Construction

Four types of insulation were tested—89 mm (3.5 in.) of fiberglass batt, fiberglass blowing wool, cellulose, and vermiculite. For each type of insulation and crack thickness, the pressure and flow rate data were analyzed as described previously in Eq 4, yielding values of C and n.

The data may be conveniently plotted as flow versus pressure difference for each slit width and each insulating material (see, for example, Fig. 2). Another way to compare the different materials is to look at the flows induced at a specific pressure difference. A value of 4 Pa has

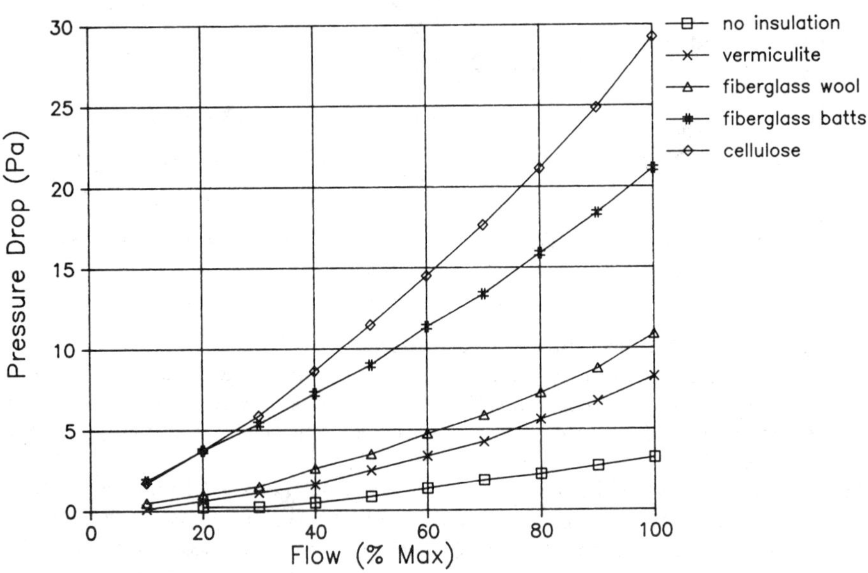

FIG. 2—*Pressure drop versus air flow through various insulating materials covering a 1.6-mm (1/16-in.) slit.*

been considered typical of winter-weather-induced pressure differences in houses [3]. The flows at 4 Pa, computed from C and n values for each configuration, are shown in Table 1.

Both the figures and the data in Table 1 lead to the same conclusions: Cellulose insulation appears to be the most effective in reducing the air leakage, with fiberglass batts a close second. Vermiculite and fiberglass blowing wool were generally comparable and significantly lower in effectiveness than cellulose and fiberglass batts in reducing leakage. However, even these choices are significantly better than the "no insulation" case.

When the flow rates at 4 Pa with different crack widths are compared, we see that without any insulation, or with vermiculite and fiberglass wool, the leakage greatly increases from smaller to larger cracks, as expected. But with fiberglass batts and cellulose, the leakage actually levels off and does not depend on the size of the opening. For instance, there was no difference in leakage with cellulose over the 1.6-mm slit and the 3.2-mm slit. With fiberglass batts, the flow rate does not change monotonically with slit width. The flow rate is comparable for the three slits and actually somewhat higher with the narrowest slit. This suggests that the leakage through fiberglass was perhaps more susceptible to differences in installation. For the narrow slit, cellulose was installed at the two different packing densities. The tightly packed case showed only a slightly lower leakage. These results suggest that beyond a certain crack width, the flow for these two materials is no longer limited by the size of the opening but by the insulation material and its ability to limit flow.

Second Application of Insulation, Representing Retrofit

In addition to single materials, some typical retrofit configurations were tested. Because of their ease of application, blown-in cellulose and blown-in fiberglass are popular materials for retrofit attic insulation. Our retrofits involved adding 150 mm (6 in.) of cellulose over the 89 mm (3.5 in.) of each material tested previously. The retrofit tests were all performed with the 1.6-mm (1/16-in.) crack. One additional retrofit involved spraying a fine coat of water over the 150 mm of cellulose placed on top of the fiberglass batt. The water dries and forms a crusty coating, which is claimed to reduce air flow through the material.

The addition of 150 mm (6 in.) of cellulose over the various insulating materials produced significant reductions in the 4-Pa flow rate. Leakage dropped by 33, 27, 21, and 13% for cellulose blown over fiberglass batts, fiberglass blowing wool, vermiculite, and cellulose, respectively. In general, as would be expected, the poorest flow-retarding materials benefit the most from the addition of the cellulose. The water-sprayed cellulose fiberglass batts yielded an 18% decrease in flow from the unsprayed condition. These results are shown in Table 2. The addition

TABLE 1—*Leakage flow at 4 Pa for three slit widths.*[a]

Insulation Type	Slit Width, mm (in.)		
	0.8 (1/32)	1.6 (1/16)	3.2 (1/8)
No insulation	63.9	109.7	...[b]
Vermiculite	50.6	66.0	104.5
Fiberglass blowing wool	28.7	53.1	93.3
Fiberglass batts	28.3	22.2	26.5
Cellulose (loose packed)	7.1	21.3	21.4
Cellulose (tightly packed)	6.0	...	...

[a]The leakage flow is expressed as a percentage of the full scale.
[b]With this large slit width, the flow rate was beyond our measurement range.

TABLE 2—*Flow for retrofit configurations for a slit 1.6 mm (1/16 in.).*[a]

Base Insulating Material	Flow at 4 Pa	Flow (Relative to Base Insulation Only)
Fiberglass batts	15	0.67
Cellulose	19	0.87
Fiberglass blowing wool	39	0.73
Fiberglass blowing wool + water-sprayed cellulose	32	0.60
Vermiculite	52	0.79

[a]Flows are again expressed as percentages of the full scale. Retrofits consist of adding 150 mm (6 in.) of cellulose above 89 mm (3.5 in.) of various insulating materials.

of 150 mm of the same insulating material would, of course, reduce air flow as well, but air flows in those cases were not measured in our experiment.

Field Measurements

A number of field measurements were taken to determinc thc effect of retrofit insulation on overall air leakage of houses. The fan pressurization method was used to measure the airtightness of houses in terms of a set of pressure versus air flow measurements [see ASTM Test for Determining Air Leakage Rate by Fan Pressurization (E 779-81)]. Equations 3 and 4 were again used to determine the values of the flow parameters, n and C.

The houses were chosen based on the schedule of a local cellulose insulation contractor. The airtightness of the houses were monitored before and after the addition of wall or attic insulation, or both. The contractor interrupted his procedures so that measurements could be made to separate the effects of adding wall and attic insulation.

The results for wall insulation retrofits on three of these houses are summarized in Table 3, and a more detailed tabulation for one of the houses is shown in Table 4. In two of these houses, both of ballon frame construction, air leakage was significantly reduced after the wall insulation retrofit. Including other houses we have tested, the air leakage reduction at 50 Pa (0.2 in. of water) ranges from virtually zero to above 40% and is typically in the 10 to 30% range.

The more detailed data on House S, shown in Table 4, provide some insight into the effect of cellulose wall insulation on wall leakage and why it is likely to be more pronounced in balloon frame houses. Consider the leakage (in Table 4) expressed as air changes per hour (ACH) at 50 Pa—$q(50)$. Before insulation, its value is much larger when the door connecting the living space to the basement is open. This implies that the basement has large leakage sites. In balloon frame houses, the wall framing on the house perimeter creates vertical passages that connect basement and attic. The process of adding cellulose wall insulation blocks these air passages and greatly reduces the basement air leakage. In our test house, $q(50)$ is lowered by 17 ACH with the basement door open versus 11 ACH with the basement door closed. In platform frame houses, vertical passages connecting basement and attic are far less pronounced, so that there is likely to be much less air leakage reduction following retrofit wall insulation.

Wall insulation also reduced living space air leakage through the walls and in the perimeter of the space between floors. Ceiling insulation was added subsequently but had a much smaller effect—0.7 to 0.8 ACH reduction in $q(50)$. The equivalent leakage area (ELA), also indicated

TABLE 3—*Reduction in conductive and air leakage heat loss from a wall insulation retrofit.*

House	50-Pa Leakage, ACH			Wall Framing Type	Net Wall Area, ft^2	Equivalent *UA* Reduction, Btu/h/°F[a]		*UA* Change Ratio, Air Leakage to Conduction
	Before, ACH	After, ACH	Reduction, %			Conduction[b]	Leakage[c]	
C	17.5	11.7	33.2	balloon	1224	224.4	120.1	0.54
G	13.1	12.7	3.6	platform	734	134.6	1.9	0.01
S	35.5	18.8	47.0	balloon	1306	239.4	125.5	0.52

[a] 1 Btu/h/°F = 0.527 W/°C.
[b] *R* values of the wall assembly before and after insulation are assumed to be *R*-4 and *R*-15, respectively.
[c] The average air infiltration rate is assumed to be 1/20th of the average of the pressurization and depressurization leakage flow rates measured at an inside-outside pressure difference of 50 Pa (0.2 in. of water).

TABLE 4—*Leakage data from House S.*[a]

	Basement Door Closed		Basement Door Open	
	$q(50)$, ACH	ELA, cm^2	$q(50)$, ACH	ELA, cm^2
Preretrofit	27.5	1424	35.5	1425
After wall retrofit	16.1	735	18.8	780
After ceiling retrofit	15.3	708	18.1	853

[a]Leakage data are the average of the pressurization and depressurization measurements, which differed by less than 15% in each case:

$q(50)$ = air leakage rate at an interior-exterior pressure difference of 50 Pa, expressed in air changes per hour (ACH), and
ELA = equivalent Leakage Area, calculated from house air leakage at 4 Pa, using the Lawrence Berkeley Laboratory method [*3*].

in Table 4, shows large drops following the wall insulation retrofit, comparable to the reduction in $q(50)$. However, the ELA actually increased following the ceiling insulation retrofit in one case (with the basement door open). ELAs are calculated from extrapolations of the leakage flow data to a pressure difference far below the measured values. The 50-Pa leakage is essentially a direct measurement. Thus, ELA is less well determined than $q(50)$ and occasionally increases after an airtightness retrofit.

Table 3 shows the estimated reduction in wall conductance (UA) values corresponding to the wall insulation retrofit. The R value is assumed to increase from 0.70 to 2.64 m$^2 \cdot$ °C/W (4 to 15 ft$^2 \cdot$ h $\cdot$ °F/Btu). An equivalent UA value for air infiltration can be expressed as the product of the volumetric heat capacity of air, the house volume, and the air infiltration rate (ACH). The equivalent UA value for air leakage reduction is then given by

$$\Delta UA\,(\text{equivalent to } AI) = C_p V \Delta AI$$

where

C_p = the volumetric heat capacity of air,
V = the house volume, and
ΔAI = the reduction in natural air infiltration rate (air change per hour).

The air infiltration rate was not directly measured. There are various models for estimating the natural air infiltration rate from pressurization leakage data [*3,4*]. In the simplest empirical relationship, which generally characterizes the winter average air infiltration rate to within ±25%

$$AI = \frac{q(50)}{20}$$

where $q(50)$ is the air change rate (house volumes per hour) at 50 Pa. Using this relationship, we estimated the equivalent UA reduction from airtightness associated with wall insulation; these are also shown in Table 3. The ratio of equivalent UA reduction from airtightness to the UA reduction from reduced conduction, shown in the last column of Table 3, is an indication of the relative effect of cellulose wall insulation retrofit on these components. In the two houses with significant air leakage reduction, the equivalent UA reduction associated with airtightness was more than half of the estimated drop in UA from reduced conductive heat loss. Thus, the energy savings from reduced air leakage following blown-in cellulose wall insulation can be significant and would affect the cost-effectiveness of wall insulation.

A reduction in air leakage, typically 10 to 30% at 50 Pa, from cellulose wall insulation retrofit also suggests that this insulation retrofit is one of the more effective means of reducing the house's air leakage. In contrast, the air leakage reduction (again at 50 Pa) of houses in the modular retrofit experiment averaged about 8% [5]. If retrofits that reduce air leakage are placed in a special category because of their potential impact on ventilation and indoor air quality, then cellulose wall insulation should be placed high on that list.

Our field experiments did not examine other retrofit wall insulating materials such as rock wool or blown-in fiberglass. Since our laboratory tests showed significant differences in the air leakage characteristics of various insulating materials, it would be inappropriate to draw conclusions about other materials from our measurements.

Conclusions and Discussions

Insulation installed over slits 0.8 to 3.2 mm (1/32 to 1/8) in. wide reduces air flow.

Air flow reduction depends strongly on the choice of insulating material. Blown-in cellulose and tightly installed fiberglass batts are much more effective than vermiculite and fiberglass blowing wool.

Adding 150 mm (6 in.) of cellulose on top of 89 mm (3.5 in.) of various base insulating materials further reduced air flow. The greatest reduction occurred for vermiculite and fiberglass blowing wool, which are by themselves least effective in blocking air flow. Adding 150 mm of another insulating material would presumably reduce air flow as well.

Pressurization data taken before and after wall insulation retrofits show widely varying air leakage reduction from virtually nothing to over 40%.

Heat loss reduction by airtightness associated with wall insulation retrofit can be more than half of the heat loss reduction through reduced conduction, and can therefore influence the cost effectiveness of wall insulation.

Cellulose wall insulation retrofits may have a major potential impact on house ventilation and indoor air quality.

Our field data on the effects of cellulose wall insulation retrofit cannot be generalized to other applications or other insulating materials. Much more data need to be collected for wall, ceiling, and floor insulation retrofits on different styles and vintages of houses, using different insulating materials. Ideally, these measurements should be combined with fuel use measurements, so that projected heat loss reductions can be compared with actual energy savings.

Acknowledgments

The authors wish to thank the Office of Building Energy Research and Development, U.S. Department of Energy, for their support of this research through Contract No. DE-AC02-77CS20062.

The authors also wish to thank Kenneth Gadsby for his contributions in the field testing of the retrofitted houses and Lee Allen and his staff at Energy Savers for allowing us entry into houses they were retrofitting and for modifying their procedures to permit us to make our measurements.

References

[1] "Ventilation and Infiltration," in *ASHRAE Handbook of Fundamentals*, American Society of Heating Refrigeration and Air Conditioning Engineers, Atlanta, GA, 1981, Chapter 22.

[2] Streeter, V. L., *Fluid Mechanics,* McGraw Hill, New York, 1966.
[3] Sherman, M. H. and Grimsrud, D. T., "Infiltration-Pressurization Correlation: Simplified Physical Modeling," *A.S.H.R.A.E. Transactions,* Vol. 86, 1980, pp. 778–807.
[4] Persily, A. K., "Understanding Air Infiltration in Homes," Report 129, Princeton University Center for Energy and Environmental Studies, Princeton, NJ, 1982.
[5] Dutt, G. S., Jacobson, D. I., and Socolow, R. H., "Air Leakage Reduction and Energy Savings in the Modular Retrofit Experiment," Report 113, Princeton University Center for Energy and Environmental Studies, Princeton, NJ, 1983.

Research

Ronald P. Tye,[1] Karl G. Coumou,[1] André O. Desjarlais,[1] and David M. Haines[1]

Historical Development of Large Heat Flow Meter Apparatus for Measurements of Thermal Resistance of Insulations

REFERENCE: Tye, R. P., Coumou, K. G., Desjarlais, A. O., and Haines, D. M., **"Historical Development of Large Heat Flow Meter Apparatus for Measurements of Thermal Resistance of Insulations,"** *Thermal Insulation: Materials and Systems, ASTM STP 922,* F. J. Powell and S. L. Matthews, Eds., American Society for Testing and Materials, Philadelphia, 1987, pp. 651–664.

ABSTRACT: The development and utilization of various types of heat flow meter apparatus, based on the ASTM Test for Steady-State Thermal Transmission Properties by Means of the Heat Flow Meter (C 518-76) are presented and assessed in an historical context.

Among the major factors discussed are the events and rationale that established the ultimate need for the large heat meter apparatus; improvements in heat flux transducer and other component technologies; design requirements and apparatus development; calibration requirements, including the internal development of necessary transfer standards prior to the availability of National Bureau of Standards (NBS) certified samples; and the international applications and impact on test specifications.

Results utilizing the current large heat meter apparatus are presented and discussed to illustrate its effects on the stability and reproducibility of results; its accuracy and precision, as determined from interlaboratory, round-robin, and National Voluntary Laboratory Accreditation Program (NVLAP) studies; the extent of the thickness effect for a number of different low-density insulation materials; and the resolution of differences in value. Future needs and requirements are also addressed, specifically with regard to higher temperatures of operation.

KEY WORDS: calibration, guarded hot plate, heat flow meter, heat flux transducer, product variability, quality assurance, radiation, reference materials, thermal insulation, thickness effect, transfer standards

Until the early 1970s, the insulative property of thermal insulation materials was most commonly evaluated by means of the ASTM Test for Steady-State Thermal Transmission Properties by Means of the Guarded Hot Plate (C 177-76). Furthermore, measurements were usually undertaken on specimens, described as homogeneous, which were of a limited area, up to 300 mm square with thicknesses of 20 to 30 mm. The resulting property was described as a thermal conductivity, and the thermal resistance was assumed to increase linearly with thickness. In a number of cases, the ASTM Test for Steady-State Thermal Transmission Properties by Means of the Heat Flow Meter (C 518-76) was beginning to be used more widely because of its faster and simpler operation and its obvious applicability to quality assurance procedures. However, the specimen sizes were still of the same order as those for the guarded hot plate.

In the early 1970s, the impact of the energy crisis on the use of thermal insulation materials and products was significant. Not only was thermal insulation considered and used more exten-

[1]Manager of market development, vice president/general manager of the Instruments Division, manager of thermal insulation evaluation, and manager of Thermatest instruments, respectively, Dynatech Scientific Inc., Cambridge, MA 02139.

sively, particularly in buildings, but insulation thicknesses increased. This was due to the introduction of voluntary and mandatory conservation and regulatory requirements [*1*] and to expected lifetime cost savings. New and modified products and materials were developed and used in these larger thicknesses. Many, especially the blanket and loose-fill fiber type, were of low density and inhomogeneous in nature, with considerable variability [*2,3*]. In addition, as a result of improved understanding of heat transmission mechanisms in such materials [*4–6*], thermal resistance could no longer be assumed to be linear with thicknesses below 100 to 150 mm. Existing quality assurance procedures undertaken on low-density fibrous materials using specimens of small area, and extrapolation of the results on thin specimens cut from a thick sample were no longer adequate to provide representative thermal resistance values for the products [ASTM Guide for Determination of the Thermal Resistance of Low-Density Blanket-Type Mineral Fiber Insulation (C 653-83)]. The need to undertake measurements on specimens up to 150 to 200 mm thick, with areas of up to 1 m^2, was clearly established.

The existing primary method of measurement for evaluating thermal performance of insulations was the guarded hot plate. This apparatus, as mentioned earlier, was, in general, inadequate because of size limitations. In addition, the apparatus was expensive to build and operating it was time-consuming, and, thus, it was unsuited for quality assurance requirements. The alternative primary method using the guarded hot box [ASTM Test for Steady-State Thermal Performance of Building Assemblies by Means of a Guarded Hot Plate (C 236-80)] was suitable for evaluating large specimens, but was even more unsuitable for quality control because of its higher costs and longer times of operation. Existing transient methods [*7*], though their operations were rapid, were unsuited to evaluating materials in which the predominant mode of heat transmission was not solid conduction. In addition, these methods normally utilize very small volumes and require homogeneous specimens.

The most promising method, therefore, appeared to be the heat flow meter technique, which had already proved to be suitable for rapid measurements on thin specimens. The need for the development of a larger-scale version of this apparatus was clearly identified. The present paper describes the development of such an apparatus. It includes a discussion of the various improvements that needed to be developed to satisfy the changing requirements in the thermal insulation field over the past decade.

Principle of the Heat Flow Meter Method

The principle of the heat flow meter method is well known and need not be discussed in detail. Essentially a test specimen or specimens of a thickness, x, and having a central metering area, A_m, which should not be more than 25% of the total area, A_T, is sandwiched between hot and cold plates, which are maintained at different constant temperatures so that a temperature difference, ΔT, is always maintained. A thin heat flux transducer or transducers of negligible thermal resistance, in comparison with that of the specimen, is placed in series with the test specimen either at the hot or cold plate or between two specimens, all within an insulated enclosure.

At equilibrium conditions, the heat flux transducer output, $Q(V)$, is measured and the apparent thermal conductivity, λ_{app}, is determined from

$$\lambda_{app} = NQx/\Delta T \tag{1}$$

The value of N (in watts per square metre per volt) is an equipment/calibration factor for a particular set of conditions. It is obtained by calibrating the apparatus with a standard reference material or materials or with an appropriate transfer standard or standards of known thermal resistance at the same conditions at which measurements are desired on a test specimen.

It should be pointed out that N will differ from the calibration constant of the heat flux transducer normally supplied by the manufacturer of the transducer. This occurs because such calibration involves levels of heat flux and temperature difference that are usually outside the range of those used for measurements on insulations. The real advantage of the *in situ* calibration is that the test specimen may be considered somewhat self-guarding. Provided that calibration specimens of the same order of thickness or resistance, or both, as the test specimens are available, the heat losses can be considered to be eliminated for each particular condition at which a calibration is undertaken, thereby ensuring that a good level of accuracy is maintained. This factor, when combined with the relative simplicity, the speed of operation, and the requirement of a simple form of uninstrumented specimen, makes the method particularly suited to commercial instrument development. This applies not only for use in the laboratory but also for quality control and assurance purposes in the plant, where less-skilled operators are often utilized.

Instruments Development

Initial Research and Development Instrument

In the early 1960s, the heat flow meter was becoming recognized as a viable method for measuring thermal insulations, both for laboratory uses [*8*] and for quality control purposes [*9*]. This interest culminated in the publication in 1964 of a proposed tentative version of ASTM Test C 518 (C 518-64T).

In concert with these developments, a laboratory instrument, conforming to the specifications of ASTM Test C 518-64T, was introduced to fulfill the needs at that time. These were primarily for an instrument that was basically oriented toward research and development and that was used by skilled operators on relatively small specimens. Additional requirements were for temperature capabilities not restricted to ambient room temperatures and for more rapid operation than the guarded hot plate. The apparatus is shown schematically in Fig. 1. The major features of the design at that time were the following:

(*a*) A_T = 300 by 300 mm; A_M = 150 by 150 mm;
(*b*) x = 51 mm maximum;
(*c*) λ range of up to 0.5 W/m · K;
(*d*) temperature range of −10 to 200°C mean, with ΔT variable;
(*e*) stability of temperatures at ±0.15°C, utilizing thick, high-conductivity plates and a combination of electrical heating and fluid cooling;
(*f*) heat flow up or down;
(*g*) utilization of an integrating area heat flux transducer developed by Pelanne and Bradley [*9*];
(*h*) equilibrium times of 30 to 75 min for specimens up to 51 mm thick and having different thermal diffusivities; and
(*i*) accuracy of ±5% or better and a reproducibility of ±2% or better.

The hand-made heat flux transducer design, based on a regular woven ribbon thermocouple array, was chosen because the approximately 200 sensors were distributed evenly over the measurement area. This was in contrast to the more usual commercial forms, which tend to be less homogeneous, since they consist of strips of high-thermal-conductivity sensors distributed within a low-conductivity matrix. The former design ensures that the measured heat flux is representative of that for the total area and, potentially, is not affected so much by inhomogenities in the specimen. The total thickness of the transducer was only 3 mm. The materials of construction were such that it could be cycled between −20 and 220°C without degradation and

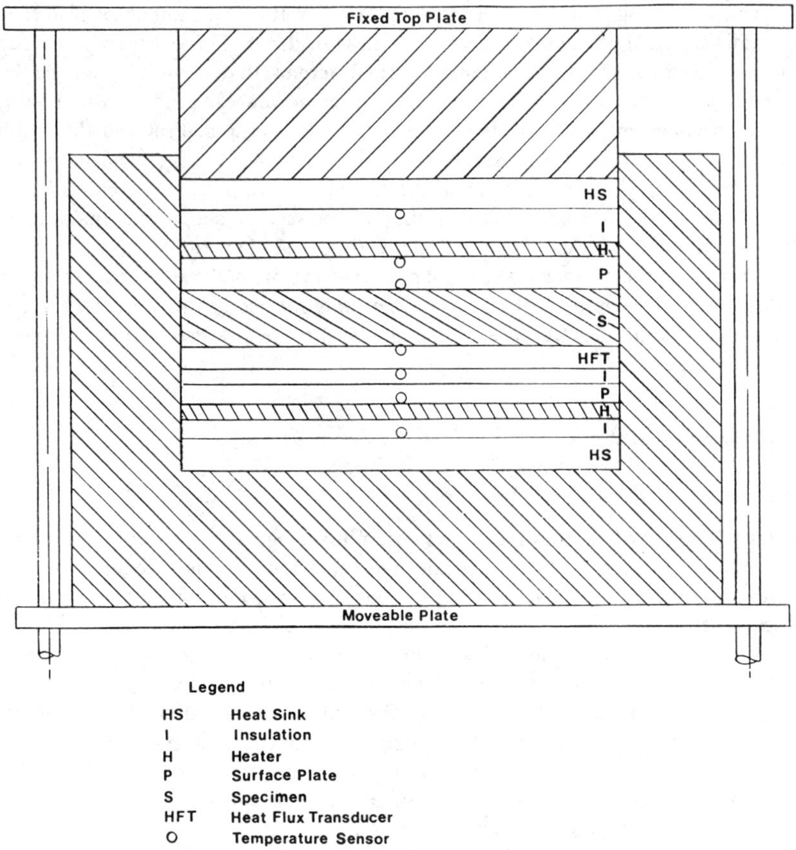

FIG. 1—*Schematic of heat flow meter instrument.*

other changes taking place. Its calibration remained stable within a ±1% range at all temperatures.

Calibration over the whole temperature range presented some problems initially. The prime calibration could be undertaken by means of 25-mm-thick high-density fiberglass transfer standards obtained from the National Bureau of Standards (NBS) for only the approximate range of −5 to 40°C, which were the limits of their apparatus. In order for the instrument to be utilized over the wider range, secondary standards were made available.

These specimens consisted of pieces cut from a lot of resin-bonded high-density fiberglass similar to that used by the NBS. The lot was characterized at Dynatech by undertaking measurements on several pieces over the complete temperature range in a standard guarded hot plate. The performance of this plate had been verified by measuring a pair of the NBS transfer standards over the more limited temperature range for which values were available.

Subsequent to this, in 1966, both the hot plate and one of the heat flow meter instruments were involved in the International Refrigeration Institute round robin test [*10*]. The interesting feature of this study was that careful guarded hot plate measurements, over a period of some 14 days, provided results that were within 1.7% of the NBS/NPL 0 to 40°C baseline data, whereas the heat flow meter results, obtained in less than 3 days, were within 0.5%. It was significant that the slopes of the two lines were identical, thus providing added confidence in the results obtained at other temperatures used in the material characterization for subsequent apparatus calibration.

This instrument, with little modification, has been widely used during the past 20 years for the whole range of thermal insulation including, for example, cellular plastics [*11*], microporous materials [*12*], and low-density concretes [*13*].

Small Quality Control Instrument

By 1970 much research and development had been undertaken within the thermal insulation community on the method and various forms of instrumentation. As a result, the 1964 tentative version of ASTM Test C 518 was revised and upgraded to full test method status (ASTM Test C 518-70). In addition, a number of manufacturers of different types of thermal insulation were using the method for in-plant quality control purposes. In particular, the method was being used extensively within the mineral wool industry. The result of this latter work was the development of ASTM Recommended Practice C 653-70 and the ASTM Recommended Practice for Determination of the Thermal Resistance of Low-Density Fibrous Loose Fill-Type Building Insulation (C 687-71) for deriving the thermal resistance of mineral fiber batts and measuring thermal resistance of fibrous loose-fill insulation, respectively. During the same time period, improvements in the performance of existing electronic components, especially in linearity of response and reduced size, had taken place.

As a result of these factors, a separate modified version of the original instrument was developed primarily for quality control applications. The basic objective was to make the instrument simpler to use, make its operation faster, and have direct digital display of the results.

The instrument and its operation have been described in detail [*14*]. The following were the major modifications in specification:

(*a*) A_T = 300 by 300 mm; A_T = 100 mm diameter;
(*b*) x = 51 mm;
(*c*) λ up to 0.085 W/m · K; and
(*d*) T and ΔT fixed at preselected values in a limited range of −5 to 100°C. These were normally T = 24°C and ΔT = 22 or 27°C, respectively, for most North American applications, and T = 10°C and ΔT = 20°C elsewhere, especially in Europe.

This instrument has been used by all sectors of the insulation industry worldwide since 1970. Various interlaboratory comparisons and round robins tests—for example a recent one on cellular plastics [*15*]—have illustrated that an accuracy of better than ±2% and a reproducibility of better than ±1% can be attained on a regular basis.

However, during the initial development there were indications from potential users that thick specimens would require evaluation in the future. Concern was expressed regarding the effect on the accuracy limits if such large specimens were used with only thin specimens available for calibration. After undertaking an analysis of the system, followed by an experimental verification [*14*], it was found that x could be increased to 100 mm by utilizing a reduced transducer metering area, A_M, of 50 by 50 mm without affecting the overall 5% accuracy limits. It was also apparent that the accuracy at thicknesses in excess of 50 mm could also be improved if thicker or higher-resistance transfer standards were available for initial calibration purposes. Finally, it is necessary for the apparatus to be in an environment maintained at a constant temperature (±1°C or better) to minimize errors due to differences between the test temperature and environment temperature.

Large Quality Control Instrument

The decade of the 1970s was a period of intense activity and progress in the thermal insulation field. The increased use of thermal insulation for energy conservation purposes, and especially the mandating of higher thermal resistance values for building envelope components, stimulated the use of greater thickness of all types and forms of thermal insulations. Many of these were of low density and inhomogeneous. Small specimens with metering areas on the order of 50 to 100 mm square and thicknesses of 50 to 100 mm were often not representative of a bulk material which, in use, might be up to 1 m wide with thicknesses well in excess of 100 mm.

An additional factor was the significant amount of work that had been stimulated by the increased use of these products in trying to understand the heat transfer behavior in low-density thermal insulations. This was particularly true for the radiation mode and the impact of the "thickness effect" in low-density materials [*16*]. This subsequently well-researched topic [*5,6,18,19*] indicated that, although radiation heat transfer between parallel surfaces is independent of distance, the measurement of λ_{app} cannot be undertaken without regard to the specimen thickness, since radiation absorption by the material is thickness dependent.

Thus, calculation of a thermal resistance for a thick specimen cannot be carried out by using results for these specimens with the assumption that the resistance is directly proportional to the thickness. Therefore, measurements have to be carried out on specimens of increasing thickness up to a value at which the variation in the measured value is found to be equal to or less than the uncertainty in the measurement. This is the so-called representative thickness for a material. This differs with the type of material [*20*] and is dependent on optical properties [*21*]; however, from a practical point of view, it appears that, for current products, 100 to 150 mm is thick enough for a measurement to be made with an uncertainty of better than 2%.

Each of these issues had an impact on the test method documents, so that the 1976 versions of both ASTM Test C 177 (guarded hot plate) and ASTM Test C 518 (heat flow meter standards) were revised significantly to reflect the changing needs. Furthermore, similar standardization at the International Organization for Standardization was also started during this time period because of corresponding worldwide needs for comparability and reliability.

Finally, regulations which mandated that advertised *R*-values for products were accurate and reproducible were being promulgated. Thus, higher accuracies of both the measurement and its reproducibility were now necessary in the quality control area.

The need for reliable, simple to operate, accurate, reproducible, larger-scale heat flow meter instruments for thick specimens was clearly established. The essential requirements for such an apparatus were the following:

(*a*) adequate areal dimensions,

(*b*) hot and cold plates having improved uniformity and stability of temperature over the total area,

(*c*) a total area integrating a heat flux transducer of a sufficient central area to provide an adequate representation of the heat flow through the specimen, and

(*d*) sufficient reference materials and transfer standards to provide a suitable calibration of the whole apparatus for the range of thicknesses (thermal resistances) and products to be evaluated.

Until 1975, the only large-area heat flux transducers available were those fabricated from a number of smaller commercial strip forms set in a square array. However, because of the efforts of Klarsfeld and DeGenne at Centre de Recherche of St. Gobain Industries, a new type of large, stable, thin, fast-response heat flux transducer was developed. This design is based on the electrodeposition of a very uniformly distributed array of copper and nickel junctions [*22*] over a large area. A central smaller metering area can then be selected within the total area, thus ensuring that the unit is both of uniform thermal resistance and self-guarding.

A major obstacle to stable heat flux measurements in inhomogeneous materials was therefore alleviated by this development. Since that time the authors have improved on their initial designs so that they are now thinner, even faster in response, and have a much higher sensitivity [*23*]. Similar types of transducers have since been developed in Japan [*24*] and in several eastern European countries.[2]

As a result, a large-area heat meter instrument, designated the R-Matic, was developed. This is shown schematically in Fig. 2. The basic features were the following:

(*a*) A_T = 600 by 600 mm; A_M = 250 by 250 mm;
(*b*) x = 200 mm;
(*c*) λ range of up to 0.45 W/m · K; R = 0.5 to 3.7 $m^2 \cdot K/W$;
(*d*) temperature range of 0 to 40°C, usually set at 24°C mean;
(*e*) temperature stability ±0.1°C;
(*f*) equilibrium time of 15 to 75 min, depending on the thickness and thermal diffusivity of the specimen;
(*g*) accuracy of ±3% or better; reproducibility ±1% or better; and
(*h*) self-contained with digital display.

Although the instrument is similar in its basic design to the other models, several important changes were incorporated to satisfy the new criteria and requirements. These include platinum resistance thermometers, thermistors, and improved electronic components to provide better temperature stability of better than ±0.1°C, improved linearity of thickness measurement devices, and the use of an additional guard heater in the main hot plate heater. Two special versions were also developed for particular requirements.

The first of these, shown also in Fig. 2, was a modification in the form of a single-sided guarded hot plate. This had a stabilized direct-current (D-C) power supply to provide a steady power input so that the instrument could be used in an absolute mode. A special guard (or bucking) heater was installed beneath the main heater, with a second heat flux transducer separating them. The output of this transducer regulated power to the heater by means of a null-balance controller. The instrument could therefore be calibrated in an absolute mode without the use of transfer standards with proper evaluation of all uncertainties. If the voltage, V, and current, I, to the heater are measured for the given A_M, then

$$\lambda = NQ\frac{X}{\Delta T} = \frac{EI}{A_m}\frac{X}{\Delta T} \tag{2}$$

Thus

$$N = \frac{EI}{QA_m} \tag{3}$$

The instrument can then be operated in the relative mode with the set point controller regulating the power to the heater to maintain the temperature constant regardless of the specimen.

The second special version was necessary in order to satisfy the needs of other testing communities, particularly those in Europe, and also for areas of very high humidity. In the former, the

[2]Kovács, T., National Institute of Measurement, Budapest, Hungery, personal communication, 1984.

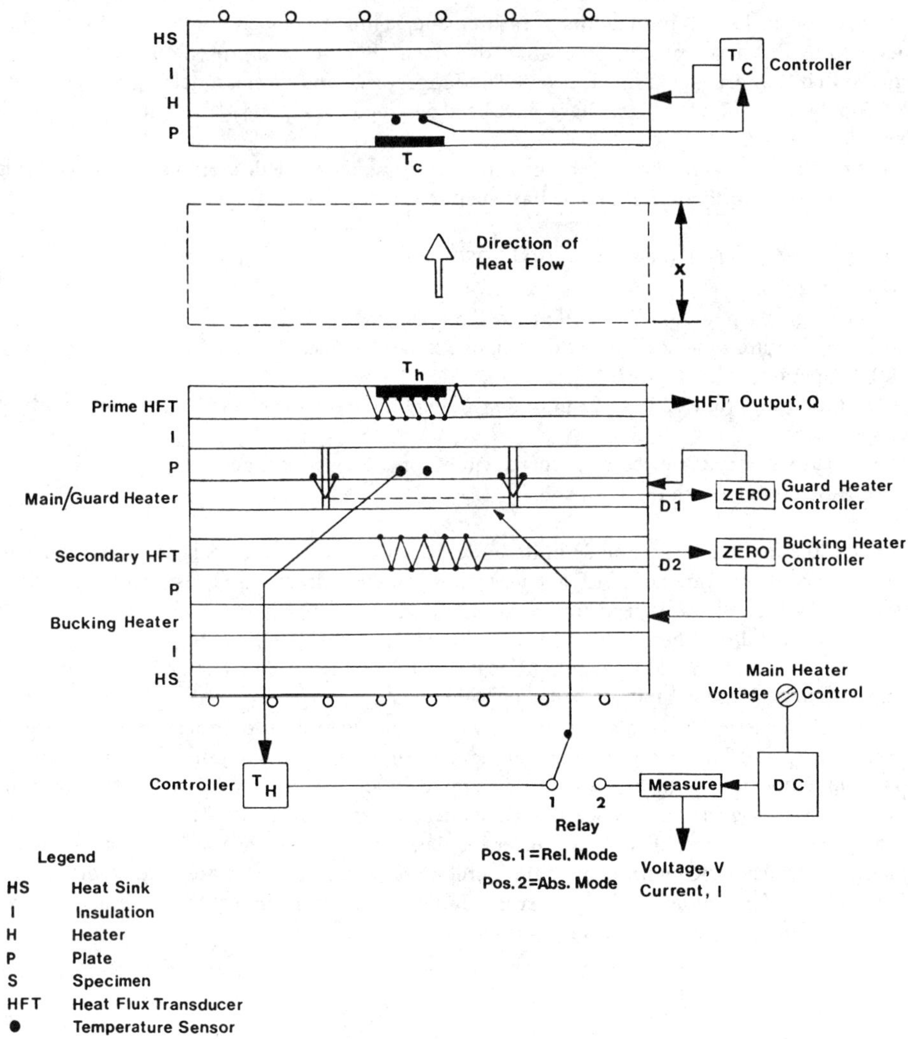

FIG. 2—*Schematic of large heat flow meter instrument.*

nominal test temperature conditions are 10°C mean and a ΔT of 20°C. There was a need to avoid condensation of moisture on the cold plate operating at about 0°C or at other temperatures in high-humidity areas. This was accomplished by a design of a test stack and chamber that included better seals, combined with the forced circulation of a current of internally dried and temperature-regulated air around the complete stack.

Calibration

The biggest problem faced in the developments in 1974–75 was the lack of transfer standards for calibration. The only general material available, and which could be provided to the user, was the NBS high-density fiberglass board (now SRM 1450b) or its equivalent, mentioned earlier. This was only available in 25.4-mm-thick pieces.

It would be possible to stack several of these in series to give different thicknesses. However, equilibrium times for 37 mm of this material were found to be on the order of 1.5 h and in excess of 24 h for 150 mm. In addition, contact resistance problems could arise in stacking because of the nature of the hard-bonded surfaces of this material. Finally, the heat transmission mechanisms in such materials are different from those in low-density materials. There is reason to believe that, for high-accuracy calibration purposes, it is necessary that transfer standards should not only have the same order of thermal resistance but also have similar heat transmission mechanisms to those of the test material [*5*].

In order to investigate the extent of this problem, secondary calibration materials were obtained or developed by Dynatech for in-house use. The first of these was a measured pair of fiberglass materials of approximately 50 kg/m^3 and a thickness of approximately 37 mm obtained from Laboratoire D'Essais in France. The second was a pair of ten-year-old aged polystyrene specimens approximately 48 kg/m^3 and 88 mm thick. These had been evaluated in the 600-mm-square guarded hot plate mentioned earlier [*14*].

Figure 3 illustrates the best-fit linear calibration curve, which was obtained using these and the NBS 1450b materials singly and in pairs to cover the approximate range of 25 to 170 mm. Overall, the results show that, if the instrument is calibrated with 25 or 50-mm specimens, measurements on specimens of 150 mm would be inaccurate to the order of 3 to 4% because of this calibration problem. From a practical point of view, this was still within the overall accuracy of the basic method but not totally suitable for the needs of the community. However, the reproducibility of the instrument at a particular thickness or condition was still of the order of 1%, and thus accurate results could be obtained comparatively.

Since the development of individual secondary calibration specimens of this type was both very time-consuming and expensive, it was impractical commercially to supply them to the community. Thus, the instrument could only be provided with a high-density fiberglass specimen for calibration, together with information on the likely extent of the change over the thickness range.

Similar calibration exercises, with similar results, were carried out by other organizations with materials and guarded hot plate facilities to produce secondary transfer standards. However, calibration remained a serious problem in the United States because reference materials or transfer standards from a national laboratory were required for use for certification purposes. It was not until 1981, after the development of a large hot plate, that the National Bureau of Standards could make available individually measured pairs of a low-density fiberglass for calibration and verification purposes. There is clearly still an outstanding need for further reference materials or transfer standards to be made available to cover the approximate range of 25 to 170 mm.

Some information on the overall accuracy of the instrument is provided by the recent publication [*25*] of the results of three interlaboratory, intermethod studies on both thick and thin specimens of mineral fiber materials. These three different studies involved both guarded hot plate and heat flow meter instrumentation. The Dynatech R/D Co. R-Matic, situated in its Thermophysics Laboratory, was one such heat flow meter instrument in this program.

Overall, the results were all in very good agreement to the order of $\pm 2.2\%$. There appeared to be no definite trend of results due to type or size of apparatus or of the basic method. Thus, the claim that the basic instrument is within $\pm 3\%$ has been verified. Comparisons of measurements by the R-Matic with those by the 600-mm guarded hot plate on some of the mineral fiber specimens used in the older study [*2*] indicated agreement of $\pm 1.5\%$ or better. Further confirmation of this accuracy has been obtained during the course of proficiency testing in the National Voluntary Laboratory Accreditation Program (NVLAP) program on thermal insulation [*26*].

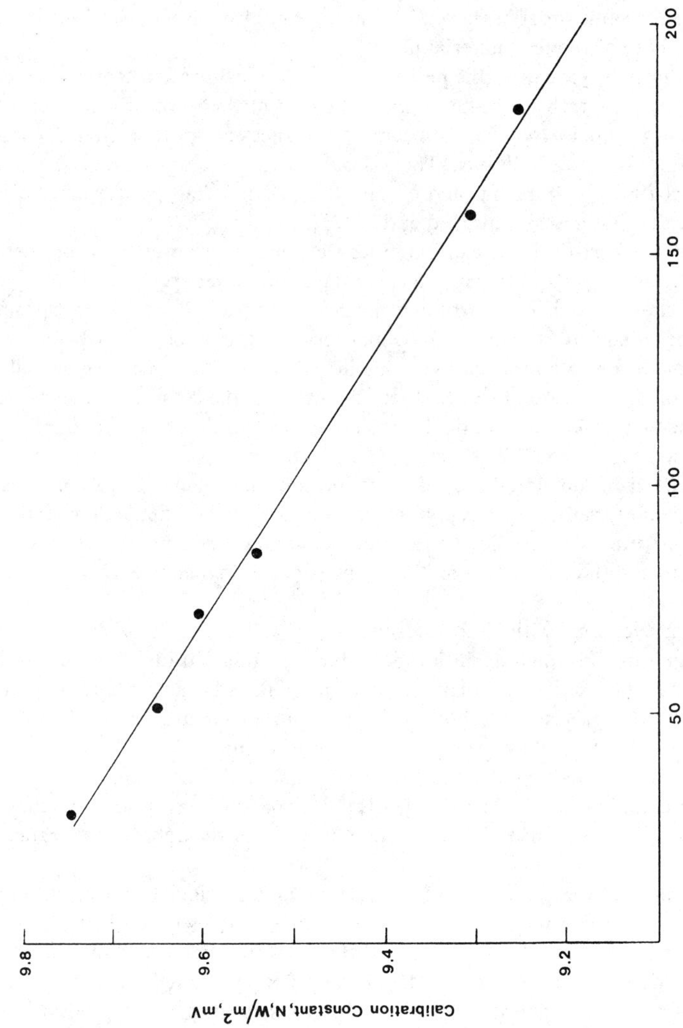

FIG. 3—*Calibration constant versus thickness for the R-Matic instrument.*

Uses of the Instrument

Product Performance

One of the major advantages of using larger area testing is the fact that dispersion in test results can be reduced significantly because the test area is more representative. This is illustrated in Fig. 4. Specimens cut from an inhomogeneous material were measured using both small-area and large-area heat flow meter apparatus. The densities of the actual heat metered sections of the specimens were also measured. The results show both that the number of tests is reduced and that the results from the larger area testing more closely represent the performance characteristics of the product.

Thickness Effect

Although the instrument was designed primarily as an accurate quality control capability, its accuracy and reproducibility make it ideal for more basic studies. To this end, it has been utilized by organizations for investigations of the thickness effect [*27*]. In order to evaluate this effect, possible instrument error due to effects of lateral heat loss as thickness increases had to be evaluated first. Although calibration at different thicknesses should eliminate these effects, small errors can still be obtained, and any effect needs to be quantified.

The most appropriate way to investigate this issue in low-density materials is to measure increasing thicknesses of an insulation with and without paper septa introduced between the layers [*27*]. Measurements on specimens without septa should provide results that are independent of thickness. If there is any increase in value as the thickness increases, this is the extent of the apparatus error. This value should be subtracted from the results obtained on the specimens measured without the septa.

Table 1 shows the results of one such investigation on specimens of a low-density polystyrene. They indicate that the real thickness effect between 25 and 100 mm is of the order of 6% rather

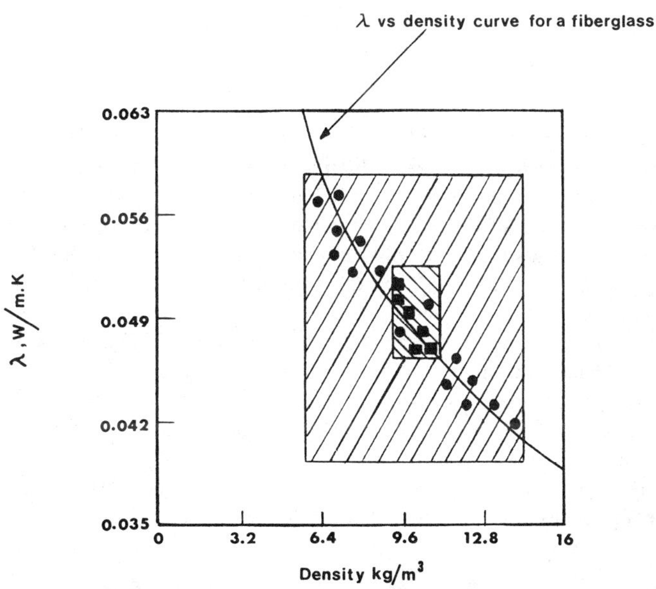

FIG. 4—*Comparison of results on inhomogeneous material using large and small metering areas.*

TABLE 1—*Extent of thickness effect for several low-density materials.*

Material	25 mm	50 mm	75 mm	100 mm	150 mm
Molded polystyrene, 12.8 kg/m³					
With septa	0.0373	0.0375	0.0378	0.0383	0.0387
Without septa	0.0375	0.0393	0.0403	0.0407	0.0412
Fiberglass, 9 kg/m³	0.0452	0.0464	0.0477[a]	...	0.0481
Cellulose, 30 kg/m³	0.0389	0.0394	0.0396	...	0.0396
Urea formaldehyde, 11 kg/m³	0.0292	0.0317	0.0328	...	0.034

[a]At 89 mm.

than 9%. Table 1 contains results based on this correction for several other low-density insulations.

Resolution of Differences

As a result of the need for energy-conserving materials to be made available for building applications, various products or materials have been introduced into the marketplace, often having inflated *R*-value claims.

A specific example of one product in this category is spray-applied cellulose. Recently, claims by two manufacturers of λ_{app} of the order of 0.032 to 0.036 W/m · K (0.212 to 0.222 Btu · in./h · ft² · °F) have been made [*28*]. These were for high-density products of the order of 85 to 90 kg/m³. Such values are quite contrary to previous findings for spray-applied cellulose in this density range [*29*]. In fact, the values are well below those accepted for blown cellulose fiber at much lower densities of use.

As part of an interlaboratory study to compare results obtained by different types of test apparatus and test methods,[3] measurements have been made on thick specimens of the preceding products at the appropriate density. In each case, values of 0.045 W/m · K (0.31 Btu · in./h · ft² · °F) were obtained. This value is within 5% of that which could be derived from the relationship given in Ref *29*.

Future Needs

Sufficient work has now been carried out worldwide on the basic heat flow meter method to illustrate that it is both accurate and reproducible, while being considerably more simple and less time-consuming than an absolute method. However, the majority of uses of the method have been at ambient temperatures or for a restricted range. Limited work has been carried out on its use at elevated temperatures [*8*], but only by an organization that has the means to develop secondary transfer standards [*30*] and fabricate special heat flux transducers of sufficient size and stability and suitable sensitivity.

These factors are the two main obstacles to development of higher-temperature heat flow meter instruments. Such equipment is most necessary for undertaking the same types of studies on industrial insulations. This is particularly relevant since heat transmission phenomena seen in materials at ambient conditions will be magnified considerably at higher temperatures and for materials of higher density [*31*].

The two immediate needs, therefore, are the development of stable high-temperature heat flux transducers of a measuring area considerably greater than those currently available. These

[3]Yarbrough, D. W., Graves, R. W., and McElroy, D. L., Oak Ridge National Laboratory, Oak Ridge, TN, personal communication, 1985.

are generally of the order of 20 to 30 mm in diameter, which is not sufficient for the applications relating to thermal insulations. The second need is to have available several high-temperature reference materials of different thicknesses having sufficient stability over a wide temperature range that appropriate transfer standards can be made available.

A high-temperature version of the single-sided guarded hot plate has been developed already. It is somewhat similar in basic design to that described for one version of the R-Matic. It utilizes a 300 by 300-mm specimen with a metering area of 150 by 150-mm and can be used at temperatures up to 700°C. This could readily be converted to the heat flow meter form once a heat flux transducer of the order of 100 by 100 mm and one or more transfer standards were made available.

Summary

The developments leading to the construction of reliable forms of heat flow meter apparatus for undertaking thermal resistance measurements on thick specimens of thermal insulations are described. Illustrations of its accuracy and versatility have been included. The two basic needs for extending the use of heat flow meter methodology to higher temperatures are discussed.

References

[*1*] "Labelling and Advertising of Home Insulation," Trade Regulation Rule, 16 CFR, Part 460, Federal Trade Commission, Washington, DC, 1979.

[*2*] Tye, R. P., Desjarlais, A. O., Yarbrough, D. Y., and McElroy, D. L., "An Experimental Study of Thermal Resistance Values of Low Density Mineral Fiber Building Insulation Batts Commercially Available in 1977," ORNL/TM-7266, Oak Ridge National Laboratory, Oak Ridge, TN, April 1980.

[*3*] Sherman, M., *Thermal Insulation Performance,* American Society for Testing and Materials, Philadelphia, 1980, pp. 298–306.

[*4*] *Heat Transmission Measurements and Thermal Insulations, ASTM STP 544,* American Society for Testing and Materials, Philadelphia, 1974.

[*5*] *Heat Transmission Measurements of Insulation, ASTM STP 660,* American Society for Testing and Materials, Philadelphia, 1978.

[*6*] *Thermal Insulation Performance, ASTM STP 718,* American Society for Testing and Materials, Philadelphia, 1980.

[*7*] Pratt, A. W., in *Thermal Conductivity,* Vol. 1, R. P. Tye, Ed., Academic Press, London, 1969.

[*8*] Verschoor, J. D. and Wilber, A., *Transactions of the American Society of Heating and Ventilating Engineers,* Vol. 60, 1954, p. 329.

[*9*] Pelanne, C. M. and Bradley, C. B., *Materials Research and Standards,* Vol. 2, No. 7, 1962, p. 549.

[*10*] *Thermal Insulating Measurements,* Commission II Report, Institut International du Froid, Paris, France, 1968.

[*11*] Tye, R. P., *Proceedings,* XIII International Congress on Refrigeration, Institut International du Froid, Paris, 1971, pp. 371–378.

[*12*] Tye, R. P., *Proceedings,* Ninth Thermal Conductivity Conference, Conference 691002, Physics (TLD 4500), U.S. Atomic Energy Commission, 1968, pp. 341–351.

[*13*] Tye, R. P. and Spinney, S. C., *Annex 1976-2 Bulletin I.T.R.,* Institut International du Froid, Paris, 1976, pp. 119–127.

[*14*] Howard, J. F., Coumou, K. G., and Tye, R. P., *Journal of Cellular Plastics,* Vol. 9, 1973, p. 266.

[*15*] Tye, R. P. and Baker, C. F., "Development of Experimental Data on Cellular Plastic Insulations Under Simulated Winter Conditions," in this publication, pp. 000–000.

[*16*] Poltz, H., *Allgemeine Wärmetechnik,* Vol. 11, 1962, p. 64.

[*17*] Jones, T. T., *Proceedings of the Seventh Thermal Conductivity Conference,* NBS Special Publication 302, National Bureau of Standards, Washington, DC, 1967, pp. 737–748.

[*18*] Lao, B. Y. and Skochdopole, R. E., *Proceedings,* Fourth International Cellular Plastics Conference, Society of Plastics Industry, New York, 1976, pp. 175–182.

[*19*] Cammerer, W. F., *Proceedings,* ITF-IIR Commission B1, Zurich, Switzerland, Institut International du Froid, 1974, pp. 189–200.

[*20*] Tye, R. P., *Proceedings,* 24th SPI Annual Urethane Division Technical Conference, Atlanta, GA, Society of Plastics Industry, New York, 1978, pp. 129–134.

[21] Wang, K. Y. and Tien, C. L., *Journal of Quantitative Spectroscopy and Radiative Transfer*, Vol. 30, 1983, p. 213.

[22] Degenne, M., Klarsfeld, K., and Barthe, M.-P., *Thermal Transmission Measurements of Insulations, ASTM STP 660*, American Society for Testing and Materials, Philadelphia, 1978, pp. 130–144.

[23] DeGenne, M. and Klarsfeld, S., *Building Applications of Heat Flux Transducers, ASTM STP 885*, American Society for Testing and Materials, Philadelphia, 1985, pp. 163–171.

[24] Miyake, Y. and Eguchi, K., *Guarded Hot Plate and Heat Flow Meter Technology, ASTM STP 879*, American Society for Testing and Materials, Philadelphia, 1985, pp. 171–179.

[25] Hust, J. G. and Pelanne, C. M., "Round Robin on the Apparent Thermal Conductivity of Low Density Glass Fiber Insulations Using Guarded Hot Plate and Heat Flow Meter Apparatus," NBSIR 85-3026, National Bureau of Standards, Boulder, CO, 1985.

[26] Kirkpatrick, D. and Horlick, J., *Thermal Conductivity 17*, J. G. Hust, Ed., Plenum Press, New York, 1983, pp. 497–506.

[27] Pelanne, C. M., *Thermal Insulation Performance, ASTM STP 718*, American Society for Testing and Materials, Philadelphia, 1980, pp. 322–334.

[28] *Energy Design Update*, J. D. Nisson, Ed., Vol. 3, No. 7, July 16, 1985.

[29] Tye, R. P., *Journal of Testing and Evaluation*, Vol. 2, 1974, p. 176.

[30] Mitchell, M., "Development of a Refractory Fiber Insulation for Use as a High Temperature Thermal Transmission Calibration Sample," *Thermal Conductivity 19*, Plenum Press, New York, in press.

[31] Tye, R. P., Desjarlais, A. O., and Smith, S. E., *Thermal Conductivity 18*, T. Ashworth and D. R. Smith, Eds., Plenum Press, New York, 1985, pp. 317–326.

Sorïn Klarsfeld,[1] *Jean Boulant,*[2] *and Catherine Langlais*[2]

Thermal Conductivity of Insulants at High Temperature: Reference Materials and Standards

REFERENCE: Klarsfeld, S., Boulant, J., and Langlais, C., **"Thermal Conductivity of Insulants at High Temperature: Reference Materials and Standards,"** *Thermal Insulation: Materials and Systems, ASTM STP 922,* F. J. Powell and S. L. Matthews, Eds., American Society for Testing and Materials, Philadelphia, 1987, pp. 665–676.

ABSTRACT: Extensive work has been done on insulants at ambient temperature: standards, reference materials, and national and international interlaboratory measurements (ISO/TC 163 round robin test). The situation is more complex at high temperatures, and this paper in an attempt to clarify this complexity, taking into account the new International Organization for Standardization (ISO) draft proposal on guarded hot plate apparatus.

Using a simple model describing heat transfer through fibrous insulants, and introducing the notion of optical thickness or opacity, the study proposes a quantitative criterion to minimize the radiative transfer, in order to obtain measurement results independent of the experimental conditions (the thickness of the specimen and emissivities of the limiting surfaces).

This criterion shows the importance of optical parameters and should, from a practical point of view, help in choosing the reference materials used for checking and in making interlaboratory comparisons of guarded hot plate apparatuses.

KEY WORDS: fibrous insulants, guarded hot plate, high-temperature optical parameters, optical thickness, porous material, radiative heat transfer, reference material, thermal conductivity, extinction coefficient, scattering coefficient, absorption coefficient, thermal insulation

The study presented here is directly related to the new International Organization for Standardization (ISO)/Technical Committee (TC) 163 Draft International Standard (DIS) 8302 concerning the "Determination of Steady-State Thermal Resistance and Related Properties by the Guarded Hot Plate Apparatus" [*1*]. In this draft Section 0.2 states that heat transfer within light-density porous materials "can involve a complex combination of different contributions of radiation, conduction, and convection (in some operating conditions)" and that, as a consequence, "thermal conductivity . . . may be not an intrinsic property of the material itself." Further, one can read that "after completion of new apparatus, it is recommended that it be verified by undertaking tests on one or more of the various reference materials."

These quotations clearly show that the choice of appropriate reference materials is very critical. This problem is similar to the radiative-conduction heat transfer through a still air layer presented by Jaouen and Klarsfeld in a paper earlier in this volume [*2*], but is extended here to the case of a porous material. The dependence of the measurement results on the experimental conditions—the layer thickness and emissivities of the limiting surfaces observed at ambient temperature for air—is indeed closely related to what happens at high temperatures on relatively heavy porous materials because of the radiative component of the total heat transfer. The

[1] Research engineer, Saint Gobain Recherche, 93304 Aubervilliers, France.

[2] Engineering supervisor and heat transfer laboratory manager, respectively, Isover Saint-Gobain, Centre de Recherches Industrielles, 60290 Rantigny, France.

problem that is raised is the need to define the thermal conductivity of reference materials independent of the experimental conditions. There are two possible ways to do this:

(*a*) to separate the radiative component of the total heat transfer energy, as in the case of the still air layer, and
(*b*) to minimize the radiative component by using practically "opaque materials."

The authors preferred to use this second approach, and the purpose of this present paper is to propose a quantitative criterion for the choice of such materials.

Starting from the description of the combined radiation-conduction heat transfer in a still air layer, equations are derived for fibrous materials by simply introducing an optical characteristic of the medium. This approach allows one to derive a simple heat transfer model in the case of a weak coupling between conduction and radiation. The computations are rather straightforward, the model simply requiring thermal conductivity measurement methods and results, in particular, the guarded hot plate method.

Combined Radiation-Conduction Heat Transfer

In the case of the still air layer, we have seen [2] how to express the radiation heat flux density, ϕ_r, and how to introduce an "equivalent radiative thermal conductivity," λ_r,[3]

$$\phi_r = \frac{\sigma(T_1^4 - T_2^4)}{\dfrac{1}{\epsilon_1} + \dfrac{1}{\epsilon_2} - 1} \simeq \frac{4\sigma\bar{T}^3\Delta T}{\dfrac{2}{\epsilon} - 1} = \lambda_r \frac{\Delta T}{L} \tag{1}$$

and

$$\lambda_r = \frac{4\sigma\bar{T}^3 \cdot L}{\dfrac{2}{\epsilon} - 1} \tag{2}$$

where

σ = Stefan-Boltzmann constant,
T_1, T_2 = hot and cold side temperatures,
$\epsilon = \epsilon_1 \simeq \epsilon_2$ emissivities of the hot and cold sides,
$\bar{T} = (T_1 + T_2)/2$,
L = air layer thickness, and
$\Delta T = T_1 - T_2$.

Let us imagine that between the two parallel, isothermal plane surfaces of temperatures T_1 and T_2, we introduce a divided solid phase, optically representative of an absorbing-scattering medium, such as a porous or fibrous material. In steady state, the radiation heat flux density, ϕ_r', through the layer of thickness, L, will be lower than ϕ_r.

It is possible to express this decrease by introducing a dimensionless parameter, **F**, at the denominator of the ratio giving ϕ_r. This parameter simply expresses the "optical presence" of

[3] $T_1^4 - T_2^4 \simeq 4\bar{T}^3\Delta T$. This approximation is better than 1% if $\Delta T < 50°C$.

the porous layer, and will vary with the temperature and material in question. The radiation heat flux density, ϕ_r', can be written as

$$\phi_r' = \frac{4\sigma\bar{T}^3 \cdot \Delta T}{\left(\frac{2}{\epsilon} - 1\right) + \mathbf{F}} = \lambda_r' \frac{\Delta T}{L} < \phi_r \qquad (1')$$

One can introduce the notion of a mass extinction coefficient, $\tilde{\sigma}_{m\,\mathrm{EXT}}$ (in square metres per kilogram), expressing the total developed area of the solid phase per unit mass of this solid-phase absorbing and scattering medium, contributing to the decrease of the radiative energy through the layer. One can suppose that the **F** parameter will be proportional to $\tilde{\sigma}_{m\,\mathrm{EXT}}$ as well as to the quantity of mass per unit of surface of the layer, that is, the areal mass, M_s (in kilograms per square metre), of the porous medium.

$$\mathbf{F} = \tilde{\sigma}_{m\,\mathrm{EXT}} \cdot M_s$$

If the extinction is expressed by a volume coefficient, $\tilde{\sigma}_{v\,\mathrm{EXT}}$ (in square metres per cubic metres), **F** can also be written as

$$\mathbf{F} = \tilde{\sigma}_{v\,\mathrm{EXT}} \cdot L = \tilde{\sigma}_{m\,\mathrm{EXT}} \cdot \rho \cdot L = \tilde{\sigma}_{m\,\mathrm{EXT}} \cdot M_s \qquad (3)$$

$$\left[\left(\frac{\mathrm{m}^2}{\mathrm{m}^3}\right)(\mathrm{m})\right] = \left[\left(\frac{\mathrm{m}^2}{\mathrm{kg}}\right)\left(\frac{\mathrm{kg}}{\mathrm{m}^3}\right)(\mathrm{m})\right] = \left[\left(\frac{\mathrm{m}^2}{\mathrm{kg}}\right)\left(\frac{\mathrm{kg}}{\mathrm{m}^2}\right)\right]$$

where ρ is the density of the layer of thickness, L.

Taking into account Eq 1′ and the three possible formulations of **F**, one can associate three different expressions of the equivalent radiative thermal conductivity, λ_r', to the considered layer of porous medium.

$$\lambda_r' = \frac{4\sigma\bar{T}^3 \cdot L}{\left(\frac{2}{\epsilon} - 1\right) + \tilde{\sigma}_{v\,\mathrm{EXT}} \cdot L} \qquad (2'a)$$

$$\lambda_r' = \frac{4\sigma\bar{T}^3 \cdot L}{\left(\frac{2}{\epsilon} - 1\right) + \tilde{\sigma}_{m\,\mathrm{EXT}} \cdot \rho \cdot L} \qquad (2'b)$$

$$\lambda_r' = \frac{4\sigma\bar{T}^3 \cdot L}{\left(\frac{2}{\epsilon} - 1\right) + \tilde{\sigma}_{m\,\mathrm{EXT}} \cdot M_s} \qquad (2'c)$$

According to Siegel and Howell, "A relatively simple idea to obtain the combined energy transfer by radiation and conduction is to assume that the interaction between the two transfer processes is so weak that each process can be considered to act independently. Then the conduction and radiation transfers are each formulated as if the other mechanism were not present" [*3*].

The assumption of a weak interaction between the different modes of heat transfer, as in the case of a still air layer, can be justified for fibrous materials, which constitute our main interest in this study. For such materials, the solid phase indeed represents a very small proportion of the total volume. The porosity of fibrous materials is extremely high, the fraction of solid phase varying between 0.3 and 8.0% for densities from 8 to 200 kg/m^3.

In this case, the total heat flux density, ϕ_T, can be written

$$\phi_T = \phi_g + \phi_s + \phi_r' = \lambda^* \frac{\Delta T}{L} \tag{4}$$

Or, introducing the corresponding equivalent thermal conductivities, λ^*, λ_g, λ_s, and λ_r', we have

$$\lambda^* = \lambda_g + \lambda_s + \lambda_r' \tag{5}$$

where λ_g, λ_s, and λ_r' are, respectively, the gaseous, solid, and radiative thermal conductivities.

If we substitute λ_r' in Eq 5 by one of the expressions given by Eqs 2′, and if we introduce a formula previously established for λ_g and λ_s [*4*], we have

$$\lambda^* = 0.2572\bar{T}^{0.81} + 0.0527\rho^{0.91}\left[1 + 0.13\frac{\bar{T}}{100}\right] + \frac{4\sigma\bar{T}^3L}{\left(\frac{2}{\epsilon} - 1\right) + \bar{\sigma}_{m\mathrm{EXT}} \cdot \rho \cdot L} \tag{6}$$

Equation 6 is a semiempiric relationship giving, however, excellent agreement with experimental data up to a density of 100 kg/m^3. For higher densities, the equation should be somewhat modified as the solid conduction term increases. Equation 6 holds for products manufactured under the same conditions. Indeed, $\bar{\sigma}_{m\mathrm{EXT}}$ is a parameter characteristic of the material, depending on the following:

(*a*) the nature of the glass,
(*b*) the specific surface,
(*c*) the product "texture" (the fibers' orientation and repartition), and
(*d*) the temperature.

Validation of the Model

The coefficient $\bar{\sigma}_{m\mathrm{EXT}}$ should be determined by optical means. At the present time, for a given manufacturing process and a given temperature, we compute this coefficient from Eq 6 using a large number of thermal conductivity measurements[4] performed on products of the considered fabrication for the temperature of interest and at a low mean density.

An example is given on Fig. 1, which shows an excellent agreement between the experimental points and the theoretical curve.

Figure 2 indicates how the thermal conductivity varies with temperature. At $\bar{T} = 400°C$, the agreement between experience and theory is not as good as at ambient temperature but is still rather satisfactory.

[4]These measurements have been achieved on performant guarded hot plate or heat flow meter apparatus; in measurement accuracy is approximately 2%, and the precision is better than ±1%.

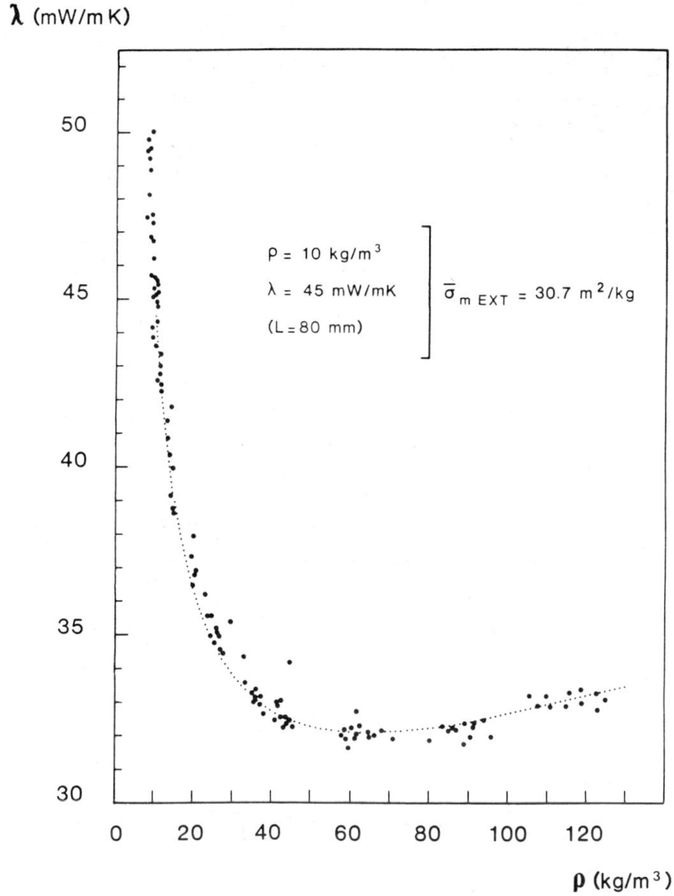

FIG. 1—*Thermal conductivity versus density: a comparison between theoretical (. . .) and experimental results (●).*

Optical Characteristics

The "optical" behavior of the fibrous material layer is described in the proposed model by the mass extinction coefficient, $\bar{\sigma}_{m\mathrm{EXT}}$, which must be considered here as a total hemispherical value. The porous medium being an absorbing-scattering material, it is necessary to define also a mass scattering coefficient, $\bar{\sigma}_{mS}$ (in square metres per kilogram), and a mass absorption coefficient, $\bar{\sigma}_{mA}$ (in square metres per kilogram). The three coefficients satisfying the relation [5]

$$\bar{\sigma}_{m\mathrm{EXT}} = \bar{\sigma}_{mS} + \bar{\sigma}_{mA} \tag{7}$$

can be computed from infrared hemispherical transmission measurements through "cold" fiber layers of different thicknesses using the Schuster-Schwarzchild approximation, as described in Refs *6* through *8*. The radiation source is a blackbody working at different temperatures.

The spatial integration of the transmitted radiation in a solid angle, π, is made by using a gold-coated integrating sphere. In Table 1, one can find the results derived from these measure-

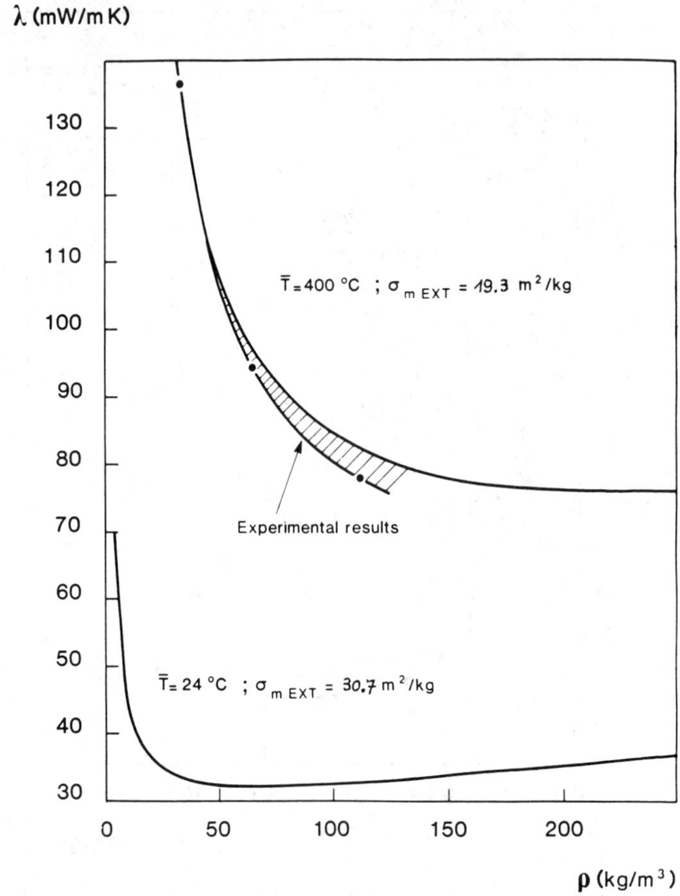

FIG. 2—*Thermal conductivity versus density for two mean temperatures, 24 and 400°C.*

TABLE 1—*Extinction scattering and absorption mass coefficients for different source temperatures,* T_s *(optical measurements).*[a]

T_s, °C	$\bar{\sigma}_{m\mathrm{EXT}}$, m²/kg	$\bar{\sigma}_{mS}$, m²/kg	$\bar{\sigma}_{mA}$, m²/kg
500	33.8	32.7	1.4
600	30.1	29.1	1.0
700	24.2	22.8	1.4
800	20.7	19.2	1.5
900	17.7	15.9	1.8
1000	14.4	11.8	2.6

[a] See Fig. 3.

ments, for a "standard" fiberglass material (mean diameter ~4 μm) at different blackbody temperatures, T_s. Two remarks can be made based on the data in this table:

1. Extinction and scattering coefficients are very close; consequently, the main mechanism of radiation transfer is scattering, while absorption, and therefore reemission, are low.[5]

$$\tilde{\sigma}_{m\,\mathrm{EXT}} \approx \tilde{\sigma}_{mS} \gg \tilde{\sigma}_{mA} \tag{7'}$$

2. Looking at the variation of $\tilde{\sigma}_{mS}$ with the source temperature, T_s, one can see that fibrous insulants do not behave like a graybody in the range of the considered temperatures. The lower limit of 500°C is imposed by the sensitivity of the sensor used in the transmission measurements.

Another way of determining the fibrous material mass scattering coefficient, $\tilde{\sigma}_{mS}$ ($\approx \tilde{\sigma}_{m\,\mathrm{EXT}}$), is to use the model derived in Eq 6. In this case, the starting point is the measure of thermal conductivity as a function of temperature with the guarded hot plate apparatus. We can say it is a calorimetric determination of an optical characteristic. The advantage of this last method, in the particular case of strongly scattering materials with a weak interaction between the different modes of heat transfer, is the possibility of determining the coefficients at temperatures below 500°C. The results, $\tilde{\sigma}_{mS}(T)$, obtained by the two methods on the same type of fiber are shown in Fig. 3. One can note a linear variation of $\tilde{\sigma}_{mS}$ with $\overline{T}$, but with different slopes for the two methods. On the other hand, the results obtained from thermal conductivity measurements fall on a same line as those obtained from optical measurements for $\overline{T} > 800$°C. One can imagine

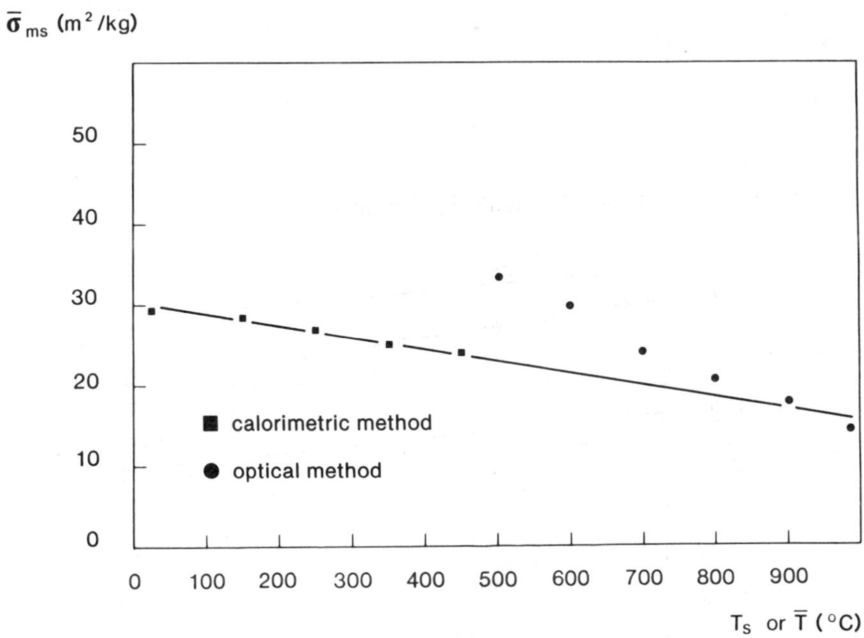

FIG. 3—*Mass scattering coefficient (total hemispherical) versus temperature for a "standard" glass fiber insulant (mean diameter ~4 μm).*

[5] According to recent results [*10*], this remark must be reconsidered.

that the accuracy of optical measurement is not as good with decreasing temperature, and the authors preferred to take into consideration the calorimetric determination of the optical characteristic, $\bar{\sigma}_{mS}$.

The morphology of the two types of specimens used for the optical and thermal measurements may also explain the differences between the results obtained with the two methods. Figure 4 shows other correlations, $\bar{\sigma}_{mS}(T) = A\bar{T} + B$, obtained on other fibrous materials of different natures and different morphology of the solid matrix.

Optical Thickness of the Reference Material

To describe the transparency of the reference material, we will introduce the spectral optical thickness, τ_λ, defined by the relation

$$\tau_\lambda(L) = \int_0^L \sigma_{v\text{EXT},\lambda}(s)ds \tag{8}$$

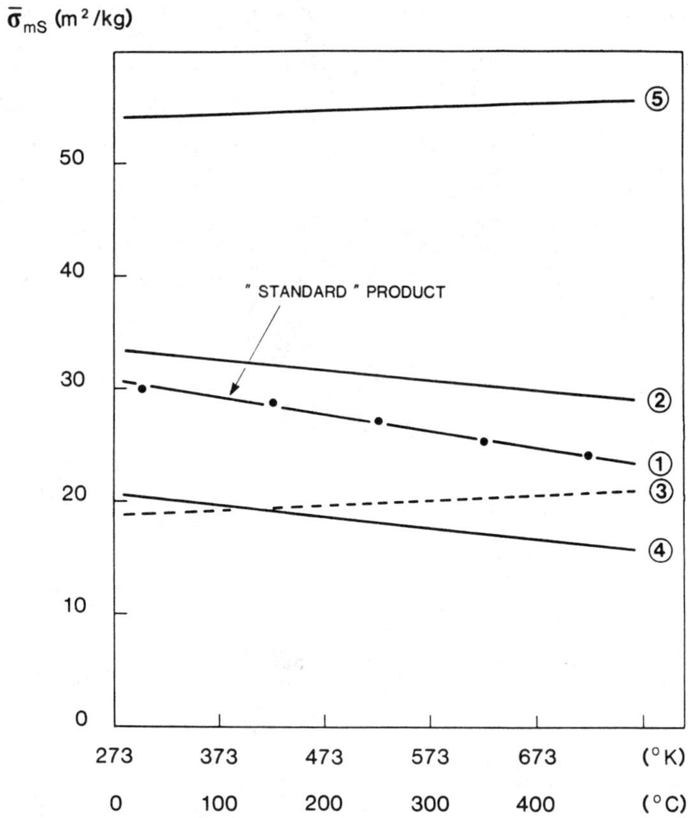

FIG. 4—*Mass scattering coefficient versus temperature for different types of fiber insulants.*

where

$\sigma_{v\text{EXT},\lambda}$ = spectral volume extinction coefficient, m^{-1},
L = characteristic length, such as the reference material thickness, m, and
s = the position of the considered point.

The spectral optical thickness or opacity is a measure of the ability of a given path length of material to attenuate radiation of a given wavelength. A large optical thickness means large attenuation. The quantity τ_λ is therefore a convenient dimensionless parameter. For instance, if $\tau_\lambda \gg 1$, then the medium is optically thick, that is to say, the mean penetration distance is quite small compared with the characteristic dimension of the medium. If the material is considered to be homogeneous, isotropic, and isothermal in its whole thickness, it becomes

$$\sigma_{v\text{EXT},\lambda}(s) \simeq \sigma_{v\text{EXT},\lambda}[T(s)] \simeq \text{constant}$$

and Eq 8 can be written

$$\tau_\lambda = \sigma_{v\text{EXT},\lambda} \cdot L \tag{8'}$$

For the total spectrum of the wavelengths constituting the incident radiation, the optical thickness will be

$$\tau = \bar{\sigma}_{v\text{EXT}} \cdot L \tag{9}$$

Introducing the photon mean path, $l_p = 1/\bar{\sigma}_{v\text{EXT}}$, produces

$$\tau = \frac{L}{l_p} \tag{9'}$$

Depending on the τ values, the specimens can be grouped into three categories [9]:

(*a*) $\tau \ll 1$—the specimen is optically thin;
(*b*) $\tau \gtrapprox 1$—the specimen is optically thick or dense; and
(*c*) $\tau > 5$—the specimen is optically opaque.

Case *c* is clearly the case we are interested in. To apply this condition to fibrous layers, we can express $\bar{\sigma}_{v\text{EXT}}$ and L as

$$\bar{\sigma}_{v\text{EXT}} \simeq \bar{\sigma}_{vS} = \bar{\sigma}_{mS} \cdot \rho \tag{10a}$$

$$L = L_s = F_s \cdot L_T = \frac{\rho}{\rho_s} \cdot L_T \tag{10b}$$

In Eq (10*b*), the characteristic length, L, corresponds to the thickness of the solid phase only, L_s, which is derived by assuming that the fluid phase is practically transparent. L_s is obtained by multiplying the total thickness of the specimen, L_T, by the solid fraction of the fibrous material, $F_s = \rho/\rho_s$, where ρ is the "apparent" density and ρ_s is the solid-phase glass density. Equation 10*a* is simply derived from the definitions introduced in the preceding section.

In the particular case of fibrous specimens, the opacity criterion, $\tau > 5$, then becomes

$$\frac{\bar{\sigma}_{mS}(T)}{\rho_s} \cdot \rho^2 \cdot L_T \geq 5 \tag{11}$$

When Relation 11 is verified, one can consider that ϕ_r is small or, in other words, that the heat transfer is mainly a conduction heat transfer.

Opacity Criteria—Diagrams

The opacity criterion expressed by Relation 11 can be "controlled" by playing on three terms characterizing the specimen:

(*a*) the total thickness, L_T, in metres,
(*b*) the "apparent" density, ρ, in kilograms per cubic metre, and
(*c*) the material mass scattering coefficient, linear function of the mean temperature of use

$$\bar{\sigma}_{mS}(T) = aT + b \qquad (\text{m}^2/\text{kg})$$

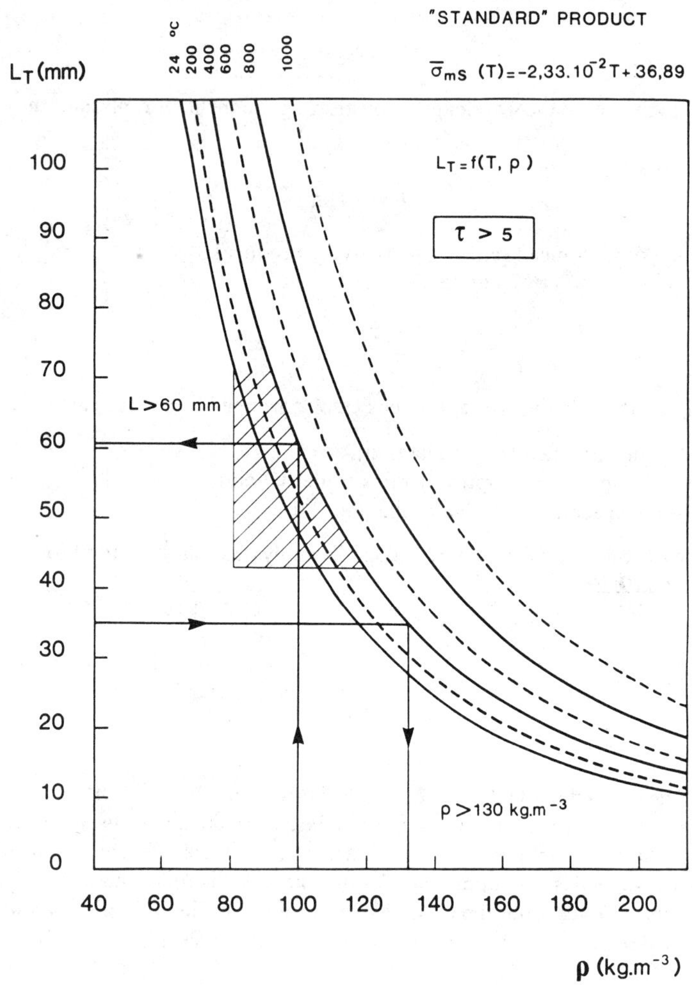

FIG. 5—*Correlation between thickness, density, and temperature, for $\tau \geq 5$ "standard" glass fiber insulants.*

For a given fibrous material, characterized by the function $\bar{\sigma}_{mS}(T)$, determined experimentally (see the section on Optical Characteristics and Fig. 4), we have represented the correlation $L_T = f(\rho; \bar{T})$ for $\tau = 5$ in Fig. 5. For instance, if the reference material has a density of 100 kg/m^3, and if the maximum temperature of use is $\bar{T} = 400°C$, Fig. 5 indicates that the total thickness of the specimen must be equal to or greater than 60 mm to fulfill the condition $\tau > 5$. Now, if we imagine that for the same material and the same range of temperatures, the measurement conditions of the thermal conductivity (plates dimensions in particular) impose a maximum specimen thickness of 35 mm, the diagram in Fig. 5 indicates that the density must be increased; this density must be equal to or greater than 130 kg/m^3.

The possible influence of the "optical behavior" of the material, that is, of the mass scattering coefficient, $\bar{\sigma}_{mS}(T)$, is shown in Fig. 6. For a given density—for instance, 100 kg/m^3—the specimen thickness can be decreased by increasing $\bar{\sigma}_{mS}(T)$, or at a given thickness, the density

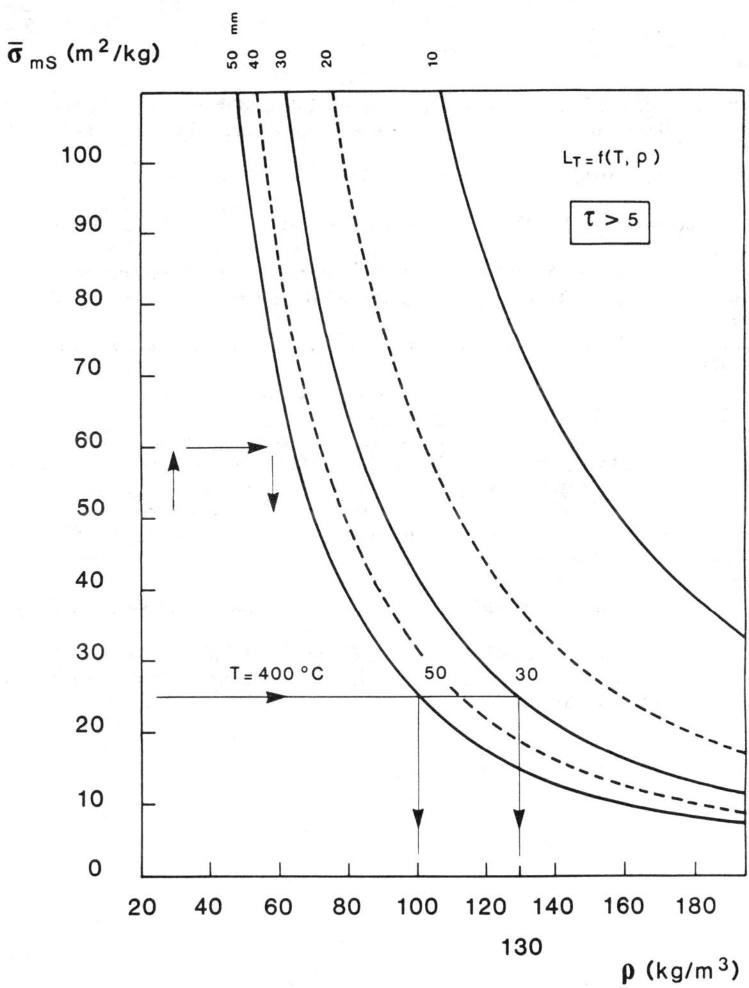

FIG. 6—*Correlation between the mass scattering (total hemispherical) coefficient, density, and thickness, for $\tau \geq 5$.*

can be decreased. From Relation 11, other representative curves can be imagined and drawn as long as the function $\bar{\sigma}_{mS}(T)$ is known.

Conclusions

Introducing the notion of "optical thickness" or "opacity" for fibrous material, it has been possible to propose a quantitative criterion allowing the choice of a type and size of material, taking into account the range of temperature in order to minimize the radiative heat transfer.

The use of this criterion shows the importance of the optical parameters characterizing the fibrous material, in particular, the mean total hemispherical mass scattering coefficient. This coefficient has to be determined experimentally (by optical or calorimetric measurements).

This approach, in addition to its theoretical assumptions, which contribute to a better understanding of the results of the measurements, should be helpful, from a practical point of view, in choosing the reference materials needed for high-temperature thermal conductivity measurements.

References

[*1*] "Determination of Steady-State Thermal Resistance and Related Properties by the Guarded Hot Plate Apparatus," ISO/TC 163/DIS 8302, International Organization for Standardization, Geneva, Switzerland.

[*2*] Jaouen, J.-L. and Klarsfeld, S., "Heat Transfer Through a Still Air Layer," in this publication, pp. 283–294.

[*3*] Siegel, R. and Howell, J. R., *Thermal Radiation Heat Transfer,* 2nd ed., McGraw-Hill, New York, 1981, p. 687.

[*4*] Langlais, C. and Klarsfeld, S., "Modelisation des phénomènes de transfert dans les isolants fibreux et optimisation de leurs performances," paper presented at the 20th Colloquium of CSTB-CFI, Paris, 5 Dec. 1984.

[*5*] Özisik, M. N., *Radiative Transfer and Interaction with Conduction and Connection,* Wiley, New York, 1973.

[*6*] Larkin, B. K. and Churchill, S. W., "Heat Transfer by Radiation through Porous Insulations," *A.I.Ch.E. Journal,* Vol. 5, No. 5, 1959, pp. 467–474.

[*7*] Linford, R. M. F., Schmitt, R. J., and Hughes, T. A., "Radiative Contribution to the Thermal Conductivity of Fibrous Insulations," *Heat Transmission Measurements in Thermal Insulation, ASTM STP 544,* American Society for Testing and Materials, Philadelphia, 1974, p. 68.

[*8*] Hyrien, M., Langlais, C., Guilbert, G., and Klarsfeld, S., "Correlation between Thermal Resistance and Structure Parameters of Fibrous and Cellular Media, in Relation with the Heat Transfer by Radiation," paper presented at the Seventh European Thermophysical Properties Conference, Antwerpen, Belgium, 30 June–4 July 1980.

[*9*] Sacadura, J. F., "Initiation aux tranferts thermiques," *Technique et Documentation Lavoisier,* Paris, 1978.

[*10*] Guilbert, G., Langlais, C., Jeandel, G., Morlot, G., and Klarsfeld, S., "Optical Characteristics of Semitransparent Porous Media," *Proceedings,* European Thermophysical Properties Conference, Rome, Italy, 22–26 Sept. 1986.

Reports

Claes G. Bankvall[1]

Thermal Performance of the Building Envelope as Influenced by Workmanship

REFERENCE: Bankvall, C. G., "**Thermal Performance of the Building Envelope as Influenced by Workmanship,**" *Thermal Insulation: Materials and Systems, ASTM STP 922,* F. J. Powell and S. L. Matthews, Eds., American Society for Testing and Materials, Philadelphia, 1986, pp. 679–684.

ABSTRACT: The thermal resistance of the building envelope depends on thermal insulation and design solutions. In practice, however, workmanship may be just as important in the total thermal performance. In the building envelope, certain areas are more susceptible to deficiencies in workmanship than others. This fact should be considered in the design stage.

An ideally installed thermal insulation in a structure will have a well-known thermal resistance. Deviations from this situation will increase the *U* value. The influence is most important on walls and roofs with low *U* values, where even small deficiencies may seriously increase the heat loss.

The wind and pressure situation around a building and the air permeability of the structure may lead to air movements in the envelope. This may increase the heat loss. The importance of proper installation of wind protection in an insulated crossbar wall is shown. The increase in *U* value, especially in combination with deficiencies in the thermal insulation installation, is given. Airtightness of a structure is usually accomplished with a vapor barrier and specially designed joints between different building elements. Large air flows through a wall usually indicate cracks or openings in the vapor barrier and the inside. The increase in heat loss resulting from such flaws is shown. The influence on a building envelope with high thermal resistance is very large.

The results and discussions are related to building envelopes with high thermal resistance and low heat losses. This situation is typical of today's building design in Scandinavia.

KEY WORDS: thermal insulation, building envelope, workmanship, wind protection, airtightness

The thermal resistance of the building envelope depends on the thermal insulation and the building design. In practice, however, the result will be influenced by workmanship. This is obvious for the installation of the insulation, since the thermal resistance of the insulating material is often more than ten times greater than the resistance of the air space that the material fills. Air movements and resulting heat loss through the structure and the insulation will, to a large degree, depend on the vapor barrier, joints, and wind protection. Good workmanship is essential in all these areas. It may be crucial to the thermal resistance experienced in practice.

A sketch of a typical Scandinavian one-family house is shown in Fig. 1. Structural sections with known problems in airtightness and thermal insulation are indicated. There are several reasons for these difficulties. One is susceptibility to strong climatic conditions—for example, high wind loads and pressure differences. Another reason is the difficulty of designing the structure so that its construction can be easily carried out in practice. A third reason may be the building process itself, which makes good workmanship in some areas more difficult. In this paper, the importance of workmanship in three areas will be discussed and exemplified. These areas are the installation of thermal insulation, the application of wind protection, and the airtightness of the structure. For further information on the theoretical analysis and details of measurements, Refs *1–3* should be consulted.

[1]Professor, Swedish National Testing Institute, S-501 15, Borås, Sweden.

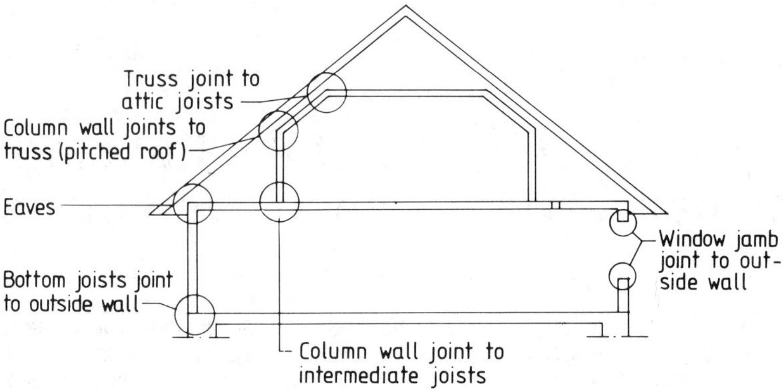

FIG. 1—*Thermal insulation and airtightness-sensitive sections on a one-family house.*

Thermal Insulation Installation

Ideally, installed thermal insulation normally leads to a well-known thermal resistance and *U* value. In practice, deviations from this are common.

Deficiencies in the installation can often be attributed to the insulation not completely filling the space to be insulated. Cracks or air spaces may occur around the thermal insulation. The consequences of this are illustrated in Fig. 2. The diagram, based on calculations and laboratory measurements, illustrates deficiencies in an insulated space, where the insulation thickness is intended to be 0.15 m, giving a nominal *U* value of 0.25 $W/m^2 \cdot °C$. The diagram is valid for normal building envelope temperatures [*3*].

At one of the crossbars there may be an air crack going from the warm to the cold side. This noticeably increases the *U* value. Also, air spaces at one side of the insulation will increase the *U* value, but to a lesser degree. The situation giving the largest, and most difficult to assess, increase in *U* value is a combination of cracks and spaces around the thermal insulation.

These difficulties will be more serious as the nominal thermal resistance of the structure is increased. Even small openings will degrade the thermal performance, especially if conditions are close to those in which natural convection may take place inside a fully insulated space [*4*]. This is due to the possibility of interaction of air movements in the material and the openings.

Similar deficiencies in other parts of a building envelope—for example, in the roof—will have similar consequences [*5,6*].

Wind Protection

The pressure field around the building envelope and the air permeability of the different materials and structural parts may lead to air flow in the structure. Such movements will increase the heat loss from the building [*7*].

The face wall will give some protection to the insulation in the structure. In this case, the design of the inlets and outlets to the normally ventilated space behind the face wall is of importance because it affects the pressure changes found at the insulated part of the structure. In a careful design for high thermal resistance, a special wind protection is often used for the insulated space (see Fig. 2). A good installation of the wind protection will give a higher thermal resistance to the structure than a poor one [*7*].

As was pointed out in the previous section, deficiencies in installation of the thermal insulation will influence the *U* value of the building envelope. It is obvious that the heat loss will increase further if air interchange can take place between the thermal insulation and the outside

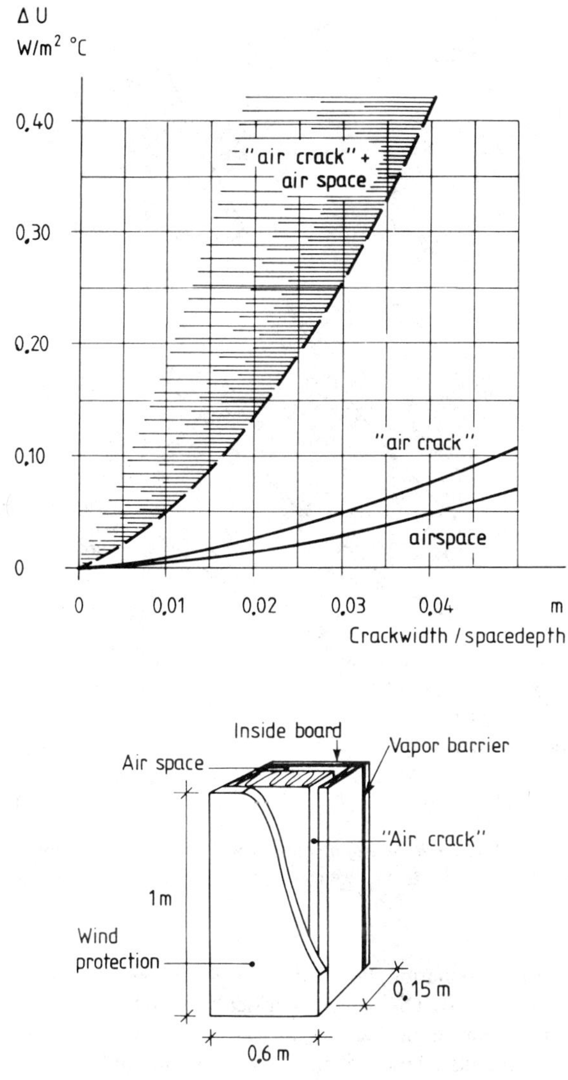

FIG. 2—*Increase in thermal transmittance from a wall section with deficiencies in the installation of the thermal insulation.*

air. The wind protection is especially important in such a case. This is illustrated in Fig. 3, which shows measurements on the wall section in Fig. 2. The figure shows the increase in U value with a pressure change of 1 Pa/m in the air space along the insulation [*7*], for both wind protection and no wind protection conditions, when there is an insulation deficiency in the form of a crack (of the indicated width) in the insulated space. In these measurements, the inside and outside air temperatures were +20 and −15°C. The insulation was a mineral wool of about 30 kg/m^3.

For the asphalt-impregnated porous board, which was carefully installed, the bottom of the marked area in Fig. 3 indicates the result due to the "built-in" crack. The additional increases in U value resulting from the air flow in the outside air space are small (top of the marked area).

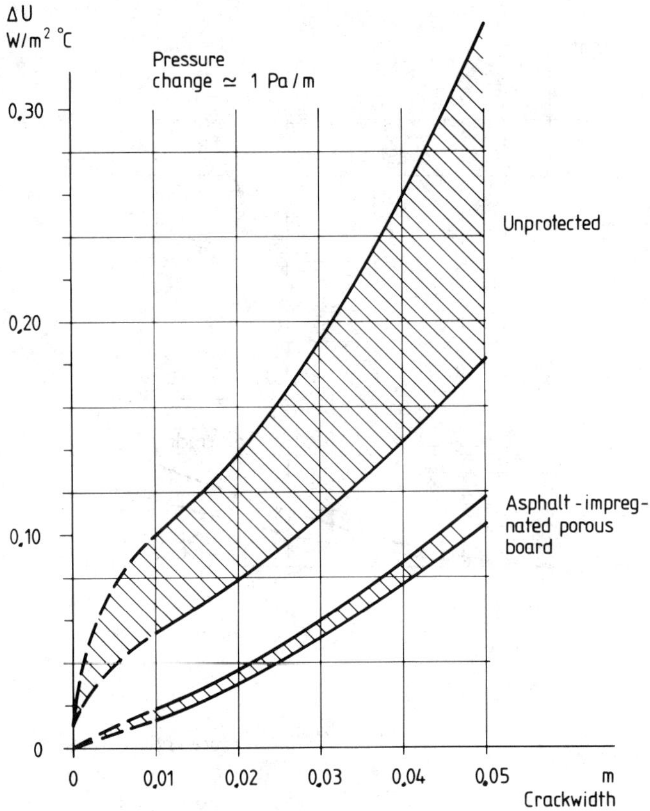

FIG. 3—*Increase in thermal transmittance of a crossbar wall due to air flow along the insulation and a vertical crack along one of the crossbars (see Fig. 2).*

For the unprotected thermal insulation, however, these effects, especially combined with installation deficiencies, are very great. They are also difficult to analyze. The situation is relatively more serious for structures with high thermal resistance. The figure shows the importance of good workmanship both for installation of thermal insulation and for application of a wind protection.

Airtightness

The airtightness of a multilayer structure is usually designed into the structure by use of a vapor barrier and by specially designed joints between different materials and building elements. If unwanted air infiltration is to be avoided, these measures must be well designed and realized by good workmanship. The importance of the vapor barrier as the main airtightness component has been shown before [7].

Even in modern designs some penetrations will occur in the building envelope. This may be caused by some of the electrical installations. Other problems related to the outside building envelope are openings for windows, doors, ventilation, and so on.

The presence of large air flows through a wall area usually presupposes deficiencies in the vapor barrier and the inside of the wall. This is illustrated in Fig. 4, which shows an example of

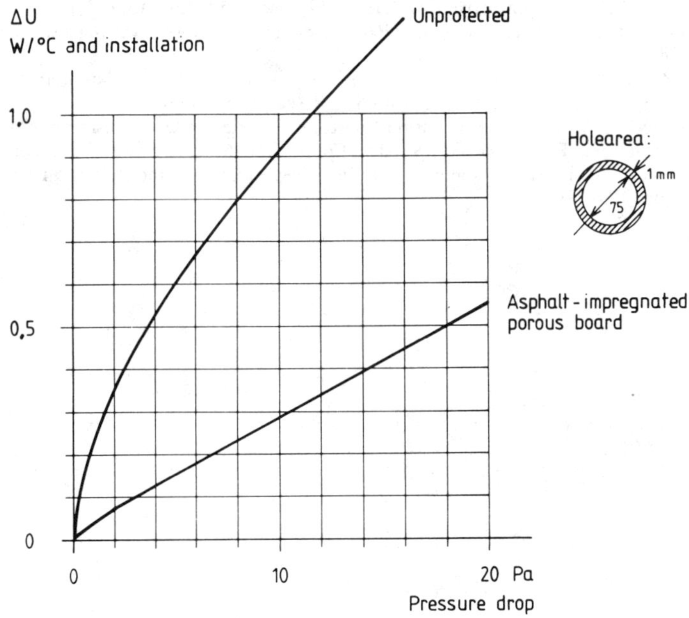

FIG. 4—*Increase in thermal transmittance due to air leakage around an electrical outlet.*

deficiencies that occur when mounting an electrical outlet in the building envelope. It is assumed that this outlet will lead to a 1-mm-wide crack around the socket, which goes through the vapor barrier and the inside board. This will allow air infiltration and, depending on the pressure difference between inside and outside [*7*], will lead to an increase in heat loss from the building. This heat loss per installation can be compared with the original *U* value of the wall section. The figure shows that both with no wind protection and with conventional wind protection the heat loss is quite large. This additional heat loss is, in most cases, independent of the thermal insulation [*7*]. The example is intended to illustrate the influence of deficiencies on the main airtightness function of the building envelope.

Conclusion

In practice, workmanship will be just as decisive for the thermal performance of the building envelope as the insulating material and the chosen thermal design. This is especially true for structures with high thermal resistance, that is, low *U* value.

Three situations are of great importance: the installation of the thermal insulation, its wind protection, and the airtightness of the envelope.

References

[*1*] Bankvall, C. G., "Forced Convection. Practical Thermal Resistance Under the Influence of Wind and Workmanship," Swedish National Testing Institute, Borås, Sweden, 1979.

[*2*] Bankvall, C. G., "Forced Convection. Practical Thermal Conductivity in an Insulated Structure Under the Influence of Workmanship and Wind," *Thermal Transmission Measurements of Insulation, ASTM STP 660,* American Society for Testing and Materials, Philadelphia, 1978.

[*3*] Bankvall, C. G., "Byggnadskonstruktioners värmeisoleringsförmåga," T18:1981, Swedish Council for Building Research, Stockholm, Sweden.

[4] Bankvall, C. G., "Natural Convective Heat Transfer in Permeable Insulation," in *Thermal Transmission Measurements of Insulation, ASTM STP 660,* American Society for Testing and Materials, Philadelphia, 1978.
[5] Pettersson, B. and Axén, B., "Thermography. Testing of the Thermal Insulation and Airtightness of Buildings," D5:1980, Swedish Council of Building Research, Stockholm.
[6] Elmroth, A. and Levin, P., "Air Infiltration Control in Housing, a Guide to International Practice," D2:1983, International Energy Agency, Swedish Council for Building Research, Stockholm.
[7] Bankvall, C. G., "Air Movements and the Thermal Performance of the Building Envelope," in this publication, pp. 124–131.

David Ober[1]

Final Report on ASTM Committee C-8 Thermal Conductivity Ceramic Fiber Round Robin

REFERENCE: Ober, D., "**Final Report on ASTM Committee C-8 Thermal Conductivity Ceramic Fiber Round Robin,**" *Thermal Insulation: Materials and Systems, ASTM STP 922,* F. J. Powell and S. L. Matthews, Eds., American Society for Testing and Materials, Philadelphia, 1987, pp. 685–703.

ABSTRACT: A 1978 round robin analysis of guarded hot plate and calorimeter apparatuses for thermal testing of ceramic fiber insulation between temperatures of 260 and 1093°C (500 and 2000°F) is reported. A significant difference between laboratories was observed in the results, even though the results of individual laboratories were reproducible. Factors to be considered in a round robin using this material are discussed, and some of the possible sources for the differences between laboratories are reviewed. The major result of the round robin was the large range of measured thermal conductivities that different laboratories are obtaining for this material.

KEY WORDS: thermal insulation, ceramic fiber insulation, insulation, testing, test methods, high-temperature tests, thermal conductivity, round robin, interlaboratory comparison

Nomenclature

σ = Standard deviation of a Stefan-Boltzmann constant
B_1 = Backscattering parameter, m^2/kg ($ft^3/in. \cdot lb$)
B_2 = Gas conduction coefficient, $W/m \cdot K$ ($Btu \cdot in./h \cdot ft^2 \cdot {}^\circ F$)
B_3 = Solid conduction coefficient, $W \cdot m^2/kg \cdot K$ ($Btu \cdot in. \cdot ft^3/h \cdot ft^2 \cdot {}^\circ F \cdot lb$)
D = Density, kg/m^3 (lb/ft^3)
R^2 = Goodness-of-fit parameter
L = Test thickness, m (in.)
T = Temperature, K (R)
K = Thermal conductivity, $W/m \cdot K$ ($Btu \cdot in./h \cdot ft^2 \cdot {}^\circ F$)
S = Fractional shot content
E = Emissivity

Subscripts

c = Cold plate
h = Hot plate
m = Mean temperature
b = Bulk density including shot
f = Bulk density excluding shot

[1]Senior research engineer, Manville Service Corp., Research and Development Center, Denver, CO 80217.

ASTM Subcommittee C08.05 on Heat Transfer and Insulation, a subcommittee of ASTM Committee C-8 on Refractories, initiated a round robin analysis for ceramic fiber insulation material in March 1978. The purpose of the round robin was to gain information on the comparability of two techniques for determining the thermal conductivity of these materials. Since several methods are currently in use, the participants were free to select the apparatus and method of their choice. This would provide an evaluation of the variability currently existing among the methods and techniques employed. Industrial, commercial, and university laboratories were involved in the round robin.

Apparatus

Two types of test apparatus were used by the participants in the round robin: the guarded hot plate and the water calorimeter.

The guarded hot plate (GHP) is covered by the ASTM Test for Steady-State Thermal Transmission Properties by Means of the Guarded Hot Plate (C 177-76). It consists of a central hot plate that has a center metering section separated from a guard ring by an insulated gap. The power is measured in the central section while the guard is maintained at the same temperature as the meter section to prevent lateral heat flow in the plate. A test specimen is placed on each side of the hot plate between it and a cold plate. Sufficient insulation is placed around the assembly to isolate it from the environment. Thermocouples measure the temperatures of the hot and cold plates and may be mounted either in the plates or between the specimens and the plates. The test procedure consists of establishing steady but different temperatures on the hot and cold plates and measuring the power input to the metering section. The distance between the hot and cold plates is maintained by gage blocks.

The water calorimeter (CAL) is covered by the ASTM Test for Thermal Conductivity of Refractories [C 201-68 (1979)]. It is a one-sided apparatus that consists of a copper plate divided into a central meter section, an inner guard, and an outer guard through which water flows. The specimen is placed between the copper plate and a carborundum plate above which electric heater elements reside to provide power. Thermocouples are placed between the carborundum plate and the test specimen, halfway between the two plates within the 50.8-mm (2-in.) test specimen, and between the test specimen and the copper plate. The heat flow is determined by measuring the water flow rate and the temperature rise between the inlet and outlet of the central metering section. Gage blocks are used to maintain the spacing between the copper and carborundum plates. The specimen is brought to steady-state test conditions, and the temperatures and heat flow are measured.

The guarded hot plate and water calorimeter are similar in some respects. Both are steady-state apparatuses designed to establish planar, one-dimensional heat flow in the metering area of the test specimen. The guarded hot plate uses two test specimens [usually each 25.4 mm (1 in.) thick] on each side of the hot plate to ensure that all of the heat input goes through the specimens. Thus, the average conductivity of the two specimens is obtained and care must be taken to select nearly identical specimens. The calorimeter uses only one specimen, usually 50.8 mm (2 in.) in thickness. Since the test specimen covers only one side of the copper plate, where the heat flow is measured, the heat transfer between the back side of the copper plate and the room air must be minimized to obtain accurate results.

The temperatures of the hot and cold plates of the guarded hot plate are independently adjustable over the operating range of the instrument; thus, the temperature difference can be constant throughout the test. In the water calorimeter, the copper (cold) plate cannot be above 100°C (212°F) and is usually around 38°C (100°F). The hot side temperature varies between 260 and 1316°C (500 and 2400°F), which results in a variable temperature difference across the test specimen and a maximum mean temperature of 704°C (1300°F) for the 50.8-mm (2-in.) specimen. To obtain higher mean temperatures, a thermocouple is placed midway in the specimen, in effect creating an upper and a lower specimen. The mean temperature of the upper specimen is then of the order of 927°C (1700°F). A critical assumption of this method is that the radiative properties of the thermocouple are the same as the insulation so that the thermocouple can measure the temperature correctly. It should be noted that the calorimeter was originally designed for measuring high-conductivity (high-density) refractory bricks and cements resulting in higher heat flows than those encountered with ceramic fiber insulation. For refractory bricks, the ASTM Test for Thermal Conductivity of Refractory Brick [C 202-71 (1977)] provides a standard test method.

Material Selection

The test specimens were aluminosilicate fibers manufactured in a blanket form to a nominal 25.4-mm (1-in.) thickness. Five consecutive rolls 4.6 m^2 (50 ft^2) of material were obtained. Sufficient specimens were cut for each apparatus to allow selection of specimens with the same density. This was necessitated by the density variation of the material. The average density was found to be 110 kg/m^3 (6.9 lb/ft^3).

This material was selected purposely to introduce the effects of high radiative heat transfer on the thermal conductivity results. Although a higher-density material (which would be easier to handle) could have been selected, it would have compared the apparatuses only on a conduction heat transfer basis rather than including the radiation transfer.

Round Robin Procedure

Each participant received enough material (already cut to size) for two tests using the apparatuses involved in the round robin with specimens selected to be as close as possible to the nominal density. The two tests would provide information on the intralaboratory precision. Two 25.4-mm (1-in.) specimens were provided for each guarded hot plate test (one for each side of the apparatus). Two 25.4-mm (1-in.) specimens were also provided for each calorimeter test. In this case, the specimens were stacked to provide a 50.8-mm (2-in.) specimen.

For each test, four ascending (in temperature) points and one descending point were obtained to provide information on reproducibility. The nominal temperatures were as follows: for the guarded hot plate—mean temperatures of 260, 538, 816, and 1093°C (500, 1000, 1500, and 2000°F); for the calorimeter—hot side temperatures of 260, 649, 982, and 1316°C (500, 1200, 1800, and 2400°F). After testing, the data and specimens were returned to Manville Service Corp., where the metering section density and shot (unfiberized) content were determined.

Test Results

The original intent was to select specimens with a density of 128 kg/m^3 (8.0 lb/ft^3), but because of material variability this was not possible. Most of the tests were run at the 110 kg/m^3 (6.9-lb/ft^3) density. However, some participants performed the tests at 128 kg/m^3 (8 lb/ft^3) by compressing the specimens. Another problem occurred in maintaining a constant test thickness. The thicknesses before and after testing (where reported) are shown in Table 1. The laboratory is indicated by a letter and the individual test by a number. Thus, B2 indicates the replicate test by Laboratory B. As shown in the table, significant changes occurred in Tests A1, C1, and C2, and it is difficult to ascertain which thickness should be used in the data analysis. The method used to measure the thickness before and after testing and the method for maintaining the specimen thickness during the test were left to the discretion of the participants and consequently are a potential source of variability between laboratories. The cause of the thickness changes is unknown, but these changes may be caused by some laboratories not using gage blocks and relying on the material to support the carborundum plate.

TABLE 1*a*—*Comparison of thickness measurements, in inches,*[a] *by calorimeter, before and after testing, as reported by participants.*

	Overall		Bottom Layer		Top Layer	
Test Code	Before	After	Before	After	Before	After
A1	1.4[b]	1.16	...	...	...	...
B1	2.00[b]	...	1.00[b]	...	1.00[b]	...
B2	1.88[b]	...	0.94[b]	...	0.94[b]	...
C1	2.198	2.000[b]	...	0.950	...	1.050
C2	2.198	2.000[b]	...	0.950	...	1.050
F1	1.70	1.766[b]	0.85 est[c]	0.970[b]	0.85 est	0.796[b]
F2	1.70	1.83[b]	0.85 est	1.02[b]	0.85 est	0.81[b]

[a]1 in. = 25.4 mm.
[b]Thickness used in calculations by participants for calculating thermal conductivity and density.
[c]Est = estimated as one half of total thickness.

TABLE 1*b*—*Comparison of thickness measurements, in inches,*[a] *by guarded hot plate before and after testing, as reported by participants.*

	Side 1		Side 2	
Test Code	Before	After	Before	After
G1	0.873[b]	...	0.900[b]	...
D1	0.963[b]	0.960	0.963[b]	0.957
D2	1.052[b]	1.034	1.052[b]	1.041
H1	1.05[b]	...	1.05	...

[a]1 in. = 25.4 mm.
[b]Thickness used in calculations by participants for calculating thermal conductivity and density.

Table 1 indicates which thickness was used by the participant to calculate the thermal conductivity and which was used in this analysis. Table 2 shows the test densities and also indicates which density was used in the data analysis. The densities are based on the appropriate thickness from Table 1. The unfiberized (Table 3) portion was determined after the test by Manville, since this test destroys the specimen. Tables 4 and 5 show the test results reported by the participants for the GHP and CAL tests. Some of the planned duplicate tests were not performed because of time limitations; each test takes approximately one week.

Factors Affecting the Test Results

In any round robin, it is necessary to eliminate all sources of variability except those under investigation. In this round robin, the purpose of the analysis was to determine the variability between laboratories using two types of test apparatuses to measure the thermal conductivity of ceramic insulation. Thus, the main objective was to determine the variability between laboratories and between different types of apparatuses. Since three GHP and four CAL apparatuses were used, interlaboratory precision of the GHPs and CALs could be determined separately, and some idea of systemic differences could be obtained.

Ideally, a single specimen (or identical specimens) would have been tested by each laboratory under identical test conditions (that is, identical in the test conditions known to affect the mea-

TABLE 2—*Density, in pounds per cubic foot, of specimens measured by participants.*

Test Code	Before Testing	After Testing	Meter Section
A1	8.1[b]	10.1	...
B1	6.90[b]	...	7.14
B2	6.96[b]	...	6.85
C1	...	6.92[b]	...
C2	...	6.84[b]	...
D1	6.90[b]	...	...
D2	6.90[b]	...	...
F1	7.7	...	8.4[b]
F2	8.2	...	8.3[b]
G1	8.0[b]	...	...
H1	7.0[b]	...	...

[a]1 lb/ft^3 = 16.0 kg/m^3.
[b]Density used in analysis.

TABLE 3—*Comparison of fractional shot content and mean fiber diameter.*[a]

Test Code	Shot Content, %	Mean Fiber Diameter, 10^{-6} *m*
D1	32.2	3.7
D2	28.5	3.8
G1	30.1	3.8
H1	28.1	3.8
A1	[b]	[b]
B1	27.1	3.3
B2	[b]	[b]
C1	31.0	3.9
C2	30.4	3.9
F1	29.3	4.1
F2	25.7	3.5
Average	29.1	3.8

[a]All measurements were made by Manville Service Corp.
[b]Not measured.

sured k). Since this was not possible, it was necessary to consider the effects caused by both the material characteristics and the apparatus/test method experimental characteristics that affect the measured thermal conductivity.

Specifically, the material characteristics are those that affect the heat transfer through ceramic fiber insulations. Heat is transferred through fiber insulation primarily by radiation and gas conduction, with lesser contributions from solid conduction and convection. The gas conduction is primarily a function of temperature. The radiation heat transfer is a function of the number of effective scattering centers (bulk density less the unfiberized portion), the scattering center optical properties (index of refraction and scattering parameter), and the scattering center size (fiber diameter). The gas and solid conduction are approximately linear functions of

TABLE 4—*Calorimeter test data as reported by participants.*[a]

Test Code	Test Point	Temperature, °F Hot	Middle	Cold	Thermal Conductivity, Hot to Cold, Btu · in./h · ft^2 · °F
A1	1	820	534	95	0.313
	2	1445	1008	177	0.468
	3	2087	1537	309	0.734
	4	2369	1807	409	1.005
	5	2032	1526	319	0.785
B1	1	468	291	75	0.31
	2	967	627	82	0.44
	3	1471	1012	98	0.63
	4	1954	1399	128	0.85
	5	1470	1008	99	0.62
B2	1	455	286	75	0.33
	2	960	629	83	0.43
	3	1464	1016	96	0.62
	4	1950	1409	117	0.86
	5	1463	1015	97	0.62
C1	1	773	...	75	0.99
	2	1403	...	110	1.01
	3	1990	...	156	1.04
	4	2240	...	191	1.24
	5	1994	...	168	1.15
C2	1	800	...	80	0.69
	2	1454	...	93	0.71
	3	2045	...	134	0.94
	4	2231	...	151	1.08
	5	2051	...	136	.94
F2	1	605	379	72	0.352
	2	1181	787	91	0.524
	3	1787	1284	122	0.836
	4	2380	1787	175	1.269
F1	1	615	398	74	0.361
	2	1187	818	91	0.532
	3	1782	1326	123	0.841
	4	2380	1836	193	1.288
	5	1779	1328	136	0.861

[a]Conversions:

$$°C = \frac{°F - 32}{1.8}$$

1 W/m · °C = 6.938 Btu · in./h · ft^2 · °F

TABLE 5—*Guarded hot plate test data as reported by participants.*[a]

Test Code	Test Point	Temperatures, °F				Thermal Conductivity, Btu · in./h · ft² · °F
		Hot	Cold	ΔT	Mean	
D1	1	598	400	198	499	0.414
	2	902	701	201	802	0.615
	3	1197	1000	198	1099	0.903
	4	1401	1200	201	1300	1.148
	5	1502	1301	201	1402	1.291
	6	1801	1598	204	1700	1.732
	7	998	799	199	898	0.729
D2	1	600	401	199	500	0.395
	2	902	700	201	801	0.600
	3	1000	799	200	899	0.684
	4	1200	999	201	1100	0.878
	5	1398	1200	198	1299	1.110
	6	1785	1556	229	1670	1.618
	7	998	799	199	899	0.697
G1	1	312	112	200	212	0.313
	2	500	300	200	400	0.421
	3	1000	800	200	900	0.752
	4	1400	1200	200	1300	1.093
	5	1800	1600	200	1700	1.491
H1	1	510	310	200	410	0.45
	2	1011	811	200	911	0.83
	3	1420	1220	200	1320	1.35
	4	1819	1619	200	1719	1.91
	5	1037	837	200	937	0.86

[a]Conversions:

$$°C = \frac{°F - 32}{1.8}$$

$$1 \text{ W/m} \cdot °C = 6.938 \text{ Btu} \cdot \text{in./h} \cdot \text{ft}^2 \cdot °F$$

temperature and thickness. However, the radiation is a nonlinear function of these parameters and, additionally, depends on the density, shot content, and fiber diameter.

The important test characteristics are the temperature level (mean temperature), temperature difference, thickness, and emittance of the hot and cold plates. It is important to emphasize that if two specimens are tested at the same mean temperature, density, shot content, and fiber diameter but at either different thicknesses or different temperature differences, or both, two different thermal conductivities will be obtained. For ceramic fiber material this result is primarily due to the radiative heat transfer through the material. These facts are important since the test conditions for the GHP and CAL are different. The GHP uses a 25.4-mm (1-in.) test thickness and a 93°C (200°F) temperature difference whereas the CAL uses a 50.8-mm (2-in.) test thickness and a variable temperature difference. Thus, even if the GHP and CAL were perfectly accurate and the same specimen were tested in both, different thermal conductivities would be obtained, and one would erroneously conclude that a difference existed between the apparatuses. It should also be emphasized that present data reduction techniques, which only calculate a thermal conductivity for the specific test conditions in the respective test methods, fail to provide any means for dealing with these factors.

It is apparent that any data reduction technique must account for (1) material variability in the specimens tested by each laboratory and (2) differences in the test conditions between the GHP and CAL. A mathematical model that models the radiation, gas conduction, and solid conduction through the insulation was selected. This model has been used by Striepens [*1*] and by Linford et al [*2*] for fibrous insulations and is a compromise between more complicated models, which are very difficult to use, and simpler models, which do not account for as many factors. The model is shown in the Appendix.

With this model, the procedure to analyze the data is as follows:

1. The appropriate parameters are fitted to each data set by nonlinear least squares (NLLS).
2. The test results are then compared at identical specimen and test conditions—that is, density, shot content, thickness, mean temperature, and temperature difference.

In short, this is an analytical procedure to reduce the variability in the test results due to nonidentical specimen and test conditions.

Least Squares Fit to the Data

Since the model is quite flexible, a wide choice of parameters could be selected to be fitted to the data. However, since there were only four data points for most of the tests, the parameters had to be kept to under four, and, if any statistics were to be obtained, the number of parameters had to be less than four. Thus, the most important parameters were selected, and values were assigned for the less important parameters. The two most important terms are the radiation and gas terms, and the most important parameters in these are B_1 and B_2. Except as noted in the following section, these terms were fit to the test data by the NLLS routine. The other four terms were chosen as follows.

The emittance of the plates was assumed to be 0.85 and independent of temperature. The GHP method calls for an emittance of 0.82 or greater. The CAL has no specification and presents some problems. For the cold side, the copper plate is initially polished but becomes oxidized during the test with the emittance starting low (0.1) and ending high (0.8). The carborundum plate has an e of 0.86. Thus, a constant emittance for the CAL is questionable. This is not a serious problem if the emittance is greater than 0.5, since the emittance term is a small fraction of the denominator of the radiation term. However, if the emittance is 0.1 throughout the test, this would be one third of the term and thus significant. Since the emittance for the CAL is unknown, it was assumed to be 0.85. An obvious improvement in the CAL test would be to fix emittance to a known constant. A high emittance is desirable since it then has little effect on the denominator and thus does not have to be known precisely. This could be easily accomplished by painting the surface black.

The gas conduction exponent was determined by fitting thermal conductivity data for air from Ref *3* to

$$k = B_2\left[\frac{T_m}{492}\right]^a \tag{1}$$

Coefficient B_2 determined in this way was 0.0244 W/m · °C (0.169 Btu · in./h · ft^2 · °F), and the exponent was found to be 0.742. The value of B_2 found by fitting the gas term to the thermal conductivity of air can be compared with the B_2 term found by fitting the model to the test data on the round robin material. The solid conduction coefficient was set at 4.5 (10^{-5}) W · m^2/°C · kg (0.005 Btu · in./h · ft^2 · °F) and was extracted from Ref *4*. Since this term has a small effect on the total k, its value does not need to be known precisely. All other values (T_h, T_c, L, D_b, S) were determined from the specimen and test conditions. For the CAL, the hot and cold surface temperatures (see Table 4) were used in the analysis. Except as noted, the middle temperature data were not used.

For all of the laboratories except Laboratory C, coefficients B_1, B_2 were fit to the data (Table 6). Coefficient B_3 was set to 4.5 (10^{-5}) W · m²/°C · kg (0.005 Btu · in. · ft³/h · ft² · °F · lb) [4]. For Laboratory C, all three coefficients were used to achieve an adequate fit to the data. The poor ability of the model to fit these data is due to the low slope of the k versus T curve at low temperatures. This slope is not supported by the other data. In addition, this laboratory showed poor intralaboratory reproducibility. This is probably because this was the first time the CAL apparatus had been used by the laboratory, and without a specific test method the laboratory had not gained the experience necessary to test this material successfully. The coefficients, the standard deviation of the coefficients, the standard error of estimate, and the coefficient of determination resulting from the NLLS fits to each data set are given in Table 6. For Laboratory C, physically unrealistic B_2 coefficients were obtained. For all of the other laboratories the B_2 coefficients were between 0.017 and 0.035 W/m · °C (0.12 and 0.24 Btu · in./h · ft² · °F). This compares favorably with the coefficient of 0.0244 W/m · °C (0.169 Btu · in./h · ft² · °F) obtained from fitting Eq 1 to thermal conductivity data for air.

A final issue for the model fits was how to handle the extra data generated by the middle thermocouple for the CAL tests. As previously mentioned, this method generated data at higher mean temperatures, but its accuracy was questionable because of uncertainties in the thermocouple location within the specimen and the effect of the high radiative flux on the accuracy of

TABLE 6—*Model parameters determined by least squares.*[a,b]

Test Code	Parameter: Radiation		Gas Coefficient		Solid Coefficient		Standard Error of Estimate, Btu · in./h · ft² · °F	R^2[c]
	B_1	σ	B_2	σ	B_3	σ		
A1	14.956	2.08	0.1208	0.025	0.005	N/A[d]	0.041	0.9876
B1	15.861	0.81	0.1903	0.0063	0.005	N/A	0.010	0.9987
B2	15.566	1.07	0.1920	0.0084	0.005	N/A	0.014	0.9977
C1	11.804	9.14	−0.5054	0.560	0.353	0.18	0.077	0.8502
C2	8.006	0.46	−0.6473	0.062	0.327	0.02	0.008	0.9995
F1	10.736	0.21	0.1843	0.0043	0.005	N/A	0.008	0.9998
F2	10.360	0.19	0.1802	0.0041	0.005	N/A	0.007	0.9998
D1	12.257	0.39	0.1836	0.0084	0.005	N/A	0.019	0.9988
D2	12.189	0.37	0.1768	0.0071	0.005	N/A	0.017	0.9987
G1	15.756	1.40	0.2379	0.016	0.005	N/A	0.037	0.9957
H1	11.825	0.93	0.2450	0.022	0.005	N/A	0.044	0.9968

[a]Conversions:

1 W/m · °C = 6.938 Btu · in./h · ft² · °F
1 m²/kg = 0.406 ft³/lb · in.
1 W · m²/°C · kg = 111 Btu · in. · ft³/h · ft² · °F · lb

[b]Units:

B_1 = ft³/lb · in.
B_2 = Btu · in./h · ft² · °F
B_3 = Btu · in. · ft³/h · ft² · °F · lb

[c]R^2 = coefficient of determination.
[d]N/A = only B_1 and B_2 were fit; B_3 = 0.005 was fixed.

the temperature measurement. These factors pointed toward using only the hot plate to cold plate (H/C) data. To investigate this effect, fits were made using the H/C data (four points) and all of the data generated using the middle thermocouple (twelve points) for each CAL test. Laboratory C's test could not be fit to the twelve points (because of nonconvergence of the NLLS) and was not included. As can be seen from Table 7, the two fits were very close for Tests B1, B2, and F1 but different for A1 and F2. The differences between the two fits for A1 may be due to the large change in thickness that occurred during the test. The differences for Test F2 are unknown, but may be due to the greater change in thickness during the test than F1 experienced. It is encouraging that for three of the tests the model using four data points agreed with that using twelve points, thus indicating that with some techniques the middle thermocouple can generate useful data. However, the poor agreement of Tests A1 and F2 using the two fits shows that more work is needed to determine if this method is indeed valid. For these reasons, only the H/C data were used in subsequent analysis.

The objective of fitting the model to each set of test data was to be able to compare the tests on a common basis as if identical specimens had been tested by all the laboratories under identical

TABLE 7—*Comparison of calorimeter analysis using four and twelve data points.*[a]

Laboratory	Mean Temperature, °F	Thermal Conductivity, Btu · in./h · ft² · °F		Difference, %
		4 Points	12 Points	
A1	500	0.305	0.266	−14.7
	1000	0.582	0.579	−0.5
	1500	1.05	1.15	8.6
	2000	1.79	2.06	13.1
B1	500	0.415	0.420	1.2
	1000	0.722	0.723	0.0
	1500	1.21	1.20	−0.8
	2000	1.95	1.91	−2.1
B2	500	0.419	0.420	0.2
	1000	0.731	0.735	0.5
	1500	1.23	1.24	0.8
	2000	1.98	1.99	0.5
F1	500	0.441	0.452	2.5
	1000	0.835	0.831	−0.5
	1500	1.50	1.46	−2.7
	2000	2.53	2.42	−4.5
F2	500	0.438	0.468	6.4
	1000	0.840	0.825	−1.8
	1500	1.52	1.40	−8.6
	2000	2.58	2.26	−14.2

[a]Conversions:

$$°C = \frac{°F - 32}{1.8}$$

$1\ W/m \cdot °C = 6.938\ Btu \cdot in./h \cdot ft^2 \cdot °F$

test conditions. Table 8 shows the results at selected temperatures for each laboratory for the following common test conditions:

D_b = 110 kg/m³ (6.9 lb/ft³),
S = 29%,
L = 25.4 mm (1 in.), and
ΔT = 93.3°C (200°F).

TABLE 8—*Comparison of individual thermal conductivity tests (in Btu · in./h · ft² · °F) at common test conditions (D = 6.9 lb/ft³; S = 29%; L = 1 in.; ΔT = 200°F).*[a]

Apparatus	Test Code	Mean Temperature, °F			
		500	1000	1500	2000
CAL	A1	0.305	0.583	1.05	1.79
	B1	0.415	0.722	1.21	1.95
	B2	0.419	0.731	1.23	1.98
	C1	1.00	0.960	1.19	1.79
	C2	0.690	0.679	1.07	1.98
	F1	0.441	0.835	1.50	2.53
	F2	0.438	0.840	1.52	2.58
	x[b]	0.530	0.764	1.25	2.09
	σ	0.238	0.124	0.188	0.331
	CV[c]	90	32	30	32
GHP	D1	0.426	0.785	1.38	2.30
	D2	0.415	0.772	1.37	2.28
	G1	0.493	0.831	1.35	2.11
	H1	0.530	0.935	1.58	2.56
	x	0.465	0.830	1.42	2.31
	σ	0.0549	0.0739	0.107	0.186
	CV	24	18	15	16
All data	x	0.506	0.788	1.31	2.17
	σ	0.190	0.109	0.179	0.299
	CV	75	28	27	28
	minimum	0.305	0.583	1.05	1.79
	maximum	1.00	0.960	1.58	2.58
	ratio[d]	137	48	40	36

[a]Conversions:

$$°C = \frac{°F - 32}{1.8}$$

1 W/m · °C = 6.938 Btu · in./h · ft² · °F

[b]x = average.

[c]$CV = \frac{2}{x} \sigma 100$ coefficient of variation.

[d]$\text{Ratio} = \frac{\text{maximum} - \text{minimum}}{x} 100.$

Plots of the results are shown in Figs. 1 and 2 for the GHP and CAL, respectively. The first comparison (Table 8) includes all of the data, since the results of Tests C1 and C2 were very different from each other. The average, standard deviation, and coefficient of variation (which normalizes the standard deviation) have been calculated for the CAL and GHP separately and then for all of the data combined. Also included for all of the data are the minimum, maximum, and range expressed as a percentage of the overall average. The largest variation between tests was at 260°C (500°F), at which temperature the range was 137% of the average. At higher temperatures, the range was 48% at 538°C (1000°F), 40% at 816°C (1500°F), and 36% at 1093°C (2000°F). The GHP apparatuses also appear to show less scatter, although this may be due to fewer of them participating in the round robin. There is a significant difference between tests.

The next comparison was of the ability of each laboratory to reproduce results on the two test specimens. This includes the precision of the apparatus plus interspecimen variability. For the laboratories that tested both specimens, the reproducibility (Table 9 and Fig. 3) is very good except for Laboratory C. Thus, reproducible results can be obtained for this material. Furthermore, the variability in the data is due to systematic differences between the laboratories. Any approach to improving the test results must address these systematic differences. In the development of a test method, it is not sufficient for one laboratory to show repeatability, more serious problems exist between laboratories than within them.

To exclude the undue influence of replicate tests by each laboratory on the results that exist in Table 8, Table 10 has been prepared by averaging the results of the replicate tests for Laboratories B, C, F, and D. For the CAL the coefficient of variation is between 31 and 94%, and for the GHP it is between 18 and 23%. For all of the data combined, the coefficient of variation is between 28 and 69%. The range is between 35 and 110% for all of the data. For both the coefficient of variation and the range, the largest scatter is at 260°C (500°F) because of Laboratory C's results. From 538°C to 1093°C (1000°F to 2000°F) the scatter is much smaller.

The CAL and GHP are compared in Table 11. The first comparison is without Laboratory C and the second is with Laboratory C. Without Laboratory C, the difference between the averages of the GHP laboratories and that of the CAL laboratories is 10 to 24%, with the GHP being higher. When Laboratory C is included, the CAL average is raised, and the difference is reduced. The result without Laboratory C is probably more representative. Although a systematic difference may exist between the CAL and GHP, it is more meaningful to note that because of the wide range of the results, given any two laboratories, either the GHP or CAL could be higher (or lower). Before any conclusions could be drawn, both methods need to be more reproducible between laboratories.

Conclusions

The round robin demonstrated that there is a difference between the averages of the CAL and GHP test methods, although for particular combinations of laboratories, either method may appear to give higher (or lower) results. Fundamentally, there should be no difference between the two types of apparatus if they are properly constructed and operated. The results of this round robin indicate the following:

1. The thermal conductivity of ceramic fiber insulation showed a range of 141 to 137% between laboratories at temperatures from 260 to 1093°C (500 to 2000°F), indicating that standardized test results are not being obtained using current test procedures.
2. The intralaboratory precision appears to be ±2% (except as noted), showing that systematic differences exist between laboratories.
3. While the GHP appears to be more precise than the CAL, the precision of both within these temperature ranges could be improved.

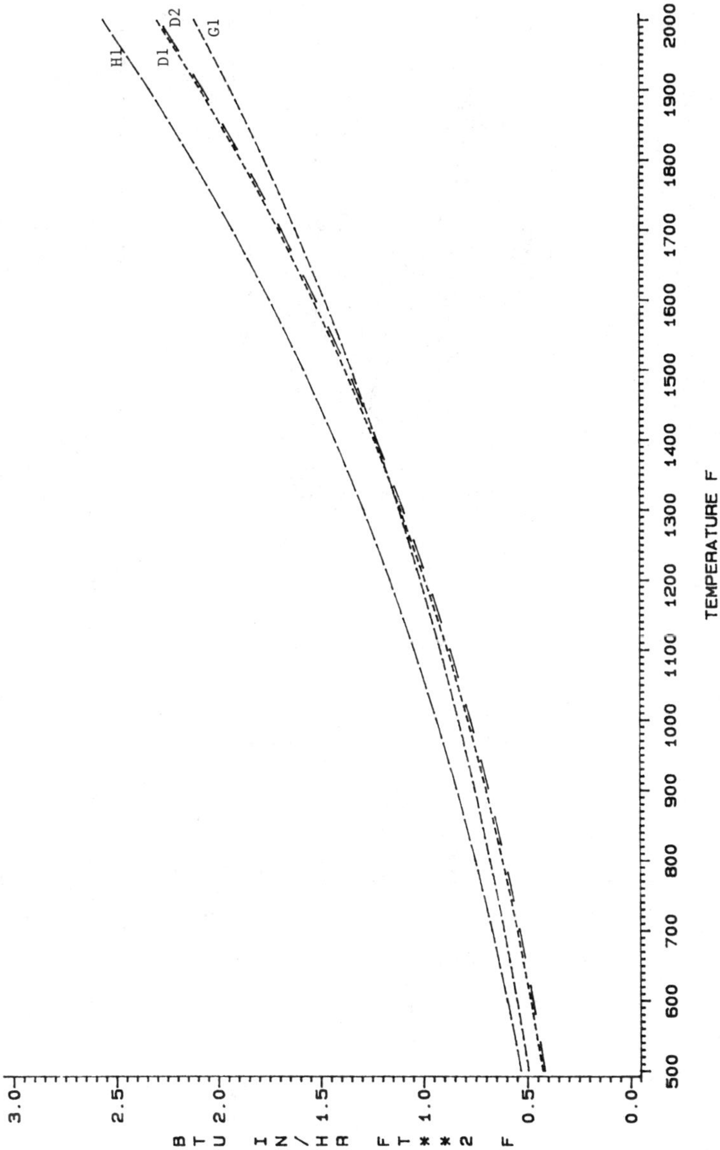

FIG. 1—*Results of GHP tests—thermal conductivity versus mean temperature.*

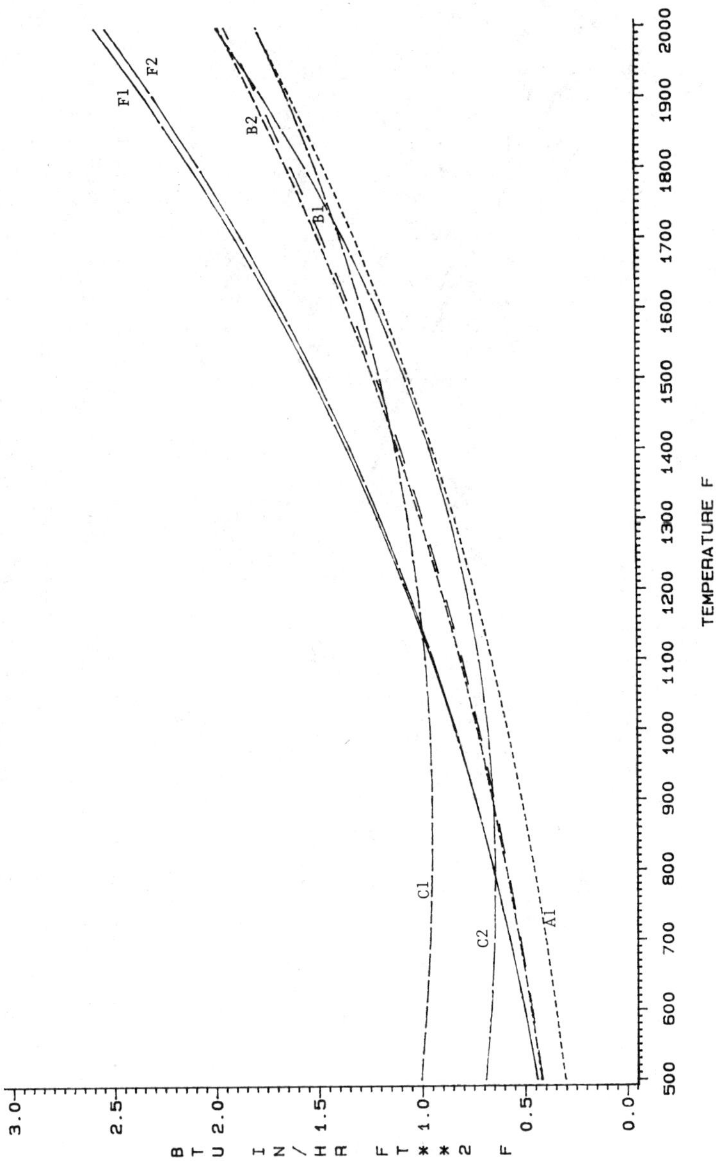

FIG. 2—*Results of calorimeter tests—thermal conductivity versus mean temperature.*

TABLE 9—*Intralaboratory reproducibility (*D = *6.9 lb/ft*^{3}*;* S = *29%;* L = *1 in.;* ΔT = *200°F).*[a]

Laboratory	Mean Temperature	Thermal Conductivity, Btu · in./(h · ft^2 · °F) Test 1	Test 2	Difference from Test 1, %
B	500	0.4146	0.4187	+1.0
	1000	0.7222	0.7310	+1.2
	1500	1.210	1.227	+1.4
	2000	1.946	1.976	+1.5
C	500	1.003	0.6901	−31
	1000	0.9588	0.6795	−29
	1500	1.194	1.072	−10
	2000	1.789	1.985	+11
D	500	0.4259	0.4152	−2.6
	1000	0.7851	0.7718	−1.7
	1500	1.379	1.365	−1.0
	2000	2.296	2.282	−0.6
F	500	0.4408	0.4381	−0.6
	1000	0.8349	0.8398	+0.6
	1500	1.498	1.520	+1.5
	2000	2.528	2.581	+2.1

[a]Conversions:

$$°C = \frac{°F - 32}{1.8}$$

1 W/m · °C = 6.938 Btu · in./h · ft^2 · °F

4. A systematic difference may exist between the GHP and CAL methods as they are currently implemented, however, the results are not conclusive.
5. The CAL method could be improved by specifying a known, constant emissivity for the copper plate.
6. A standardized data reduction technique would be helpful in eliminating biases introduced by the different test conditions in the GHP and CAL.
7. The use of the thermocouple in the center of the CAL specimen needs to be thoroughly investigated to determine its validity.
8. The test methods need to be improved to eliminate differences between laboratories that may be due to different implementations of the types of test apparatus.
9. There is a lack of standard specimens.

Future round robins could be improved by the following means:

(*a*) ensuring that all participants test the material at the same density to reduce the adjustments in the test data to common conditions;

(*b*) standardizing the test procedures (this was intentionally avoided in this round robin);

(*c*) obtaining more specific information on the test apparatus (such as thermocouple calibration, placement, installation, thickness measurement, heat flow measurement, area determination, control, and measurement of the guard to the meter heat flow);

(*d*) obtaining detailed information from each laboratory on all calculations; and

(*e*) having each laboratory perform replicate tests.

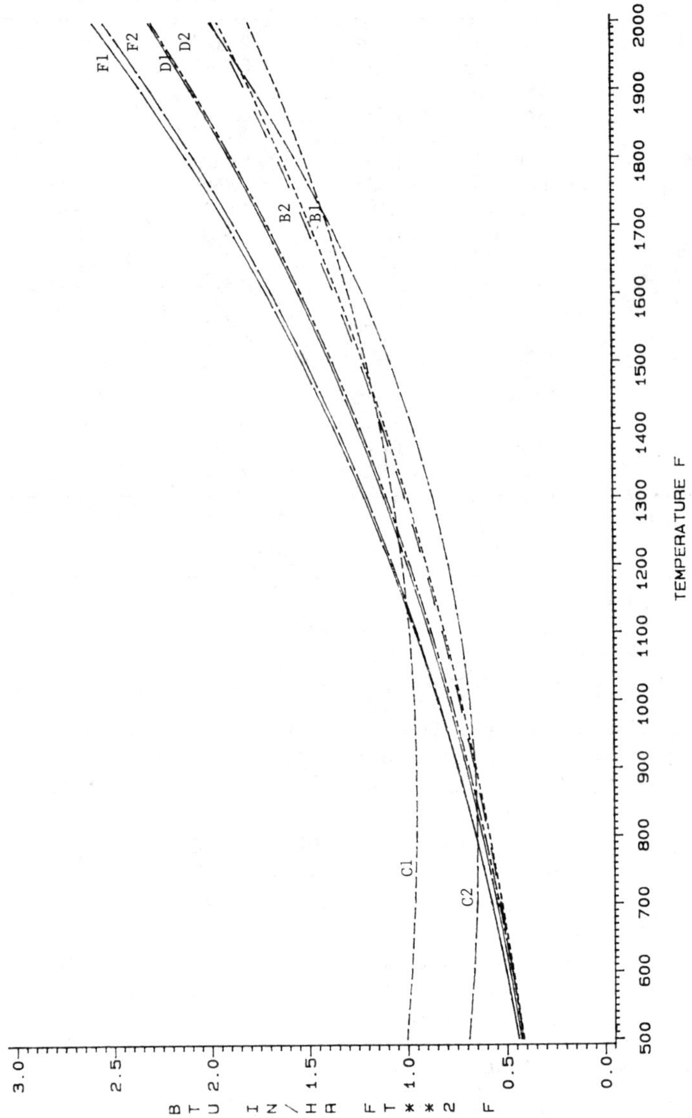

FIG. 3—*Intralaboratory reproducibility—thermal conductivity versus mean temperature.*

TABLE 10—*Thermal conductivity results (in Btu · in./(h · ft² · °F) at common test conditions using averages for each laboratory (D = 6.9 lb/ft³; S = 29%; L = 1 in.; Δ T = 200°F).*[a]

Apparatus	Laboratory	Mean Temperature, °F 500	1000	1500	2000
CAL	A	0.305	0.583	1.05	1.79
	B	0.417	0.726	1.22	1.97
	F	0.440	0.837	1.51	2.55
	C	0.845	0.820	1.13	1.89
	x	0.502	0.742	1.23	2.05
	σ	0.236	0.116	0.201	0.341
	CV[b]	94	31	33	33
GHP	D	0.421	0.779	1.37	2.29
	G	0.493	0.831	1.35	2.11
	H	0.530	0.935	1.58	2.56
	x	0.481	0.848	1.44	2.32
	σ	0.0554	0.079	0.122	0.226
	CV	23	19	18	20
All laboratories	x	0.493	0.787	1.32	2.17
	σ	0.170	0.110	0.194	0.310
	CV	69	28	29	29
	minimum	0.305	0.583	1.05	1.79
	maximum	0.845	0.935	1.58	2.56
	ratio[c]	110	45	40	35

[a]Conversions:

$$°C = \frac{°F - 32}{1.8}$$

$$1\ W/m \cdot °C = 6.938\ Btu \cdot in./h \cdot ft^2 \cdot °F$$

[b]$CV = \frac{2}{x} \sigma (100).$

[c]$Ratio = \frac{maximum - minimum}{x} 100.$

As a final note, it must be remembered that most of the data on these materials generated to date have not been adjusted for differences in the two test methods, namely, the different temperature differences and the different thicknesses. This could cause difficulty in comparing data that have been generated with the CAL and GHP.

Acknowledgments

ASTM Subcommittee C08.05 on Heat Transfer and Insulation, a subcommittee of ASTM Committee C-8 on Refractories, initiated this round robin.

TABLE 11—*Comparison of the averages of all CAL and GHP tests.*[a,b]

Average of Laboratories	Mean Temperature, °F			
	500	1000	1500	2000
CAL A, B, F	0.387	0.715	1.26	2.10
GHP D, G, H	0.481	0.848	1.44	2.32
Percentage GHP over CAL	24	19	14	10
CAL A, B, C, F	0.502	0.742	1.23	2.05
GHP D, C, H	0.481	0.848	1.44	2.32
Percentage GHP over CAL	−4	14	17	13

[a]Thermal conductivity in Btu · in./h · ft² · °F.

[b]Conversions:

$$°C = \frac{°F - 32}{1.8}$$

$$1\ W/m \cdot °C = 6.938\ Btu \cdot in./h \cdot ft^2 \cdot °F$$

APPENDIX

Mathematical Model

The model used to reduce the data contains a radiation term, gas conduction term, and solid conduction term. It can account for differences in mean temperature, temperature difference, thickness, density, and plate emittance. It cannot account for fiber diameter explicitly. Thus, differences in fiber diameters between specimens were not considered. As seen in Table 3, the variation in fiber diameter was small.

$$k = \frac{\sigma(T_h^4 + T_c^4)L}{(T_h - T_c)\left[\dfrac{1}{E_h} + \dfrac{1}{E_c} - 1 + D_f B_1 L\right]} \text{ (radiation conductivity)}$$

$$+ B_2\left[\frac{T_m}{492}\right]^{0.742} \text{ (gas conductivity)} + B_3 D_f \text{ (solid conductivity)}$$

where

$$D_f = (1 - S)D_b$$

References

[1] Striepens, A. H., "Heat Transfer in Refractory Fiber Insulations," *Thermal Transmission Measurements of Insulations, ASTM STP 660,* R. P. Tye, Ed., American Society for Testing and Materials, Philadelphia, 1978, pp. 293–309.

[2] Linford, R. M. F., Schmitt, R. J., and Hughes, T. A. in *Heat Transmission Measurements in Thermal Insulations, ASTM STP 544,* American Society for Testing and Materials, Philadelphia, 1974, pp. 68–84.

[3] Bolz, R. E. and Tuue, G. L., *Handbook of Tables for Applied Engineering Science,* Chemical Rubber Co., Cleveland, OH, 1970.

[4] Verschoor, J. D. and Greebler, P., "Heat-Transfer by Gas Conduction and Radiation in Fibrous Insulations," *Transactions of the Institute of Mining, Metallurgical, and Petroleum Engineers,* August 1952.

John R. Mumaw[1]

Human Protection from Burns by Heated Surfaces—The Problem and Solution

REFERENCE: Mumaw, J. R., **"Human Protection from Burns by Heated Surfaces—The Problem and Solution,"** *Thermal Insulation: Materials and Systems, ASTM STP 922,* F. J. Powell and S. L. Matthews, Eds., American Society for Testing and Materials, Philadelphia, 1987, pp. 704-712.

ABSTRACT: Insulation in some form has been used in the home for centuries to protect the occupants against extremes of heat and cold. With the Industrial Revolution, insulation was applied to industrial processes to conserve energy and, more important, to protect the process products during manufacture. As technology expanded, bringing more complex processes, higher temperatures, and more worker exposure situations, an increased awareness of the need for worker safety led to the use of insulation for personnel protection. Unfortunately, it was not until after World War II that adequate medical data were available to aid in the definition of a "safe" heated surface. To this day, the definition of "safe conditions" for heated surfaces varies widely among consultants, appliance manufacturers, insulation producers, and regulatory agencies. In 1983, ASTM Committee C-16 on Thermal Insulation began work toward developing standards to resolve these issues. This paper outlines the approach taken by the working task groups to establish (1) a standard definition of the human burn mechanism for heated surfaces and (2) the tools required to determine if a surface presents a hazard.

KEY WORDS: human burn protection, heated surfaces, standards, burns, personnel protection, hyperthermia, thermal insulation

The Problem

Since the discovery of fire in prehistoric times, man has faced the potential of skin burns from touching hot surfaces. At that time, he was concerned more with treating the injury than with developing some means of preventing its occurrence. As civilization advanced, man developed crude insulation forms to control the extremes of heat to which he was exposed. The greatest improvement to these systems has come since the Industrial Revolution, during which the use of high-temperature power and process systems dictated the development of modern insulation systems. These systems had not only to conserve energy but also to protect process products during manufacture. As technology expanded to include higher temperatures, more complex processes, and thus more worker exposure situations, worker organizations and, later, governmental agencies demanded the increased use of insulation for protection of personnel.

At the same time the workplace was becoming more hazardous, the increased development of consumer products that heated, steamed, and cooked was increasing the potential hazards found in consumer products. Higher operating temperatures and increased efficiency demands for consumer products forced the use of more insulation and protection for the operator. One very good example of increased hazard is the development of self-cleaning ranges, in which internal temperatures can exceed 427°C (800°F) during the clean cycle, in comparison with the 205°C (400°F) needed for most cooking operations. Personal protection is now required every-

[1]Senior engineer, Research and Development Division, Owens-Corning Fiberglas Corp., Technical Center, Granville, OH 43023.

where for consumer products. Examples include curling irons, ranges, irons, dryers, dish washers, light fixtures, furnace and heating fixtures, to name only a few.

Most readers quickly recognize the hazard potential involved with heated surfaces. The obvious solution is simply to insulate the heated part, and thus isolate the hazard from the user. Unfortunately, the random application of insulation without detailed analysis can sometimes disrupt the process, for example, overheating in instances when some loss is desired, or produce an economic handicap to the overall cost of the project. To date, most applications of insulation to heated process systems are made on the basis of trade-offs between the cost of the installed insulation and the cost of the energy lost. Using this criterion or the more common rule-of-thumb approach—that is, "put on about an inch like we always do"—can create exposed surface temperatures that exceed even the shortest-term human exposure limits. Thus, to protect both operators and casual visitors in an area, an analysis of the exposed surfaces must be undertaken to identify those having temperatures capable of causing burns.

Once consumer product and industrial system designers recognized the need to design for personnel safety, they established what they felt were safe operating limits for exposed surfaces. Since few studies were available before 1950, many industries chose to establish their own standards for maximum surface temperatures based on combinations of available research results and personal experience; this remains the current method for evaluating surface hazards. Every industry regulatory agency or consumer group has its own specification for determining hazard exposure. In 1983, ASTM Committee C-16 on Thermal Insulation undertook the study of a proposal to establish a "standard" criterion for evaluating burn hazard potential. This standard was to be well documented and easy to use. As an adjunct to this effort, a second standard was proposed to establish a means for evaluating existing or proposed systems for hazard level by either physical measurement or mathematical modeling.

The remainder of this paper deals with the definition of human burns through an overview of the existing medical research and, finally, a summary of the approach ASTM has taken in developing standards for burn hazard potential and its evaluation.

Physiological Mechanism of a Burn—Background

Before World War II, little, if any, research had been performed in developing an understanding of the physiology of burns to the human body. With the increased destruction potential of more powerful weapons, burn injuries became a common battle problem, and the military began to support research to study the relationships between burn damage and the severity of exposure. At that time, little was known about the mechanism by which hyperthermia (high temperature exposure) leads to irreversible damage. The chemical reactions occurring within the skin cells upon exposure and the relationships between exposure temperature and duration on the transfer of heat into the skin were also subjects of research.

The first significant research on the subject was conducted by Henriques and Moritz at the Harvard Medical School [*1–6*]. The results were released for publication in the open literature in 1946 through 1948. This research, performed primarily on swine (which happen to have skin properties similar to those of humans), with some human subjects added later, helped define the significant parameters controlling the flow of heat into the skin. Later, the relationship between temperature and duration of exposure to the extent of damage observed was established to serve as a guide for future work. Some of the significant results of their initial work [*1*] are as follows:

1. The burning of human skin occurs as a complex, nonsteady heat transfer between a contacted medium, that is, the hot surface, and the surface of the skin. The rate of heating depends on the temperature and heating capacity of the source and the heat capacity and thermal conductivity of the skin layers (see Fig. 1). Neglected in these studies were the flow of blood to carry

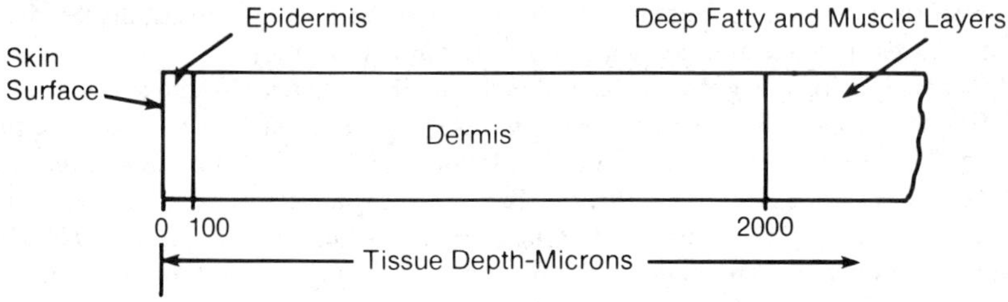

FIG. 1—*Cross section of human tissue.*

heat away and the physiological changes in skin properties as the damaged zone traverses the outer skin layers.

2. Factors that increase the complexity of the problem include (1) site variations with respect to the thickness of the different skin layers; (2) variations of initial conditions within the skin with respect to time, position, and physical condition of the subject; (3) the unknown average rate of blood flow through the skin layers and variations within the layers with respect to location and ambient temperatures (that is, warm ambient temperature causes increased flow near the surface and cold ambient temperature results in less flow near the surface); and (4) the appearance of watery fluids in variable quantities after exposure that result in alternations of skin density, heat capacity, thickness, and thermal conductivity.

3. Analysis of the experimental results showed that it was possible to assume average conditions and to develop an approximate first-order Fourier's law equation to describe the transient heat flow in the contact problem. The modeling work by Henriques neglected the influence of contact resistance and blood flow and assumed that both the skin and the touched surface could be treated as semiinfinite. Succeeding experiments showed that the assumption of semiinfinite solids and a neglecting of blood flow were valid for the time/temperature conditions of interest. The experiments performed at Harvard used a direct contact water bath, which avoided the issue of contact resistance.

After their initial work was complete, Moritz and Henriques extended their work to include the effects on human skin of hyperthermia of varying duration and varying degrees of intensity. These studies [2] led to a clearer definition of the degree of burning. Several additional conclusions were forthcoming from that research and are outlined here:

1. The pain reaction to prolonged hyperthermal exposure first occurs as a stinging sensation at between 47.5° and 48.5°C. The level of discomfort does not always correlate with the level of damage sustained or with intensity between subjects or in the same subject on different days.

2. The lowest temperature at which significant epidermal (outside skin layer) damage occurs is 44°C when exposure is sustained for 6 h.

3. As the temperature of contact increases above 44°C, the time to damage is shortened by approximately 50% for each 1-degree-Celsius rise in temperature up to about 51°C.

4. Testing showed that increasing the pressure of contact within an expected range was not sufficient to collapse the blood vessels and increase the vulnerability of the epidermis to thermal injury.

5. At temperatures above 70°C, the rate of injury from a high-capacity surface exceeds the body reaction time (that is, it takes less than 1 s to produce complete death of epidermis cells), so that the blood vessel location or flow has little effect on the level of burn.

6. The level of skin damage can be related to the duration and intensity of surface contact by the curve in Fig. 2. Exposures below the lower curve should not produce permanent injury in normal humans. Exposures between the curves are described as second-degree burns and have

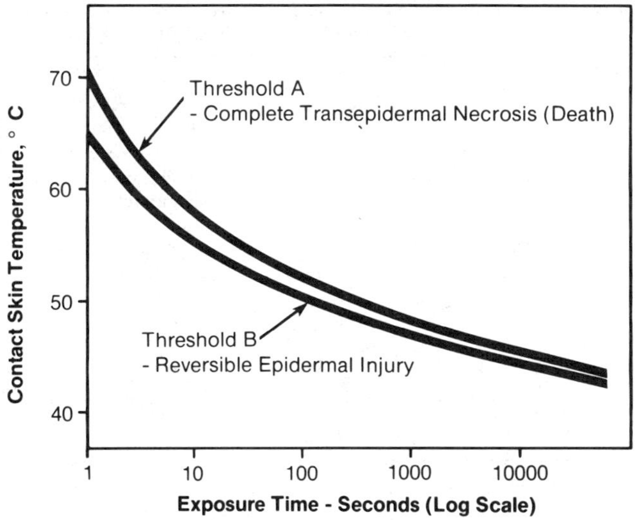

FIG. 2—*Temperature = time relationship for burns.*

intermediate levels of cell damage. Exposures at levels above the top line are defined as third-degree burns, which will cause deep, permanent cell damage and scaring.

After the initial research just described, several other researchers studied the same problems to extend the knowledge of burns to more realistic situations. Most significant here are problems with contact resistance and source surfaces having noninfinite thermal inertias. Wu [*7*] took the analysis developed by Moritz one step further by adding the heat transfer reaction for a source of high energy. His treatment, assuming contact between the two semiinfinite bodies of finite thermal inertia (as measured by the square root of thermal diffusivity) at different temperatures, showed that sources of low inertia—for example, wood, insulation, and some plastics—cause a slower rise in skin temperature than a source of high thermal inertia—for example, steel and aluminum—at the same temperature. In short, this is explained by observing that high thermal inertia materials can make more energy available at the surface in a given time than those of lower thermal inertia. Wu also pointed out that cell death (necrosis) is a result of irreversible thermal denaturation of the protein present within the cell. This denaturation is a rate process having a very high temperature coefficient, which corresponds to a very high activation energy. In short, this means that the higher the temperature of exposure, the faster damage occurs. This explanation confirms the results of Henriques and Moritz.

Wu also developed the information presented in Table 1, which outlines the relationship between the pain sensation, exposed skin color, tissue temperature at a 80-μm depth, and the cell process.

Stoll [*8*], on the other hand, looked at the relationship between pain, reaction times, and injury. Stoll found approximately a $\pm 10\%$ day-to-day variation in pain thresholds for individual human subjects. This research established that the minimum time to sense and react to the pain at any temperature is 0.3 s. For those situations in which pain was reached after 0.3 s, Stoll found that complete epidermal necrosis occurred at a time approximately 2.5 times longer than the time for the initial pain sensation.

Several years after his initial work, Wu [*9*] proposed a third model composed of three layers (see Fig. 3), so that the properties of the surface layer and the substrate could be different. This model describes a case identical to that of an insulation covered by a jacket material. The equa-

TABLE 1—*Thermal sensations and associated effects throughout the range of temperatures compatible with tissue life.*

Sensation	Skin Color	Tissue Temperature, °C (°F)	Process	Injury
Numbness	white	72 (162)	protein coagulation	irreversible
	mottled	68	thermal	possibly
	red and white	64	inactivation	reversible
Maximum pain	bright red	60 (140)	of tissue	
		56	contents	
Severe pain		52		
Threshold pain	light red	48		
		44 (111)		
Hotness	flushed	40	normal	none
Warmth		36	metabolism	
		32 (90)		

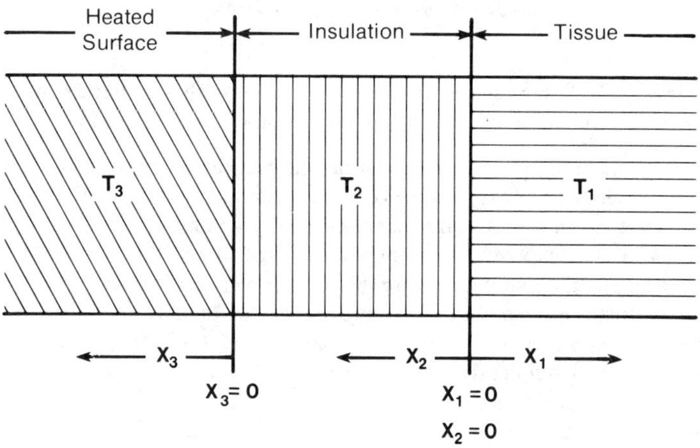

FIG. 3—*Schematic of the heat transfer model.*

tions he developed are a basis for establishing an extrapolation of Moritz's work to real insulated systems.

Wu also recommended that a 1-min exposure limit be used for design purposes for persons who have slow reactions (infants, the elderly, or the infirm) or who "freeze" under severe hazard conditions.

The remaining problem of contact resistance has yet to be resolved. Hatton et al [*10*] demonstrated that the results of Stoll on pain and blistering times were better correlated if the contact resistance was included in the model. He defined pain as the point at which the interface between the epidermis and dermis reaches a temperature of 44°C. His improved correlations were accomplished using a surface coefficient of 1000 $W/m^2 \cdot K$. He points out, however, that depending on skin conditions, this coefficient could be as low as 10 $W/m^2 \cdot K$.

Finally, McChesney [*11*] added a final point to the understanding of burn prevention when he suggested that some factor be included in the analysis to account for the heating wave continuing to penetrate the skin for some time after the contact is removed. He did not, however, guess what that factor should be, since it would depend on the method of cooling the contact location on the skin.

Current ASTM Standardization Work

In 1983, several members of ASTM Committee C-16 on Thermal Insulation requested that a task group be established within the existing subcommittee structure to study the problem of burn hazard evaluation. The initial task group was established within ASTM Subcommittee C16.24 on Health and Safety Hazard Potentials with the charter to establish "a guide for the determination of safe surface operating conditions for heated systems." The scope of this work included establishing (1) a uniform definition of the human burn hazard and (2) a usable practice for design and evaluation of heated systems to prevent serious injury on contact with exposed surfaces.

After initial review of the scope and objectives for this first task group, a second area was identified that was necessary to support the work of the first group. At the fall 1983 ASTM Committee C-16 meeting, a second task group within ASTM Subcommittee C16.30 on Thermal Measurement was established with the objective of developing the analytical tools necessary for evaluating the contact burn potential of heated surfaces either on existing equipment or during design. These tools, when used with the guide established by the first group, are intended to provide the user, designer, or manufacturer with the procedures needed to evaluate the relative safeness of a piece of hardware or system.

Although only approximately 18 months has passed since this work began within ASTM, significant progress has been made toward accomplishing the original objectives. First and foremost in this effort was a survey of available literature to establish the state of the art on the subject and to determine what standards were already in place. The information in the background section of this report briefly summarizes some of the significant work done to date in this area. More than 25 separate technical papers have been identified that relate to burn hazard evaluation and associated medical research. The list of references at the end of this paper gives a sampling of these reports. After review of the available literature, the task groups felt that adequate knowledge was available to develop the standards desired, and work then continued.

In April 1984, the task groups presented the first drafts of the proposed standards in their respective areas. The standard developed by the ASTM Subcommittee C16.24 task group establishes the definitions of burn hazards and a guide for evaluating the combinations of time of exposure, surface temperature, and surface composition that make up a hazardous system. This standard has been approved as the ASTM Guide for Heated System Surface Conditions That Produce Contact Burn Injuries (C 1055-86). The second task group has identified two tools for the evaluation of specific systems for hazardous conditions. The first tool, intended for existing systems, is a device called the thermesthesiometer. Developed by Marzetta [*12–14*] at the National Bureau of Standards (NBS), this device simulates the thermophysical reaction of the human skin to touch contact with a heated surface. Although this device is relatively accurate and easy to use, it has the drawback of requiring an existing system for testing and cannot be used during the design phase. The second tool identified combines the previously established ASTM Recommended Practice for Determination of Heat Gain or Loss, and Surface Temperature of Insulated Pipe and Equipment Systems by the Use of a Computer Program (C 680-82) for surface temperature prediction with the modeling work of Dissan [*14*] to predict, for a given design, the expected contact temperature for the system. This standard has been approved by ASTM and is designated as ASTM Practice for Determination of Skin Contact Temperature From Heated Surfaces Using a Mathematical Model and Thermesthesiometer (C 1057-86). The following sections briefly describe the significant portions of the two standards.

Summary of the Burn Hazard Evaluation Standard Guide

The initial step in the use of ASTM Guide C 1055-86, which was prepared by ASTM Subcommittee C16.24 on Health and Safety Hazard Potentials (a subcommittee of ASTM Committee C-16 on Thermal Insulation), is to establish the operating system parameters for the system in question. This step provides input information for the analysis and may preclude any further work concerning burn hazard. The following are items that need to be identified:

(*a*) system description—shape, size, insulation material and thickness, jacketing material, thickness, and surface emissivity;
(*b*) operation conditions—temperatures of heated system, times of year, cycle, and so on;
(*c*) ambient temperature—worst-case design for burn hazards would be summer design dry bulb or, for inside conditions, the room ambient air temperature; and
(*d*) ambient air velocity—in metres per second or still.

Note: The worst-case design conditions for burn hazard evaluation may be different from those used for heat loss analysis. For example, the highest ambient temperature is used for burn hazard analysis and the lowest for heat loss.

The second step is to determine the surface temperature of the jacket material at the "worst" design condition. This is accomplished by inserting the system dimensions, material and jacket properties, and operating conditions into an analysis technique conforming to ASTM Recommended Practice C 680-82. This technique must be used during design or where the system jacket temperatures cannot be physically measured at worst-case conditions. If the system can be monitored at worst case, direct contact thermometry (thermocouple or resistance device) or infrared noncontact thermometry can be used. Care must be taken in attaching measurement devices on hot systems since burns can result.

Now, if the surface temperature is below 44°C, no short-term (that is, less than 6 h) hazard exists, and further analysis can be ignored. If not, the user enters the graph (see Fig. 2), using the measurement or estimation of surface temperature for the system in question, and checks whether the intersection of the operating temperature and the appropriate time of contact (5 s is usually recommended) falls below the threshold (Line B) temperature (for example, 57°C at 5 s). This is a worst-case assumption.

If the operating surface condition is below the point on the threshold (Line B) curve, then the system will be safe (no permanent cell damage) for the exposure prescribed. If, however, the point falls above the threshold, the system as configured may meet the safety requirements—that is, no burn worse than second-degree level—only if certain combinations of insulation and jacketing are used. Analysis procedures for the jacketing/insulating effects, as outlined in the analytical test standard in the next section, must be used to evaluate the system further.

If, after analysis using the analytical model, the system temperature still fails to meet the maximum level at design, then increasing the insulation thickness, changing the jacketing material or emissivity, or both, or other means must be used to lower the surface temperature. ASTM Recommended Practice C 680-82 will be helpful in determining the levels required.

Once a new combination of jacket and insulation is determined, the analysis just described should be repeated to confirm that operating conditions are safe.

An alternative to calculation of the hazard level is available for those systems already operating. A probe, simulating the workings of a human finger, has been developed [*15*] to measure the same phenomena as the computer method. The following "tools" standard outlines the correct procedures for use of this device for surface hazard evaluation. The output from the thermesthesiometer is the contact temperature of the skin and can be related to Fig. 2 with no corrections for jacketing type needed. As with the calculation, if the contact temperature exceeds the limit threshold in Fig. 2, the owner/user can assume that his system exceeds an acceptable limit and should undertake corrective action.

Summary of the Practice Describing the Analytical Tools

ASTM Practice C 1057-86, which was prepared by ASTM Subcommittee C16.30 on Thermal Measurement, a subcommittee of ASTM Committee C-16 on Thermal Insulation, actually provides two procedures for evaluation of the contact burn potential from heated surfaces. The two methods, a mathematical model and the physical measurement, are both useful depending on the availability of the system—that is, is it operating or is it in the design state—and on the ambient conditions in which it operates. The first step in using this practice is to determine which procedure is to be used. Unless the system of interest is operating at worst-case design conditions, that is, high system temperatures and high ambient temperature, the calculation procedure is recommended. On the other hand, for an operating system, if the question is whether it is safe at the present conditions, the thermesthesiometer provides a quick measurement with no auxiliary calculations. Once the method of testing is selected, the procedures follow the steps summarized here.

Calculation Procedure (Method A)—In this method, the surface temperature of the insulated system has been determined either by direct measurement, using either thermocouples, thermistors, or infrared noncontact techniques, or by modeling of the system using ASTM Recommended Practice C 680-82. Once the surface temperature is known, the designer uses the equation set included in the method to determine the estimated epidermal contact temperature at the permitted time (usually 5 s) after contact with the surface. This temperature is a function of surface temperature and composition of both the jacketing material and insulation substrate. The designer than refers to the standard by ASTM Subcommittee C16.24 to determine the safety of the surface.

Thermesthesiometer (Method B)—In this procedure, the operator places the sensor probe face against the heated surface for a predetermined time (usually 5 s). The device directly reads the contact surface temperature of the probe. This temperature can then be used directly with the standard by ASTM Subcommittee C16.24 to determine the relative safeness of the surface. For a detailed description of the thermesthesiometer, refer to the Marzetta technical paper [*15*].

Conclusions

Injury of personnel resulting from contact with heated surfaces can be prevented by proper design of insulation systems or other protective measures. ASTM Committee C-16 on Thermal Insulation, through its Subcommittees C16.24 on Health and Safety Hazard Potentials and C16.30 on Thermal Measurement, has developed consensus standards to define surface burn hazards. The work of these committees established a guide for what constitutes safe surface conditions and will standardize the tools by which proposed or existing systems can be examined for potential burn hazard. These standards, supported by significant research into both the physical and medical processes involved, provide the designer with the tools he needs to balance the expected exposure times, operating conditions, and system geometry to obtain the safest yet most economical systems.

References

[*1*] Moritz, A. R. and Henriques, F. C., "Studies of Thermal Injury, Part I: The Conduction of Heat to and Through Skin and the Temperatures Attained Therein—A Theoretical and Experimental Investigation," *American Journal of Pathology*, Vol. 23, 1947, pp. 531–549.

[*2*] Moritz, A. R. and Henriques, F. C., "Studies of Thermal Injury, Part II: The Relative Importance of Time and Surface Temperature in the Causation of Cutaneous Burns," *American Journal of Pathology*, Vol. 23, 1947, pp. 695–720.

[*3*] Moritz, A. R., Henriques, F. C., Dutra, F. R., and Weisiger, J. R., "Studies of Thermal Injury, Part

IV: Exploration of Casualty Producing Attributes of Conflagrations—The Local and Systematic Effects of Generalized Cutaneous Exposure to Excessive Circumambient (Air) and Circumradiant Heat of Varying Duration and Intensity," *Archives of Pathology,* Vol. 43, 1947, pp. 466–488.

[*4*] Moritz, A. R. and Henriques, F. C., "Studies in Thermal Injury, Part III: The Relative Importance of Time and Surface Temperature in the Causation of Cutaneous Burns," *American Journal of Pathology,* 1947.

[*5*] Henriques, F. C., "Studies in Thermal Injury V, The Predictability and Significance of Thermally Induced Rate Processes Leading to Irreversible Epidermal Injury, *Archives of Pathology,* Vol. 43, 1947, pp. 489–502.

[*6*] Henriques, F. C., "Studies in Thermal Injury, Part VIII: Automatic Recording Calorie Applicator and Skin Tissue and Skin Surface Thermocouples," *Revised Scientific Instruments,* 1947.

[*7*] Wu, Yung-Chi, "Material Properties Criteria for Thermal Safety," *Journal of Materials,* JMLSA, Vol. 1, No. 4, Dec. 1972, pp. 573–579.

[*8*] Stoll, A. M., Chianta, M. A., and Piergallini, J. R., "Thermal Conduction Effects in Human Skin," *Aviation, Space and Environmental Medicine,* Vol. 50, No. 8, Aug. 1979, pp. 778–787.

[*9*] Wu, Yung-Chi, "Control of Thermal Impact for Thermal Safety," *A.I.A.A. Journal,* Vol. 15, No. 5, 1977, pp. 674–680.

[*10*] Hatton, A. P. and Halfdanarson, "Role of Contact Resistance in Skin Burns," *Journal of Biomedical Engineering,* Vol. 4, April 1982, pp. 97–102.

[*11*] McChesney, M. and McChesney, P., "Preventing Burns from Insulated Pipes," *Chemical Engineering,* 27 July 1981, pp. 58–64.

[*12*] Marzetta, L. A., "A Thermesthesiometer—An Instrument for Burn Hazard Measurement," *I.E.E.E. Transactions on Biomedical Engineering,* Sept. 1974, pp. 452–427.

[*13*] Marzetta, L. A., "Thermesthesiometer," U.S. Patent No. 3,878,728, 22 April 1975.

[*14*] Marzetta, L. A., *Engineering and Construction Manual for an Instrument to Make Burn Hazard Measurements in Consumer Products,* NBS Technical Note 816, National Bureau of Standards, Washington, DC, Feb. 1974.

[*15*] Dussan, B. I. and Weiner, R. I., "Study of Burn Hazard in Human Tissue and Its Implication on Consumer Product Design," *A.S.M.E. Journal,* 1972. Paper No. 71-WA/HT-39, presented at the Winter Meeting, 28 Nov.–2 Dec.

Charles C. Roberts, Jr.[1]

Building Component Test Development Using Module Simulators

REFERENCE: Roberts, C. C., Jr., **"Building Component Test Development Using Module Simulators,"** *Thermal Insulation: Materials and Systems, ASTM STP 922,* F. J. Powell and S. L. Matthews, Eds., American Society for Testing and Materials, Philadelphia, 1987, pp. 713-719.

ABSTRACT: In certain instances consensus standard test methods may not model building responses to environmental influences. Researchers in the past have utilized module simulators, which are model buildings for comparing the effects of specific building components and interactions between several building components. Examples of module tests are presented. A draft ASTM comparative module test method is also presented; this test method may be useful both in analyzing complex phenomena not addressed by present consensus standard test methods and as a basis for the development of a consensus standard test.

KEY WORDS: module testing, test development, test method proposal, thermal insulation

Renewed interest in building science stimulated by increased energy costs has resulted in many new and innovative products that may be useful in reducing energy consumption in buildings. Some products work very well and save energy while others do not. It is therefore necessary to evaluate product performance to verify the claimed beneficial characteristics. Physical testing of the new building product, such as a new type of insulation, is a common way of validating its energy performance. Physical tests usually involve placing the building component (insulation, sheathing, wall covering, and so on) in a standard test instrument and determining such physical properties as R value. Physical testing is differentiated from theoretical modeling in which a concept is tested through mathematical methods. ASTM has developed several test methods for evaluating building component characteristics. These test methods are recommended as evaluation procedures since they are standardized and well recognized.

However, some ASTM test methods may not be applicable when evaluating certain building products. There is the possibility that a particular ASTM test method may not demonstrate energy-saving characteristics that may have been observed in the field. There is also the possibility that an ASTM test method has not been developed to demonstrate characteristics of some newly discovered phenomenon.

When this situation exists, it is interesting to note that some investigators have utilized module simulators to initiate their testing programs. Figure 1*a* shows the concept of the module simulator. It is essentially a model building with sides and roof of a configuration similar to a building but significantly reduced in size. A model heater and thermostat are placed in the module, and energy usage is recorded. The test material is placed in the module in its normal configuration. Usually two test modules are analyzed, one with standard building materials (such as fiberglass insulation) and the other with the new product. If the module with the new product shows less energy consumption than that of the standard, one may make an initial suggestion that the new product has more favorable energy-saving characteristics. Figure 1*b* shows a typical comparative module test while Fig. 1*c* shows associated instrumentation. In Fig.

[1]Consulting engineer, Warrenville, IL 60555.

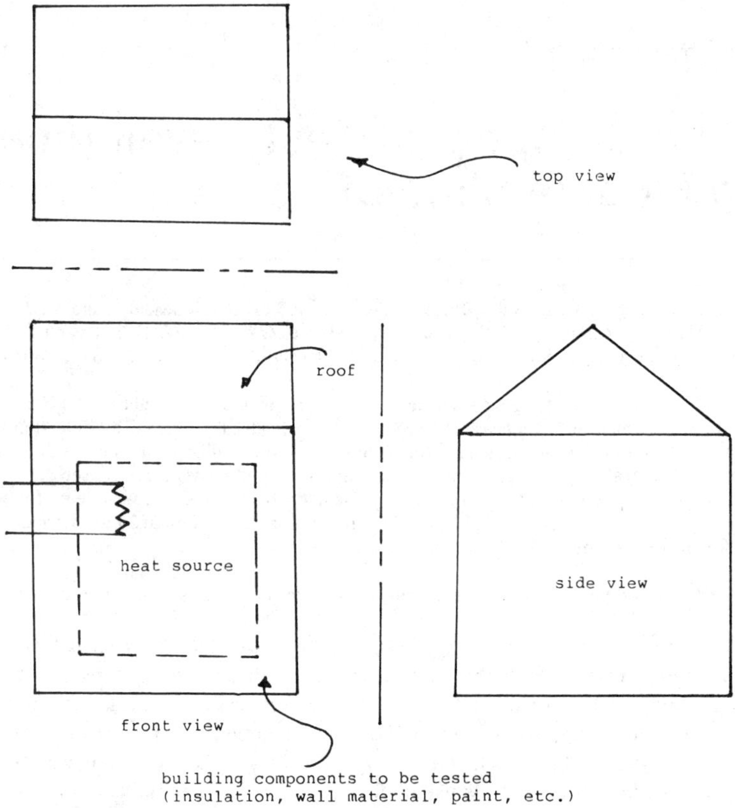

FIG. 1*a*—*Module simulator concept.*

1*c*, a thermocouple recorder is shown that tracks various temperatures in the modules, such as the inside ambient air temperature, mean radiant temperature, inside wall temperature, and exterior wall temperature. The small timers to the left of the recorder monitor the time the heating system is operational, thereby gaging the comparative energy consumption of the modules.

This paper reviews several examples of comparative module tests that have been performed by researchers. It is interesting to note the similarity between the tests. A possible ASTM test method utilizing the comparative module simulation technique is discussed.

Department of Energy Solar Test Modules

Los Alamos National Laboratory (LANL) has been active in the development of solar energy systems. Figure 2 shows a typical LANL test module used to evaluate passive solar performance of various wall combinations. References *1* and *2* demonstrate the feasibility of this concept. In Fig. 2 we see the module configuration composed of walls, glass panel, and thermal storage wall. This research found that the test modules accurately predicted the energy consumption of solar buildings constructed from similar materials. It was proposed in Ref *2* that test modules be located at a building site prior to building construction to obtain data for refining the solar design of the building to be constructed. The test modules were an economical way of evaluating several passive solar system concepts. The module program paved the way for more detailed

FIG. 1*b*—*Two typical modules.*

FIG. 1*c*—*Heating system running time meters (*left*) and thermocouple strip chart recorder (*right*).*

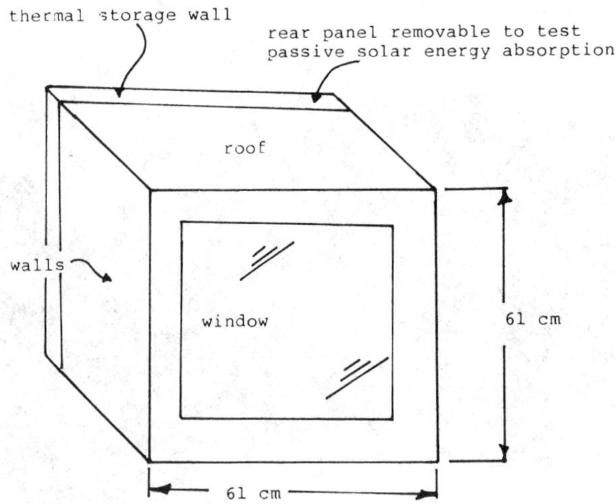

FIG. 2—*U.S. Department of Energy passive solar test box [1,2].*

analysis on full-sized test rooms. Test modules were used in this study because no ASTM test method for performing this type of test exists presently.

Vapor Retarder Paint Tests

Vapor retarders help prevent the transport of water vapor from the inside of a building to the insulation. The water vapor that condenses in the insulation may reduce its effectiveness. A test was performed to determine the energy-saving characteristics of a vapor retarder paint.[2] The questions to be answered were how much insulation effectiveness was lost and which building component suffered the most R value reduction as a result of moisture in the insulation. The module test of Fig. 3 was conducted. Two modules of identical construction were placed in a cold chamber to simulate severe weather environments. The modules were constructed of wood framing and fiberglass batt insulation. A humidity and heat source was placed in the module. Heat flow transducers monitored envelope performance. One module had internal gypsum board painted with a vapor retarder paint [average permeability according to the ASTM Test for Water Vapor Transmission of Materials (E 96-80) was 0.6 perms]. The other module was painted with a standard non-vapor retarder paint. Energy consumption was monitored throughout the test period, and at the end of the test, moisture tests on the insulation were performed. ASTM testing (ASTM Test E 96-80) was used to determine the perm rating of the paint; however, module testing was used to illustrate the complex thermal environment in buildings and the performance characteristics of the moisture retarder paint. The module internal temperatures were set at 20°C, the relative humidity at 50%, and external ambient temperature at 3°C. Approximately 25% more energy was required to maintain the non-vapor retarder module compared with the vapor retarder module.

Thermal Insulation Module Tests

Figure 4 shows a module test used to evaluate an attic insulation.[2] In this case it was claimed that the insulation performed best when air was free to move over the insulation in the attic. The

[2]Unpublished research, C. Roberts Consulting Engineers, Warrenville, IL 60555.

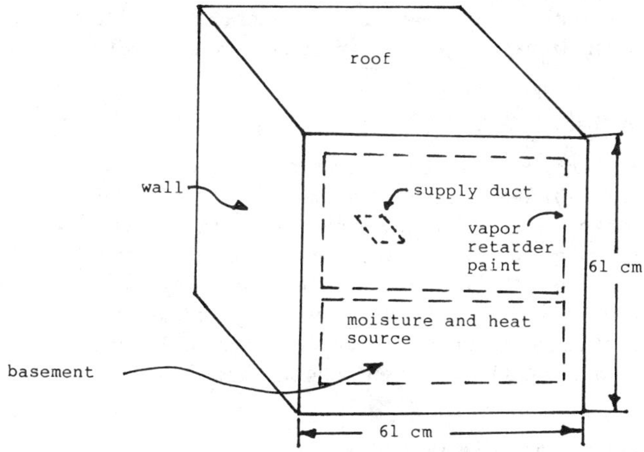

FIG. 3—*Vapor retarder paint test.*

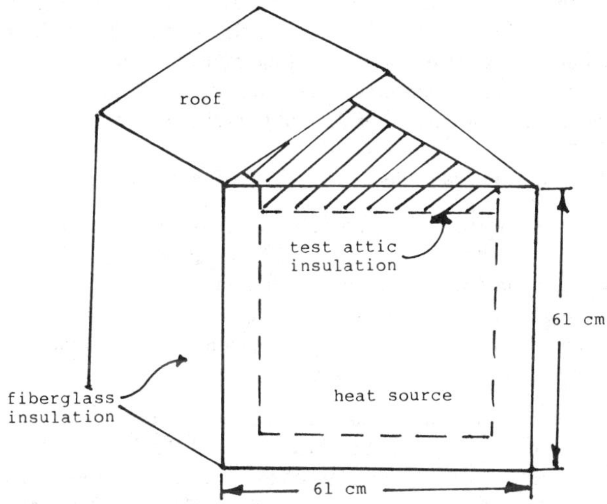

FIG. 4—*Attic insulation test module.*

modules were constructed of standard building materials, which, in this case, were gypsum board, 8.9-cm fiberglass batt insulation, and plywood siding. The tops or simulated attic of the modules were removable so that standard materials and test materials could be used. A module constructed of expanded perlite in the attic and fiberglass batt as wall insulation was used as a comparison standard. The attic insulation in the test module was expanded perlite coated with sodium silicate.

During winter heating simulation, no difference was noted between the standard perlite insulation module and the sodium silicate-coated insulation module. Adsorption and absorption of water in the sodium silicate during summer diurnal cycles resulted in a 17% lower air conditioning energy consumption in the silicate-coated perlite module compared with the uncoated perlite module. The author is not aware of any consensus standard test method for evaluating convective latent heat transport by absorption and adsorption in attic insulations.

Proposed Test Method

In the previous examples, it has been shown that comparative module testing was a viable technique for demonstrating energy-related characteristics of building components. In each case, an ASTM or another consensus test method had not as yet been developed. The development of an ASTM test method based on the module simulation method would be very helpful to researchers working on materials that cannot be tested using existing ASTM or other standards. The comparative module testing would give insight into the new phenomenon and aid in the development of tests germane to the new phenomenon. The following is an outline of a possible standard that could be developed along the lines of comparative testing.

Standard Test Method for Comparative Thermal Performance of Building Components

Scope—This method covers the use of comparative module simulators in the comparative measurement of thermal performance of building components or systems.

Significance—This thermal performance test method is comparative in nature. It shows only the difference between two building components or systems and not an absolute physical quantity. The method is useful in demonstrating thermal characteristics of building components or systems that cannot be characterized by other test methods. Interactive aspects between various building components that are influenced by environmental conditions such as orientation, unsteady-state temperatures, solar radiation, humidity cycles, and diurnal cycles can be investigated.

Definitions—The following are terms used in this test method:

Building components—any part of a building that affects energy usage in the building, such as insulation, sheathing, windows, furnace, or air conditioner.

Building system—any combination of building components that affect energy usage, such as heating system, cooling system, or building envelope.

Module—a test fixture used to simulate energy response in a building and compare energy performance of different building components or systems.

Apparatus—Two or more modules must be constructed of identical materials. Provisions must be made so that the component to be tested can easily be fixed to all modules. The modules must be instrumented identically and placed in the same environment, such as a test chamber or out of doors. Care should be taken in establishing nearly identical positioning of the modules out of doors. For instance, one module should not be shielded from the wind by neighboring obstructions while the other modules are exposed. Instrumentation should be installed to monitor the performance of the modules and characteristics of their environment.

Procedure—At the beginning of the test period, each module is set up with the identical building components in order to determine the similarity of the modules. Energy is applied to each module, and data recording equipment is started. For steady-state testing, data can be taken after transients have disappeared. For unsteady-state testing, data can be taken after three cycles. The module performances (energy usage and so on) should then be compared. If the performances differ by more than 10%, the modules must be corrected so that the comparative performance does not differ by more than 10%. (The 10% value is arbitrarily chosen. Perhaps module testing at accuracies of 3, 5, and 10% may be characterized as Class A, Class B, and Class C module tests, respectively.)

After it has been demonstrated that the modules perform within 10% of each other, the comparative testing can begin. The new components to be analyzed should be placed in their respective modules. Standard, well-documented components are placed in at least one module to act as the standard.

For steady-state testing, the test period is arbitrary as long as transient factors have disappeared. For cyclic testing, at least three cycles should constitute a minimum test period. Data shall be recorded for all test periods. The remaining test periods are identical to the first, with the exception that the building components in each module are switched to account for construction differences in the modules.

Report—The report shall include drawings or photographs of the modules, experimental data (temperature, heat flow, moisture transfer, and so on), and the comparative performance of the modules.

Commentary on Proposed Test Method

The test can demonstrate building component performance differences, especially the effects of several types of components and their complex interactions in a given environment. The test can serve as a basis for the development of other tests that specifically address the unusual phenomenon demonstrated. The test does not yield a material property or characteristic; it demonstrates a difference between a known material and another material.

Conclusions

Module tests have been utilized by some researchers for several years. Module testing is an easily accomplished and often inexpensive way of identifying building component systems that can save energy. Module tests are not intended to replace ASTM test methods. Module testing is a preliminary investigation and should be used to gather background information that leads to the use of, or development of, specific ASTM tests methods.

References

[*1*] Collier, R. K. and Grimmer, D. P., "The Experimental Evaluation of Phase Change Material Building Walls Using Small Passive Test Boxes," Third National Passive Solar Conference, San Jose, CA, 14 June 1979, U.S. Department of Energy, Washington, DC, 1979.

[2] Grimmer, D. P., "The Use of Small Passive-Solar Test-Boxes to Model the Thermal Performance of Passively Solar Heated Building Designs," ASES/ISES 1977, U.S. Energy Research and Development Administration, Washington, DC, 1977.

R. Gerry Miller,[1] James A. Berry,[1] and Morton Sherman[1]

Thermal Performance of Metal Furred/Foam Board Insulated Wall Systems

REFERENCE: Miller, R. G., Berry, J. A., and Sherman, M., **"Thermal Performance of Metal Furred/Foam Board Insulated Wall Systems,"** *Thermal Insulation: Materials and Systems, ASTM STP 922,* F. J. Powell and S. L. Matthews, Eds., American Society for Testing and Materials, Philadelphia, 1987, pp. 720–728

ABSTRACT: System thermal resistance, *R*, values were measured in full-scale guarded hot box tests [according to the ASTM Test for Steady-State Thermal Performance of Building Assemblies by Means of a Guarded Hot Box (C 236-80)] on a series of foam board insulated wall systems. Tests were run on Hi-Hat, Temp Guard insulation furring (TGIF), and Z-furring securement systems at a mean temperature of approximately 24°C (75°F).

System *R* values were calculated for these insulated constructions by means of the series/parallel and zone methods recommended by the American Society of Heating, Refrigerating, and Air-Conditioning Engineers (ASHRAE). Results of comparisons between measured and calculated *R* values are reported in this paper. Thermal effects of the various metal furring systems are also discussed.

KEY WORDS: thermal insulation, metal furred wall system, guarded hot box

The prescriptive sections of the three major model building codes in the United States[2] require the securement of a 15-min-rated[3] thermal barrier interior surface finish, such as regular 1.3-cm (1/2-in.) gypsum board, when constructing new exterior walls insulated with code-qualified combustible cellular plastic (foam) thermal insulation. Architect, engineer, and builder preferences for these higher thermal efficiency insulation materials on the interior sides of concrete or masonry exterior walls have resulted in the commercial development of alternative metal furring securement systems for installing foam insulation board products and the thermal barrier interior finish required to meet these code requirements. The competitive systems most often specified in the construction marketplace for use on the interior sides of insulated concrete or masonry walls include the following: Hi-Hat metal furring channel, a conventionally available drywall installation product; Temp Guard insulation furring (TGIF), a proprietary metal furring strip manufactured by Judkins Associates Inc.; Thermo-Stud furring channel, a proprietary metal furring strip manufactured by W. R. Grace and Co.; and Z-configuration metal furring, another conventionally available metal furring product.

In this paper, a laboratory comparison of the system thermal resistance, *R*, values for Hi-Hat, TGIF, and Z-furring wall constructions were made at a mean temperature of 24°C (75°F)

[1]Research associate, senior research civil engineer, and senior research associate, Jim Walter Research Corp., St. Petersburg, FL 33702.

[2]Building codes: Building Officials and Code Administrators International Inc., Section 1319.0; International Conference of Building Officials, Torrance, CA, Section 1712; and Southern Building Code Congress International, Birmingham, AL, Section 717.

[3]Uniform Building Code Standard 17.3, "Test Method for the Evaluation of Thermal Barriers (Thermal Barrier Index)," a standard of the International Conference of Building Officials.

in a full-scale guarded hot box facility. The overall test specimen size was 2.4 by 2.4 m (8 by 8 ft), and the metered area was 1.8 by 1.8 m (6 by 6 ft). This facility is accredited by the National Voluntary Laboratory Accreditation Program (NVLAP) and has been fully described by Miller et al [*1*].

Masonry Securement Systems

This section of the paper provides a description of each of the 14 constructions tested in the guarded hot box.

Because of the difficulties in testing a masonry wall (moisture content, a long stabilization time, and other problems), the authors decided to simulate the concrete block/stucco component of the wall with a less massive material of similar *R* value. Since the tests were conducted under steady-state conditions and the results were to be compared with each other, this substitution would not affect the overall results.

Using Table 3A of the *ASHRAE Handbook of Fundamentals,* published by the American Society of Heating, Refrigeration, and Air-Conditioning Engineers (ASHRAE) [*2*], the *R* value of a 20.3-cm (8-in.) concrete block/stucco combination was 0.23 $K \cdot m^2/W$ (1.31 $h \cdot ft^2 \cdot {}^\circ F/Btu$). From this same table, 2.5-cm (1-in.) plywood has an *R* value of 0.22 $K \cdot m^2/W$ (1.25 $h \cdot ft^2 \cdot {}^\circ F/Btu$). This difference in *R* value was considered negligible when viewed from the overall *R* values of the systems. Therefore, 2.5-cm (1-in.) plywood boards were used in all the system constructions to simulate the exterior concrete block/stucco component.

Test Construction 1

Test Construction (TC) 1 consisted of two nominal 1.9-cm (3/4-in.) by 1.2-m (4-ft) by 2.4-m (8-ft) aluminum-foil-faced, glass-fiber-reinforced polyisocyanurate foam boards. The boards were installed in the test frame by taping (with aluminum tape) along the perimeter on both the hot and cold surfaces and double taping over the center joint.

Test Construction 2

The boards from TC 1 were used with the TGIF system for TC 2. The TGIF strip is a 4.1-cm (15/8-in.) galvanized steel furring strip with 3-mm (1/8-in.) ribs that give it flexural strength and keep the fastener heads from touching the drywall. Every 12.7 cm (5 in.), small tab-like "darts" on the strip point downward to penetrate the foam board and support the strip until it is mechanically fastened. With the TGIF strip, thermal bridging is due only to the fasteners through the insulation and the small amount of penetration into the foam by the darts.

Figure 1 shows a schematic of the TGIF system. From the exterior to the interior side, TC 2 consisted of 2.5-cm (1-in.) plywood with the center joint caulked; the nominal 1.9-cm (3/4-in.) aluminum-foil-faced, glass-fiber-reinforced polyisocyanurate foam boards from TC 1 with the center joint taped; TGIF strips attached vertically to the foam boards on 0.6-m (2-ft) centers and fastened to the plywood boards with 4.1-cm (15/8-in.) drywall screws, 25.4 cm (10 in.) on center; and 1.3-cm (1/2-in.) gypsum boards fastened to the TGIF strips with 2.5-cm (1-in.) drywall screws, 30.5 cm (12 in.) on center, and a center joint finished with joint compound and tape. All the TGIF strips contained a 6-mm (1/4-in.) thermal break gap at the meter/guard interface. The perimeter of the construction was caulked.

Test Construction 3

A new pair of nominal 1.9-cm (3/4-in.) aluminum-foil-faced, glass-fiber-reinforced polyisocyanurate foam boards was tested as TC 3. They were installed as in TC 1.

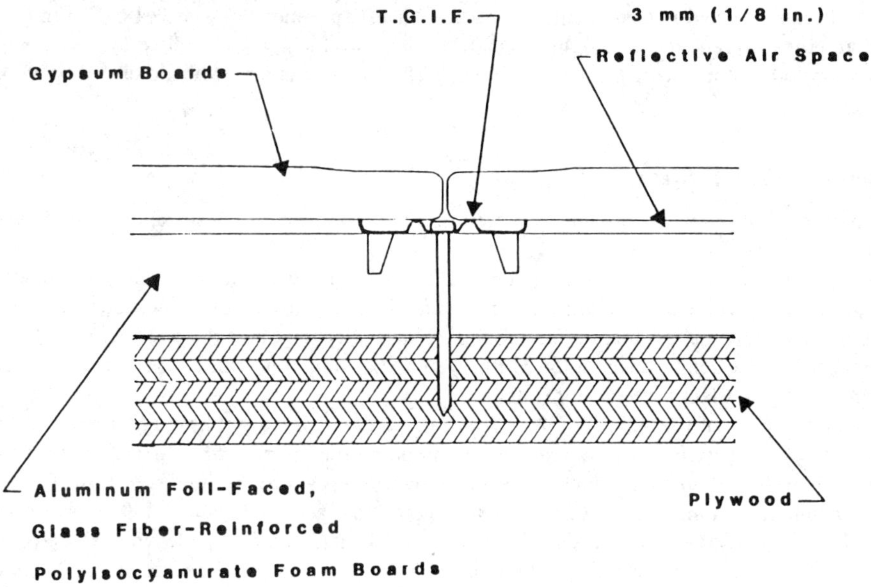

FIG. 1—*Test construction for the TGIF system.*

Test Construction 4

The boards from TC 3 were used with the Hi-Hat system for TC 4. The Hi-Hat strip is a galvanized steel furring strip which, when viewed in cross section, resembles a high hat profile, hence its name. With the Hi-Hat strip, thermal bridging is due only to the fasteners through the insulation.

Test Construction 4 (Fig. 2) consisted of, from exterior to interior, 2.5-cm (1-in.) plywood with the center joint caulked; the foam boards from TC 3 with the center joint taped; Hi-Hat strips located vertically on 0.6-m (2-ft) centers and fastened to the plywood boards with 3.8-cm (1½-in.) drywall screws, 25.4 cm (10 in.) on center; and 1.3-cm (½-in.) gypsum boards fastened to the Hi-Hat strips with 2.5-cm (1-in.) drywall screws, 30.5 cm (12 in.) on center, and a center joint finished with joint compound and tape. All the Hi-Hat strips contained a 6-mm (¼-in.) thermal break gap at the meter/guard interface. The perimeter of the construction was caulked.

Test Construction 5

TC 5 consisted of a pair of nominal 2.5-cm (1-in.) aluminum-foil-faced, glass-fiber-reinforced polyisocyanurate foam boards installed as in TC 1.

Test Construction 6

TC 6 was identical to TC 2 except that the nominal 2.5-cm (1-in.) foam boards from TC 5 were used.

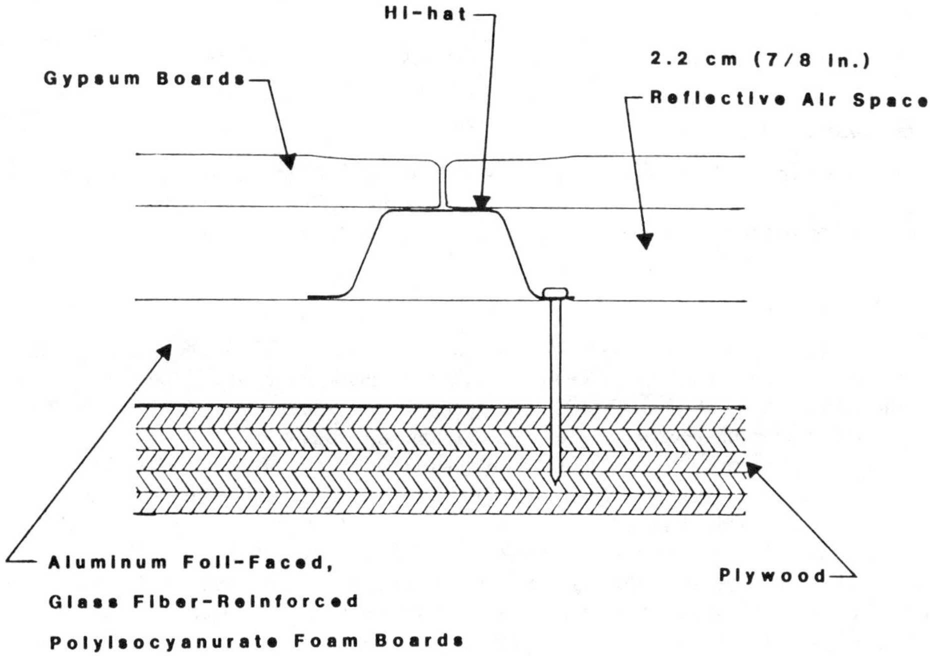

FIG. 2—*Test construction for the Hi-Hat system.*

Test Construction 7

A new pair of nominal 2.5-cm (1-in.) aluminum-foil-faced, glass-fiber-reinforced polyisocyanurate foam boards, installed as in TC 1, was tested as TC 7.

Test Construction 8

TC 8 was identical to TC 4 except that the insulation boards from TC 7 were used.

Test Construction 9

TC 9 consisted of a pair of nominal 3.2-cm (1¼-in.) aluminum-foil-faced, glass-fiber-reinforced polyisocyanurate foam boards.

Test Construction 10

This construction was identical to TC 2 except that it contained the foam boards from TC 9.

Test Construction 11

A new pair of nominal 3.2-cm (1¼-in.) aluminum-foil-faced, glass-fiber-reinforced polyisocyanurate foam boards, installed as in TC 1, was tested as TC 11.

Test Construction 12

TC 12 was identical to TC 4 except that the boards from TC 11 were used.

Test Construction 13

TC 13 consisted of four nominal 5.1-cm (2-in.) by 0.6-m (2-ft) by 2.4-m (8-ft) extruded polystyrene foam boards. The boards were taped along the butt joints and along the perimeter on both the hot and cold surfaces.

Test Construction 14

The boards from TC 13 were used with the Z-furring system for TC 14. The Z-furring strip is a galvanized steel strip with a cross-section profile in the shape of the letter Z. These strips form a continuous metal path through the insulation joint from the hot to the cold surfaces. Therefore, thermal bridging occurs because of the Z-furring strips themselves.

Figure 3 is a schematic of the Z-furring construction, consisting of the foam boards in TC 13 installed over 2.5-cm (1-in.) plywood utilizing the Z-furring strips vertically on 0.6-m (2-ft) centers and fastened to the plywood with 1.3-cm (1/2-in.) drywall screws, 25.4 cm (10 in.) on center, and 1.3-cm (1/2-in.) gypsum boards fastened to the Z-furring strips with 2.5-cm (1-in.) drywall screws, 30.5 cm (12 in.) on center, and a center joint finished with joint compound and tape. All the Z-furring strips contained a 6-mm (1/4-in.) thermal break gap at the meter/guard interface. The perimeter of the construction was caulked.

Guarded Hot Box Test Results

Fourteen guarded hot box tests were conducted, according to the ASTM Test for Steady-State Thermal Performance of Building Assemblies by Means of a Guarded Hot Box (C 236-80), at a mean temperature of approximately 24°C (75°F). Of the 14 tests, 7 were on the masonry securement systems; the other 7 were on the insulation boards used in the respective

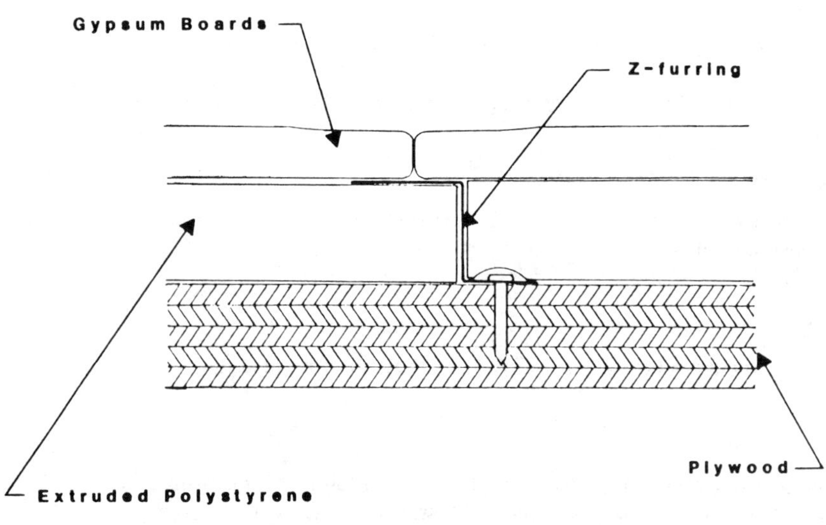

FIG. 3—*Test construction for the Z-furring.*

systems. It should be noted that the 2.5-cm (1-in.) plywood boards, used to simulate the concrete block/stucco component of the wall, were a constant in all the system tests. The insulation board and the securement system itself (Hi-Hat, TGIF, and Z furring) were the variables.

As walls were being tested, the heat flow direction was horizontal. For the system tests, the gypsum board was considered the interior surface and was exposed to the hot chamber.

Figures 1 through 3 show the three securement systems that were tested: TGIF, Hi-Hat, and Z-furring, respectively. Three thicknesses, 1.9, 2.5, and 3.2 cm (3/4, 1, and 1 1/4 in.), of the foil-faced, glass-fiber-reinforced polyisocyanurate foam boards were used in both the TGIF and Hi-Hat securement systems. The 5.1-cm (2-in.) extruded polystyrene boards were used in the Z-furring system. The results of the 14 tests are shown on Table 1.

Table 2 shows the *R* value comparisons between the three securement systems, as well as those between the systems and boards alone. The Hi-Hat securement system outperformed the TGIF system; both systems outperformed the Z-furring system. When comparing the board *R* value with its respective system *R* value, it was expected that the system *R* value would be greater because of the additional resistances from the gypsum boards, plywood, and, in two cases, reflective air spaces. This was the case for the Hi-Hat and TGIF systems. The last column on Table 2 shows the ΔR, the difference between the system and the board *R* values.

The comparison between the Z-furring system *R* value, 1.67 K · m^2/W (9.49 h · ft^2 · °F/Btu), and the extruded polystyrene *R* value, 1.73 K · m^2/W (9.80 h · ft^2 · °F/Btu), showed just the opposite. In this case, the additional resistances from the gypsum board and plywood were negated by the metal Z-furring, which formed a continuous metal path (thermal short) through the insulation from the hot side to the cold side.

Comparison Between the Calculated and Measured System *R* Values

The ASHRAE series/parallel method [*3*] was used to calculate the *R* values for the Hi-Hat and TGIF systems. The ASHRAE zone method [*3*] was used for the Z-furring system. In each of these calculation procedures, the *R* values of the systems' components are needed. For the gypsum and plywood boards, the *R* values were obtained from the *ASHRAE Handbook of Funda-*

TABLE 1—*Test Results for the wall systems.* [a]

Test Construction No.	Hot Surface Temperature		Cold Surface Temperature		Mean Temperature		Surface to Surface Thermal Resistance	
	°C	(°F)	°C	(°F)	°C	(°F)	K · m^2/W	(h · ft^2 · °F/Btu)
1	41.0	(105.8)	6.3	(43.4)	23.7	(74.6)	1.06	(6.04)
2	38.6	(101.5)	10.2	(50.3)	24.4	(75.9)	1.45	(8.26)
3	40.3	(104.6)	7.1	(44.8)	23.8	(74.8)	1.02	(5.80)
4	40.1	(104.1)	6.9	(44.5)	23.5	(74.3)	1.69	(9.58)
5	46.7	(116.0)	1.2	(34.1)	23.9	(75.0)	1.40	(7.94)
6	40.9	(105.7)	6.2	(43.1)	23.6	(74.4)	1.79	(10.16)
7	44.9	(112.9)	2.3	(36.1)	23.6	(74.5)	1.30	(7.41)
8	42.4	(108.3)	5.3	(41.5)	23.8	(74.9)	1.94	(11.03)
9	42.9	(109.3)	4.1	(39.3)	23.5	(74.3)	2.03	(11.52)
10	43.9	(111.1)	3.6	(38.5)	23.8	(74.8)	2.31	(13.10)
11	43.4	(110.1)	4.5	(40.1)	23.9	(75.1)	2.03	(11.53)
12	43.6	(110.5)	4.1	(39.3)	23.8	(74.9)	2.51	(14.23)
13	40.3	(104.5)	7.3	(45.1)	23.8	(74.8)	1.73	(9.80)
14	40.1	(104.1)	8.0	(46.4)	24.1	(75.3)	1.67	(9.49)

[a] The original measurements were made in English units.

TABLE 2—R *value comparisons for the boards and systems.*

Test Construction No.	Board		Z-Furring		TGIF		Hi-Hat		ΔR ($R_{system} - R_{bound}$)	
	$m^2 \cdot K/W$	($h \cdot ft^2 \cdot °F/Btu$)	$m^2 \cdot K/W$	($h \cdot ft^2 \cdot °F/Btu$)	$m^2 \cdot K/W$	($h \cdot ft^2 \cdot °F/Btu$)	$m^2 \cdot K/W$	($h \cdot ft^2 \cdot °F/Btu$)	$m^2 \cdot K/W$	($h \cdot ft^2 \cdot °F/Btu$)
1	1.06	(6.04)	...		...		...		...	
2	...		...		1.45	(8.26)	...		0.39	(2.22)
3	1.02	(5.80)	...		...		...		...	
4	...		...		...		1.69	(9.58)	0.67	(3.78)
5	1.40	(7.94)	...		...		...		...	
6	...		...		1.79	(10.16)	...		0.39	(2.22)
7	1.30	(7.41)	...		...		...		...	
8	...		...		...		1.94	(11.03)	0.64	(3.62)
9	2.03	(11.52)	...		...		...		...	
10	...		...		2.31	(13.10)	...		0.28	(1.58)
11	2.03	(11.53)	...		...		...		...	
12	...		...		...		2.51	(14.23)	0.48	(2.70)
13	1.73	(9.80)	...		...		...		...	
14	...		1.67	(9.49)	...		...		−0.06	(−0.31)

mentals [2]. The R values for the insulation boards were measured in this study (Table 1). The R value of the reflective air space in the TGIF system was calculated as follows

$$R = \frac{1}{h_c + Eh_r} \tag{1}$$

where Eh_r is the radiation coefficient

$$Eh_r \cong 0.00686E \left[\frac{t_m + 460}{100}\right]^3 \tag{2}$$

and h_c is assumed to represent the heat transfer by conduction alone through the air and is calculated by [4]

$$h_c = \frac{0.159(1 + 0.0016t_m)}{l} \tag{3}$$

where t_m is the mean temperature in degrees Fahrenheit, assumed to be 38°C (100°F), in these systems and l is the air space thickness (in inches). In the TGIF system l = 0.32 cm (1/8 in.).

Using Eqs 1 through 3, the reflective air space R value was 0.11 m^2 · K/W (0.65 h · ft^2 · °F/Btu). This value was used in the series/parallel method calculation for each TGIF system.

To determine the R value of the reflective air space in the Hi-Hat system, the surface temperatures bounding the air space were measured, and the calculation procedure outlined in Ref 5 was followed. Table 3 contains the average surface temperatures and calculated reflective air space R values for the three Hi-Hat constructions. These R values were used in the series/parallel method calculations for the three Hi-Hat constructions.

The two calculation procedures (the series/parallel and zone methods) neglect the heat loss through the fasteners. This would tend to produce a calculated R value higher than if the fastener heat loss was included.

Table 4 contains the comparison between the guarded hot box measured R values and the ASHRAE calculated R values. For both the TGIF and Hi-Hat systems, the series/parallel method calculated R values were higher than the measured R values. The average difference was +2.6% for the TGIF system (Test Constructions 2, 6, and 10) and +9.8% for the Hi-Hat system (Test Constructions 4, 8, and 12).

TABLE 3—*Calculated reflective air space* R *values for the Hi-Hat system.*[a,b]

Test Construction No.	t_h		t_c		t_m		Δt		R	
	°C	(°F)	°C	(°F)	°C	(°F)	°C	(°F)	m^2 · K/W	(h · ft^2 · °F/Btu)
4	38.9	(102.0)	29.9	(85.9)	34.4	(94.0)	9.0	(16.1)	0.59	(3.37)
8	41.2	(106.1)	32.2	(90.0)	36.7	(98.0)	9.0	(16.1)	0.59	(3.36)
12	42.6	(108.7)	35.4	(95.7)	39.0	(102.2)	7.2	(13.0)	0.61	(3.46)

[a]Calculated using the methodology outlined in Housing Research Paper No. 32 [5].
[b]Key to symbols:
t_h = hot surface temperature.
t_c = cold surface temperature.
t_m = mean temperature, $(t_h + t_c)/2$.
Δt = temperature difference across reflective air space, $t_h - t_c$.

TABLE 4—*Comparison of measured and calculated* R *values.*

Test Construction No.	Measured *R* m²·K/W	Measured *R* (h·ft²·°F/Btu)	Calculated *R* m²·K/W	Calculated *R* (h·ft²·°F/Btu)	Difference As Percent of Measured *R*
1	1.06	(6.04)	. . .		. . .
2	1.45	(8.26)	1.47	(8.36)	+1.21
3	1.02	(5.80)	. . .		. . .
4	1.69	(9.58)	1.81	(10.31)	+7.62
5	1.40	(7.94)	. . .		. . .
6	1.79	(10.16)	1.81	(10.26)	+0.98
7	1.30	(7.41)	. . .		. . .
8	1.94	(11.03)	2.10	(11.93)	+8.16
9	2.03	(11.52)	. . .		. . .
10	2.31	(13.10)	2.44	(13.84)	+5.65
11	2.03	(11.53)	. . .		. . .
12	2.51	(14.23)	2.85	(16.16)	+13.56
13	1.73	(9.80)	. . .		. . .
14	1.67	(9.49)	1.53	(8.71)	−8.22

The ASHRAE zone method calculation gave an *R* value 8.2% *lower* than the measured *R* value, which was already lower than the *R* value of the insulation boards used in the construction. This is further evidence of the substantial effect of the thermal shorting caused by the Z-furring strips in the constructions.

Conclusions

Tests were run on full-scale Hi-Hat, TGIF, and Z metal furring systems, combined with foam board insulation in a wall construction, using the guarded hot box, according to the requirements of ASTM Test C 236-80. The mean temperature for the tests was approximately 24°C (75°F). The results showed that the Hi-Hat system performed best. It was also shown that the *R* value of the Z-furring system was less than the *R* value of the insulation boards used in the construction.

The ASHRAE series/parallel method was used to calculate the system *R* values for the Hi-Hat and TGIF constructions. For the Z-furring system, the ASHRAE zone method was used. The calculated *R* values were higher than the measured *R* values for the Hi-Hat and TGIF systems but were lower for the Z-furring system. On the average, the calculations predicted the measured *R* values within ±10%.

References

[1] Miller, R. G., Perrine, E. L., and Lineham, P. W. in *Thermal Transmission Measurements of Insulation, ASTM STP 660,* R. P. Tye, Ed., American Society for Testing and Materials, Philadelphia, 1978, pp. 329–341.

[2] *ASHRAE Handbook of Fundamentals,* American Society of Heating, Refrigeration, and Air-Conditioning Engineers, Atlanta, GA, 1981, pp. 23.14–23.17.

[3] *ASHRAE Handbook of Fundamentals,* American Society of Heating, Refrigeration, and Air-Conditioning Engineers, Atlanta, GA, 1981, pp. 23.2–23.4.

[4] *ASHRAE Handbook of Fundamentals,* American Society of Heating, Refrigeration, and Air-Conditioning Engineers, Atlanta, GA, 1981, p. 23.13.

[5] Robinson, H. E., Powlitch, F. J., and Dill, R. S., "The Thermal Insulating Value of Air Spaces," Housing Research Paper No. 32, Housing and Home Finance Agency, Washington, DC, 1954.

Indexes

Author Index

Subject Index

B

C

G

I

Q

R

S

T

U

V

Y

Z